Teacher Edition

SpringBoard®
Mathematics

Algebra 1

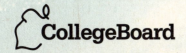

About the College Board

The College Board is a mission-driven not-for-profit organization that connects students to college success and opportunity. Founded in 1900, the College Board was created to expand access to higher education. Today, the membership association is made up of more than 5,900 of the nation's leading educational institutions and is dedicated to promoting excellence and equity in education. Each year, the College Board helps more than seven million students prepare for a successful transition to college through programs and services in college readiness and college success—including the SAT® and the Advanced Placement Program®. The organization also serves the education community through research and advocacy on behalf of students, educators, and schools.

For further information, visit www.collegeboard.org.

ISBN: 1-4573-0158-X
ISBN: 978-1-4573-0158-2

Copyright © 2014 by the College Board. All rights reserved.

College Board, Advanced Placement Program, AP, AP Central, AP Vertical Teams, CollegeEd, Pre-AP, SpringBoard, connect to college success, SAT, and the Acorn logo are registered trademarks of the College Board. College Board Standards for College Success, connect to college success, English Textual Power, and SpringBoard are trademarks owned by College Board. PSAT/NMSQT is a registered trademark of the College Board and the National Merit Scholarship Corporation. Microsoft and PowerPoint are registered trademarks of Microsoft Corporation. All other products and services may be trademarks of their respective owners.

2 3 4 5 6 7 8   14 15 16 17 18 19
Printed in the United States of America

Acknowledgments

The College Board gratefully acknowledges the outstanding work of the classroom teachers and writers who have been integral to the development of this revised program. The end product is testimony to their expertise, understanding of student learning needs, and dedication to rigorous but accessible mathematics instruction.

Michael Allwood
Brunswick School
Greenwich, Connecticut

Shawn Harris
Ronan Middle School
Ronan, Montana

Dr. Roxy Peck
California Polytechnic Institute
San Luis Obispo, California

Floyd Bullard
North Carolina School of Science and Mathematics
Durham, North Carolina

Marie Humphrey
David W. Butler High School
Charlotte, North Carolina

Katie Sheets
Harrisburg School
Harrisburg, South Dakota

Marcia Chumas
East Mecklenburg High School
Charlotte, North Carolina

Brian Kotz
Montgomery College
Monrovia, Maryland

Andrea Sukow
Mathematics Consultant
Nashville, Tennessee

Kathy Fritz
Plano Independent School District
Plano, Texas

Chris Olsen
Prairie Lutheran School
Cedar Rapids, Iowa

Stephanie Tate
Hillsborough School District
Tampa, Florida

Product Development

Betty Barnett
Executive Director, SpringBoard

Robert Sheffield
Sr. Director, SpringBoard Implementation

Allen Dimacali
Editorial Director, Mathematics
SpringBoard

Judy Windle
Sr. Mathematics Instructional Specialist
SpringBoard

John Nelson
Editor, SpringBoard

Alex Chavarry
Sr. Director, SpringBoard Strategic Accounts

Acknowledgments continued

Research and Planning Advisors

We also wish to thank the members of our SpringBoard Advisory Council and the many educators who gave generously of their time and their ideas as we conducted research for both the print and online programs. Your suggestions and reactions to ideas helped immeasurably as we planned the revisions. We gratefully acknowledge the teachers and administrators in the following districts.

ABC Unified
Cerritos, California

Albuquerque Public Schools
Albuquerque, New Mexico

Amarillo School District
Amarillo, Texas

Baltimore County Public Schools
Baltimore, Maryland

Bellevue School District 405
Bellevue, Washington

Charlotte Mecklenburg Schools
Charlotte, North Carolina

Clark County School District
Las Vegas, Nevada

Cypress Fairbanks ISD
Houston, Texas

District School Board of Collier County
Collier County, Florida

Denver Public Schools
Denver, Colorado

Frisco ISD
Frisco, Texas

Gilbert Unified School District
Gilbert, Arizona

Grand Prairie ISD
Grand Prairie, Texas

Hillsborough County Public Schools
Tampa, Florida

Houston Independent School District
Houston, Texas

Hobbs Municipal Schools
Hobbs, New Mexico

Irving Independent School District
Irving, Texas

Kenton County School District
Fort Wright, Kentucky

Lee County Public Schools
Fort Myers, Florida

Newton County Schools
Covington, Georgia

Noblesville Schools
Noblesville, Indiana

Oakland Unified School District
Oakland, California

Orange County Public Schools
Orlando, Florida

School District of Palm Beach County
Palm Beach, Florida

Peninsula School District
Gig Harbor, Washington

Polk County Public Schools
Bartow, Florida

Quakertown Community School District
Quakertown, Pennsylvania

Rio Rancho Public Schools
Rio Rancho, New Mexico

Ronan School District
Ronan, Montana

St. Vrain Valley School District
Longmont, Colorado

Scottsdale Public Schools
Phoenix, Arizona

Seminole County Public Schools
Sanford, Florida

Southwest ISD
San Antonio, Texas

Spokane Public Schools
Spokane, Washington

Volusia County Schools
DeLand, Florida

Contents

To the Teacher ... xi

Instructional Units

UNIT 1   EQUATIONS AND INEQUALITIES
Planning the Unit ... 1a
Unit 1 Overview ... 1
Getting Ready ... 2

Activity 1   Investigating Patterns—*Cross-Country Adventures* ... **3**
Lesson 1-1 Numeric and Graphic Representations of Data ... 3
Lesson 1-2 Writing Expressions ... 8
Activity 1 Practice ... 14

Activity 2   Solving Equations—*What's My Number?* ... **15**
Lesson 2-1 Writing and Solving Equations ... 15
Lesson 2-2 Equations with Variables on Both Sides ... 19
Lesson 2-3 Solving More Complex Equations ... 22
Lesson 2-4 Equations with No Solution or Infinitely Many Solutions ... 25
Lesson 2-5 Solving Literal Equations for a Variable ... 28
Activity 2 Practice ... 31

Embedded Assessment 1   Patterns and Equations—*Of Music and Money* ... 33

Activity 3   Solving Inequalities—*Physical Fitness Zones* ... **35**
Lesson 3-1 Inequalities and Their Solutions ... 35
Lesson 3-2 Solving Inequalities ... 38
Lesson 3-3 Compound Inequalities ... 43
Activity 3 Practice ... 47

Activity 4   Absolute Value Equations and Inequalities—*Student Distances* ... **49**
Lesson 4-1 Absolute Value Equations ... 49
Lesson 4-2 Absolute Value Inequalities ... 54
Activity 4 Practice ... 59

Embedded Assessment 2   Inequalities and Absolute Value—*Diet and Exercise* ... 61

UNIT 2   FUNCTIONS
Planning the Unit ... 63a
Unit 2 Overview ... 63
Getting Ready ... 64

Activity 5   Functions and Function Notation—*Vending Machines* ... **65**
Lesson 5-1 Relations and Functions ... 65
Lesson 5-2 Domain and Range ... 71
Lesson 5-3 Function Notation ... 76
Activity 5 Practice ... 79

Activity 6   Interpreting Graphs of Functions—*Shake, Rattle, and Roll* ... **81**
Lesson 6-1 Key Features of Graphs ... 81
Lesson 6-2 More Complex Graphs ... 87
Lesson 6-3 Graphs of Real-World Situations ... 92
Activity 6 Practice ... 95

Contents continued

Activity 7 **Graphs of Functions—*Experiment Experiences*** **97**
    Lesson 7-1 The Spring Experiment 97
    Lesson 7-2 The Falling Object Experiment 101
    Lesson 7-3 The Radioactive Decay Experiment 105
    Activity 7 Practice 109

Activity 8 **Transformations of Functions—*Transformers*** **111**
    Lesson 8-1 Exploring $f(x) + k$ 111
    Lesson 8-2 Exploring $f(x + k)$ 115
    Activity 8 Practice 119

Embedded Assessment 1 Representations of Functions—*Bryce Canyon Hiking* 121

Activity 9 **Rates of Change—*Ramp It Up*** **123**
    Lesson 9-1 Slope 123
    Lesson 9-2 Slope and Rate of Change 128
    Lesson 9-3 More About Slopes 133
    Activity 9 Practice 137

Activity 10 **Linear Models—*Stacking Boxes*** **139**
    Lesson 10-1 Direct Variation 139
    Lesson 10-2 Indirect Variation 144
    Lesson 10-3 Another Linear Model 148
    Lesson 10-4 Inverse Functions 152
    Activity 10 Practice 157

Activity 11 **Arithmetic Sequences—*Picky Patterns*** **159**
    Lesson 11-1 Identifying Arithmetic Sequences 159
    Lesson 11-2 A Formula for Arithmetic Sequences 162
    Lesson 11-3 Arithmetic Sequences as Functions 166
    Lesson 11-4 Recursive Formulas 168
    Activity 11 Practice 171

Embedded Assessment 2 Linear Functions and Equations—*Text Message Plans* 173

Activity 12 **Forms of Linear Functions—*Under Pressure*** **175**
    Lesson 12-1 Slope-Intercept Form 175
    Lesson 12-2 Point-Slope Form 179
    Lesson 12-3 Standard Form 183
    Lesson 12-4 Slopes of Parallel and Perpendicular Lines 187
    Activity 12 Practice 191

Activity 13 **Equations from Data—*Pass the Book*** **193**
    Lesson 13-1 Scatter Plots and Trend Lines 193
    Lesson 13-2 Linear Regression 197
    Lesson 13-3 Quadratic and Exponential Regressions 200
    Activity 13 Practice 205

Embedded Assessment 3 Linear Models and Slope as Rate of Change—
A 10K Run 207

UNIT 3 EXTENSIONS OF LINEAR CONCEPTS

Planning the Unit — 209a
Unit 3 Overview — 209
Getting Ready — 210

Activity 14 Piecewise-Defined Linear Functions—*Breakfast for Bowser* — 211
Lesson 14-1 Function Notation and Rate of Change — 211
Lesson 14-2 Writing Functions and Finding Domain and Range — 215
Lesson 14-3 Evaluating Functions and Graphing Piecewise-Defined Linear Functions — 219
Lesson 14-4 Comparing Functions — 221
Activity 14 Practice — 225

Activity 15 Comparing Equations—*A Tale of a Trucker* — 227
Lesson 15-1 Writing Equations from Graphs and Tables — 227
Lesson 15-2 Comparing Functions with Inequalities — 231
Lesson 15-3 Writing Equations from Verbal Descriptions — 235
Activity 15 Practice — 237

Activity 16 Inequalities in Two Variables—*Shared Storage* — 239
Lesson 16-1 Writing and Graphing Inequalities in Two Variables — 239
Lesson 16-2 Graphing Inequalities in Two Variables — 242
Activity 16 Practice — 247

Embedded Assessment 1 Graphing Inequalities and Piecewise-Defined Functions—*Earnings on a Graph* — 249

Activity 17 Solving Systems of Linear Equations—*A Tale of Two Truckers* — 251
Lesson 17-1 The Graphing Method — 251
Lesson 17-2 Using Tables and the Substitution Method — 256
Lesson 17-3 The Elimination Method — 261
Lesson 17-4 Systems Without a Unique Solution — 264
Lesson 17-5 Classifying Systems of Equations — 267
Activity 17 Practice — 271

Activity 18 Solving Systems of Linear Inequalities—*Which Region Is It?* — 273
Lesson 18-1 Representing the Solution of a System of Inequalities — 273
Lesson 18-2 Interpreting the Solution of a System of Inequalities — 278
Activity 18 Practice — 281

Embedded Assessment 2 Systems of Equations and Inequalities—*Tilt the Scales* — 283

UNIT 4 EXPONENTS, RADICALS, AND POLYNOMIALS

Planning the Unit — 285a
Unit 4 Overview — 285
Getting Ready — 286

Activity 19 Exponent Rules—*Icebergs and Exponents* — 287
Lesson 19-1 Basic Exponent Properties — 287
Lesson 19-2 Negative and Zero Powers — 291
Lesson 19-3 Additional Properties of Exponents — 294
Activity 19 Practice — 297

Contents continued

Activity 20 Operations with Radicals—*Go Fly a Kite* 299
    Lesson 20-1 Radical Expressions 299
    Lesson 20-2 Adding and Subtracting Radical Expressions 304
    Lesson 20-3 Multiplying and Dividing Radical Expressions 307
    Activity 20 Practice 311

Activity 21 Geometric Sequences—*Go Viral!* 313
    Lesson 21-1 Identifying Geometric Sequences 313
    Lesson 21-2 Formulas for Geometric Sequences 316
    Activity 21 Practice 321

Embedded Assessment 1 Exponents, Radicals, and Geometric Sequences—*Taking Stock* 323

Activity 22 Exponential Functions—*Protecting Your Investment* 325
    Lesson 22-1 Exponential Functions and Exponential Growth 325
    Lesson 22-2 Exponential Decay 329
    Lesson 22-3 Graphs of Exponential Functions 333
    Activity 22 Practice 339

Activity 23 Modeling with Exponential Functions—*Growing, Growing, Gone* 341
    Lesson 23-1 Compound Interest 341
    Lesson 23-2 Population Growth 347
    Activity 23 Practice 351

Embedded Assessment 2 Exponential Functions—*Family Bonds* 353

Activity 24 Adding and Subtracting Polynomials—*Polynomials in the Sun* 355
    Lesson 24-1 Polynomial Terminology 355
    Lesson 24-2 Adding Polynomials 359
    Lesson 24-3 Subtracting Polynomials 364
    Activity 24 Practice 367

Activity 25 Multiplying Polynomials—*Tri-Com Computers* 369
    Lesson 25-1 Multiplying Binomials 369
    Lesson 25-2 Special Products of Binomials 376
    Lesson 25-3 Multiplying Polynomials 379
    Activity 25 Practice 381

Embedded Assessment 3 Polynomial Operations—*Measuring Up* 383

Activity 26 Factoring—*Factors of Construction* 385
    Lesson 26-1 Factoring by Greatest Common Factor (GCF) 385
    Lesson 26-2 Factoring Special Products 388
    Activity 26 Practice 391

Activity 27 Factoring Trinomials—*Deconstructing Floor Plans* 393
    Lesson 27-1 Factoring $x^2 + bx + c$ 393
    Lesson 27-2 Factoring $ax^2 + bx + c$ 398
    Activity 27 Practice 401

Activity 28 Simplifying Rational Expressions—*Totally Rational* 403
    Lesson 28-1 Simplifying Rational Expressions 403

	Lesson 28-2 Dividing Polynomials	406
	Lesson 28-3 Multiplying and Dividing Rational Expressions	411
	Lesson 28-4 Adding and Subtracting Rational Expressions	413
	Activity 28 Practice	417
Embedded Assessment 4	Factoring and Simplifying Rational Expressions—*Rock Star Demands*	419

UNIT 5 QUADRATIC FUNCTIONS

	Planning the Unit	421a
	Unit 5 Overview	421
	Getting Ready	422
Activity 29	**Introduction to Quadratic Functions—*Touchlines***	**423**
	Lesson 29-1 Modeling with a Quadratic Function	423
	Lesson 29-2 Graphing and Analyzing a Quadratic Function	427
	Activity 29 Practice	431
Activity 30	**Graphing Quadratic Functions—*Transformers***	**433**
	Lesson 30-1 Translations of the Quadratic Parent Function	433
	Lesson 30-2 Stretching and Shrinking the Quadratic Parent Function	440
	Lesson 30-3 Multiple Transformations of the Quadratic Parent Function	444
	Activity 30 Practice	451
Embedded Assessment 1	Graphing Quadratic Functions—*Parabolic Paths*	453
Activity 31	**Solving Quadratic Equations by Graphing and Factoring—*Trebuchet Trials***	**455**
	Lesson 31-1 Solving by Graphing or Factoring	455
	Lesson 31-2 The Axis of Symmetry and the Vertex	459
	Lesson 31-3 Graphing a Quadratic Function	462
	Activity 31 Practice	465
Activity 32	**Algebraic Methods of Solving Quadratic Equations—*Keeping It Quadratic***	**467**
	Lesson 32-1 The Square Root Method	467
	Lesson 32-2 Completing the Square	471
	Lesson 32-3 The Quadratic Formula	474
	Lesson 32-4 Choosing a Method and Using the Discriminant	477
	Lesson 32-5 Complex Solutions	480
	Activity 32 Practice	483
Activity 33	**Applying Quadratic Equations—*Rockets in Flight***	**485**
	Lesson 33-1 Fitting Data with a Quadratic Function	485
	Lesson 33-2 Interpreting Solutions of Quadratic Equations	488
	Activity 33 Practice	491
Embedded Assessment 2	Solving Quadratic Equations—*Egg Drop*	493
Activity 34	**Modeling with Functions—*Photo App***	**495**
	Lesson 34-1 Constructing Models	495
	Lesson 34-2 Comparing Models	500

Contents continued

	Lesson 34-3 Extending Models	503
	Activity 34 Practice	507
Activity 35	**Systems of Equations**—*Population Explosion*	**509**
	Lesson 35-1 Solving a System Graphically	509
	Lesson 35-2 Solving a System Algebraically	513
	Activity 35 Practice	517
Embedded Assessment 3	Solving Systems of Equations—*Sports Collector*	519

UNIT 6   PROBABILITY AND STATISTICS

	Planning the Unit	521a
	Unit 6 Overview	521
	Getting Ready	522
Activity 36	**Measures of Center and Spread**—*To Text, or Not to Text*	**523**
	Lesson 36-1 Mean, Median, Mode, and MAD	523
	Lesson 36-2 Another Measure of Variability	532
	Activity 36 Practice	536
Activity 37	**Dot and Box Plots and the Normal Distribution**—*Disturbing Coyotes*	**537**
	Lesson 37-1 Dot Plots and Box Plots	537
	Lesson 37-2 Modified Box Plots	543
	Lesson 37-3 Normally Distributed	548
	Activity 37 Practice	554
Embedded Assessment 1	Comparing Univariate Distributions—*Splitting the Bill*	557
Activity 38	**Correlation**—*What's the Relationship?*	**559**
	Lesson 38-1 Scatter Plots	559
	Lesson 38-2 Correlation Coefficient	564
	Activity 38 Practice	569
Activity 39	**The Best-Fit Line**—*Regressing Linearly*	**571**
	Lesson 39-1 Line of Best Fit	571
	Lesson 39-2 Residuals	577
	Lesson 39-3 Interpreting the Slope and Intercept of the Best-Fit Line	582
	Lesson 39-4 Plotting Residuals	588
	Activity 39 Practice	594
Activity 40	**Bivariate Data**—*Categorically Speaking*	**595**
	Lesson 40-1 Bivariate Categorical Data	595
	Lesson 40-2 Presenting Relative Frequency Data Graphically	600
	Activity 40 Practice	607
Embedded Assessment 2	Bivariate Distributions—*Dear Traveling Tooth*	609

RESOURCES   613

Formulas	614
Learning Strategies	618
Glossary	621
Academic Vocabulary Graphic Organizers	626

To the Teacher

Welcome to *SpringBoard Mathematics*, a highly engaging, student-centered instructional program. This revised edition of SpringBoard is based on the standards defined by the **College and Career Readiness Standards for Mathematics** for each course. The program may be used as a core curriculum that will provide the instructional content that students need to be prepared for future mathematical courses.

Shifts in Mathematics Instruction

With an increased emphasis on better preparing students to understand and master mathematical concepts, mathematics instruction has become a major focus of attention. Efforts at improvement center around the following points:

Greater Focus on the Content of the Standards:

- Learn more about less by spending more time on fewer concepts.
- Focus on the essential learning that helps students **develop strong foundational knowledge** and deep conceptual understanding.

Coherence to Link Major Topics:

- Connect learning within a grade and build knowledge across grades.
- Focus on learning progressions so that teachers can continue counting on students' deep conceptual understanding of core content and build on it.

Rigor with Balance:

- Develop fluency in procedural skills.
- Promote depth and mastery by connecting concepts, practice, and independent application.
- Learn and apply the mathematical practices.

College and Career Readiness

The goal of this increased focus on standards and mathematical practices is, of course, helping students be prepared for the expectations of either college or a career, or both. Students who are prepared for college or career will be able to:

- **Build on content knowledge:** Students will have a base knowledge of math concepts on which to extend their learning.
- **Use mathematical models:** Students will be able to use a variety of mathematical representations to model what they know and to justify how they are using their knowledge.
- **Communicate mathematics:** Students will communicate verbally and in writing to explain their discoveries and understanding of mathematics and how it works theoretically and in the real world.
- **Collaborate with others:** Students will participate in discourse focused on discovery and problem solving, evaluate the contributions of others, and collaborate to present and defend viable solutions.
- **Use technology:** Students will use appropriate technology to enhance their understanding of mathematics and to gain greater precision in areas where technology is appropriate.

To the Teacher continued

The implications of these student expectations are that students will need to develop greater depth of knowledge, higher-level thinking skills, and effective communication skills. What they need less of will be memorization, drills and worksheets, and "one size fits all" content.

SpringBoard's Role in Preparing Students for College and Career Success

Based on the College and Career Readiness Standards for Mathematics and current research on best instructional practices, SpringBoard uses a "back-mapping" instructional design that starts with the end in mind, namely, the skills and knowledge students need to use mathematics effectively and to demonstrate that ability through performance on various assessments.

The mathematics instruction follows a balanced approach in which concepts are presented based on the most effective instructional methods: *directed* for basic mathematics principles, including examples and practice; *guided* for concepts that need a combination of direct instruction and investigatory learning; and *investigative* activities that allow students to explore and discover mathematics concepts through a contextual setting.

Organization of the Content

Instructional content is organized into coherent units of mathematical concepts. Each unit contains multiple activities that are divided into shorter lessons. The units are structured as follows:

- Unit opener content sets the stage for what students will learn in the unit.
- Getting Ready helps teachers assess students' current skills and knowledge to determine whether they have the basic knowledge on which to build new content presented in the unit.
- Multiple lessons per activity.
- Worked-out examples as needed to help students learn and apply concepts.
- Frequent Check Your Understanding questions to help students assimilate and apply knowledge.
- Mathematical practices called out so students are reminded to apply them as they respond to problems and applications.
- Lesson Practice problems to provide the opportunity to practice new learning and to build fluency.
- Activity Practice provides additional problems for each lesson.
- Embedded Assessments give students new contexts for applying the concepts learned in the unit and give you the opportunity for regular formative assessment.

Integration of Mathematical Practices

With its process of questioning students within a lesson and asking them to think through concepts and applications, SpringBoard reinforces the actions and practices that help students build knowledge and skills. SpringBoard requires students to:

- Make sense of and connect mathematics concepts to everyday life.
- Model with mathematics to solve problems, justify solutions and their reasonableness, and communicate mathematical ideas.
- Use appropriate tools, such as number lines, protractors, technology, or paper and pencil, strategically to help solve problems.
- Communicate abstract and quantitative reasoning both orally and in writing through viable arguments and critiques.
- Analyze mathematical relationships through structure and repeated reasoning to connect ideas.
- Attend to precision in both written and oral communication.

Engaging and Interactive Online Edition

With this new edition, SpringBoard introduces an all-new interactive online experience for both students and teachers. In addition to providing all content online, the new SpringBoard Digital program:

- Allows access at any time.
- Discerns the device you're using and adjusts content to fit screens—from desktops to laptops to tablets.
- Provides exciting tools such as text marking (highlight, underline, circle, and so on), online calculators, graphing and equation tools, and handwriting recognition for entering equations, note-taking, and uploading of student papers.
- Allows teachers to edit teaching commentary, personalizing by adding their own notes and comments about lessons.
- Includes online student and teacher resources such as graphic organizers, blackline masters, mini-lessons, and other content to support instruction.

New Assessment Options

The SpringBoard program now provides the option of using the ExamView test generator program for all grades. Teachers will have multiple options for choosing pre-made tests or making their own. Options include:

- Unit tests aligned to standards and the content in each unit.
- Test banks allowing teachers to choose items and create tests for multiple needs, including benchmark tests and quarter or semester tests.
- Expanded test item types, including short response and interactive simulations and manipulatives.

What Sets SpringBoard Apart from Other Mathematics Programs?

Three key things set SpringBoard apart:

1. The expectation that students can do rigorous work with the **right** preparation.
2. Learning materials that reflect both **rigor** and the **expectations** about what students should know and be able to do.
3. Extensive teacher support through **professional development** and coaching services.

To the Teacher continued

Unique features of SpringBoard include:

- **Rigorous, standards-based instruction:** Instructional content organized around the College and Career Readiness Standards for Mathematics to provide coherent topics that build knowledge and skills throughout each course and across grade levels.
- **Mathematical practices:** Integration of the Standards for Mathematical Practice that support student learning and higher level thinking.
- **Research-based instruction:** Back-mapped instructional design gives students a learning target and scaffolds activities to develop students' knowledge and skills and prepare them to demonstrate their learning on an Embedded Assessment.
- **Student-centered, interactive, collaborative activities and lessons:** Each course is organized into short, interactive activities that are further divided into focused lessons. Lessons engage students and aid learning by having students participate in class discussions, solve problems and justify solutions, and demonstrate learning through multiple means of evaluation.
- **Integrated teaching and learning strategies:** Suggested Learning Strategies in each lesson help students use methodical approaches to learning new content, helping students take control of their own learning by identifying which strategies work best for them. Teachers also use strategies for instruction that demands a reflective and metacognitive approach to teaching and learning.
- **Embedded Differentiated Instruction:** Scaffolded activity design supports best practices for special populations and English Language Learners. Additional support for these learners is available at point of use within lessons.
- **Assessment for learning:** Multiple assessment opportunities provide a formative look at students' knowledge and skills: before starting a unit of instruction to assess prerequisite knowledge (Getting Ready), during instruction to monitor understanding (Check Your Understanding), and after instruction to evaluate knowledge of concepts and how to apply them in a variety of situations (Practice, Embedded Assessments, Unit Tests).
- **Professional development:** Unparalleled professional development builds teacher capacity to deliver challenging curriculum to meet the needs of all students while honoring the creativity and intelligence teachers bring to the classroom. Face-to-face training is supported by an online system featuring resources that include an interactive professional learning *Community* that allows peer-to-peer sharing and sustains successful teaching.

The Pathway to Advanced Placement and College Readiness

SpringBoard provides a comprehensive and systematic approach to preparing ALL students for the demands of rigorous AP courses, college classes, and other post-secondary experiences. SpringBoard prepares students through sequential, scaffolded development of the prerequisite skills and knowledge needed for success in AP Calculus and Statistics. In each unit of study, explicit AP Connections are outlined in the Planning the Unit pages of the teacher editions and are reinforced as they appear in student activities.

From Pre-AP to AP and Beyond

Beginning in middle school, students are introduced to concepts and skills that are fundamental to success in AP mathematics and statistics courses.

Grade 6 students learn to:

- Model functions in numerical, symbolic (equation), table, and graphical forms.
- Communicate mathematics in writing and verbally, justifying answers and clearly labeling charts and graphs.
- Explore and represent data in a variety of forms.
- Use multiple representations to communicate their mathematical understanding.

Grade 7 students continue to:

- Acquire an algebraic and graphical understanding of functions.
- Write, solve, and graph linear equations; recognize and verbalize patterns; and model slope as a rate of change.
- Communicate clearly to explain methods of problem solving and to interpret results.
- Investigate concepts presented visually and verbally.

Grade 8 students extend their knowledge by:

- Writing algebraic models from a variety of physical, numeric, and verbal descriptions.
- Solving equations using a variety of methods.
- Justifying answers using precise mathematical language.
- Relating constant rate of change to verbal, physical, and algebraic models.
- Using technology to solve problems.
- Reinforcing and extending the vocabulary of probability and statistics.

Algebra 1 students:

- Gain an understanding of the properties of real numbers.
- Formalize the language of functions.
- Explore the behavior of functions numerically, graphically, analytically, and verbally.
- Use technology to discover relationships, test conjectures, and solve problems.
- Write expressions, equations, and inequalities from physical models.
- Communicate mathematics understanding formally and informally.

Geometry students:

- Read, analyze, and solve right triangle and trigonometric functions within contextual situations.
- Develop area formulas necessary for determining volumes of rotational solids, solids with known cross sections, and area beneath a curve.
- Explain work clearly so that the reasoning process can be followed throughout the solution.

To the Teacher continued

Algebra 2 students:

- Develop the algebra of functions.
- Read and analyze contextual situations involving exponential and logarithmic functions.
- Work with functions graphically, numerically, analytically, and verbally.
- Learn optimization problems.
- Compare the relative rate of change of linear and exponential functions.
- Learn the concept of infinite sum as a limit of partial sums.
- Work with statistics in numerical summaries, calculations using the normal curve, and the modeling of data.

Precalculus students:

- Gain an introductory understanding of convergence and divergence.
- Collect, analyze, and draw conclusions from data.
- Solve problems in contextual situations dealing with polynomial, rational, logarithmic, and trigonometric functions.
- Model motion using parametric equations and vectors.
- Develop an intuitive understanding of limits and continuity.

The SpringBoard Mathematics Classroom

A SpringBoard classroom is an environment that supports high expectations for all students.

Collaborative Groups

The **student-centered classroom** capitalizes on collaboration. Collaborative groups provide a setting in which students feel safe to explore ideas and learn effective communication skills. Collaborative groups allow learning to be active as students engage in discussions, make conjectures, question, and discover new ideas as they fulfill tasks within the group.

Debriefing/Reflections

Frequently in a mathematics classroom, students and teachers should engage in **debriefings**. The purpose of debriefing is to allow students to reflect on their learning, correct misconceptions, identify thinking processes used during an activity, summarize information, and process what they have learned.

Interactive Word Wall

The class **Word Wall** facilitates vocabulary development and provides a reference during class and group discussions. Creating and maintaining a Word Wall is an ongoing activity. It should be an instructional tool, not just a display.

Math Notebook

Keeping a **Math Notebook** helps students learn and explore new vocabulary while also summarizing notes about math concepts and ideas. It is an intentional tool for students to expand their understanding of mathematics terms and concepts. The Math Notebook may be an online tool for students who have regular access to SpringBoard Mathematics Digital.

Unit 1 Planning the Unit

In this Unit, students recognize and generalize patterns using words, tables, expressions, and graphs. Students will also generate rules for solving simple linear equations and inequalities, as well as absolute value equations and inequalities.

Vocabulary Development

The key terms for this unit can be found on the Unit Opener page. These terms are divided into Academic Vocabulary and Math Terms. Academic Vocabulary includes terms that have additional meaning outside of math. These terms are listed separately to help students transition from their current understanding of a term to its meaning as a mathematics term. To help students learn new vocabulary:

- Have students discuss meaning and use graphic organizers to record their understanding of new words.
- Remind students to place their graphic organizers in their math notebooks and revisit their notes as their understanding of vocabulary grows.
- As needed, pronounce new words and place pronunciation guides and definitions on the class Word Wall.

Embedded Assessments

Embedded Assessments allow students to do the following:

- Demonstrate their understanding of new concepts.
- Integrate previous and new knowledge by solving real-world problems presented in new settings.

They also provide formative information to help you adjust instruction to meet your students' learning needs.

Prior to beginning instruction, have students unpack the first Embedded Assessment in the unit to identify the skills and knowledge necessary for successful completion of that assessment. Help students create a visual display of the unpacked assessment and post it in your class. As students learn new knowledge and skills, remind them that they will be expected to apply that knowledge to the assessment. After students complete each Embedded Assessment, turn to the next one in the unit and repeat the process of unpacking that assessment with students.

CollegeBoard

AP / College Readiness

Unit 1 develops concepts that engender a solid algebraic foundation by:

- Developing pattern recognition necessary for success in AP Statistics.
- Providing a constructivist approach to solving equations, enabling students to compare and evaluate multiple methods of solution.
- Allowing for an intuitive understanding of absolute value and methods for solving equations and inequalities involving absolute value.

Unpacking the Embedded Assessments

The following are the key skills and knowledge students will need to know for each assessment.

Embedded Assessment 1

Patterns and Equations, *Of Music and Money*

- Identify patterns
- Model patterns with expressions
- Use patterns to make predictions
- Write, solve, and interpret multi-step equations
- Solve literal equations for a variable

Embedded Assessment 2

Inequalities and Absolute Value, *Diet and Exercise*

- Write, solve, and graph inequalities
- Write and graph compound inequalities
- Solve and graph absolute value inequalities

Planning the Unit continued

Suggested Pacing

The following table provides suggestions for pacing using a 45-minute class period. Space is left for you to write your own pacing guidelines based on your experiences in using the materials.

	45-Minute Period	Your Comments on Pacing
Unit Overview/Getting Ready	1	
Activity 1	2	
Activity 2	5	
Embedded Assessment 1	1	
Activity 3	3	
Activity 4	2	
Embedded Assessment 2	1	
Total 45-Minute Periods	**15**	

Additional Resources

Additional resources that you may find helpful for your instruction include the following, which may be found in the eBook Teacher Resources.

- Unit Practice (additional problems for each activity)
- Getting Ready Practice (additional lessons and practice problems for the prerequisite skills)
- Mini-Lessons (instructional support for concepts related to lesson content)

Equations and Inequalities

Unit Overview
Investigating patterns is a good foundation for studying Algebra 1. You will begin this unit by analyzing, describing, and generalizing patterns using tables, expressions, graphs, and words. You will then write and solve equations and inequalities in mathematical and real-world problems.

Key Terms
As you study this unit, add these and other terms to your math notebook. Include in your notes your prior knowledge of each word, as well as your experiences in using the word in different mathematical examples. If needed, ask for help in pronouncing new words and add information on pronunciation to your math notebook. It is important that you learn new terms and use them correctly in your class discussions and in your problem solutions.

Academic Vocabulary
- consecutive

Math Terms
- sequence
- common difference
- expression
- variable
- equilateral
- equation
- solution
- formula
- literal equation
- graph of an inequality
- solution of an inequality
- compound inequality
- conjunction
- disjunction
- absolute value
- absolute value notation
- absolute value equation
- absolute value inequality

ESSENTIAL QUESTIONS

How can you represent patterns from everyday life by using tables, expressions, and graphs?

How can you write and solve equations and inequalities?

EMBEDDED ASSESSMENTS
These assessments, following Activities 2 and 4, will give you an opportunity to demonstrate what you have learned about patterns, equations, and inequalities.

Embedded Assessment 1:
Patterns and Equations p. 33

Embedded Assessment 2:
Inequalities and Absolute Value p. 61

Unit Overview
Ask students to read the unit overview and mark the text to identify key phrases that indicate what they will learn in this unit.

Key Terms
As students encounter new terms in this unit, help them to choose an appropriate graphic organizer for their word study. As they complete a graphic organizer, have them place it in their math notebooks and revisit as needed as they gain additional knowledge about each word or concept.

Essential Questions
Read the essential questions with students, and ask them to share possible answers. As students complete the unit, revisit the essential questions to help them adjust their initial answers as needed.

Unpacking Embedded Assessments
Prior to beginning the first activity in this unit, turn to Embedded Assessment 1 and have students unpack the assessment by identifying the skills and knowledge they will need to complete the assessments successfully. Guide students through a close reading of the assignment, and use a graphic organizer or other means to capture their identification of the skills and knowledge. Repeat the process for each Embedded Assessment in the unit.

Developing Math Language
As this unit progresses, help students make the transition from general words they may already know (the Academic Vocabulary) to the meanings of those words in mathematics. You may want students to work in pairs or small groups to facilitate discussion and to build confidence and fluency as they internalize new language. Ask students to discuss new academic and mathematics terms as they are introduced, identifying meaning as well as pronunciation and common usage. Remind students to use their math notebooks to record their understanding of new terms and concepts.

As needed, pronounce new terms clearly and monitor students' use of words in their discussions to ensure that they are using terms correctly. Encourage students to practice fluency with new words as they gain greater understanding of mathematical and other terms.

UNIT 1
Getting Ready

Use some or all of these exercises for formative evaluation of students' readiness for Unit 1 topics.

Prerequisite Skills

- Perform operations with fractions (Items 1, 2) 7.NS.A.1b
- Understand exponents (Item 3) 8.EE.A.1
- Perform operations with mixed numbers (Item 4) 7.NS.A.1b
- Compare and perform operations with integers (Items 5, 6) 7.NS.A.3
- Perform operations with decimals (Item 7) 6.NS.B.3
- Solve one-step equations (Item 8) 8.EE.C.7
- Simplify expressions by combining like terms (Item 9) 7.EE.A.1
- Interpret Venn diagrams (Item 10) 6.SP.B.5

Answer Key

1. $1\frac{7}{15}$
2. The fractions must have the same denominator.
3. Megan is correct.
   $4^2 \times 2^2 = (4 \times 4) \times (2 \times 2) = 16 \times 4 = 64$
4. $\frac{3}{4}$ ft or 9 in.
5. c (-10), a (-2), b (2)
6. $-8 + 3$ has a greater value.
   $-8 + 3 = -5$
   $-8 \times 3 = -24$
   $-5 > -24$
7. All three expressions are equal to 14.95.
8. d
9. a. $6x$
   b. $-12n$
   c. $9.1y - 2$
   d. $-m + 4$, or $4 - m$
10. Student B plays the piano and is in the band, because B is in the overlapping region for piano and band. Student G is in the band only, because G is in the circle for band but not in the region that overlaps the circle for piano.

UNIT 1
Getting Ready

Write your answers on notebook paper. Show your work.

1. What is $\frac{2}{3} + \frac{4}{5}$?
2. What condition must be met before you can add or subtract fractions?
3. Jennifer is checking Megan's homework. They disagree on the answer to this problem: $4^2 \times 2^2$. Jennifer says the product is 32 and Megan says it is 64. Who has the correct answer? Explain how she arrived at that correct product.
4. A piece of lumber $2\frac{1}{4}$ feet long is to be cut into 3 equal pieces. How long will each piece of cut wood be? Give the measurement in feet and in inches.
5. Arrange the following expressions in order of their value from least to greatest.
   a. $4 - 6$
   b. $-4 + 6$
   c. $-4 - 6$
6. Which expression has the greater value? Justify your answer.
   A. $-8 + 3$
   B. -8×3
7. Which of the following are equal to 14.95?
   A. 2.3×6.5
   B. $21.45 - 6.5$
   C. $8.32 + 6.63$
8. Which equation has the least solution?
   A. $x + 5 = 13$
   B. $-6x = -30$
   C. $\frac{x}{4} = 18$
   D. $x - 2 = -11$
9. Combine like terms in the following expressions.
   a. $10x - 4x$
   b. $-15n + 3n$
   c. $7.5y + 1.6y - 2$
   d. $m + 4 - 2m$
10. The Venn diagram below provides a visual representation of the students in Mr. Griffin's class who participate in music programs after school. What does the diagram tell you about the musical involvement of Student B and Student G? Explain how you reached your conclusion.

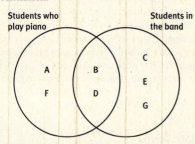

Getting Ready Practice

For students who may need additional instruction on one or more of the prerequisite skills for this unit, Getting Ready practice pages are available in the Teacher Resources at SpringBoard Digital. These practice pages include worked-out examples as well as multiple opportunities for students to apply concepts learned.

Investigating Patterns
Cross-Country Adventures
Lesson 1-1 Numeric and Graphic Representations of Data

ACTIVITY 1

Learning Targets:
- Identify patterns in data.
- Use tables, graphs, and expressions to model situations.
- Use expressions to make predictions.

SUGGESTED LEARNING STRATEGIES: Sharing and Responding, Create Representations, Discussion Groups, Look for a Pattern, Interactive Word Wall

Mizing spent his summer vacation traveling cross-country with his family. Their first stop was Yellowstone National Park in Wyoming and Montana. Yellowstone is famous for its geysers, especially one commonly referred to as Old Faithful. A geyser is a spring that erupts intermittently, forcing a fountain of water and steam from a hole in the ground. Old Faithful can have particularly long and fairly predictable eruptions. As a matter of fact, park rangers have observed the geyser over many years and have developed patterns they use to predict the timing of the next eruption.

Park rangers have recorded the information in the table below.

Length of Eruption (in minutes)	Approximate Time Until Next Eruption (in minutes)
1	46
2	58
3	70
4	82

CONNECT TO HISTORY
Yellowstone National Park was the first National Park. The park was established by Congress on March 1, 1872. President Woodrow Wilson signed the act creating the National Park Service on August 25, 1916.

1. Describe any patterns you see in the table.
   Answers will vary. The numbers in the second column increase by 12 minutes for each increase in one minute in the duration of an eruption.

2. Why might it be important for park rangers to be able to predict the timing of Old Faithful's eruptions?
   Answers will vary; crowd control or safety

DISCUSSION GROUP TIPS
Work with your peers to set rules for:
- discussions and decision-making
- clear goals and deadlines
- individual roles as needed

3. If an eruption lasts 8 minutes, about how long must park visitors wait to see the next eruption? Explain your reasoning using the patterns you identified in the table.
   130 minutes; explanations will vary.

Common Core State Standards for Activity 1

HSN-Q.A.1	Use units as a way to understand problems and to guide the solution of multi-step problems; choose and interpret units consistently in formulas; choose and interpret the scale and the origin in graphs and data displays.
HSN-Q.A.2	Define appropriate quantities for the purpose of descriptive modeling.
HSA-SSE.A.1a	Interpret parts of an expression, such as terms, factors, and coefficients.
HSF-BF.A.2	Write arithmetic and geometric sequences both recursively and with an explicit formula, use them to model situations, and translate between the two forms.*

ACTIVITY 1
Investigative

Activity Standards Focus
In previous grades, students began to develop algebraic thinking skills by writing numeric and algebraic expressions to represent situations and by using tables and graphs to examine the relationship between two quantities. In Activity 1, students build on these skills and concepts by using tables, graphs, and expressions to explore linear and nonlinear patterns in real-world situations. They also use patterns to make predictions.

Lesson 1-1

PLAN
Pacing: 1 class period
Chunking the Lesson
#1–3 #4–5 #6–7
#8–9 #10
Check Your Understanding
Lesson Practice

TEACH

Bell-Ringer Activity
Have students determine the relationship between the number of chairs in the classroom and the number of chair legs. Ask them to predict how the number of chair legs would change if 8 more chairs were brought into the room. Discuss with students how they made their predictions.

1–3 Look for a Pattern, Think-Pair-Share, Sharing and Responding
In this set of items, students identify patterns in a table and use the patterns to make predictions. As students investigate the patterns in the table, make sure they look for patterns both within each column and between the two columns. For example, most students should be able to see that the numbers in the first column increase by 1 and the numbers in the second column increase by 12. Together, these two patterns show that each increase of 1 minute in the length of the eruption results in an increase of 12 minutes in the approximate time until the next eruption. It is important for students to have an opportunity to share the patterns they recognize before moving on to the graph and the representation of the information in the table as a sequence later in the lesson. These three questions provide the teacher with valuable formative assessment information about the level of student thinking.

ACTIVITY 1 Continued

4–5 Look For A Pattern, Create Representations, Debriefing Due to the scale on the graph, students must approximate the correct location for each data point. Emphasize the importance of placing the points as accurately as possible so that it will be easier to see patterns in the data. In this case, students should see that the data points appear to lie on a line.

In Item 5, some students may extend the table to find the approximate eruption length. Others may draw a line through the points on the graph and extend it. Still other students may repeatedly add 12 minutes until they get a sum that is close to 120 minutes. It is important for the debrief of this portion of the lesson to validate each of the solution methods, and for students to recognize that any of these methods are appropriate.

TEACHER to TEACHER

Some students may not recognize that 2 hours needs to be converted to 120 minutes. Keep a watch out for this as you monitor group discussion.

Developing Math Language

Have students complete a graphic organizer like the one below to show their understanding of the word *expression*. Explain that characteristics are attributes of the word that distinguish it from related mathematical terms.

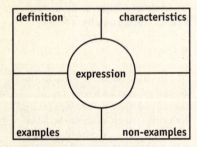

6–7 Look for a Pattern, Create Representations, Discussion Groups, Sharing and Responding In these items, students write the sequence that represents the approximate times until the next Old Faithful eruption and describe the pattern in the sequence. Attention should be drawn to the Math Terms box and the mathematical definition of a sequence, including the use of an ellipsis (. . .) to indicate that the pattern in a sequence continues. Students will describe the patterns they see in different ways. Encourage the proper use of mathematical vocabulary in their descriptions.

ACTIVITY 1 continued

Lesson 1-1
Numeric and Graphic Representations of Data

4. Graph the data from the table on the grid below.

5. **Reason quantitatively.** Mizing and his family arrived at Old Faithful to find a sign indicating they had just missed an eruption and that it would be approximately 2 hours before the next one. How long was the eruption they missed? Explain how you determined your answer.
   About 7 minutes; 2 hours is 120 minutes; 46 + 12 + 12 + 12 + 12 + 12 + 12 = 118 minutes. 118 minutes is very close to 120 minutes. Students may also extend the table from Item 1.

Patterns can be written as *sequences*.

MATH TERMS
A **sequence** is a list of numbers, and each number is called a *term* of the sequence. For example: 2, 4, 6, 8, … and 2, 5, 10, 17, … are sequences.

6. Using the table or graph above, write the approximate times until the next Old Faithful eruption as a sequence.
   46, 58, 70, 82, …

7. How would you describe this sequence of numbers?
   Answers will vary. It starts at 46 and then each subsequent term increases by 12.

---

4 SpringBoard® Mathematics **Algebra 1, Unit 1** • Equations and Inequalities

Lesson 1-1
Numeric and Graphic Representations of Data

ACTIVITY 1
continued

In the table below, 5 and 8 are *consecutive* terms. Some sequences have a *common difference* between consecutive terms. The common difference between the terms in the table below is 3.

Sequence: 5, 8, 11, 14…

Term number	Term
1	5
2	8
3	11
4	14

8. Identify two consecutive terms in the sequence of next eruption times that you created in Item 6.
   Answers will vary. 70 and 82

9. The sequence of next eruption times has a common difference. Identify the common difference.
   12 minutes

10. Each term in the sequence above can be written using the first term and repeated addition of the common difference. For example, the first term is 5, the second term is $5 + 3$, and the third term can be expressed as $5 + 3 + 3$ or $5 + 2(3)$. Similarly, the terms in the sequence of next eruption times can also be written using repeated addition of the common difference.

    a. Write the approximate waiting time for the next eruption after eruptions lasting 4 and 5 minutes using repeated addition of the common difference.
    4 minutes: $46 + 12 + 12 + 12$ or $46 + 12(3)$
    5 minutes: $46 + 12 + 12 + 12 + 12$ or $46 + 12(4)$

    b. **Model with mathematics.** Let n represent the number of minutes an eruption lasts. Write an *expression* using the *variable* n that could be used to determine the waiting time until the next eruption.
    $46 + 12(n - 1)$

    c. Check the accuracy of your expression by evaluating it when $n = 2$.
    58

    d. Use your expression to determine the number of minutes a visitor to the park must wait to see another eruption of Old Faithful after a 12-minute eruption.
    178 minutes

ACADEMIC VOCABULARY

Consecutive refers to items that follow each other in order.

MATH TIP

A common difference is also called a ***constant difference***.

MATH TERMS

An ***expression*** may consist of numbers, variables, and operations. A ***variable*** is a letter or symbol used to represent an unknown quantity.

ACTIVITY 1 Continued

8–9 Interactive Word Wall, Discussion Groups, Sharing and Responding The focus of these two items is on the understanding of the vocabulary associated with sequences. Student answers will vary in Item 8 so allowing groups to share out will show students that no matter which two consecutive terms are used, the constant difference remains the same.

10 Create Representations, Predict and Confirm, Discussion Groups, Debriefing Item 10 includes the first use of variables in this course. Students use repeated addition to facilitate the transition from the pattern they have observed in the table and graph to an algebraic representation. In this way, students should recognize that the number multiplied by 12 is one minute less than the eruption time and represent this as $(n - 1)$. Students then confirm the validity of their expression by substituting a value for the variable and checking for the correct output based on data given in the problem. Students can use this method of checking the accuracy of expressions throughout this course. The whole group debriefing should focus on the idea of the use of variables to represent quantities and the appropriate use of the variable n and the expression $n - 1$ in this situation. Students should be reminded to define variables no matter the situation.

Differentiating Instruction

Support students who need to be reminded of the order of operations by reviewing the order:
1. Parentheses
2. Exponents
3. Multiply and divide in order from left to right.
4. Add and subtract in order from left to right.

Some students forget "in order from left to right" in steps 3 and 4.

Activity 1 • Investigating Patterns 5

ACTIVITY 1 Continued

Check Your Understanding

Debrief students' answers to these items to ensure that they understand how to represent the relationship between two quantities using tables, graphs, and expressions. The items also assess whether students can write a sequence and determine its common difference.

Answers

11.

Number of Gigabytes Used	Total Data Charge
1	$20
2	$40
3	$60
4	$80
5	$100

12.

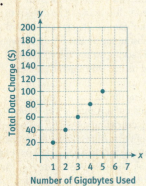

13. 20, 40, 60, 80, …
14. $20
15. $20 + 20(n - 1)$ or $20n$
16. $200

Lesson 1-1
Numeric and Graphic Representations of Data

Check Your Understanding

SB-Mobile charges $20 for each gigabyte of data used on any of its smartphone plans.

11. Copy and complete the table showing the charges for data based on the number of gigabytes used.

Number of Gigabytes Used	Total Data Charge
1	
2	
3	
4	
5	

12. Graph the data from the table. Be sure to label your axes.
13. Write a sequence to represent the total price of a data plan.
14. The sequence you wrote in Item 13 has a common difference. Identify the common difference.
15. Let n represent the number of gigabytes used. Write an expression that can be used to determine the total data charge for the phone plan.
16. Use your expression to calculate the total data charge if 10 gigabytes of data are used.

Lesson 1-1
Numeric and Graphic Representations of Data

LESSON 1–1 PRACTICE

Travis owns stock in the SBO Company. After the first year of ownership the stock is worth $45 per share. Travis estimates that the value of a share will increase by $2.80 per year.

17. Copy and complete the table showing the value of the stock over the course of several years.

Year	Share Value
1	$45
2	
3	
4	
5	

18. Write a sequence to show the increase in the stock value over the course of several years.
19. **Make use of structure.** The sequence you wrote in Item 18 has a common difference. Identify the common difference.
20. Let n represent the number of years that have passed. Write an expression that can be used to determine the value of one share of SBO stock.
21. Use your expression to calculate the value of one share of stock after 20 years.

ACTIVITY 1 Continued

ASSESS

Students' answers to lesson practice problems will provide the teacher with a formative assessment of student understanding of the lesson concepts and their ability to apply their learning.

See the Activity Practice for additional problems for this lesson. You may assign the problems here or use them as a culmination for the activity.

LESSON 1-1 PRACTICE

17.

Year	Share Value
1	$45
2	$47.80
3	$50.60
4	$53.40
5	$56.20

18. $45, $47.80, $50.60, $53.40, $56.20
19. $2.80
20. $45 + 2.8(n - 1)$
21. $98.20

ADAPT

Check students' answers to the Lesson Practice to ensure that they understand linear patterns and are ready to transition to exploring and representing nonlinear patterns. Students who may need more practice with the concepts in this lesson will have additional opportunities to work with linear patterns in Lesson 1-2.

ACTIVITY 1 Continued

Lesson 1-2

PLAN

Pacing: 1 class period
Chunking the Lesson
#1–4 #5–8 #9–10
#11–13 #14–17 #18–21
Check Your Understanding
Lesson Practice

TEACH

Bell-Ringer Activity
Have students identify a geometric pattern in the classroom, such as a pattern of shapes on a t-shirt or a color pattern in the floor tiles. Ask them to describe the pattern as completely as possible, first in writing and then by sharing aloud with the class.

1–4 Look for a Pattern, Create Representations, Think-Pair-Share, Debriefing In these items, students extend and describe a geometric pattern and then use it to make a prediction. Before students draw the next two figures in the geometric pattern, ask students to discuss with their partners the similarities and differences among the given figures. Identifying how each figure differs from the previous figure can help them decide what to keep the same and what to change when extending the pattern. As students complete the table that describes the geometric pattern, some will need to continue to draw the figures while some will begin to identify and use patterns in the other rows of the table to find the number of squares in these figures. Both methods should be encouraged and validated.

Differentiating Instruction

Support students who miscount the number of squares in the figures by suggesting that they make a mark on each square when counting. This step may help them to avoid skipping a square or counting a square more than once.

ACTIVITY 1 continued

Lesson 1-2
Writing Expressions

Learning Targets:
- Use patterns to write expressions.
- Use tables, graphs, and expressions to model situations.

SUGGESTED LEARNING STRATEGIES: Look for a Pattern, Create Representations, Think-Pair-Share, Discussion Groups, Sharing and Responding

Mizing and his family also visited Mesa Verde National Park in Colorado. As Mizing investigated the artifacts on display from the ancestral Pueblo people who once called the area home, Mizing began to notice that the patterns used to decorate pottery, baskets, and textiles were geometric.

Mizing found a pattern similar to the one below particularly interesting.

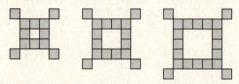

CONNECT TO HISTORY
Mesa Verde National Park was created by President Theodore Roosevelt in 1906 as the first National Park designated to preserve "the works of man." The park protects nearly 5000 known archeological sites and 600 cliff dwellings, offering a look into the lives of the ancestral Pueblo people who lived there from 600–1300 AD.

1. **Reason abstractly.** Draw the next two figures in the pattern.

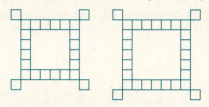

2. Create a table showing the relationship between the figure number and the number of small squares in each figure.

Figure Number	Number of Squares
1	12
2	16
3	20
4	24
5	28
6	32
7	36

Lesson 1-2
Writing Expressions

continued

3. Use the variable *n* to represent the figure number. Write an expression that could be used to determine the number of small squares in any figure number.
   $12 + 4(n - 1)$

4. Use your expression to determine the number of small squares in the 12th figure.
   56

Mizing noticed that many times the centers of the figures in the pattern were filled in with small squares of the same size as the outer squares but in a different color.

5. Fill in the centers of the diagrams with small colored squares.

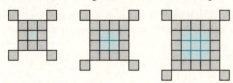

6. Draw the next two figures in the pattern. Be sure to include the inner colored squares.

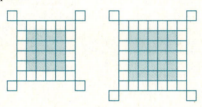

7. Copy the first two columns of the table you created in Item 2 and add a column to show the relationship between the figure number and the number of inner colored squares.

Figure Number	Number of Outer Squares	Number of Inner Colored Squares
1	12	1
2	16	4
3	20	9
4	24	16
5	28	25
6	32	36
7	36	49

ACTIVITY 1 Continued

1–4 (continued) After a few minutes of time to work and think individually, ask students to share with their shoulder partner and then their whole group. Students should recognize that the sequence of the numbers of small squares has a constant difference of 4. As a result, they can use repeated addition of the constant difference to extend the sequence. Students can use steps similar to the ones they used in Lesson 1-1 to write an algebraic expression for the sequence. Be sure to debrief these items with students to make sure that they have identified and represented the geometric pattern correctly.

Differentiating Instruction

Extend learning by having students use properties of real numbers to write two other expressions that are equivalent to the one they wrote in Item 3, such as $4n + 8$ or $8 + 4n$. These representations should be shared during the whole class debrief and validated as equivalent to the expression $12 + 4(n - 1)$.

5–8 Look for a Pattern, Create Representations, Discussion Groups, Sharing and Responding
In these items, students extend and describe a different geometric pattern, but in this case, the pattern in the number of small squares is not linear. Students may or may not recognize this by examining only the table but should be able to describe other patterns that will lead to this understanding as they progress through the next few items. Students should begin to realize that the number of inner colored squares is always a perfect square and that the perfect square is equal to the figure number times itself.

ACTIVITY 1 Continued

9–10 Create Representations, Think-Pair-Share, Sharing and Responding In these items, students use graphs to compare a linear pattern and a non-linear pattern. When drawing the graphs, students must choose an appropriate scale for the *y*-axis. If needed, allow students to generate ideas about how to choose an appropriate scale. For the first graph, the greatest number of outer squares listed in the table is 36. As a result, the *y*-axis needs to show values from 0 to at least 36. Using a scale of 1 on the *y*-axis would result in values only from 0 to 9; using a scale of 4, by contrast, would result in values from 0 to 36. So, a scale of 4 or greater would include all of the data values. Use a similar discussion of scale for the second graph.

TEACHER TO TEACHER

Oftentimes, more than one scale is appropriate for an axis on a graph. An appropriate scale should include the entire range of the data set within a reasonably sized graph. Patterns in the data values can affect the choice of scale. For example, if all of the values in a data set are multiples of 5, the graph will be easier to read if a scale of 5 is used for the corresponding axis rather than a scale of 2.

9–10 (continued) When students compare the graphs, make sure that they compare patterns formed by the points. Some students may need to use a straightedge to determine whether the patterns are linear.

11–12 Think-Pair-Share, Sharing and Responding At this point students are required to verbalize the patterns they have observed and use the patterns to make predictions within the problem situation before moving on to an algebraic representation.

ACTIVITY 1 continued

Lesson 1-2
Writing Expressions

My Notes

8. Describe any numerical patterns you see in the table.
   Answers will vary. The numbers in column 3 are the squares of the numbers in column 1.

9. Write the numbers of inner colored squares as a sequence.
   1, 4, 9, 16, 25, 36, 49, …

10. Does the sequence of numbers of inner colored squares have a common difference? If so, identify it. If not, explain.
    No; the difference between consecutive terms is not the same for each pair of consecutive terms.

11. **Model with mathematics.** Graph the data from the table on the appropriate grid. Be sure to label an appropriate scale on the *y*-axis.

    a.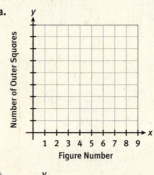

    Students' graphs will vary based on their chosen scale for the *y*-axis. Check that points are graphed at (1, 12), (2, 16), (3, 20), (4, 24), (5, 28), (6, 32), and (7, 36).

    b.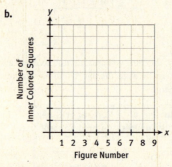

    Students' graphs will vary based on their chosen scale for the *y*-axis. Check that points are graphed at (1, 1), (2, 4), (3, 9), (4, 16), (5, 25), (6, 36), and (7, 49).

Lesson 1-2
Writing Expressions

12. Compare the graphs.
    Answers will vary. The first graph is a line and the second is not.

13. **Reason quantitatively.** Use the patterns you have described to predict the number of inner colored squares in the 10th figure of the pattern.
    100

14. How is the number of inner squares related to the figure number?
    It is the figure number squared.

15. Use the variable n to represent the figure number. Write an expression that could be used to determine the number of inner colored squares in any figure number.
    n^2

16. Use your expression to determine the number of inner colored squares in the 17th figure.
    289

Mizing discovered another pattern in the artifacts. He noticed that when triangles were used, the triangles were all *equilateral* and often multicolored.

      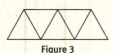
Figure 1   Figure 2   Figure 3

17. **Attend to precision.** Determine the perimeter of each figure in the pattern if each side of one triangle measures 1 cm.
    Figure 1: 3 cm; Figure 2: 5 cm; Figure 3: 7 cm

MATH TERMS
In an **equilateral** triangle, all three sides have the same measure.

ACTIVITY 1 Continued

13–17 Look for a Pattern, Create Representations, Discussion Groups, Debriefing In these items, students represent a non-linear pattern in the geometric figures using an algebraic expression. They also use the non-linear pattern to make predictions. Some students may try to draw the 10th and 17th figures of the pattern to answer Items 13 and 16. If so, use this as an opportunity to discuss why identifying patterns can be useful: Using patterns can help you make predictions. If you identify a pattern, it is not always necessary to draw each figure or calculate each number in the sequence. Patterns can be expressed by writing a rule or an expression, which can be used to make predictions. It is important that the debrief for this section of the lesson include a discussion on the differences between a linear and non-linear function as they are compared in a table, on a graph, and in an expression.

Differentiating Instruction

Some students may use the expression $2n$ to represent the number of inner colored squares in any figure of the pattern, incorrectly reasoning that the figure number times itself is $2n$.

Support students who make this error by having them evaluate their expressions for $n = 3$ and checking whether the result gives the correct number of inner colored squares for the third figure in the pattern. This will help students see the expression is incorrect. Point out that $2n$ is equivalent to $n + n$, not $n \times n$. Remind students that they need to use an exponent to show repeated multiplication of the same factor: $n \times n = n^2$.

Developing Math Language

This lesson contains the vocabulary term *equilateral*. Encourage students to mark the text by drawing a sketch of an equilateral triangle next to the definition of this term. Point out that the prefix *equi-* means "equal." So, an *equilateral* triangle is one that has *equal* sides, or sides with the same length. Add the word to your class Word Wall.

Activity 1 • Investigating Patterns 11

ACTIVITY 1 Continued

18–20 Look for a Pattern, Create Representations, Debriefing In this chunk of items, students explore a pattern involving equilateral triangles and represent it using a table, a sequence, and a graph. In Item 18, students are given an algebraic expression that represents the pattern and are asked to use it to complete a table. Be sure students understand that the variable n in the expression represents the figure number and that the value of the expression represents the perimeter in centimeters of figure number n. Use Item 20 as an opportunity to discuss the importance of labeling the axes of a graph. Debrief this portion of the lesson by asking students to connect the common difference they identified in the sequence written in Item 19 to the expression given in Item 18, and the graph created in Item 20.

TEACHER to TEACHER

In Item 18, students are given an algebraic expression that represents the perimeter of the figures in the pattern. Once students have noted that the sequence representing the perimeters has a common difference, you may want to have students use the common difference to write an alternate expression for the pattern. They can follow the steps they used for the linear patterns earlier in this lesson and in Lesson 1-1. In this case, students can observe that the perimeter is equal to 3 plus the product of 2 and some number. The number multiplied by 2 is one less than the figure number. So, $3 + 2(n - 1)$ is another expression that gives the perimeter in centimeters of figure number n.

ACTIVITY 1 continued

My Notes

Lesson 1-2
Writing Expressions

18. Mizing found that he could determine the perimeter of any figure in the pattern using the expression $2n + 1$. Use Mizing's expression to calculate the perimeters of the next three figures in the pattern. Use the table below to record your calculations.

Figure Number	Perimeter (cm)
1	3
2	5
3	7
4	9
5	11
6	13
7	15

19. Create a sequence to represent the perimeters of the figures in the pattern. Does the sequence have a common difference? If so, identify it. If not, explain.

    3, 5, 7, 9, 11, 13, 15, …; yes; common difference: 2

20. Represent the relationship between the figures in the pattern and their perimeters as a graph. Be sure to label your axes and the scale on the y-axis.

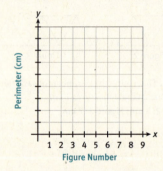

Students' graphs will vary based on their chosen scale for the y-axis. Check that points are graphed at (1, 3), (2, 5), (3, 7), (4, 9), (5, 11), (6, 13), and (7, 15).

Lesson 1-2
Writing Expressions

Check Your Understanding

A pattern of small squares is shown below. Use the pattern to respond to the following questions.

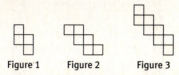

Figure 1   Figure 2   Figure 3

21. Create a table to show the number of small squares in the first through the fifth figures, assuming the pattern continues.
22. Write the number of small squares in each figure as a sequence. Does the sequence have a common difference? If so, identify it. If not, explain.
23. How many small squares would be in the 10th figure? Justify your response using the sequence or the table.
24. Use the variable n to write an expression that could be used to determine the number of small squares in any figure in the pattern.
25. Use your expression to determine the number of small squares in the 20th figure.

LESSON 1-2 PRACTICE

A toothpick pattern is shown below. Use the pattern for Items 26–29.

Figure 1   Figure 2   Figure 3

26. Create a table to show the number of toothpicks in the first through the fifth figures, assuming the pattern continues.
27. Write the number of toothpicks in each figure as a sequence. Does the sequence have a common difference? If so, identify it. If not, explain.
28. **Express regularity in repeated reasoning.** How many toothpicks would be in the 15th figure? Justify your response using the sequence or the table.
29. Use the variable n to write an expression that could be used to determine the number of toothpicks in any figure in the pattern.

LESSON 1-2 PRACTICE

26.

Figure Number	Number of Toothpicks
1	3
2	5
3	7
4	9
5	11

27. 3, 5, 7, 9, 11, …; common difference: 2
28. 31
29. $3 + 2(n-1)$ or $2n + 1$

ACTIVITY 1 Continued

ACTIVITY PRACTICE

1. Answers will vary. The numbers in the last column are all 3's.
2. 3, 6, 9, 12, …; common difference: 3
3.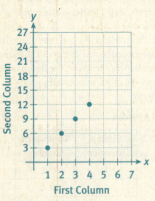
4. $3 + 3(n - 1)$ or $3n$
5. Substitute 25 for n. Subtract 1 and then multiply by 3. Lastly, add 3. The value in the 25th row would be 75.
6.
7. Figure 1: 5 in.; Figure 2: 8 in.; Figure 3: 11 in.; Figure 4: 14 in.; Figure 5: 17 in.; Figure 6: 20 in.
8. 32 inches; Justifications will vary. Sample justification: You begin with a perimeter of 5 inches and add 3 nine times.
9. $5 + 3(n - 1)$ or $3n + 2$
10. 152 inches
11. 3
12. 120
13. With only the information given, both Nancy and Richard could be correct. Nancy assumed that the number of tiles was the square of the figure number, and Richard assumed the number of tiles increased by 3 each time. You need information about more figures in the pattern to determine which is correct.

ADDITIONAL PRACTICE
If students need more practice on the concepts in this activity, see the eBook Teacher Resources for additional practice problems.

ACTIVITY 1 continued

Investigating Patterns
Cross-Country Adventures

ACTIVITY 1 PRACTICE
Write your answers on notebook paper.
Show your work.

Lesson 1-1

1	3	3	3
2	6	12	3
3	9	27	3
4	12	48	3

1. Describe any patterns you notice in the table of values.
2. Write the second column of the table as a sequence. Identify the common difference.
3. Create a graph to show the relationship between the first two columns of the table.
4. Write an expression to represent the relationship between the first two columns of the table.
5. Explain how you could use your expression to determine the value that would appear in the 25th row of the second column.

Lesson 1-2
Use the visual pattern below for Items 6–10.

Figure 1

Figure 2

Figure 3

6. Draw the next three figures in the pattern.
7. Determine the perimeter of each of the first six figures. Assume each side of each pentagon measures 1 inch.
8. What is the perimeter of the 10th figure? Justify your response.
9. Use the variable n to write an expression that could be used to determine the perimeter of any figure in the pattern.
10. Use your expression to calculate the perimeter of the 50th figure. Be sure to include units.
11. Evaluate $3x - 12$ for $x = 5$.
12. Evaluate $45 + 15(n - 1)$ for $n = 6$.

MATHEMATICAL PRACTICES
Construct Viable Arguments and Critique the Reasoning of Others

13. Figure 1 of a mosaic pattern contains one tile and figure 2 contains four tiles. Nancy and Richard were asked to predict the number of tiles in figure 3. Nancy wrote 9, and Richard wrote 7. Who is correct? Explain your reasoning.

Solving Equations

ACTIVITY 2

What's My Number?
Lesson 2-1 Writing and Solving Equations

Learning Targets:
- Use the algebraic method to solve an equation.
- Write and solve an equation to model a real-world situation.

SUGGESTED LEARNING STRATEGIES: Guess and Check, Create Representations, Discussion Groups, Identify a Subtask, Note Taking

Let's play "What's My Number?"

1. **Make sense of problems.** Determine the number, and explain how you came up with the solution.

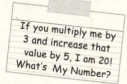

If you multiply me by 3 and increase that value by 5, I am 20! What's My Number?

5; explanations may vary. Use guess and check or write an equation.

2. You may have used an *equation* to answer Item 1. Write an equation that could be used to represent the problem in Item 1.
   $3x + 5 = 20$

3. For each equation, tell whether the given value of x is a *solution*. Explain.
   a. $2 + 4(x - 1) = 22$; $x = 6$
      Yes; $2 + 4(6 - 1) = 2 + 4(5) = 2 + 20 = 22$
   b. $12 - \frac{x}{4} = 8$; $x = 24$
      No; $12 - \frac{24}{4} = 12 - 6 = 6 \neq 8$
   c. $3.8 + 6x = 8.6$; $x = 0.9$
      No; $3.8 + 6(0.9) = 3.8 + 5.4 = 9.2 \neq 9.6$

One way to solve an equation containing a variable is to use the algebraic method. This method is also called the *symbolic method* or *solving equations using symbols*.

DISCUSSION GROUP TIPS

Work with your peers to set rules for:
- discussions and decision-making
- clear goals and deadlines
- individual roles as needed

MATH TERMS

An **equation** is a mathematical statement that shows that two expressions are equal. A **solution** is any value that makes an equation true when substituted for the variable.

Common Core State Standards for Activity 2

HSN-Q.A.1	Use units as a way to understand problems and to guide the solution of multi-step problems; choose and interpret units consistently in formulas; choose and interpret the scale and the origin in graphs and data displays.
HSN-Q.A.2	Define appropriate quantities for the purpose of descriptive modeling.
HSA-REI.A.1	Explain each step in solving a simple equation as following from the equality of numbers asserted at the previous step, starting from the assumption that the original equation has a solution. Construct a viable argument to justify a solution method.
HSA-REI.B.3	Solve linear equations and inequalities in one variable, including equations with coefficients represented by letters.

ACTIVITY 2 Guided

Activity Standards Focus

In Activity 2, students use properties of equality and inverse operations to write and solve linear equations in one variable, including multi-step equations and equations with variables on both sides. Students also explore linear equations that have no solution or infinitely many solutions, and they solve literal equations for a given variable. Throughout this activity, stress that the equal sign in an equation indicates that the expressions on either side have the same value. If the same operation is performed on both sides of the equation, the two sides will remain equal.

Lesson 2-1

PLAN

Pacing: 1 class period

Chunking the Lesson
#1–3   Example A
Check Your Understanding
#6–10
Check Your Understanding
Lesson Practice

TEACH

Bell-Ringer Activity

As a review of content from previous courses, have students solve the following one-step equations. Discuss with students the methods they used to find the values of the variables.
1. $15x = 60$ $[x = 4]$
2. $t - 23 = 6$ $[t = 29]$
3. $n + 12 = -18$ $[n = -30]$

1–3 Guess and Check, Create Representations, Discussion Groups

In this set of items, students first use their own methods to solve a problem involving an unknown number. Then they are led to see that they can model the problem by using an equation in which a variable represents the unknown number. Have students discuss the methods they used to solve the problem in Item 1. Some may have used the strategy of guess and check, and some may have worked backward. The game of "What's My Number?" can be played several times by changing the numbers and operations. For example: "If you multiply 4 times the sum of 3 and me and then subtract 8 from that value, I am 12! What's My Number?" [2] Have students give reasons why it might be useful to represent problems such as those in "What's My Number?" with an equation. As students check solutions to the equations in Item 3, be sure they replace x with the given value.

Activity 2 • Solving Equations   15

ACTIVITY 2 Continued

Developing Math Language
This lesson contains the vocabulary terms *equation, solution,* and *properties of equality*. In this lesson, the word *solution* refers to the solution of an equation, but as students progress through the course, they will encounter this word in other contexts, as when finding the solution of a system of equations. Have students add the words to their math notebooks. They can also make a vocabulary organizer in their notebooks for the properties of equality along with the Commutative, Associative, and Distributive Properties. For each property, students can list the name of the property, a verbal description, and an example of the property.

Example A Activating Prior Knowledge, Identify a Subtask, Note Taking
In this example, students review the algebraic method of solving an equation. As the teacher works through each step, the purpose of each step should be pointed out. For instance, the Commutative Property is used in the first step to switch the terms 90 and $2x$. The purpose of this step is to put the like terms $3x$ and $2x$ together so that they can be combined. Work with students to develop a general process, such as the one given below, for solving the equations in this lesson.

1. Simplify each side of the equation by combining like terms if possible.
2. Isolate the term containing the variable.
3. Isolate the variable.

TEACHER to TEACHER
Emphasize that when students check solutions to equations by using substitution, they should always substitute the value of the variable into the *original* equation, not one of the equations derived during the solution process. Doing so will make it more likely that students will catch mistakes. Substituting into the original equation will also be important when students need to identify extraneous solutions in future math courses.

Check Your Understanding
Debrief students' answers to these questions to ensure that they understand the definition of a solution of an equation as well as how to apply the properties of equality.

Answers
4. No; substituting 4 for x in $2(x - 3) + 7$ gives 9, not 23.
5. Multiplication Property of Equality; you would multiply both sides by 5.

ACTIVITY 2 continued

Lesson 2-1 Writing and Solving Equations

MATH TIP

The **Properties of Equality** state that you can perform the same operation on both sides of an equation without affecting the solution.

Addition Property of Equality
If $a = b$, then $a + c = b + c$.

Subtraction Property of Equality
If $a = b$, then $a - c = b - c$.

Multiplication Property of Equality
If $a = b$, then $ac = bc$.

Division Property of Equality
If $a = b$, then $a \div c = b \div c$, when $c \neq 0$.

Other important properties are:

Symmetric Property of Equality
If $a = b$, then $b = a$.

Commutative Property of Addition/Multiplication
$a + b = b + a$
$ab = ba$

Associative Property of Addition/Multiplication
$(a + b) + c = a + (b + c)$
$(ab)c = a(bc)$

Distributive Property
$a(b + c) = ab + ac$

Example A
Solve the equation $3x + 90 + 2x = 360$ using the algebraic method, showing each step. List a property or provide an explanation for each step. Check your solution.

$3x + 90 + 2x = 360$	Original equation
$3x + 2x + 90 = 360$	Commutative Property of Addition
$5x + 90 = 360$	Combine like terms.
$5x + 90 - 90 = 360 - 90$	Subtraction Property of Equality
$5x = 270$	Combine like terms.
$\frac{5x}{5} = \frac{270}{5}$	Division Property of Equality
$x = 54$	Simplify.

Check by substitution:
$3(54) + 90 + 2(54) = 360$
$162 + 90 + 108 = 360$
$360 = 360$

Solution: $x = 54$

Try These A
a. Solve the "What's My Number?" problem using the equation you wrote in Item 2 and the algebraic method. **5**

Solve each equation using the algebraic method, showing each step. List a property or provide an explanation for each step.

b. $-5x - 6 = 1$
$x = -\frac{7}{5}$

c. $12d + 2 - 3d = 5$
$d = \frac{1}{3}$

d. $7.4p - 9.2p = -2.6 + 5.3$
$p = -1.5$

e. $\frac{x}{4} + 7 = 12$
$x = 20$

f. $20x - 3 + 5x = 22$
$x = 1$

g. $8 = \frac{3}{2}w - 12 + 8$
$w = 8$

Check Your Understanding

4. Is $x = 4$ a solution of $2(x - 3) + 7 = 23$? How do you know?
5. Which property of equality would you use to solve the equation $\frac{x}{5} = 13$? Explain your answer.

Common Core State Standards for Activity 2 (continued)

HSA-SSE.A.1	Interpret expressions that represent a quantity in terms of its context.*
HSA-SSE.A.1a	Interpret parts of an expression, such as terms, factors and coefficients.*
HSA-SSE.A.1b	Interpret complicated expressions by viewing one or more of their parts as a single entity.*
HSA-CED.A.1	Create equations and inequalities in one variable and use them to solve problems. Include equations arising from linear and quadratic functions and simple rational and exponential functions.
HSA-CED.A.4	Rearrange formulas to highlight a quantity of interest, using the same reasoning as in solving equations.

Lesson 2-1
Writing and Solving Equations

Julio has 5 more dollars than Dan. Altogether, Julio and Dan have 19 dollars. How much money does each young man have?

6. Let d represent the amount of money, in dollars, that Dan has. Use d to write an expression that represents the amount of money that Julio has.
   $d + 5$

7. Write an equation to represent the problem situation.
   $d + 5 + d = 19$

8. In the space below, solve your equation from Item 7, showing each step. State a property or provide an explanation for each step. Check your solution.

$d + 5 + d = 19$	Original equation
$d + d + 5 = 19$	Commutative Property of Addition
$2d + 5 = 19$	Combine like terms.
$2d + 5 - 5 = 19 - 5$	Subtraction Property of Equality
$2d = 14$	Combine like terms.
$\frac{2d}{2} = \frac{14}{2}$	Division Property of Equality
$d = 7$	Simplify.

   Check: $7 + 5 + 7 = 19$
   $12 + 7 = 19$
   $19 = 19$

9. Interpret your solution to Item 8 within the context of the problem.
   Dan has $7. Julio has $7 + $5 = $12.

10. Verify the reasonableness of your solution by checking that your answer to Item 9 matches the information given in the original problem situation at the top of the page.
    If Dan has $7 and Julio has $12, then Julio has 5 more dollars than Dan, and together they have $12 + $7 = $19. This matches the information given in the problem situation.

MATH TIP

When using the Addition or Subtraction Property of Equality, you may find it helpful to add or subtract vertically on both sides of the equation.

MINI-LESSON: Algebra Tiles and Flowchart Methods of Solving Equations

Students who have difficulty using the algebraic method to solve equations may benefit from first practicing with algebra tiles. This mini-lesson provides students with an opportunity to explore the algebra-tiles method of solving equations.

Another mini-lesson explains an alternative to the algebraic method of solving equations. In the flowchart method, students create a flowchart to show what happens to the variable in an equation. They then work backward to solve the equation by performing the inverse operations.

See SpringBoard's eBook Teacher Resources for student pages for these mini-lessons.

ACTIVITY 2 Continued

6–10 Create Representations, Think-Pair-Share, Debriefing In this set of items, students write an equation in one variable to represent a real-world situation. Then they solve the equation algebraically and check the reasonableness of their answer. Check that students have written a correct equation in Item 7 before they move on to solve it. To answer Item 9, students will need to be able to interpret the solution to the equation. Students have already established that the variable d represents the amount of money in dollars that Dan has. So, a solution of $d = 7$ means that Dan has $7.

TEACHER to TEACHER

Be sure to point out that when students define a variable, they should state not only the quantity that the variable represents, but also the units used to measure the quantity. For example, if students use the variable h to represent a height, they need to state whether the height is measured in feet, inches, meters, or some other unit.

Differentiating Instruction

Some students may apply the properties of equality incorrectly by performing an operation on one side of the equation and then performing the inverse operation on the other side of the equation. Show examples of some equations being solved vertically:

Support students who make this mistake by suggesting that they use colored pencils or highlighters when applying properties of equality. Emphasize that the text they write in color on one side of the equation (such as -5) must exactly match the text in color on the other side (-5).

Activity 2 • Solving Equations

ACTIVITY 2 Continued

Check Your Understanding
Debrief students' answers to these questions to ensure that they understand how to define a variable and represent a situation with an equation. Discuss with students if it matters whether the variable in Item 12 is used to represent Elaine's age or Tyler's age.

Answers
11. a. $j - 5$
    b. $j - 5 + j = 19$; $j = 12$; Julio has $12.
    c. Yes; you would have gotten that Julio has $12 and Dan has $12 − 7 = $5.
12. Let e represent Elaine's age; $e + (e + 8) = 34$; $e = 13$; Elaine is 13, and Tyler is 21.

ASSESS

Students' answers to lesson practice problems will provide you with a formative assessment of their understanding of the lesson concepts and their ability to apply their learning.

See the Activity Practice for additional problems for this lesson. You may assign the problems here or use them as a culmination for the activity.

ADAPT

Check students' answers to the Lesson Practice to ensure that they understand how to write and solve linear equations in one variable. Students who may need more practice with the concepts in this lesson may benefit from solving equations by using alternative methods, such as the flowchart method or the algebra tiles method as explained in the mini-lesson on page 17.

ACTIVITY 2 continued

My Notes

Lesson 2-1
Writing and Solving Equations

Check Your Understanding

11. In Item 6, the variable j could have been defined as the amount of money, in dollars, that Julio has.
    a. Use j to write an expression to represent the amount of money, in dollars, that Dan has. Describe the similarities and differences between this expression and the one you wrote in Item 6.
    b. Write and solve an equation using the variable j to represent the problem situation. Interpret the solution within the context of the problem.
    c. Does the definition of the variable in a problem situation change the solution to the problem? Explain your reasoning.
12. Elaine is 8 years younger than her brother Tyler. The sum of their ages is 34. Define a variable and then write and solve an equation to find Elaine's and Tyler's ages.

LESSON 2-1 PRACTICE

13. **Attend to precision.** Justify each step in the solution of $5x + 15 = 0$ below by stating a property or providing an explanation for each step.

    $$5x + 15 = 0$$
    $$5x + 15 - 15 = 0 - 15$$
    $$5x = -15$$
    $$\frac{5x}{5} = \frac{-15}{5}$$
    $$x = -3$$

Solve the equations below using the algebraic method. State a property or provide an explanation for each step. Check your solutions.

14. $3x - 24 = -6$ 15. $\frac{4x+1}{3} = 11$

Define a variable for each problem. Then write and solve an equation to answer the question. Check your solutions.

16. Last week, Donnell practiced the piano 3 hours longer than Marcus. Together, Marcus and Donnell practiced the piano for 11 hours. For how many hours did each young man practice the piano?
17. Olivia ordered 24 cupcakes and a layer cake. The layer cake cost $16, and the total cost of the order was $52. What was the price of each cupcake?

LESSON 2-1 PRACTICE

13. $5x + 15 = 0$ Original equation
     $5x + 15 - 15 = 0 - 15$ Subtr. Prop. of Eq.
     $5x = -15$ Combine like terms.
     $\frac{5x}{5} = \frac{-15}{5}$ Div. Prop. of Eq.
     $x = -3$ Simplify.

14. $3x - 24 = -6$ Original equation
     $3x - 24 + 24 = -6 + 24$ Add. Prop. of Eq.
     $3x = 18$ Combine like terms.
     $\frac{3x}{3} = \frac{18}{3}$ Div. Prop. of Eq.
     $x = 6$ Simplify.

15. $\frac{4x+1}{3} = 11$ Original equation
     $3\left(\frac{4x+1}{3}\right) = 3(11)$ Mult. Prop. of Eq.
     $4x + 1 = 33$ Simplify.
     $4x + 1 - 1 = 33 - 1$ Subtr. Prop. of Eq.
     $4x = 32$ Combine like terms.
     $\frac{4x}{4} = \frac{32}{4}$ Div. Prop. of Eq.
     $x = 8$ Simplify.

16. Let d represent the number of hours that Donnell practiced piano; $d + (d - 3) = 11$. Donnell practiced 7 hours and Marcus practiced 4 hours.
17. Let c represent the cost of a cupcake; $24c + 16 = 52$. Each cupcake cost $1.50.

Lesson 2-2
Equations with Variables on Both Sides

ACTIVITY 2
continued

Learning Targets:
- Write and solve an equation to model a real-world situation.
- Interpret parts of an expression in terms of its context.

SUGGESTED LEARNING STRATEGIES: Close Reading, KWL Chart, Create Representations, Discussion Groups, Construct an Argument

The Future Engineers of America Club (FEA) wants to raise money for a field trip to the science museum. The club members will hold an engineering contest to raise money. They are deciding between two different contests, the Straw Bridge contest and the Card Tower contest.

The Straw Bridge contest will cost the club $5.50 per competitor plus $34.60 in extra expenses. The Card Tower contest will cost $4.25 per competitor plus $64.60 in extra expenses.

To help decide which contest to host, club members want to determine how many competitors they would need for the costs of the two contests to be the same.

1. Write an equation that sets the costs of the two contests equal.
   $5.50x + 34.60 = 4.25x + 64.60$

2. Solve the equation from Item 1 by using the algebraic method, showing each step. List a property of equality or provide an explanation for each step.

Equations	Properties/Explanations
$5.50x + 34.60 = 4.25x + 64.60$	Original equation
$550x + 3460 = 425x + 6460$	Multiplication Property of Equality: Multiply each side by 100.
$550x - 425x + 3460 = 425x - 425x + 6460$	Subtraction Property of Equality: Subtract $425x$ from each side.
$125x + 3460 = 6460$	Combine like terms.
$125x + 3460 - 3460 = 6460 - 3460$	Subtraction Property of Equality: Subtract 3460 from each side.
$125x = 3000$	Combine like terms.
$\frac{125x}{125} = \frac{3000}{125}$	Division Property of Equality: Divide each side by 125.
$x = 24$	Solution

MINI-LESSON: Properties Used in Solving Equations

If students need additional review of the properties of equality, a mini-lesson is available to provide more practice. This mini-lesson also includes review of the Commutative, Associative, and Distributive Properties along with the Additive and Multiplicative Inverse Properties.

See SpringBoard's eBook Teacher Resources for a student page for this mini-lesson.

ACTIVITY 2 Continued

Lesson 2-2

PLAN

Pacing: 1 class period
Chunking the Lesson
#1–3   #4–6
#7   #8
Check Your Understanding
Lesson Practice

TEACH

Bell-Ringer Activity
Write the equation $6x + 4 = 3x + 1$ on the board, and ask students to consider the steps and properties they would use to solve the equation. Discuss with students how one of the key steps would be to get the variable terms on one side of the equation.

1–3 Close Reading, Marking the Text, Create Representations, KWL Chart (Know, Want to Know, Learn), Debriefing In these items, students write and solve an equation with variables on both sides. After reading the introduction, but before students begin work on this chunk of the lesson, complete the Know and Want to Know columns of a KWL chart for solving equations with variables on both sides. Students can complete their charts as they work through the lesson. Make sure students understand that the expression on one side of the equation represents the cost of one contest and the expression on the other side represents the cost of the other contest. As these items are debriefed, be sure students understand that the variable represents the number of competitors. Therefore, the solution to the equation represents the number of competitors that will make the cost of hosting the two different contests equal. The variable in the equation should represent the number of competitors.

TEACHER to TEACHER

In Item 2, students start to solve the equation by multiplying each side by 100. The purpose of this step is to eliminate the decimals in the equation, leaving only whole-number constants and coefficients. Many students find it easier to work with whole numbers than with decimals, so eliminating decimals from an equation can be a useful first step. To do so, students can multiply both sides of the equation by an appropriate power of 10.

Activity 2 • Solving Equations

ACTIVITY 2 Continued

4–6 Discussion Groups, Construct an Argument, Create Representations, Debriefing In these items, students are asked to make a recommendation based on the situation. They also write and solve an equation to determine the break-even point for the club's fundraiser. Discuss vocabulary such as cost, profit, revenue and break-even before students begin this chunk of the lesson. Emphasize the information in the Connect to Business text box. Remind students to consider the fact that the club is trying to raise money before they begin their discussions. Students need to determine which contest is cheaper when the number of competitors is greater than 30. If needed, suggest that students choose a number greater than 30 and then evaluate the expressions for the costs of the contests for that number of competitors.

TEACHER to TEACHER

Use Item 4 as an opportunity to discuss how mathematics can help people to make and justify decisions in everyday life.

In Item 6b, students will find that the solution of the equation is a decimal. Because the variable in the equation represents a number of competitors, the solution must be rounded up or down to the nearest whole number. Discuss whether it would be better to round up or down in this situation. Include in the debrief of this chunk of the lesson a discussion to be sure students realize that if they round the number of competitors down to 11, the club's costs will be greater than its revenue and it will have a loss of $1.35. If students round the number of competitors up to 12, the club's revenue will be greater than its costs and it will have a profit of $4.40. These results show that the club needs to have at least 12 competitors to break even and not have a loss. So, in this case, it is more appropriate to round the solution up to 12.

7 Identify a Subtask, Discussion Groups, Create Representations, Debriefing In this item, students use equations to solve problems involving the club's revenue, costs, and profit. To answer Item 7, students can break the problem into subtasks: first finding the club's revenue, next finding its costs, and finally subtracting the costs from the revenue to find the profit.

ACTIVITY 2 continued

Lesson 2-2
Equations with Variables on Both Sides

3. **Model with mathematics.** Interpret the meaning of the value of x in the context of the problem.

   The solution $x = 24$ represents the number of competitors that would make the cost of hosting the Straw Bridge contest the same as the cost of hosting the Card Tower contest.

4. The FEA club estimates they will have more than 30 competitors in their contest. Make a recommendation to the club explaining which contest would be the better choice and why.

   Recommendation: The Card Tower contest is the better choice. Explanations may vary. Choose a value greater than 30 and substitute it into the cost expression for each company. For 35 competitors, the Straw Bridge contest will cost $5.50(35) + 34.60$ dollars, or $227.10. For 35 competitors, the Card Tower contest will cost $4.25(35) + 64.60$ dollars, or $213.35. So, the cost to the club will be less for the Card Tower contest.

5. The FEA club will charge each competitor $10 to enter the engineering contest. Write an expression for the club's revenue if x competitors enter the contest.

   $10x$

CONNECT TO BUSINESS

Revenue is the amount of money made selling a product or service.
Profit is earnings after costs are subtracted from the revenue.

Profit = Revenue − Cost

The *break-even point* occurs when revenue equals cost.

Revenue = Cost

6. a. Write an equation to find the break-even point for the fundraiser using the contest you recommended to the FEA club.

   $10x = 4.25x + 64.60$

   b. Solve the equation. State a property of equality or provide an explanation for each step. How many competitors does the club need to break even?

$10x = 4.25x + 64.60$	Original equation
$1000x = 425x + 6460$	Multiplication Property of Equality
$1000x - 425x = 425x - 425x + 6460$	Subtraction Property of Equality
$575x = 6460$	Combine like terms.
$x \approx 11.23$	Division Property of Equality

   There cannot be 11.23 competitors, so round up to 12. The club needs 12 competitors to break even.

7. How much profit will the FEA club earn from 32 competitors if they use the contest you recommended?

   $R = 10(32) = \$320$
   $C = 4.25(32) + 64.60 = \$200.60$
   $P = R - C = 320 - 200.60 = \$119.40$

MINI-LESSON: Connect to Business

If students need additional work with problems involving revenue, costs, and profit, a mini-lesson is available to provide more practice.

See SpringBoard's eBook Teacher Resources for a student page for this mini-lesson.

20 SpringBoard® Mathematics Algebra 1, Unit 1 • Equations and Inequalities

Lesson 2-2
Equations with Variables on Both Sides

8. The Future Engineers of America Club treasurer was going back through the fundraising records. On Monday, the club made revenue of $140 selling contest tickets at $10 each. One person sold 8 tickets, but the other person selling that day forgot to write down how many she sold. Write and solve an equation to determine the number of tickets the other person sold.
   $10(8 + x) = 140$; $x = 6$; the other person sold 6 tickets.

Check Your Understanding

9. How can you use the Multiplication Property of Equality to rewrite the equation $0.6x + 4.8 = 7.2$ so that the numbers in the problem are integers and not decimals?
10. When writing an expression or equation to represent a real-world situation, why is it important to be able to describe what each part of the expression or equation represents?

LESSON 2-2 PRACTICE

On-the-Go Phone Company has two monthly plans for their customers. The EZ Pay Plan costs $0.15 per minute. The 40 to Go Plan costs $40 per month plus $0.05 per minute.

11. **a.** Write an expression that represents the monthly bill for x minutes on the EZ Pay Plan.
    b. Write an expression that represents the monthly bill for x minutes on the 40 to Go Plan.
12. Write an equation to represent the point at which the monthly bills for the two plans are equal.
13. Solve the equation, showing each step. List a property of equality or provide an explanation for each step.
14. Interpret the solution of the equation within the context of the problem.
15. **Construct viable arguments.** Which plan should you choose if you want only 200 minutes per month? Justify your response.

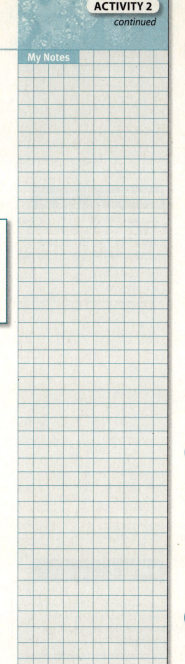

ACTIVITY 2 Continued

8 Create Representations, Think-Pair-Share, Debriefing After students have written an equation to represent the problem in Item 8, have them explain what each part of the equation represents. Doing so can help students check that their equation correctly models the situation.

Check Your Understanding
Debrief this lesson by asking students to explain how the properties of equality can be used to solve equations with variables on both sides. Discuss with students why it might be useful to rewrite an equation containing decimals so that the numbers in the equation are integers.

Answers
9. Answers may vary. Each number in the equation is a decimal to the tenths place. The equation can be rewritten without decimals by multiplying each side by 10: $10(0.6x + 4.8) = 10(7.2)$, which simplifies to $6x + 48 = 72$.
10. Answers may vary. If you can describe what each part of the expression or equation represents, you can be fairly certain that you have modeled the real-world situation correctly.

ASSESS

Students' answers to lesson practice problems will provide you with a formative assessment of their understanding of the lesson concepts and their ability to apply their learning.

See the Activity Practice for additional problems for this lesson. You may assign the problems here or use them as a culmination for the activity.

ADAPT

Check students' answers to the Lesson Practice to ensure that they understand how to write and solve equations with variables on both sides. If students have not yet mastered the solution process, it may help to present them with several equations with variables on both sides, and ask them to explain the steps they would perform to solve the equations and why they would perform each step.

LESSON 2-2 PRACTICE
11. **a.** $0.15x$
    b. $40 + 0.05x$
12. $0.15x = 40 + 0.05x$.
13. $x = 400$; Check students' work.
14. The solution $x = 400$ represents the number of minutes that will make the monthly bill for the EZ Pay Plan equal to the monthly bill for the 40 to Go Plan.
15. You should choose the EZ Pay Plan. The monthly bill for 200 minutes on EZ Pay is $30. The monthly bill for 200 minutes on 40 to Go is $50.

Activity 2 • Solving Equations

ACTIVITY 2 Continued

Lesson 2-3

PLAN

Pacing: 1 class period
Chunking the Lesson
#1–2   #3
Check Your Understanding
Lesson Practice

TEACH

Bell-Ringer Activity
Present students with the equation $2(x + 4) = 6$, and ask them to solve it. Then discuss students' solution methods. Some students may have applied the Distributive Property as a first step; others may have started by dividing both sides of the equation by 2. Students that subtract 4 from each side may need to review the order of operations.

1–2 Create Representations, Construct an Argument, Think-Pair-Share In these items, students work with equations that require multiple steps to solve. These steps include applying the Distributive Property, combining like terms, and gathering variable terms on one side of the equation. Work with students to develop a general procedure, such as the one below, to solve these types of equations.

1. Simplify each side if possible by applying the Distributive Property and/or combining like terms.
2. Gather the variable terms on one side of the equation.
3. Isolate the term that contains the variable.
4. Isolate the variable.

Help students understand that when they list a property or provide an explanation for each step as they solve an equation, they are constructing an argument that shows that their solution is correct. Make sure students understand that in this case, an *argument* is a logical progression of ideas leading to a conclusion that makes sense.

Differentiating Instruction

Support students who make careless errors when solving equations by reminding them of the importance of using substitution to check their answers, especially when the equations are complicated. If their answer does not make the equation true, they should review their work carefully to find the mistake.

ACTIVITY 2 continued

My Notes

MATH TIP
If an equation includes the product of a number and an expression in parentheses, you can simplify by applying the Distributive Property.
Distributive Property:
$a(b + c) = ab + ac$

MATH TIP
You can eliminate the fractions in an equation by multiplying both sides of the equation by the least common denominator of the fractions.

Lesson 2-3
Solving More Complex Equations

Learning Targets:
- Solve complex equations with variables on both sides and justify each step in the solution process.
- Write and solve an equation to model a real-world situation.

SUGGESTED LEARNING STRATEGIES: Create Representations, Construct an Argument, Think-Pair-Share, Create a Plan

Some equations require multiple steps to solve them efficiently.

1. The equation $3x - 2(x + 3) = 5 - 2x$ is solved in the table below. Complete the table by stating a property or providing an explanation for each step.

Equations	Properties/Explanations
$3x - 2(x + 3) = 5 - 2x$	Original equation
$3x - 2x - 6 = 5 - 2x$	Distributive Property
$x - 6 = 5 - 2x$	Combine like terms.
$x + 2x - 6 = 5 - 2x + 2x$	Addition Property of Equality: Add 2x to both sides.
$3x - 6 = 5$	Combine like terms.
$3x - 6 + 6 = 5 + 6$	Addition Property of Equality: Add 6 to both sides.
$3x = 11$	Combine like terms.
$\frac{3x}{3} = \frac{11}{3}$	Division Property of Equality: Divide both sides by 3.
$x = 3\frac{2}{3}$	Solution

Solution: $x = 3\frac{2}{3}$

2. Solve the following equations. State a property of equality or provide an explanation for each step.

   a. $5x + 8 = 3x - 3$
      $x = -\frac{11}{2}$

   b. $2(4y + 3) = 16$
      $y = \frac{5}{4}$

   c. $\frac{2}{3}p + \frac{1}{5} = \frac{4}{5}$
      $p = \frac{9}{10}$

   d. $\frac{3}{4}a - \frac{1}{6} = \frac{2}{3}a + \frac{1}{4}$
      $a = 5$

Lesson 2-3
Solving More Complex Equations

3. **Model with mathematics.** Bags of maple granola cost $2 more than bags of apple granola. The owner of a restaurant ordered 6 bags of maple granola and 5 bags of apple granola. The total cost of the order was $56.
   a. Let m represent the cost of a bag of maple granola. Write an expression for the cost of 6 bags of maple granola.
   $6m$

   b. Use the variable m to write an expression for the cost of a bag of apple granola.
   $m - 2$

   c. Write an expression for the cost of 5 bags of apple granola.
   $5(m - 2)$

   d. Write an equation to show that the cost of 6 bags of maple granola and 5 bags of apple granola was $56.
   $6m + 5(m - 2) = 56$

   e. Solve your equation to find the cost per bag of each type of granola.
   Maple: $6; apple: $4

Check Your Understanding

4. Suppose you are asked to solve the equation $\frac{3}{4}x - \frac{2}{3} = \frac{1}{6}x$.
   a. What number could you multiply both sides of the equation by so that the numbers in the problem are integers and not fractions?
   b. What property allows you to do this?

5. Explain how the Commutative Property of Addition could help you solve the equation $-6x + 10 + 8x = 12 - 4x$.

ACTIVITY 2 Continued

ASSESS

Students' answers to lesson practice problems will provide you with a formative assessment of their understanding of the lesson concepts and their ability to apply their learning.

See the Activity Practice for additional problems for this lesson. You may assign the problems here or use them as a culmination for the activity.

LESSON 2-3 PRACTICE

6. $x = 7$
7. $x = -40$
8. $x = -\frac{1}{6}$
9. $x = 1$
10. $x = \frac{3}{5}$
11. $x = -7$
12. a. Let j represent the price of a pair of jeans; $4(j - 3) + 8 = 92$
    b. $24
13. The student found the solution $x = 2.6$, which is close to 3. Using mental math, you can see that substituting 3 for x on the left side of the equation gives 0, while substituting 3 for x on the right side of the equation gives 20. Since 0 is not close to 20, $x = 2.6$ is not the correct answer.

ADAPT

Students that can write a correct equation but have difficulty arriving at the correct answer may require review on the order of operations. Check that they can evaluate the side of the equation with the variable for certain values of the variable. Also, students should check their answers in terms of *the context*, not just in terms of the equation.

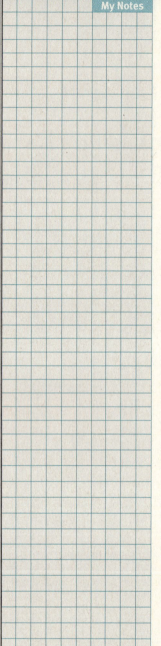

ACTIVITY 2 continued

My Notes

Lesson 2-3
Solving More Complex Equations

LESSON 2-3 PRACTICE

Solve the following equations, and explain each step.

6. $6x + 3 = 5x + 10$
7. $6 + 0.10x = 0.15x + 8$
8. $5 - 4x = 6 + 2x$
9. $9 - 2x = 7x$
10. $2(x - 4) + 2x = -6x - 2$
11. $\frac{1}{2}x - 3 = \frac{3}{2}x + 4$
12. Ben bought 4 pairs of jeans and a T-shirt that cost $8. He had a coupon for $3 off the price of each pair of jeans and spent a total of $92 before tax.
    a. Define a variable and write an equation to represent the situation.
    b. What was the price of each pair of jeans?
13. **Use appropriate tools strategically.** A student solved the equation $3(2x - 6) = 4x + 8$ and found the solution to be $x = 2.6$. How could you use estimation and mental math to show that the student's solution is incorrect?

Lesson 2-4
Equations with No Solution or Infinitely Many Solutions

Learning Targets:
- Identify equations that have no solution.
- Identify equations that have infinitely many solutions.

SUGGESTED LEARNING STRATEGIES: Marking the Text, Think-Pair-Share, Predict and Confirm, Construct an Argument, Sharing and Responding

Remember that a solution of an equation with one variable is a value of the variable that makes the equation true.

1. The set of numbers $\left\{\frac{1}{2}, 3, 6, 17, 0, 11\right\}$ contains possible solutions to the following equations. Determine which of these numbers are solutions to each of the following equations.

   a. $9x + 5 = 4(x + 2) + 5x$
   None of the numbers is a solution.

   b. $7x - 10 = 3x + 14$
   6

   c. $3x - 12 = 3(x + 1) - 15$
   All of the numbers are solutions.

An equation has *no solution* if there is no value of the variable that will create a true mathematical statement. An equation has *infinitely many solutions* if there are an unlimited number of values of the variable that will create a true mathematical statement.

2. Laura, Nia, and Leo solved the following three equations as shown. Identify each of the equations as having *one solution, no solution,* or *infinitely many solutions.* Justify your responses.

Laura	Nia	Leo
$-(5x + 3) - 4x = 8 + 6x$	$3(x + 4) - 10 = 3x + 2$	$2(4x + 3) - 3x = 17 + 5x$
$-5x - 3 - 4x = 8 + 6x$	$3x + 12 - 10 = 3x + 2$	$8x + 6 - 3x = 17 + 5x$
$-9x - 3 = 8 + 6x$	$3x + 2 = 3x + 2$	$5x + 6 = 17 + 5x$
$+9x \qquad +9x$	$-3x \qquad -3x$	$-5x \qquad -5x$
$-3 = 8 + 15x$	$2 = 2$	$6 = 17$
$-8 \qquad -8$		
$-11 = 15x$		
$\frac{-11}{15} = \frac{15x}{15}$		
$-\frac{11}{15} = x$		

Laura's equation has one solution, $x = -\frac{11}{15}$. Nia's equation has infinitely many solutions because $2 = 2$ is always true. Leo's equation has no solutions because $6 = 17$ is always false.

MATH TIP

An equation is true when both sides of the equation have the same value. Otherwise, the equation is false.

$2 + 3 = 5$ is a true mathematical statement because $2 + 3$ has the same value as 5.

$2 + 3 = 6$ is a false mathematical statement because $2 + 3$ does not have the same value as 6.

ACTIVITY 2 Continued

Check Your Understanding
Debrief by having students describe what the solutions should look like for equations with infinitely many solutions and equations with no solution. Then verify that students understand concepts related to these types of equations by asking them to share their answers to the Check Your Understanding questions.

Answers
3. $z = 0$
4. all real numbers
5. no solutions

6 Think-Pair-Share, Predict and Confirm, Construct an Argument, Debriefing Student reasoning may vary so students should be encouraged to compare processes and results with other groups during the debrief of this item. The debriefing discussion should include the fact that, in Item 6b, both sides of the equation are identical; no matter what number you substitute for x, the equation will be true. Therefore, the value of x can be any real number.

7–10 Discussion Groups, Critique Reasoning, Construct an Argument, Sharing and Responding As students work through Items 7–10 they evaluate each situation to help them build a deeper understanding of the three types of equations and the accompanying solution.

TEACHER to TEACHER

An equation that is true for all values of the variable, such as $2x = 2x$, is called an *identity*. An equation that is false for all values of the variable, such as $x = x + 4$, is called a *contradiction*.

Differentiating Instruction

Support students who are visual learners by having them make a graphic organizer that lists characteristics and gives examples of equations with one solution, with infinitely many solutions, and with no solution.

ACTIVITY 2 continued

Lesson 2-4
Equations with No Solution or Infinitely Many Solutions

Check Your Understanding

Determine solutions to each of the following equations.
3. $3(2z + 4) = 6(5z + 2)$
4. $3(x + 1) + 1 + 2x = 2(2x + 2) + x$
5. $8b + 3 - 10b = -2(b - 2) + 3$

Some equations are true for all values of the variable. This type of equation has all real numbers as solutions. This means the equation has *infinitely many solutions*. Other equations are false for all values of the variable. This type of equation has *no solutions*.

6. Which of the following equations has no solutions and which has all real numbers as solutions? Explain your reasoning.
   a. $3x + 5 = 3x$
   There are no solutions. A number cannot be equal to itself plus 5.
   b. $4r - 2 = 4r - 2$
   The solution is all real numbers. The expression $4r - 2$ is equal to itself for all values of r.

7. **Critique the reasoning of others.** A student claims that the equation $2x + 6 = 4x + 4$ has no solutions because when you substitute 0 for x, the left side has a value of 6 and the right side has a value of 4. Is the student's reasoning correct? Explain.
   No; just because $x = 0$ is not a solution does not mean that the equation has no solutions. Solving the equation for x shows that $x = 1$, so the equation has a solution.

8. Explain why the equation $x + 3 = x + 2$ has no solution.
   Answers may vary. Subtracting x from both sides of the equation results in the false statement $3 = 2$.

9. **Reason quantitatively.** For what value of a does the equation $3x + 5 = ax + 5$ have infinitely many solutions? Explain.
   $a = 3$; if $a = 3$, the equation becomes $3x + 5 = 3x + 5$. Both sides of the equation are the same, so the equation is always true.

10. Consider the equation $nx - 4 = 6$.
    a. What are the solutions if the value of n is 0? Explain.
    No solutions; when you substitute 0 for n in the equation and simplify, you get $-4 = 6$, which is false.
    b. What if the value of n is 2? What is the solution of the equation? How do you know?
    $x = 5$; when you substitute 2 for n, the equation is $2x - 4 = 6$. Solving shows that $x = 5$.

26 SpringBoard® Mathematics **Algebra 1, Unit 1** • Equations and Inequalities

Lesson 2-4
Equations with No Solution or Infinitely Many Solutions

Check Your Understanding

Make use of structure. Create an equation that will have each of the following as its solution.

11. One solution
12. No solution
13. Infinitely many solutions
14. A solution of zero

LESSON 2-4 PRACTICE

Solve each equation. If an equation has no solutions or if an equation has infinitely many solutions, explain how you know.

15. $3x - x - 5 = 2(x + 2) - 9$
16. $7x - 3x + 7 = 3(x - 4) + 20$
17. $-2(x - 2) - 4x = 3(x + 1) - 9x$
18. $5(x + 2) - 3 = 3x - 8x + 7$
19. $4(x + 3) - 4 = 8x + 10 - 4x$
20. $3(x + 2) + 4x - 5 = 7(x + 1) - 6$
21. **Construct viable arguments.** Justify your response for each of the following.
    a. Write an equation with no solutions that has the expression $3x + 6$ on the left side of the equal sign. Demonstrate that your equation has no solutions.
    b. Write an equation with infinitely many solutions that has the expression $3x + 6$ on the left side of the equal sign. Demonstrate that your equation has infinitely many solutions.

ACTIVITY 2 Continued

Lesson 2-5

PLAN

Pacing: 1 class period
Chunking the Lesson
Example A
Check Your Understanding
Example B   #3
Check Your Understanding
Lesson Practice

TEACH

Bell-Ringer Activity
Ask students to name some of the formulas they have learned in previous mathematics or science classes and to explain what each variable in the formulas represent.

Developing Math Language
This lesson contains the vocabulary terms *literal equation* and *formula*. Have students add these terms to their math notebooks. Include the terms on the class Interactive Word Wall. Some students may be familiar with the definition of *literal* meaning "true or actual," as in the sentence, "I'm not exaggerating; I literally did 4 hours of math homework last night." However, point out that *literal* can also mean "composed or consisting of letters." Thus, a literal equation is one that includes more than one letter or variable.

Example A Activating Prior Knowledge, Identify a Subtask, Note Taking In this example, students solve literal equations for a given variable. In Step 2 of the example, students divide both sides of the equation by 4 to isolate the variable. Be sure that they understand that the entire expression on the right side needs to be divided by 4, not just the constant 12 or the variable b. Remind students that an expression such as $\frac{12-b}{4}$ can also be written as $\frac{12}{4} - \frac{b}{4}$.

TEACHER to TEACHER

Refer back to the mini-lesson on using the flowchart method to solve equations in Lesson 2-1, page 17. This method can be useful for solving literal equations for a given variable. Students can make a flowchart to show the operations that are applied to the given variable and the order in which these operations are applied. Students can then work backward to undo these operations in reverse order.

ACTIVITY 2 continued

My Notes

MATH TERMS

A **literal equation** has more than one variable, and the equation can be solved for a specific variable.

Formulas are examples of literal equations. A **formula** is an equation written using symbols that describes the relationship between different quantities.

Lesson 2-5
Solving Literal Equations for a Variable

Learning Targets:
- Solve literal equations for a specified variable.
- Use a formula that has been solved for a specified variable to determine an unknown quantity.

SUGGESTED LEARNING STRATEGIES: Identify a Subtask, Close Reading, Work Backward, Create Representations, Discussion Groups

A *formula* describes how two or more quantities are related. Formulas are important in many disciplines; geometry, physics, economics, sports, and medicine are just a few examples of fields in which formulas are widely used.

A formula is an example of a **literal equation**. A literal equation contains more than one variable. Literal equations and formulas can be solved for a specific variable using the same procedures as equations containing one variable.

Example A
Solve the equation $4x + b = 12$ for x.

Step 1: Isolate the term that contains x by subtracting b from both sides.

$4x + b = 12$ Original equation
$4x + b - b = 12 - b$ Subtraction Property of Equality
$4x = 12 - b$ Combine like terms.

Step 2: Isolate x by dividing both sides by 4.

$\frac{4x}{4} = \frac{12-b}{4}$ Division Property of Equality

$x = \frac{12-b}{4}$ Simplify.

Solution: $x = \frac{12-b}{4}$, or $x = 3 - \frac{b}{4}$

Try These A
Solve each equation for x.

a. $ax + 7 = 3$
$x = -\frac{4}{a}$

b. $cx - 10 = -5$
$x = \frac{5}{c}$

c. $-3x + d = -9$
$x = 3 + \frac{d}{3}$

---

28 SpringBoard® Mathematics **Algebra 1, Unit 1** • Equations and Inequalities

Lesson 2-5
Solving Literal Equations for a Variable

Check Your Understanding

1. Is the equation $2x + 4 = 5x - 6$ a literal equation? Explain.
2. Describe the similarities and differences between solving an equation containing one variable and solving a literal equation for a variable.

Example B
The equation $v = v_0 + at$ gives the velocity in meters per second of an object after t seconds, where v_0 is the object's initial velocity in meters per second and a is its acceleration in meters per second squared.
a. Solve the equation for a.
b. Determine the acceleration for an object whose velocity after 15 seconds is 25 meters per second and whose initial velocity was 15 meters per second.

a.
$v = v_0 + at$	Original equation
$v - v_0 = v_0 - v_0 + at$	Subtraction Property of Equality: Subtract v_0 from both sides.
$v - v_0 = at$	Combine like terms.
$\frac{v - v_0}{t} = \frac{at}{t}$	Division Property of Equality: Divide both sides by t.
$\frac{v - v_0}{t} = a$	Simplify.
$a = \frac{v - v_0}{t}$	Symmetric Property of Equality

b. To determine the acceleration for an object whose velocity after 15 seconds is 25 meters per second and whose initial velocity was 15 meters per second, substitute 25 for v, 15 for v_0, and 15 for t.

$$a = \frac{25 - 15}{15} = \frac{10}{15} = \frac{2}{3} \text{ m/s}^2$$

Try These B
The equation $t = 13p + 108$ can be used to estimate the cooking time t in minutes for a stuffed turkey that weighs p pounds. Solve the equation for p. Then find the weight of a turkey that requires 285 minutes to cook.

$p = \frac{t - 108}{13}$; about 13.6 pounds

CONNECT TO PHYSICS
An object's *velocity* is its speed in a particular direction. Its *acceleration* is the rate of change in velocity. If an object has a positive acceleration, it is speeding up; if it has a negative acceleration, it is slowing down.

READING MATH
Sometimes a variable may include a subscript. A subscript is a small number or letter written to the lower right of a variable. For example, the variable v_0 has the subscript 0. The subscript 0 is often read "nought." (*Nought* is another word for 0.) A variable with a subscript of 0 usually indicates an initial value. So, v_0 indicates the initial value of the velocity, or the velocity when the time $t = 0$.

ACTIVITY 2 Continued

3 Work Backward, Create Representations, Discussion Groups In this item, students solve formulas for a given variable. Monitor group discussions of this item to ensure that all members are participating. Point out that students do not need to understand the scientific vocabulary used in the table when working with the formulas, but encourage them to ask for clarification of unfamiliar terms within their group discussions.

Check Your Understanding

Verify that students understand concepts related to literal equations by reviewing their answers to the Check Your Understanding questions. Ask students to explain the steps they used to solve the equation in Item 5 for the variable c.

Answers

4. $c = \dfrac{s}{w+1}$

5. Being able to solve a literal equation for a variable allows you to write a formula for any quantity in the literal equation. For example, if you know that the perimeter of a rectangle is given by $P = 2(l + w)$, you can write a formula for the length of a rectangle $\left(l = \dfrac{P}{2} - w\right)$ or the width of a rectangle $\left(w = \dfrac{P}{2} - l\right)$.

ASSESS

Students' answers to lesson practice problems will provide you with a formative assessment of their understanding of the lesson concepts and their ability to apply their learning.

See the Activity Practice for additional problems for this lesson. You may assign the problems here or use them as a culmination for the activity.

ADAPT

Check students' answers to the Lesson Practice to ensure that they understand how to solve literal equations for an indicated variable. If students have not yet mastered this skill, suggest substituting numbers for each of the variables in a literal equation with the exception of the variable they are solving for. Students can then solve the equation, recording each step. They can use these steps as a model to go back and solve the literal equation.

LESSON 2-5 PRACTICE

6. $d = \dfrac{W}{F}$
7. $W = Pt$
8. $t = \dfrac{W}{P}$
9. $k = \dfrac{on + r}{a}$
10. a. $E = \dfrac{9R}{I}$ Original equation
    $IE = 9R$ Multiply both sides by I.
    $I = \dfrac{9R}{E}$ Divide both sides by E.
    b. 225 innings

Lesson 2-5
Solving Literal Equations for a Variable

3. **Reason abstractly.** Solve for the indicated variable in each formula.

Name	Formula		Solve for
Distance	$d = rt$, where d is the distance an object travels, r is the average rate of speed, and t is the time traveled	r	$r = \dfrac{d}{t}$
Pressure	$p = \dfrac{F}{A}$, where p is the pressure on a surface, F is the force applied, and A is the area of the surface	F	$F = pA$
Kinetic energy	$k = \dfrac{1}{2}mv^2$, where k is the kinetic energy of an object, m is its mass, and v is its velocity	m	$m = \dfrac{2k}{v^2}$
Gravitational energy	$U = mgh$, where U is the gravitational energy of an object, m is its mass, g is the acceleration due to gravity, and h is the object's height	h	$h = \dfrac{U}{mg}$
Boyle's Law	$p_1V_1 = p_2V_2$, where p_1 and V_1 are the initial pressure and volume of a gas and p_2 and V_2 are the final pressure and volume of the gas when the temperature is kept constant	V_2	$V_2 = \dfrac{p_1V_1}{p_2}$

Check Your Understanding

4. Solve the equation $w + i = \dfrac{s}{c}$ for c.
5. Why do you think being able to solve a literal equation for a variable would be useful in certain situations?

LESSON 2-5 PRACTICE

Solve each equation for the indicated variable.

6. $W = Fd$, for d
7. $P = \dfrac{W}{t}$, for W
8. $P = \dfrac{W}{t}$, for t
9. $ak - r = on$, for k
10. **Reason quantitatively.** In baseball, the equation $E = \dfrac{9R}{I}$ gives a pitcher's earned run average E, where R is the number of earned runs the player allowed and I is the number of innings pitched.
    a. Solve the equation for I. State a property or provide an explanation for each step.
    b. Last season, a pitcher had an earned run average of 2.80 and allowed 70 earned runs. How many innings did the pitcher pitch last season?

30 SpringBoard® Mathematics Algebra 1, Unit 1 • Equations and Inequalities

Solving Equations
What's My Number?

ACTIVITY 2
continued

ACTIVITY 2 PRACTICE
Write your answers on notebook paper.
Show your work.

Lesson 2-1

1. Which of the following shows the Addition Property of Equality?
   A. $\frac{4x}{4} = \frac{10}{4}$
   $x = 2.5$
   B. $16x - 2 + 2 = 18 + 2$
   $16x = 20$
   C. $8x + 2x = 10x$
   D. $4(3x + 1) = 8$
   $12x + 4 = 8$

Solve each equation.

2. $3z - 2 = 13$
3. $-5y - 10 = -60$
4. $4x - 5 + 2x = -2$
5. $\frac{x-8}{3} = 3$
6. Which equation has the solution $x = -4$?
   A. $2(x - 4) = 16$
   B. $3x + 6 - 2x = -2$
   C. $\frac{1}{4}x + 4 = 3$
   D. $(x + 6)3 = -6$
7. Is 7 a solution of $5x - 3 = 12$? Justify your answer.
8. The perimeter of triangle ABC is 54. The triangle has side lengths $AB = 3x$, $BC = 4x$, and $AC = 5x$. Find the length of each side.

A group of 19 students want to see the show at the planetarium. Tickets cost $11 for each student who is a member of the planetarium's frequent visitor program and $13 for each student who is not a member. The total cost of the students' tickets is $209. Use this information for Items 9–16.

9. Let x represent the number of students in the group who are members of the frequent visitor program. Write an expression in terms of x for the number of students in the group who are not members.

10. Write an equation to determine the value of x. Explain what each part of your equation represents.

11. Solve the equation you wrote in Item 10. List a property of equality or provide an explanation for each step.

12. How many students in the group are members of the frequent visitor program? How many are not members? Explain how you know.

Lesson 2-2

13. Joining the frequent visitor program at the planetarium costs $5 per year. Write an equation that can be used to determine n, the number of visits per year for which the cost of being a member of the frequent visitor program is equal to the cost of not being a member.

14. Solve your equation from Item 13. List a property of equality or provide an explanation for each step.

15. Explain the meaning of the solution of the equation.

16. Nash plans to visit the planetarium twice in the next year. Should he join the frequent visitor program? Explain.

For Items 17–24, solve the equations, and explain each step.

17. $8x + 5 = 3x + 15$
18. $3x + 11 = 2x - 5$
19. $6x - 9 = 8x + 11$
20. $0.5x - 3.5 = 0.2x - 0.5$

Lesson 2-3

21. $6 - 2(x + 6) = 3x + 4$
22. $3x + 2(x - 1) = 9x + 4$
23. $5(x - 2) + x = 6(x + 3) - 4x$
24. $2 - 3(4 - x) = 5(2 - x) + 4x$

ACTIVITY 2 Continued

ACTIVITY PRACTICE
Explanations may vary in Items 10 and 16.

1. B
2. $z = 5$
3. $y = 10$
4. $x = 0.5$
5. $x = 17$
6. C
7. No. $5(7) - 3 = 35 - 3 = 32$, not 12
8. 13.5, 18, 22.5
9. $19 - x$
10. $11x + 13(19 - x) = 209$. The equation shows that the cost of tickets for members plus the cost of the tickets for nonmembers is equal to the total cost of $209. The cost of tickets for members is equal to the cost of one ticket, $11, times the number of members, x. The cost of tickets for nonmembers is equal to the cost of one ticket, $13, times the number of nonmembers, $19 - x$.
11. $x = 19$; Check students' work.
12. All 19 students in the group are members. The solution $x = 19$ shows that 19 students in the group are members. Because there are only 19 students in all, there aren't any students who are not members.
13. $11n + 5 = 13n$
14. $n = 2.5$; Check students' work.
15. The solution shows that you would have to visit the planetarium 2.5 times per year for the total cost of being a member to equal the total cost of being a nonmember. So, you would need to visit at least 3 times per year to make it worth being a member of the frequent visitor program.
16. Nash should not join the frequent visitor program. Substitute 2 for n in the cost expressions for being a member and being a nonmember. For 2 visits, being a member will cost $11(2) + 5$ dollars, or $27. For 2 visits, being a nonmember will cost $13(2)$ dollars, or $26. So, the total cost will be less for being a nonmember.
17. $x = 2$
18. $x = -16$
19. $x = -10$
20. $x = 10$
21. $x = -2$
22. $x = -\frac{3}{2}$
23. $x = 7$
24. $x = 5$

Activity 2 • Solving Equations

ACTIVITY 2 Continued

Explanations may vary in Items 28, 30, 32, and 33.

25. C
26. (1) Original equation
    (2) Distributive Property
    (3) Combine like terms.
    (4) Addition Property of Equality
    (5) Combine like terms.
    (6) Addition Property of Equality
    (7) Combine like terms.
    (8) Division Property of Equality
    (9) Solution
27. **a.** Let w represent the hours Tyrell spent walking;
    $4w + 12(w - 1) = 36$
    b. 3 hours walking, 2 hours biking
28. $x = $ all real numbers. When you simplify both sides of the equation, you get $3x + 8 = 3x + 8$. When you subtract $3x$ from both sides, you get $8 = 8$, which is true for any value of the variable. So, the original equation has infinitely many solutions.
29. $x = 1$
30. No solutions. When you simplify both sides of the equation, you get $-2x + 5 = -2x + 3$. When you add $2x$ to both sides, you get $5 = 3$, which is a false statement. So, the original equation has no solutions.
31. $x = 0$
32. No solutions. When you simplify both sides of the equation, you get $8x + 8 = 8x - 8$. When you subtract $8x$ from both sides, you get $8 = -8$, which is a false statement. So, the original equation has no solutions.
33. $x = $ all real numbers. When you eliminate the fractions and simplify both sides of the equation, you get $6x + 9 = 6x + 9$. When you subtract $6x$ from both sides, you get $9 = 9$, which is true for any value of the variable. So, the original equation has infinitely many solutions.
34. D
35. $m = \dfrac{w}{g}$
36. $P = 2(Q - 15)$
37. $R = \dfrac{V}{I}$
38. $m = \dfrac{y - b}{x}$
39. **a.** $t = f - d - e$
    b. 16.175

ADDITIONAL PRACTICE

If students need more practice on the concepts in this activity, see the eBook Teacher Resources for additional practice problems.

ACTIVITY 2 continued

25. Which equation has the greatest solution?
    A. $0.4x - 2.5 = 1.3x + 4.7$
    B. $2.4(x - 6) = -5.6x + 8$
    C. $\dfrac{3}{4}x + 5 = \dfrac{1}{2}x + 9$
    D. $\dfrac{1}{3}x + \dfrac{2}{3} = \dfrac{1}{4}(x + 3)$

26. Provide a reason for each step in solving the equation shown below.
    $2(x - 1) - 3(x + 2) = 8 - 4x$
    $2x - 2 - 3x - 6 = 8 - 4x$
    $-1x - 8 = 8 - 4x$
    $-1x + 4x - 8 = 8 - 4x + 4x$
    $3x - 8 = 8$
    $3x - 8 + 8 = 8 + 8$
    $3x = 16$
    $\dfrac{3x}{3} = \dfrac{16}{3}$
    $x = 5\dfrac{1}{3}$

27. Tyrell exercised this week both by walking and by biking. He walked at a rate of 4 mi/h and biked at a rate of 12 mi/h. The total distance he covered both walking and biking was 36 miles, and Tyrell spent one more hour walking than biking.
    a. Define a variable and write an equation for this situation.
    b. How many hours did Tyrell spend on each activity?

Lesson 2-4

Solve each equation. If an equation has no solutions, or if an equation has infinitely many solutions, explain how you know.

28. $3(x + 4) - 4 = 2(x + 4) + x$
29. $4(2x + 6) = 5(x + 5) + 2$
30. $6x - 8x + 5 = -2(x + 2) + 7$
31. $0.3x + 1.8 = 0.4(x + 5) - 0.2$
32. $6(x + 2) + 2x - 4 = 8(x - 2) + 8$
33. $\dfrac{1}{2}(x + 1) + \dfrac{1}{4} = \dfrac{2}{3}x - \dfrac{1}{6}x + \dfrac{3}{4}$

40. No, the equation is not solved correctly. The student made a mistake in the last step when applying the Division Property of Equality. Correct solution:

$2(x - 4) - 4x = -6x + 9x + 4$	Original equation
$2x - 8 - 4x = -6x + 9x + 4$	Distrib. Prop.
$-2x - 8 = 3x + 4$	Combine like terms.
$-2x + 2x - 8 = 3x + 2x + 4$	Add. Prop. of Eq.
$-8 = 5x + 4$	Combine like terms.
$-8 - 4 = 5x + 4 - 4$	Subtr. Prop. of Eq.
$-12 = 5x$	Combine like terms.
$-\dfrac{12}{5} = \dfrac{5x}{5}$	Div. Prop. of Eq.
$-\dfrac{12}{5} = x$	Solution

Solving Equations
What's My Number?

Lesson 2-5

34. Which shows the equation $c = a\left(\dfrac{w}{150}\right)$ correctly solved for the variable w?
    A. $w = a(150c)$
    B. $w = a\left(\dfrac{c}{150}\right)$
    C. $w = \dfrac{150}{ac}$
    D. $w = \dfrac{150c}{a}$

Solve each equation for the indicated variable.

35. $w = gm$, for m
36. $Q = \dfrac{1}{2}P + 15$, for P
37. $I = \dfrac{V}{R}$, for R
38. $y = mx + b$, for m
39. The equation $f = d + e + t$ can be used to find an athlete's final score f in an Olympic trampoline event, where d is the difficulty score, e is the execution score, and t is the time of flight score.
    a. Solve the equation for t.
    b. An athlete's final score is 55.675. His difficulty score is 14.6, and his execution score is 24.9. What is the athlete's time of flight score?

MATHEMATICAL PRACTICES
Construct Viable Arguments and Critique the Reasoning of Others

40. A student solved the equation $2(x - 4) - 4x = -6x + 9x + 4$ as shown below. Did the student solve the equation correctly? If so, list a property or explanation for each step. If not, solve the equation correctly, and list a property or explanation for each step.

    $2(x - 4) - 4x = -6x + 9x + 4$
    $2x - 8 - 4x = -6x + 9x + 4$
    $-2x - 8 = 3x + 4$
    $-2x + 2x - 8 = 3x + 2x + 4$
    $-8 = 5x + 4$
    $-8 - 4 = 5x + 4 - 4$
    $-12 = 5x$
    $-60 = x$

Patterns and Equations
OF MUSIC AND MONEY

Embedded Assessment 1
Use after Activity 2

As part of a social studies class project on economics, Annette and Jeff are researching the benefits of membership in an online music club

1. Yearly membership with the online music club costs $48. Members pay $0.99 per song to download music. Nonmembers may download songs for $1.29 each.
   a. Copy and complete the tables below to represent the yearly cost to download songs for members and nonmembers.

Members	
Number of Songs	Total Cost
0	
1	
2	
3	
4	

Nonmembers	
Number of Songs	Total Cost
0	
1	
2	
3	
4	

   b. Describe any patterns you notice in the tables.
   c. Represent the total cost of songs purchased for members and nonmembers as sequences. Tell whether each sequence has a common difference, and if so, identify it.
   d. Use the variable n to write expressions for the total cost of downloading n songs for members and for nonmembers.
   e. Use your expressions to determine the total cost of 8 songs for members and for nonmembers.

2. To determine whether becoming a member of the online music club is cost effective, Annette and Jeff must know at what point the costs for members and nonmembers are equal.
   a. Write an equation to represent the point at which the total cost of downloading songs as a club member is equal to the total cost of downloading songs as a nonmember. Then solve your equation and interpret your solution within the context of the problem.
   b. Assume you download 4 songs per week. Would it be beneficial for you to become a member of the online music club? Justify your response. (*Remember: There are 52 weeks in a year.*)

3. Members pay a $48 membership fee each year. The literal equation $c = 48y + 0.99n$ represents the total cost c for a person who is a member of the music club for y years and who downloads n songs. Solve the equation for y.

Embedded Assessment 1

Assessment Focus
- Identifying patterns
- Modeling patterns with expressions
- Using patterns to make predictions
- Writing, solving, and interpreting multi-step equations
- Solving literal equations for a variable

TEACHER to TEACHER
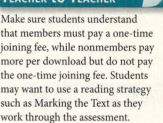
Make sure students understand that members must pay a one-time joining fee, while nonmembers pay more per download but do not pay the one-time joining fee. Students may want to use a reading strategy such as Marking the Text as they work through the assessment.

Answer Key
1. a. See student page.
   b. Members have a starting cost of $48, and then the cost increases by $0.99 for each song downloaded. Nonmembers have no starting cost, but the cost increases by $1.29 for each song downloaded.
   c. Members: {48, 48.99, 49.98, 50.97, 51.96,…}; yes; common difference = 0.99; Nonmembers: {0, 1.29, 2.58, 3.87, 5.16,…}; yes; common difference = 1.29
   d. Members: $48 + 0.99n$; Nonmembers: $1.29n$
   e. Members: $48 + 0.99(8) = \$55.92$; Nonmembers: $1.29(8) = \$10.32$
2. a. $48 + 0.99n = 1.29n$; $n = 160$; The total cost will be the same for members and nonmembers if 160 songs are bought in a year.
   b. Sample answer: Four songs per week is equivalent to 208 songs per year. For more than 160 songs per year, it is more cost effective to be a member of the online music club.
3. $y = \dfrac{c - 0.99n}{48}$

Common Core State Standards for Embedded Assessment 1

HSN-Q.A.1	Use units as a way to understand problems and to guide the solution of multi-step problems; choose and interpret units consistently in formulas; choose and interpret the scale and the origin in graphs and data displays.
HSN-Q.A.2	Define appropriate quantities for the purpose of descriptive modeling.
HSA-CED.A.1	Create equations and inequalities in one variable and use them to solve problems. Include equations arising from linear and quadratic functions and simple rational and exponential functions.
HSA-CED.A.4	Rearrange formulas to highlight a quantity of interest, using the same reasoning as in solving equations.

Embedded Assessment 1

4. Annette: In Step 4, Annette should have simplified $1500 - 1000$ to 500, not -500.
   Jeff:
   Step 1: Original equation
   Step 2: Distributive Property
   Step 3: Comm. Property of Addition
   Step 4: Combine like terms.
   Step 5: Add. Property of Equality
   Step 6: Combine like terms.
   Step 7: Subtr. Property of Equality
   Step 8: Combine like terms.
   Step 9: Div. Property of Equality
   Step 10: Simplify.
   Step 11: Symmetric Property of Eq.

Teacher to Teacher

You may wish to read through the scoring guide with students and discuss the differences in the expectations at each level. Check that students understand the terms used.

Unpacking Embedded Assessment 2

Once students have completed this Embedded Assessment, turn to Embedded Assessment 2 and unpack it with them. Use a graphic organizer to help students understand the concepts they will need to know to be successful on Embedded Assessment 2.

Embedded Assessment 1
Use after Activity 2

Patterns and Equations
OF MUSIC AND MONEY

4. The class is also working on creating family budgets. A sample monthly mortgage payment M can be represented by the equation $3250 - M = 2(M - 500) + 1500$. Jeff and Annette each solved the equation, but they disagree on the solution. Decide who is correct. For the correct solution, justify each step by writing a property or an explanation. For the incorrect solution, identify the error in the solution process.

Annette

$3250 - M = 1500 + 2(M - 500)$
$3250 - M = 1500 + 2M - 1000$
$3250 - M = 2M + 1500 - 1000$
$3250 - M = 2M - 500$
$3250 - M + M = 2M + M - 500$
$3250 = 3M - 500$
$3250 + 500 = 3M - 500 + 500$
$\dfrac{3750}{3} = \dfrac{3M}{3}$
$1250 = M$
$M = \$1250$

Jeff

$3250 - M = 1500 + 2(M - 500)$
$3250 - M = 1500 + 2M - 1000$
$3250 - M = 1500 - 1000 + 2M$
$3250 - M = 500 + 2M$
$3250 - M + M = 500 + 2M + M$
$3250 = 500 + 3M$
$3250 - 500 = 500 - 500 + 3M$
$2750 = 3M$
$\dfrac{2750}{3} = \dfrac{3M}{3}$
$916.67 = M$
$M = \$916.67$

Scoring Guide	Exemplary	Proficient	Emerging	Incomplete
	The solution demonstrates the following characteristics:			
Mathematics Knowledge and Thinking (Items 1a–e, 2a, 3, 4)	• Fluent use of patterns, sequences, and tables to write expressions and equations • Accuracy in solving a literal equation	• Adequate understanding of how to use patterns, sequences, and tables to write expressions and equations • Correct solution of a literal equation	• Partial understanding of how to use patterns, sequences, and tables to write expressions and equations • Partially solved literal equation	• Inaccurate or incomplete understanding of how to use patterns, sequences, and tables to write expressions and equations • No attempt to solve a literal equation
Problem Solving (Items 1e, 2a, 2b, 4)	• Appropriate and efficient strategy that results in a correct answer • Correct identification of an error in a solution process	• Strategy that may include unnecessary steps but results in a correct answer. • Correct identification of an error, but with an incorrect reason given	• Strategy that results in some incorrect answers • Correct identification of an error with no reason given	• No clear strategy when solving problems • No identification of an error in a solution process
Mathematical Modeling / Representations (Items 1a, 1d, 2a)	• Clear and accurate creation of a table to describe a real-world scenario • Effective understanding of how to write expressions and equations to represent a real-world scenario	• Little difficulty creating a table to describe a real-world scenario • Functional understanding of how to write expressions and equations to represent a real-world scenario	• Partially accurate table to describe a real-world scenario • Partial understanding of how to write expressions and equations to represent a real-world scenario	• Inaccurate or incomplete table to describe a real-world scenario • Little or no understanding of how to write expressions and equations to represent a real-world scenario
Reasoning and Communication (Items 1b, 2b, 4)	• Precise use of appropriate math terms and language to describe patterns and to justify each step in the solution of an equation • Clear and accurate conclusion drawn from an equation	• Adequate description of patterns and justification of each step in the solution of an equation • Reasonable conclusion drawn from an equation	• Confusing description of patterns and/or justification of the steps in the solution of an equation • Partially correct conclusion drawn from an equation	• Incomplete or inaccurate description of patterns and/or justification of the steps in the solution of an equation • Incomplete or inaccurate conclusion drawn from an equation

Common Core State Standards for Embedded Assessment 1 *(cont.)*

HSA-REI.A.1 Explain each step in solving a simple equation as following from the equality of numbers asserted at the previous step, starting from the assumption that the original equation has a solution. Construct a viable argument to justify a solution method.

HSA-REI.B.3 Solve linear equations and inequalities in one variable, including equations with coefficients represented by letters.

HSF-LE.A.1b Recognize situations in which one quantity changes at a constant rate per unit interval relative to another.

Solving Inequalities
Physical Fitness Zones
Lesson 3-1 Inequalities and Their Solutions

ACTIVITY 3

Learning Targets:
- Understand what is meant by a solution of an inequality.
- Graph solutions of inequalities on a number line.

SUGGESTED LEARNING STRATEGIES: Levels of Questions, Think-Pair-Share, Interactive Word Wall, Construct an Argument, Quickwrite

Spartan Middle School students participate in Physical Education testing each semester. In order to pass, 12- and 13-year-old girls have to do at least 7 push-ups and 4 modified pull-ups. They also have to run one mile in 12 minutes or less.

You can use an inequality to express the passing marks in each test.

	Push-Ups, p	Modified Pull-Ups, m	One-Mile Run, r
Verbal	At least 7 push-ups	At least 4 pull-ups	12 minutes or less
Inequality	$p \geq 7$	$m \geq 4$	$r \leq 12$
Graph	(number line 5–11, dotted, from 7)	(number line 2–8, dotted, from 4)	(number line 7–13, solid, to 12)

MATH TERMS

The **graph of an inequality** in one variable is all the points on a number line that make the inequality true.

1. Why do you think the graphs of push-ups and pull-ups are dotted but the graph of the mile run is a solid ray?
   Push-ups and pull-ups are whole numbers. Running time does not have to be a whole number of minutes.

2. **Reason quantitatively.** Jamie ran one mile in 12 minutes 15 seconds, did 8 push-ups, and did 4 modified pull-ups. Did she pass the test? Explain.
   Jamie didn't pass. She didn't run one mile in less than 12 minutes.

3. Karen did 7 push-ups.
   a. Is this a passing number of push-ups? Which words in the verbal description indicate this? Explain.
   Yes; "at least" indicates that 7 is a passing number of push-ups.

   b. How is this represented in the inequality $p \geq 7$?
   The symbol $\geq$ is used to indicate that 7 is a solution.

WRITING MATH

Other phrases that are equivalent to "at least" are "no less than" and "no fewer than." These phrases are also represented by the inequality symbol $\geq$.

The **solution of an inequality** in one variable is the set of numbers that make the inequality true. To verify a solution of an inequality, substitute the value into the inequality and simplify to see if the result is a true statement.

Common Core State Standards for Activity 3

HSA-CED.A.1 Create equations and inequalities in one variable and use them to solve problems. Include equations arising from linear and quadratic functions and simple rational and exponential functions.

HSA-CED.A.3 Represent constraints by equations or inequalities and by systems of equations and/or inequalities and interpret solutions as viable or non-viable options in a modeling context.

HSA-REI.B.3 Solve linear equations and inequalities in one variable, including equations with coefficients represented by letters.

ACTIVITY 3 Guided

Activity Standards Focus

In Activity 3, students write and solve linear inequalities in one variable, including multi-step inequalities and inequalities with variables on both sides. They graph solutions of inequalities on number lines and explore how inequalities can represent constraints in real-world situations. They also solve and graph compound inequalities. Throughout this activity, emphasize the importance of paying attention to the inequality sign and the circumstances in which it should be reversed.

Lesson 3-1

PLAN

Pacing: 1 class period
Chunking the Lesson
#1–3   #4–7   #8–9
Check Your Understanding
Lesson Practice

TEACH

Bell-Ringer Activity
As a review of inequality signs and how to compare numbers, have students complete each of the following statements with $<$ or $>$.

1. $3.05 \;\square\; 3.204$ [$<$]
2. $-12 \;\square\; -8$ [$<$]
3. $\dfrac{5}{6} \;\square\; \dfrac{3}{4}$ [$>$]

Discuss with students the methods they used to compare the numbers.

1–3 Shared Reading, Levels of Questions, Activating Prior Knowledge, Think-Pair-Share, Sharing and Responding In this set of items, students explore different ways of representing inequalities: verbal descriptions, symbolic forms, and graphs. They then determine whether given values meet the constraints represented by the inequalities. Make sure students understand that each column of the table shows different ways to represent the same information. Emphasize that when the graph of an inequality is a ray, the ray includes all real values on the number line up to (and perhaps including) its endpoint. Check students' understanding by asking them whether 10.25 minutes is a valid result for the one-mile run. Students should reply yes. However, note that 10.25 push-ups would not be a valid result because the number of push-ups must be counted in whole numbers.

Activity 3 • Solving Inequalities 35

ACTIVITY 3 Continued

Developing Math Language
This lesson contains the vocabulary terms *graph of an inequality* and *solution of an inequality*. Have students add the terms to their math notebooks, and encourage them to include an example of each term. For instance, they can sketch the graph of $x \geq 2$ to illustrate the graph of an inequality, and then list the *x*-values 3 and 4 as examples of solutions of the inequality. Use the class Word Wall to keep the new terms in front of students as they work on this activity.

4–7 Interactive Word Wall, Construct an Argument, Debriefing In these items, students determine whether given values are solutions of an inequality. They also compare solutions of an inequality with the solution of the related equation. When students complete the table in Item 4, make sure they respond "yes" or "no" as well as showing their work. Based on their experience with linear equations in one variable, some students may assume that linear inequalities have a single solution. Use these items as formative assessment opportunities to determine whether students understand that inequalities generally have infinitely many solutions.

8–9 Think-Pair-Share, Quickwrite, Debriefing In these items, students compare and contrast the graphs of two inequalities and draw conclusions about when the endpoint should be an open circle or a solid circle and when the ray should point to the right or the left. The questions in Item 9 point out critical differences students may have missed in Item 8. Students should not associate the graph as a result of the inequality symbol only, but instead should check values to see the inequality results in a true statement. This is because the inequalities $x < 3$ and $3 > x$ are equivalent.

ACTIVITY 3 continued

Lesson 3-1
Inequalities and Their Solutions

4. Use the table below to figure out which *x*-values are solutions to the equation and which ones are solutions to the inequality. Show your work in the rows of the table.

x-values	Solution to the equation? $2x + 3 = 5$	Solution to the inequality? $2x + 3 > 5$
−1	$2(-1) + 3 \neq 5$; no	$2(-1) + 3$ is not > 5; no
0	$2(0) + 3 \neq 5$; no	$2(0) + 3$ is not > 5; no
1	$2(1) + 3 = 5$; yes	$2(1) + 3$ is not > 5; no
2	$2(2) + 3 \neq 5$; no	$2(2) + 3 > 5$; yes
8.5	$2(8.5) + 3 \neq 5$; no	$2(8.5) + 3 > 5$; yes

5. How many solutions are there to the equation $2x + 3 = 5$? Explain.
   There is only one solution. Only one value will make the equation true.

6. Which numbers in the table are solutions to the inequality $2x + 3 > 5$? Are these the only solutions to the inequality? Explain.
   2 and 8.5; no. Any number greater than 1 is a solution.

7. Would 1 be a solution to the inequality $2x + 3 \geq 5$? Explain.
   Yes. The symbol $\geq$ includes "equal to."

Here are the number line graphs of two different inequalities.

$x < 3$ 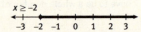 $x \geq -2$

8. Use the graphic organizer to compare and contrast the two inequalities and graphs that are shown above.
   Answers may vary.

Similarities	Differences
Some of the solutions are the same for these inequalities.	The arrows on the number lines go in different directions indicating different ranges of values.
The graphs both show a range of values.	The graph on the left shows an open circle, which does not include the value of 3; the graph on the right shows a filled-in circle, which does include the value of −2.
	The inequalities have different signs, one less than and the other greater than or equal to.

MINI-LESSON: Verifying Solutions to Inequalities

If students need additional help with determining whether given values are solutions of an inequality, a mini-lesson is available to provide practice.

See SpringBoard's eBook Teacher Resources for a student page for this mini-lesson.

Lesson 3-1
Inequalities and Their Solutions

9. Think about why the graphs are different.
   a. Why is one of the graphs showing a solid ray going to the left and the other graph showing a solid ray going to the right?
   The first graph has a solid ray going to the left because numbers that are less than 3 are to the left of 3 on a number line. The second graph has a solid ray going to the right because numbers that are greater than 2 are to the right of 2 on the number line.
   b. Why does one graph have an open circle and the other graph a filled-in circle?
   An open circle indicates that the value at that point is not included in the solutions. A filled-in circle indicates that the value at that point is included in the solutions.

Check Your Understanding

10. Write an inequality to represent each statement.
    a. x is less than 12.
    b. m is no greater than 35.
    c. Your height h must be at least 42 inches for you to ride a theme park ride.
    d. A child's age a can be at most 12 for the child to order from the children's menu.

11. How are the graphs of $x > -4$ and $x \geq -4$ alike and how are they different?

LESSON 3-1 PRACTICE

Graph each inequality on a number line.

12. $x < -2$
13. $x \geq 5$
14. $x < 4$
15. $x > -2$
16. $x \geq \frac{1}{2}$

17. Write a real-world statement that could be represented by the inequality $x \leq 6$.

18. **Attend to precision.** Consider the inequalities $x \leq -3$ and $x \geq -3$.
    a. Graph $x \leq -3$ and $x \geq -3$ on the same number line.
    b. Describe any overlap in the two graphs.
    c. Describe the combined graphs.

WRITING MATH

An open circle represents $<$ or $>$ inequalities, and a filled-in circle represents $\leq$ or $\geq$ inequalities.

LESSON 3-1 PRACTICE

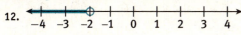
12.

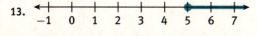
13.

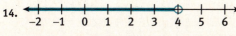
14.

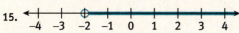
15.

16.

17. Possible answer: There can be at most 6 people on a team.

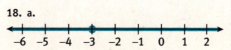
18. a.

b. The overlap is the value of -3.
c. The combined graphs form the real number line.

ACTIVITY 3 Continued

Lesson 3-2

PLAN

Pacing: 1 class period
Chunking the Lesson
#1–4
Check Your Understanding
#7–10   #11–13
Example A
Check Your Understanding
Lesson Practice

TEACH

Bell-Ringer Activity
Write the equation $-4x + 5 = -2x - 1$ on the board. Have students discuss the steps they would use to solve this equation. Use the discussion as an opportunity to review the properties of equality before students begin working with the properties of inequality in this lesson.

1–4 Marking the Text, Create Representations, Guess and Check, Activating Prior Knowledge, Debriefing In these items, students write an inequality and then use two different methods to solve it: first by guess and check and then algebraically. Be sure to debrief Item 1 to make sure that all students have written the correct inequality before moving on to solve it. Have students discuss the advantages and disadvantages of each method of solving the inequality.

TEACHER to TEACHER

As students begin to solve the inequality algebraically, you may want to introduce the Addition and Subtraction Properties of Inequality. (The Multiplication and Division Properties can wait until later in the lesson.) The Addition Property of Inequality states that if $a > b$, then $a + c > b + c$. Likewise, the Subtraction Property of Inequality states that if $a > b$, then $a - c > b - c$. These properties are also true for inequalities that include the signs $<, \geq,$ and $\leq$.

ACTIVITY 3 continued

Lesson 3-2 Solving Inequalities

Learning Targets:
- Write inequalities to represent real-world situations.
- Solve multi-step inequalities.

SUGGESTED LEARNING STRATEGIES: Create Representations, Guess and Check, Look for a Pattern, Think-Pair-Share, Identify a Subtask

1. **Make sense of problems.** Chloe and Charlie are taking a trip to the pet store to buy some things for their new puppy. They know that they need a bag of food that costs $7, and they also want to buy some new toys for the puppy. They find a bargain barrel containing toys that cost $2 each.
   a. Write an expression for the amount of money they will spend if the number of toys they buy is t.
      $2t + 7$

   b. Chloe has $30 and Charlie has one-third of this amount with him. Use this information and the expression you wrote in Part (a) to write an inequality for finding the number of toys they can buy.
      Answers may vary. $40 \geq 2t + 7$

There are different methods for solving the inequality you wrote in the previous question. Chloe suggests that they guess and check to find the number of new toys that they could buy.

2. Use Chloe's suggestion to find the number of new puppy toys that Chloe and Charlie can buy with their combined money.
   Since the number of toys must be a whole number, students will probably use whole numbers as they guess and check. They will find that 16 is a solution of the inequality but 17 is not. So Chloe and Charlie may buy 16 or fewer toys.

Charlie remembered that they could use algebra to solve inequalities. He imagined that the inequality symbol was an equal sign. Then he used equation-solving steps to solve the inequality.

3. Use Charlie's method to solve the inequality you wrote in Item 1b.

$$40 \geq 2t + 7$$
$$\underline{-7 \quad\quad -7}$$
$$\frac{33}{2} \geq \frac{2t}{2}$$
$$16.5 \geq t$$

Lesson 3-2
Solving Inequalities

4. Did you get the same answer using Charlie's method as you did using Chloe's method? Explain.
   In the guess and check method used by Chloe, students probably found whole number solutions. Because Charlie's method allows for a result like 16.5, the answer needs to be reviewed in the real-world situation. It needs to be modified to read, "$16 \geq t$ for whole number values of t" because Charlie and Chloe cannot buy part of a toy.

Check Your Understanding

5. How would you graph the solution to Charlie and Chloe's inequality?
6. Jaden solved an inequality as shown below. Describe and correct any errors in his work.
$$3x + 5 + 6x > 23$$
$$9x + 5 > 23$$
$$9x > 28$$
$$x > 7\tfrac{1}{9}$$

Chloe liked the fact that Charlie's method for solving inequalities did not involve guess and check, so she asked him to show her the method for the inequality $-2x - 4 > 8$.

Charlie showed Chloe the work below to solve $-2x - 4 > 8$.
$$-2x - 4 > 8$$
$$-2x - 4 + 4 > 8 + 4$$
$$-2x > 12$$
$$\frac{-2x}{-2} > \frac{12}{-2}$$
$$x > -6$$

When Chloe went back to check the solution by substituting a value for x back into the original inequality, she found that something was wrong.

7. Confirm or disprove Chloe's conclusion by substituting values for x into the original inequality.
   Chloe could confirm that Charlie's answer was incorrect by choosing a number greater than -6 to substitute into the original inequality.

Lesson 3-2
Solving Inequalities

Chloe tried the problem again but used a few different steps.

$$-2x - 4 > 8$$
$$-2x + 2x - 4 > 8 + 2x$$
$$-4 > 8 + 2x$$
$$-4 - 8 > 8 - 8 + 2x$$
$$-12 > 2x$$
$$\frac{-12}{2} > \frac{2x}{2}$$
$$-6 > x$$

Chloe concluded that $x < -6$.

8. Is Chloe's conclusion correct? Explain.

   Chloe's conclusion is correct. Substituting any number less than -6 in the inequality results in a true statement.

9. Explain what Chloe did to solve the inequality.

   Chloe added 2x to both sides of the equation to avoid having a negative coefficient that would require dividing by a negative number.

Charlie looked back at his work. He said that he could easily fix his work by simply switching the inequality sign.

10. **Critique the reasoning of others.** What do you think about Charlie's plan? Explain.

    Charlie could just switch the inequality sign when he divides by -2. Any inequality that involves multiplication and/or division by a negative number can be solved correctly by switching the sign at the point of the multiplication or division.

Although all of these methods worked, Charlie and Chloe wanted to know why they were working.

Here is an experiment to discover what went wrong with Charlie's first method. Look at what happens when you multiply or divide by a negative number.

Directions	Numbers	Inequality
Pick two different numbers.	2 and 4	$2 < 4$
Multiply both numbers by 3.	2(3) and 4(3)	$6 < 12$
Multiply both numbers by -3.	$2(-3)$ and $4(-3)$	$-6 > -12$
Divide both numbers by 2.	$2 \div 2$ and $4 \div 2$	$1 < 2$
Divide both numbers by -2.	$2 \div (-2)$ and $4 \div (-2)$	$-1 > -2$

11. Try this experiment again with two different numbers. Record your results in the *My Notes* section of this page. Compare your results to the rest of your class.

    Answers may vary but should show the same results as the table above.

Lesson 3-2
Solving Inequalities

12. Express regularity in repeated reasoning. What happens each time you multiply each side of an inequality by a negative number? What happens each time you divide each side of an inequality by a negative number?

When multiplying by a negative number, the values on each side of the inequality change in relationship to each other. For example, 2 is less than 4, but when the inequality is multiplied by -3, -6 is greater than -12. Similarly, when dividing each side by a negative number, the relationship between the numbers on either side of the inequality sign also changes.

13. How does this affect how you solve an inequality?

When you multiply or divide by a negative number, the inequality sign must be reversed so that it reflects the change in the relationship of the numbers on either side of the inequality.

Example A
Solve and graph: $-5x + 8 \leq -2x + 23$

Step 1: Subtract 8 from both sides.
$$-5x + 8 - 8 \leq -2x + 23 - 8$$
$$-5x \leq -2x + 15$$

Step 2: Add $2x$ to both sides.
$$-5x + 2x \leq -2x + 2x + 15$$
$$-3x \leq 15$$

Step 3: Divide both sides by -3.
Remember to reverse the inequality sign.
$$\frac{-3x}{-3} \geq \frac{15}{-3}$$
$$x \geq -5$$

Solution: $x \geq -5$

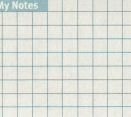

Try These A
Solve and graph each inequality.

a. $3 - 4x \leq 11$
   $x \geq -2$

b. $6 - 3(x + 2) > 15$
   $x < -5$
   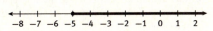

c. $2(x + 5) < 8(x - 3)$
   $x > 5\frac{2}{3}$

MATH TIP
Substitute some sample answers back into the original inequality to check your work.

ACTIVITY 3 Continued

Check Your Understanding

Debrief this lesson by asking students to explain how solving multi-step inequalities is similar to and different from solving multi-step equations. Ask students to share their answers to the Check Your Understanding questions. Discuss with students why more than one inequality can have the same solution and the same graph.

Answers

14. Sample answers: $x < 3$, $3x + 2 < 11$
15. Sample answer: When you multiply or divide both sides of an inequality by a negative number, the relationship of the numbers on each side of the inequality sign changes order, so the inequality sign reverses.

ASSESS

Students' answers to lesson practice problems will provide you with a formative assessment of their understanding of the lesson concepts and their ability to apply their learning.

See the Activity Practice for additional problems for this lesson. You may assign the problems here or use them as a culmination for the activity.

LESSON 3-2 PRACTICE

16. $x > -1$

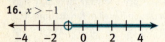

17. $x < 1$

18. $x > 2$

19. $x \geq -\frac{2}{3}$

ADAPT

Check students' answers to the Lesson Practice to ensure that they understand how to write, solve, and graph the solutions of multi-step inequalities. Watch for students who reverse the inequality sign when it is not necessary or who do not reverse the inequality sign when it is necessary. Reinforce student's understanding of the properties of inequality as you discuss when the inequality sign should and should not be reversed.

ACTIVITY 3 continued

My Notes

MATH TIP

Remember to substitute some sample answers back into the original inequality to check your work.

Lesson 3-2
Solving Inequalities

Check Your Understanding

14. Write two different inequalities that have the solution graphed on the number line below.

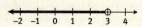

15. Explain why you reverse the inequality sign when you multiply or divide both sides of an inequality by a negative number.

LESSON 3-2 PRACTICE

Solve and graph each inequality.

16. $5 < 3x + 8$
17. $5 < -3x + 8$
18. $3x - 8 + 4x > 6$
19. $-5x + 2 \geq -8x$
20. $4 - 2(x + 1) < 18$
21. $-6x - 3 \leq -4x + 1$
22. $3(x + 7) \geq 2(2x + 8)$
23. $-2x - 3 + 8 < -3(3x + 5)$

24. **Model with mathematics.** Riley and Rhoda plan to buy several bags of dog food and a dog collar. Each bag of dog food costs $7, and the dog collar costs $5.
    a. Use the information above to write an expression for the amount of money they will spend if they buy b bags of food.
    b. Riley and Rhoda have $30. Use this information and the expression you wrote for Part (a) to write an inequality for finding the number of bags of food they could buy.
    c. Solve the inequality and graph the solutions. Check your answer in the original situation.

25. In Example A, $-5x + 8 \leq -2x + 23$ was solved by dividing each side of the inequality by -3 in the last step. Is there another way to solve this inequality so that you can avoid dividing by a negative number? Explain.

---

20. $x > -8$

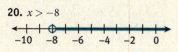

21. $x \geq -2$

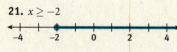

22. $x \leq 5$

23. $x < -\frac{20}{7}$

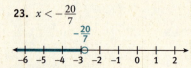

24. a. $7b + 5$
    b. $7b + 5 \leq 30$
    c. $b \leq 3\frac{4}{7}$

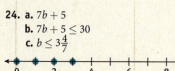

25. Yes; -23 and $5x$ could be added to both sides, resulting in the inequality $-15 \leq 3x$. This inequality can be solved by dividing by 3.

Lesson 3-3
Compound Inequalities

ACTIVITY 3 continued

My Notes

Learning Targets:
- Graph compound inequalities.
- Solve compound inequalities.

SUGGESTED LEARNING STRATEGIES: Vocabulary Organizer, Look for a Pattern, Create Representations, Think-Pair-Share, Note Taking

Compound inequalities are two inequalities joined by the word *and* or by the word *or*. Inequalities joined by the word *and* are called **conjunctions**. Inequalities joined by the word *or* are **disjunctions**. You can represent compound inequalities using words, symbols, or graphs.

1. Complete the table. The first two rows have been done for you.

Verbal Description	Some Possible Solutions	Inequality	Graph
all numbers from 3 to 8, inclusive	3.5, 4, $4\frac{1}{3}$, 5, 6, 7.9, 8	$x \geq 3$ and $x \leq 8$	number line from 2 to 9, closed circles at 3 and 8
all numbers less than 5 or greater than 10	-2, 0, 3, 4, 4.8, $10\frac{3}{4}$, 11	$x < 5$ or $x > 10$	number line from 3 to 11, open circles at 5 and 10
all numbers greater than -1 and less than or equal to 4	Answers may vary. 0, 0.25, 1, 2, 3.5	$x > -1$ and $x \leq 4$	number line from -3 to 5, open circle at -1, closed at 4
all numbers less than or equal to 3 or greater than 6	Answers may vary. 0, 0.5, 1.6, 2, 6.05, 7, 9	$x \leq 3$ or $x > 6$	number line from 0 to 8, closed circle at 3, open at 6

2. Use the graphic organizer below to compare and contrast the graphs for conjunctions and disjunctions.
   Answers may vary.

Similarities	Differences
Both graphs show ranges of numbers on the number line. **Both graphs have points at the ends that represent the numbers that are or are not solutions to the inequalities.**	**The graphs for the "and" statements have one segment between two numbers, and the graphs for the "or" statements have two rays going in opposite directions.**

ACTIVITY 3 Continued

Example A Create Representations, Levels of Questions, Note Taking In this example, students write and graph compound inequalities to represent real-world situations. The table in this example is a good opportunity to teach the Levels of Questions strategy to students. For instance, students can start by asking literal questions about the table, such as, "What do the numbers 10–20 mean in the boys' push-up column?" As they work through the example, they can ask interpretative questions, such as "Why are individual points drawn on the number line in Step 2 instead of a continuous segment from 10 to 20?" After completing the example, they can move on to more universal questions, such as "In what circumstances are the solutions of an inequality restricted to whole numbers?"

CONNECT TO AP

Inequalities can be expressed in interval notation, which is often used in more advanced math courses. In interval notation, a parenthesis indicates that the following or preceding number is *not* included in the solution set. A bracket indicates that the following or preceding number *is* included in the solution set. Thus, the interval notation for $x > 3$ is $(3, \infty)$, and the interval notation for $x \leq -8$ is $(-\infty, -8]$. The interval notation for $-3 \leq x \leq 7$ is $[-3, 7]$, and the interval notation for $-6 < x \leq 9$ is $(-6, 9]$.

3 Activating Prior Knowledge, Think-Pair-Share, Debriefing This item provides another opportunity to remind students to consider the solutions of an inequality in the context of the situation. Refer back to earlier discussions in this activity about how the number of push-ups and pull-ups must each be a whole number while the time needed to run a mile can be measured in partial units.

ACTIVITY 3 continued

Lesson 3-3
Compound Inequalities

WRITING MATH

The inequality $p \geq 10$ is equivalent to the inequality $10 \leq p$. So the compound inequality "$p \geq 10$ and $p \leq 20$" can also be written as "$10 \leq p \leq 20$."

CONNECT TO AP

In upper-level mathematics classes, inequalities are expressed in interval notation. The interval notation for $x > 3$ is $(3, \infty)$.

Example A

Spartan Middle School distributes this chart to students each year to show what students must be able to do to pass the fitness test.

Age	Mile Run (min:sec)		Push-Ups		Modified Pull-Ups	
	Boys	Girls	Boys	Girls	Boys	Girls
12	8:00–10:30	9:00–12:00	10–20	7–15	7–20	4–13
13	7:30–10:00	9:00–12:00	12–25	7–15	8–22	4–13

Write and graph a compound inequality that describes the push-up range for 12-year-old boys.

Step 1: Choose a variable.
Let p represent the number of push-ups for 12-year-old boys.

Step 2: Determine the range and write an inequality.
The range is whole numbers p such that $p \geq 10$ and $p \leq 20$.

Solution: The compound inequality is $10 \leq p \leq 20$.

9 10 11 12 13 14 15 16 17 18 19 20 21

Try These A

Write and graph a compound inequality for each range or score.

a. the push-up range for 13-year-old boys
 $12 \geq x$ and $x \leq 25$, which can be written as $12 \leq x \leq 25$
 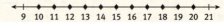
 11 12 13 14 15 16 17 18 19 20 21 22 23 24 25 26

b. the pull-up range for 13-year-old girls
 $x \geq 4$ and $x \leq 13$, which can be written as $4 \leq x \leq 13$

 3 4 5 6 7 8 9 10 11 12 13 14

c. the mile run range for 12-year-old girls
 $x \geq 9$ and $x \leq 12$, which can be written as $9 \leq x \leq 12$

 6 7 8 9 10 11 12 13 14 15

d. the mile run range for 13-year-old boys
 $x \geq 7.5$ and $x \leq 10$, which can be written as $7.5 \leq x \leq 10$

 5 6 7 8 9 10 11 12

e. a score outside the healthy fitness zone for girls' push-ups
 $x < 7$ or $x > 15$

 3 5 7 9 11 13 15 17 19

3. **Attend to precision.** Why are individual points used in the graphs for Example A and some of the graphs in Try These A?
 The individual points are shown because push-ups and pull-ups are counted in units of one. A student cannot count a part of a push-up or a pull-up.

Lesson 3-3
Compound Inequalities

The solution of the conjunction will be *the solutions that are common to both parts*.

Example B
Solve and graph the conjunction: $3 < 3x - 6 < 8$

Step 1: Break the compound inequality into two parts.
$3 < 3x - 6$ and $3x - 6 < 8$

Step 2: Solve and graph $3 < 3x - 6$.
$$3 < 3x - 6$$
$$3 + 6 < 3x - 6 + 6$$
$$9 < 3x$$
$$3 < x \text{ or } x > 3$$

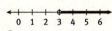

Step 3: Solve and graph $3x - 6 < 8$.
$$3x - 6 < 8$$
$$3x - 6 + 6 < 8 + 6$$
$$3x < 14$$
$$x < 4\tfrac{2}{3}$$

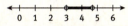

Step 4: Determine what is common to the solutions of each part. In the inequalities and graphs in Steps 2 and 3, the points between 3 and $4\tfrac{2}{3}$ are in common.

Solution: $3 < x < 4\tfrac{2}{3}$

Try These B
Solve and graph each conjunction.

a. $-1 < 3x + 5 < 6$
   $-2 < x < \tfrac{1}{3}$

b. $2 < \tfrac{x}{3} - 5 < 6$
   $21 < x < 33$

c. $3 < 2(x + 2) - 7 \leq 13$
   $3 < x \leq 8$

d. $-2 < 3(x + 6) < 18$
   $-6\tfrac{2}{3} < x < 0$

MATH TIP

Remember to substitute some sample answers back into the original inequality to check your work.

ACTIVITY 3 Continued

Example C Note Taking, Think-Pair-Share, Create Representations In this example, students solve disjunctions and graph their solutions. Make sure students understand how to combine the solutions of the two parts of the compound inequality. The combined solutions include all of the solutions of the first part of the disjunction as well as all of the solutions of the second part.

Check Your Understanding

Debrief this lesson by having students compare and contrast simple inequalities with compound inequalities. Then verify that students understand concepts related to compound inequalities by asking them to share their answers.

Answers

4. $0 < x < 5$
5. The graph of the conjunction is the overlap of the graphs of $x > 2$ and $x < 10$, which is all the points between the points 2 and 10; the graph of the disjunction shows all of the points of both $x > 2$ and $x > 10$.

ASSESS

Students' answers to lesson practice problems will provide you with a formative assessment of their understanding of the lesson concepts and their ability to apply their learning.

See the Activity Practice for additional problems for this lesson. You may assign the problems here or use them as a culmination for the activity.

LESSON 3-3 PRACTICE

6. $x \geq 96$ and $x \leq 110$
7. $x > -2.5$ and $x < 3$

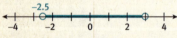

8. $x < 2$ or $x > 3$

9. Sample answer: The price p of a hat is greater than $7.50 but at most $18.50.

ADAPT

Check students' answers to the Lesson Practice to ensure that they understand how to write, solve, and graph the solutions of compound inequalities. Reinforce students' understanding of compound inequalities by reviewing the differences between conjunctions and disjunctions. You can have the students create a chart similar to the one in Item 2 to summarize all the similarities and differences, in addition to those that are graph-related.

46 SpringBoard® Mathematics **Algebra 1, Unit 1** • Equations and Inequalities

---

ACTIVITY 3 continued

My Notes

MATH TIP

Remember to substitute some sample answers back into the original inequality to check your work.

Lesson 3-3
Compound Inequalities

The solution of a disjunction will be *all the solutions from both its parts*.

Example C

Solve and graph the compound inequality: $2x - 3 < 7$ or $4x - 4 \geq 20$.

Step 1: Solve and graph $2x - 3 < 7$.
$$2x - 3 + 3 < 7 + 3$$
$$2x < 10$$
$$x < 5$$

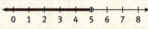

Step 2: Solve and graph $4x - 4 \geq 20$.
$$4x - 4 \geq 20$$
$$4x - 4 + 4 \geq 20 + 4$$
$$4x \geq 24$$
$$x \geq 6$$

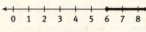

Step 3: Combine the solutions.

Solution: $x < 5$ or $x \geq 6$

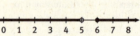

Try These C

Solve and graph each compound inequality.

a. $5x + 1 > 11$ or $x - 1 < -4$
  $x > 2$ or $x < -3$

b. $-5x > 20$ or $x - 2 \geq -7$
  $x < -4$ or $x \geq -5$

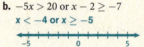

Check Your Understanding

4. The solutions of a conjunction are graphed below. What is the inequality?

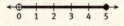

5. Describe the difference in the graph of the conjunction "$x > 2$ and $x < 10$" and the graph of the disjunction "$x > 2$ or $x > 10$."

LESSON 3-3 PRACTICE

6. **Reason quantitatively.** A Category 2 hurricane has wind speeds of at least 96 miles per hour and at most 110 miles per hour. Write the wind speed of a Category 2 hurricane as two inequalities joined by the word *or* or *and*.

Solve and graph each compound inequality on a number line.

7. $-2x + 3 < 8$ and $3(x + 4) - 11 < 10$
8. $-3x + 5 > -1$ or $2(x + 4) > 14$
9. Write a real-world statement that could be represented by the compound inequality $7.50 < p \leq 18.50$.

© 2014 College Board. All rights reserved.

Solving Inequalities
Physical Fitness Zones

ACTIVITY 3 PRACTICE
Write your answers on notebook paper.
Show your work.

Lesson 3-1

1. Describe the similarities and differences in the solutions of $2x - 7 = 15$ and $2x - 7 \leq 15$.

2. For the equation $-3x + 2 = 8$ and the inequality $-3x + 2 > 8$, which x-values indicated below are solutions of the equation and which are solutions of the inequality?
   A. -3
   B. -2
   C. -1
   D. 0

3. Describe the graph of $x > -2$.

4. Describe the graph of $x \leq -2$.

5. Graph $x < 1\frac{1}{2}$ on a number line.

6. Graph $x \geq -3$ on a number line.

Lesson 3-2

7. Mayumi plans to buy pencils and a notebook at the school store. A pencil costs $0.15, and a notebook costs $1.59. Mayumi has $5.00. Which inequality could she use to find the number of pencils she can buy?
   A. $5.00 < 0.15x + 1.59$
   B. $5.00 \geq 0.15x + 1.59$
   C. $5.00 < 0.15x - 1.59$
   D. $5.00 \geq 0.15x - 1.59$

8. Which values of a and b disprove the statement below?
   $$\text{If } a > b, \text{ then } a^2 > b^2.$$
   A. $a = 2, b = 0$
   B. $a = 4, b = 1$
   C. $a = 3, b = -5$
   D. $a = \frac{3}{4}, b = \frac{1}{2}$

Solve the inequality and graph the solutions on a number line. Check your answers.

9. $5x - 4 > -4$
10. $8 > 6 + \frac{2}{5}x$
11. $5 - 3x \leq 8$
12. $3x - 4 \geq 6x + 11$
13. $x - 2 \geq -8x + 16$
14. $3x - 7 < 2(2x - 1)$
15. $2(\frac{1}{2}x - 4) < -(x - 5)$
16. $\frac{2x-11}{3} \geq -x - 2$

Write an inequality that requires more than one step to solve and that has the given solution.

17. $x < -3$
18. $x > 1$

19. Roy is attending his cousin's graduation ceremony in another town. Roy has already driven 30 miles and the ceremony starts in 2 hours.
   a. Let r represent Roy's driving speed in miles per hour. Write an expression to show the total distance Roy will have driven in 2 hours.
   b. The graduation is 150 miles from Roy's home. Use this information and your expression from Part (a) to write an inequality showing the possible speeds Roy could drive to make it to the ceremony on time.
   c. Solve your inequality and graph the solutions. What does your solution mean in the context of the problem?

ACTIVITY 3 Continued
ACTIVITY PRACTICE

1. $2x - 7 = 15$ is an equation with one solution, which is 11. $2x - 7 \leq 15$ is an inequality with infinitely many solutions; its solutions are all real numbers less than or equal to 11.

2.
	Equation	Inequality
a. -3	no	yes
b. -2	yes	no
c. -1	no	no
d. 0	no	no

3. The graph of $x > -2$ has an open circle at -2 and a ray that starts at -2 and goes to the right.

4. The graph of $x \leq -2$ has a closed circle at -2 and a ray that starts at -2 and goes to the left.

5. [number line showing open circle at $1\frac{1}{2}$ with ray to the left]

6. [number line showing open circle at -3 with ray to the left]

7. B
8. C
9. $x > 0$ [number line]
10. $x < 5$ [number line]
11. $x \geq -1$ [number line]
12. $x \leq -5$ [number line]
13. $x \geq 2$ [number line]
14. $x > -5$ [number line]
15. $x < 6.5$ [number line]
16. $x \geq 1$ [number line]
17. Sample answer: $2 + 4x < -10$
18. Sample answer: $3(x + 5) > 18$
19. a. $30 + 2r$
    b. $30 + 2r \geq 150$
    c. $r \geq 60$ [number line]
    Roy must drive at least 60 miles per hour.

ACTIVITY 3 Continued

20. $75 \leq x \leq 97$
21. $20 \leq x \leq 36$
22. **a.** Anna correctly solved the inequality $x + 6 > 11$, but she forgot to include the requirement that $x < 11 + 6$.
    b. $5 < x < 17$
23. Answers may vary. Any value of n such that $n \leq 0$ is correct.
24. $-2 < x < 9$

25. $2 < x < 3$
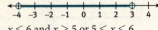
26. $x < 3$ and $x \geq -4$ or $-4 \leq x < 3$
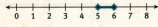
27. $x \leq 6$ and $x \geq 5$ or $5 \leq x \leq 6$
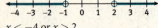
28. $x < -1$ or $x > 2$
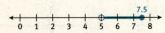
29. $x < -4$ or $x > 2$

30. $5 < x \leq 7.5$

31. $-1 \leq x < 5$

32. $x < -10$ or $x > -5$

33. Adding 5 to a number cannot result in a value that is less than adding 4 to the number.

ADDITIONAL PRACTICE

If students need more practice on the concepts in this activity, see the eBook Teacher Resources for additional practice problems.

ACTIVITY 3 continued

Lesson 3-3

Use the table for Items 20 and 21.

Average High and Low Monthly Temperatures		
	January	August
Austin, TX	41–62°F	75–97°F
Columbus, OH	20–36°F	63–83°F

20. Write a compound inequality for the range of temperatures for Austin, TX, in August.
21. Write a compound inequality for the range of temperatures for Columbus, OH, in January.
22. The sum of the lengths of any two sides of a triangle must be greater than the length of the third side.

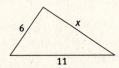

   a. For the triangle shown, Anna said that any value of x greater than 5 is possible. Explain Anna's error.
   b. Write a compound inequality that represents all possible values of x.
23. Find a value for n so that the compound inequality $-n < x < n$ has no solutions.

Solving Inequalities
Physical Fitness Zones

Solve each compound inequality and graph the solutions on a number line. Check your answers.

24. $-2 < 3x + 4 < 31$
25. $1 < 5x - 9 < 6$
26. $-2x + 7 > 1$ and $4x + 3 \geq -13$
27. $0 \leq \frac{6-x}{9}$ and $-2x \leq -10$
28. $5x - 2 < -7$ or $2x + 1 > 5$
29. $7x - 2 < -30$ or $4x + 5 > 13$
30. $\frac{1}{2} < \frac{2x-9}{2} \leq 3$
31. $3 \leq 2(x + 4) - 3 < 15$
32. $-2(x + 2) - 7 > 9$ or $3(x + 3) > -6$

MATHEMATICAL PRACTICES
Construct Viable Arguments and Critique the Reasoning of Others

33. The inequality $x + 5 < x + 4$ has no solutions. Explain why.

Absolute Value Equations and Inequalities
Student Distances
Lesson 4-1 Absolute Value Equations

ACTIVITY 4

Learning Targets:
- Understand what is meant by a solution of an absolute value equation.
- Solve absolute value equations.

SUGGESTED LEARNING STRATEGIES: Paraphrasing, Create Representations, Think-Pair-Share, Note Taking, Identify a Subtask

Ms. Patel is preparing the school marching band for the homecoming show. She has the first row of band members stand in positions along a number line on the floor of the band room. The students' positions match the points on a number line as shown.

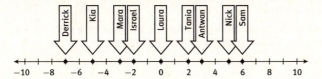

1. Use the number line to write each student's distance from 0 next to their name. For example, Tania is 2 units away from 0. Israel's distance from 0 is also 2 units even though he is at −2.

Derrick	7	Laura	0
Kia	5	Israel	2
Mara	3	Antwan	3
Tania	2	Nick	5
Sam	6		

The *absolute value* of a number is the distance from 0 to the number on a number line. Using *absolute value notation,* Mara's distance is $|-3|$ and Antwan's distance is $|3|$. Since Mara and Antwan are each 3 units from 0, $|-3| = 3$ and $|3| = 3$.

2. **Attend to precision.** Write each person's distance from 0 using absolute value notation.
   Derrick: $|-7|$; Kia: $|-5|$; Mara: $|-3|$; Israel $|-2|$; Laura $|0|$; Tania: $|2|$; Antwan $|3|$; Nick $|5|$; Sam: $|6|$

Absolute value equations can represent distances on a number line.

3. The locations of the two students who are 5 units away from 0 are the solutions of the absolute value equation $|x| = 5$. Which two students represent the solutions to the equation $|x| = 5$?
   Kia and Nick

READING MATH

Read $|-3|$ as "the absolute value of negative three."

MATH TERMS

An **absolute value equation** is an equation involving the absolute value of a variable expression.

Common Core State Standard for Activity 4

HSA-CED.A.1 Create equations and inequalities in one variable and use them to solve problems. Include equations arising from linear and quadratic functions and simple rational and exponential functions.

ACTIVITY 4
Guided

Activity Standards Focus
In Activity 4, students use absolute value equations and absolute value inequalities to solve problems. Students apply the definition of absolute value to write an absolute value equation as two separate equations, which they can solve using the algebraic method. They use a similar process when solving and graphing the solutions of absolute value inequalities.

Lesson 4-1

PLAN

Pacing: 1 class period
Chunking the Lesson
#1–2   #3–4
#5–9   #10–11
Example A   Example B
Check Your Understanding
Lesson Practice

TEACH

Bell-Ringer Activity
Present students with this problem: *Jack is walking on a trail and comes to mile marker 5. If he continues walking for another mile, what is the next mile marker he will come to?* Discuss how this problem has two possible solutions. Depending on which direction Jack is walking, he will come next to either mile marker 4 or mile marker 6.

1–2 Activating Prior Knowledge, Paraphrase, Interactive Word Wall, Create Representations In this set of items, students determine the distance of numbers from 0 on a number line and then connect these distances to the concept of absolute value. Have students paraphrase the instructions for Item 1 to determine whether they understand that distance is a nonnegative quantity, even for the people standing at points to the left of 0.

3–4 Think Pair, Share, Create Representations, Debriefing These items transition students from finding the absolute value of a number to finding values that make an absolute value equation true. Be sure to discuss why the equation $|x| = 5$ has two solutions. Be sure students understand that there is both a positive number that is 5 units from 0 and a negative number that is 5 units from 0. Both 5 and −5 make the equation true when substituted for *x*.

Activity 4 • Absolute Value Equations and Inequalities   49

ACTIVITY 4 Continued

Developing Math Language
This lesson contains the vocabulary terms *absolute value*, *absolute value notation*, and *absolute value equation*. Set aside time for students to record the terms in their math notebooks. Many students will recall that absolute value is never negative, but this is the time to help them refine their definition of absolute value as it relates to distance on a number line.

Differentiating Instruction

Support students who are kinesthetic learners by having them model the situation in Items 1–9 by assuming the roles of the students in the marching band. A taped number line on the floor with equal space between consecutive integers may be helpful.

5–9 Think-Pair-Share, Create Representations, Work Backward, Debriefing In these items, students consider how they can write and solve an absolute value equation that models distance from a point other than 0 on a number line. They start by determining the numbers that are 4 units from 1. They can use these numbers (5 and −3) to work backward and check that they have written the correct equations in Item 7. While students are being expected to solve algebraically, they should be reminded of the conceptual interpretation as well.

Lesson 4-1
Absolute Value Equations

4. You can create a graph on a number line to represent the solutions of an absolute value equation. Graph the solutions of the equation $|x| = 5$ on the number line below. Then use the graph to help you explain why it makes sense that the equation $|x| = 5$ has two solutions.
   There are two solutions because there are two points at a distance of 5 units from 0.

   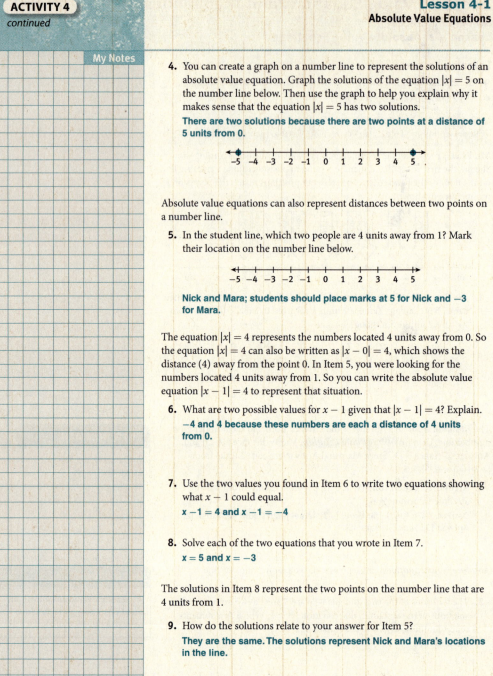

   Absolute value equations can also represent distances between two points on a number line.

5. In the student line, which two people are 4 units away from 1? Mark their location on the number line below.

   Nick and Mara; students should place marks at 5 for Nick and −3 for Mara.

   The equation $|x| = 4$ represents the numbers located 4 units away from 0. So the equation $|x| = 4$ can also be written as $|x - 0| = 4$, which shows the distance (4) away from the point 0. In Item 5, you were looking for the numbers located 4 units away from 1. So you can write the absolute value equation $|x - 1| = 4$ to represent that situation.

6. What are two possible values for $x - 1$ given that $|x - 1| = 4$? Explain.
   −4 and 4 because these numbers are each a distance of 4 units from 0.

7. Use the two values you found in Item 6 to write two equations showing what $x - 1$ could equal.
   $x - 1 = 4$ **and** $x - 1 = -4$

8. Solve each of the two equations that you wrote in Item 7.
   $x = 5$ **and** $x = -3$

   The solutions in Item 8 represent the two points on the number line that are 4 units from 1.

9. How do the solutions relate to your answer for Item 5?
   They are the same. The solutions represent Nick and Mara's locations in the line.

In future math courses, students may be asked to give solutions to equations and inequalities in set notation. A mini-lesson is provided for the teacher to extend this lesson for more advanced students by introducing set notation now.

See SpringBoard's eBook Teacher Resources for a student page for this mini-lesson.

50 SpringBoard® Mathematics **Algebra 1, Unit 1** • Equations and Inequalities

Lesson 4-1
Absolute Value Equations

10. Draw a number line to show the answer to each question. Then write an absolute value equation to represent the points described.

    a. Which two points are 2 units away from 0?

    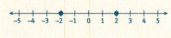

    $|x| = 2$

    b. Which two points are 5 units away from -2?

    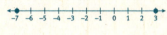

    $|x - (-2)| = 5$

    c. Which two points are 3 units away from 4?

    $|x - 4| = 3$

11. Solve these absolute value equations.

    a. $|x| = 10$
    $-10, 10$

    b. $|x| = -3$
    no solution

    c. $|x + 2| = 7$
    $-9, 5$

    d. $|2x - 1| = 5$
    $-2, 3$

In the equation $|x - 3| = 7$, the 7 indicates that the distance between x and 3 is 7 units. There are two points that are 7 units from 3. These can be found by solving the equations $x - 3 = 7$ and $x - 3 = -7$. Rewriting an absolute value equation as two equations allows you to solve the absolute value equation using algebra.

MATH TIP

If there are no values of x that make an equation true, the equation has **no solution**.

Example A

Solve the equation $|x - 3| = 7$.

Step 1: Rewrite the equation as two equations that do not have absolute value symbols.

$x - 3 = 7 \qquad x - 3 = -7$

Step 2: Solve $x - 3 = 7$.
$x - 3 + 3 = 7 + 3$
$x = 10$

Step 3: Solve $x - 3 = -7$.
$x - 3 + 3 = -7 + 3$
$x = -4$

Solution: $10, -4$

ACTIVITY 4 Continued

Example B Note Taking, Identify a Subtask, Discussion Groups, Debriefing This example shows students how to solve a more complex absolute value equation, one in which the absolute value expression must be isolated before the equation can be rewritten as two equations with absolute value symbols. After students have worked on the Try These problems individually, have them explain their solutions and steps in their discussion groups. Monitor group discussions to ensure that all members of the group are participating and that each member understands the language and terms used in the discussion.

TEACHER to TEACHER

Some students may wonder whether it is necessary to isolate the absolute value expression before rewriting an absolute value equation as two separate equations. Use the equation $|x - 1| + 2 = 4$ as an example. If students solve the equation by first isolating the absolute value expression, they will get the correct solution of $x = 3, x = -1$. Students who do not first isolate the absolute value expression might incorrectly rewrite the equation as $x - 1 + 2 = 4$ and $x - 1 + 2 = -4$. Solving these equations gives an incorrect solution of $x = 3, x = -5$.

ACTIVITY 4 continued

My Notes

Lesson 4-1
Absolute Value Equations

Try These A
Solve each absolute value equation.
a. $|x - 5| = 1$
   4, 6
b. $|x + 5| = 2$
   $-7, -3$
c. $|2x + 3| = 11$
   $-7, 4$
d. $|3x - 4| = 8$
   $-1\frac{1}{3}, 4$

You may need to first isolate the absolute value expression to solve an absolute value equation.

Example B
Solve the equation $6|x + 2| = 18$.

Step 1: Divide both sides of the equation by 6 so that the absolute value expression is alone on one side of the equation.
$$\frac{6|x+2|}{6} = \frac{18}{6}$$
$$|x + 2| = 3$$

Step 2: Rewrite the equation as two equations that do not have absolute value symbols.
$x + 2 = 3$ and $x + 2 = -3$

Step 3: Solve $x + 2 = 3$.
$x + 2 - 2 = 3 - 2$
$x = 1$

Step 4: Solve $x + 2 = -3$.
$x + 2 - 2 = -3 - 2$
$x = -5$

Solution: $1, -5$

Try These B
Solve each absolute value equation.
a. $3|x - 1| = 12$
   5, -3
b. $|x| - 14 = 6$
   20, -20
c. $|x + 4| + 5 = 8$
   $-1, -7$
d. $3|x + 6| - 7 = 20$
   3, -15

MATH TIP
Check your answers by substituting the solutions in the original equation.
$6|1 + 2| = 18$
$6|3| = 18$
$6(3) = 18$
$18 = 18$

$6|-5 + 2| = 18$
$6|-3| = 18$
$6(3) = 18$
$18 = 18$

Lesson 4-1
Absolute Value Equations

> **Check Your Understanding**
>
> 12. Tell whether each statement is true or false. Explain your answers.
>   a. For $x > 0$, $|x| = x$.
>   b. For $x < 0$, $|x| = -x$.
> 13. Kate says that the opposite of $|-6|$ is 6. Is she correct? Explain.

LESSON 4–1 PRACTICE

Draw a number line to show the answer for each question. Then write an absolute value equation that has the numbers described as solutions.

14. Which two numbers are 3 units away from 0?
15. Which two numbers are 4 units away from -1?
16. Which two numbers are 3 units away from 3?

Solve each equation. Check your answers.

17. $|x - 5| = 8$
18. $|-2(x + 2)| = 1$
19. $|-(x - 5)| = 8.5$
20. $|3(x + 1)| = 15$
21. $2|x - 7| = -4$
22. $-2|x - 7| = -4$
23. **Make sense of problems.** Use the equations $|x - 3| = 7$ and $|x| - 3 = 7$ to answer the following questions.
    a. Describe the similarities and differences between the equations.
    b. Which of the following values are solutions of each equation: $-10, -4, 10$?
    c. Are the equations $|x - 3| = 7$ and $|x| - 3 = 7$ equivalent? Explain.
    d. Are the equations $|x| - 3 = 7$ and $|x| = 10$ equivalent? Explain.

---

17. $x = -3, x = 13$
18. $x = -2.5, x = -1.5$
19. $x = -3.5, x = 13.5$
20. $x = -6, x = 4$
21. no solution
22. $x = 5, x = 9$

23. a. Similarities: Both equations use the same numbers, and in both the value on the right is 7. Differences: In $|x - 3| = 7$, the expression on the left side of the equation has 3 inside of the absolute value bars. In $|x| - 3 = 7$, the expression on the left side of the equation has 3 outside of the absolute value bars.
    b. $|x - 3| = 7$ has solutions $x = -4$, $x = 10$; $|x| - 3 = 7$ has solutions $x = -10, x = 10$.
    c. The equations are not equivalent because their solutions are not the same.
    d. The equations are equivalent because their solutions are the same.

ACTIVITY 4 Continued

Lesson 4-2

PLAN

Pacing: 1 class period
Chunking the Lesson
#1–2   #3–6
#7–10 #11–12
Example A
Check Your Understanding
Lesson Practice

TEACH

Bell-Ringer Activity
Present students with this scenario: *The temperature of a heated swimming pool varies from 78°F by no more than 2°F.* Ask students to name the least and greatest possible temperatures of the pool. Then have them write a compound inequality to represent all of the pool's possible temperatures. In this lesson, students will learn how to represent situations similar to this one by using absolute value inequalities.

1–2 Activating Prior Knowledge, Role Play, Create Representations
In these items, students identify numbers that are within a given distance from 0 on a number line. As in Lesson 4-1, students can take the roles of students in the marching band and line up on a taped number line to model the problems. In Item 2, make sure students shade all values on the number line between −3 and 3, not just the integers.

Developing Math Language

Ask students what they think the term *absolute value inequality* means based on what they know about the terms *absolute value* and *inequality*. Then have students add the new term to their math notebooks.

3–6 Interactive Word Wall, Discussion Groups, Guess and Check, Activating Prior Knowledge, Create Representations, Construct an Argument, Debriefing In these items, students identify solutions of an absolute value inequality and then write it as a compound inequality. Be sure students explain how they identified which numbers in Item 3 are solutions. The two decimal values can be used to emphasize that the solutions of the inequality are not restricted to integers. Items 5 and 6 activate students' prior knowledge about compound inequalities. Be sure to debrief the inequalities that are written. This should serve as a formative assessment of student understanding of "and" and "or" inequalities.

ACTIVITY 4 continued

My Notes

Lesson 4-2
Absolute Value Inequalities

Learning Targets:
- Solve absolute value inequalities.
- Graph solutions of absolute value inequalities.

SUGGESTED LEARNING STRATEGIES: Role Play, Visualization, Create Representations, Guess and Check, Think-Pair-Share

Here is the marching band line-up once again.

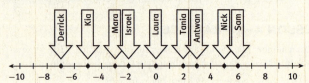

1. Which people in the line up are 3 or fewer units from 0?
   Mara, Israel, Laura, Tania, Antwan

2. Show the portion of the number line that includes numbers that are 3 or fewer units from 0.

The graph you created in Item 2 can be represented with an **absolute value inequality**. The inequality $|x| \leq 3$ represents the numbers on a number line that are 3 or fewer units from 0.

3. Circle the numbers below that are solutions of $|x| \leq 3$. Explain why you chose those numbers.

      3   4   −3.1

   Explanations may vary. I circled the numbers that are 3 or fewer units from 0 on a number line.

4. **Reason abstractly.** How many solutions does the inequality $|x| \leq 3$ have?
   Infinitely many

5. If you were to write a compound inequality for the graph of $|x| \leq 3$ that you sketched in Item 2, would it be a conjunction ("and" inequality) or a disjunction ("or" inequality)? Explain.
   Conjunction. Explanations may vary. The values in common to the two parts are the solutions.

54 SpringBoard® Mathematics Algebra 1, Unit 1 • Equations and Inequalities

Lesson 4-2
Absolute Value Inequalities

6. Write a compound inequality to represent the solutions to $|x| \leq 3$.
   $x \leq 3$ and $x \geq -3$, which can also be written as $-3 \leq x \leq 3$

7. What numbers are more than 4 units away from 3 on a number line? Show the answer to this question on the number line.

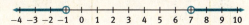

The absolute value inequality $|x - 3| > 4$ represents the situation in Item 7. A "greater than" symbol indicates that the distances are greater than 4.

8. Circle the numbers below that are solutions to the inequality $|x - 3| > 4$. Explain why you chose those numbers.

   (7.1)   7   0   (−2)   6.9   (100)

   Explanations may vary. I circled the numbers that are more than 4 units away from 3.

9. **Construct viable arguments.** If you were to write a compound inequality for the graph of $|x - 3| > 4$ that you sketched in Item 7, would it be a conjunction or disjunction? Explain.
   Disjunction; the solution includes all values of x that make x − 3 farther and farther from 4. This would resemble the two parts of an "or" compound inequality.

10. To solve $|x - 3| > 4$ for x, you need to write the absolute value inequality as a compound inequality.
    a. Based on the graph from Item 7, the expression $x - 3$ is either greater than 4 or less than -4. Write this statement as a compound inequality.
       $x − 3 > 4$ or $x − 3 < −4$

    b. Solve each of the inequalities you wrote in Part (a). Graph the solution.
       $x > 7$ or $x < −1$

ACTIVITY 4 Continued

11–12 Create Representations, Guess and Check, Think-Pair-Share In these items, students write and solve absolute value inequalities by using number lines. These items give students more practice in connecting absolute value inequalities with a pictorial representation before they move on to solve these types of inequalities algebraically. Have students share their inequalities and graphs, both to serve as a learning tool and as a formative assessment.

Example A Note Taking, Identify a Subtask, Think-Pair-Share, Debriefing This example shows students how to solve an absolute value inequality using the algebraic method. Make sure students understand how to write the absolute value inequality as a compound inequality once the absolute value expression is isolated: If the inequality includes $<$ or $\leq$, the relationship is a conjunction and involves "and". If the inequality includes $>$ or $\geq$, the relationship is a disjunction and involves "or". After solving individually, students should share solutions to the Try These items with their partners and then their groups. Remind students to check their work by using substitution when solving the Try These problems.

Teacher to Teacher

In Example A, the definition of absolute value can be used to to rewrite the inequality $|2x| > 6$ as $2x > 6$ or $-(2x) > 6$. Dividing both sides of the second inequality by -1 results in $2x < -6$. Thus, the compound inequality equivalent to $|2x| > 6$ is the disjunction $2x > 6$ or $2x < -6$, as stated in Step 2.

ACTIVITY 4 continued

Lesson 4-2
Absolute Value Inequalities

11. Make a graph that represents the answer to each question. Then write an absolute value inequality that has the solutions that are graphed. Finally, write each absolute value inequality as a compound inequality.

   a. What numbers are less than 2 units from 0?

   $|x| < 2;\ -2 < x < 2$

   b. What numbers are 4 or more units away from 0?
   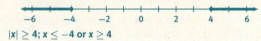
   $|x| \geq 4;\ x \leq -4\ \text{or}\ x \geq 4$

   c. What numbers are 4 or fewer units away from -2?
   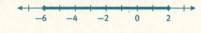
   $|x - (-2)| \leq 4;\ -4 \leq x - (-2) \leq 4$

12. Describe the absolute value inequalities $|x| < 3$ and $|x| > 3$ as conjunctions or disjunctions and justify your choice in each case.
    $|x| < 3$ **indicates the points between -3 and 3. So it is a conjunction.**
    $|x| > 3$ **indicates the points that are farther than 3 units from 0. So it is a disjunction.**

You can solve absolute value inequalities algebraically.

Example A
Solve the inequality $|2x| + 3 > 9$. Graph the solutions.

Step 1: Subtract 3 from both sides of the inequality so that the absolute value expression is alone on one side.
$|2x| + 3 - 3 > 9 - 3$
$|2x| > 6$

Step 2: Rewrite the equation as a compound inequality. Determine if the relationship is "and" or "or."
$2x > 6$ or $2x < -6$

Step 3: Solve the inequalities.
$2x > 6$ or $2x < -6$
$x > 3$ or $x < -3$

Solution: $x > 3$ or $x < -3$

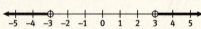

56 SpringBoard® Mathematics **Algebra 1, Unit 1** • Equations and Inequalities

Lesson 4-2
Absolute Value Inequalities

Try These A

Solve each absolute value inequality and graph the solutions.

a. $|2x - 7| > 3$

   $x < 2$ or $x > 5$

   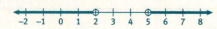

b. $|3x + 8| < 5$

   $-4\frac{1}{3} < x < -1$

   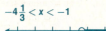

c. $|4x| - 3 \geq 5$

   $x \leq -2$ or $x \geq 2$

d. $2|x| + 7 \leq 11$

   $-2 \leq x \leq 2$

e. $-3|x - 9| \geq -21$

   $x \geq 2$ and $x \leq 16$

f. $-5|2x - 8| < -20$

   $x < 2$ or $x > 6$

   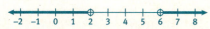

MATH TIP

Remember that when multiplying or dividing each side of an inequality by a negative number, you must reverse the inequality symbol.

ACTIVITY 4 Continued

Check Your Understanding

Debrief this lesson by having students describe the relationship between absolute value inequalities and compound inequalities. Then verify that students understand concepts related to solving an absolute value inequality by asking them to share their answers.

Answers

13. Similarities: For both situations, it is necessary to break up the absolute value equation or inequality into two new equations/inequalities that do not have absolute value symbols. Differences: When solving an absolute value inequality, it is also necessary to determine whether the compound inequality is a conjunction ("and") or a disjunction ("or").

ASSESS

Students' answers to lesson practice problems will provide you with a formative assessment of their understanding of the lesson concepts and their ability to apply their learning.

See the Activity Practice for additional problems for this lesson. You may assign the problems here or use them as a culmination for the activity.

LESSON 4-2 PRACTICE

14. a.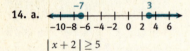
    $|x + 2| \geq 5$
    b.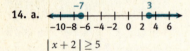
    $|x + 2| \leq 5$
15. $2 \leq x \leq 6$
    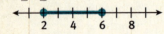
16. $x < 2$ or $x > 8$

17. $x < -13.5$ or $x > -1.5$

ADAPT

Check students' answers to the Lesson Practice to ensure that they understand how to write, solve, and graph the solutions of absolute value inequalities. Reinforce students' understanding by reviewing how to determine when an absolute value inequality can be rewritten as a conjunction and when it can be rewritten as a disjunction.

18. $-3 \leq x \leq 6\frac{1}{3}$

19. $-7 \leq x \leq 7$

20. $x < -4$ or $x > 4$

ACTIVITY 4 continued

My Notes

Lesson 4-2
Absolute Value Inequalities

Check Your Understanding

13. Describe the similarities and differences between solving an absolute value equation and solving an absolute value inequality.

LESSON 4-2 PRACTICE

14. Graph the following and then write an absolute value inequality that represents each question.
    a. What numbers are 5 units or more away from -2 on a number line?
    b. What numbers are 5 units or fewer from -2 on a number line?

Solve each inequality and graph the solutions.

15. $|x - 4| \leq 2$
16. $|x - 5| > 3$
17. $\left|\frac{2}{3}x + 5\right| > 4$
18. $\left|\frac{3x - 5}{2}\right| \leq 7$
19. $|x| - 3 \leq 4$
20. $-3|x| < -12$
21. $6 \leq |2x - 9|$
22. $-5|x + 12| > -35$
23. **Critique the reasoning of others.** Isabelle was asked to write an absolute value inequality to represent the numbers that are less than 1 unit away from 7 on a number line. Isabelle wrote $|x - 1| < 7$. Explain and correct Isabelle's error.

21. $x \leq 1.5$ or $x \geq 7.5$
    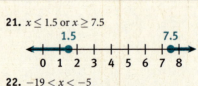
22. $-19 < x < -5$
    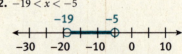
23. Isabelle switched the positions of 1 and 7 in her inequality. She wrote an inequality that represents the numbers that are less than 7 units away from 1 on a number line. The correct inequality is $|x - 7| < 1$.

58 SpringBoard® Mathematics Algebra 1, Unit 1 • Equations and Inequalities

Absolute Value Equations and Inequalities
Student Distances

ACTIVITY 4 PRACTICE
Write your answers on notebook paper.
Show your work.

Lesson 4-1

1. Use a number line to show the numbers that are 3 units from −1 on a number line. Then write an absolute value equation to describe the graph.

2. Explain why $|x| = -5$ does not have a solution.

3. Which graph shows the solutions of $|x - 11| = 7$?

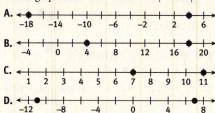

4. Suppose that x is negative and $|x| = n$. What conclusion can you draw?
   A. x and n are opposites.
   B. x and n are equal.
   C. x is greater than n.
   D. None of the above

For Items 5–17, solve each absolute value equation. Check your answers.

5. $|x| = 7$
6. $|x - 2| = -2$
7. $|x - (-2)| = 5$
8. $|3(x - 1)| = 15$
9. $\left|\frac{2}{5}x\right| = 4$
10. $|2(x - 3)| = 10$
11. $|3(x - 2)| = x$
12. $|4(x + 2)| + 9 = 15$
13. $|-2x + 3| = 7$
14. $|-3(x - 7)| = 21$
15. $-5|x - 2| = -20$
16. $-3|x + 5| + 7 = 4$
17. $\left|\frac{2x - 5}{7}\right| = 3$

Lesson 4-2

For Items 18–21, graph the solutions. Then write an absolute value inequality that represents each question.

18. What numbers are more than 3 units from −1 on a number line?

19. What numbers are less than 3 units from −1 on a number line?

20. What numbers are 5 or fewer units away from 3?

21. What numbers are 3 or more units away from 5?

For Items 22–25, graph the solutions of each absolute value inequality and write compound inequalities for the solutions.

22. $|x| > 3$
23. $|x| < 3$
24. $|x - 4| \geq 7$
25. $|x - 4| \leq 7$

26. Which describes the solutions of $|6x - 3| > 21$?
    A. all numbers greater than 4
    B. all numbers greater than −3 and greater than 4
    C. all numbers between −3 and 4
    D. all numbers less than −3 and greater than 4

27. Without solving, match each absolute value equation or inequality with its number of solutions. Justify your answers.

    $|x - 7| < -2$ one solution
    $|x| = 0$ no solutions
    $|x + 1| > -5$ infinitely many solutions

ACTIVITY 4 Continued

28. D
29. Graphic organizers may vary. Organizer may show two points associated with the equation, a section of a number line between two points for the ≤ inequality, and two rays in opposite directions for the > inequality.
30. $x < -1$ or $x > 5$
31. $3 < x < 7$
32. $x \leq -6$ or $x \geq -1$
33. $-4\frac{1}{3} \leq x \leq 3$
34. $-1.8 < x < 3$
35. $x < -3$ or $x > 5$
36. no solution
37. $-10 < x < 0$
38. $-5 \leq x \leq 3$
39. $x \leq -6\frac{1}{3}$ or $x \geq 7$
40. $x \leq \frac{1}{3}$ or $x \geq 2\frac{1}{3}$
41. Answers will vary. Note that any inequality in which the absolute value expression is multiplied by a negative number will work, such as $-3|x| < -5$.

ADDITIONAL PRACTICE
If students need more practice on the concepts in this activity, see the eBook Teacher Resources for additional practice problems.

ACTIVITY 4
continued

Absolute Value Equations and Inequalities
Student Distances

28. The solutions to which absolute value inequality are shown in the graph below?

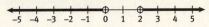

A. $|x + 1| < 1$
B. $|x + 1| > 1$
C. $|x - 1| < 1$
D. $|x - 1| > 1$

29. Create a graphic organizer that compares and contrasts the following equation and inequalities.
$|x - 3| = 6$
$|x - 3| \leq 6$
$|x| - 3 > 6$

For Items 30–40, solve each absolute value inequality and graph the solutions.

30. $|x - 2| > 3$
31. $|x - 5| < 2$
32. $|2x + 7| \geq 5$
33. $|3x + 2| \leq 11$
34. $\left|\dfrac{5x - 3}{2}\right| < 6$
35. $|4(x - 1)| > 16$
36. $|x - 7| + 3 < 2$
37. $|x + 5| - 2 < 3$
38. $|2(x + 1)| - 7 \leq 1$
39. $\left|\dfrac{3x - 1}{4}\right| \geq 5$
40. $-2|3x - 4| \leq -6$
41. Marty said that all absolute value inequalities that contain the symbol $<$ are conjunctions. Give a counterexample to show that Marty's statement is incorrect.

MATHEMATICAL PRACTICES
Model with Mathematics

42. According to some medical websites, normal body temperature can be as much as one degree above or below 98.6°F. Write a compound inequality that shows the range of normal body temperatures, t. Then write an absolute value inequality that shows the same information. Explain how you wrote the absolute value inequality, and include a number line graph in your explanation.

---

42. $97.6 \leq t \leq 99.6$; The graph of the compound inequality is shown below:

It shows all of the points whose distance from 98.6 is less than or equal to 1. The absolute value inequality that represents this situation is $|t - 98.6| \leq 1$.

Inequalities and Absolute Value
DIET AND EXERCISE

Embedded Assessment 2
Use after Activity 4

The table below shows ranges for the daily calorie needs of 15-year-old males according to the United States Department of Agriculture. Use the table for Items 1 and 2.

Daily Calories for 15-Year-Old Males	
Sedentary	no more than 2200 calories
Moderately Active	2400 to 2800 calories
Highly Active	2800 to 3200 calories

1. Write and graph an inequality for the daily number of calories that are recommended for a sedentary 15-year-old male.

2. Use the information for a moderately active 15-year-old male.
   a. Draw a graph on a number line for the daily calorie requirements.
   b. Write a compound inequality for the graph.

It is recommended that teenagers between 14 and 18 years old consume at least 46 grams of protein per day. The table shows the amounts of protein present in various foods. Use the table for Items 3 and 4.

Food	Amount of Protein
Milk	8 grams per cup
Chicken	7 grams per ounce
Beans	16 grams per cup
Yogurt	11 grams per cup

3. Darwin is 15 years old. So far today, he has consumed a total of 25 grams of protein. He plans to eat chicken at dinner.
   a. Let c represent the number of ounces of chicken Darwin eats at dinner. Write an inequality to show how much chicken Darwin can eat and meet the minimum requirement for daily protein.
   b. Solve your inequality from Part (a) and graph the solutions. How many ounces of chicken must Darwin eat?

4. Describe at least two other foods or combinations of foods from the table that Darwin could eat at dinner and meet the minimum requirement for daily protein. Justify your answers.

5. Darwin also keeps track of his target heart rate when he is exercising. The range for the number of heart beats per minute R for someone his age is $|R - 136| \leq 20$. Determine the solutions to the inequality and graph them on a number line.

Embedded Assessment 2

Assessment Focus
- Writing, solving, and graphing inequalities
- Writing and graphing compound inequalities
- Solving and graphing absolute value inequalities

TEACHER to TEACHER
Encourage students to use reading strategies such as Marking the Text and Close Reading as they work through the assessment.

Answer Key

1. $C \leq 2200$

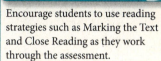

2. a.

   b. $2400 \leq C \leq 2800$

3. a. $25 + 7c \geq 46$
   b. $c \geq 3$

   Darwin must eat at least 3 ounces of chicken.

4. Darwin needs to consume at least $46 - 25 = 21$ grams of protein. 2 cups of beans and 1 cup of milk = $2(16) + 8 = 40$ grams; 1 cup of beans and 1 cup of yogurt = $16 + 11 = 27$ grams.

5. $116 \leq R \leq 156$

Common Core State Standards for Embedded Assessment 2

HSA-CED.A.1	Create equations and inequalities in one variable and use them to solve problems. Include equations arising from linear and quadratic functions and simple rational and exponential functions.
HSA-CED.A.3	Represent constraints by equations or inequalities and by systems of equations and/or inequalities and interpret solutions as viable or non-viable options in a modeling context.
HSA-REI.B.3	Solve linear equations and inequalities in one variable, including equations with coefficients represented by letters.

Embedded Assessment 2

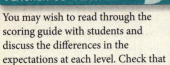

TEACHER to TEACHER

You may wish to read through the scoring guide with students and discuss the differences in the expectations at each level. Check that students understand the terms used.

Embedded Assessment 2
Use after Activity 4

Inequalities and Absolute Value
DIET AND EXERCISE

Scoring Guide	Exemplary	Proficient	Emerging	Incomplete
	The solution demonstrates these characteristics:			
Math Knowledge and Thinking (Items 1, 2a, 2b, 3a, 3b, 5)	• Clear and accurate understanding of how to solve and graph inequalities, including compound and absolute value inequalities	• Largely correct understanding of how to solve and graph inequalities, including compound and absolute value inequalities	• Partial understanding of how to solve and graph inequalities, including compound and absolute value inequalities	• Inaccurate or incomplete understanding of how to solve and graph inequalities, including compound and absolute value inequalities
Problem Solving (Items 3b, 4, 5)	• Appropriate and efficient strategy that results in a correct answer	• Strategy that may include unnecessary steps but results in a correct answer	• Strategy that results in some incorrect answers	• No clear strategy when solving problems
Mathematical Modeling / Representations (Items 1, 2a, 2b, 3a, 5)	• Clear and accurate understanding of how to write and graph inequalities, including compound and absolute value inequalities, to represent real-world data or a real-world scenario	• Largely correct understanding of how to write and graph inequalities, including compound and absolute value inequalities, to represent real-world data or a real-world scenario	• Partial understanding of how to write and graph inequalities, including compound and absolute value inequalities, to represent real-world data or a real-world scenario	• Little or no understanding of how to write and graph inequalities, including compound and absolute value inequalities, to represent real-world data or a real-world scenario
Reasoning and Communication (Items 3b, 4)	• Clear and accurate conclusions drawn from an inequality and a table of data	• Reasonable conclusions drawn from an inequality and a table of data	• Partially correct conclusions drawn from an inequality and a table of data	• Incomplete or inaccurate conclusions drawn from an inequality and a table of data

Unit 2 Planning the Unit

In this unit, students study functions and function concepts, including domain, range, slope as rate of change, and intercepts. Students write linear functions given a point and a slope, two points, a table of values, an arithmetic sequence, or a graph. They collect and model data with linear, quadratic, or exponential functions.

Vocabulary Development

The key terms for this unit can be found on the Unit Opener page. These terms are divided into Academic Vocabulary and Math Terms. Academic Vocabulary includes terms that have additional meaning outside of math. These terms are listed separately to help students transition from their current understanding of a term to its meaning as a mathematics term. To help students learn new vocabulary:

- Have students discuss meaning and use graphic organizers to record their understanding of new words.
- Remind students to place their graphic organizers in their math notebooks and revisit their notes as their understanding of vocabulary grows.
- As needed, pronounce new words and place pronunciation guides and definitions on the class Word Wall.

Embedded Assessments

Embedded Assessments allow students to do the following:

- Demonstrate their understanding of new concepts.
- Integrate previous and new knowledge by solving real-world problems presented in new settings.

They also provide formative information to help you adjust instruction to meet your students' learning needs.

Prior to beginning instruction, have students unpack the first Embedded Assessment in the unit to identify the skills and knowledge necessary for successful completion of that assessment. Help students create a visual display of the unpacked assessment and post it in your class. As students learn new knowledge and skills, remind them that they will be expected to apply that knowledge to the assessment. After students complete each Embedded Assessment, turn to the next one in the unit and repeat the process of unpacking that assessment with students.

CollegeBoard

AP / College Readiness

Unit 2 continues to hone student understanding of function by:

- Formalizing the language of functions.
- Making the connection that the slope of a line represents a constant rate-of-change.
- Exploring functions and linear functions and their behavior in a variety of ways: numerical, graphical, analytical, and verbal.
- Collecting data and modeling with a linear, quadratic, or exponential function.

Unpacking the Embedded Assessments

The following are the key skills and knowledge students will need to know for each assessment.

Embedded Assessment 1

Representations of Functions,
Bryce Canyon Hiking

- Identify functions and use function notation
- Interpret key features of graphs

Embedded Assessment 2

Linear Functions and Equations,
Text Message Plans

- Model with, write, and use linear functions
- Identify a direct variation

Embedded Assessment 3

Linear Models and Slope as Rate of Change,
A 10K Run

- Make a scatter plot and perform a linear regression
- Interpret slope in a real-world context

Planning the Unit continued

Suggested Pacing

The following table provides suggestions for pacing using a 45-minute class period. Space is left for you to write your own pacing guidelines based on your experiences in using the materials.

	45-Minute Period	Your Comments on Pacing
Unit Overview/Getting Ready	1	
Activity 5	3	
Activity 6	3	
Activity 7	3	
Activity 8	2	
Embedded Assessment 1	1	
Activity 9	3	
Activity 10	4	
Activity 11	4	
Embedded Assessment 2	1	
Activity 12	4	
Activity 13	3	
Embedded Assessment 3	1	
Total 45-Minute Periods	**33**	

Additional Resources

Additional resources that you may find helpful for your instruction include the following, which may be found in the eBook Teacher Resources.

- Unit Practice (additional problems for each activity)
- Getting Ready Practice (additional lessons and practice problems for the prerequisite skills)
- Mini-Lessons (instructional support for concepts related to lesson content)

Functions

2

Unit Overview
In this unit, you will build linear models and use them to study functions, domain, and range. Linear models are the foundation for studying slope as a rate of change, intercepts, and direct variation. You will learn to write linear equations given varied information and express these equations in different forms.

Key Terms
As you study this unit, add these and other terms to your math notebook. Include in your notes your prior knowledge of each word, as well as your experiences in using the word in different mathematical examples. If needed, ask for help in pronouncing new words and add information on pronunciation to your math notebook. It is important that you learn new terms and use them correctly in your class discussions and in your problem solutions.

Academic Vocabulary
- causation

Math Terms
- relation
- function
- vertical line test
- independent variable
- dependent variable
- continuous
- discrete
- y-intercept
- relative maximum
- relative minimum
- extrema
- x-intercept
- parent function
- absolute value function
- direct variation
- constant of variation
- indirect variation
- inverse function
- one-to-one
- arithmetic sequence
- explicit formula
- recursive formula
- slope-intercept form
- point-slope form
- standard form
- scatter plot
- trend line
- correlation
- line of best fit
- linear regression
- quadratic regression
- quadratic function
- exponential regression
- exponential function

ESSENTIAL QUESTIONS

 How can you show mathematical relationships?

 Why are linear functions useful in real-world settings?

EMBEDDED ASSESSMENTS
This unit has three embedded assessments, following Activities 8, 11, and 13. They will give you an opportunity to demonstrate what you have learned.

Embedded Assessment 1:
Representations of Functions — p. 121

Embedded Assessment 2:
Linear Functions and Equations — p. 173

Embedded Assessment 3:
Linear Models and Slope as Rate of Change — p. 207

63

Unit Overview
Ask students to read the unit overview and mark the text to identify key phrases that indicate what they will learn in this unit.

Materials
- paper cups
- rubber bands
- paper clips
- measuring tapes
- same-sized washers
- butcher paper or tag board
- rulers
- markers
- graph paper
- colored pencils
- graphing calculators

Key Terms
Read through the vocabulary list with students. Assess prior knowledge by asking students to identify and give a description of any terms they feel they are already familiar with. As students encounter new terms in this unit, help them to choose an appropriate graphic organizer for their word study. As they complete a graphic organizer, have them place it in their math notebooks and revisit as needed as they gain additional knowledge about each word or concept.

Essential Questions
Read the essential questions with students and ask them to share possible answers. As students complete the unit, revisit the essential questions to help them revise their initial responses as needed.

Unpacking Embedded Assessments
Prior to beginning the first activity in this unit, turn to Embedded Assessment 1 and have students unpack the assessment by identifying the skills and knowledge they will need to complete the assessments successfully. Guide students through a close reading of the assessment, and use a graphic organizer or other means to capture their identification of the skills and knowledge. Repeat the process for each Embedded Assessment in the unit.

Developing Math Language
As this unit progresses, help students make the transition from general words they may already know (the Academic Vocabulary) to the meanings of those words in mathematics. You may want students to work in pairs or small groups to facilitate discussion and to build confidence and fluency as they internalize new language. Ask students to discuss new academic and mathematical terms as they are introduced, identifying meaning as well as pronunciation and common usage. Remind students to use their math notebooks to record their understanding of new terms and concepts.

As needed, pronounce new terms clearly and monitor students' use of words in their discussions to ensure that they are using terms correctly. Encourage students to practice fluency with new words as they gain greater understanding of mathematical and other terms.

UNIT 2
Getting Ready

Use some or all of these exercises for formative evaluation of students' readiness for Unit 2 topics.

Prerequisite Skills

- Identify and extend patterns (Item 1) 4.OA.C.5; 5.OA.B.3
- Solve and interpret inequalities (Item 2) 6.EE.B.5
- Evaluate algebraic expressions (Item 3) 6.EE.A.2c
- Graph points on the coordinate plane (Items 4, 5 and 8) 6.NS.C.8
- Represent data using an equation (Item 6) 8.F.B.4
- Solve linear equations (Item 7) 7.EE.B.4a; 8.EE.C.7b

Answer Key

1. Column one, row six: 14; Column two, row four: 17
2. $-3, -2, -1, 0, 1, 2, 3$
3. a. 1
   b. 6
4. a. S
   b. R
   c. T
5. Sample answer: Begin at the origin. Count three units to the right and then 4 units down.
6. B
7. D
8. A

UNIT 2
Getting Ready

Write your answers on notebook paper. Show your work.

1. Copy and complete the table of values.

-1	-1
2	5
5	11
8	
11	23
	29

2. List the integers that make this statement true.
   $$-3 \leq x < 4$$

3. Evaluate for $a = 3$ and $b = -2$.
   a. $2a - 5$ b. $3b + 4a$

4. Name the point for each ordered pair.
   a. $(-3, 0)$ b. $(-1, 3)$ c. $(2, -2)$

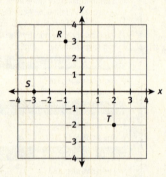

5. Explain how you would plot $(3, -4)$ on a coordinate plane.

6. Which of the following equations represents the data in the table?

x	1	3	5	7
y	2	8	14	20

   A. $y = 2x - 1$ B. $y = 3x - 1$
   C. $y = x + 1$ D. $y = 2x + 1$

7. If $2x + 6 = 2$, what is the value of x?
   A. 4   B. 2   C. 0   D. -2

8. Which of the following are the coordinates of a point on this line?

   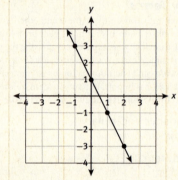

   A. $(-1, 3)$ B. $(1, -3)$
   C. $(-1, -3)$ D. $(1, 3)$

Getting Ready Practice

For students who may need additional instruction on one or more of the prerequisite skills for this unit, Getting Ready practice pages are available in the Teacher Resources at SpringBoard Digital. These practice pages include worked-out examples as well as multiple opportunities for students to apply concepts learned.

Functions and Function Notation
Vending Machines
Lesson 5-1 Relations and Functions

ACTIVITY 5

Learning Targets:
- Represent relations and functions using tables, diagrams, and graphs.
- Identify relations that are functions.

SUGGESTED LEARNING STRATEGIES: Visualization, Create Representations, Think-Pair-Share, Interactive Word Wall, Paraphrasing

Use this machine to answer the questions below.

DVD Vending Machine

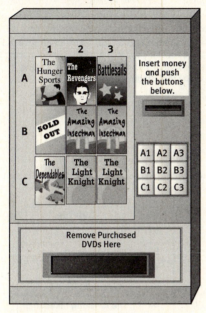

1. What DVD would you receive if you inserted your money and pressed:
   a. A1?
   The Hunger Sports

   b. C2?
   The Light Knight

   c. B3?
   The Amazing Insectman

2. Assuming the machine were filled properly, describe what would happen if you pressed the same button twice.
   You should receive two identical DVDs.

Common Core State Standards for Activity 5

HSF-IF.A.1   Understand that a function from one set (called the domain) to another set (called the range) assigns to each element of the domain exactly one element of the range. If f is a function and x is an element of its domain, then $f(x)$ denotes the output of f corresponding to the input x. The graph of f is the graph of the equation $y = f(x)$.

HSF-IF.A.2:   Use function notation, evaluate functions from inputs in their domains, and interpret statements that use function notation in terms of a context.

ACTIVITY 5
Investigative

Activity Standards Focus
Prior to this activity, students begin to express relationships between quantities, using tables, graphs, and equations. In Activity 5, students learn to recognize a function, identify the domain and range of a function and to use function notation.

Lesson 5-1

PLAN

Pacing: 1 class period
Chunking the Lesson
#1–3   #4   #5
Check Your Understanding
#7–8   #9–12
Check Your Understanding
Lesson Practice

TEACH

Bell-Ringer Activity
On the board write the following:
1 or 2: pencil   3, 4, or 5: eraser
6: excused from homework
Have students pretend that if they roll a number cube, they will win a prize as shown on the board. Ask questions such as, "If you roll a 2, what will you win?" and "Suppose you won an eraser. What number did you roll?"

1–3 Visualization, Activating Prior Knowledge, Think-Pair-Share Use these items and other questions to help students connect their experiences with vending machines to the idea that a specific input results in a specific output.

ACTIVITY 5 Continued

1–3 (continued) To help students understand Item 3, encourage them to explain what "having no output" means in terms of the vending machine. If students are having difficulty with this have pairs identify a button that "has no output" or does not result in a DVD.

Developing Math Language
This lesson has several key math terms. Help students develop the idea of a *mapping* by having them map students' names to their birth months or their heights to their shoe sizes. Compare the use of a table and a mapping to show the same information. Note that when both the input and output consist of numbers, the mapping shows a *relation* and can be written as *ordered pairs*. Have students describe how they can graph ordered pairs.

4 Marking the Text, Create Representations, Sharing and Responding
Students may not be familiar with writing mappings without numerical content, but it is important for students to understand that both representations relay the same information. Be sure students include button B1 and the fact that there is no output in the table.

ACTIVITY 5 continued

My Notes

MATH TERMS
A **mapping** is a visual-representation of a relation in which an arrow associates each input with its output.

MATH TERMS
An **ordered pair** shows the relationship between two elements, written in a specific order using parentheses notation and a comma separating the two values.

MATH TERMS
A **relation** is information that can be represented by a set of ordered pairs.

**Lesson 5-1
Relations and Functions**

Each time you press a button, an **input**, you may receive a DVD, an **output**.

3. In the DVD vending machine situation, does every input have an output? Explain your response.
   No. Button B1 does not have an output since it is sold out.

4. Each combination of input and output can be expressed as a **mapping** written *input → output*. For example, B2 → The Amazing Insectman.
   a. Write as mappings each of the possible combinations of buttons pushed and DVDs received in the vending machine.
      A1 → The Hunger Sports; A2 → The Revengers; A3 → Battlesails;
      B2 → The Amazing Insectman; B3 → The Amazing Insectman;
      C1 → The Dependables; C2 → The Light Knight;
      C3 → The Light Knight

   b. Create a table to illustrate how the inputs and outputs of the vending machine are related.

Button	DVD Title
A1	The Hunger Sports
A2	The Revengers
A3	Battlesails
B1	(no output)
B2	The Amazing Insectman
B3	The Amazing Insectman
C1	The Dependables
C2	The Light Knight
C3	The Light Knight

Mappings that relate values from one set of numbers to another set of numbers can be written as **ordered pairs**. A **relation** is a set of ordered pairs.

Relations can have a variety of representations. Consider the relation {(1, 4),

Lesson 5-1
Relations and Functions

Relations can have a variety of representations. Consider the relation {(1, 4), (2, 3), (6, 5)}, shown here as a set of ordered pairs. This relation can also be represented in these ways.

Table	Mapping	Graph
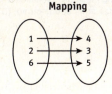	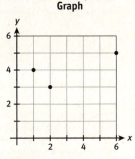	

5. Write the following numerical mappings as ordered pairs.

Input		Output	Ordered Pairs
1	→	−2	(1, −2)
2	→	1	(2, 1)
3	→	4	(3, 4)
4	→	7	(4, 7)

Check Your Understanding

6. A vending machine at the Ocean, Road, and Air show creates souvenir coins. You select a letter and a number and the machine creates a souvenir coin with a particular vehicle imprinted on it. The graph shows the vending machine letter/number combinations for the different coins.

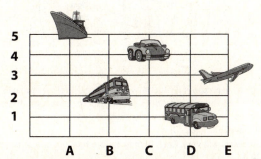

a. Make a table showing each coin's letter/number combination.
b. Write the letter/number combinations as a set of ordered pairs.
c. Write the letter/number combinations in a mapping diagram.

ACTIVITY 5 Continued

Developing Math Language
The concept of a function is important throughout the study of algebra and higher levels of mathematics. After reading the definition of a *function*, ask students to explain it using their own words. Draw a mapping of (2, 3), (1, 4), and (3, 4) as well as one of (1, 5), (3, 4), and (1, 2). Have students describe why the first mapping shows a function but the second does not.

7–8 Paraphrasing, Interactive Word Wall, Vocabulary Organizer Have students relate the concept of a function to the context of each problem. Students should be able to describe why each situation is or is not similar to a function. A vocabulary organizer can be used to show the connection between the meaning of the word *function* and the different representations (ordered pairs, graphs, tables, and mappings) of a function.

ACTIVITY 5 continued

My Notes

Lesson 5-1
Relations and Functions

A *function* is a relation in which each input is paired with exactly one output.

7. Compare and contrast the DVD Vending Machine with a function.
   Answers may vary. The vending machine resembles a function because each input (button pressed) has at most one output (DVD).

8. Suppose when pressing button C1 on the vending machine both "The Dependables" and "The Light Knight" come out. Describe how this vending machine resembles or does not resemble a function.
   Such a vending machine would not resemble a function because one input (C1) yields two different outputs (The Dependables and The Light Knight), and a function cannot have an input with more than one corresponding output.

9. Imagine a machine where you input an age and the machine gives you the name of anyone who is that age. Compare and contrast this machine with a function. Explain by using examples and create a representation of the situation.
   Answers may vary. This would not be a function because each input can be paired with multiple outputs because there are many people who are the same age. Sample representation:

   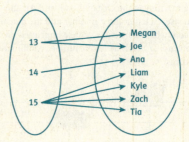

10. Create an example of a situation (math or real-life) that behaves like a function and another that does not behave like a function. Explain why you chose each example to fit the category.
    a. Behaves like a function:
       Answers will vary. Each item in a store has just one price assigned to it.
    b. Does not behave like a function:
       Answers will vary; {(2, 3), (2, 5), (3, 1)}. The input of 2 gives two possible outputs, 3 and 5.

Lesson 5-1
Relations and Functions

11. Determine whether the ordered pairs and equations represent functions. Explain your answers.
    a. {(5, 4), (6, 3), (7, 2)}
       Yes; every input is matched with exactly one output.

    b. {(4, 5), (4, 3), (5, 2)}
       No; the input of 4 can either have an output of 5 or 3.

    c. {(5, 4), (6, 4), (7, 4)}
       Yes; every input is matched with exactly one output.

    d. $y = 3x - 5$, where x represents input values and y represents output values
       Yes; every value of x will result in exactly one value for y.

    e. $y = -x + 4$, where x represents input values and y represents output values
       Yes; every value of x will result in exactly one value for y.

12. **Attend to precision.** Using positive integers, write two relations as lists of ordered pairs below, one that is a function and one that is not a function.

    Function: **Answers will vary; {(3, 5), (5, 8), (1, 9)}**

    Not a function: **Answers will vary; {(3, 5), (3, 8), (1, 9)}**

Check Your Understanding

13. Does the mapping shown represent a function? Explain.

    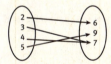

14. Does the graph shown represent a function? Explain.

    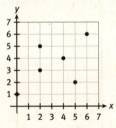

ACTIVITY 5 Continued

9–12 Create Representations, Think-Pair-Share, Debriefing
Students create a visual representation to demonstrate their understanding of a function. They also describe relations that behave like functions. Students should work together to determine when a set of ordered pairs is or is not a function. Debrief the lesson by having students summarize how to recognize a function in each of the different representations.

TEACHER to TEACHER

Make sure students understand that a function is a special type of relation. Note that all functions are relations but not all relations are functions. If students refer to different types of sets of ordered pairs as *functions* or *relations*, remind them that they are all relations and the categories they should be using are *functions* or *not functions*.

Check Your Understanding

Remind students that they must explain their reasoning for each answer. Debrief students' answers to these items to make sure they understand that in a function, each input is paired with exactly one output, although an output may be paired with more than one input.

Answers
13. Yes; because every input is matched with exactly one output.
14. No; because the input value 2 has more than one output.

ACTIVITY 5 Continued

ASSESS

Students' answers to lesson practice problems will provide you with a formative assessment of their understanding of the lesson concepts and their ability to apply their learning.

See the Activity Practice for additional problems for this lesson. You may assign the problems here or use them as a culmination for the activity.

LESSON 5-1 PRACTICE

15. 23
16. G4
17. No; N3 is a free space so it has no numerical output.
18. a. The input value 1 has two outputs.
    b. The input value 17 has two outputs.
    c. There can be more than one output value for a value of x. For example, for the input value 16, the output value could be 4 or -4 because $(4)^2 = 16$ and $(-4)^2 = 16$.

ADAPT

Check students' answers to the Lesson Practice to ensure that they understand that a relation relates an input value to an output value. They should also be able to explain why a relation is or is not a function. Students need a basic understanding of functions before they work with functions in many different contexts.

ACTIVITY 5 continued

Lesson 5-1
Relations and Functions

LESSON 5-1 PRACTICE

For the Bingo card below, suppose that a combination of a column letter and a row number, such as B1, represents an input and the number at that location, such as 7, represents an output. Use this information for Items 15–17.

15. What output corresponds to I2?
16. What input corresponds to 54?
17. Does every input have a numerical output? Explain.
18. **Construct viable arguments.** Explain why each of the following is **not** a function.

    a.
    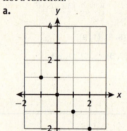

    b.

x	y
12	−8
17	3
−4	9
17	−5

    c. $y^2 = x$, where x represents input values and y represents output values.

Lesson 5-2
Domain and Range

Learning Targets:
- Describe the domain and range of a function.
- Find input-output pairs for a function.

SUGGESTED LEARNING STRATEGIES: Quickwrite, Create Representations, Discussion Groups, Marking the Text, Sharing and Responding

The set of all inputs for a function is known as the *domain* of the function. The set of all outputs for a function is known as the *range* of the function.

1. Consider a vending machine where inserting 25 cents dispenses one pencil, inserting 50 cents dispenses 2 pencils, and so forth up to and including all 10 pencils in the vending machine.

   a. Identify the domain in this situation.
   {0, 0.25, 0.50, 0.75, 1.00, 1.25, 1.50, 1.75, 2.00, 2.25, 2.50}

   b. Identify the range in this situation.
   {0, 1, 2, 3, 4, 5, 6, 7, 8, 9, 10}

2. For each function below, identify the domain and range.

   a.
input	output
7	6
3	−2
5	1

   Domain: {7, 3, 5}
   Range: {6, −2, 1}

   b. Domain: {2, 6, 8}
   Range: {4, −3}

   c.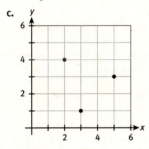

   Domain: {2, 3, 5}
   Range: {1, 3, 4}

   d. {(−7, 0), (9, −3), (−6, 2.5)}
   Domain: {−7, 9, −6}
   Range: {0, −3, 2.5}

WRITING MATH

The **domain** and **range** of a function can be written using set notation.

For example, for the function {(1, 2), (3, 4), (5, 6)}, the domain is {1, 3, 5} and the range is {2, 4, 6}.

ACTIVITY 5 Continued

1–3 (continued) In Item 3, students should be able to explain whether each answer makes sense in terms of the situation. Include in the debriefing an opportunity for students to examine the table, mapping, and graph and describe how they can use each to visually recognize the domain and the range. Include in the discussion why the data points are not connected in part c. This discussion of discrete data will lead to Item 4.

Differentiating Instruction

Support students who do not understand the context of the problem. Sketch a change machine on the board. Draw a box with an arrow pointing to the input slot. Label it *What you can put in the machine*. Have students reread the problem and discuss what is allowed to go into the machine. List the dollar amounts in the box. Draw another box with an arrow pointing to the output slot and label it *What comes out of the machine*. Ask, "If you put $1 in the machine, what would come out?" Write the answer in the output box. Continue until you have an output value for every input value. Revisit the concepts of domain and range in the context of the machine. Then have students read and discuss the questions.

ACTIVITY 5 continued

Lesson 5-2
Domain and Range

3. Consider a machine that exchanges quarters for dollar bills. Inserting one dollar bill returns four quarters and you may insert up to five one-dollar bills at a time.

   a. Is 7 a possible input for the relation this change machine represents? Justify your response.
   No; you may not input more than 5 dollar bills at one time.

   b. Could 3.5 be included in the domain of this relation? Explain why or why not.
   No; you must insert a whole number of dollar bills.

   c. **Reason abstractly.** What values are **not** in the domain? Justify your reasoning.
   Any values other than 0, 1, 2, 3, 4, or 5 are not in the domain because you must insert a whole number of dollar bills and you may not insert more than five bills at one time.

   d. Is 8 a possible output for the relation this change machine represents? Justify your response.
   Yes; if you inserted 2 dollar bills, you would receive 8 quarters.

   e. Could 3 be included in the range of this relation? Explain why or why not.
   No; there is no whole number of dollar bills that is equivalent to 3 quarters.

   f. What values are **not** in the range? Justify your reasoning.
   Any values other than 0, 4, 8, 12, 16, or 20 are not in the range because these are the only numbers of quarters you could possibly receive from the machine at one time.

Lesson 5-2
Domain and Range

4. **Make sense of problems.** Each of the functions that you have seen has a *finite* number of ordered pairs. There are functions that have an *infinite* number of ordered pairs. Describe any difficulties that may exist trying to represent a function with an infinite number of ordered pairs using the four representations of functions that have been described thus far.

   Answers may vary. Each of these representations would require an infinite number of elements, making it impossible to ever complete writing the function.

MATH TERMS

A **finite** set has a fixed countable number of elements. An **infinite** set has an unlimited number of elements.

5. Sometimes, machine diagrams are used to represent functions. In the function machine below, the inputs are labeled x and the outputs are labeled y. The function is represented by the expression $2x + 5$.

   a. What is the output if the input is $x = 7$? $x = -2$? $x = \frac{1}{2}$?
      19; 1; 6

   b. **Express regularity in repeated reasoning.** Is there any limit to the number of input values that can be used with this expression? Explain.
      There is no limit. There are infinitely many possible values to use as inputs.

Consider the function machine below.

6. Use the diagram to find the (input, output) ordered pairs for the following values.
   a. $x = -5$
      (−5, 18)
   b. $x = \frac{3}{5}$
      $\left(\frac{3}{5}, 4\frac{14}{25}\right)$
   c. $x = -10$
      (−10, 83)

ACTIVITY 5 Continued

8 Think-Pair-Share, Create Representations Students move from the function machine diagram to an equation in terms of *x* and *y*. Students may still use the words *input* and *output* but should recognize the relationship between *x* and the *input* and *y* and the *output* of a function. Students should be able to evaluate using the order of operations. The teacher may wish to add zero to the list of domain values for the function in part b. This would yield an input that does not have an output, much like the B1 button in the original vending machine situation.

Differentiating Instruction

Extend students thinking by providing a function machine with a function rule and several output values. Challenge them to find the input value that results in each output value.

Check Your Understanding

Debrief this lesson by having students describe how they can identify the domain and range of functions that are presented using various formats. Have students explain how they know which set of values forms the domain and which set forms the range.

Answers

9. Domain: {3, −1, 2, 0};
   Range: {5, 2, −1}
10. a. Domain: {−1, 0, 1, 2};
       Range: {1, 0, −1, −2}
    b. Domain: {12, 17, −4};
       Range: {−8, 3, 9}

ACTIVITY 5 continued

Lesson 5-2
Domain and Range

7. Make a function machine for the expression $10 - 5x$. Use it to find ordered pairs for $x = 3$, $x = -6$, $x = 0.25$, and $x = \dfrac{3}{4}$.

$(3, -5)$; $(-6, 40)$; $(0.25, 8.75)$; $\left(\dfrac{3}{4}, \dfrac{25}{4}\right)$

Creating a function machine can be time consuming and awkward. The function represented by the diagram in Item 5 can also be written algebraically as the equation $y = 2x + 5$.

8. For each function, find ordered pairs for $x = -2$, $x = 5$, $x = \dfrac{2}{3}$, and $x = 0.75$. Create tables of values.

a. $y = 9 - 4x$

x	y
−2	17
5	−11
$\dfrac{2}{3}$	$\dfrac{19}{3}$
0.75	6

b. $y = \dfrac{1}{x}$

x	y
−2	$-\dfrac{1}{2}$
5	$\dfrac{1}{5}$
$\dfrac{2}{3}$	$\dfrac{3}{2}$
0.75	1.333 or $\dfrac{4}{3}$

Check Your Understanding

9. The set {(3, 5), (−1, 2), (2, 2), (0, −1)} represents a function. Identify the domain and range of the function.

10. Identify the domain and range for each function.

a.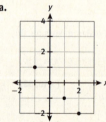

b.
x	y
12	−8
17	3
−4	9

Lesson 5-2
Domain and Range

LESSON 5-2 PRACTICE

Identify the domain and range.

11.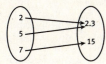

12.
x	y
1.5	4
−0.3	8
$\frac{1}{6}$	3

13. **Model with mathematics.** At an arcade, there is a machine that accepts game tokens and returns tickets that can be redeemed for prizes. Inserting 5 tokens returns 3 tickets and inserting 10 tokens returns 8 tickets. You must insert tokens in multiples of 5 or 10, and you have a total of 20 tokens.
    a. Identify the domain in this situation.
    b. Identify the range in this situation.

14. For the function machine shown, copy and complete the table of values.

x	y
−1	
0	
$\frac{1}{2}$	
1.2	

15. For each function below, find ordered pairs for $x = -1$, $x = 3$, $x = \frac{1}{2}$, and $x = 0.4$. Write your results as a set of ordered pairs.
    a. $y = 4x$
    b. $y = 2 - x^2$

ACTIVITY 5 Continued

Lesson 5-3

PLAN

Pacing: 1 class period
Chunking the Lesson
#1–3   #4–7
Check Your Understanding
Lesson Practice

TEACH

Bell-Ringer Activity
Have students write the Order of Operations on a sheet of paper. Then have them simplify these expressions:
$3 + 4(2)$ $-2^3 + (-2)^3$
$4 - 2(3) + 8$ $3(4)^2 + 7(2)$

Have students compare results. When students' answers differ, have them talk through the processes they used until they can justify which answer is correct.

1–3 Shared Reading, Think Aloud, Create Representations Students have the opportunity to use function notation and to recognize its inherent advantages. Allow ample time for the concepts to develop for students; additional examples may be necessary. After the introduction of function notation, students are brought back to the multiple representations of functions. Students must be constantly aware of the different representations.

TEACHER to TEACHER

When introducing function notation, it is essential that students do not confuse the name of the function with a variable. Students often perceive $f(x)$ as "f multiplied by x" instead of as function notation. Emphasize the Math Tip on the student page and the relationship between the notation $f(x)$ and the value of y.

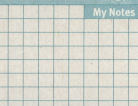

Lesson 5-3
Function Notation

Learning Targets:
- Use and interpret function notation.
- Evaluate a function for specific values of the domain.

SUGGESTED LEARNING STRATEGIES: Create Representations, Discussion Groups

When referring to the functions in Item 8 in Lesson 5-2, it can be confusing to distinguish among them since each begins with "$y =$." Function notation can be used to help distinguish among different functions.

For instance, the function $y = 9 - 4x$ in Item 8a can be written:

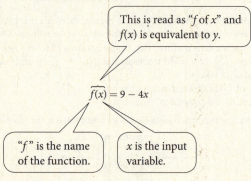

$f(x) = 9 - 4x$

This is read as "f of x" and $f(x)$ is equivalent to y.

"f" is the name of the function.

x is the input variable.

MATH TIP
It is important to recognize that $f(x)$ **does not** mean f multiplied by x.

MATH TIP
Notice that $f(x) = y$. For a **domain** value x, the associated **range** value is $f(x)$.

1. To distinguish among different functions, it is possible to use different names. Use the name h to write the function from Item 8b using function notation.

   $h(x) = \dfrac{1}{x}$

Function notation is useful for evaluating functions for multiple input values. To evaluate $f(x) = 9 - 4x$ for $x = 2$, you substitute 2 for the variable x and write $f(2) = 9 - 4(2)$. Simplifying the expression yields $f(2) = 1$.

2. Use function notation to evaluate $f(x) = 9 - 4x$ for $x = 5$, $x = -3$, and $x = 0.5$.

   $f(5) = -11; f(-3) = 21; f(0.5) = 7$

76 SpringBoard® Mathematics Algebra 1, Unit 2 • Functions

Lesson 5-3
Function Notation

3. Use the values for x and $f(x)$ from Item 2. Display the values using each representation.

   a. list of ordered pairs
   {(5, −11), (−3, 21), (0.5, 7)}

   b. table of values

x	y
5	−11
−3	21
0.5	7

   c. mapping

   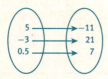

   d. graph

   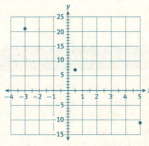

4. Given the function $f(x) = 9 - 4x$ as shown above, what value of x results in $f(x) = 1$?
   x = 2

5. Evaluate each function for $x = -5$ and $x = \frac{4}{3}$.

   a. $f(x) = 2x - 7$
   $f(-5) = -17; f\left(\frac{4}{3}\right) = -\frac{13}{3}$

   b. $g(x) = 6x - x^2$
   $g(-5) = -55; g\left(\frac{4}{3}\right) = \frac{56}{9}$

   c. $h(x) = \frac{2}{x^2}$
   $h(-5) = \frac{2}{25}; h\left(\frac{4}{3}\right) = \frac{9}{8}$

6. **Reason quantitatively.** Recall the money-changing machine from Item 3 in Lesson 5-2, in which customers can insert up to five one-dollar bills at a time and receive an equivalent amount of quarters. The function $f(x) = 4x$ represents this situation. What does x represent? What does $f(x)$ represent?
   x represents the number of dollar bills inserted; $f(x)$ represents the number of quarters received.

ACTIVITY 5 Continued

Check Your Understanding

Debrief students' answers to these items to ensure that they understand how to write and evaluate functions using function notation. This lesson prepares students to use function notation as a tool when analyzing functions and when using functions to solve problems.

Answers

8. a. −1, 14, 26
   b. 14, 0, 0, 14
9. a. −3
   b. 9

ASSESS

Students' answers to lesson practice problems will provide you with a formative assessment of their understanding of the lesson concepts and their ability to apply their learning.

See the Activity Practice for additional problems for this lesson. You may assign the problems here or use them as a culmination for the activity.

LESSON 5-3 PRACTICE
10. $f(x) = x^2 − 3x − 4$
11. $f(−2) = 6$
12. 0 or 3
13. 2

ADAPT

Check students' answers to the Lesson Practice to ensure that they understand how to use function notation. Reinforce students' understanding of function notation as just one of the ways to express functions. Note that it is an efficient way to express a function when you want to evaluate it for many input values.

Lesson 5-3
Function Notation

A function whose domain is the set of positive consecutive integers forms a sequence. The terms of the sequence are the range values of the function. For the sequence 4, 7, 10, 13, …, $f(1) = 4$, $f(2) = 7$, $f(3) = 10$, and $f(4) = 13$.

7. Consider the sequence −4, −2, 0, 2, 4, 6, 8, ….
   a. What is $f(3)$?
      0

   b. What is $f(7)$?
      8

Check Your Understanding

8. Evaluate the functions for the domain values indicated.
   a. $p(x) = 3x + 14$ for $x = −5, 0, 4$
   b. $h(t) = t^2 − 5t$ for $t = −2, 0, 5, 7$
9. Consider the sequence −7, −3, 1, 5, 9, ….
   a. What is $f(2)$?
   b. What is $f(5)$?

LESSON 5-3 PRACTICE

Use the function $y = x^2 − 3x − 4$ for Items 10–12.

10. Write the function in function notation.
11. Evaluate the function for $x = −2$. Express your answer in function notation.
12. **Make use of structure.** For what value of x does $f(x) = −4$?
13. Consider the sequence $\frac{1}{2}$, 1, $\frac{3}{2}$, 2, $\frac{5}{2}$, 3, …. What is $f(4)$?

Functions and Function Notation
Vending Machines

ACTIVITY 5 continued

ACTIVITY 5 PRACTICE
Write your answers on notebook paper.
Show your work.

Lesson 5-1

Use the Beverage Vending Machine to answer Items 1–6.

1. List all of the possible inputs.
2. List all of the possible outputs.
3. Which output results from an input of 2C?
   A. Juice
   B. Iced tea
   C. Latte
   D. Cocoa
4. Which number/letter combination would you input if you wanted the machine to output juice?
   A. 2A
   B. 1B
   C. 2B
   D. 1D
5. In a mapping of the relation shown by the vending machine, what drink would 1D map to?
6. In a table of the relation shown by the vending machine, what number/letter combination would correspond to cocoa?

For Items 7–9, two relations are given. One relation is a function and one is not. Identify each and explain.

7. $\{(5, -2), (-2, 5), (2, -5), (-5, 2)\}$
   $\{(5, -2), (-2, 5), (5, 2), (-5, 2)\}$

8.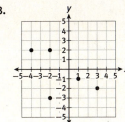

   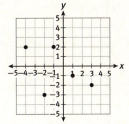

9.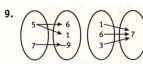

10. What value(s) of x in the relation below would create a set of ordered pairs that is not a function? Justify your answer.

    $\{(0, 5) \ (1, 5) \ (2, 6) \ (x, 7)\}$

ACTIVITY 5 Continued

ACTIVITY PRACTICE
1. A1, A2, B1, B2, C1, C2, D1, D2
2. coffee, chai, juice, cocoa, iced tea, latte, water, mocha
3. C
4. B
5. water
6. 2B
7. The first relation is a function, because each input has exactly one output. The second relation is not a function, because the input 5 has 2 outputs.
8. The first relation is not a function, because the input −2 has 2 outputs. The second relation is a function, because each input has exactly one output.
9. The first relation is not a function, because the input 5 has 2 outputs. The second relation is a function, because each input has exactly one output.
10. $x = 0, 1,$ or 2; these x-values already appear in the domain of the function, and none of them are matched with a range value of 7. Therefore, if x were to equal one of these numbers, then there would be a domain value with more than one range value.

ACTIVITY 5 Continued

11. Yes; because every x-value (input) is matched with exactly one y-value (output).
12. {−4, −1, 0, 4}
13. {−5}
14. Yes; because every x-value (input) is matched with exactly one y-value (output).
15. $f(x) = x^2 - 5x + 8$
16. 22
17. 0
18. 1
19. 1
20. 5
21. The mapping diagram represents a function because each domain value is paired with exactly one range value. It doesn't matter that each domain value is paired with the same range value, as long as no domain value is paired with more than one range value.

ADDITIONAL PRACTICE

If students need more practice on the concepts in this activity, see the eBook Teacher Resources for additional practice problems.

ACTIVITY 5 continued

11. Does the graph shown represent a function? Explain.

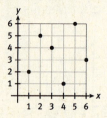

Lesson 5-2

Use the graph for Items 12–14.

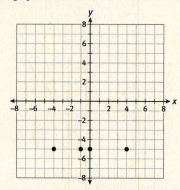

12. Identify the domain of the relation represented in the graph.
13. Identify the range of the relation represented in the graph.
14. Does the relation shown in the graph represent a function? Explain.

Functions and Function Notation
Vending Machines

Lesson 5-3

Use the function machine for Items 15–17.

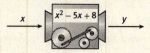

15. How would you write the function shown in the function machine in function notation?
16. What is the value of $f(-2)$?
17. What value(s) of x results in $f(x) = 8$?
18. Given the function $f(x) = -2x - 5$, determine the value of $f(-3)$.

The first seven numbers in the Fibonacci sequence are: 0, 1, 1, 2, 3, 5, 8. Use this information for Items 19 and 20.

19. What is $f(2)$?
20. What is $f(6)$?

MATHEMATICAL PRACTICES
Construct Viable Arguments and Critique the Reasoning of Others

21. Dora said that the mapping diagram below does not represent a function because each value in the domain is paired with the same value in the range. Explain the error in Dora's reasoning.

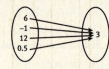

Interpreting Graphs of Functions
Shake, Rattle, and Roll
Lesson 6-1 Key Features of Graphs

ACTIVITY 6

Learning Targets:
- Relate the domain and range of a function to its graph.
- Identify and interpret key features of graphs.

SUGGESTED LEARNING STRATEGIES: Marking the Text, Visualization, Interactive Word Wall, Discussion Groups

Roller coasters can be scary but fun to ride. Below is the graph of the heights reached by the cars of the Thunderball Roller Coaster over its first 1250 feet of track. The graph displays a function because each input value has one and only one output value. You can see this visually using the *vertical line test*. Study this graph to determine the domain and range.

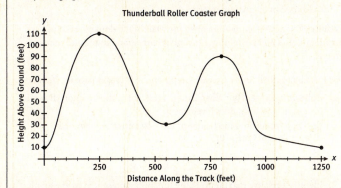

Thunderball Roller Coaster Graph

The domain gives all values of the *independent variable*: in this case, the distance along the track in feet. The domain values are graphed along the horizontal or *x*-axis. The domain of the function above can be written in set notation as:

{all real values of *x*: $0 \leq x \leq 1250$}

Read this notation as: *the set of all real values of x, between 0 and 1250, inclusive.*

The range gives the values of the *dependent variable*: in this case, the height of the roller coaster above the ground in feet. The range values are graphed on the vertical or *y*-axis. The range of the function above can be written in set notation as:

{all real values of *y*: $10 \leq y \leq 110$}

Read this notation as: *the set of all real values of y, between 10 and 110, inclusive.*

MATH TERMS

The **vertical line test** is a visual check to see if a graph represents a function. For a function, every vertical line drawn in the coordinate plane will intersect the graph in at most one point. This is equivalent to having each domain element associated with one and only one range element.

MATH TERMS

An **independent variable** is the variable for which input values are substituted in a function.
A **dependent variable** is the variable whose value is determined by the input or value of the independent variable.

ACTIVITY 6
Directed

Activity Standards Focus
In Activity 6 students determine the domain and range of various relations and identify relative maxima and minima. Students extend their thinking to real-world situations by interpreting key features of graphs within a context and by determining a reasonable domain and range for the problem situation.

Lesson 6-1

PLAN
Pacing: 1 class period
Chunking the Lesson
#1–5   #6–10
Check Your Understanding
#15–21
Check Your Understanding
Lesson Practice

TEACH

Bell-Ringer Activity
Have each student sketch a graph of his or her distance from school, from just leaving home to arriving at school. Students can compare graphs and consider the meaning of the attributes of the graphs.

Developing Math Language
This lesson contains multiple vocabulary terms that are used to describe graphs of functions. When possible, call on students' prior knowledge to help them understand the new vocabulary. For example, have students share their understanding of the terms *vertical*, *independent*, *dependent*, *intercept*, *maximum*, *minimum* and *continue*. Help them connect those understandings with the terms *vertical line test*, *independent variable*, *dependent variable*, *y-intercept*, *relative maximum*, *relative minimum* and *continuous*. Note that some math terms such as *discrete* are more difficult to connect to an everyday meaning of the word.

Common Core State Standards for Activity 6

HSF-IF.B.4:	For a function that models a relationship between two quantities, interpret key features of graphs and tables in terms of the quantities, and sketch graphs showing key features given a verbal description of the relationship. Key features include: intercepts; intervals where the function is increasing, decreasing, positive, or negative; relative maximums and minimums; symmetries; end behavior and periodicity.
HSF-IF.B.5:	Relate the domain of a function to its graph and, where applicable, to the quantitative relationship it describes.
HSF-IF.C.7:	Graph functions expressed symbolically and show key features of the graph, by hand in simple cases and using technology for more complicated cases.
HSF-IF.C.7a:	Graph linear and quadratic functions and show intercepts, maxima, and minima.

Activity 6 • Interpreting Graphs of Functions   81

ACTIVITY 6 Continued

1–5 Shared Reading, Marking the Text, Interactive Word Wall Discuss the important ideas in each paragraph. Encourage students to refer to the marked text and the Word Wall as they answer the questions involving the math terms in this lesson. Make sure students are using the vocabulary correctly.

CONNECT TO AP

The *absolute maximum* and *absolute minimum* of a function are important points on the graph of the function. In higher levels of algebra and in calculus, they provide useful information about functions.

ACTIVITY 6 continued

My Notes

CONNECT TO AP

The **absolute maximum** of a function $f(x)$ is the greatest value of $f(x)$ for all values in the domain. The **absolute minimum** of a function $f(x)$ is the least value of $f(x)$ for all values in the domain. Unlike relative maximums and relative minimums, absolute maximums and absolute minimums may correspond to the endpoints of graphs.

MATH TIP

An open interval is an interval whose endpoints are not included. For example, $0 < x < 5$ is an open interval, but $0 \leq x \leq 5$ is not.

Lesson 6-1
Key Features of Graphs

The graph above shows data that are **continuous**. The points in the graph are connected, indicating that domain and range are sets of real numbers with no breaks in between. A graph of **discrete** data consists of individual points that are not connected by a line or curve.

Many other useful pieces of information about a function can be determined by looking at its graph.

- The ***y-intercept*** of a function is the point at which the graph of the function intersects the y-axis. The y-intercept is the point at which $x = 0$.
- A ***relative maximum*** of a function $f(x)$ is the greatest value of $f(x)$ for values in a limited open domain interval.
- A ***relative minimum*** of a function $f(x)$ is the least value of $f(x)$ for values in a limited open domain interval.

Because they must occur within open intervals of the domain, relative maximums and relative minimums cannot correspond to the endpoints of graphs.

Use the Thunderball Roller Coaster Graph on the previous page for Items 1–5.

1. **Reason abstractly.** What is the y-intercept of the function shown in the graph, and what does it represent?
   (0, 10); it represents the height of the roller coaster when the ride begins (at time = 0).

2. Identify a relative maximum of the function represented by the graph.
   110 or 90

3. Identify the absolute maximum of the function represented by the graph. Interpret its meaning in the context of the situation.
   110; the greatest height reached by the roller coaster

4. Identify a relative minimum of the function represented by the graph.
   30

5. Identify the absolute minimum of the function represented by the graph. Interpret its meaning in the context of the situation.
   10; the least height reached by the roller coaster

Lesson 6-1
Key Features of Graphs

Suppose you got on a roller coaster called Cougar Mountain that immediately started climbing the track in a linear fashion, as shown in the graph.

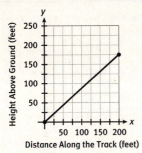

6. Identify the domain and range of the function.
   Domain: {all real values of x: $0 \leq x \leq 200$}; range: {all real values of y: $0 \leq y \leq 175$}

7. Identify the y-intercept of the function.
   (0, 0)

8. Identify the absolute maximum and minimum of the function.
   Maximum: 175; minimum: 0

9. Does the function have any relative maximum or minimum values? Explain.
   No; there is no open domain interval that has maximum or minimum values. The absolute maximum and minimum occur at the endpoints, so they cannot be the relative maximum and minimum.

10. How are the *extrema* different on this linear graph versus the nonlinear graph for the Thunderball Roller Coaster?
    On the linear graph, the extrema occur at the endpoints of the interval. On the nonlinear graph, they occur in the interior as well as the endpoints of the interval.

MATH TERMS

Extrema refers to all maximum and minimum values.

ACTIVITY 6 Continued

Check Your Understanding

Debrief students' answers to these items to ensure that they understand the mathematical terminology used to describe the graphs of functions. This lesson prepares students to study functions and their graphs with proper mathematical vocabulary.

Answers
11. No, the points are not connected. Therefore, it is a discrete function.
12. 28
13. Sometimes; It is possible that a relative minimum is not the absolute minimum.
14. Sometimes; This is not true if the absolute minimum corresponds to an endpoint, for example.

15–21 Interactive Word Wall, Group Discussion, Think Aloud, Debriefing

Before answering the questions, go over the words on the Word Wall from this activity. Have students relate each word to the graph shown. As students discuss the answers to the questions, have them address both the mathematical features as well as its meaning of each in the context of the problem.

Lesson 6-1
Key Features of Graphs

Check Your Understanding

11. The graph below shows five points that make up the function h. Is the function h continuous? Explain.

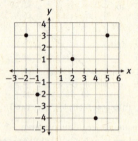

12. A function has three relative maximums: -2, 10.3, and 28. One of the relative maximums is also the absolute maximum. What is the absolute maximum?

Tell whether each statement is sometimes, always, or never true. Explain your answers.

13. A relative minimum is also an absolute minimum.
14. An absolute minimum is also a relative minimum.

Tom hiked along a circular trail known as the Juniper Loop. The graph shows his distance d from the starting point after t minutes.

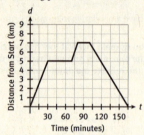

15. Identify the domain and range of the function shown in the graph.
    Domain: $0 \leq t \leq 165$; Range: $0 \leq d \leq 7$

16. Identify the absolute minimum of the function. What does it represent?
    0; Tom's least distance from the starting point

84 SpringBoard® Mathematics **Algebra 1, Unit 2** • Functions

Lesson 6-1
Key Features of Graphs

17. In this function, the absolute minimum corresponds to two points on the graph. What are the two points? What do they represent in this context?
    (0, 0) and (165, 0); Tom's least distance from the starting point (0 km) occurred both at the beginning and at the end of his hike.

18. Identify the absolute maximum of the function. What does it represent?
    7; Tom's greatest distance from the starting point (7 km)

19. What points on the graph correspond to the absolute maximum? What does this mean in the context of Tom's hike?
    All of the points along the graph from $t = 80$ to $t = 100$; Tom was 7 km from the starting point between these times (probably he stopped to rest).

20. Identify any relative minimums for the function shown in the graph.
    5

21. Identify any relative maximums for the function shown in the graph.
    5 or 7

Check Your Understanding

22. What are the independent and dependent variables for the function representing Tom's hike?
23. Explain how to determine the maximum and minimum values of a function by examining its graph.
24. Is it possible for a function to have more than one absolute maximum or absolute minimum value? Explain.

ACTIVITY 6 Continued

ASSESS

Students' answers to Lesson Practice problems will provide you with a formative assessment of their understanding of the lesson concepts and their ability to apply their learning.

See the Activity Practice for additional problems for this lesson. You may assign the problems here or use them as a culmination for the activity.

LESSON 6-1 PRACTICE

25. The independent variable is t, the number of minutes since the bath began, and the dependent variable is d, the depth of the bathwater, since the depth of the water depends on how many minutes the water has been running.
26. Domain: {all real values of t: $0 \leq t \leq 12$}; Range: {all real values of d: $0 \leq d \leq 8$}
27. Continuous; The function includes all real values of t between 0 and 12, inclusive, and all real values of d between 0 and 8, inclusive.
28. (0, 0); the depth of the bathwater at time 0
29. Relative maximum: 8; no relative minimum
30. Absolute maximum: 8; greatest depth; absolute minimum: 0; least depth

ADAPT

Check students' answers to the Lesson Practice to ensure that they understand the terminology used to describe features of the graph. Help students internalize the vocabulary by modeling the proper use of the terms throughout the unit.

ACTIVITY 6 continued

My Notes

Lesson 6-1
Key Features of Graphs

LESSON 6-1 PRACTICE

Model with mathematics. Use the graph below for Items 25–30.

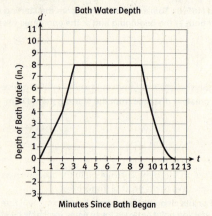

25. What are the independent and dependent variables? Explain.
26. Use set notation to write the domain and range of the function.
27. Is the function discrete or continuous? Explain.
28. What is the y-intercept? Interpret the meaning of the y-intercept in this context.
29. Identify any relative maximums or minimums of the function.
30. Identify the absolute maximum and absolute minimum values. Interpret their meanings in this context.

Lesson 6-2
More Complex Graphs

Learning Targets:
- Relate the domain and range of a function to its graph and to its function rule.
- Identify and interpret key features of graphs.

SUGGESTED LEARNING STRATEGIES: Marking the Text, Levels of Questions, Think Aloud, Create Representations, Summarizing

Examine the graph of the function $f(x) = \dfrac{1}{(x-2)^2}$, graphed below.

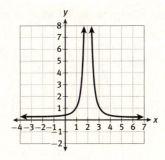

1. Describe how this graph is different from the graphs in Lesson 6-1.

   Answers will vary. Important characteristics for students to notice are that this graph consists of two distinct branches and that it continues on beyond the portion shown, as indicated by the arrowheads on the curves.

Example A

Give the domain and range of the function $f(x) = \dfrac{1}{(x-2)^2}$.

Then find the *y*-intercept, the absolute maximum, and the absolute minimum.

To find the domain and range:

Step 1: Study the graph.
The sketch of this graph is a portion of the function represented by the equation $f(x) = \dfrac{1}{(x-2)^2}$.

Step 2: Look for values for which the domain causes the function to be undefined. Look how the graph behaves near $x = 2$.

Solution: The domain and range of $f(x) = \dfrac{1}{(x-2)^2}$ can be written:

   Domain: {all real values of $x : x \neq 2$}
   Range: {all real values of $y : y > 0$}

MATH TIP

Notice the result when $x = 2$ is substituted into $f(x)$.

$$f(2) = \dfrac{1}{(2-2)^2} = \dfrac{1}{0}$$

Division by zero is undefined in mathematics.

ACTIVITY 6 Continued

Try These A Think-Pair-Share In this set of questions, students are asked to identify the key features of the graph of an important type of function (without identifying it as such): a quadratic function. Before students answer the questions, ask students to discuss with their partners the similarities and differences between the graph of this function and the graph of the function in Example A. Identifying how this function differs from the previous function should help students identify this function's key features.

As students identify these features, some students may make use of the graph to identify key features of this function, while others may use the function rule. Encourage students to share their strategies, noting that they get the same results.

Differentiating Instruction

Support students who have difficulty remembering new terminology through regular references to the words in the text as well as words you and students place on the classroom Word Wall.

Lesson 6-2
More Complex Graphs

To determine the y-intercept and identify any maximums or minimums:

Study the graph. We can see that the function intersects the y-axis at $(0, 0.25)$. The value of $f(x)$ keeps getting larger as x approaches 2 from both sides. The value of $f(x)$ approaches, but never reaches, 0 as x gets further from 2 on both sides.

Solution: The y-intercept is $(0, 0.25)$. The function does not have an absolute maximum or minimum.

Try These A

The function $f(x) = 8 + 2x - x^2$ is graphed below.

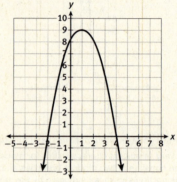

a. Identify the domain and range of the function.
   Domain: {all real numbers}

   Range: {all real values of y : $y \leq 9$}

b. Identify the y-intercept.
   $(0, 8)$

c. Identify any relative or absolute minimums of the function.
   No minimums exist.

d. Identify any relative or absolute maximums of the function.
   9

Lesson 6-2
More Complex Graphs

2. The equation $y = 2x - 1$ is graphed below.

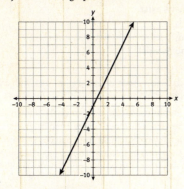

   a. Identify the domain and range.
      Domain: {all real numbers}
      Range: {all real numbers}

   b. What is the y-intercept of $y = 2x - 1$?
      (0, −1)

   c. Identify any relative or absolute minimums of $y = 2x - 1$.
      No minimums exist.

   d. Identify any relative or absolute maximums of $y = 2x - 1$.
      No maximums exist.

   e. **Construct viable arguments.** Explain whether this equation represents a function and how you determined this.
      Answers may vary. Because the graph of this equation passes the vertical line test, this equation and its graph represent a function.

3. The function $y = 2^x$ is graphed below.

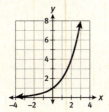

   a. Identify the domain and range.
      Domain: {all real numbers}
      Range: {$y > 0$}

   b. What is the y-intercept of the function $y = 2^x$?
      (0, 1)

ACTIVITY 6 Continued

2–3 Think-Pair-Share, Look for a Pattern, Summarizing, Debriefing In this set of items students are asked to identify the key features of the graph of two important types of functions (without identifying them as such): a linear function and an exponential function. Again, students should discuss with their partners the similarities and differences between the graphs of these function rules and the graphs of the previous function rules. Identifying how these function rules and their graphs differ from the previous functions should help students identify the key features of these functions.

Once students have finished discussing their work, go through the key terms in this activity. Call on students to describe what each term means and to identify examples on the graph of the function.

ACTIVITY 6 Continued

TEACHER to TEACHER

The purpose of Item 4 is for students to identify domain and range, *not* to graph functions. If your students do not have graphing calculators, display the graphs on an overhead or on the board and have students identify the domain and range.

4 Create Representations, Discussion Groups All of the equations can be viewed in the standard window where $-10 < x < 10$ and $-10 < y < 10$. Depending on how much practice students have had using the graphing calculator, you may need to assist students in setting the window and in entering the equations into the calculators. Functions 7, 8 and 9 may need to be written out by keystrokes if students are not familiar with the calculator nomenclature.

After students have had the opportunity to complete the chart and to compare answers with another group, debrief the activity by asking students to share what they noticed about the results and the behavior of the graphs.

Differentiating Instruction

Extend this activity by having students graph other linear, quadratic and absolute value functions. Discuss how restrictions on the domain and range of the new graphs compare with the restrictions on the previous set of functions.

ACTIVITY 6 continued

**Lesson 6-2
More Complex Graphs**

MATH TIP

The domain is restricted to avoid situations where division by zero or taking the square root of a negative number would occur.

c. Identify any relative or absolute minimums of $y = 2^x$.
   No minimums exist.

d. Identify any relative or absolute maximums of $y = 2^x$.
   No maximums exist.

4. If you have access to a graphing calculator, work with a partner to graph the equations listed in the table below. Each equation is a function.
   a. Using the graphs you create, determine the domain and range for each function from the possibilities listed below the chart.
   b. Select the appropriate domain from choices 1–6 and record your answer in the Domain column. Then select the appropriate range from choices a–f and record the appropriate range in the Range column.
   c. When the chart is complete, compare your answers with those from another group.

Function	Domain	Range
$y = -3x + 4$	1	a
$y = x^2 - 6x + 5$	1	c
$y = 9x - x^2$	1	e
$y = |x + 1|$	1	d
$y = 3 + \sqrt{x}$	5	f
$y = \dfrac{4}{x}$	3	b

Possible Domains
1) all real numbers
2) all real x, such that $x \neq -2$
3) all real x, such that $x \neq 0$
4) all real x, such that $x \neq 2$
5) all real x, such that $x \geq 0$
6) all real x, such that $x \leq 0$

Possible Ranges
a) all real numbers
b) all real y, such that $y \neq 0$
c) all real y, such that $y \geq -4$
d) all real y, such that $y \geq 0$
e) all real y, such that $y \leq 20.25$
f) all real y, such that $y \geq 3$

Lesson 6-2
More Complex Graphs

Check Your Understanding

5. How can you determine from a function's graph whether the function has any maximum or minimum values?
6. How can you determine the domain of a function by examining its graph? By examining its function rule?
7. Give an example of a function that has a restricted domain. Justify your answer.

LESSON 6-2 PRACTICE

The function $f(x) = 2x^2 - 3$ is graphed below.

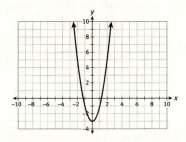

8. Give the domain, range, and y-intercept.
9. Identify any relative or absolute minimums.
10. Identify any relative or absolute maximums.
11. **Attend to precision.** Examine the graphs below. Explain why one function has an absolute minimum and an absolute maximum and the other function does not. Identify the absolute minimum and maximum values of the function for which they exist.

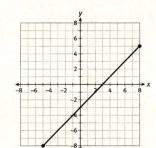   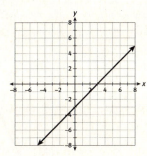

ACTIVITY 6 Continued

Lesson 6-3

PLAN

Pacing: 1 class period
Chunking the Lesson
#1, Example A, #2   #3
Check Your Understanding
Lesson Practice

TEACH

Bell-Ringer Activity
Have students work in small groups to brainstorm real-world situations in which only positive numbers can be used, situations in which both positive and negative numbers can be used, situations in which only whole numbers can be used, and situations in which fractions and decimals can be used.

1, Example A, 2 Shared Reading, Think Aloud, Visualization Students should explain why both the domain and range in Item 1 are made up of all real numbers. Students should suggest real-world situations in which the set of all real numbers would be unreasonable for the domain and/or range. As you read through the example, help students picture the situation so that they understand why it would not be reasonable for x to be negative.

Lesson 6-3
Graphs of Real-World Situations

Learning Targets:
- Identify and interpret key features of graphs.
- Determine the reasonable domain and range for a real-world situation.

SUGGESTED LEARNING STRATEGIES: Visualization, Discussion Groups, Look for a Pattern

The function $f(x) = 3 + 2x$ is graphed below.

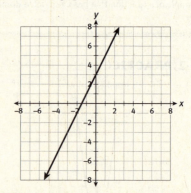

1. What are the domain and range of the function?
   Domain: {all real numbers}
   Range: {all real numbers}

In many real-world situations, not all values make sense for the domain and/or range. For example, distance cannot be negative; number of people cannot be a decimal or a fraction. In such situations, the values that make sense for the domain and range are called the *reasonable* domain and range.

Example A
A taxi ride costs an initial rate of $3.00, which is charged as soon as you get in the cab, plus $2 for each mile traveled. The cost of traveling x miles is given by the function $f(x) = 3 + 2x$. What are the reasonable domain and range?

Step 1: Sketch a graph of the function.

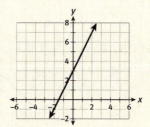

MATH TIP

Graph a function by substituting several values for x and generating ordered pairs. You can organize the ordered pairs in a table. There are infinitely many other solutions because the graph has infinitely many points.

x	f(x) = 3 + 2x	(x, y)
0	3	(0, 3)
1	5	(1, 5)
2	7	(2, 7)

92 SpringBoard® Mathematics Algebra 1, Unit 2 • Functions

Lesson 6-3
Graphs of Real-World Situations

ACTIVITY 6
continued

My Notes

Step 2: Determine the reasonable domain. Think about what the variable x represents. What values make sense?

The variable x represents the number of miles, so it does not make sense for x to be negative.

The reasonable domain is $\{x: x \geq 0\}$.

Step 3: Use the reasonable domain and the graph to determine the reasonable range.

From the graph, all y-values corresponding to the reasonable domain values are greater than or equal to 3. The reasonable range is $\{y: y \geq 3\}$.

Solution: The reasonable domain is $\{x: x \geq 0\}$. The reasonable range is $\{y: y \geq 3\}$.

Try These A

a. A banquet hall charges $15 per person plus a $100 setup fee. The cost for x people is given by the function $f(x) = 100 + 15x$. What are the reasonable domain and range?

Domain: {all whole numbers}; range: {all whole numbers $y: y \geq 100$}

b. Eight Ball Billiards charges $5 to rent a table plus $10 per hour of game play, rounded to the nearest whole hour. The cost of playing billiards for x hours is given by the function $f(x) = 5 + 10x$. What are the reasonable domain and range?

Domain: {all whole numbers}; range: {5, 15, 25, 35, …}

2. Reason quantitatively. Are the domain and range of $f(x) = 3 + 2x$ that you found in Item 1 the same as the reasonable domain and range of $f(x) = 3 + 2x$ found in Example A? Explain.

No; when the function does not model a real-world situation, its domain and range are all real numbers. However, not all real numbers make sense for the domain and range for the real-world situation modeled in Example A. The reasonable domain and range are restricted to only those values that make sense.

ACTIVITY 6 Continued

1, Example A, 2 (continued) As students discuss each of the Try These items, suggest that students first find the reasonable domain for each situation and then use that domain to generate the range.

Differentiating Instruction

Support students who do not distinguish between continuous and concrete data by presenting an example such as this: At Store A, 2 bananas cost $1.00; at Store B, 2 pounds of bananas cost $1.00. Have students describe the difference between the two price structures. Point out that each situation can be described by a function: $C = 2b$ represents the price for bananas at Store A, and $C = 2p$ represents the price for bananas based on the number of pounds of bananas at Store B. Elicit from students the fact that you cannot buy a fraction of a banana, but you can buy a fraction of a pound of bananas. After identifying a reasonable domain for each function, use it to find the corresponding ranges.

Activity 6 • Interpreting Graphs of Functions 93

ACTIVITY 6 Continued

3 Look for a Pattern, Discussion Groups, Debriefing Make sure that groups consider not only the individual value of each point but also the pattern made by the points (a straight line). Have groups consider the type of function represented by this pattern (a linear function). This will help students come up with scenarios that could be represented by this graph.

Check Your Understanding
Debrief students' answers to these questions to make sure they understand how the features of a function relate to the situation it represents. Students should use correct mathematical language when describing the features. Students will use functions to model and solve real-world problems throughout their study of mathematics and must be able to describe and interpret functions clearly and accurately.

Answers
4. Positive real numbers; The height cannot be zero or a negative number.
5. Whole numbers; The number of people cannot be a fraction, or a negative number.

ASSESS
Students' answers to Lesson Practice problems will provide you with a formative assessment of their understanding of the lesson concepts and their ability to apply their learning.

See the Activity Practice for additional problems for this lesson. You may assign the problems here or use them as a culmination for the activity.

LESSON 6-3 PRACTICE
6. Independent: number of minutes; dependent: total cost; The total cost depends on the number of minutes.
7. Domain: $\{x \geq 0\}$ because number of minutes cannot be negative; Range: $\{y \geq 20\}$ because the least possible y-value is the cost for 0 minutes, and that cost is $20.

ADAPT
Check students' answers to the Lesson Practice to ensure they can interpret the features of a function that represents a real-world situation.

ACTIVITY 6 continued

Lesson 6-3
Graphs of Real-World Situations

3. The graph below represents a real-world situation.

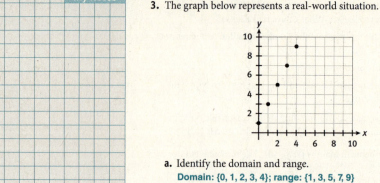

a. Identify the domain and range.
   Domain: {0, 1, 2, 3, 4}; range: {1, 3, 5, 7, 9}

b. Describe a real-world situation that matches the graph. Your answers to Part (a) should be the reasonable domain and range for your situation.
   Answers will vary. A bowling alley charges $1 to rent a pair of shoes and $2 per game for up to 4 games.

c. Identify the independent and dependent variables in your real-world situation.
   Answers will vary. Independent: number of games played; dependent: total amount charged

Check Your Understanding

4. For a function that models a real-world situation, the dependent variable y represents a person's height. What is a reasonable range? Explain.

5. A tour company charges $25 to hire a tour director plus $75 per tour member. The total cost for a group of x people is given by $f(x) = 25 + 75x$. What is the reasonable domain? Explain.

LESSON 6-3 PRACTICE
Talk the Talk Cellular charges a base rate of $20 per month for unlimited texts plus $0.15/minute of talk time. The monthly cost for x minutes is given by $f(x) = 20 + 0.15x$.

6. **Make sense of problems.** What is the independent variable and what is the dependent variable? Explain how you know.

7. What are the reasonable domain and range? Explain.

Interpreting Graphs of Functions
Shake, Rattle, and Roll

ACTIVITY 6 continued

ACTIVITY 6 PRACTICE
Write your answers on notebook paper.
Show your work.

Lesson 6-1

Use the graph below for Items 1–5.

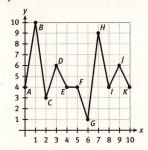

1. Which point corresponds to the absolute maximum of the function?
    A. B
    B. D
    C. G
    D. H

2. Which represents the range of the function shown in the graph?
    A. $\{0 \leq x \leq 10\}$
    B. $\{1 \leq x \leq 10\}$
    C. $\{0 \leq y \leq 10\}$
    D. $\{1 \leq y \leq 10\}$

3. Which point does **not** correspond to a relative minimum?
    A. B
    B. C
    C. E
    D. I

4. Is the function represented by the graph discrete or continuous? Explain.

5. What is the y-intercept of the function shown in the graph?

6. a. Give the domain and range for the function graphed below. Explain why this graph represents a function.

    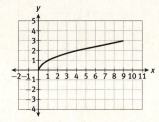

    b. What is the y-intercept of the function shown in the graph?
    c. Identify any extrema of the function shown in the graph.

Jeff walks at an average rate of 125 yards per minute. Mark's house is located 2000 yards from Jeff's house. The graph below shows how far Jeff still needs to walk to reach Mark's house. Use the graph for Items 7–10.

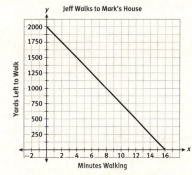

7. Identify the independent and dependent variables.

8. Identify the absolute minimum and absolute maximum values. What do these values represent?

9. Identify any relative maximums or minimums.

10. What is the y-intercept? What does it represent?

ACTIVITY 6 Continued

ACTIVITY PRACTICE
1. A
2. D
3. A
4. All the points are connected, with no breaks, so it is continuous.
5. (0, 4)
6. a. Domain: $\{x: 0 \leq x \leq 9\}$; range: $\{y: 0 \leq y \leq 3\}$; It passes the vertical line test.
   b. (0, 0)
   c. Absolute minimum: 0; absolute maximum: 3
7. Independent: minutes walking; dependent: yards left to walk
8. Absolute minimum: 0; the least distance Jeff will have to walk to Mark's house; This distance is 0 when Jeff arrives. Absolute maximum: 2000; the greatest distance Jeff has to walk to Mark's house, which occurs after 0 minutes of walking
9. None exist.
10. (0, 2000); the distance Jeff has to walk to Mark's house at time = 0

ACTIVITY 6 Continued

11. $\{x: x \neq 0\}$
12. $\{y: y \neq 0\}$
13. There is no y-intercept.
14. (0, 1)
15. 1
16. -9 and -3.8
17. the money raised
18. the amount that will be donated
19. $\{x \geq 0\}$; An amount of money raised cannot be negative.
20. $\{y \geq 250\}$; The least amount the organization will donate is $250 (if they raise no money from the event).
21. Answers may vary. An after-school job pays $5 per hour minus a one-time $3 fee for a storage locker.
22. The absolute maximum and absolute minimum values are the same, because the y-value of every point on the line is the same.

ADDITIONAL PRACTICE

If students need more practice on the concepts in this activity, see the eBook Teacher Resources for additional practice problems.

ACTIVITY 6 continued

Interpreting Graphs of Functions
Shake, Rattle, and Roll

Lesson 6-2

Use the graph for Items 11–13.

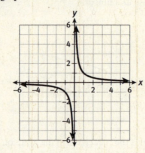

11. What is the domain of the function shown in the graph?
12. What is the range of the function shown in the graph?
13. What is the y-intercept of the function shown in the graph?

Use the graph below for Items 14–16.

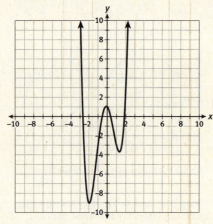

14. What is the y-intercept of the function shown in the graph?
15. Identify any relative maximums.
16. Identify any relative minimums.

Lesson 6-3

A fundraising organization will donate $250 plus half of the money it raises from a charity event. Use this information for Items 17–20.

17. What is the independent variable?
18. What is the dependent variable?
19. What is the reasonable domain? Explain.
20. What is the reasonable range? Explain.
21. Describe a real-world situation that matches the graph shown.

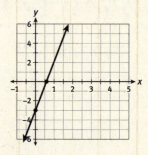

MATHEMATICAL PRACTICES
Look For and Make Use of Structure

22. The graph of a function is a horizontal line. What is true about the absolute maximum and absolute minimum values of this function? Explain.

Graphs of Functions

Experiment Experiences

Lesson 7-1 The Spring Experiment

ACTIVITY 7

Learning Targets:
- Graph a function given a table.
- Write an equation for a function given a table or graph.

> **SUGGESTED LEARNING STRATEGIES:** Discussion Groups, Look for a Pattern, Sharing and Responding, Think-Pair-Share, Create Representations, Construct an Argument

For the following experiment, you will need a paper cup, a rubber band, a paper clip, a measuring tape, and several washers.

A. Punch a small hole in the side of the paper cup, near the top rim.
B. Use the bent paper clip to attach the paper cup to the rubber band as shown in the diagram in the *My Notes* section.

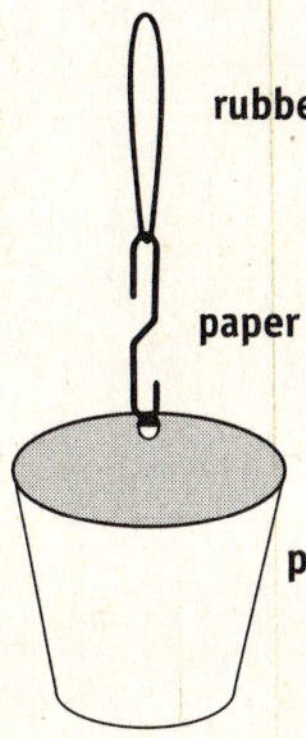

1. What is the length of the rubber band?
   Answers will vary.

Drop washers one at a time into the cup. Each time you add a washer, measure the length of the rubber band. Subtract the original length you recorded in Item 1 to find the distance that the rubber band has stretched.

2. Make a table of your data. **Check students' tables.**

Number of Washers x	Length of Stretch from Original Length y
1	
2	
3	
4	
5	

3. What patterns do you notice that might help you determine the relationship between the number of washers in the cup and the length of the rubber band stretch?
   Answers will vary. In general, the length of the stretch should be approximately a constant multiple of the number of washers.

Common Core State Standards for Activity 7

HSA-REI.D.10:	Understand that the graph of an equation in two variables is the set of all its solutions plotted in the coordinate plane, often forming a curve (which could be a line).
HSF-IF.B.5:	Relate the domain of a function to its graph and, where applicable, to the quantitative relationship it describes.
HSF-IF.C.7:	Graph functions expressed symbolically and show key features of the graph, by hand in simple cases and using technology for more complicated cases.
HSF-IF.C.7a:	Graph linear and quadratic functions and show intercepts, maxima, and minima.
HSF-IF.C.7e:	Graph exponential functions, showing intercepts and end behavior.

ACTIVITY 7
Investigative

Activity Standards Focus
In this Activity, students will use functions to explore real-world relationships. They will use equations, graphs and features of functions as tools to describe and understand the relationships. The activity will reinforce what students have already learned about functions as well as use functions to model natural phenomena.

Lesson 7-1

PLAN

Materials:
- paper cups
- rubber bands
- paper clips
- measuring tapes
- same-sized washers
- butcher paper or tag board
- rulers
- markers

Pacing: 1 class period

Chunking the Lesson
#1–7   #8–10   #11–16   #17–19
Check Your Understanding
Lesson Practice

TEACH

Bell-Ringer Activity
Have students imagine they are going to collect data concerning the length and weight of various animals. Have them list several ways to record and display the data. Have them discuss the advantages of each method

1–7 Discussion Groups, Look for a Pattern, Sharing and Responding
As students do the experiment, be sure they are measuring and recording the length of stretch from the original length and not the total length of the rubber band. Once they have their data collected, students should discover patterns that can be used to help them predict the lengths of the stretch for other numbers of washers.

ACTIVITY 7 Continued

1–7 (continued) Monitor the groups to make sure that all students are participating. Students may try to connect the points on their graphs and/or force the data to be perfectly linear. As group discussions are monitored, encourage students to think about whether their graphs should be discrete or continuous. Be sure to include a discussion of why the data is discrete in the whole-class Sharing and Responding after Item 7. In addition, students may need to be reminded that because their data may not be perfectly linear, they may need to generalize a little to write their equation. For example, the rate of change may be "about 1 inch per washer" instead of exactly 1 inch per washer. Include the fact that we are using an equation to "model" a real-world situation in the whole-class Sharing and Responding after Item 7.

Differentiating Instruction

Support students who have difficulty translating a pattern into an equation. Have students describe the pattern verbally. Have them consider which operation they would use as part of the pattern. Help them write a word equation using that operation. Then have them define the variables and translate the word equation into an algebraic equation. Have them test the equation with various ordered pairs in their table to verify that it describes the pattern.

ACTIVITY 7 continued

My Notes

CONNECT TO SCIENCE

What you have revealed with your experiment is an example of Hooke's Law. Hooke's Law states that the distance d that a spring (in this case the rubber band) is stretched by a hanging object varies directly with the object's weight w.

Lesson 7-1
The Spring Experiment

4. Use your table to make a graph. Be sure to label an appropriate scale and the units on the y-axis. **Check students' graphs.**

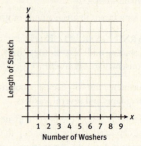

5. Describe your graph.
   The graph should resemble a line.

6. **Model with mathematics.** Use your graph and any patterns you described in Item 3 to write an equation that describes the relationship between the number of washers and the length of the stretch.
   Answers will vary.

7. Use your graph or your equation to predict the length of the stretch for 8 washers and for 10 washers.
   Answers will vary; check students' work.

A group of students performed a similar experiment with a spring and various masses. The data they collected is shown in the table below.

Mass (g)	Spring Stretch (cm)
2	6
4	12
6	18
8	24
10	30
12	36

8. Make a graph of the data in the table. **Check students' graphs.**

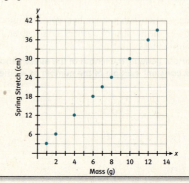

---

98 SpringBoard® Mathematics Algebra 1, Unit 2 • Functions

Lesson 7-1
The Spring Experiment

ACTIVITY 7
continued

My Notes

9. **Reason quantitatively.** How much does the spring stretch for each additional gram of mass added? Explain how you found your answer.
   3 cm; you can see from the table and from the graph that each time the mass increases by 2 g, the stretch increases by 6 cm, which is equivalent to 3 cm for each additional gram.

10. **Reason abstractly.** Use the students' data to write an equation that gives the distance d that the spring will stretch in terms of the mass m. Explain your equation.
    The stretch is always 3 times the mass, so $d = 3m$.

11. Use the equation or the graph to determine the length of the stretch for a mass of 1 gram. Graph the outcome on your graph.
    3 cm; see graph above.

12. Use the equation or the graph to determine the length of the stretch for a mass of 7 grams. Graph the outcome on your graph.
    21 cm; see graph above.

13. Use the equation or the graph to determine the length of the stretch for a mass of 13 grams. Graph the outcome on your graph.
    39 cm; see graph above.

14. **a.** What do you notice about the points you graphed in Items 11–13?
    They lie on the same line as the points graphed from the table.

    b. How could you represent the set of all possible masses and corresponding stretches?
    Connect all the points to form a line.

15. What is the y-intercept of the graph? What does it represent?
    The y-intercept is (0, 0). It represents the length that the spring stretches with no mass attached to it.

16. What is the reasonable domain? Explain.
    Domain: $\{m \geq 0\}$; m represents mass and mass cannot be negative.

ACTIVITY 7 Continued

8–10 Discussion Groups, Look for a Pattern, Create Representations, Construct an Argument Students should identify a constant increase of 6 inches per 2 grams of mass added. Some may identify this from the table, while some may use the graph. Groups should determine that the stretch distance is 3 times the mass and represent this relationship using an equation. Emphasize the need for students to explain the relationship both algebraically and verbally.

11–16 Activating Prior Knowledge, Think-Pair-Share Students should find that each of the points they determine lies on the same line as the points from the table. The understanding that the masses must not necessarily be discrete in this situation is important. Students should realize that they can connect the points on the graph to represent masses such as 2.4 ounces.

ACTIVITY 7 Continued

17–19 Think-Pair-Share, Create Representations,
Debriefing Students should notice the discrepancy in the change of weights (i.e., from 5 to 8, 3 ounces, and from 8 to 10, 2 ounces) they see in the table and explain how this will or will not affect how they answer each of the items that follow. It is important that students determine how far their spring stretches for each additional ounce of weight before attempting to write an equation. The debrief should include a discussion of any patterns students may have noticed relating the rate of change to the coefficient of the variable in each of the equations they have created in this activity. Students have not labeled the rate of change as of yet but should be beginning to see a connection.

Check Your Understanding

As you debrief with students, have them explain how they found the amount of change in the length of the spring for each 0.5- or 1-pound change in the weight. This lesson prepares students for the study of slope later in this unit.

Answers

20. 4 inches; The spring stretched 12 more inches when 3 pounds of weight were added, which is equivalent to 4 inches per pound.

ASSESS

Students' answers to Lesson Practice problems will provide you with a formative assessment of their understanding of the lesson concepts and their ability to apply their learning.

See the Activity Practice for additional problems for this lesson. You may assign the problems here or use them as a culmination for the activity.

LESSON 7-1 PRACTICE
21.

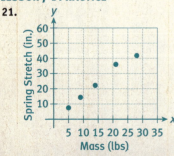

22. Shanice, because her point is not on the line formed by the other points
23. (21, 31.5)
24. $d = 1.5m$

ADAPT

Check students' answers to the Lesson Practice to ensure that they understand how to use tables, graphs and equations to model linear functions. Reinforce students' understanding of how the different representations are related.

Lesson 7-1
The Spring Experiment

Mr. Hardiff's class conducts an experiment with a spring and a set of weights. They record their data, but some of the information is missing.

Weight (oz)	Spring Stretch (in.)
5	12.5
8	20
10	25
12	30
15	37.5
16	40

17. How much does the spring stretch for each additional ounce of weight?
    2.5 inches
18. Describe how to use your answer to Item 17 to write an equation for the data in the table.
    The stretch is 2.5 inches per ounce, so the equation is $d = 2.5w$.
19. Use your equation from Item 18 to complete the table.
    See above.

Check Your Understanding

20. A 4.5-pound weight stretches a spring 18 inches and a 7.5-pound weight stretches the same spring 30 inches. How much does the spring stretch for each additional pound of weight? Explain how you found your answer.

LESSON 7-1 PRACTICE

Jeremy and his classmates conduct an experiment with a set of weights and a spring. They record their results in the table. Use the table to answer Items 21–24.

Student	Mass (lb)	Spring Stretch (in.)
Jeremy	5	7.5
Adele	8	12
Roberto	14	21
Shanice	21	36
Guillaume	28	42

21. Make a graph of the data.
22. **Critique the reasoning of others.** Which student made a mistake when taking their turn at the experiment? Explain how you know.
23. If the mistake in Item 22 were corrected, what would the correct data point be?
24. Write an equation to describe the students' data, using the corrected data point you identified in Item 23.

Lesson 7-2
The Falling Object Experiment

Learning Target:
- Graph a function describing a real-world situation and identify and interpret key features of the graph.

SUGGESTED LEARNING STRATEGIES: Discussion Groups, Look for a Pattern, Construct an Argument, Think-Pair-Share, Summarizing, Sharing and Responding

1. The Empire State Building in New York City is 1454 feet tall. How long do you think it will take a penny dropped from the top of the Empire State Building to hit the ground?
   Answers will vary.

In 1589, the mathematician and scientist Galileo conducted an experiment to answer a question much like the one in Item 1. Galileo dropped balls from the top of the Leaning Tower of Pisa in Italy and determined the time it took them to reach the ground. Galileo used several balls identical in shape but differing in mass. Because the balls all reached the ground in the same amount of time, he developed the theory that all objects fall at the same rate.

Galileo's findings can be represented with the equation $h(t) = 1600 - 16t^2$, where $h(t)$ represents the height in feet of an object t seconds after it has been dropped from a height of 1600 feet.

2. Make a table of values for Galileo's function $h(t) = 1600 - 16t^2$.

t (seconds)	$h(t)$ (feet)
0	1600
1	1584
2	1536
3	1456
4	1344
5	1200
6	1024
7	816
8	576
9	304
10	0

ACTIVITY 7 Continued
Lesson 7-2

PLAN

Pacing: 1 class period
Chunking the Lesson
#1   #2–9   #10–15
Check Your Understanding
Lesson Practice

TEACH

Bell-Ringer Activity
Ask students to sketch a nonlinear function on a coordinate plane, labeling any relative maxima and minima as well as the x- and y-intercepts. Ask students to also state the domain and the range.

1 Think-Pair-Share, Construct an Argument Give students a few minutes to think on their own and share with a partner before allowing several students to quickly share with the whole class to generate interest.

2–9 Shared Reading, Marking the Text, Summarizing, Discussion Groups, Activating Prior Knowledge, Construct an Argument, Sharing and Responding Students use the graph and the description of the problem situation to determine a reasonable domain and range and to relate those to the x- and y-intercepts. It is important for students to identify the x-intercept as the point identifying the length of time the ball was in the air before it hit the ground and the y-intercept as the height of the ball when it was dropped. Students should be able to connect this information with the idea that a negative domain would be inappropriate. Monitor groups carefully to ensure students' graphs reflect this understanding.

Activity 7 • Graphs of Functions 101

Lesson 7-2
The Falling Object Experiment

3. **Construct viable arguments.** Why would negative domain values not be appropriate in this context?
   t represents the number of seconds after an object is dropped, so negative values do not make sense.

4. Using your table of values, graph Galileo's function.

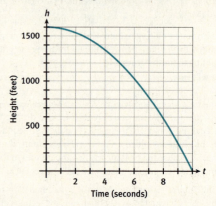

5. What is the reasonable domain of the function represented in your graph? What is the reasonable range?
   Domain: $\{0 \leq t \leq 10\}$; Range: $\{0 \leq h \leq 1600\}$

6. What is the y-intercept?
   (0, 1600)

7. What does the y-intercept represent?
   The height of the object before it is dropped (at 0 seconds)

8. What is the **x-intercept**? What does the x-intercept represent?
   (10, 0); the number of seconds it takes the object to reach the ground (a height of 0)

9. Identify any extrema of the function shown in the graph. What do the extrema represent?
   Maximum: 1600; the greatest height of the object (before it is dropped); minimum: 0; the least height of the object (when it hits the ground)

"Your homework assignment is to graph this function," your math teacher says. She then points to the following function on the board:

$$f(x) = x^2 - 2x$$

In this case, the function is not limited by a real-world situation. Therefore, it is important to use different types of domain values as you prepare to graph.

MATH TERMS

The **x-intercept** is the point where a graph crosses the x-axis. The y-coordinate of the x-intercept is 0.

Lesson 7-2
The Falling Object Experiment

10. Using various values for x, make a table of values for $f(x) = x^2 - 2x$.
    Answers will vary.

x	f(x)
−2	8
−1	3
0	0
1	−1
2	0
3	3

11. Using your table of values, graph the function.

    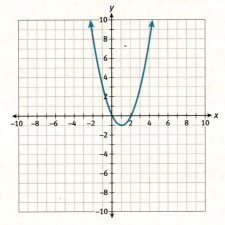

12. Describe the differences between the domain of $f(x) = x^2 - 2x$ and the domain of Galileo's function.

    For $f(x) = x^2 - 2x$, x can assume any values, including negative ones; therefore the domain is all real numbers. The domain of Galileo's function is limited to values of x that make sense in the real-world situation (non-negative numbers).

13. State the range of $f(x) = x^2 - 2x$.

    {y: y ≥ −1}

14. Identify the y-intercept of $f(x) = x^2 - 2x$.

    (0, 0)

15. What is the absolute maximum of $f(x) = x^2 - 2x$? What is the absolute minimum?

    There is no absolute maximum value of f(x) because the graph continues upward indefinitely; absolute minimum: −1

ACTIVITY 7 Continued

Check Your Understanding

Debrief this lesson by having students revisit the opening question in this lesson. As students offer their opinions, encourage them to refer to Galileo's equation and methods they used in this lesson.

Answers

16. Galileo's equation is for an object dropped from a height of 1600 feet. This is slightly greater than the height of the Empire State Building, so the amount of time it takes a penny dropped from the top of the Empire State Building to reach the ground will be close to, but less than, the amount of time Galileo's equation predicts, which is 10 seconds.

ASSESS

Students' answers to Lesson Practice problems will provide you with a formative assessment of their understanding of the lesson concepts and their ability to apply their learning.

See the Activity Practice for additional problems for this lesson. You may assign the problems here or use them as a culmination for the activity.

LESSON 7-2 PRACTICE

17.

w	1	2	3	4	5	6	7	8	9
f(w)	9	16	21	24	25	24	21	16	9

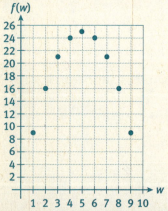

ADAPT

If students have difficulty understanding what the questions are asking for, review the terms on the Word Wall. Remind students to consider each answer in the context of the situation.

18. Domain: {all whole numbers w such that $1 \leq w \leq 9$}; w must be a whole number, and the width of a rectangle cannot be 0, so the least value for w is 1. If $w = 10$, then the area of the rectangle becomes 0, and that is not possible. Therefore, the greatest value for w is 9.
19. Because the width of a rectangle can never be 0, there is no y-intercept.
20. Absolute maximum: 25; absolute minimum: 9

21. Tables will vary.

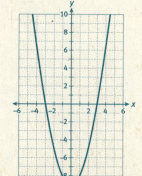

22. Domain: {all real values of x}; range: {$y: y \geq 9$}
23. y-intercept: $(0, -9)$; absolute maximum: does not exist; absolute minimum: -9

My Notes

Lesson 7-2
The Falling Object Experiment

Check Your Understanding

16. Revisit your answer to Item 1 and revise it if necessary. About how long do you think it will take a penny dropped from the top of the Empire State Building to hit the ground? How can you use Galileo's equation to help you answer this question?

LESSON 7-2 PRACTICE

The area of a rectangle with a perimeter of 20 units is given by $f(w) = 10w - w^2$, where w is the width of the rectangle. Assume that w is a whole number. Use this function to answer Items 17–20.

17. Make a table of values and a graph of the function.
18. **Attend to precision.** Give a reasonable domain for the function in this context. Explain your answers.
19. Identify the y-intercept of the function. What does the y-intercept represent within this context?
20. What is the absolute maximum of the function? What is the absolute minimum?

For Items 21–23, use the function $f(x) = x^2 - 9$.

21. Make a table of values and a graph of the function.
22. What are the domain and range?
23. Identify the y-intercept, the absolute maximum, and the absolute minimum.

Lesson 7-3
The Radioactive Decay Experiment

Learning Targets:
- Given a verbal description of a function, make a table and a graph of the function.
- Graph a function and identify and interpret key features of the graph.

SUGGESTED LEARNING STRATEGIES: Discussion Groups, Look for a Pattern, Construct an Argument, Paraphrasing, Marking the Text, Think-Pair-Share

In the late nineteenth century, the scientist Marie Curie performed experiments that led to the discovery of radioactive substances.

A radioactive substance is a substance that gives off radiation as it decays. Scientists describe the rate at which a radioactive substance decays as its *half-life*. The half-life of a substance is the amount of time it takes for one-half of the substance to decay.

1. Radium has a half-life of 1600 years. How much radium will be left from a 1000-gram sample after 1600 years?
   500 grams

2. How much radium will be left after another 1600 years?
   250 grams

3. Suppose a radioactive substance has a half-life of 1 second and you begin with a sample of 4 grams. Complete the table of values.

Time (seconds)	Amount Remaining (grams)
0	4
1	**2**
2	**1**
3	**0.5**
4	**0.25**
5	**0.125**

CONNECT TO SCIENCE

How much is half a life?
The half-life of a radioactive substance can be as little as 0.0018 seconds for Polonium-215 and as much as 4.5 billion years for Uranium-238.

ACTIVITY 7 Continued

3–4 Look for a Pattern, Create Representations, Discussion Groups Monitor groups carefully to ensure students are creating graphs that do not intersect the *x*-axis. If groups are having trouble with this idea, encourage the addition of extra units of time to the table to identify a pattern.

5–6 Construct an Argument, Discussion Groups Students should use the real-world situation and the idea that a radioactive substance will never completely decay to explain why the domain of the function modeled by the graph includes 0 while the range cannot include 0.

7–8 Discussion Groups, Sharing and Responding Students connect the real-world situation to the mathematical model by interpreting the meaning of the *y*-intercept and maximum within the context. It is important that they make the connection between the range of the function and the fact that there can be no absolute minimum. This idea lays the foundation for the identification of horizontal asymptotes and end behavior.

Differentiating Instruction

Extend student learning by asking students to research the half-lives of two different radioactive substances. Assuming they begin with 8 grams of each, students should create a table and a graph comparing the amount of each substance remaining over a period of time.

ACTIVITY 7 continued

Lesson 7-3
The Radioactive Decay Experiment

4. Graph the data from the table on the grid below.

5. **Make use of structure.** Will the amount of the substance that remains ever reach 0? Explain.
   No; the amount divides in half every second, and there is no amount for which one-half of that amount is 0.

6. What are the reasonable domain and range of the function represented in the graph? Explain.
   Domain: $\{x: x \geq 0\}$; x represents time and so cannot be negative.
   Range: $\{y: 0 < y \leq 4\}$; the greatest amount is 4 grams at time 0, and the amount remaining gets smaller and smaller but never reaches 0.

7. What is the *y*-intercept and what does it represent?
   (0, 4); there are 4 grams of the substance at time = 0.

8. Identify the absolute maximum and minimum of the function represented in the graph, and tell what they represent in the context.
   Absolute maximum: 4; the greatest amount of the substance (at time = 0). There is no absolute minimum value.

106 SpringBoard® Mathematics Algebra 1, Unit 2 • Functions

Lesson 7-3
The Radioactive Decay Experiment

ACTIVITY 7
continued

The function that describes the substance's decay is $f(x) = 4\left(\frac{1}{2}\right)^x$. The graph of this function when it does not model a real-world situation is shown below.

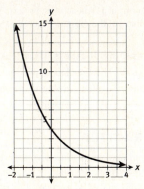

9. What are the domain and range of the function?
   Domain: {all real values of x}; Range: { $y : y > 0$ }

10. How is this graph different from your graph in Item 4?
    The graph in Item 4 appears in the first quadrant only, because x is restricted to nonnegative values. In the graph above, x can be any value, and so the graph continues on the left side of the y-axis.

11. How do the values of y change as the values of x increase?
    The values of y decrease; they approach, but do not reach, 0.

12. How do the values of y change as the values of x decrease?
    The values of y increase.

13. Identify the absolute maximum and absolute minimum of the function.
    They do not exist.

ACTIVITY 7 Continued

9–13 Think-Pair-Share, Debriefing
Students should recognize that the domain and range can change depending on the situation or the lack of a real-world situation. It is especially important that students understand why this function, without a real-world context, has no maximum or minimum value. Debrief this portion of the activity by again discussing why the function never yields a y-value of 0. Talk to students about exponential decay and how it relates to this Activity.

TEACHER to TEACHER

The functions in this lesson are examples of *exponential functions* because the variable is an exponent. Exponential functions are written in the form $y = a \cdot b^x$, where $a \neq 0$ and either $0 < b < 1$ or $b > 1$. In this lesson, $0 < b < 1$ for all the functions, and all the graphs have similar shapes. When $b > 1$ and $a \neq 0$, the graphs increase rather than decrease from left to right.

Activity 7 • Graphs of Functions 107

ACTIVITY 7 Continued

Check Your Understanding
Debrief student answers to these items to determine their level of understanding of exponential functions. At this point, student understanding will be basic but should indicate an awareness of the similarities and differences between exponential and linear functions.

Answers
14. $\frac{g}{2}$, or $\frac{1}{2}g$
15. The function never reaches 0, so it cannot be the absolute minimum.

ASSESS

Students' answers to Lesson Practice problems will provide you with a formative assessment of their understanding of the lesson concepts and their ability to apply their learning.

See the Activity Practice for additional problems for this lesson. You may assign the problems here or use them as a culmination for the activity.

LESSON 7-3 PRACTICE
16. A value is reduced by half after every fixed time period.
17.

Time (years)	Value ($)
0	20,000
1	10,000
2	5,000
3	2,500
4	1,250
5	625

18. 6 years

ADAPT

Check students' answers to the Lesson Practice to ensure that they understand exponential equations and are able to discuss the features of their graphs. Continue to expect students to use mathematical terminology when discussing the functions and their graphs.

ACTIVITY 7 continued

Lesson 7-3
The Radioactive Decay Experiment

Check Your Understanding

14. A scientist has g grams of a radioactive substance. Write an expression that shows the amount of the substance that remains after one half-life.

15. **Critique the reasoning of others.** Dylan looked at the function $f(x) = 4\left(\frac{1}{2}\right)^x$ and said, "This function is always greater than 0, so 0 is the absolute minimum." Explain why Dylan is incorrect.

LESSON 7-3 PRACTICE

Suppose the value of your new car is reduced by half every year that you own it. You paid $20,000 for your new car.

16. Describe how this situation is similar to the half-life of a radioactive substance.

17. Copy and complete the table below.

Time (years)	Value ($)
0	20,000
1	
2	
3	
4	
5	

18. **Make sense of problems.** For insurance purposes, a vehicle is considered scrap when its value falls below $500. After how many years will your new car be considered scrap?

Graphs of Functions
Experiment Experiences

ACTIVITY 7 continued

ACTIVITY 7 PRACTICE
Write your answers on notebook paper.
Show your work.

Lesson 7-1
A weight of 15 ounces stretches a spring 10 inches. A weight of 24 ounces stretches the same spring 16 inches. Use this information to answer Items 1–4.

1. How many inches does the spring stretch per ounce of additional weight?
   A. $\frac{2}{3}$ inch
   B. $\frac{3}{2}$ inches
   C. 25 inches
   D. 150 inches

2. Write an equation to describe the relationship between the distance d that the spring stretches and the weight w that is attached to it.

3. How much will the spring stretch for a weight of 9 ounces?

4. The spring is stretched 14 inches. How many ounces is the weight that is attached to it?

A spring stretches 2.5 inches for each ounce of weight. Use this information for Items 5–7.

5. Determine a function that represents this situation.

6. If you were to graph the function represented by this situation, what would be the reasonable domain? Explain.

7. Which of the following data points would **not** lie on the graph representing this function?
   A. (0, 0)
   B. (1, 2.5)
   C. (2.5, 1)
   D. (10, 25)

Lesson 7-2
Suppose that the height of an object after x seconds is given by $f(x) = 100 - 4x^2$, as shown in the graph below.

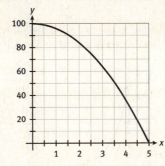

Use the function or the graph for Items 8–14.

8. What is the reasonable domain of the function?
9. What is the reasonable range of the function?
10. Identify the y-intercept of the function.
11. What does the y-intercept represent?
12. Identify the x-intercept of the function.
13. What does the x-intercept represent?
14. Loni says that because of the negative sign in front of $4x^2$, the reasonable domain for this function is only negative values. Is her reasoning correct? Explain.

ACTIVITY 7 Continued

ACTIVITY PRACTICE
1. A
2. $d = \frac{2}{3}w$
3. 6 inches
4. 21 ounces
5. $f(w) = 2.5w$
6. $\{w \geq 0\}$ because you can't have a weight less than 0.
7. C
8. $\{0 \leq x \leq 5\}$
9. $\{0 \leq y \leq 100\}$
10. (0, 100)
11. the height from which the ball is dropped
12. (5, 0)
13. the time, in seconds, it takes the ball to hit the ground
14. No; the domain represents the number of seconds the ball has been in the air. A negative number of seconds is impossible.

Activity 7 • Graphs of Functions

ACTIVITY 7 Continued

19.

Number of Cuts, x	Area of Remaining Piece, y
0	150
1	75
2	37.5
3	18.75
4	9.375

20. A value is repeatedly being reduced by one-half.

21. No; the number of cuts must be a whole number.

22. In this situation, the reasonable domain consists of whole numbers only because the domain values represent a number of cuts. In a radioactive decay situation, the reasonable domain consists of all nonnegative numbers, because the domain values represent time.

23. (0, 150); the area of the original piece of paper (after 0 cuts)

24. 150; the area of the original piece of paper, which is the piece that has the greatest area

25. Yes; Possible justification: Maude will spend her birthday money in 15 days, as shown in the table.

Day	Money Remaining
0	$100
1	$50
2	$25
3	$12.50
4	$6.25
5	$3.13
6	$1.57
7	$0.79
8	$0.40
9	$0.20
10	$0.10
11	$0.05
12	$0.03
13	$0.02
14	$0.01
15	$0.00

Because Maude cannot spend half cents, she will have to round as she spends half of her money each day (for example, on days 5–7). On day 14, she cannot spend half a cent, so she will have to spend the entire remaining cent, leaving her with no more money.

ADDITIONAL PRACTICE

If students need more practice on the concepts in this activity, see the eBook Teacher Resources for additional practice problems.

ACTIVITY 7 continued

Lesson 7-3

15. The half-life of a radioactive substance is 1 hour. If you begin with 100 ounces of the substance, how many hours does it take for 12.25 ounces to remain?

The graph below represents a radioactive decay situation. Use this graph for Items 16–18.

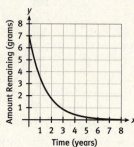

16. What is the original amount of the radioactive substance? Explain how you know.

17. What are the reasonable domain and range?

18. Identify the absolute maximum and absolute minimum values of the function. What do these values represent?

Barry has a piece of paper whose area is 150 square inches. He cuts the paper in half and discards one of the pieces. He repeats this procedure several times. Use this information for Items 19–24.

19. Copy and complete the table below to show the area of the remaining piece of paper after x cuts.

Number of Cuts, x	Area of Remaining Piece, y
0	150
1	
2	
3	
4	

20. Describe how this situation is similar to the half-life of a radioactive substance.

21. If you were to graph the points from the table, would you connect the points? Explain.

22. Describe how the reasonable domain in this situation is different from the reasonable domain in a radioactive decay situation.

23. Identify the y-intercept. What does it represent?

24. Identify the absolute maximum value. What does it represent?

MATHEMATICAL PRACTICES
Construct Viable Arguments and Critique the Reasoning of Others

25. Maude receives $100 for her birthday. "I am going to spend half of my birthday money each day until none is left," she decides. Is it reasonable for her to believe that she will eventually spend all of the money? Justify your answer.

---

15. 4 hours (Note: after 3 hours, there is 12.5 oz left, so an additional hour is needed to reach 12.25 oz)

16. 7 grams; The graph shows that this is the amount when time = 0.

17. Domain: $\{x: x \geq 0\}$; range: $\{y: 0 < y \leq 7\}$

18. Absolute maximum: 7; the greatest amount of the substance (at time = 0); There is no absolute minimum value. The amount of substance will never be 0, and the curve will never reach the x-axis.

Transformations of Functions

Transformers

Lesson 8-1 Exploring $f(x) + k$

Learning Targets:
- Identify the effect on the graph of replacing $f(x)$ by $f(x) + k$.
- Identify the transformation used to produce one graph from another.

SUGGESTED LEARNING STRATEGIES: Look for a Pattern, Interactive Word Wall, Think-Pair-Share, Create Representations, Discussion Groups

The equation and the graph of $y = x$ or $f(x) = x$ are referred to as the linear *parent function*. The graph of $f(x) = x$ is shown below.

MATH TERMS

A **parent function** is the most basic function of a particular category or type.

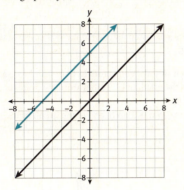

1. Complete the table for $g(x) = x + 5$.

x	$f(x) = x$	$g(x) = x + 5$
−3	−3	2
−2	−2	3
−1	−1	4
0	0	5
1	1	6
2	2	7
3	3	8

2. **Make use of structure.** How do the *y*-values for $g(x)$ compare to the *y*-values for $f(x)$? Make a conjecture about the graph of $g(x)$. As you share your ideas with your group, be sure to use mathematical terms and academic vocabulary precisely. Make notes to help you remember the meaning of new words and how they are used to describe mathematical concepts.

   The y-values for g(x) are 5 greater than the corresponding y-values for f(x). Answers may vary. The graph of g(x) will be 5 units above the graph of f(x).

Common Core State Standards for Activity 8

HSF-BF.B.3: Identify the effect on the graph of replacing $f(x)$ by $f(x) + k$, $kf(x)$, $f(kx)$ and $f(x + k)$ for specific values of k (both positive and negative); find the value of k given the graphs. Experiment with cases and illustrate an explanation of the effects on the graph using technology.

ACTIVITY 8
Investigative

Activity Standards Focus

In this activity students build on their knowledge of functions and their graphs by exploring vertical and horizontal translations of graphs produced by the addition of a constant, k, to a function.

Lesson 8-1

PLAN

Materials
- graph paper
- colored pencils
- graphing calculators

Pacing: 1 class period

Chunking the Lesson

#1–3 #4–6 #7–8
Check Your Understanding
#11–14
Check Your Understanding
Lesson Practice

TEACH

Bell-Ringer Activity

Students should use graph paper and colored pencils. Instruct students to sketch a small, simple shape in the center of the graph paper, indicating several points on the figure. Direct students to translate or slide the shape 12 units up and redraw it in a different color. Request that students also translate the original shape 12 units to the left, 12 units down, and 12 units to the right, redrawing the shape in a different color each time.

Developing Math Language

This activity introduces students to the concept of a *parent function*, the basic building block of a family of functions. As students explore translations of functions, use the term *parent functions* when referring to these basic functions.

1–3 Close Reading, Interactive Word Wall, Think-Pair-Share, Look for a Pattern, Visualization, Predict and Confirm Students need to spend a few minutes becoming familiar with the term *parent function* using the lesson introduction and Math Terms box. Add the word to the classroom word wall and encourage students to identify parent functions as they are used in the activity. As students complete the table, they should recognize that the values of $g(x)$ are 5 greater than the corresponding values of $f(x)$. They should connect this observation with the equation of $g(x)$. Completing the table for both functions enables students to visualize the differences between the graphs.

Activity 8 • Transformations of Functions **111**

ACTIVITY 8 Continued

1–3 (continued) Be sure students understand how to enter functions into a graphing calculator. The calculator should be set to view graphs in the standard window, $-10 < x < 10$ and $-10 < y < 10$.

4–6 Predict and Confirm, Create Representations, Discussion Groups
Students should identify $f(x) = x^3$ as the parent function as they compare the equations of the functions $f(x) = x^3$ and $g(x) = x^3 - 4$. Encourage students to label at least a few points on both graphs to facilitate the identification of the effect of adding the constant -4 to the parent function. Once students have confirmed or revised their conjectures about the graph of $g(x)$, they may be ready to generalize about the difference between the graphs of $f(x)$ and $g(x) = f(x) + k$.

Differentiating Instruction

Support students in using language to describe the differences between the graphs of related functions. Terms such as *up*, *down*, *left*, *right*, *same* and *different* will help some students communicate verbally what they comprehend visually from the graphs or conceptually from the equations.

ACTIVITY 8 continued

My Notes

MATH TIP
The *x*-coordinate of the *x*-intercept is called a zero of the function. You will learn more about zeros of functions when you study quadratic functions later in this course.

Lesson 8-1
Exploring $f(x) + k$

3. Test your conjecture by using a graphing calculator to graph $g(x) = x + 5$. Graph this on the grid in Item 1. **See graph.**
   a. What is the *y*-intercept of the parent function?
      (0, 0)
   b. What is the *y*-intercept of $g(x)$?
      (0, 5)
   c. What is the *x*-intercept of the parent function? What is the zero of the function $f(x)$?
      (0, 0); 0
   d. What is the *x*-intercept of $g(x)$? What is the zero of the function?
      (−5, 0); −5
   e. Revisit your original conjecture in Item 2 and revise it if necessary. How does the graph of $g(x)$ differ from the graph of the parent function, $f(x) = x$?
      The graph of $g(x)$ is 5 units above the graph of the parent function.

The graph of $f(x) = x^3$ is shown below.

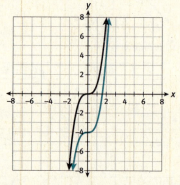

4. Make a conjecture about the graph of $g(x) = x^3 - 4$.
   Answers may vary. It will be 4 units below the graph of $f(x)$.

5. Graph both $f(x)$ and $g(x)$ on a graphing calculator. Sketch the graph of $g(x)$ on the grid above. Label a few points on each graph.
   See above graph.

6. Revisit your original conjecture in Item 4 about the graph of $g(x)$ and revise it if necessary. How does the graph of $g(x)$ differ from the graph of $f(x)$?
   The graph of $g(x)$ is 4 units below the parent graph.

7. **Express regularity in repeated reasoning.** How does the value of k in the equation $g(x) = f(x) + k$ change the graph of $f(x)$?
   The graph of $f(x)$ is shifted up k units when k is positive and down $|k|$ units when k is negative.

Exploring $f(x) + k$

A change in the position, size, or shape of a graph is a **transformation**. The changes to the graphs in Items 1–6 are examples of a transformation called a *vertical translation*.

8. In the figure, the graphs of $g(x)$ and $h(x)$ are vertical translations of the graph of $f(x) = 2^x$.

   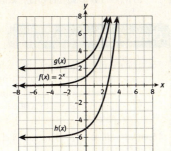

   a. Write the equation for $g(x)$.
   $g(x) = 2^x + 2$

   b. Write the equation for $h(x)$.
   $h(x) = 2^x - 6$

MATH TERMS

A **vertical translation** of a graph shifts the graph up or down. A vertical translation preserves the shape of the graph.

Check Your Understanding

9. Without graphing, describe the transformation from the graph of $f(x) = x^2$ to the graph of $g(x) = x^2 + 7$.

10. Suppose $f(x) = x - 2$. Describe the transformation from the graph of $f(x)$ to the graph of $g(x) = x + 3$. Use a graphing calculator to check your answer.

Ray's Gym charges an initial sign-up fee of $25.00 and a monthly fee of $15.00.

11. **Reason abstractly.** Write a function that describes the gym's total membership fee for x months.   **$f(x) = 15x + 25$**

12. Graph the function you wrote in Item 11 on the grid below. Label several points on the graph.

    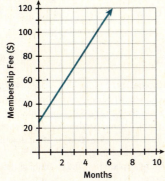

13. Identify the y-intercept. What does the y-intercept represent?
    (0, 25); the initial sign-up fee

ACTIVITY 8 Continued

11–14 (continued) As students consider the change in the sign-up fee, they should identify the parts of the equation and the graph that represent the original sign-up fee. They can then adjust the equation and the graph accordingly.

Check Your Understanding

Debrief the lesson by asking students to compare two related functions in terms of their equations and graphs, including the y-intercept. Upon completion of this lesson, students should understand that adding a constant k to the equation of a function results in a vertical translation of the graph. The effect of k should be generalized not only in terms of direction (positive = translate up, negative = translate down), but also in terms of magnitude (the greater the absolute value of k, the greater the distance translated).

Answers

15. a. The monthly fee is the same, but the sign-up fee at Gina's Gym is $20, as compared to $25 at Ray's Gym.
b. The graph of $g(x)$ would be a translation 5 units down of the graph of $f(x)$.
16. $(0, b + k)$

ASSESS

Students' answers to lesson practice problems will provide you with a formative assessment of their understanding of the lesson concepts and their ability to apply their learning.

See the Activity Practice for additional problems for this lesson. You may assign the problems here or use them as a culmination for the activity.

ADAPT

Check students' answers to the Lesson Practice to ensure that students understand vertical translations and are ready to move on to horizontal translations. Students having difficulty predicting the effect on the graph of the parent function directly from the equation may create a table of values, as in Item 1, and compare to the values for the parent function before attempting to describe the graph.

ACTIVITY 8 continued

My Notes

Lesson 8-1
Exploring $f(x) + k$

14. How would the function change if the initial sign-up fee were increased by $5.00? How would the graph change? **The function would change to $f(x) = 15x + 30$; the graph would be translated up 5 units.**

Check Your Understanding

15. The membership fee at Gina's Gym is given by the function $g(x) = 15x + 20$, where x is the number of months.
  a. How do the fees at Gina's Gym compare to those at Ray's Gym?
  b. Without graphing, describe how the graph of $g(x)$ compares to the graph of $f(x)$.

16. The y-intercept of a function $f(x)$ is $(0, b)$. What is the y-intercept of $f(x) + k$?

LESSON 8-1 PRACTICE

Identify the transformation from the graph of $f(x) = x^2$ to the graph of $g(x)$. Then graph $f(x)$ and $g(x)$ on the same coordinate plane.

17. $g(x) = x^2 - 7$ **18.** $g(x) = x^2 + 10$

Write the equation of the function described by each of the following transformations of the graph of $f(x) = x^3$.

19. Translated up 9 units **20.** Translated down 5 units

Each graph shows a vertical translation of the graph of $f(x) = x$. Write an equation to describe each graph.

21. 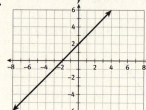 **22.**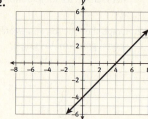

23. Model with mathematics. Orange Taxi charges $2.75 as soon as you step into the taxi and $2.50 per mile. Magenta Taxi charges $3.25 as soon as you step into the taxi and $2.50 per mile.
  a. Write a function $f(x)$ that describes the total cost of a ride of x miles with Orange Taxi. Write a function $g(x)$ that describes the total cost of a ride of x miles with Magenta Taxi.
  b. Without graphing, explain how the graph of $g(x)$ compares to the graph of $f(x)$.
  c. Check your answer to Part (b) by graphing the functions.

LESSON 8-1 PRACTICE

17. translation 7 units down
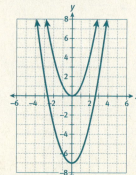

18. translation 10 units up
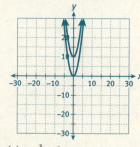

19. $g(x) = x^3 + 9$
20. $g(x) = x^3 - 5$
21. $g(x) = x + 2$

22. $g(x) = x - 4$
23. a. $f(x) = 2.5x + 2.75$; $g(x) = 2.5x + 3.25$
b. The graph of $g(x)$ is a translation 0.5 units up of the graph of $f(x)$.
c.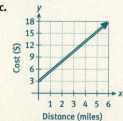

114 SpringBoard® Mathematics **Algebra 1, Unit 2 • Functions**

Lesson 8-2
Exploring $f(x + k)$

Learning Targets:
- Identify the effect on the graph of replacing $f(x)$ by $f(x + k)$.
- Identify the transformation used to produce one graph from another.

SUGGESTED LEARNING STRATEGIES: Predict and Confirm, Look for a Pattern, Create Representations, Think-Pair-Share, Discussion Groups

The function $f(x) = |x|$ is graphed below.

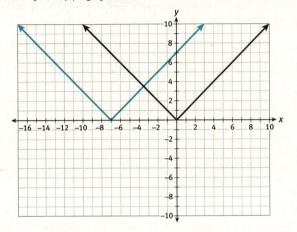

1. Write a new function, $g(x)$, by replacing x with $x + 7$.
   $g(x) = |x + 7|$

2. Graph both $f(x) = |x|$ and $g(x)$ on a graphing calculator. Sketch the graph of $g(x)$ on the grid above, labeling at least a few points on each graph.
   See above graph.

3. What is the x-intercept of $f(x) = |x|$?
   $(0, 0)$

4. What is the x-intercept of $g(x)$?
   $(-7, 0)$

5. Describe the transformation from the graph of $f(x) = |x|$ to the graph of $g(x)$.
   The graph of $f(x)$ is shifted 7 units to the left.

Note that the function $g(x)$ can be written as $f(x + 7)$. This means that x is replaced with $x + 7$ in the function $f(x)$.

MATH TERMS

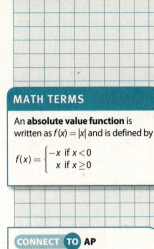

An **absolute value function** is written as $f(x) = |x|$ and is defined by
$$f(x) = \begin{cases} -x & \text{if } x < 0 \\ x & \text{if } x \geq 0 \end{cases}$$

CONNECT TO AP

The vertex of an absolute value function is an example of a **cusp** in a graph. A graph has a cusp at a point where there is an abrupt change in direction.

Activity 8 • Transformations of Functions 115

ACTIVITY 8 Continued

6–10 Discussion Groups, Predict and Confirm, Debriefing In Item 6, students should explain the reasoning behind their conjectures before they use a graphing calculator to confirm in Item 7. Encourage students to label at least a few points on both graphs to facilitate the identification of the effect of adding the constant -3 inside the parent function.

In Item 10, students should identify how the equation for $g(x)$ differs from the equation for $f(x)$. They can then discuss how to alter the graph to reflect this difference. Once students have confirmed or revised their conjectures, they may be ready to generalize about the difference between the graphs of $f(x)$ and $g(x) = f(x + k)$. Debrief by allowing students to have a quick discussion comparing and contrasting the translations of the parent function $k(x) = x$ for $f(x) = x + 8$ and $g(x) = (x + 8)$.

Differentiating Instruction

Support students who have difficulty determining the direction of a horizontal translation by connecting the equation and the graph. The graphs of the parent functions in this lesson all contain the point $(0, 0)$. Have students examine $f(x + k)$ and find the value of x for which $f(x + k) = 0$. For example, for $(x + 2)^2$, this value of x is -2 and so the graph contains the point $(-2, 0)$. This can help students see that all of the points have been translated 2 units to the left.

ACTIVITY 8 continued

My Notes

MATH TERMS

A **horizontal translation** of a graph shifts the graph left or right. Like a vertical translation, a horizontal translation preserves the shape of the graph.

Lesson 8-2
Exploring $f(x + k)$

The graph of $f(x) = x^3$ is shown below.

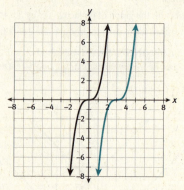

6. Make a conjecture about the graph of $g(x) = (x - 3)^3$.
   Answers may vary. The graph of $f(x)$ will be 3 units to the right of the graph of $f(x)$.

7. Graph both $f(x)$ and $g(x)$ on a graphing calculator. Sketch the graph of $g(x)$ on the grid above, labeling at least a few points on each graph.
   See above graph.

8. Revisit your original conjecture in Item 6 about the graph of $g(x)$ and revise it if necessary. How does the graph of $g(x)$ differ from the graph of $f(x)$?
   The graph of $g(x)$ is 3 units to the right of the graph of $f(x)$.

9. How does the value of k in the equation $g(x) = f(x + k)$ change the graph of the function $f(x)$?
   The graph of $f(x)$ is shifted left k units when k is positive and right $|k|$ units when k is negative.

The changes to the graphs in Items 1–8 are examples of a transformation called a **horizontal translation**.

10. The figure shows the graph of the function $f(x) = 2^x$.
    a. Without using a graphing calculator, sketch the graph of $g(x) = f(x + 8) = 2^{x+8}$ on the grid. **See graph.**

    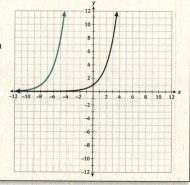

    b. Use a graphing calculator to check your graph in Part (a). Revise your graph if necessary.
    Answers will vary.

116 SpringBoard® Mathematics **Algebra 1, Unit 2** • Functions

Lesson 8-2
Exploring $f(x + k)$

Check Your Understanding

11. Without graphing, describe the transformation from the graph of $f(x) = x^2$ to the graph of $g(x)$.
    a. $g(x) = (x + 4)^2$
    b. $g(x) = f(x - 7)$
    c. $g(x) = (x - 2)^2 + 5$
    d. $g(x) = (x + 9)^2 - 1$

12. The function $f(x) = x^2$ and another function, $g(x)$, are graphed below. Write the equation for $g(x)$. Explain how you found your answer.

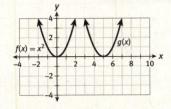

13. **Make sense of problems.** Julio went to a theme park in July. He paid $15 to enter the park and $3.00 for each ride. He went on x rides.
    a. Write a function that describes the total cost of Julio's trip to the theme park.
    $f(x) = 3x + 15$

    b. Julio went back to the theme park in September. The entrance fee was the same and each ride still cost $3.00. However, this time Julio went on 5 more rides. Use your function from Part (a) to describe Julio's second trip.
    $f(x + 5) = 3(x + 5) + 15$

    c. How does the equation for Julio's second trip to the park change the graph of the first trip?
    It moves the graph 5 units to the left.

    d. What kind of transformation describes the change from the first graph to the second graph?
    a horizontal translation

    e. Julio went to the park again in October and went on 8 fewer rides than he did in July. Use your function from Part (a) to describe Julio's third trip. How does this change the initial graph?
    $f(x - 8) = 3(x - 8) + 15$; the graph will move 8 units to the right.

ACTIVITY 8 Continued

Check Your Understanding
It is essential for students to understand the relationship between the symbolic and graphic representations. Debrief by having students describe how each function's equation is different from $f(x) = x^2$ and how those differences affect the graphs. Note that the graph of a function can be translated both vertically and horizontally, as in Items 11c and 11d.

Answers
11. a. translation 4 units to the left
    b. translation 7 units to the right
    c. translation 2 units right and 5 units up
    d. translation 9 units left and 1 unit down
12. $g(x) = (x - 5)^2$; the graph of $g(x)$ is located 5 units to the right of the graph of $f(x)$, so I used the rule for translating a function to the right, which is to replace x with $x - 5$.

13 Create Representations, Discussion Groups, Debriefing
In parts b and e, be sure students understand that they need to write expressions involving x that represent the number of rides Julio went on in September and in October. Students may need support using function notation to write $f(x)$, $f(x + 3)$, and $f(x - 8)$. Debrief by encouraging students to discuss how a change in the entrance fee affects the equation of the function differently than a change in the number of rides does.

Activity 8 • Transformations of Functions 117

ACTIVITY 8 Continued

Check Your Understanding

When debriefing, make sure students understand translations in terms of the equations, the graphs and the intercepts. Students should understand that the graph of $f(x + k)$ is a horizontal translation of the graph of $f(x)$. The effect of k should be generalized not only in terms of direction (positive = translate left, negative = translate right), but also in terms of magnitude (the greater the absolute value of k, the greater the distance translated).

Answers
14. $a - k$
15. The graph of $y = (x - 4)^3$ is a translation 8 units to the right of the graph of $y = (x + 4)^3$.

ASSESS

Students' answers to lesson practice problems will provide you with a formative assessment of their understanding of the lesson concepts and their ability to apply their learning.

See the Activity Practice for additional problems for this lesson. You may assign the problems here or use them as a culmination for the activity.

LESSON 8-2 PRACTICE
16. translation 1 unit right

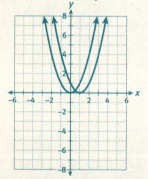

ADAPT

If students have difficulty with the direction of horizontal translations, have them continue to practice predicting the graph of a given function, and then confirming with a graphing calculator.

Lesson 8-2
Exploring $f(x + k)$

f. Julio goes to the park again in November. Now it is the off-season and the entrance fee is $10 less than it was in July. He goes on the same number of rides as he did in July. Write a function to describe Julio's fourth trip. How does the graph of the initial trip change with this new situation?

$f(x) = 3x + 5$; the graph will shift down 10 units.

Check Your Understanding

14. The x-intercept of the function $f(x)$ is $(a, 0)$. What is the x-intercept of the function $f(x + k)$?
15. Without graphing, explain how the graph of $y = (x - 4)^3$ is related to the graph of $y = (x + 4)^3$.

LESSON 8-2 PRACTICE

Identify the transformation from the graph of $f(x) = x^2$ to the graph of $g(x)$. Then graph $f(x)$ and $g(x)$ on the same coordinate plane.

16. $g(x) = (x - 1)^2$
17. $g(x) = (x + 3)^2$

Write the equation of the function described by each of the following transformations of the graph of $f(x) = x^3$.

18. Translated 7 units to the left
19. Translated 8 units to the right
20. Each graph shows a horizontal translation of the graph of $f(x) = x$. Write an equation to describe each graph.

a. 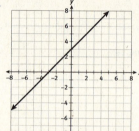 b.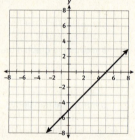

c. **Critique the reasoning of others.** Molly said that the graphs above are also vertical translations of the graph of $f(x) = x$. Is Molly correct? Explain.

21. How does the graph of $h(x) = |x - 4|$ compare with the graph of $f(x) = |x|$?

17. translation 3 units left

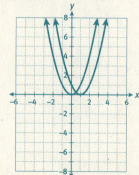

18. $g(x) = (x + 7)^3$
19. $g(x) = (x - 8)^3$
20. a. $g(x) = x + 3$
    b. $g(x) = x - 5$
    c. Yes; For the function $y = x$, a horizontal translation of k units left/right is equivalent to a vertical translation of k units up/down.

Transformations of Functions
Transformers

ACTIVITY 8 continued

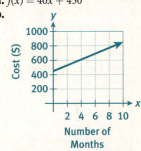

ACTIVITY 8 PRACTICE
Write your answers on notebook paper.
Show your work.

Lesson 8-1
In Items 1–4, identify the transformation from the graph of $f(x) = x^3$ to the graph of $g(x)$.

1. $g(x) = x^3 + 11$
2. $g(x) = x^3 - 4$
3. $g(x) = x^3 + 0.1$
4. $g(x) = -2 + x^3$
5. The graph of $f(x) = x^2$ is translated 9 units down to create the graph of $g(x)$. Which of the following is the equation for $g(x)$?
   A. $g(x) = x^2 + 9$
   B. $g(x) = x^2 - 9$
   C. $g(x) = (x + 9)^2$
   D. $g(x) = (x - 9)^2$

In Items 6 and 7, each graph shows a vertical translation of the graph of $f(x) = x$. Write an equation to describe the graph. Identify the zeros of each function.

6.
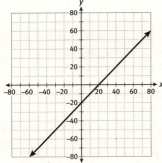

7.
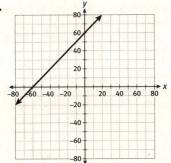

For Items 8 and 9, determine the equation of the function described by each of the following transformations of the graph of $f(x) = 3^x$.

8. Translated 15 units down
9. Translated 2.1 units up
10. An air conditioner costs $450 plus $40 per month to operate.
    a. Write a function that describes the total cost of buying and operating the air conditioner for x months.
    b. Use your calculator to graph the function.
    c. What is the y-intercept? What does it represent?
    d. How would the function change if the price of the air conditioner were reduced to $425? How would the graph change?

Given that $g(x) = f(x) + k$, with $k \neq 0$, determine whether each statement is always, sometimes, or never true.

11. The graph of $g(x)$ is a vertical translation of the graph of $f(x)$.
12. The graphs of $f(x)$ and $g(x)$ are both lines.
13. The graph of $f(x)$ has the same y-intercept as the graph of $g(x)$.
14. Caitlin drew the graph of $f(x) = x^2$. Then she translated the graph 6 units up to get the graph of $g(x)$. Next, she translated the graph of $g(x)$ 8 units down to get the graph of $h(x)$. Which of these is an equation for $h(x)$?
    A. $h(x) = x^2 + 14$
    B. $h(x) = x^2 + 2$
    C. $h(x) = x^2 - 2$
    D. $h(x) = x^2 - 14$

ACTIVITY 8 Continued

ACTIVITY PRACTICE
1. translation 11 units up
2. translation 4 units down
3. translation 0.1 unit up
4. translation 2 units down
5. B
6. $g(x) = x - 20$; 20
7. $g(x) = x + 60$; -60
8. $g(x) = 3^x - 15$
9. $g(x) = 3^x + 2.1$
10. a. $f(x) = 40x + 450$
    b.
    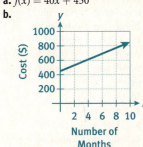
    c. (0, 450); The cost of the air conditioner itself is $450.
    d. The function would be $g(x) = 40x + 425$; the graph of $f(x)$ would be translated down 25 units.
11. always
12. sometimes
13. never
14. C

ACTIVITY 8 Continued

15. translation 3 units down
16. translation 3 units right
17. translation 4 units up
18. translation 4 units left
19. C
20. $g(x) = x^3 + 7$
21. $g(x) = x^3 - 4$
22. $g(x) = (x - 2)^3$
23. $g(x) = x^3 - 5$
24. $g(x) = (x + 3)^3$
25. $g(x) = (x - 5)^4 - 8$
26. translation 7 units right, 1 unit up
27. translation 4 units left
28. translation 9 units left, 0.2 unit down
29. translation 2 units right, 3 units down
30. B
31. a. $f(x) = 4x + 12$
    b. $g(x) = 4x + 15$
    c. translation 3 units up
    d. translation 3 units down

ADDITIONAL PRACTICE

If students need more practice on the concepts in this activity, see the eBook Teacher Resources for additional practice problems.

ACTIVITY 8 continued

Lesson 8-2

In Items 15–18, identify the transformation from the graph of $f(x) = 2^x$ to the graph of $g(x)$.

15. $g(x) = 2^x - 3$
16. $g(x) = 2^{(x-3)}$
17. $g(x) = 2^x + 4$
18. $g(x) = 2^{(x+4)}$

19. The graph of which function is a translation of the graph of $f(x) = x^2$ five units to the right?
    A. $g(x) = x^2 - 5$
    B. $g(x) = (x + 5)^2$
    C. $g(x) = (x - 5)^2$
    D. $g(x) = x^2 + 5$

Write the equation of the function described by each of the following transformations of the graph of $f(x) = x^3$.

20. Translated 7 units up
21. Translated 4 units down
22. Translated 2 units right
23. Translated 5 units down
24. Translated 3 units left
25. The figure shows the graph of $f(x) = x^4$ and the graph of $g(x)$. Write an equation for the graph of $g(x)$.

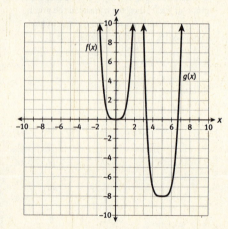

Without graphing, describe the transformation from the graph of $f(x) = x^2$ to the graph of $g(x)$.

26. $g(x) = (x - 7)^2 + 1$
27. $g(x) = f(x + 4)$
28. $g(x) = (x + 9)^2 - 0.2$
29. $g(x) = f(x - 2) - 3$

30. The graph of $f(x)$ is shown below. Which of the following is a true statement about the graph of $g(x) = f(x + 3)$?

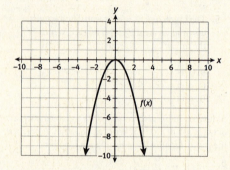

A. The x-intercept of $g(x)$ is $(3, 0)$.
B. The x-intercept of $g(x)$ is $(-3, 0)$.
C. The y-intercept of $g(x)$ is $(0, 3)$.
D. The y-intercept of $g(x)$ is $(0, -3)$.

MATHEMATICAL PRACTICES
Model with Mathematics

31. In 2011, the ticket price for entrance to a state fair was $12. Each ride had an additional $4.00 fee. In 2012, the entrance ticket cost $15 and the rides remained $4.00 each.
    a. Write a function $f(x)$ for the cost of visiting the fair and riding x rides in 2011.
    b. Write a function $g(x)$ for the cost of visiting the fair and riding x rides in 2012.
    c. What transformation could you use to obtain the graph of $g(x)$ from the graph of $f(x)$?
    d. What transformation could you use to obtain the graph of $f(x)$ from the graph of $g(x)$?

Representations of Functions
BRYCE CANYON HIKING

Embedded Assessment 1
Use after Activity 8

While on vacation, Jorge and Jackie traveled to Bryce Canyon National Park in Utah. They were impressed by the differing elevations at the viewpoints along the road. The graph describes the elevations for several viewpoints in terms of the time since they entered the park.

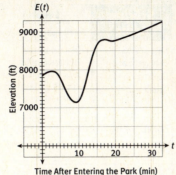

1. The graph represents a function $E(t)$. Describe why the graph represents a function. Identify the domain and range of the function.
2. Is this discrete or continuous data? Explain.
3. What is the y-intercept? Interpret the meaning of the y-intercept in the context of the problem.
4. Identify a relative maximum of the function represented by the graph.
5. What is the absolute maximum of the function represented by the graph? What does it represent?
6. Identify a relative minimum of the function represented by the graph.
7. What is the absolute minimum of the function represented by the graph? What does it represent?

While at Bryce Canyon National Park, Jorge and Jackie hiked at an average speed of about 2 miles per hour.

8. Copy and complete the table below to show the distance hiked by a person whose constant speed is 2 miles per hour.

Time (hours)	Distance (miles)
0	0
1	2
2	
3	
4	
5	

9. Write a function $f(x)$ to describe the data in the table. What are the reasonable domain and range?
10. Create a graph of the function.
11. How long will it take this person to hike 5 miles? Justify your answer.
12. On the same coordinate grid that you used in Item 9, create a graph of another function by translating the graph 5 units up.
13. Write a function to describe the graph you created in Item 12. Explain how you determined your answer.

Common Core State Standards for Embedded Assessment 1

HSF-IF.A.1: Understand that a function from one set (called the domain) to another set (called the range) assigns to each element of the domain exactly one element of the range. If f is a function and x is an element of its domain, then $f(x)$ denotes the output of f corresponding to the input x. The graph of f is the graph of the equation $y = f(x)$.

HSF-IF.A.2: Use function notation, evaluate functions from inputs in their domains, and interpret statements that use function notation in terms of a context.

HSF-IF.B.4: For a function that models a relationship between two quantities, interpret key features of graphs and tables in terms of the quantities, and sketch graphs showing key features given a verbal description of the relationship. Key features include: intercepts; intervals where the function is increasing, decreasing, positive, or negative; relative maximums and minimums; symmetries; end behavior and periodicity.

Embedded Assessment 1

Assessment Focus
- Functions, range and domain
- Graphs of functions and their key features
- Writing and using equations of functions
- Transforming functions

Materials
- graph paper

TEACHER to TEACHER
Remind students to use reading strategies such as Marking the Text as they work to reinforce the vocabulary being used in this Assessment.

Answer Key

1. $E(t)$ is a function because it passes the vertical line test.
   Domain: $0 \leq t \leq 33$;
   Range: $7800 \leq E(t) \leq 9300$
2. Answers may vary. The data are continuous because the domain and range include all real numbers between given values.
3. $(0, 7800)$; Their beginning elevation when they entered the park (at time = 0) was 7800 feet.
4. 8000 or 8800
5. 9300; their greatest elevation since entering the park
6. 7200 or 8700
7. 7200; their least elevation since entering the park
8.

Time (hours)	Distance (miles)
0	0
1	2
2	4
3	6
4	8
5	10

9. $f(x) = 2x$; Domain: $x \geq 0$; range: $y \geq 0$
10.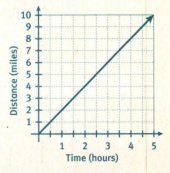

Unit 2 • Functions 121

Embedded Assessment 1

11. 2.5 hours; Students may use the table, graph or equation to justify their response.

12.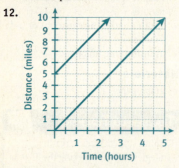

13. $g(x) = 2x + 5$; The graph is a vertical translation 5 units up from the original graph, so add 5 to the function rule.

TEACHER to TEACHER

You may wish to read through the scoring guide with students and discuss the differences in the expectations at each level. Check that students understand the terms used.

Unpacking Embedded Assessment 2

Once students have completed this Embedded Assessment, turn to Embedded Assessment 2 and unpack it with them. Use a graphic organizer to help students understand the concepts they will need to know to be successful on Embedded Assessment 2.

Embedded Assessment 1
Use after Activity 8

Representations of Functions
BRYCE CANYON HIKING

Scoring Guide	Exemplary	Proficient	Emerging	Incomplete
	The solution demonstrates the following characteristics:			
Mathematics Knowledge and Thinking (Items 1, 3–7)	• Clear and accurate identification of key features of the function and its graph, including domain, range, y-intercept, maximums, and minimums	• Correct identification of most of the key features of the function and its graph, including domain, range, y-intercept, maximums, and minimums	• Partially correct identification of some of the key features of the function and its graph, including domain, range, y-intercept, maximums, and minimums	• Inaccurate or incomplete identification of key features of the function and its graph, including domain, range, y-intercept, maximums, and minimums
Problem Solving (Item 11)	• Appropriate and efficient strategy that results in a correct answer	• Strategy that may include unnecessary steps but results in a correct answer	• Strategy that results in some incorrect answers	• No clear strategy when solving problems
Mathematical Modeling / Representations (Items 8–10, 12, 13)	• Effective understanding of how to complete a table of real-world data, and how to write, graph, and interpret the associated function • Fluency in translating a graph and writing the associated function	• Largely correct understanding of how to complete a table of real-world data, and how to write, graph, and interpret the associated function • Little difficulty translating a graph and writing the associated function	• Partial understanding of how to complete a table of real-world data, and how to write, graph, and interpret the associated function • Some difficulty translating a graph and writing the associated function	• Inaccurate or incomplete understanding of how to complete a table of real-world data, and how to write, graph, and interpret the associated function • Significant difficulty translating a graph and writing the associated function
Reasoning and Communication (Items 1–3, 5, 7, 13)	• Precise use of appropriate math terms and language to describe key features of a graph and to explain how a function rule was determined from a translated graph • Clear and accurate interpretations of the graph of a function	• Adequate description of key features of a graph • Reasonable interpretations of the graph of a function • Adequate explanation of how a function rule was determined from a translated graph	• Confusing description of key features of a graph • Partially correct interpretations of the graph of a function • Confusing explanation of how a function was determined from a translated graph	• Incomplete or inaccurate description of key features of a graph • Incomplete or inaccurate interpretation of the graph of a function • Incomplete or inaccurate explanation of how a function was determined from a translated graph

Common Core State Standards for Embedded Assessment 1 (cont.)

HSF-IF.B.5: Relate the domain of a function to its graph and, where applicable, to the quantitative relationship it describes.

HSF-BF.B.3: Identify the effect on the graph of replacing $f(x)$ by $f(x) + k$, $kf(x)$, $f(kx)$ and $f(x + k)$ for specific values of k (both positive and negative); find the value of k given the graphs. Experiment with cases and illustrate an explanation of the effects on the graph using technology.

Rates of Change
Ramp it Up
Lesson 9-1 Slope

ACTIVITY 9

Learning Targets:
- Determine the slope of a line from a graph.
- Develop and use the formula for slope.

> **SUGGESTED LEARNING STRATEGIES:** Close Reading, Summarizing, Sharing and Responding, Discussion Groups, Construct an Argument, Identify a Subtask

Margo's grandparents are moving in with her family. The family needs to make it easier for her grandparents to get in and out of the house. Margo has researched the specifications for building stairs and wheelchair ramps. She found the government website that gives the Americans with Disabilities Act (ADA) accessibility guidelines for wheelchair ramps and discovered the following diagram:

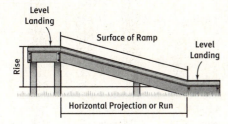

Then, Margo decided to look for the requirements for building stairs and found the following diagram:

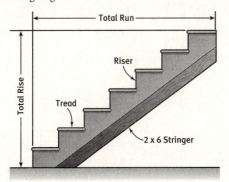

Review with your group the background information that is given as you solve the following items.

1. What do you think is meant by the terms *rise* and *run* in this context?
   Answers may vary. Rise represents the vertical movement, and run represents the horizontal movement.

CONNECT TO SOCIAL SCIENCE

The table gives information from the ADA website about the slope of wheelchair ramps.

Slope	Maximum Rise		Maximum Run	
	in.	mm	ft	m
$\frac{1}{16} < m \le \frac{1}{12}$	30	760	30	9
$\frac{1}{20} \le m < \frac{1}{16}$	30	760	40	12

Common Core State Standards for Activity 9

HSF-IF.B.6	Calculate and interpret the average rate of change of a function (presented symbolically or as a table) over a specified interval. Estimate the rate of change from a graph.
HSF-LE.A.1	Distinguish between situations that can be modeled with linear functions and with exponential functions.
HSF-LE.A.1a:	Prove that linear functions grow by equal differences over equal intervals; and that exponential functions grow by equal factors over equal intervals.
HSF-LE.A.1b:	Recognize situations in which one quantity changes at a constant rate per unit interval relative to another.

ACTIVITY 9
Investigative

Activity Standards Focus
In this activity, students explore slope as a rate of change by using the ratio of vertical change to horizontal change and developing the slope formula. Students make connections between linear functions and the idea of a constant rate of change. They develop an understanding of when the slope of a line is positive, negative, zero, or undefined.

Lesson 9-1

PLAN

Pacing: 1 class period
Chunking the Lesson
Introduction, #1 #2–6 #7–9 #10–11
Check Your Understanding
Lesson Practice

TEACH

Bell-Ringer Activity
Ask students to imagine they are helping to design a playground. They should think about the types of straight slides they would design for preschoolers, elementary-age children, and middle school–age students. Then sketch a slide for each age group. Students then describe in writing how the three slides are different.

Introduction, 1 Close Reading, Summarizing, Paraphrasing, Sharing and Responding Have students examine the table and the diagram of the ramp. Ask them to explain in their own words how the information in the table applies to the dimensions of the ramp. Discuss similarities and differences in the diagrams of the ramp and the stairs. Allow students to share responses to Item 1. All responses should be valued.

ACTIVITY 9 Continued

2–6 Discussion Groups, Vocabulary Organizer, Interactive Word Wall, Debriefing Monitor groups carefully to be sure students study the graph carefully and understand the difference between horizontal and vertical. Be sure students are counting the number of squares to find the distance from the beginning of a segment to the end of a segment. In Item 4, students should write the ratios in lowest terms to facilitate the idea that the slope between any two points on a line is constant. Add the math term *slope* to the classroom word wall. Create with students a spider vocabulary organizer with the term *slope* in the center. Each spoke of the organizer should show a different representation of slope and may be added to at any time during the activity or unit as new representations are explored. Students may choose to add diagrams or pictures to help with memory.

Developing Math Language

A solid understanding of the term *slope* is crucial for students as they continue their study of mathematics. Draw lines with various slopes. Have students brainstorm words that describe the differences among the lines: slant, incline, pitch, tilt, and so on. Invite students who speak other languages to add words from their languages to the list.

Explain to students that mathematics uses the word *slope* to describe the differences among the lines. Ask students to draw three lines with different slopes.

ACTIVITY 9 continued

Lesson 9-1
Slope

MATH TERMS

Slope is a measure of the amount of decline or incline of a line. The variable *m* is often used to represent slope.

Consider the line in the graph below:

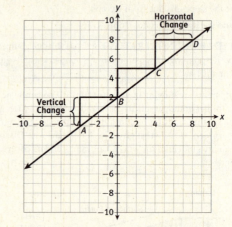

Vertical change can be represented as a *change in y*, and horizontal change can be represented by a *change in x*.

2. What is the vertical change between:
   a. points *A* and *B*? b. points *A* and *C*? c. points *C* and *D*?
   3 units   **6 units**   **3 units**

3. What is the horizontal change between:
   a. points *A* and *B*? b. points *A* and *C*? c. points *C* and *D*?
   4 units   **8 units**   **4 units**

The ratio of the vertical change to the horizontal change determines the *slope* of the line.

$$\text{slope} = \frac{\text{vertical change}}{\text{horizontal change}} = \frac{\text{change in } y}{\text{change in } x} = \frac{\Delta y}{\Delta x}$$

4. Find the slope of the segment of the line connecting:
   a. points *A* and *B*   b. points *A* and *C*   c. points *C* and *D*
   $\frac{3}{4}$ $\frac{6}{8} = \frac{3}{4}$ $\frac{3}{4}$

5. What do you notice about the slope of the line in Items 4a, 4b, and 4c?
   The slopes are all the same.

6. What does your answer to Item 5 indicate about points on a line?
   The slope between any two points on a line is always the same.

124 SpringBoard® Mathematics Algebra 1, Unit 2 • Functions

Lesson 9-1
Slope

7. Slope is sometimes referred to as $\frac{rise}{run}$. Explain how the ratio $\frac{rise}{run}$ relates to the ratios for finding slope mentioned above.
   Answers may vary. Rise refers to the vertical change and run refers to the horizontal change.

8. **Reason quantitatively.** Would the slope change if you counted the run (horizontal change) before you counted the rise (vertical change)? Explain your reasoning.
   Answers may vary. The slope should not change as long as you remember to use the run value as the denominator and the rise value as the numerator.

9. Determine the slope of the line graphed below.

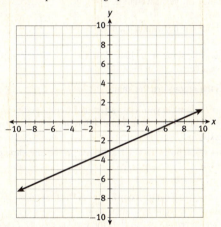

   The slope is $\frac{3}{7}$.

WRITING MATH
In mathematics the Greek letter Δ (delta) represents a change or difference between mathematical values.

MATH TIP
Select two points on the line and use them to compute the slope.

ACTIVITY 9 Continued

10–11 Discussion Groups, Identify a Subtask, Construct an Argument, Create Representations, Debriefing Students develop the slope formula by relating the horizontal and vertical change to the differences between the x and y coordinates of points on a line. As students write the formula for determining the slope of a line, some may use the expressions $y_2 - y_1$ and $x_2 - x_1$, while some may use $y_1 - y_2$ and $x_1 - x_2$. Both orders are correct as long as students are consistent in the order used for both the change in y and the change in x. Include in the debriefing a discussion, and possibly an example, to show why this is true.

Students must find the slope using its definition and two points on the line. Students may choose any two points on the line, but encourage them to use points with whole-number coordinates. Once they find the slope and compare it with the answer to Item 10, have them compare the processes of using the slope formula and using the definition of slope.

Differentiating Instruction

Support students who have trouble correctly substituting into the slope formula. Write the formula on the board, using red for the numerator and blue for the denominator. Have students copy the formula including the color coding. Then have them list two ordered pairs, circling the x-coordinates with blue and the y-coordinates with red. Have them substitute the values into the formula and simplify it. Caution students that when they choose a y to use as y_1, they must use the x from the same point as x_1.

ACTIVITY 9 continued

Lesson 9-1
Slope

Although the slope of a line can be calculated by looking at a graph and counting the vertical and horizontal change, it can also be calculated numerically.

10. Recall that the slope of a line is the ratio $\frac{\text{change in } y}{\text{change in } x}$.
    Answers may vary.

   a. Identify two points on the graph above and record the coordinates of the two points that you selected.

	x-coordinate	y-coordinate
1st point	0	−3
2nd point	7	0

   b. Which coordinates relate to the vertical change on the graph?
      The y-coordinates, 0 and −3
   c. Which coordinates relate to the horizontal change on a graph?
      The x-coordinates, 7 and 0
   d. Determine the vertical change.
      0 − (−3) = 3
   e. Determine the horizontal change.
      7 − 0 = 7
   f. Calculate the slope of the line. How does this slope compare to the slope that you found in Item 10?
      $\frac{3}{7} = \frac{3}{7}$; the slopes are the same.
   g. If other students in your class selected different points for this problem, should they have found different values for the slope of this line? Explain.
      No. The slope between any two points on the line should remain the same.

11. It is customary to label the coordinates of the first point (x_1, y_1) and the coordinates of the second point (x_2, y_2).
    a. Write an expression to calculate the vertical change, $\triangle y$, of the line through these two points.
       $y_2 - y_1$ or $y_1 - y_2$

    b. Write an expression to calculate the horizontal change, $\triangle x$, of the line through these two points.
       $x_2 - x_1$ or $x_1 - x_2$

    c. Write an expression to calculate the slope of the line through these two points.
       $\frac{y_2 - y_1}{x_2 - x_1}$ or $\frac{y_1 - y_2}{x_1 - x_2}$

Lesson 9-1
Slope

Check Your Understanding

12. Use the slope formula to determine the slope of a line that passes through the points (4, 9) and (−8, −6).
13. Use the slope formula to determine the slope of the line that passes through the points (−5, −3) and (9, −10).
14. Explain how to find the slope of a line from a graph.
15. Explain how to find the slope of a line when given two points on the line.

LESSON 9-1 PRACTICE

16. Find $\triangle x$ and $\triangle y$ for the points (7, −2) and (9, −7).
17. **Critique the reasoning of others.** Connor determines the slope between (−2, 4) and (3, −3) by calculating $\frac{4-(-3)}{-2-3}$. April determines the slope by calculating $\frac{3-(-2)}{-3-4}$. Explain whose reasoning is correct.
18. When given a table of ordered pairs, you can find the slope by choosing any two ordered pairs from the table. Determine the slope represented in the table below.

x	5	7	9	11
y	5	3	1	−1

19. Determine the slope of the given line.

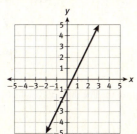

ACTIVITY 9 Continued

Lesson 9-2

PLAN

Pacing: 1 class period
Chunking the Lesson
#1   #2   #3
Check Your Understanding
Lesson Practice

TEACH

Bell-Ringer Activity
Write the expressions $5x$, $0.3x$, and $20x$ on the board. Ask students to write three real-world situations that could be represented by each expression.

1 Marking the Text, Discussion Groups, Create Representations, Look for a Pattern, Vocabulary Organizer, Interactive Word Wall, Sharing and Responding This item adds context to the process of graphing a linear function and determining the slope. Students are given the opportunity to choose their own input values. Discuss the choice of negative inputs and why they may or may not be needed in the context of the problem. Before they begin, ask students to identify the dependent and independent variables in the situation and to relate these terms to the rate of change. In parts e and f, students should make the connection between the coefficient of x, the slope of the line, and the cost of each piece of wood. Add *rate of change* to the classroom word wall and to student vocabulary organizers for slope begun in the previous lesson.

Developing Math Language
Build on students' understanding of rate to help them understand *rate of change*. Have students give real-world examples of rates and explain each context to make sure their classmates understand it. For each example, ask students to identify the independent and dependent variables. For instance, if a student says *50 miles per hour*, he or she should identify time as the independent variable and distance traveled as the dependent variable because the distance traveled depends on the time elapsed. Students should understand the expression as a rate of change because every time the number of hours increases by 1, the number of miles increases by 50.

ACTIVITY 9 continued

Lesson 9-2
Slope and Rate of Change

My Notes

Learning Targets:
- Calculate and interpret the rate of change for a function.
- Understand the connection between rate of change and slope.

SUGGESTED LEARNING STRATEGIES: Discussion Groups, Create Representations, Look for a Pattern, Think-Pair-Share

The **rate of change** for a function is the ratio of the change in y, the dependent variable, to the change in x, the independent variable.

1. Margo went to the lumberyard to buy supplies to build the wheelchair ramp. She knows that she will need several pieces of wood. Each piece of wood costs $3.

   a. **Model with mathematics.** Write a function $f(x)$ for the total cost of the wood pieces if Margo buys x pieces of wood.
   $f(x) = 3x$

   b. Make an input/output table of ordered pairs and then graph the function.

Pieces of Wood, x	Total Cost, $f(x)$
0	0
1	3
2	6
3	9
4	12
5	15

   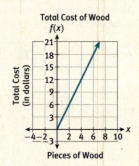

   c. What is the slope of the line that you graphed?
   3

   d. By how much does the cost increase for each additional piece of wood purchased?
   $3

128 SpringBoard® Mathematics **Algebra 1, Unit 2** • Functions

Lesson 9-2
Slope and Rate of Change

e. How does the slope of this line relate to the situation with the pieces of wood?
   The slope is the same as the cost for each piece of wood.

f. Is there a relationship between the slope of the line and the equation of the line? If so, describe that relationship.
   Yes. The slope is the coefficient of the variable x in the equation of the line.

2. Margo is going to work with a local carpenter during the summer. Each week she will earn $10.00 plus $2.00 per hour.
   a. Write a function $f(x)$ for Margo's total earnings if she works x hours in one week.
   $f(x) = 2x + 10$

   b. Make an input/output table of ordered pairs and then graph the function. Label your axes.

Hours, x	Earnings, $f(x)$ (dollars)
0	10
1	12
2	14
3	16
4	18
5	20
6	22

Margo's Summer Earnings

ACTIVITY 9 Continued

3 Discussion Groups, Create Representations, Look for a Pattern, Debriefing Before they begin working, ask students to consider whether the amount of money Margo has will increase or decrease as the amount of wood she buys increases. Again, monitor group discussions carefully to ensure choice of input values appropriate to the context. Debrief by asking students to compare and contrast this problem situation with the previous two. Students should make the connection between money being spent, a negative rate of change, and a line with a negative slope.

ACTIVITY 9 continued

Lesson 9-2
Slope and Rate of Change

My Notes

c. How much will Margo's earnings change if she works 6 hours instead of 2? If she works 4 hours instead of 3? How much do Margo's earnings change for each additional hour worked?
   Margo will earn $22 − $14 = $8 more if she works 6 hours instead of 2. She will earn $18 − $16 = $2 more if she works 4 hours instead of 3 hours. She earns $2 more for each additional hour worked.

d. Does the function have a constant rate of change? If so, what is it?
   Yes; 2

e. What is the slope of the line that you graphed?
   Coordinates used may vary, but students should get slope = 2.

f. Describe the meaning of the slope within the context of Margo's job.
   The slope of this line tells how much money Margo earns each hour.

g. Describe the relationship between the slope of the line, the rate of change, and the equation of the line.
   The slope is the same as the coefficient of x in the equation of the line. The slope is also equal to the rate of change.

h. How much will Margo earn if she works for 8 hours in one week?
   $26.00

3. By the end of the summer, Margo has saved $375. Recall that each of the small pieces of wood costs $3.
   a. Write a function $f(x)$ for the amount of money that Margo still has if she buys x pieces of wood.
   $f(x) = 375 − 3x$

Lesson 9-2
Slope and Rate of Change

b. Make an input/output table of ordered pairs and then graph the function.

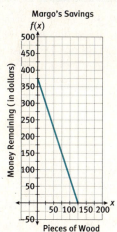
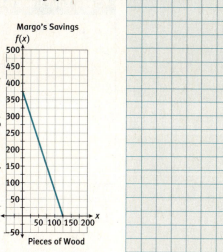

Pieces of Wood, x	Money Remaining, f(x) (dollars)
0	375
25	300
50	225
75	150
100	75
125	0

c. How much will the amount Margo has saved change if she buys 100 instead of 25 pieces of wood? If she buys 50 instead of 0 pieces of wood? For each additional piece of wood? Explain.

The amount of money in her savings changes by $75 − $300 = −$225. The amount of money in her savings changes by $225 − $375 = −$150. The amount of money in her savings changes by −$3 for each additional piece of wood she buys.

d. Does the function have a constant rate of change? If so, what is it?
Yes; −3

e. What is the slope of the line that you graphed?
Coordinates used may vary, but students should get slope = −3.

f. How are the rate of change of the function and the slope related?
They are equal.

g. Describe the meaning of the slope within the context of Margo's savings.
The slope of this line tells how much the amount of Margo's savings changes for each piece of wood purchased.

h. How does this slope differ from the other slopes that you have seen in this activity?
Answers may vary. This slope is negative, and the others were positive.

ACTIVITY 9 Continued

Differentiating Instruction

Extend students' understanding of slope by having them compare the graphs of functions with different slopes. Give students the example of different-sized beverages:

Beverages
Small: $0.50 *Medium:* $1 *Large:* $2

For each size of beverage, have students write a function modeling how much x beverages would cost. Have them create a table and a graph for each. Finally, ask students to describe in writing how the graphs of $f(x) = 0.5x$ and $h(x) = 2x$ are each different from the graph of the parent function, $g(x) = x$.

Activity 9 • Rates of Change 131

ACTIVITY 9 Continued

Check Your Understanding

To debrief this lesson, have students use the definition of slope to justify their answers. Pay close attention to their answers to Item 5. Students should grasp the idea that the amount of change from one row to another does not have to be constant, but the ratios of the changes in y and the changes in x must be constant.

Answers

4. The graph of the function goes down 5 units for each 1 unit the line moves from left to right.
5. Yes, there is a constant rate of change of $\frac{5}{1}$, and the slope is also $\frac{5}{1}$.

ASSESS

Students' answers to Lesson Practice problems will provide you with a formative assessment of their understanding of the lesson concepts and their ability to apply their learning.

See the Activity Practice for additional problems for this lesson. You may assign the problems here or use them as a culmination for the activity.

LESSON 9-2 PRACTICE

6. a. $f(x) = 50 + 15x$
   b.

x	f(x)
3	95
7	155
9	185
12	230

   Check students' graphs.
   c. $\frac{15}{1}$; $m = 15$
   d. The slope of the line is the same as the cost of each trip to the museum, $15.
7. The slope for a line through $(-2, 7)$ and $(3, 0)$ is $\frac{0-7}{3-(-2)} = \frac{-7}{5}$. The slope of a line through $(2, 1)$ and $(12, -13)$ is $\frac{-13-1}{12-2} = \frac{-14}{10} = -\frac{7}{5}$. The slopes are equal.

ADAPT

If students have difficulty with the idea of a rate of change and its relationship to slope, review the concept of slope as the ratio of the change in y to the change in x using a table such as the one in Check Your Understanding Item 5. Ask students to calculate the rate of change using the table. Then have them graph the points to create a line and determine the slope of the line using slope triangles.

ACTIVITY 9 continued

My Notes

Lesson 9-2
Slope and Rate of Change

Check Your Understanding

4. The constant rate of change of a function is -5. Describe the graph of the function as you look at it from left to right.
5. Does the table represent data with a constant rate of change? Justify your answer.

x	y
2	−5
4	5
7	20
11	40

LESSON 9-2 PRACTICE

6. The art museum charges an initial membership fee of $50.00. For each visit the museum charges $15.00.
   a. Write a function $f(x)$ for the total amount charged for x trips to the museum.
   b. Make a table of ordered pairs and then graph the function.
   c. What is the rate of change? What is the slope of the line?
   d. How does the slope of this line relate to the number of museum visits?
7. **Critique the reasoning of others.** Simone claims that the slope of the line through $(-2, 7)$ and $(3, 0)$ is the same as the slope of the line through $(2, 1)$ and $(12, -13)$. Prove or disprove Simone's claim.

Lesson 9-3
More About Slopes

ACTIVITY 9
continued

Learning Targets:
- Show that a linear function has a constant rate of change.
- Understand when the slope of a line is positive, negative, zero, or undefined.
- Identify functions that do not have a constant rate of change and understand that these functions are not linear.

SUGGESTED LEARNING STRATEGIES: Look for a Pattern, Think-Pair-Share, Construct an Argument, Sharing and Responding, Summarizing

You have seen that for a linear function, the rate of change is constant and equal to the slope of the line. This is because linear functions increase or decrease by equal differences over equal intervals. Look at the graph below.

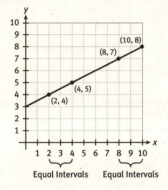

1. Over the interval 2 to 4, by how much does the function increase? Explain.
   The function increases by 1, from $y = 4$ to $y = 5$.

2. Over the equal interval 8 to 10, by how much does the function increase? Explain.
   The function again increases by 1, from $y = 7$ to $y = 8$.

"Equal differences over equal intervals" is an equivalent way of referring to constant slope. "Differences" refers to $\triangle y$, and "intervals" refers to $\triangle x$. "Equal differences over equal intervals" means $\frac{\triangle y}{\triangle x}$, which represents the slope, will always be the same.

ACTIVITY 9 Continued

3 Marking the Text, Discussion Groups, Construct an Argument, Sharing and Responding Students should examine the coordinates in the table to determine the ratios of the changes in the *x*-values to the changes in the *y*-values. Once they determine that the rate of change is not constant, they should make the connection that the function is not linear. Discuss as a whole group why a function without a constant rate of change is not linear, using a graph of the function if necessary.

4–5 Look for a Pattern, Think-Pair-Share, Sharing and Responding As students complete Items 4–5, they make connections between the rise and fall of the graph of a line and the slope of the line. Students should notice that the graph of a line with a positive slope rises from left to right and that the graph of a line with a negative slope falls from left to right.

ACTIVITY 9 continued

My Notes

**Lesson 9-3
More About Slopes**

3. The table below represents a function.

x	y
−8	62
−6	34
−1	−1
1	−1
5	23
7	47

a. Determine the rate of change between the points (−8, 62) and (−6, 34).

$$\frac{\Delta y}{\Delta x} = \frac{34 - 62}{-6 - (-8)} = \frac{-28}{2} = -14$$

b. Determine the rate of change between the points (−1, −1) and (1, −1).

$$\frac{\Delta y}{\Delta x} = \frac{-1 - (-1)}{1 - (-1)} = \frac{0}{2} = 0$$

c. **Construct viable arguments.** Is this a linear function? Justify your answer.

No; the rate of change is not constant.

4. a. Determine the slopes of the lines shown.

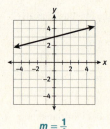
$m = \frac{1}{4}$

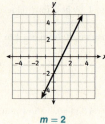
$m = 2$

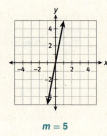
$m = 5$

b. **Express regularity in repeated reasoning.** Describe the slope of any line that rises as you view it from left to right.

The slope of the line will be positive.

134 SpringBoard® Mathematics Algebra 1, Unit 2 • Functions

Lesson 9-3
More About Slopes

5. **a.** Determine the slopes of the lines shown.

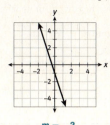 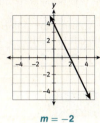 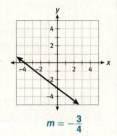

$m = -3$ $m = -2$ $m = -\frac{3}{4}$

b. Express regularity in repeated reasoning. Describe the slope of any line that falls as you view it from left to right.
The slope of the line will be negative.

6. **a.** Determine the slopes of the lines below.

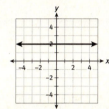 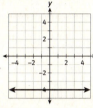 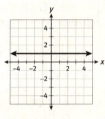

$m = 0$ for all three graphs.

b. What is the slope of a horizontal line?
The slope of any horizontal line is 0.

7. **a.** Determine the slopes of the lines shown.

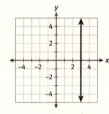 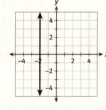 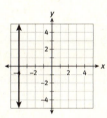

The slope of each graph is undefined.

b. What is the slope of a vertical line?
The slope of any vertical line is undefined.

ACTIVITY 9 Continued

8 Think-Pair-Share, Summarizing, Vocabulary Organizer After students have summarized their findings in the chart, allow time to add the information to students' vocabulary organizers for slope from the previous lessons. Encourage students to add visual representations along with verbal descriptions on their vocabulary organizers.

Check Your Understanding
As you debrief this lesson, focus on the idea that, for a linear function, the ratios of the change in y to the change in x between any two points on the line are equivalent. Students should also be able to describe lines with positive, negative, zero, and undefined slopes.

Answers
9. If the rate of change or the slope is not constant, the function is not linear.
10. Look at the graph from left to right. If the line goes up, it has a positive slope. If the line goes down, it has a negative slope.
11. A vertical line has an undefined slope. It is undefined because, for a vertical line, all points have the same x-coordinate. This means that the denominator in the slope formula $x_2 - x_1$ will always be 0 and division by 0 is undefined.

ASSESS
Students' answers to Lesson Practice problems will provide you with a formative assessment of their understanding of the lesson concepts and their ability to apply their learning.

See the Activity Practice for additional problems for this lesson. You may assign the problems here or use them as a culmination for the activity.

LESSON 9-3 PRACTICE
12. a.
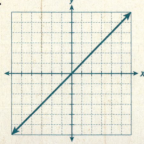

ADAPT
If students have difficulty determining the slope of a line, review the use of the formula, making connections to graphs and tables. Assign additional practice as needed.

Lesson 9-3
More About Slopes

8. Summarize your findings in Items 4–7. Tell whether the slopes of the lines described in the table below are positive, negative, 0, or undefined.

Up from left to right	Down from left to right	Horizontal	Vertical
positive	negative	0	undefined

Check Your Understanding

9. Suppose you are given several points on the graph of a function. Without graphing, how could you determine whether the function is linear?
10. How can you tell from a graph if the slope of a line is positive or negative?
11. Describe a line having an undefined slope. Why is the slope undefined?

LESSON 9-3 PRACTICE
12. **Make use of structure.** Sketch a line for each description.
    a. The line has a positive slope.
    b. The line has a negative slope.
    c. The line has a slope of 0.
13. Does the table represent a linear function? Justify your answer.

x	y
1	−1
4	9
7	19
11	29

14. Are the points (12, 11), (2, 7), (5, 9), and (1, 5) part of the same linear function? Explain.

b.
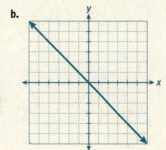

c.
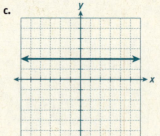

13. No, the rate of change between the points is not constant; it is $\frac{10}{3}$ between the first and third points, and $\frac{5}{2}$ between the third and fourth.
14. No, the rate of change between the points is not constant; it is $\frac{2}{5}$ between the first two points, $\frac{2}{3}$ for the next two, and 1 for the last two.

Rates of Change
Ramp it Up

ACTIVITY 9 continued

ACTIVITY 9 PRACTICE
Write your answers on notebook paper.
Show your work.

Lesson 9-1

1. Find $\triangle x$ and $\triangle y$ for each of the following pairs of points.
   a. $(2, 6), (-6, -8)$
   b. $(0, 9), (4, -8)$
   c. $(-3, -3), (7, 10)$

For Items 2 and 3, use the table to calculate the slope.

2.
x	y
-5	-1
0	2
5	5
10	8

3.
x	y
-4	20
-3	14
0	-4
2	-16

4. Two points on a line are $(-10, 1)$ and $(5, -5)$. If the y-coordinate of another point on the line is -3, what is the x-coordinate?

For Items 5–7, determine the slope of the line that passes through each pair of points.

5. $(-4, 11)$ and $(1, -9)$
6. $(-10, -3)$ and $(-5, 1)$
7. $(-2, -7)$ and $(-8, -4)$

8. Are the three points $(2, 3)$, $(5, 6)$, and $(0, -2)$ on the same line? Explain.

9. Which of the following pairs of points lies on a line with a slope of $-\frac{3}{5}$?
   A. $(4, 0), (-2, 10)$
   B. $(4, 2), (10, 4)$
   C. $(-4, -10), (0, -2)$
   D. $(10, -2), (0, 4)$

For Item 10, determine the slope of the line that is graphed.

10.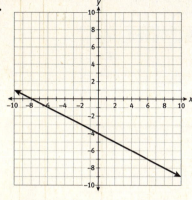

Lesson 9-2

11. Juan earns $7 per hour plus $20 per week making picture frames.
    a. Write a function $g(x)$ for Juan's total earnings if he works x hours in one week.
    b. Without graphing the function, determine the slope.
    c. Describe the meaning of the slope within the context of Juan's job.

12. The graph shows the height of an airplane as it descends to land.

    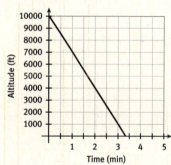

    a. Does the function have a constant rate of change? If so, what is it?
    b. What is the slope of the line?
    c. How are the rate of change and the slope of the line related?
    d. Describe the meaning of the slope within the context of the situation.

ACTIVITY 9 Continued

ACTIVITY PRACTICE

1. a. $\triangle x = 8$ or -8, $\triangle y = 14$ or -14
   b. $\triangle x = -4$ or 4, $\triangle y = 17$ or -17
   c. $\triangle x = 10$ or -10, $\triangle y = 13$ or -13
2. $m = \frac{3}{5}$
3. $m = -6$
4. $x = 0$
5. -4
6. $0.8 = \frac{4}{5} = m$
7. $m = -\frac{1}{2}$
8. No; the slope between points $(2, 3)$ and $(5, 6)$ is $\frac{6-3}{5-2} = \frac{3}{3} = 1$. The slope between points $(5, 6)$ and $(0, -2)$ is $\frac{-2-6}{0-5} = \frac{-8}{-5} = \frac{8}{5}$. Since the slopes are not equal, the three points are not on the same line.
9. D
10. $m = -\frac{1}{2}$
11. a. $g(x) = 7x + 20$
    b. 7
    c. The slope of this line tells how much money Juan earns each hour.
12. a. Yes, the function has a constant rate of change, which is 3000.
    b. 3000
    c. The rate of change and the slope are the same.
    d. The plane's altitude changes by -3000 feet each minute.

Activity 9 • Rates of Change 137

ACTIVITY 9 Continued

13. No, it is not linear because the rate of change between the first two points is −20 and the rate of change between the next two points is −5, so there is not constant slope.
14. Yes, it is linear because the rate of change, $-\frac{1}{5}$, is constant and there is a constant slope.
15. Yes, it is linear because the rate of change, −8, is constant and there is a constant slope.
16. Answers will vary. Check students' responses.
17. B
18. undefined
19. negative
20. Sample answer: The y-coordinate of any point must be 4 so as to have a slope 0 and pass through (−3, 4).
21. Sample answer: Count grid squares, use delta notation, or use the formula. Check students' explanations.

ADDITIONAL PRACTICE

If students need more practice on the concepts in this activity, see the eBook Teacher Resources for additional practice problems.

ACTIVITY 9 continued

Rates of Change
Ramp it Up

Lesson 9-3

For Items 13–15, tell whether the function is linear. Justify your response.

13.

x	y
−3	44
−1	4
0	−1
1	4

14.

x	y
−5	−7
0	−8
5	−9
10	−10

15.

x	y
4	−30
6	−46
8	−62
9	−70

16. One point on the line described by $y = -2x + 3$ is shown below. Use your knowledge of slope to give the coordinates of three more points on the line.

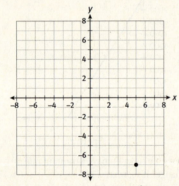

17. Which of the following is **not** a linear function?
    A. (4, −6), (7, −12), (8 −14), (10, −18), (2, −2)
    B. (−2, −6), (1, 0), (4, −30), (0, 2), (7, −96)
    C. (−4, 9), (0, 7), (2, 6), (6, 4), (8, 3)
    D. (2, 18), (6, 50), (−3, −22), (0, 2), (3, 26)

For Items 18 and 19, identify the slope of the line in each graph as positive, negative, 0, or undefined.

18.
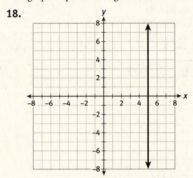

19.
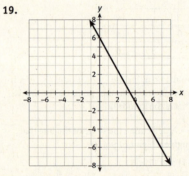

20. The slope of a line is 0. It passes through the point (−3, 4). Identify two other points on the line. Justify your answers.

MATHEMATICAL PRACTICES
Look For and Make Use of Structure

21. Describe three different ways to determine the slope of a line and the similarities and differences between the methods.

Linear Models
Stacking Boxes
Lesson 10-1 Direct Variation

ACTIVITY 10

Learning Targets:
- Write and graph direct variation.
- Identify the constant of variation.

> **SUGGESTED LEARNING STRATEGIES:** Create Representations, Interactive Word Wall, Marking the Text, Sharing and Responding, Discussion Groups

You work for a packaging and shipping company. As part of your job there, you are part of a package design team deciding how to stack boxes for packaging and shipping. Each box is 10 cm high.

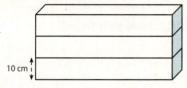

1. Complete the table and make a graph of the data points (number of boxes, height of the stack).

Number of Boxes	Height of the Stack (cm)
0	0
1	10
2	20
3	30
4	40
5	50
6	60
7	70

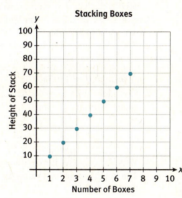

2. Write a function to represent the data in the table and graph above.

   $f(x) = 10x$, or $y = 10x$

WRITING MATH
Either y or $f(x)$ can be used to represent the output of a function.

3. What is a reasonable and realistic domain for the function? Explain.

   The domain is the set of integers that are greater than or equal to 0. The domain cannot include values less than 0 because there is no such thing as fewer than 0 boxes.

4. What is a reasonable and realistic range for the function? Explain.

   The range is 0 and positive multiples of 10. The range cannot include values less than 0 because the height of a stack cannot be less than 0. You cannot have a fraction of a box, so the range cannot include values that are not multiples of 10.

Common Core State Standards for Activity 10

HSN-Q.A.3	Choose a level of accuracy appropriate to limitations on measurement when reporting quantities.
HSA-CED.A.1	Create equations and inequalities in one variable and use them to solve problems. Include equations arising from linear and quadratic functions and simple rational and exponential functions.
HSF-IF.B.5	Relate the domain of a function to its graph and, where applicable, to the quantitative relationship it describes.
HSF-BF.A.1	Write a function that describes a relationship between two quantities.
HSF-BF.A.1a	Determine an explicit expression, a recursive process, or steps for calculation from a context.

ACTIVITY 10
Guided

Activity Standards Focus

In this activity students solve problems by gathering real-world data, recording the results in tables, representing results with graphs, and writing function equations. They also learn to write and use inverse functions. These concepts and skills will be important to students as they use functions in increasingly complex mathematical contexts.

Lesson 10-1

PLAN

Pacing: 1 class period

Chunking the Lesson
#1–6   #7–8   #9–10   #11–12
Check Your Understanding
Lesson Practice

TEACH

Bell-Ringer Activity
Show students the ratios $\frac{2}{3}$, $\frac{9}{15}$, $\frac{15}{21}$ and $\frac{15}{25}$. Ask them to identify the two equivalent ratios. Students should then write five other ratios that are equivalent to those two ratios.

1–6 Marking the Text, Discussion Groups, Create Representations, Look for a Pattern, Sharing and Responding Ordered pairs in the table should correspond to discrete points on students' graphs since they are being used to represent the number of and heights of boxes. As students respond to Items 3 and 4, be sure students are choosing appropriate values for the given problem situation. As students share responses with the class, focus the discussion on why there are no negative values for x or y. Students should understand that there cannot be a negative number of boxes nor a negative height for the stack. Students should identify patterns focusing on the constant difference of 10 as it appears in both the table and the graph.

ACTIVITY 10 Continued

Developing Math Language
Refer to the table as you discuss the term *directly proportional*. Discuss with students that when quantities have *direct variation*, the two values are directly proportional. All the ratios of $y : x$ are equivalent in a situation. Relate direct variations to the concept of slope in which the ratios along a line of $\triangle y : \triangle x$ are equivalent. Note that the *constant of variation* is the ratio $\frac{y}{x}$ and is represented by k. Demonstrate how to rewrite the equation $\frac{y}{x} = k$ as $y = kx$.

7–8 Marking the Text, Activating Prior Knowledge, Discussion Groups, Interactive Word Wall
Monitor student discussions carefully to be sure students understand the new vocabulary and how the words are interrelated. Direct proportionality is a concept that is familiar to students. By connecting to prior knowledge, students will better understand the new concept. Add the terms *directly proportional* and *direct variation* to the classroom word wall.

9–10 Think-Pair-Share, Construct an Argument, Interactive Word Wall, Sharing and Responding In these items students derive the equation for determining the constant of variation. Add new vocabulary to the classroom word wall. Be sure students understand before they begin that the constant of variation, k, represents a ratio, and that the ratio may be a whole number or a fraction.

TEACHER to TEACHER
The variable k is also sometimes called the *constant of proportionality*. Students should understand that they may see it called this as well as *constant of variation*.

ACTIVITY 10 continued

Lesson 10-1
Direct Variation

MATH TERMS
A **direct proportion** is a relationship in which the ratio of one quantity to another remains constant.

5. What do $f(x)$, or y, and x represent in your equation from Item 2?
   $f(x)$ represents the height of the stack and x represents the number of boxes.

6. Describe any patterns that you notice in the table and graph representing your function.
   Answers may vary. Each time a box is added to the stack, the height increases by 10 cm.

7. The number of boxes is **directly proportional** to the height of the stack. Use a proportion to determine the height of a stack of 12 boxes.
   $\frac{1 \text{ box}}{10 \text{ cm}} = \frac{12 \text{ boxes}}{120 \text{ cm}}$; the height is 120 cm.

When two values are directly proportional, there is a **direct variation**. In terms of stacking boxes, the height of the stack *varies directly* as the number of boxes.

8. Using variables x and y to represent the two values, you can say that y varies directly as x. Use your answer to Item 6 to explain this statement.
   Answers may vary. y is the height of the stack and x is the number of boxes, so "y varies directly as x" means that the height of the stack varies as the number of boxes changes.

9. Direct variation is defined as $y = kx$, where $k \neq 0$ and the coefficient k is the **constant of variation**.
   a. Consider your answer to Item 2. What is the constant of variation in your function?
      10
   b. Why do you think the coefficient is called the constant of variation?
      The height *constantly varies* by 10 with each addition of another box.
   c. **Reason quantitatively.** Explain why the value of k cannot be equal to 0.
      Answers may vary. If $k = 0$, then $y = 0x = 0$, which means that the value of y will always be 0 and will not vary.
   d. Write an equation for finding the constant of variation by solving the equation $y = kx$ for k.
      $k = \frac{y}{x}$

Common Core State Standards for Activity 10 (continued)

HSF-BF.B.4	Find inverse functions.
HSF-BF.B.4a	Solve an equation of the form $f(x) = c$ for a simple function f that has an inverse and write an expression for the inverse.
HSF-LE.B.5	Interpret the parameters in a linear or exponential function in terms of a context.

Lesson 10-1
Direct Variation

10. a. Interpret the meaning of the point (0, 0) in your table and graph.
   A stack of 0 boxes has no height.

b. True or False? Explain your answer. "The graphs of all direct variations are lines that pass through the point (0, 0)."
   True; explanations may vary. In an equation of the form $y = kx$, when $x = 0$ then $y = 0$.

c. Identify the slope and y-intercept in the graph of the stacking boxes.
   $m = 10$; y-intercept $= (0, 0)$

d. Describe the relationship between the constant of variation and the slope.
   The constant of variation and the slope are equal.

Direct variation can be used to answer questions about stacking and shipping your boxes.

11. The height y of a different stack of boxes varies directly as the number of boxes x. For this type of box, 25 boxes are 500 cm high.

a. Find the value of k. Explain how you found your answer.
   $k = 20$; use the equation $k = \frac{y}{x}$ to find the constant of variation; $k = \frac{y}{x} = \frac{500 \text{ cm}}{25 \text{ boxes}} = 20$

b. Write a direct variation equation that relates y, the height of the stack, to x, the number of boxes in the stack.
   $y = 20x$

c. How high is a stack of 20 boxes? Explain how you would use your direct variation equation to find the height of the stack.
   400 cm; replace x with 20 in the equation, $y = 20x$, and solve for y. $y = 20(20) = 400$

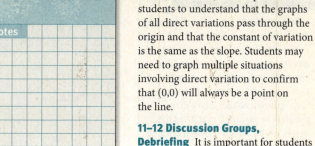

ACTIVITY 10 Continued

9–10 (continued) It is important for students to understand that the graphs of all direct variations pass through the origin and that the constant of variation is the same as the slope. Students may need to graph multiple situations involving direct variation to confirm that (0,0) will always be a point on the line.

11–12 Discussion Groups, Debriefing
It is important for students to understand that y, or $f(x)$, represents the height of the stack, and that x represents the number of boxes. The constant of variation represents the rate at which the stack height increases with the addition of each box. In this case, it is also the height of each box.

ACTIVITY 10 Continued

Differentiating Instruction

Extend students' understanding of direct variations by having them examine a linear function that is not a direct variation. Change the problem about the boxes by saying that the boxes are stacked on a platform that is 5 centimeters high. Have them generate an equation that gives the height of the stack of boxes including the platform. Instruct them to make a table and a graph to show the height with various numbers of boxes. Then have them analyze the table and graph to determine whether or not it is a direct variation. They should consider both the ratio of height to boxes as well as the y-intercept of the equation.

Check Your Understanding

As you debrief the lesson, continue to emphasize that, in a linear function, the ratio of $y : x$ is constant, and the graph passes through the point $(0, 0)$. Use the vocabulary introduced in this lesson as you discuss the problems.

Answers

13. **a.** No; the graph is not a line through the origin.
    b. Yes; the graph is a line through the origin.
    c. Yes; the values can be described by the equation $y = 4x$, which is in the form $y = kx$, where $k = 4$; also, the graph is a line through the origin.
    d. No; the graph is not a line.
    e. Yes; the equation is in the form $y = kx$.
    f. No; the equation cannot be written in the form $y = kx$.

Lesson 10-1
Direct Variation

12. At the packaging and shipping company, you get paid each week. One week you earned $48 for 8 hours of work. Another week you earned $30 for 5 hours of work.

   a. Write a direct variation equation that relates your wages to the number of hours you worked each week. Explain the meaning of each variable and identify the constant of variation.

   $y = 6x$ where y equals your wages, x is the number of hours you work in a week, and the constant of variation is 6.

   b. How much would you earn if you worked 3.5 hours in one week?

   $21.00

Check Your Understanding

13. Tell whether the tables, graphs, and equations below represent direct variations. Justify your answers.

   a. 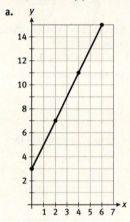   **b.**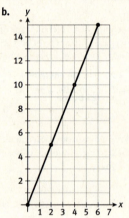

   c.
x	y
2	12
4	24
6	36

   d.
x	y
2	8
4	12
6	16

   e. $y = 20x$ **f.** $y = 3x + 2$

Lesson 10-1
Direct Variation

LESSON 10-1 PRACTICE

14. In the equation $y = 15x$, what is the constant of variation?
15. In the equation $y = 8x$, what is the constant of variation?
16. The value of y varies directly with x and the constant of variation is 7. What is the value of x when $y = 63$?
17. The value of y varies directly with x and the constant of variation is 12. What is the value of y when $x = 5$?
18. **Model with mathematics.** The height of a stack of boxes varies directly with the number of boxes. A stack of 12 boxes is 15 feet high. How tall is a stack of 16 boxes?
19. Jan's pay is in direct variation to the hours she works. Jan earns $54 for 12 hours of work. How much will she earn for 18 hours work?

ACTIVITY 10 Continued

ASSESS

Students' answers to Lesson Practice problems will provide you with a formative assessment of their understanding of the lesson concepts and their ability to apply their learning.

See the Activity Practice for additional problems for this lesson. You may assign the problems here or use them as a culmination for the activity.

LESSON 10-1 PRACTICE
14. 15
15. 8
16. 9
17. 60
18. 20 feet
19. $81

ADAPT

Check students' answers to the Lesson Practice to ensure that they understand how to write and use equations of direct variation. Reinforce the correct use of *directly proportional*, *direct variation*, and *constant of variation* as you discuss the problems.

Activity 10 • Linear Models

ACTIVITY 10 Continued

Lesson 10-2

PLAN

Pacing: 1 class period
Chunking the Lesson
#1–4 #5 #6–7
#8-9 #10–11
Check Your Understanding
Lesson Practice

TEACH

Bell-Ringer Activity
Ask students to determine the volumes of rectangular prisms with these dimensions: 3 in. × 4 in. × 5 in.; 15 cm × 7 cm × 11 cm; 2 ft × 4 ft × 8 ft. Challenge them to find all possible whole-number dimensions for rectangular prisms with volumes of 12 ft^3, 30 in.^{3}, and 60 cm^3.

1–4 Marking the Text, Create Representations, Discussion Groups, Look for a Pattern, Sharing and Responding Students should observe that the product of the length and width is 40. Students should also notice that the graph is decreasing but not linear. This chunk of items provides the opportunity for rich discussion about distinguishing between a constant rate of change and a nonconstant rate of change. In addition, it is important for students to understand that both x and y can come very close to zero, but they can never be equal to it. Have students discuss any patterns they see in the table and graph.

ACTIVITY 10 continued

My Notes

MATH TIP
The volume of a rectangular prism is found by multiplying length, width, and height: $V = lwh$.

Lesson 10-2
Indirect Variation

Learning Targets:
- Write and graph indirect variations.
- Distinguish between direct and indirect variation.

SUGGESTED LEARNING STRATEGIES: Create Representations, Marking the Text, Sharing and Responding, Think-Pair-Share, Discussion Groups

When packaging a different product, your team at the packaging and shipping company determines that all boxes for this product will have a volume of 400 cubic inches and a height of 10 inches. The lengths and the widths will vary.

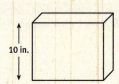

1. To explore the relationship between length and width, complete the table and make a graph of the points.

Width (x)	Length (y)
1	40
2	20
4	10
5	8
8	5
10	4
20	2

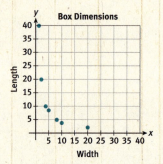

2. How are the lengths and widths in Item 1 related? Write an equation that shows this relationship.
   Answers may vary. The product of the length and width must be 40, so look for pairs of factors that have a product of 40; $xy = 40$.

3. Use the equation you wrote in Item 2 to write a function to represent the data in the table and graph above.
   $f(x) = \frac{40}{x}$, or $y = \frac{40}{x}$

4. Describe any patterns that you notice in the table and graph representing your function.
   Answers may vary. As the width increases, the length decreases.

Lesson 10-2
Indirect Variation

In terms of box dimensions, the length of the box varies indirectly as the width of the box. Therefore, this function is called an **indirect variation**.

5. Recall that direct variation is defined as $y = kx$, where $k \neq 0$ and the coefficient k is the constant of variation.

   a. How would you define indirect variation in terms of y, k, and x?
   $y = \dfrac{k}{x}$

   b. Are there any limitations on these variables as there are on k in direct variation? Explain.
   $k \neq 0, x \neq 0, y \neq 0$. Answers may vary. If $k = 0$, then x can be any number except 0 and y will always be 0. From the graph in Item 1, the values of x and y can get closer and closer to 0 but never equal 0.

   c. Write an equation for finding the constant of variation by solving for k in your answer to Part (a).
   $k = xy$

6. **Reason abstractly.** Compare and contrast the equations of direct and indirect variation.
   Answers may vary. With direct variation you multiply k by x to find y. With indirect variation you divide k by x to find y.

7. Compare and contrast the graphs of direct and indirect variation.
   Answers may vary. With direct variation, as x increases, y also increases by a given constant of variation, k. With indirect variation, as x increases, y decreases because the constant of variation, k, is divided by x.

8. Use your function in Item 3 to determine the following measurements for your company.
   a. Find the length of a box whose width is 80 inches.
      0.5 inches

   b. Find the length of a box whose width is 0.4 inches.
      100 inches

MATH TIP
Indirect variation is also known as **inverse variation**.

ACTIVITY 10 Continued

5 Marking the Text, Interactive Word Wall, Think-Pair-Share, Sharing and Responding It is important for students to understand that as the width of the box increases, the length of the box decreases. Monitor pair discussions carefully and refer students who are having difficulty back to their tables and graphs in Item 1. Note that in an indirect variation, the product of the two quantities represented by the variables is always a constant. As students share responses to part b with the class, be sure the reasons that x, y and k cannot equal 0 are highlighted.

6–7 Quickwrite Students should be able to use real-world language to compare and contrast these new terms. Provide real-world contexts for students who are struggling.

Developing Math Language
Review the mathematical definition of an *indirect variation* as a relationship between two quantities in which the value of one decreases as a result of an increase in the other. Ask students to provide real-world examples of indirect variation relationships. A possible example of direct variation is the ratio of minutes a person types to the number of pages typed. An example of indirect variation is the ratio of the height of a candle to the length of time it has burned.

8–9 Discussion Groups, Create Representations Monitor group discussions carefully to be sure students understand that, in Item 9, the constant of variation is found by solving the inverse variation equation for k and then substituting a known ordered pair (x,y).

Lesson 10-2
Indirect Variation

9. The time, y, needed to load the boxes on a truck for shipping varies indirectly as the number of people, x, working. If 10 people work, the job is completed in 20 hours.

   a. Explain how to find the constant of variation. Then find it.
      To find the constant of variation multiply the number of people working by the number of hours worked; $k = 10(20) = 200$.

   b. Write an indirect variation equation that relates the time to load the boxes to the number of people working.
      $y = \frac{200}{x}$

   c. How long does it take 8 people to finish loading the boxes? Use your equation to answer this question.
      $y = \frac{200}{8} = 25$ hours

   d. On the grid below, make a graph to show the time needed for 2, 4, 5, 8, 10, and 25 people to load the boxes on the truck.

   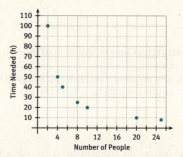

10. The cost for the company to ship the boxes varies indirectly with the number of boxes being shipped. If 25 boxes are shipped at once, it will cost $10 per box. If 50 boxes are shipped at once, the cost will be $5 per box.

    a. Write an indirect variation equation that relates the cost per box to the number of boxes being shipped.
       $y = \frac{250}{x}$

    b. How much would it cost to ship only 10 boxes?
       $25

11. Is an indirect variation function a linear function? Explain.
    No, an indirect function $y = \frac{k}{x}$ is not a linear function because the graph of an indirect variation is not a line.

MINI-LESSON: Recognizing Variation Equations

Give each student three index cards. On the first card they should write *Direct Variation* and an equation in the form $y = kx$. On the second card they should write *Indirect Variation* and an equation in the form $y = \frac{k}{x}$. On the third card they should write *Neither Direct nor Indirect* and write an equation that is neither a direct nor indirect variation. On the back of each card, students should create a table of values that corresponds to the equation on the front of the card. Collect and shuffle the cards. Distribute three cards to each student, table side up. Students should use the table to decide if it is a direct variation, an indirect variation or neither. If it is a variation, they should determine the constant of variation. They can turn over the card to check the answer.

Lesson 10-2
Indirect Variation

Check Your Understanding

12. Identify the following graphs as direct variation, indirect variation, neither, or both.

a.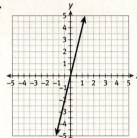

b.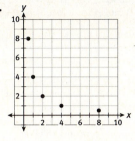

c.

13. Which equations are examples of indirect variation? Justify your answers.

   A. $y = 2x$
   B. $y = \dfrac{x}{2}$
   C. $y = \dfrac{2}{x}$
   D. $xy = 2$

14. In the equation $y = \dfrac{80}{x}$, what is the constant of variation?

LESSON 10-2 PRACTICE

15. Graph each function. Identify whether the function is an indirect variation.

a.
x	−4	−2	−1	1	2	4
y	−3/4	−3/2	−3	3	3/2	3/4

b.
x	−4	−2	−1	1	2	4
y	11	5	3	−4	−7	−13

16. **Make sense of problems.** For Parts (a) and (b) below, y varies indirectly as x.
   a. If $y = 6$ when $x = 24$, find y when $x = 16$.
   b. If $y = 8$ when $x = 20$, find the value of k.

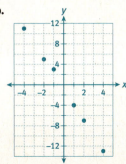

It is not an indirect variation.

16. a. 9
    b. 160

ACTIVITY 10 Continued

Lesson 10-3

PLAN

Materials
- ruler or measuring tape marked with centimeters
- at least 6 each of two sizes of paper cups for each student

Pacing: 1 class period

Chunking the Lesson
#1–2 #3–6 #7–8
#9–10 #11–12 #13
Check Your Understanding
Lesson Practice

TEACH

Bell-Ringer Activity
Give each student a centimeter ruler or measuring tape. Ask students to find several measurements accurate to the nearest centimeter. You might have them find the length, width and height of their notebooks or other books, the lengths of pens or pencils, or the length of their thumb or hand span. Assist students who need help measuring accurately.

1–2 Shared Reading, Marking the Text, Use Manipulatives, Discussion Groups Have students work with their groups to collect the data for Item 1. Some students may mistakenly want to measure along the slant height. Differences between heights of stacks with consecutive numbers of cups should be nearly the same, although individual measurements may yield imperfect data. Allow students to discuss the patterns they find in their data with the whole class.

ACTIVITY 10 continued

Lesson 10-3
Another Linear Model

CONNECT TO GEOMETRY

The carton will be a right rectangular prism. A **rectangular prism** is a closed, three-dimensional figure with three pairs of opposite parallel faces that are congruent rectangles.

Learning Targets:
- Write, graph, and analyze a linear model for a real-world situation.
- Interpret aspects of a model in terms of the real-world situation.

SUGGESTED LEARNING STRATEGIES: Marking the Text, Discussion Groups, Create Representations, Guess and Check, Use Manipulatives

Your design team at the packaging and shipping company has been asked to design a cardboard box to use when packaging paper cups for sale. Your supervisor has given you the following requirements.

- All lateral faces of the container must be rectangular.
- The base of the container must be a square, just large enough to accommodate one cup.
- The height of the container must be given as a function of the number of cups the container will hold.
- All measurements must be in centimeters.

To help discover which features of the cup affect the height of the stack, collect data on two types of cups found around the office.

1. **Use appropriate tools strategically.** Use two different types of cups to complete the tables below.

   Answers will vary depending on cup dimensions.

CUP 1		CUP 2	
Number of Cups	Height of Stack	Number of Cups	Height of Stack
1		1	
2		2	
3		3	
4		4	
5		5	
6		6	

2. **Express regularity in repeated reasoning.** What patterns do you notice that might help you figure out the relationship between the height of the stack and the number of cups in that stack?

   Answers may vary. The height increased by the same amount each time a new cup was added to the stack.

---

148 SpringBoard® Mathematics Algebra 1, Unit 2 • Functions

Lesson 10-3
Another Linear Model

Use your data for Cup 1 to complete Items 3–13.

3. Make a graph of the data you collected.
   Answers will vary depending on cup dimensions. Graphs should reflect tabular data for Cup 1 and be discrete.

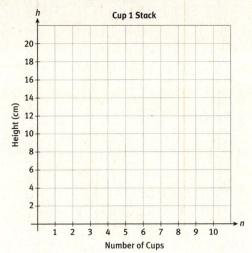

4. Predict, without measuring, the height of a stack of 16 cups. Explain how you arrived at your prediction.
   Answers will vary depending on cup dimensions, but should be based on the pattern students observed in Items 1 and 2.

5. Predict, without measuring, the height of a stack of 50 cups. Explain how you arrived at your prediction.
   Answers will vary depending on cup dimensions, but should be based on the pattern students observed in Items 1 and 2.

6. Write an equation that gives the height of a stack of cups, h, in terms of n, the number of cups in the stack.
   Answers should be equivalent to $h = d \cdot n + (c - d)$ or $h = c + (n - 1)d$ where d is the difference in height between consecutive entries on the table and c is the height of one cup. Values of c and d in students' solutions will depend on cup dimensions.

7. Use your equation from Item 6 to find h when $n = 16$ and when $n = 50$. Do your answers to this question agree with your predictions in Items 4 and 5?
   Answers will vary.

ACTIVITY 10 Continued

3–6 Discussion Groups, Create Representations, Look for a Pattern, Guess and Check, Debriefing

Students may want to connect the points on the graph. Help students realize that the graphs should consist of only discrete collinear points. If students connect the points, encourage self-editing by asking if packing a fraction of a cup makes sense. Students will probably not extend the graph to accommodate 50 cups in Item 7. It is more likely they will use the pattern in the table. Students who are unable to see the additive pattern may have difficulty connecting it to linear functions.

TEACHER to TEACHER

Some students may extend the grid, draw a line through the points, and extend the line to $x = 16$ to estimate the height of the stack. Others might use the additive pattern in the table. A common error is to add 16 times the differences in the table to 1 cup. Another common error is to determine the height of 8 cups and then double it.

Some students may require significant assistance when writing the equation in Item 6. Students will create different forms of the equation. Allow time during debriefing for a whole-class discussion about the equation forms and their relevance in the situation.

Activity 10 • Linear Models 149

ACTIVITY 10 Continued

7–8 Create Representations, Discussion Groups, Guess and Check Correct equations should yield values close to the predictions made in the beginning of the lesson. Again, students may try to graph the equation as a solid line. The graph should contain only discrete points because the input values are numbers of cups.

9–10 Activating Prior Knowledge, Think-Pair-Share Students' graphs will be similar. The equation should give ordered pairs very close to, if not exactly the same as, the ordered pairs found in the collected data.

11–12 Think-Pair-Share, Sharing and Responding The length and width of the carton are required to complete the design. Students must realize that these dimensions depend on the diameter of the top of a cup. In Item 12, students should evaluate the equation for $n = 25$ and identify that value as the height of the carton. Emphasize units.

Differentiating Learning

Support the acquisition of academic vocabulary for English Language Learners during this hands-on activity. As you use clear academic language when describing and discussing the task, support it with nonverbal cues. Ask students to repeat or paraphrase what you say. To facilitate continued acquisition of academic language during group work, pair students with strong English skills with those who need more practice with academic vocabulary.

ACTIVITY 10 continued

My Notes

Lesson 10-3
Another Linear Model

8. Sketch the graph of your equation from Item 6.
   Answers will vary depending on the values of c and d. Graphs should be discrete and include values of $n > 0$.

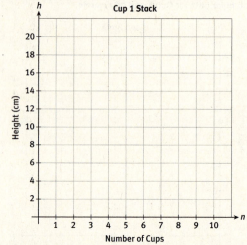

9. How are the graphs you made in Items 3 and 8 the same? How are they different?
   Answers may vary. Both graphs are discrete but this new graph includes more points. The graph in Item 3 contains only data that was collected from measuring a finite number of cups. The y-values may be slightly different because the graph from Item 3 was based on measuring the height of the stack and not the equation.

10. Do the graphs in Items 3 and 8 represent direct variation, indirect variation, or neither? Explain.
    Neither; it is not a direct variation because there is an initial height for the first cup so the graph does not pass through (0, 0). It is not an indirect variation because the graph appears to be linear.

11. Remember that you are designing a container with a square base. What dimension(s), other than the height of the stack, do you need to design your cup container? Use Cup 1 to find this/these dimension(s).
    Answers may vary. The diameter of the wider end of the cup. The actual diameter will depend on the dimensions of Cup 1.

12. Find the dimensions of a container that will hold a stack of 25 cups.
    Answers will vary depending on the dimensions of Cup 1.

150 SpringBoard® Mathematics **Algebra 1, Unit 2** • Functions

Lesson 10-3
Another Linear Model

13. Your team has been asked to communicate its findings to your supervisor. Write a report to her that summarizes your findings about the cup container design. Include the following information in your report.
    - The equation your team discovered to find the height of the stack of Cup 1 style cups
    - A description of how your team discovered the equation and the minimum number of cups needed to find it
    - An explanation of how the numbers in the equation relate to the physical features of the cup
    - An equation that could be used to find the height of the stack of Cup 2 style cups

    Answers will vary.

MATH TIP

When writing your answer to Item 13, you can use a RAFT.
- Role—team leader
- Audience—your boss
- Format—a letter
- Topic—stacks of cups

Check Your Understanding

14. A group of students performed the cup activity described in this lesson. For their Cup 1, they found the equation $h = 0.25n + 8.5$, where h is the height in inches of a stack of cups and n is the number of cups.
    a. What would be the height of 25 cups? Of 50 cups?
    b. Graph this equation. Describe your graph.

LESSON 10-3 PRACTICE

15. **Reason quantitatively.** A group of students performed the cup activity in this lesson using plastic drinking cups. Their data is shown below.

CUP 1	
Number of Cups	Height of Stack
1	14.5 cm
2	16 cm
3	17.5 cm
4	19 cm
5	20.5 cm

CUP 2	
Number of Cups	Height of Stack
1	10.5 cm
2	11.75 cm
3	13 cm
4	14.25 cm
5	15.5 cm

For each cup, write and graph an equation. Describe your graphs.

16. A consultant earns a flat fee of $75 plus $50 per hour for a contracted job. The table shows the consultant's earnings for the first four hours she works.

Hours	0	1	2	3	4
Earnings	$75	$125	$175	$225	$275

The consultant has a 36-hour contract. How much will she earn?

ACTIVITY 10 Continued

13 Shared Reading, Marking the Text, RAFT, Prewriting, Self Revision/Peer Revision Read this aloud so all students know what is expected. The minimum number of cups needed to determine the equation is 2. Note that students use Cup 2 data for their second equation. Students with a thorough understanding of linear equations and their relationship to this context should be able to replicate the process they followed for Cup 1. Encourage the use of correct mathematical vocabulary as students begin to write.

Check Your Understanding

Debrief this lesson by having students explain how to use the given equation, including how to determine a reasonable domain for the situation. Model and encourage correct use of mathematical vocabulary as students describe their graphs.

Answers

14. a. 14.75 inches; 21 inches
    b.
    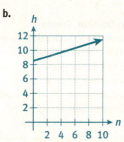
    The graph is a line.

ASSESS

Students' answers to Lesson Practice problems will provide you with a formative assessment of their understanding of the lesson concepts and their ability to apply their learning.

See the Activity Practice for additional problems for this lesson. You may assign the problems here or use them as a culmination for the activity.

ADAPT

Check students' answers to the Lesson Practice to assess whether they can write and graph an equation that represents a set of linear data. If students have difficulty writing the equations, have them describe each pattern verbally. Then help them translate their words into an equation.

Differentiating Learning

Extend the lesson to make a connection to arithmetic sequences. The height equations could be expressed as $a_n = d(n-1) + a_1$. Using sequences is appropriate here because the domain is the set of counting numbers.

LESSON 10-3 PRACTICE

15. Cup 1: $h = 1.5n + 13$
    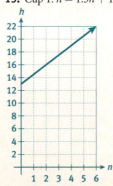
    Cup 2: $h = 1.25n + 9.25$
    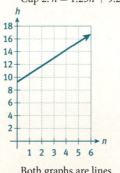
    Both graphs are lines.
16. $1,875

ACTIVITY 10 Continued

Lesson 10-4

PLAN

Pacing: 1 class period
Chunking the Lesson
#1–4   #5–8
Check Your Understanding
#12–13
Example A   #14–16   #17–18
Check Your Understanding
Lesson Practice

TEACH

Bell-Ringer Activity
Have students consider the concept of an opposite or inverse by having them tell how to undo each of the following:
- climb up 3 stairs
- travel east 4 miles
- put on socks, then put on shoes
- close the door, then lock the door

Be sure students note that in the compound actions, they not only use opposite actions, but also reverse the order of the actions.

1–4 Visualization, Create Representations, Activating Prior Knowledge, Think-Pair-Share, Sharing and Responding These items provide a good opportunity for formative assessment. It is important that students are able to interpret the constants 12.5 and 0.5 as the height of one cup minus the difference and 0.5 as the difference to give a cup height of 13 cm.

ACTIVITY 10 continued

Lesson 10-4
Inverse Functions

Learning Targets:
- Write the inverse function for a linear function.
- Determine the domain and range of an inverse function.

> **SUGGESTED LEARNING STRATEGIES:** Visualization, Create Representations, Think-Pair-Share, Discussion Groups, Construct an Argument

After reading your report, your supervisor was able to determine the equation for the height of the stack for the specific cup that the company will manufacture. The company will use the function $S(n) = 0.5n + 12.5$.

1. What do S, n, and $S(n)$ represent?
   S is the height of a stack, n is the number of cups, and $S(n)$ is the height of a stack that has n cups.

2. What do the numbers 0.5 and the 12.5 in the function S tell you about the physical features of the cup?
   12.5 is the height of one cup minus the difference, and 0.5 is the height increase when a new cup is added to the stack.

3. Evaluate $S(1)$ to find the height of a single cup.
   $S(1) = 0.5(1) + 12.5 = 13$ cm

4. How tall is a stack of 35 cups? Show your work using function notation.
   $S(35) = 0.5(35) + 12.5 = 30$ cm

5. If you add 2 cups to a stack, by how much does the height of the stack increase?
   The height increases by 1 cm.

6. If you add 20 cups to a stack, by how much does the height of the stack increase?
   The height increases by 10 cm.

7. **Critique the reasoning of others.** A member of one of the teams stated: "If you double the number of cups in a stack, then the height of the stack is also doubled." Is this statement correct? Explain.
   Answers may vary. The height of 1 cup is 13 cm and the height of a stack of 2 cups is 13.5 cm. The height of a stack of 3 cups is 14 cm and the height of a stack of 6 cups is 15.5 cm. These two examples show that doubling the number of cups does not double the height of the stack.

152 SpringBoard® Mathematics **Algebra 1, Unit 2** • Functions

Lesson 10-4
Inverse Functions

8. If you were to graph the function $S(n) = 0.5n + 12.5$, you would see that the points lie on a line.

   a. What is the slope of this line?
   0.5

   b. Interpret the slope of the line as a rate of change that relates a change in height to a change in the number of cups.
   Each additional cup adds 0.5 cm to the height of the stack.

Check Your Understanding

Use this table for Items 9 and 10.

x	1	2	3	4	5
y	1	5	9	13	17

9. Write an equation for y in terms of x.

10. Explain how the numbers in your equation relate to the numbers in the table.

11. Evaluate the function you wrote in Item 9 for each of the following values of x.

    a. $x = 8$ b. $x = 12$ c. $x = 15$ d. $x = 0$

12. a. The supervisor wanted to increase the height of a container by 5 cm. How many more cups would fit in the container?
    16 cups

    b. If the supervisor wanted to increase the height of a container by 6.4 cm, how many more cups would fit in the container?
    The container would fit 12 more cups, which would use 6 cm of the additional 6.4 cm.

    c. How many cups fit in a container that is 36 cm tall?
    47 cups; $0.5n + 12.5 = 36$ when $n = 47$

    d. How many cups fit in a container that is 50 cm tall?
    75 cups; $0.5n + 12.5 = 50$ when $n = 75$

ACTIVITY 10 Continued

5–8 Discussion Groups, Look for a Pattern, Construct an Argument, Debriefing These items address slope as a rate of change, $\frac{\Delta \text{height}}{\Delta \text{cups}}$. The implication is that for each cup added, the height increases by 0.5 cm, or that for each two cups added, the height increases 1 cm. The most direct way for students to address Item 7 is by using a counterexample. Some students may use words or pictures. For example, a diagram of cups stacked on top, but not inside of, each other is a powerful argument. Debrief by allowing students to discuss responses to Item 7, using manipulatives as models if necessary.

Check Your Understanding

Debrief students' answers to make sure they know how to write a linear equation from a table of coordinates of points on a line and how to use the equation to find y-values when given x-values. This portion of the lesson prepares students to extend their understanding of functions to include the inverse of functions.

Answers

9. $y = 4x - 3$

10. Answers may vary. Sample answer: The y-values increase by 4 as the x-values increase by 1, so the coefficient of x in the equation is 4. Comparing the x-value to the corresponding y-value, the only expression in terms of x that gives y is $4x - 3$.

11. a. $y = 29$
    b. $y = 45$
    c. $y = 57$
    d. $y = 3$

12–13 Discussion Groups, Create Representations, Work Backward, Debriefing Students may use proportions to solve using the rate of change $\frac{\Delta \text{height}}{\Delta \text{cups}} = \frac{1}{2}$. In 12b, the numerical answer comes to 12.8. The number of cups must be an integer, so the maximum number of cups is 12. Some students may be tempted to round 12.8 to 13. Be sure students understand that 13 cups would exceed the height in question.

Activity 10 • Linear Models

ACTIVITY 10 Continued

12–13 (continued) In Item 13, some students may immediately recognize that solving the equation for *n* is all that is necessary. Others may find the equation of the line by using ordered pairs of the form (S, n). What students may not realize is that they are finding the inverse of the original function. In the inverse function, the input values are heights *S* and output values are numbers of cups *n*. Debrief these items by asking students to discuss what it means to solve an equation for a variable. Have students discuss the advantages and disadvantages of solving the equation for the dependent variable before trying to find the independent variable. Ask students to compare the slope of the new equation to the slope of the original equation.

Developing Math Language

The concept of an *inverse function* is important in algebra. Have students highlight it and then add it to their math notebooks. As you add it to the classroom Word Wall, have students explain it in their own words. You might note that the function $f(x)$ is useful when you know a value of *x* and want to find the corresponding value of *y*. The inverse function $f^{-1}(x)$ is useful when you know a value of *y* and want to find the corresponding value of *x*. Note that $f^{-1}(x)$ means the inverse of $f(x)$, not $\frac{1}{f(x)}$.

Example A Note Taking Observe that it is not difficult to find the inverse of a function, but that there are several steps that must be completed in order. Use the Try These to have students demonstrate how to complete each step in the context of an actual function. Ask students to explain the process to their shoulder partner before moving on.

ACTIVITY 10
continued

My Notes

READING MATH

$f^{-1}(x)$ is read as "*f* inverse of *x*". It does **not** mean "*f* to the negative one power."

Lesson 10-4
Inverse Functions

13. The function $S(n) = 0.5n + 12.5$ describes the height *S* in terms of the number of cups *n*.

   a. Solve this equation for *n* to describe the number of cups *n* in terms of the height *S*.
   $n = 2S - 25$

   b. How many cups fit in a carton that is 85 cm tall? Compare your method of answering this question to your method used in Items 12c and 12d.
   $n = 2(85) - 25 = 145$. Comparisons may vary.

   c. What is the slope of the line represented by your equation in Part (a)? Interpret it as a rate of change and compare it to the rate of change found in Item 8b.
   The slope is $2 = \frac{2}{1} = \frac{\Delta \text{ cups}}{\Delta \text{ height}}$. For each additional 1 cm of height, there are two additional cups.

An *inverse function* is a function that interchanges the independent and dependent variables of another function. In Item 13, you found the inverse function for $S(n)$. In general, the inverse function for $f(x)$ is $f^{-1}(x)$.

Example A

Use the table below to fill in the steps to find the inverse function for $f(x) = 2x + 3$.

Write the function, replacing $f(x)$ with y.	$y = 2x + 3$
Switch x and y.	$x = 2y + 3$
Solve for y in terms of x.	$x - 3 = 2y$ $\frac{x-3}{2} = y$ or $y = \frac{1}{2}x - \frac{3}{2}$
Replace y with $f^{-1}(x)$.	$f^{-1}(x) = \frac{1}{2}x - \frac{3}{2}$

Try These A

Determine the inverses of each of the following of functions.

a. $f(x) = -4x - 5$
$f^{-1}(x) = -\frac{1}{4}x - \frac{5}{4}$

b. $f(x) = \frac{2}{3}x + 2$
$f^{-1}(x) = \frac{3}{2}x - 3$

c. $f(x) = -\frac{1}{2}x + 4$
$f^{-1}(x) = -2x + 8$

Lesson 10-4
Inverse Functions

Only those functions that are *one-to-one* functions have an inverse function. Functions that are not one-to-one must have their domain restricted for an inverse function to exist.

14. Is $S(n) = 0.5n + 12.5$ a one-to-one function? Explain.
    Yes; each value of n is paired with a different value of $S(n)$.

15. Do the following graphs of functions show one-to-one functions? Justify your answers.

    a.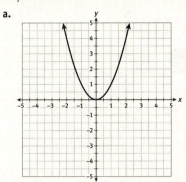

    No; there are points such as $(2, 4)$ and $(-2, 4)$ where two different x-values are paired with the same y-value.

    b.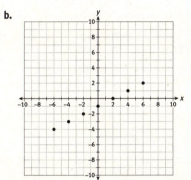

    Yes; each value of x is paired with a different value of y.

A visual test for a one-to-one function is the horizontal line test. If you can draw a horizontal line that intersects the graph of a function in more than one place, that function is not one-to-one.

16. **Construct viable arguments.** Are linear functions one-to-one functions? Justify your response.
    Most linear functions are one-to-one; their graphs pass the horizontal line test. However, linear functions whose graphs are horizontal lines do not pass the horizontal line test and are therefore not one-to-one.

MATH TERMS

For a function to be **one-to-one** means that no two values of x are paired with the same value of y.

ACTIVITY 10 Continued

17–18 Marking the Text, Activating Prior Knowledge, Think-Pair-Share
Discuss the fact that the domain and range of a function are reversed for its inverse function. Ask students to explain in their own words why this makes sense.

Check Your Understanding
As you debrief this lesson, model and encourage the use of mathematical language. Discuss when the function is most useful to solve a problem and when the inverse function is most useful.

Answers
19. $18.50
20. domain = {all real numbers greater than 0}, range = {all real numbers greater than 3.5}
21. $f^{-1}(x) = \frac{x - 3.5}{2.5}$; domain = {all real numbers greater than 3.5}, range = {all real numbers greater than 0}
22. x represents cost of the cab ride.
23. $f^{-1}(46) = \frac{46 - 3.5}{2.5} = 17$ miles

ASSESS

Students' answers to Lesson Practice problems will provide you with a formative assessment of their understanding of the lesson concepts and their ability to apply their learning.

See the Activity Practice for additional problems for this lesson. You may assign the problems here or use them as a culmination for the activity.

LESSON 10-4 PRACTICE
24. $f^{-1}(x) = \frac{x + 5}{3}$
25. $f^{-1}(x) = -\frac{1}{2}x + 5$
26. $f^{-1}(x) = \frac{3}{7}x + \frac{1}{14}$

ADAPT

If students have difficulty writing inverse functions, review the steps presented in the lesson while making visual connections to graphs, tables and mappings. Assign additional practice as needed.

My Notes

17. A function is defined by the ordered pairs {(−3, −1), (−1, 0), (1, 1), (3, 2), (5, 3)}. What are the domain and range of the function?
    Domain = {−3, −1, 1, 3, 5}; Range = {−1, 0, 1, 2, 3}

Because inputs and outputs are switched when writing the inverse of a function, the domain of a function is the range of its inverse function, and the range of a function is the domain of its inverse function.

18. What are the domain and range of the inverse function for the function in Item 17?
    Domain = {−1, 0, 1, 2, 3}; Range = {−3, −1, 1, 3, 5}

Check Your Understanding

The function $f(x) = 2.5x + 3.5$ gives the cost $f(x)$ of a cab ride of x miles.
19. What is the cost of a 6-mile ride?
20. What are the reasonable domain and range of the function?
21. Write the inverse function, $f^{-1}(x)$. What are the domain and range of $f^{-1}(x)$?
22. What does x represent in the inverse function?
23. A cab ride costs $46. Show how to use the inverse function to find the distance of the cab ride in miles.

LESSON 10-4 PRACTICE

Make use of structure. Find the inverse function, $f^{-1}(x)$, for the functions in Items 24–26.

24. $f(x) = 3x - 5$
25. $f(x) = -2x + 10$
26. $f(x) = \frac{7x}{3} - \frac{1}{6}$
27. The yearly membership fee for the Art Museum is $75. After paying the membership fee, the cost to enter each exhibit is $7.50.
    a. Write a function for the total cost of a member for one year of attending the art museum.
    b. What is the total cost for a member who sees 12 exhibits?
    c. What are the domain and range for the function?
    d. What is $f^{-1}(x)$? What are the domain and range for $f^{-1}(x)$?
    e. What does x represent in $f^{-1}(x)$?
    f. How many exhibits can a member see in a year for a total of $210, including the membership fee?

27. a. $f(x) = 75 + 7.5x$
    b. $165
    c. Domain is the natural numbers; range is real numbers greater than or equal to 75.
    d. $f^{-1}(x) = \frac{2}{15}x - 10$; Domain is real numbers greater than or equal to 75, and the range is the natural numbers.
    e. x represents the total cost for a member for one year.
    f. 18 exhibits

156 SpringBoard® Mathematics **Algebra 1, Unit 2** • Functions

Linear Models
Stacking Boxes

ACTIVITY 10 PRACTICE
Write your answers on notebook paper.
Show your work.

Lesson 10-1

1. The value of *y* varies directly as *x* and $y = 125$ when $x = 25$. What is the value of *y* when $x = 2$?

2. Which is the graph of a direct variation?

   A.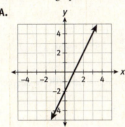

   B.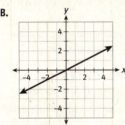

   C.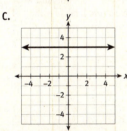

   D.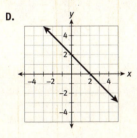

3. Which equation **does not** represent a direct variation?
   A. $y = \frac{x}{3}$
   B. $y = \frac{2}{5}x$
   C. $y = \frac{3}{x}$
   D. $y = \frac{5x}{2}$

4. The value of *y* varies directly as *x* and $y = 9$ when $x = 6$. What is the value of *y* when $x = 15$?

5. The tailor determines that the cost of material varies directly with the amount of material. The cost is $42 for 14 yards of material. What is the cost for 70 yards of material?

Lesson 10-2

6. The value of *y* varies indirectly as *x* and $y = 4$ when $x = 20$. What is the value of *y* when $x = 40$?
   A. $y = 2$
   B. $y = 8$
   C. $y = 50$
   D. $y = 80$

7. The temperature varies indirectly as the distance from the city. The temperature equals 3°C when the distance from the city is 40 miles. What is the temperature when the distance is 20 miles from the city?

8. The amount of gas left in the gas tank of a car varies indirectly to the number of miles driven. There are 9 gallons of gas left after 24 miles. How much gas is left after the car is driven 120 miles?

ACTIVITY 10 Continued

ACTIVITY PRACTICE
1. 10
2. B
3. C
4. 10
5. $210
6. A
7. 6°
8. 1.8 gallons

ACTIVITY 10 Continued

9. 3
10. 33
11. Possible answer: Direct; Explanations may vary. If the numbers 0 through 11 are assigned to January through December, the relationship can be modeled by $y = 3x$, a direct variation. If, however, the numbers 1 through 12 are assigned to the months, then the equation is $y = 3x + 3$, which is not a direct or indirect variation.
12. 3; 3 new stores open per month.
13. $h(n) = 3n + 27$
14. 11 chairs; Justifications may vary. 5 feet is equal to 60 inches. A stack of 11 chairs will be $3(11) + 27 = 60$ inches tall.
15. $f^{-1}(x) = -\frac{1}{8}x + \frac{1}{2}$
16. $f^{-1}(x) = 4x + 12$
17. $f^{-1}(x) = \frac{1}{8}x + \frac{15}{8}$
18. $f^{-1}(x) = x - 1$
19. $f^{-1}(x) = -x + 1$
20. 212°F
21. $\frac{9}{5}$
22. 10°
23. $C = \frac{5}{9}(F - 32)$
24. Answers may vary. Working backwards and using an inverse function are similar in that you are starting with what is usually the output of a problem and finding the input. They differ in that working backwards undoes the problem using the output and undoing the operations. Inverse functions are equations that have already swapped the input and output and are solved as any other equation.

ADDITIONAL PRACTICE

If students need more practice on the concepts in this activity, see the eBook Teacher Resources for additional practice problems.

ACTIVITY 10 continued

Linear Models
Stacking Boxes

Lesson 10-3

The Pete's Pets chain of pet stores is growing. The table below shows the number of stores in business each month. Use the table for Items 9–12.

Month	Stores
January	0
February	3
March	6
April	9
May	12
June	15

9. According to the table, how many new stores open per month?
10. How many stores will be in business by December?
11. Are the Pete's Pets data an example of indirect variation, direct variation, or neither? Explain your reasoning.
12. What is the slope of this function? Interpret the meaning of the slope.

Jeremy collected the following data on stacking chairs. Use the data for Items 13 and 14.

Number of Chairs	Height (in.)
1	30
2	33
3	36
4	39
5	42
6	45

13. Write a linear function that models the data.
14. Chairs cannot be stacked higher than 5 feet. What is the maximum number of chairs Jeremy can stack? Justify your answer.

Lesson 10-4

Write the inverse function for each of the following.

15. $f(x) = -8x + 4$
16. $f(x) = \frac{1}{4}x - 3$
17. $f(x) = 8x - 15$
18. $f(x) = x + 1$
19. $f(x) = -x + 1$

The formula to convert degrees Celsius C to degrees Fahrenheit F is $F = \frac{9}{5}C + 32$. Use this formula for Items 20–23.

20. Use the formula to convert 100°C to degrees Fahrenheit.
21. What is the slope?
22. The temperature is 50°F. What is the temperature in degrees Celsius?
23. Solve for C to derive the formula that converts degrees Fahrenheit to degrees Celsius.

MATHEMATICAL PRACTICES
Look for and Make Use of Structure

24. Describe the similarities and differences between finding the inverse of a function and working backward to solve a problem.

Arithmetic Sequences
Picky Patterns
Lesson 11-1 Identifying Arithmetic Sequences

ACTIVITY 11

Learning Targets:
- Identify sequences that are arithmetic sequences.
- Use the common difference to determine a specified term of an arithmetic sequence.

> **SUGGESTED LEARNING STRATEGIES:** Look for a Pattern, Create Representations, Discussion Groups, Marking the Text, Use Manipulatives

1. Use toothpicks to make the following models.

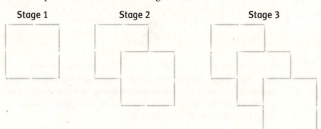

Stage 1   Stage 2   Stage 3

2. Continue to Stage 4 and Stage 5. Draw your models below.

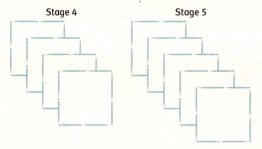

Stage 4   Stage 5

3. Complete the table for the number of toothpicks used for each stage of the models up through Stage 5.

Stage	Number of Toothpicks
1	8
2	14
3	20
4	26
5	32

My Notes

Common Core State Standards for Activity 11

- **HSF-IF.A.3:** Recognize that sequences are functions, sometimes defined recursively, whose domain is a subset of the integers.
- **HSF-IF.B.5:** Relate the domain of a function to its graph and, where applicable, to the quantitative relationship it describes.
- **HSF-BF.A.2:** Write arithmetic and geometric sequences both recursively and with an explicit formula, use them to model situations and translate between the two forms.
- **HSF-LE.A.2:** Construct linear and exponential functions, including arithmetic and geometric sequences, given a graph, a description of a relationship or two input-output pairs (include reading these from a table).

ACTIVITY 11
Guided

Activity Standards Focus
In Activity 11, students learn to identify arithmetic sequences of numbers and to express them as functions whose domains are subsets of the integers. Students learn to describe arithmetic sequences algebraically using both explicit and recursive formulas.

Lesson 11-1

PLAN

Materials:
- toothpicks

Pacing: 1 class period

Chunking the Lesson
#1–4   #5–6
Check Your Understanding
#9–11
Check Your Understanding
Lesson Practice

TEACH

Bell-Ringer Activity
Have students write a sequence of numbers that represents the number of students in each row (or column) of the class. Then ask the following questions: Could the same number be in every row? Could each row increase by the same number of students?

1–4 Discussion Groups, Look for a Pattern, Create Representations, Use Manipulatives, Activating Prior Knowledge Give each group about 50 toothpicks to create the models shown in the diagram. Monitor group discussions carefully to be sure students are using words like *domain*, *discrete*, and *slope* to describe the visual representations, tables, and graphs. Students should realize that

ACTIVITY 11 Continued

1–4 (continued) their graphs must be discrete because a portion of a toothpick cannot be used to create the diagram.

5–6 Shared Reading, Close Reading, Marking the Text, Debriefing Read the information about sequences and the ways to refer to terms of sequence. Be sure students are able to distinguish between the term number and the term itself. Focus on the use of indexed variables to represent specific terms in a sequence.

Developing Math Language

Use the sequence, 2, 6, 10, 14, ... Ask students to use the words *sequence* and *term* to describe the patterns they observe. Have students identify the third and fourth terms and label the first six terms as a_1, a_2, a_3, a_4, etc., by writing the indexed variables above each term.

$$a_1 \quad a_2 \quad a_3 \quad a_4$$
$$2, \; 6, \; 10, \; 14, \ldots$$

Model how to read the notation "a sub 1 is 2 and a sub 2 is 6." Add the words *sequence* and *term* to the classroom word wall.

Check Your Understanding

Debrief students' answers to these items to be sure they understand how to identify the terms in an arithmetic sequence before they study sequences in more depth.

Answers
7. 1
8. 17

9–11 Shared Reading, Marking the Text, Interactive Word Wall, Discussion Groups, Debriefing
Monitor group discussions carefully to be sure students understand the concept of a common difference. It may help to have students write the data they collected in Item 3 as a sequence before they respond to these question items. Debrief this portion of the lesson by asking students to share the connections they have noticed between the common difference of the sequence and the slope and the rate of change of the function represented by the table and the graph.

ACTIVITY 11 continued

My Notes

> **READING MATH**
> Read a_n as "a sub n."

> **READING MATH**
> A common difference may also be called a **constant difference**.

Lesson 11-1
Identifying Arithmetic Sequences

4. Model with mathematics. Use the following grid to make a graph of the data in the table.

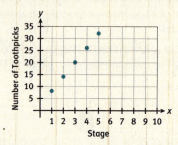

a. Is your graph discrete or continuous? Explain your answer.
Discrete; stages can be represented by whole numbers only. Also, because toothpicks are individual objects and the models include complete toothpicks only, the number of toothpicks is also represented by whole numbers only.

b. Is your graph the graph of a linear function? Explain your answer.
Yes; it is the graph of a linear function, because the points lie on a line.

An ordered list of numbers is called a **sequence**. The numbers in a sequence are **terms**. To refer to the *n*th term in a sequence, you can use either function notation, $f(n)$, or the indexed variable a_n.

The toothpick data form a sequence. The numbers of toothpicks at each stage are the terms of the sequence.

5. What are the first four terms of the toothpick sequence?
$a_1 = $ **8** $\quad a_2 = $ **14** $\quad a_3 = $ **20** $\quad a_4 = $ **26**

6. In a sequence, what is the distinction between a term and a term number?
A term is a number in a sequence; a term number gives a term's position in the sequence.

Check Your Understanding

7. For the sequence 7, −5, −3, 1, 1, . . . , what is a_4?

8. For the sequence 1, 5, 9, 13, 17, . . . , what is a_5?

An **arithmetic sequence** is a sequence in which the difference between terms is constant. The difference between consecutive terms in an arithmetic sequence is called the **common difference**.

9. Explain why the toothpick sequence is an arithmetic sequence.
The toothpick sequence is arithmetic because there is a common difference of 6 between consecutive terms.

Lesson 11-1
Identifying Arithmetic Sequences

10. What is the rate of change for the toothpick data?
    The number of toothpicks increases by 6 in each successive stage.

11. Look back at the graph in Item 4.
    a. Determine the slope between any two points on the graph.
       6
    b. Describe the connections between the slope, the rate of change, and the common difference.
       They are all equal.

Check Your Understanding

Tell whether each sequence is an arithmetic sequence. For each arithmetic sequence, find the common difference.

12. 9, 16, 23, 30, 37, ...
13. $-24, -20, -14, -10, -4, 0, \ldots$
14. $-2.8, -2.2, -1.6, -1.0, \ldots$
15. 3, 5, 8, 12, 17, ...
16. $\frac{1}{4}, \frac{1}{2}, \frac{3}{4}, 1, \ldots$
17. **Reason abstractly.** Can the common difference in an arithmetic sequence be negative? If so, give an example. If not, explain why not.

LESSON 11-1 PRACTICE

Tell whether each sequence is arithmetic. If the sequence is arithmetic, identify the common difference and find the indicated term.

18. $-9, -4, 1, 6, 11, \ldots; a_7 = ?$
19. $2, 4, 7, 11, 16, 22, \ldots; a_9 = ?$
20. $-7, -1, 5, 11, 17, \ldots; a_6 = ?$
21. $1.2, 1.9, 2.6, 3.3, 4.0, \ldots; a_8 = ?$
22. $3, \frac{5}{2}, \frac{3}{2}, -\frac{3}{2}, \ldots; a_7 = ?$
23. Write an arithmetic sequence in which the last digit of each term is 4. What is the common difference for your sequence?
24. **Critique the reasoning of others.** Jim said that the terms in an arithmetic sequence must always increase, because you must add the common difference to each term to get the next term. Is Jim correct? Justify your reasoning.

ACTIVITY 11 Continued

Lesson 11-2

PLAN

Pacing: 1 class period
Chunking the Lesson
#1–3 #4–5 Example A
#6–10 #11–12 #13–16
Check Your Understanding
Lesson Practice

TEACH

Bell-Ringer Activity
Tell students that some people think of multiplication as repeated addition.
Ask students to explain the statement by writing a paragraph that includes examples.

1–3 Look for a Pattern, Think-Pair-Share, Activating Prior Knowledge These items provide students an opportunity to revisit the toothpick sequence from Lesson 1 and to consider why it might be difficult to determine the values of certain terms in the sequence.

4–5 Shared Reading, Marking the Text, Interactive Word Wall, Look for a Pattern, Create Representations After students realize the need for a formula to generate requested terms, these items allow students to develop a general formula that can be used to determine the nth term of any arithmetic sequence. Some students may recall following a similar process to determine the number of small squares in any figure number in Lesson 1-2.

Developing Math Language

Review the words that make up the term *explicit formula*. Have students state formulas they know. Discuss the meaning of the word *formula*, steering the discussion to the idea that a formula is a mathematical rule written using symbols that describes a relationship between quantities. Note that a formula for a sequence gives the relationship between the term and the term number. Have students look up the word *explicit* to find that it means something fully and clearly expressed. Have students combine the definitions to explain what an explicit formula is.

ACTIVITY 11 continued

Lesson 11-2
A Formula for Arithmetic Sequences

My Notes

Learning Targets:
- Develop an explicit formula for the nth term of an arithmetic sequence.
- Use an explicit formula to find any term of an arithmetic sequence.
- Write a formula for an arithmetic sequence given two terms or a graph.

SUGGESTED LEARNING STRATEGIES: Look for a Pattern, Create Representations, Interactive Word Wall, Predict and Confirm, Think-Pair-Share

1. Rewrite the terms of the toothpick sequence and identify the common difference.
   8, 14, 20, 26, 32; common difference = 6

2. Find the next three terms in the sequence without building toothpick models. Explain how you found your answers.
   38, 44, 50; add 6 to 32 to get 38; add 6 again to get 44; and add 6 again to get 50.

3. Why might it be difficult to find the 100th term of the toothpick sequence using repeated addition of the common difference?
   You would have to find all 100 terms of the sequence by repeatedly adding 6, which would be time-consuming.

MATH TERMS

An **explicit formula** for an arithmetic sequence describes any term in the sequence using the first term and the common difference.

An *explicit formula* for a sequence allows you to compute any term in a sequence without computing all of the terms before it.

4. Develop an explicit formula for the toothpick sequence using the first term and the common difference.

   The first term of the sequence is $a_1 = 8$.

   The second term is $a_2 = 8 + 6$.

   The third term is $a_3 = 8 + 6 + 6$, or $a_3 = 8 + 2(6)$.

   a. Write an expression for the fourth term using the value of a_1 and the common difference.
   $a_4 =$ 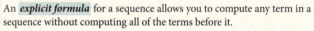 **$26 = 8 + 6 + 6 + 6$, or $8 + 3(6)$**

   b. **Express regularity in repeated reasoning.** Use the patterns you have observed to determine the 15th term. Justify your reasoning.
   $a_{15} =$ **$8 + 14(6) = 92$; the pattern is to multiply 6, the common difference, times one less than the term number and then add to 8, the first term.**

   c. Write an expression that can be used to find the nth term of the toothpick sequence.
   $a_n =$ **$8 + (n - 1)(6) = 2 + 6n$**

162 SpringBoard® Mathematics **Algebra 1, Unit 2** • Functions

Lesson 11-2
A Formula for Arithmetic Sequences

The formula you wrote in Item 4c is the explicit formula for the toothpick sequence.

For any arithmetic sequence, a_1 refers to the first term and d refers to the common difference.

5. Write an explicit formula for finding the nth term of any arithmetic sequence.

   $a_n = a_1 + (n - 1)d$

Example A
Write the explicit formula for the arithmetic sequence 3, −3, −9, −15, −21, ... Then use the formula to find the value of a_{10}.

Step 1: Find the common difference.

3, −3, −9, −15, −21
   −6 −6 −6 −6

The common difference is −6.

Step 2: Write the explicit formula and simplify.
$a_n = a_1 + (n - 1)d$
$a_n = 3 + (n - 1)(-6) = 9 - 6n$

Step 3: Use the formula to find a_{10} by substituting for n.
$a_{10} = 9 - 6(10) = -51$

Solution: The explicit formula is $a_n = 9 - 6n$ and $a_{10} = -51$.

Try These A
For the following arithmetic sequences, find the explicit formula and the value of the indicated term.

a. 2, 6, 10, 14, 18, . . . ; a_{21}
   $a_n = 2 + (n - 1)(4) = -2 + 4n$; 82

b. −0.6, −1.0, −1.4, −1.8, −2.2, . . . ; a_{15}
   $a_n = -0.6 + (n - 1)(-0.4) = -0.4n - 0.2$; −6.2

c. $\frac{1}{3}$, 1, $\frac{5}{3}$, $\frac{7}{3}$, . . . ; a_{37}
   $a_n = \frac{1}{3} + \frac{2}{3}(n - 1) = \frac{2}{3}n - \frac{1}{3}$; $24\frac{1}{3}$, or $\frac{73}{3}$

An arithmetic sequence can be graphed on a coordinate plane. In the ordered pairs the term numbers (1, 2, 3, . . .) are the x-values and the terms of the sequence are the y-values.

6. Look back at the sequence in Example A. Make a prediction about its graph.
   Answers will vary.

ACTIVITY 11 Continued

6–12 Discussion Groups, Create Representations, Predict and Confirm, Look for a Pattern, Sharing and Responding As students begin to consider graphing sequences, refer them back to the table created from the toothpick data in Lesson 11-1 and ask them to make connections to the sequence, terms of the sequence, and the term numbers. Monitor group discussions to be sure students understand that the term numbers are the values of the independent variable, x, and that the terms of the sequence are the values of the dependent variable, y. Student graphs should show discrete linear functions. Students should realize the relationship between the common difference and the slope of the graphs. Have students share with the whole class their observations about the connections between sequences, linear functions, and their graphs.

ACTIVITY 11 continued

My Notes

Lesson 11-2
A Formula for Arithmetic Sequences

7. On the grid below, create a graph of the arithmetic sequence in Example A. Revise your prediction in Item 6 if necessary.

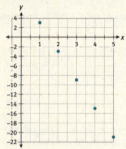

8. Determine the slope between any two points on your graph in Item 7. How does the slope compare to the common difference of the sequence?
   −6; the slope and the common difference are the same.

9. The first three terms of the arithmetic sequence 2, 5, 8, . . . are graphed below. Determine the common difference. Then graph the next three terms.

   Common difference = 3

10. Determine the slope between any two points you graphed in Item 9. How does the slope compare to the common difference of the sequence?
    3; the slope and the common difference are the same.

11. Write the explicit formula for the sequence graphed in Item 9.
    $a_n = 2 + 3(n - 1) = 3n - 1$

12. If you are given a graph of an arithmetic sequence, how do you find the explicit formula?
    First, find the ordered pair whose x-coordinate is 1. The y-coordinate of that point is a_1. The slope of the graph is d. Use these values to write the explicit formula.

The 11th term of an arithmetic sequence is 59 and the 14th term is 74.

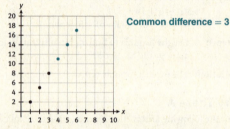

13. **Reason quantitatively.** How could you determine the value of d? What is the value of d?
    From the illustration above, you can see that $74 = 59 + 3d$. Solve this equation to find $d = 5$.

Lesson 11-2
A Formula for Arithmetic Sequences

14. How could you determine the value of a_1? What is this value?

    $a_{11} = 59$ and $d = 5$. Substitute these values into the explicit formula: $59 = a_1 + 10(5)$. Solve this equation to find $a_1 = 9$.

15. Write the explicit formula for the nth term of the sequence.

    $a_n = 9 + (n - 1)5$, or $4 + 5n$

16. Determine a_{30}, the 30th term of the sequence.

    $a_{30} = 9 + (29)5 = 154$

Check Your Understanding

17. The explicit formula for the nth term of an arithmetic sequence is $a_n = a_1 + (n - 1)d$. What does each variable in the explicit formula represent?

18. What is a_{50} of the arithmetic sequence 45, 40, 35, 30, ...?

19. The 8th term of an arithmetic sequence is 12.5 and the 13th term is 20. What is a_{25}?

20. **Construct viable arguments.** Could an arithmetic sequence also be a direct variation? Justify your answer.

21. Why is the graph of an arithmetic sequence made up of discrete points?

LESSON 11-2 PRACTICE

For the following arithmetic sequences, find the explicit formula and the value of the term indicated.

22. 2, 11, 20, 29, ...; a_{30}

23. 0.5, 0.75, 1, ...; a_{18}

24. $\frac{1}{6}, 0, -\frac{1}{6}, -\frac{1}{3}, ...; a_{42}$

25. The 3rd term of an arithmetic sequence is -1 and the 7th term is -13. Find the explicit formula for this sequence. What is a_{22}?

26. **Make use of structure.** Write the explicit formula for each arithmetic sequence graphed below. Then find the 25th term.

a.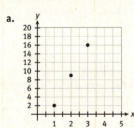
b.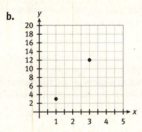
c.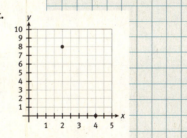

ACTIVITY 11 Continued

Lesson 11-3

PLAN

Pacing: 1 class period
Chunking the Lesson
#1–3   #4–7
Check Your Understanding
Lesson Practice

TEACH

Bell-Ringer Activity
Write $f(x) = 3x - 2$ and $g(x) = 7 - 2x$ on the board. Say that they are examples of functional notation and have students explain what that means. Then have students determine the values of $f(5)$, $f(-3)$, $f(0)$, $g(-2)$, $g(5)$, and $g(3)$ and represent the input and output values of each as ordered pairs.

1–3 Shared Reading, Marking the Text, Look for a Pattern, Create Representations, Discussion Groups, Sharing and Responding
Some students may need to create a table to help as they write a function in Item 3. If student responses are equivalent but not identical with the exception of the a^n and $f(n)$ in Items 2 and 3, encourage groups to simplify by distributing or combining like terms. It is important for students to notice that the equations for both the sequence and the function are identical with the exception of the notation.

4–7 Discussion Groups, Look for a Pattern, Debriefing, Group Presentation In these items, students are asked to make the connection between the common difference and the slope of the linear function. It is also important for students to realize that while sequences can be represented by linear functions, the domain must be restricted to the set of positive integers because the input values are the term numbers. If students struggle to describe the graph of $f(n)$, encourage them to use the My Notes section to draw a graph before responding to Item 7. Debrief this lesson by asking student groups to create their own arithmetic sequence and then represent the sequence using an explicit formula, an equation written in function notation, and a graph. They should identify the first term of the sequence, the y-intercept of the graph, the common difference of the sequence, and the slope of the linear function and graph. Groups should then share with the whole class.

166 SpringBoard® Mathematics **Algebra 1, Unit 2** • Functions

---

ACTIVITY 11
continued

Lesson 11-3
Arithmetic Sequences as Functions

Learning Targets:
- Use function notation to write a general formula for the nth term of an arithmetic sequence.
- Find any term of an arithmetic sequence written as a function.

SUGGESTED LEARNING STRATEGIES: Look for a Pattern, Create Representations, Discussion Groups, Sharing and Responding, Group Presentation

An arithmetic sequence is a special case of a linear function. The terms of the sequence are the functional values $f(1), f(2), f(3), \ldots, f(n)$ for some n.

1. Fill in the next three terms of the arithmetic sequence.

   $a_1 = 7 = f(1)$
   $a_2 = 10 = f(2)$
   $a_3 = 13 = f(3)$
   $a_4 = \underline{\ 16\ } = f(4)$
   $a_5 = \underline{\ 19\ } = f(5)$
   $a_6 = \underline{\ 22\ } = f(6)$

2. What is the nth term of the sequence?
   $a_n = 7 + 3(n - 1)$, or $4 + 3n$

3. What function f could be used to describe the sequence?
   $f(n) = 7 + 3(n - 1)$, or $4 + 3n$

4. What is the common difference of the sequence? How is the common difference related to the function you wrote in Item 3?
   3; it is the coefficient of n.

5. **Attend to precision.** Describe the domain of f using set notation. (*Hint*: What values are used as inputs for f?)
   The domain is the term numbers, which is the set of positive integers $\{1, 2, 3, \ldots\}$.

6. What ordered pair represents the nth term of the sequence?
   $(n, 4 + 3n)$

7. Describe the graph of f. How is the common difference related to the graph?
   The graph consists of discrete points that lie on a line. The slope of the line is the common difference, 3.

Lesson 11-3
Arithmetic Sequences as Functions

Check Your Understanding

For Items 8–11, use the arithmetic sequence −5, 1, 7, 13, ….

8. What is $f(1)$?
9. What is $f(4)$?
10. Write a function to describe the sequence.
11. Use your function to find $f(14)$.

LESSON 11-3 PRACTICE

Write a function to describe each arithmetic sequence.

12. 10, 14, 18, 22, …
13. 8.5, 10.3, 12.1, 13.9, …
14. $\frac{1}{3}, \frac{7}{12}, \frac{5}{6}, \frac{13}{12}, …$
15. −7, −4.5, −2, 0.5, …
16.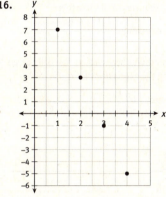

The 1st term of an arithmetic sequence is 5, and the common difference is 1.5. Use this information for Items 17–19.

17. What is $f(3)$?
18. Write a function to describe this arithmetic sequence.
19. Determine the 25th term of the sequence. Use function notation in your answer.
20. **Make sense of problems.** The 3rd term of an arithmetic sequence is −2, and the 8th term is −32. Write a function to describe this sequence.

ACTIVITY 11 Continued

Check Your Understanding

As you debrief students' answers, use mathematical terminology to discuss the reasoning behind each. This lesson helps students to connect sequences and functions and to see that an arithmetic sequence is a special type of function.

Answers
8. −5
9. 13
10. $f(n) = −5 + 6(n − 1) = 6n − 11$
11. $f(14) = 73$

ASSESS

Students' answers to Lesson Practice problems will provide you with a formative assessment of their understanding of the lesson concepts and their ability to apply their learning.

See the Activity Practice for additional problems for this lesson. You may assign the problems here or use them as a culmination for the activity.

LESSON 11-3 PRACTICE

12. $f(n) = 10 + 4(n − 1) = 4n + 6$
13. $f(n) = 8.5 + 1.8(n − 1) = 1.8n + 6.7$
14. $f(n) = \frac{1}{3} + \frac{1}{4}(n − 1) = \frac{1}{4}n + \frac{1}{12}$
15. $f(n) = −7 + 2.5(n − 1) = 2.5n − 9.5$
16. $f(n) = 7 − 4(n − 1) = −4n + 11$
17. 8
18. $f(n) = 5 + 1.5(n − 1) = 1.5n + 3.5$
19. $f(25) = 41$
20. $f(n) = 10 − 6(n − 1) = −6n + 16$

ADAPT

Check students' answers to ensure they know how to write arithmetic sequences as functions. If they have difficulty, review how to write explicit formulas for sequences and then relate the formulas to functions.

ACTIVITY 11 Continued

Lesson 11-4

PLAN

Materials:
- markers
- poster board

Pacing: 1 class period

Chunking the Lesson
Introduction, #1   #2
Check Your Understanding
#7–10
Check Your Understanding
Lesson Practice

TEACH

Bell-Ringer Activity
Have students sketch an equilateral triangle and write down the number of sides (3). Then have students draw another same-size triangle so it shares one side of the first triangle (the new shape will be a parallelogram) and have them write down the number of sides the shape now has by adding 2 to 3. Have students continue to add triangles to the shape (it will alternate between a parallelogram and a trapezoid) and to determine the number of sides in the figure. Ask students how to determine the number of sides in the next figure provided they know the number of sides in a given figure in the pattern.

Introduction, 1 Shared Reading, Close Reading, Look for a Pattern
Guide students through examination of the examples of recursion in the opening using Close Reading. Students should determine that in each case, 6 is added to the previous term. As they compare sequence notation to function notation, help students understand the connections between the two.

2 Shared Reading, Paraphrasing, Marking the Text, Think-Pair-Share
Have students explain the general recursive formulas in their own words. Students are given the opportunity to represent the toothpick data using a recursive formula and function. As students share, discuss how these are similar to and different from the explicit formula and function.

Check Your Understanding
Debrief these items by asking students to identify the number they started with and the amount that was added each time. Have them relate those two answers to their formulas. It is important for students to be comfortable with writing recursive formulas and functions before they compare them with explicit formulas and functions.

ACTIVITY 11 continued

Lesson 11-4
Recursive Formulas

My Notes

MATH TIP
In a sequence, $f(n-1) = a_{n-1}$ refers to the term before $f(n) = a_n$. Item 1 is asking "For any value of n, how can you find the term before $f(n) = a_n$?"

Learning Targets:
- Write a recursive formula for a given arithmetic sequence.
- Use a recursive formula to find the terms of an arithmetic sequence.

SUGGESTED LEARNING STRATEGIES: Look for a Pattern, Create Representations, Close Reading, Marking the Text, Discussion Groups

In a sequence, the term before $f(n) = a_n$ is $f(n-1) = a_{n-1}$.

The first four terms of the toothpick sequence can be written as

$a_1 = 8$ $f(1) = 8$
$a_2 = 14 = a_1 + 6$ $f(2) = 14 = f(1) + 6$
$a_3 = 20 = a_2 + 6$ $f(3) = 20 = f(2) + 6$
$a_4 = 26 = a_3 + 6$ $f(4) = 26 = f(3) + 6$

1. For any value of n, how can you find the value of $f(n-1) = a_n - 1$?
   Subtract the common difference from a_n.

A **recursive formula** can be used to represent an arithmetic sequence. Recursion is the process of choosing a starting term and repeatedly applying the same process to each term to arrive at the following term.

A recursive formula for an arithmetic sequence looks like this:

$$\begin{cases} a_1 = \text{1st term} \\ a_n = a_{n-1} + d \end{cases}, \text{ or in function notation: } \begin{cases} f(1) = \text{1st term} \\ f(n) = f(n-1) + d \end{cases}$$

2. The recursive formulas for the toothpick sequence are partially given below. Complete them by writing the expressions for a_n and $f(n)$.

$$\begin{cases} a_1 = 8 \\ a_n = \mathbf{a_{n-1} + 6} \end{cases} \text{ and } \begin{cases} f(1) = 8 \\ f(n) = \mathbf{f(n-1) + 6} \end{cases}$$

Check Your Understanding

Write the recursive formula for the following arithmetic sequences. Include the recursive formula in function notation.

3. $2, 4, 6, 8, \ldots$
4. $-2, -5, -8, -11, \ldots$
5. $-3, -\frac{3}{2}, 0, \frac{3}{2}, \ldots$
6. Suppose that $a_{n-1} = -4$.
   a. Find the value of a_n for the arithmetic sequence with the recursive formula $\begin{cases} a_1 = 6 \\ a_n = a_{n-1} + (-5) \end{cases}$.
   b. What term did you find? (In other words, what is n equal to?)

Answers

3. $\begin{cases} a_1 = 2 \\ a_n = a_{n-1} + 2 \end{cases}$ or

   $\begin{cases} f(1) = 2 \\ f(n) = f(n-1) + 2 \end{cases}$

4. $\begin{cases} a_1 = -2 \\ a_n = a_{n-1} + (-3) \end{cases}$ or

   $\begin{cases} f(1) = -2 \\ f(n) = f(n-1) + (-3) \end{cases}$

5. $\begin{cases} a_1 = -3 \\ a_n = a_{n-1} + 1\frac{1}{2} \end{cases}$ or

   $\begin{cases} f(1) = -3 \\ f(n) = f(n-1) + 1\frac{1}{2} \end{cases}$

6. a. -9
   b. 4

168 SpringBoard® Mathematics Algebra 1, Unit 2 • Functions

Lesson 11-4
Recursive Formulas

7. An arithmetic sequence has the recursive formula below.
$$\begin{cases} f(1) = \frac{1}{2} \\ f(n) = f(n-1) + 2 \end{cases}$$

   a. Determine the first five terms of the sequence.

   $\frac{1}{2}, \frac{5}{2}, \frac{9}{2}, \frac{13}{2}, \frac{17}{2}$

   b. Write the explicit formula for the sequence using function notation.

   $f(n) = \frac{1}{2} + 2(n-1) = 2n - \frac{3}{2}$

8. An arithmetic sequence has the explicit formula $a_n = 3n - 8$.
   a. What are the values of a_1 and a_2?

   $a_1 = 3(1) - 8 = -5; a_2 = 3(2) - 8 = -2$

   b. How can you use the values of a_1 and a_2 to find d? What is d?

   You add 3 to get from a_1 to a_2 so $d = 3$.

   c. Use your answers to Parts (a) and (b) to write the recursive formula for the sequence.

   $\begin{cases} a_1 = -5 \\ a_n = a_{n-1} + 3 \end{cases}$

In the 12th century, Leonardo of Pisa, also known as Fibonacci, first described a sequence known as the **Fibonacci sequence**. The sequence can be described by the recursive formula below.

$$\begin{cases} a_1 = 1 \\ a_2 = 1 \\ a_n = a_{n-1} + a_{n-2}, \text{ for } n > 2 \end{cases}$$

Notice that the first two terms of the sequence are 1 and that the expression describing a_n applies to those terms after the 2nd term.

9. Use the recursive formula to determine the first 10 terms of the Fibonacci sequence.

   1, 1, 2, 3, 5, 8, 13, 21, 34, 55

10. Is the Fibonacci sequence an arithmetic sequence? Justify your response.

    No; it is not an arithmetic sequence because it does not have a common difference between consecutive terms.

ACTIVITY 11 Continued

Developing Math Language
Pronounce the name *Fibonacci* and have students repeat it. Throughout the lesson, model the correct pronunciation of Fibonacci and help students pronounce it correctly.

Check Your Understanding
After debriefing the activity, have students discuss the advantages of each representation of an arithmetic sequence. Then have them tell which representation they prefer and give reasons for their preference.

Answers
11. Using the explicit formula, you can find a specific term without having to know the previous term, as you do with the recursive formula.
12. The sum of the first two terms, which are both 1, is 2, which is the third term; add the second and third terms to get 3 for the fourth term, and so on.

ASSESS

Students' answers to Lesson Practice problems will provide you with a formative assessment of their understanding of the lesson concepts and their ability to apply their learning.

See the Activity Practice for additional problems for this lesson. You may assign the problems here or use them as a culmination for the activity.

LESSON 11-4 PRACTICE

13. $\begin{cases} a_1 = 1 \\ a_n = a_{n-1} + 5 \end{cases}$ or $\begin{cases} f(1) = 1 \\ f(n) = f(n-1) + 5 \end{cases}$

14. $\begin{cases} a_1 = 1 \\ a_n = a_{n-1} + 3 \end{cases}$ or $\begin{cases} f(1) = 1 \\ f(n) = f(n-1) + 3 \end{cases}$

15. $\begin{cases} a_1 = 8 \\ a_n = a_{n-1} - 3 \end{cases}$ or $\begin{cases} f(1) = 8 \\ f(n) = f(n-1) - 3 \end{cases}$

ADAPT

If students have difficulty with these questions, have them first list the beginning number and the common difference of each sequence. Then have them describe how to use those numbers to find successive terms. Finally, have them describe how they could use that information to write a recursive formula.

16. $\begin{cases} a_1 = \frac{2}{5} \\ a_n = a_{n-1} + \frac{3}{20} \end{cases}$ or $\begin{cases} f(1) = \frac{2}{5} \\ f(n) = f(n-1) + \frac{3}{20} \end{cases}$

17. $\begin{cases} a_1 = 18 \\ a_n = a_{n-1} - 7 \end{cases}$ or $\begin{cases} f(1) = 18 \\ f(n) = f(n-1) - 7 \end{cases}$

18. 1.5
19. The sequences are similar to the Fibonacci sequence because after the first two terms, subsequent terms are found by adding the previous two terms.
Next two terms of sequence a: 52, 84
Next two terms of sequence b: 26, 42

Lesson 11-4
Recursive Formulas

Check Your Understanding

11. **Attend to precision.** Compare and contrast the explicit and recursive formulas for an arithmetic sequence.
12. Explain how to find any term of the Fibonacci sequence.

LESSON 11-4 PRACTICE
Write the recursive formula for each arithmetic sequence. Include the recursive formula in function notation.

13. 1, 6, 11, 16, ...
14. 1, 4, 7, 10, 13, ...
15. $a_n = 11 - 3n$
16. $a_n = \frac{1}{4} + \frac{3}{20}n$
17.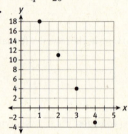

18. Given $f(n - 1) = 1.2$, use the recursive formula below to find $f(n)$.
$\begin{cases} f(1) = -0.6 \\ f(n) = f(n-1) + 0.3 \end{cases}$

19. **Reason quantitatively.** Describe how each sequence is similar to the Fibonacci sequence. Then find the next two terms.
    a. 4, 4, 8, 12, 20, 32, ...
    b. 2, 2, 4, 6, 10, 16, ...

170 SpringBoard® Mathematics Algebra 1, Unit 2 • Functions

Arithmetic Sequences
Picky Patterns

ACTIVITY 11 continued

ACTIVITY 11 PRACTICE
Write your answers on notebook paper.
Show your work.

Lesson 11-1

For Items 1–3, refer to the toothpick pattern shown below.

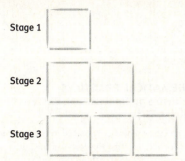

Stage 1
Stage 2
Stage 3

1. Copy and complete the table below.

Stage	Number of Toothpicks
1	
2	
3	
4	
5	

2. Write the number of toothpicks as a sequence.

3. Is the sequence you wrote in Item 2 an arithmetic sequence? If so, determine the common difference. If not, explain why not.

4. Which of the following is **not** an arithmetic sequence?
    A. $\frac{1}{2}, 1, \frac{3}{2}, 2 \ldots$
    B. 11, 14, 17, 20, …
    C. 2, 4, 8, 16, …
    D. 5, 2, −1, −4, …

For Items 5 and 6, find the common difference for each arithmetic sequence.

5. −2.3, −1.1, 0.1, 1.3, …

6. −9, −13, −18, −23, …

7. What are the next three terms in the arithmetic sequence −6, −10, −14, … ?

8. Write an arithmetic sequence in which some of the terms are whole numbers and the common difference is $\frac{3}{4}$.

Lesson 11-2

For Items 9 and 10, determine the explicit formula for each arithmetic sequence.

9. 3, −3, −9, −15, …

10. 1, 6, 11, 16, …

11. What is the 20th term of the arithmetic sequence: 7, 4, 1, … ?

12. The 9th and 10th terms of an arithmetic sequence are −24 and −30, respectively. What is the 30th term?

13. The 15th and 21st terms of an arithmetic sequence are −67 and −97, respectively. What is the 30th term?

14. The 9th and 14th terms of an arithmetic sequence are 23 and 33, respectively. What is the 1st term?

15. For the sequence graphed below, write the explicit formula. Then find the 57th term of the sequence.

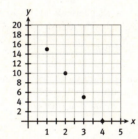

Lesson 11-3

16. Write a function to describe the arithmetic sequence graphed below. Then find the 5th term of the sequence. Use function notation in your answer.

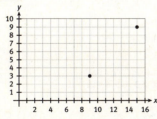

ACTIVITY 11 Continued

ACTIVITY PRACTICE

1.

Stage	Number of Toothpicks
1	4
2	7
3	10
4	13
5	16

2. 4, 7, 10, 13, 16, …
3. yes; common difference = 3
4. C
5. 1.2
6. −5
7. −18, −22, −26
8. Sample answer:
   $\frac{1}{4}, 1, \frac{7}{4}, \frac{5}{2}, \frac{13}{4}, 4, \ldots$
9. $a_n = 3 - 6(n-1) = -6n + 9$
10. $a_n = 1 + 5(n-1) = 5n - 4$
11. −50
12. −150
13. −142
14. 7
15. $a_n = 15 - 5(n-1) = -5n + 20$; −265
16. $f(n) = -5 + 1(n-1) = n - 6$; −1

ACTIVITY 11 Continued

17. D
18. 4, 1, −2, −5, −8
19. 5, 9, 13, 17, 21
20. $\begin{cases} f(1) = -1\frac{1}{2} \\ f(n) = f(n-1) + \frac{1}{2} \end{cases}$
21. $f(n) = -4n + 1.5$
22. $\begin{cases} a_1 = 5 \\ a_n = a_{n-1} + 3 \end{cases}$ or $\begin{cases} f(1) = 5 \\ f(n) = f(n-1) + 3 \end{cases}$
23. $\begin{cases} a_1 = -1 \\ a_n = a_{n-1} - \frac{3}{4} \end{cases}$ or $\begin{cases} f(1) = -1 \\ f(n) = f(n-1) - \frac{3}{4} \end{cases}$
24. a. The first addends, 1, 1, 2, 3, start with the first terms in the Fibonacci sequence. The second addends start with the third term of the Fibonacci sequence.
    b. $a_6 = 5 + 13 = 18$
       $a_7 = 8 + 21 = 29$

ADDITIONAL PRACTICE

If students need more practice on the concepts in this activity, see the eBook Teacher Resources for additional practice problems.

ACTIVITY 11 continued

Arithmetic Sequences
Picky Patterns

17. For an arithmetic sequence, $f(1) = \frac{4}{5}$ and the common difference is −1. What is $f(20)$?
    A. $18\frac{1}{5}$
    B. $-\frac{1}{5}$
    C. $-20\frac{1}{5}$
    D. $-18\frac{1}{5}$

18. An arithmetic sequence is described by the function $f(n) = -3n + 7$. Determine the first five terms of this sequence.

Lesson 11-4

19. What are the first five terms in the arithmetic sequence with the recursive formula below?
    $$\begin{cases} a_1 = 5 \\ a_n = a_{n-1} + 4 \end{cases}$$

20. What is the recursive formula for the arithmetic sequence described by the function below?
    $$f(n) = \frac{1}{2}n - 2$$

21. What is the explicit formula for the arithmetic sequence that has the recursive formula below?
    $$\begin{cases} f(1) = -2.5 \\ f(n) = f(n-1) - 4 \end{cases}$$

For Items 22 and 23, write the recursive formula for the arithmetic sequence that is graphed. Include the recursive formula in function notation.

22.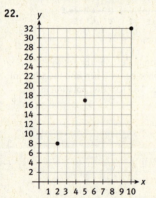

23.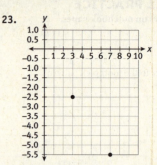

MATHEMATICAL PRACTICES
Make Sense of Problems and Persevere in Solving Them

24. The Lucas sequence is related to the Fibonacci sequence. The first five terms of the Lucas sequence are given below.
    $a_1 = 1$
    $a_2 = 1 + 2 = 3$
    $a_3 = 1 + 3 = 4$
    $a_4 = 2 + 5 = 7$
    $a_5 = 3 + 8 = 11$

    a. Beginning with a_2, what do you observe about the first addends in each sum? What do you observe about the second addends?
    b. What are the next two terms of the Lucas sequence? Explain how you determined your answer.

Linear Functions and Equations
TEXT MESSAGE PLANS

Embedded Assessment 2
Use after Activity 11

Pedro is planning to add a text messaging feature to his cell phone plan. He has gathered information about the two different plans offered by his wireless phone company.

Plan A: $4.00 per month plus 4 cents for each message
Plan B: 5 cents per message

1. Use the mathematics you have been studying in this unit to provide Pedro with the following information for each plan.
   a. Plan A
      - a table of data
      - a graph of the data
      - the linear function that fits this plan
      - the domain and range of the function
   b. Plan B
      - a table of data
      - a graph of the data
      - the linear function that fits this plan
      - the domain and range of the function

2. If Pedro sends 360 messages on average each month, which plan would you recommend that he choose? Support your recommendation using mathematical evidence.

3. If Pedro knows that his average usage is going to increase to 500 text messages per month, should he change to a different plan? Explain and justify your reasoning.

4. Explain whether either of the plans represents a direct variation.

5. Pedro's friend Chenetta is considering another text messaging plan that advertises the following: "A one-time joining fee of $3.00 and $0.08 per message."
   a. Write an explicit formula for the text messaging plan.
   b. Chenetta knows that she sends and receives about 1800 text messages per month. Use an example and other mathematical evidence to let Chenetta know if you think this plan would be a good deal for her.

Embedded Assessment 2

Assessment Focus
- Modeling with tables, graphs and linear functions
- Analyzing linear models

Materials
- graph paper
- rulers

Answer Key
1. Plan A

Messages	Charges
100	$8.00
200	$12.00
300	$16.00
400	$20.00

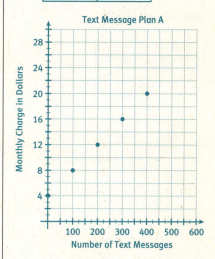

Linear function: $f(x) = 0.04x + 4$

Domain: {whole numbers x: $x \geq 0$}

Range: {Rational numbers y: $y \geq 4 +$ multiples of 0.04}

Common Core State Standards for Embedded Assessment 2

HSF-IF.B.5	Relate the domain of a function to its graph and, where applicable, to the quantitative relationship it describes.
HSF-BF.A.1	Write a function that describes a relationship between two quantities.
HSF-LE.A.2	Construct linear and exponential functions, including arithmetic and geometric sequences, given a graph, a description of a relationship, or two input-output pairs (include reading these from a table).

Unit 2 • Functions 173

Embedded Assessment 2

Teacher to Teacher

You may wish to read through the scoring guide with students and discuss the differences in the expectations at each level. Check that students understand the terms used.

Unpacking Embedded Assessment 3

Once students have completed this Embedded Assessment, turn to Embedded Assessment 3 and unpack it with them. Use a graphic organizer to help students understand the concepts they will need to know to be successful on Embedded Assessment 3.

Answer Key (cont.)

Plan B

Messages	Charges
100	$5.00
200	$10.00
300	$15.00
400	$20.00

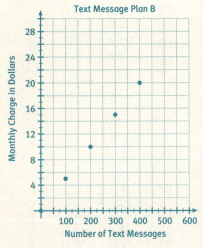

Text Message Plan B

Linear function: $f(x) = 0.05x$

Domain: {whole numbers $x: x \geq 0$}

Range: {rational numbers $y: y$ is a multiple of 0.05}

2. If Pedro's average number of text messages per month is 360, then he should choose Plan B, because 360 messages on Plan B costs $0.05(360) = $18.00, while Plan A costs 40 cents more since $4 + $0.04(360) = $18.40.

3. If Pedro's average number of text messages per month is 500, then he should choose Plan A, because it will cost him $1.00 per month less. Plan A's cost calculates out to $4 + $0.04(500) = $24 while Plan B works out to $0.05(500) = $25.

Embedded Assessment 2
Use after Activity 11

Linear Functions and Equations
TEXT MESSAGE PLANS

Scoring Guide	Exemplary	Proficient	Emerging	Incomplete
	The solution demonstrates the following characteristics:			
Mathematics Knowledge and Thinking (Items 1, 4, 5a)	• Clear and accurate understanding of linear models, including direct variation • Effective understanding of arithmetic sequences	• Largely correct understanding of linear models, including direct variation • Adequate understanding of arithmetic sequences	• Partial understanding of linear models, including direct variation • Some difficulty with arithmetic sequences	• Inaccurate or incomplete understanding of linear models, including direct variation • Little or no understanding of arithmetic sequences
Problem Solving (Items 2, 3, 5b)	• Appropriate and efficient strategy that results in a correct answer	• Strategy that may include unnecessary steps but results in a correct answer	• Strategy that results in some incorrect answers	• No clear strategy when solving problems
Mathematical Modeling / Representations (Items 1, 5a)	• Clear and accurate tables of real-world data, graphs of the data, and linear functions to model the data, including reasonable domain and range • Fluency in writing an explicit formula to model a real-world scenario	• Correct tables of real-world data, graphs of the data, and linear functions to model the data, including reasonable domain and range • Little difficulty writing an explicit formula to model a real-world scenario	• Partially correct tables of real-world data, graphs of the data, and linear functions to model the data, including reasonable domain and range • Some difficulty writing an explicit formula to model a real-world scenario	• Inaccurate or incomplete tables of real-world data, graphs of the data, and linear functions to model the data, including reasonable domain and range • Significant difficulty writing an explicit formula to model a real-world scenario
Reasoning and Communication (Items 2–4, 5b)	• Precise use of appropriate math terms and language to make and justify a recommendation • Clear and accurate explanation of whether one of the plans represents a direct variation	• Appropriate recommendations with adequate justifications • Largely correct explanation of whether one of the plans represents a direct variation	• Misleading or confusing recommendations and/or justifications • Partially correct explanation of whether one of the plans represents a direct variation	• Incomplete or inaccurate recommendations and/or justifications • Incomplete or inaccurate explanation of whether one of the plans represents a direct variation

4. Yes, Plan B; Explanations may vary. Plan B is of the form $y = kx$. The constant of variation is 0.05 and (0, 0) is a point on the line. Every time the number of text messages x increases by 1, the cost increases by 5 cents.

5. a. $a_n = 3 + 0.08(n - 1) = 0.08n + 2.92$
   b. Answers may vary.

   Chenetta, I do not think this plan is a good deal for you. If you send and receive 1800 text messages per month, then the monthly cost of this plan will be $0.08(1800) + $2.92 = $146.92. Both of the plans that Pedro's wireless company offers are significantly less expensive than this. Under Plan A, your monthly cost would be only $4 + $0.04(1800) = $76. Under Plan B, your monthly cost would be only $0.05(1800) = $90. I think you will be able to find a better deal.

Forms of Linear Functions
Under Pressure
Lesson 12-1 Slope-Intercept Form

ACTIVITY 12

Learning Targets:
- Write the equation of a line in slope-intercept form.
- Use slope-intercept form to solve problems.

SUGGESTED LEARNING STRATEGIES: Create Representations, Think-Pair-Share, Marking the Text, Discussion Groups

When a diver descends in a lake or ocean, pressure is produced by the weight of the water on the diver. As a diver swims deeper into the water, the pressure on the diver's body increases at a rate of about 1 *atmosphere of pressure* per 10 meters of depth. The table and graph below represent the total pressure, y, on a diver given the depth, x, under water in meters.

x	y
0	1
1	1.1
2	1.2
3	1.3
4	1.4
5	1.5
6	1.6

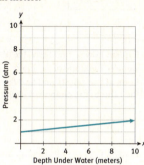

1. Write an equation describing the relationship between the pressure exerted on a diver and the diver's depth under water.
   $y = 1 + 0.1x$

2. What is the slope of the line? What are the units of the slope?
   $m = 0.1$; the units are atmospheres per meter.

3. What is the y-intercept? Explain its meaning in this context.
   (0, 1); the amount of pressure on the diver at the surface of the water (when depth = 0)

> **Slope-Intercept Form** of a **Linear Equation**
> $$y = mx + b$$
> where m is the slope of the line and $(0, b)$ is the y-intercept.

4. Identify the slope and y-intercept of the line described by the equation $y = -2x + 9$.
   slope = -2; y-intercept = (0, 9)

CONNECT TO SCIENCE

Pressure is force per unit area. *Atmospheric pressure* is defined using the unit atmosphere. 1 atm is 14.6956 pounds per square inch.

MATH TERMS

A **linear equation** is an equation that can be written in standard form $Ax + By = C$ where A, B, and C are constants and A and B cannot both be zero.

MATH TIP

Linear equations can be written in several forms.

Common Core State Standards for Activity 12

| HSA-REI.D.10 | Understand that the graph of an equation in two variables is the set of all its solutions plotted in the coordinate plane, often forming a curve (which could be a line). |
| HSF-LE.A.2 | Construct linear and exponential functions, including arithmetic and geometric sequences, given a graph, a description of a relationship or two input-output pairs (include reading these from a table). |

ACTIVITY 12 Guided

Activity Standards Focus

In Activity 12, students write linear equations in slope-intercept form, point-slope form, and standard form and use these forms to solve problems. They also write the equations of lines parallel or perpendicular to a given line. Throughout this activity, students relate the equations of lines to alternate representations, such as graphs, verbal descriptions, and tables of values.

Lesson 12-1

PLAN

Pacing: 1 class period
Chunking the Lesson
#1–3 #4 #5–9
Check Your Understanding
#14–16
Check Your Understanding
Lesson Practice

TEACH

Bell-Ringer Activity
Present the following table to students. Tell them that the relationship in the table is linear, and ask them to find the slope $\left[\text{slope} = \frac{1}{3}\right]$. Invite students to share the methods they used to determine the slope.

x	−3	0	3	6
y	3	4	5	6

1–3 Shared Reading, Marking the Text, Activating Prior Knowledge, Create Representations, Think-Pair-Share In this set of items, students activate prior knowledge by exploring a linear relationship given a table of values and a graph. Ask students to explain how they determined the slope and y-intercept of the line.

4 Think-Pair-Share This item can be used as a formative assessment of students' understanding of the definition of slope-intercept form.

ACTIVITY 12 Continued

Developing Math Language
This lesson contains the vocabulary terms *linear equation*, *y-intercept*, and *slope-intercept form*. Add the terms to the class word wall, and suggest that students create a Vocabulary Organizer for forms of linear functions. Fold a sheet of paper into fourths. In the center, write "Forms of Linear Functions." In the top left quadrant, write "slope-intercept form." Students should record the representation for slope-intercept form, labeling m and b as the slope and y-intercept. They can also include an example. Students can build on their organizers as they learn additional vocabulary related to linear equations and forms of linear functions later in this activity.

5–9 Create Representations, Discussion Groups, Debriefing In these items, students explain how to determine the slope and y-intercept of a linear relationship when given a table of values or a graph. As this portion of the lesson is debriefed, ask students to discuss the advantages and disadvantages of using each representation to determine the slope and y-intercept.

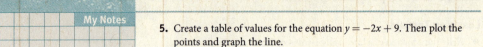

Lesson 12-1
Slope-Intercept Form

5. Create a table of values for the equation $y = -2x + 9$. Then plot the points and graph the line.

x	y
−2	13
0	9
2	5
4	1
5	−1

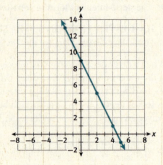

6. Explain how to find the value of the slope from the table. What is the value of the slope of the line?

   Look at two points and find the change in y and the change in x; write the ratio of the change in y over the change in x; the slope is −2.

7. Explain how to find the y-intercept from the table. What is the y-intercept?

   Find the point where $x = 0$; the y-intercept is (0, 9).

8. Explain how to find the value of the slope from the graph. What is the value of the slope?

   Find a point and move from it to another point. Find the change in y and the change in x and write it as a ratio; −2.

9. Explain how to find the y-intercept from the graph. What is the y-intercept?

   Find the point where the line intersects the y-axis. That point is the y-intercept; (0, 9).

MINI-LESSON: Using Slope-Intercept Form

A mini-lesson is available for students who need more practice writing a linear equation in slope-intercept form when given its slope and y-intercept. The mini-lesson also provides practice in identifying the slope and y-intercept of a linear equation when given its slope-intercept form.

See SpringBoard's eBook Teacher Resources for a student page for this mini-lesson.

Lesson 12-1
Slope-Intercept Form

Check Your Understanding

10. What are the slope and y-intercept of the line described by the equation $y = -\frac{4}{5}x - 10$?

11. Write the equation in slope-intercept form of the line that is represented by the data in the table.

x	−2	−1	0	1	2	3
y	9	7	5	3	1	−1

12. Write the equation, in slope-intercept form, of the line with a slope of 4 and a y-intercept of (0, 5).

13. Write an equation of the line graphed in the *My Notes* section of this page.

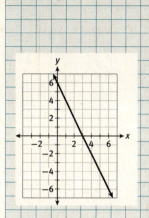

Monica gets on an elevator in a skyscraper. The elevator starts to move at a rate of −20 ft/s. After 6 seconds on the elevator, Monica is 350 feet from the ground floor of the building.

14. The rate of the elevator is negative. What does this mean in the situation? What value in the slope-intercept form of an equation does this rate represent?
    The elevator is moving down; slope, or m

15. **a.** How many feet was Monica above the ground when she got on the elevator? Show how you determined your answer.
    350 + 20(6) = 470 ft

    b. What value in the slope-intercept form does your answer to Part (a) represent?
    the y-coordinate of the y-intercept, or b

16. **Model with mathematics.** Write an equation in slope-intercept form for the motion of the elevator since it started to move. What do x and y represent?
    $y = -20x + 470$; x represents the time in seconds since Monica got on the elevator, and y represents the height of the elevator above the ground in feet.

    a. What does the y-intercept represent?
    Monica's original height above the ground in feet.

    b. Use the equation you wrote to determine, at this rate, how long it will take after Monica enters the elevator for her to exit the elevator on the ground floor. Explain how you found your answer.
    At the ground floor, y = 0. Solve the equation $0 = -20x + 470$ to find x = 23.5 seconds.

ACTIVITY 12 Continued

Check Your Understanding

Debrief students' answers to these items to make sure they understand concepts related to writing linear equations in slope-intercept form. Ask students to explain what the slope and y-intercept of the linear equation in Item 18 represent in the situation.

Answers

17. Sample answer: Subtract $3x$ from both sides: $-2y = -3x + 16$; divide both sides by -2: $y = \frac{3}{2}x - 8$.
18. **a.** $y = 0.25x + 6.5$
    b. x represents the time in days it takes to reach a height of y inches.
    c. $11.25 = 0.25x + 6.5$; $x = 19$; 19 days

ASSESS

Students' answers to Lesson Practice problems will provide the teacher with a formative assessment of student understanding of the lesson concepts and their ability to apply their learning.

See the Activity Practice for additional problems for this lesson. You may assign the problems here or use them as a culmination for the activity.

LESSON 12-1 PRACTICE

19. $m = -\frac{1}{2}$; $b = 2$
20. $y = \frac{2}{3}x - 5$
21. $y = -\frac{5}{6}x + 2$
22. **a.** $7.50; The y-intercept is $(0, -7.5)$. It means that Matt pays $7.50 even if he sells no books.
    b. $1.50; The slope is 1.5, so each time he sells one book, the earnings increase by $1.50.
    c. $y = 1.5x - 7.5$
    d. 25 books; $y = 30$, so $30 = 1.5x - 7.5$, $37.5 = 1.5x$, $x = 25$
23. **a.** ascending; crosses y-axis at $(0, 0)$
    b. ascending; crosses y-axis at $(0, 2)$
    c. descending; crosses y-axis at $(0, -5)$
    d. descending, crosses y-axis at $(0, 4)$

ADAPT

Check students' answers to the Lesson Practice to ensure that they understand how to represent and interpret a linear equation in slope-intercept form. If students need more practice working with this form of a linear equation, use the mini-lesson on page 176.

Lesson 12-1
Slope-Intercept Form

Check Your Understanding

17. Write the equation $3x - 2y = 16$ in slope-intercept form. Explain your steps.
18. A flowering plant stands 6.5 inches tall when it is placed under a growing light. Its growth is 0.25 inches per day. Today the plant is 11.25 inches tall.
    a. Write an equation in slope-intercept form for the height of the plant since it was placed under the growing light.
    b. In your equation, what do x and y represent?
    c. Use the equation to determine how many days ago the plant was placed under the light.

LESSON 12-1 PRACTICE

19. What are the slope, m, and y-intercept, $(0, b)$, of the line described by the equation $3x + 6y = 12$?
20. Write an equation in slope-intercept form for the line that has a slope of $\frac{2}{3}$ and y-intercept of $(0, -5)$.
21. Write an equation in slope-intercept form for the line that passes through the points $(6, -3)$ and $(0, 2)$.
22. Matt sells used books on the Internet. He has a weekly fee he has to pay for his website. He has graphed his possible weekly earnings, as shown.
    a. What is the weekly fee that Matt pays for his website? How do you know?
    b. How much does Matt make for each book sold? How do you know?
    c. Write the equation in slope-intercept form for the line in Matt's graph.
    d. How many books does Matt have to sell to make $30 for the week? Explain.

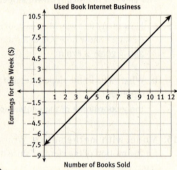

23. **Make use of structure.** Without graphing, describe the graph of each equation below. Tell whether the line is ascending or descending from left to right and where the line crosses the y-axis.
    a. $y = 3x$ **b.** $y = 5x + 2$ **c.** $y = -2x - 5$ **d.** $y = -6x + 4$

Lesson 12-2
Point-Slope Form

ACTIVITY 12 continued

Learning Targets:
- Write the equation of a line in point-slope form.
- Use point-slope form to solve problems.

SUGGESTED LEARNING STRATEGIES: Create Representations, Marking the Text, Note Taking, Think-Pair-Share, Critique Reasoning, Sharing and Responding

Another form of the equation of a line is the point-slope form. The point-slope form of the equation is found by solving the slope formula $m = \dfrac{y - y_1}{x - x_1}$ for $y - y_1$, by multiplying both sides by $x - x_1$. You may use this form when you know a point on the line and the slope.

> **Point-Slope Form** of a Linear Equation
> $$y - y_1 = m(x - x_1)$$
> where m is the slope of the line and (x_1, y_1) is a point on the line.

Example A
Write an equation of the line with a slope of $\frac{1}{2}$ that passes through the point (2, 5). Graph the line.

Step 1: Substitute the given values into point-slope form.
$$y - y_1 = m(x - x_1)$$
$$y - 5 = \tfrac{1}{2}(x - 2)$$

Step 2: Graph $y - 5 = \frac{1}{2}(x - 2)$. Plot the point (2, 5) and use the slope to find another point.

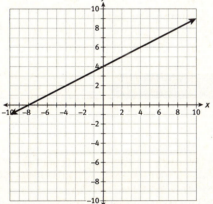

CONNECT TO AP

In calculus, the point-slope form of a line is used to write the equation of the line tangent to a curve at a given point.

MATH TIP

If you needed to express the solution to Example A in slope-intercept form, you could apply the Distributive Property and combine like terms.

MINI-LESSON: Using Point-Slope Form

A mini-lesson is available for students who need more practice writing a linear equation in point-slope form when given its slope and the coordinates of a point on the line. The mini-lesson also provides practice in using the point-slope form of a linear equation to identify the slope and the coordinates of a point on the line.

See SpringBoard's eBook Teacher Resources for a student page for this mini-lesson.

ACTIVITY 12 Continued

TEACHER to TEACHER

When using point-slope form, be sure to discuss the differences between (x, y) and (x_1, y_1). In this case, x and y are the independent and dependent variables, and x_1 and y_1 represent the coordinates of a specific point on the line.

Differentiating Instruction

Support students who make errors with signs when writing the point-slope form of an equation by emphasizing that the coordinates of the given point are *subtracted* from the variables x and y. Suggest that students circle the subtraction signs when writing the point-slope form.

Extend learning for students by having them use the Distributive Property to write their answers to the Try These problems in slope-intercept form.

1–4 Marking the Text, Think-Pair-Share, Create Representations, Debriefing In these items, students write a linear equation in point-slope form to model a real-world situation. Debrief this chunk of the lesson by asking students to share their interpretation of the meaning of the slope and the point (x_1, y_1) in this situation. Ask students to discuss how they can check that the linear equation they wrote in Item 3 correctly models the situation.

Check Your Understanding

Debrief this portion of the lesson by asking students to discuss how they would determine Violet's earnings for a week in which she sells 25 paint sets.

Answers

5. $y - 125 = -18(x - 86)$
6. $y - 2.25 = 0.75(x - 8)$

ACTIVITY 12 continued

Lesson 12-2
Point-Slope Form

Try These A
Find an equation of the line given a point and the slope.

a. $(-2, 7)$, $m = \frac{2}{3}$
$y - 7 = \frac{2}{3}(x + 2)$

b. $(6, -1)$, $m = -\frac{5}{4}$
$y + 1 = -\frac{5}{4}(x - 6)$

The town of San Simon charges its residents for trash pickup and water usage on the same bill. Each month the city charges a flat fee for trash pickup and a fee of $0.25 per gallon for water used. In January, one resident used 44 gallons of water, and received a bill for $16.

1. If x is the number of gallons of water used during a month, and y represents the bill amount in dollars, write a point (x_1, y_1).
   (44, 16)

2. What does $0.25 per gallon represent?
   The rate of change of the bill. For every additional gallon, there is an increase of $0.25.

3. **Reason abstractly.** Use point-slope form to write an equation that represents the bill cost y in terms of the number of gallons of water x used in a month.
   $y - 16 = 0.25(x - 44)$

4. Write the equation in Item 3 in slope-intercept form. What does the y-intercept represent?
   $y = 0.25x + 5$; it represents the flat fee.

Check Your Understanding

5. Determine the equation of the line given the point (86, 125) and the slope $m = -18$.

6. Violet has an Internet business selling paint sets. After an initial website fee each week, she makes a profit of $0.75 on each set she sells. If she sells 8 sets, she makes $2.25. Write an equation representing her weekly possible earnings.

180 SpringBoard® Mathematics Algebra 1, Unit 2 • Functions

Lesson 12-2
Point-Slope Form

7. **Critique the reasoning of others.** Jamilla and Ryan were asked to write the equation of the line through the points (6, 4) and (3, 5). Both Jamilla and Ryan determined that the slope was $-\frac{1}{3}$. Jamilla wrote the equation of the line as $y - 4 = -\frac{1}{3}(x - 6)$. Ryan wrote the equation of the line as $y - 5 = -\frac{1}{3}(x - 3)$.

   a. Rewrite each student's equation in slope-intercept form and compare the results.

   Jamilla's line: $y = -\frac{1}{3}x + 6$

   Ryan's line: $y = -\frac{1}{3}x + 6$

   The equations are the same.

   b. Whose equation was correct? Justify your response.

   Both equations were correct. The equations represent the same line.

8. Find the equation in point-slope form of the line shown in the graph.
   $y - 2 = 5(x - 2)$

9. Write the equation of the line in slope-intercept form.
   $y = 5x - 8$

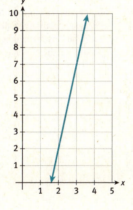

Check Your Understanding

10. Explain the process you would use to write an equation of a line in point-slope form when given two points on the line.

11. Describe the similarities and differences between point-slope form and slope-intercept form.

MINI-LESSON: Writing Point-Slope Form Given Two Points

A mini-lesson is available to provide students with practice writing the point-slope form of a linear equation when given the coordinates of two points on the line.

See SpringBoard's eBook Teacher Resources for a student page for this mini-lesson.

ACTIVITY 12 Continued

ASSESS

Students' answers to Lesson Practice problems will provide the teacher with a formative assessment of student understanding of the lesson concepts and their ability to apply their learning.

See the Activity Practice for additional problems for this lesson. You may assign the problems here or use them as a culmination for the activity.

LESSON 12-2 PRACTICE

12. $y + 8 = 0.25(x + 1)$

13. Slope $= -\frac{2}{3}$; points may vary. Sample answer: $(-3, 3)$

14. $y = \frac{1}{5}x + 3\frac{3}{5}$; the y-intercept is $3\frac{3}{5}$.

15. a. $y - 54 = 3.5(x - 5)$
    b. $y = 3.5x + 36.5$
    c. $36.50; when $x = 0$, $y = 36.5$
    d. $61
    e. 12

16. a. Rising; the slope is positive, so the height increases as time increases.
    b. 40 feet per minute
    c. 120 feet

ADAPT

Check students' answers to the Lesson Practice to ensure that they understand how to represent and interpret a linear equation in point-slope form. If students need more practice working with this form of a linear equation, use the mini-lesson on page 179.

Lesson 12-2
Point-Slope Form

LESSON 12-2 PRACTICE

12. Write an equation of the line with a slope of 0.25 that passes through the point $(-1, -8)$.

13. Find the slope and a point on the line for the line whose equation is $y = 3 - \frac{2}{3}(x + 3)$.

14. Write the equation of the line through the points $(-3, 3)$ and $(7, 5)$ in slope-intercept form. What is the y-intercept?

15. Jay pays a flat fee each month for basic cable service. He also pays $3.50 for each movie he orders during the month. Last month, he ordered 5 movies and his total bill came to $54.
    a. Write an equation in point-slope form that represents the total bill, y, in terms of the number of movies, x.
    b. Write the equation in slope-intercept form.
    c. What is the monthly fee for basic cable service? How do you know?
    d. Next month, Jay plans to order 7 movies. What will be his total bill for the month?
    e. This month, Jay's total bill is $78.50. How many movies did he order this month?

16. **Attend to precision.** The equation $y - 160 = 40(x - 1)$ represents the height in feet, y, of a hot-air balloon x minutes after the pilot started her stopwatch.
    a. Is the hot-air balloon rising or descending? Justify your answer.
    b. At what rate is the hot-air balloon rising or descending? Be sure to use appropriate units.
    c. What was the height of the balloon when the pilot started her stopwatch?

Lesson 12-3
Standard Form

Learning Targets:
- Write the equation of a line in standard form.
- Use the standard form of a linear equation to solve problems.

SUGGESTED LEARNING STRATEGIES: Create Representations, Note Taking, Discussion Groups, Think-Pair-Share, Identify a Subtask

A **linear equation** can be written in the form $Ax + By = C$ where A, B, and C are constants and A and B are not both zero.

> **Standard Form** of a Linear Equation
> $$Ax + By = C$$
> where $A \geq 0$, A and B are not both zero, and A, B, and C are integers whose **greatest common factor** is 1.

MATH TERMS
The **greatest common factor** of two or more integers is the greatest integer that is a divisor of all the integers.

1. **Reason abstractly.** You can use the coefficients of this form of an equation to find the x-intercept, y-intercept, and slope.
   a. Determine the x-intercept.
      $\left(\dfrac{C}{A}, 0\right)$
   b. Determine the y-intercept.
      $\left(0, \dfrac{C}{B}\right)$
   c. Write $Ax + By = C$ in slope-intercept form to find the slope.
      $y = -\dfrac{A}{B}x + \dfrac{C}{B}$; the slope is $-\dfrac{A}{B}$.

The definition of standard form states that both A and B are not 0. However, one of A or B may be equal to 0.

2. Write the standard form if $A = 0$.
   $By = C$
   a. Suppose $A = 0$, $B = -1$, and $C = 3$. Write the equation of the line in standard form.
      $0x - y = 3$, or $-y = 3$
   b. Graph the line on the grid in the *My Notes* section. Describe the graph. What is the slope?
      It is a horizontal line; 0

3. Write the standard form if $B = 0$.
   $Ax = C$
   a. Suppose $A = 1$, $B = 0$, and $C = -6$. Write the equation of the line in standard form.
      $x + 0y = -6$, or $x = -6$
   b. Graph the equation on the grid in the *My Notes* section. Describe the graph. What is the slope?
      It is a vertical line; undefined

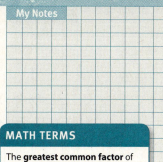

ACTIVITY 12 Continued

Developing Math Language
This lesson includes the vocabulary terms *standard form* of a linear equation and *x-intercept*. Many students may be able to determine the definition of *x-intercept* based on their previous knowledge of the term *y-intercept*. If students made a **Vocabulary Organizer** related to linear equations in Lesson 12-1, they should add "Standard Form" in the lower left fourth of the organizer and include these new terms as they relate to standard form. An example may be added to the organizer as well.

4–5 Create Representations, Think-Pair-Share, Debriefing In these items, students convert between standard form and other forms of linear equations. Ask students how they could check that the equations they wrote are equivalent to the original equations. One possible way is to substitute a value for x in the original equation and determine the corresponding value of y. Then use substitution to check that this ordered pair is also a solution of the rewritten equation. Debrief the first portion of this lesson by asking students to discuss whether the linear equation $-2x + 6y = 10$ is written in standard form. Students should conclude that the equation is not in standard form because A is negative and A, B, and C have a common factor of 2.

> **TEACHER to TEACHER**
>
> If the slope-intercept form of a linear equation contains fractions, the fractions must be eliminated before the equation can be written in standard form. To do so, students can multiply both sides of the equation by the least common denominator of the fractions.

Check Your Understanding
Ask students to discuss their answers to the Check Your Understanding questions to ensure that they understand concepts related to the standard form of a linear equation.

Answers
6. $y = \frac{3}{2}x - 8$
7. $6x + 5y = -20$
8. The graph is a horizontal line.

Lesson 12-3
Standard Form

4. Write $3x + 2y = 8$ in slope-intercept form.
   $y = -\frac{3}{2}x + 4$

5. Write the equation $y - 7 = 2(x + 1)$ in standard form.
   $2x - y = -9$

Check Your Understanding

6. Write the equation $2x + 3y = 18$ in slope-intercept form.
7. Write the equation $y = -\frac{6}{5}x - 4$ in standard form.
8. Describe the graph of any line whose equation, when written in standard form, has $A = 0$.

9. Susheila is making a large batch of granola to sell at a school fundraiser. She needs to buy walnuts and almonds to make the granola. Walnuts cost $3 per pound and almonds cost $2 per pound. She has $30 to spend on these ingredients.
   a. Write an equation that represents the different amounts of walnuts, x, and almonds, y, that Susheila can buy.
      $3x + 2y = 30$
   b. Graph the x- and y-intercepts on the coordinate plane below. Use these to help you graph the line.

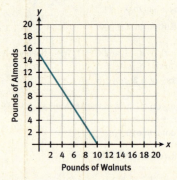

   c. If Susheila buys 4 pounds of walnuts, how many pounds of almonds can she buy?
      9 pounds

MINI-LESSON: Writing Linear Equations in Standard Form

A mini-lesson is available for students who need more practice writing a linear equation in standard form when given a different form of the equation.

See SpringBoard's eBook Teacher Resources for a student page for this mini-lesson.

Lesson 12-3
Standard Form

10. Refer to the graph you made in Item 9b. What is the x-intercept? What does it represent?

    (10, 0); if Susheila buys no almonds and spends all the money on walnuts, she can buy 10 pounds of walnuts.

11. Write an equation in standard form for the line shown.

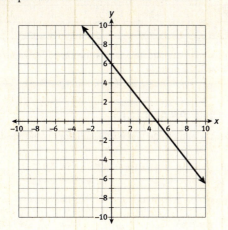

    $5x + 4y = 24$

12. **Make use of structure.** The equation $2x - 5y = 20$, the table below, and the graph below represent three different linear functions.

x	y
−3	1
−2	4
−1	7
0	10
1	13
2	16
3	19

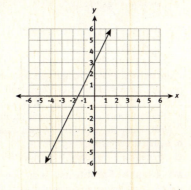

Which function represents the line with the greatest slope? Explain your reasoning.

The table; the slope of the line described in the table is 3 while the slope of the graphed line is 2 and the slope of the line described by the equation is $\frac{2}{5}$; $3 > 2 > \frac{2}{5}$.

ACTIVITY 12 Continued

Check Your Understanding

Debrief students' answers to these items to ensure that they understand how to write a linear equation in standard form. Have students explain the steps they used to write each equation.

Answers
13. $2x + y = 5$
14. $7x - y = 5$

ASSESS

Students' answers to Lesson Practice problems will provide the teacher with a formative assessment of student understanding of the lesson concepts and their ability to apply their learning.

See the Activity Practice for additional problems for this lesson. You may assign the problems here or use them as a culmination for the activity.

LESSON 12-3 PRACTICE

15. x-intercept: 7; y-intercept: -3; slope: $\frac{3}{7}$
16. a. $4x - 7y = 13$
    b. $6x + 7y = 84$
17. a. $x - y = 0$
    b. $x + y = 6$
18. a. $4x + 8y = 20$
    b.

c. 2.5; if Pedro only runs, he has to run for a total of 2.5 hours during the week.
19. a. $15.50
    b. $3.50

ADAPT

Check students' answers to the Lesson Practice to ensure that they understand how to represent and interpret a linear equation in standard form. If students need more practice working with this form of a linear equation, use the mini-lesson on page 184.

ACTIVITY 12 continued

My Notes

Check Your Understanding

13. Write an equation in standard form for the line that is represented by the data in the table.

x	-2	-1	0	1	2	3
y	9	7	5	3	1	-1

14. Write an equation in standard form for the line with a slope of 7 that passes through the point (1, 2).

LESSON 12-3 PRACTICE

15. Determine the x-intercept, y-intercept, and slope of the line described by $-3x + 7y = -21$.
16. Write each equation in standard form.
    a. $8x = 26 + 14y$
    b. $y = -\frac{6}{7}x + 12$
17. Write an equation in standard form for each line below.
    a.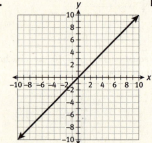
    b.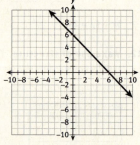

18. Pedro walks at a rate of 4 miles per hour and runs at a rate of 8 miles per hour. Each week, his exercise program requires him to cover a total distance of 20 miles with some combination of walking and/or running.
    a. Write an equation that represents the different amounts of time Pedro can walk, x, and run, y, each week.
    b. Graph the equation.
    c. What is the y-intercept? What does this tell you?

19. **Make sense of problems.** Keisha bought a discount pass at a movie theater. It entitles her to a special discounted admission price for every movie she sees. Keisha wrote an equation that gives the total cost y of seeing x movies. In standard form, the equation is $7x - 2y = -31$.
    a. What was the cost of the pass?
    b. What is the discounted admission price for each movie?

Lesson 12-3 Standard Form

Lesson 12-4
Slopes of Parallel and Perpendicular Lines

Learning Targets:
- Describe the relationship among the slopes of parallel lines and perpendicular lines.
- Write an equation of a line that contains a given point and is parallel or perpendicular to a given line.

SUGGESTED LEARNING STRATEGIES: Think-Pair-Share, Predict and Confirm, Create Representations, Look for a Pattern, Discussion Groups

Parallel lines and perpendicular lines are pairs of lines that have special relationships.

Parallel lines in a plane are equidistant from each other at all points.

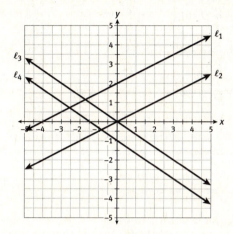

1. Consider lines l_1, l_2, l_3, and l_4 on the graph above. Determine the slope of each line.
   The slope of l_1 and l_2 is $\frac{1}{2}$. The slope of l_3 and l_4 is $-\frac{2}{3}$.

2. **Reason quantitatively.** In the graph above, l_1 is parallel to l_2 and l_3 is parallel to l_4. Write a conjecture about the slopes of parallel lines.
   The slopes of parallel lines are equal.

3. Determine the slope of a line that is parallel to the line whose equation is $y = -3x + 4$.
   -3

4. Write the equation of a line that is parallel to the line $y = \frac{3}{4}x - 1$ and has a y-intercept of $(0, 5)$.
   $y = \frac{3}{4}x + 5$

ACTIVITY 12 Continued

5–7 Marking the Text, Activating Prior Knowledge, Sharing and Responding In these items, students conclude that all horizontal lines are parallel and that all vertical lines are parallel. Check that students refer to the concept of slope in their answers to Items 5 and 6. Before students answer Item 7, you may want to have them write an equation to represent the x-axis [$y = 0$] and an equation to represent the y-axis [$x = 0$].

8 Think-Pair-Share, Create Representations In this item, students write the equation of a line through a given point that is parallel to a given line. Students are not told what form of a linear equation to use in this item, so have them discuss with their partner which would be easiest to use in this situation. Students may choose to use point-slope form because they are given a point on the line and can determine the slope from the parallel line. Others may choose to graph the line with a slope of 3 and passing through (1, 4) to determine the y-intercept. Validate both methods. Should both methods be used, debrief by discussing the advantages and disadvantages of both.

TEACHER to TEACHER

The Parallel Postulate states that given a line and a point not on the line, there is exactly one line through the point that is parallel to the given line. Students will learn more about the Parallel Postulate when they study geometry.

9–10 Activating Prior Knowledge, Discussion Groups, Predict and Confirm, Debriefing In these items, students make and confirm conjectures about the slopes of perpendicular lines. Students may need to be reminded about how to graph lines with negative slope. Suggest that they use the corner of an index card to help identify the right angles formed by the intersection of perpendicular lines on their graphs. Debrief by asking groups to share responses to Item 10. Some groups may recognize that the slopes are negative reciprocals, while others might recognize that the product of the slopes is -1. Both responses should be validated. Discuss how the two responses are related.

Lesson 12-4
Slopes of Parallel and Perpendicular Lines

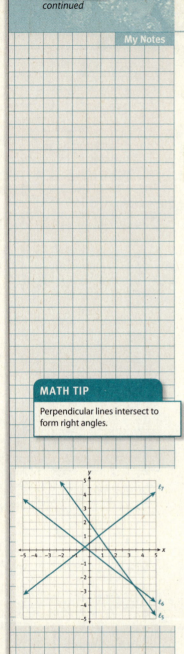

MATH TIP
Perpendicular lines intersect to form right angles.

5. Horizontal lines are described by equations of the form $y = number$. For example, the equation of the x-axis is $y = 0$, because all points on the x-axis have y-coordinate 0. Explain why any two horizontal lines are parallel.
   They both have the same slope, 0.

6. Vertical lines are described by equations of the form $x = number$. For example, the equation of the y-axis is $x = 0$, because all points on the y-axis have x-coordinate 0. Do you think that any two vertical lines are parallel? Explain why or why not.
   Yes; they both have undefined slopes.

7. Use the information in Items 5 and 6 to write the equation of a line that is
   a. parallel to the x-axis.
      Answers may vary; $y = 1$
   b. parallel to the y-axis.
      Answers may vary; $x = 1$

8. A line is parallel to $y = 3x + 2$ and passes through the point (1, 4).
   a. What is the slope of the line? Explain how you know.
      3; the line is parallel to $y = 3x + 2$, so it has the same slope.
   b. Write an equation of the line.
      $y = 3(x - 1) + 4$, or $y = 3x + 1$

9. Graph and label each line described below on the grid in the My Notes section. Which lines appear to be perpendicular?
   - l_5 has slope $-\frac{4}{3}$ and contains the point (0, 2).
   - l_6 has slope $-\frac{3}{4}$ and contains the point (0, 0).
   - l_7 has slope $\frac{3}{4}$ and contains the point $(-2, -1)$.

   l_5 and l_7 are perpendicular.

10. Write a conjecture about the slopes of perpendicular lines.
    The slopes of perpendicular lines are opposite reciprocals of each other.

Lesson 12-4
Slopes of Parallel and Perpendicular Lines

11. Use your prediction from Item 10 to write the equations of two lines that are perpendicular. On the grid in the *My Notes* section on the previous page, graph both lines and confirm that they are perpendicular.

    Answers will vary; $y = \frac{1}{4}x$ and $y = -4x + 1$

12. In the coordinate plane, what is true about a line that is perpendicular to a horizontal line?

    It is vertical.

13. Line l_1 contains the points $(0, -1)$ and $(3, 1)$. It is perpendicular to line l_2 that contains the point $(-1, 2)$.
    a. What is the slope of each line? Explain how you know.

    The slope of l_1 is $\frac{1-(-1)}{3-0} = \frac{2}{3}$. The slope of l_2 is $-\frac{3}{2}$ because it is perpendicular to l_1 so its slope is the opposite reciprocal of the slope of l_1.

    b. Write the equation of each line.

    $l_1: y = \frac{2}{3}x - 1$; $l_2: y = -\frac{3}{2}x + \frac{1}{2}$

Check Your Understanding

14. Determine whether the lines with the given slopes are parallel, perpendicular, or neither.
    a. $m_1 = -4, m_2 = \frac{1}{4}$
    b. $m_1 = -3, m_2 = 3$
    c. $m_1 = \frac{10}{12}, m_2 = -1\frac{1}{5}$
    d. $m_1 = \frac{1}{2}, m_2 = \frac{1}{2}$

15. The equation of line l_1 is $y = \frac{1}{3}x - 2$.
    a. Write the equation of a line parallel to l_1. Explain.
    b. Write the equation of a line perpendicular to l_1. Explain.

16. Write the equation of a line that is parallel to the line $3x + 4y = 4$ and contains the point $(8, 1)$.

17. Write an equation of a line that is perpendicular to the line $y = 5x + 1$ and contains the point $(-10, 2)$.

Answers
14. a. perpendicular
    b. neither
    c. perpendicular
    d. parallel
15. a. Sample answer: $y = \frac{1}{3}x$; The slopes of parallel lines are equal.
    b. Sample answer: $y = -3x - 2$; The slopes of perpendicular lines are negative reciprocals of each other.
16. $y = -\frac{3}{4}x + 7$
17. $y = -\frac{1}{5}x$

ACTIVITY 12 Continued

ASSESS

Students' answers to Lesson Practice problems will provide the teacher with a formative assessment of student understanding of the lesson concepts and their ability to apply their learning.

See the Activity Practice for additional problems for this lesson. You may assign the problems here or use them as a culmination for the activity.

LESSON 12-4 PRACTICE

18. a. neither
    b. perpendicular
    c. parallel
19. a. $-\frac{1}{2}$; 3; 0
    b. 2; $-\frac{1}{3}$; undefined
20. $m = -\frac{1}{2}$
21. Sample answer: $y = \frac{1}{4}x + 5$; The given line has a slope of $\frac{1}{4}$, and parallel lines have equal slopes.
22. $m = -\frac{4}{3}$
23. $y = \frac{5}{2}x + 23$
24. $x = 4$
25. Neither; The slope of line a is $-\frac{5}{3}$ and the slope of line b is $\frac{5}{3}$. The slopes are not equal or negative reciprocals of each other, so the lines are neither parallel nor perpendicular to each other.

ADAPT

Pair students if they need more practice with writing the equations of parallel and perpendicular lines. One student in each pair can write the equation of a line in slope-intercept form and the coordinates of a point not on the line. The student's partner can then write the equations of the lines through the point that are parallel and perpendicular to the given line. Students can confirm their work by graphing the equations by hand or with technology. Partners can then switch roles.

ACTIVITY 12 continued

Lesson 12-4
Slopes of Parallel and Perpendicular Lines

LESSON 12-4 PRACTICE

18. Determine whether the lines with the given slopes are parallel, perpendicular, or neither.
    a. $m_1 = 5, m_2 = \frac{1}{5}$
    b. $m_1 = -6, m_2 = \frac{1}{6}$
    c. $m_1 = -\frac{2}{3}, m_2 = -\frac{2}{3}$

19. The slopes of three lines are given below.
    $m_1 = -\frac{1}{2}$ $m_2 = 3$ $m_3 = 0$
    a. Determine the slope of a line that is parallel to a line with each given slope.
    b. Determine the slope of a line that is perpendicular to a line with each given slope.

20. Determine the slope of any line that is parallel to the line described by $y = -\frac{1}{2}x + 5$.

21. Write the equation of a line that is parallel to the line described by $x - 4y = 8$. Explain how you know the lines are parallel.

22. Determine the slope of any line that is perpendicular to the line described by $y = \frac{3}{4}x - 9$.

23. Write an equation of the line that is perpendicular to the line $2x + 5y = -15$ and contains the point $(-8, 3)$.

24. Determine the equation of a line perpendicular to the x-axis that passes through the point $(4, -1)$.

25. **Construct viable arguments.** A line a passes through points with coordinates $(-3, 5)$ and $(0, 0)$ and a line b passes through points with coordinates $(3, 5)$ and $(0, 0)$. Are lines a and b parallel, perpendicular, or neither? Explain your answer.

Forms of Linear Functions
Under Pressure

ACTIVITY 12
continued

ACTIVITY 12 PRACTICE
Write your answers on notebook paper. Show your work.

Lesson 12-1

1. Write the equation of a line in slope-intercept form that has a slope of -8 and a y-intercept of $(0, 3)$.

2. Write the equation of a line in slope-intercept form that passes through the point $(0, -7)$ and has a slope of $\frac{3}{4}$.

3. Find the slope and the y-intercept of the line whose equation is $-5x + 3y - 8 = 0$.

4. Which of the following is the slope-intercept form of the equation of the line in the graph?

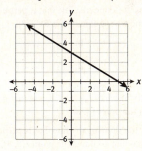

 A. $y = -\frac{5}{3}x + 3$
 B. $y = -\frac{3}{5}x + 5$
 C. $y = -\frac{3}{5}x + 3$
 D. $y = -\frac{5}{3}x + 5$

After paying an initial fee each week, Mike can sell packs of baseball cards in a sports shop. He displays his possible earnings for one week on the following graph. Use the graph for Items 5–9.

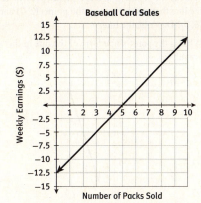

5. What is the initial fee Mike pays each week?
6. How many packs does Mike have to sell to break even?
7. What is the price of one pack of cards?
8. What is the equation in slope-intercept form for the line shown in graph?
9. How many packs of cards must Mike sell to make $40? Explain.

Lesson 12-2

10. What is the equation in point-slope form of the line that passes through $(-9, 12)$ with a slope of $\frac{5}{6}$?

11. What is the equation in slope-intercept form of the line that has a slope of 0.25 and passes through the point $(6, -8)$?

12. What is the equation in point-slope form of the line that passes through the points $(2, -3)$ and $(-5, 8)$?

ACTIVITY 12 Continued

ACTIVITY PRACTICE
1. $y = -8x + 3$
2. $y = \frac{3}{4}x - 7$
3. $m = \frac{5}{3}; (0, b) = \left(0, \frac{8}{3}\right)$
4. C
5. $12.50
6. 5 packs
7. $2.50
8. $y = 2.5x - 12.5$
9. 21 packs; use the equation in slope-intercept form and replace y with 40 to find the corresponding value of x.
10. $y - 12 = \frac{5}{6}(x + 9)$
11. $y = 0.25x - 9.5$
12. $y + 3 = -\frac{11}{7}(x - 2)$ or $y + 8 = -\frac{11}{7}(x + 5)$

ACTIVITY 12 Continued

13. $y = 3x - 10$
14. $y = 7$; a horizontal line with y-intercept (0, 7)
15. Sample answer: $y - 5 = 3(x - 5)$
16. $3x - y = 10$
17. a. $0.8x + 1.2y = 12$
    b.

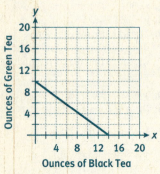

   c. (15, 0); If David buys only black tea, he can buy 15 ounces of black tea.
   d. $3\frac{1}{3}$ ounces
18. No; there is a common factor, 3, that should be divided out.
19. C
20. $5x + 3y = 15$
21. $m = -\frac{3}{5}$
22. $m = \frac{1}{2}$
23. B
24. $y = 2x - 4$
25. -1
26. No; the line $y = 2x + b$ is parallel to the line through (2, 5) and (−1, −1) for any value of b other than 1. When $b = 1$, the line is coincident with the line through (2, 5) and (−1, −1).

ADDITIONAL PRACTICE

If students need more practice on the concepts in this activity, see the eBook Teacher Resources for additional practice problems.

ACTIVITY 12 continued

Forms of Linear Functions
Under Pressure

13. Write an equation in slope-intercept form of the line that passes through the points (4, 2) and (1, −7).
14. What is the equation in slope-intercept form of the line that passes through the points (2, 7) and (6, 7)? Describe the line.
15. What is the point-slope form of the line in the graph?

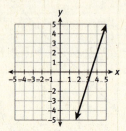

Lesson 12-3

16. Write the equation of the line in the graph from Item 15 in standard form.
17. David is ordering tea from an online store. Black tea costs $0.80 per ounce and green tea costs $1.20 per ounce. He plans to spend a total of $12 on the two types of tea.
    a. Write an equation that represents the different amounts of black tea, x, and green tea, y, that David can buy.
    b. Graph the equation.
    c. What is the x-intercept? What does it represent?
    d. Suppose David decides to buy 10 ounces of black tea. How many ounces of green tea will he buy?
18. Is the equation $6x - 15y = -12$ in standard form? Why or why not?

19. Which is a true statement about the line $x - 4y = 8$?
    A. The x-intercept of the line is (2, 0).
    B. The y-intercept of the line is (0, 2).
    C. The slope of the line is $\frac{1}{4}$.
    D. The line passes through the origin.
20. Write the equation of a line in standard form that has an x-intercept of (3, 0) and a y-intercept of (0, 5).

Lesson 12-4

21. What is the slope of a line parallel to a line whose equation is $3x + 5y = 12$?
22. What is the slope of a line perpendicular to a line whose equation is $-4x - 2y + 18 = 0$?
23. Which is the slope of a line that is perpendicular to the line whose equation is $5x - 3y = -10$?
    A. $\frac{3}{5}$ B. $-\frac{3}{5}$
    C. $\frac{5}{3}$ D. $-\frac{5}{3}$
24. What is the equation of the line that is perpendicular to $2x + 4y = 1$ and that passes through the point (6, 8)?
25. What is the slope of any line that is perpendicular to the line that contains the points (8, 8) and (12, 12)?

MATHEMATICAL PRACTICES
Construct Viable Arguments and Critique the Reasoning of Others

26. Aidan stated that for any value of b, the line $y = 2x + b$ is parallel to the line that passes through (2, 5) and (−1, −1). Do you agree with Aidan? Explain why or why not.

Equations From Data
Pass the Book
Lesson 13-1 Scatter Plots and Trend Lines

ACTIVITY 13

Learning Targets:
- Use collected data to make a scatter plot.
- Determine the equation of a trend line.

SUGGESTED LEARNING STRATEGIES: Predict and Confirm, Sharing and Responding, Create Representations, Look for a Pattern, Interactive Word Wall

How fast can you and your classmates pass a textbook from one person to the next until the book has been relayed through each person in class?

1. Suppose your entire class lined up in a row. Estimate the length of time you think it would take to pass a book from the first student in the row to the last. Assume that the book starts on a table and the last person must place the book on another table at the end of the row.

   Estimated time to pass the book: **Answers will vary.**

2. As a class, experiment with the actual time it takes to pass the book using small groups of students in your class. Use the table below to record the times.
   Answers will vary. Check students' tables.

Number of students passing the book	3	6	9	11	13	15
Time to pass the book (nearest tenth of a second)						

3. **Reason quantitatively.** Based on the data you recorded in the table above, would you revise your estimated time from Item 1? Explain the reasoning behind your answer.
   Answers will vary.

Common Core State Standards for Activity 13

HSF-IF.A.2	Use function notation, evaluate functions from inputs in their domains, and interpret statements that use function notation in terms of a context.
HSF-IF.B.4	For a function that models a relationship between two quantities, interpret key features of graphs and tables in terms of the quantities, and sketch graphs showing key features given a verbal description of the relationship.
HSF-IF.C.7	Graph functions expressed symbolically and show key features of the graph, by hand in simple cases and using technology for more complicated cases.
HSF-IF.C.7a	Graph linear and quadratic functions and show intercepts, maxima, and minima.
HSF-IF.C.7e	Graph exponential functions, showing intercepts and end behavior.

ACTIVITY 13
Investigative

Activity Standards Focus
In Activity 13, students distinguish between situations that can be modeled with linear and nonlinear functions. They make scatter plots of data sets and determine the equations of trend lines. Students also employ technology to perform regressions. They interpret key features of their function models and apply the models to make predictions.

Lesson 13-1

PLAN

Materials
- stopwatch

Pacing: 1 class period

Chunking the Lesson
#1–3   #4–6   #7
#8–10   #11–13   #14
Check Your Understanding
Lesson Practice

TEACH

Bell-Ringer Activity
Present the following scenario: *A 2-week old wolf pup weighs 3.5 pounds. At 4 weeks, the pup weighs 8.7 pounds.* Ask students to predict what the wolf will weigh at age 6 weeks, and have students discuss how they made their predictions.

1–3 Predict and Confirm, Visualization, Use Manipulatives, Sharing and Responding In this set of items, students make an initial estimate of the time it would take to pass a book down a row of students. They then collect a set of data to help them revise their predictions. Students will compare their predictions to the actual time later in the lesson. Pass the book with the numbers of students suggested in the table. However, if your class has the same number of students as one given in the table, change the numbers in the table. This will allow students to predict the time for the entire class without having already collected data for the entire class. Ask students to discuss how they used patterns in the data to revise their estimates.

Activity 13 • Equations From Data 193

ACTIVITY 13 Continued

Developing Math Language
This lesson contains the vocabulary terms *scatter plot*, *trend line*, and *correlation*. As you guide students through their learning of these mathematical terms, explain meanings in words that are accessible for your students. As possible, provide concrete examples to help students gain understanding. For example, some students may know that to *scatter* seeds means to toss them loosely on the ground rather than planting them in a precise row. You can help students make the connection that the points on a *scatter plot* can appear to have been scattered on the graph. Add these terms to the classroom word wall as they are encountered in the lesson.

4–6 Create Representations, Think-Pair-Share, Look for a Pattern, Interactive Word Wall In this set of items, students create a scatter plot of the data they collected and then look for patterns in the graph. The data will not be perfectly linear, but should display a linear trend. Students' answers to Item 5 will give insight about their understanding of what constitutes linear data. Most students should be able to use the scatter plot to determine that the data are not perfectly linear, but some students may not understand how to recognize linear data from a table. Item 6 is an open-ended question. Be very accepting of most answers, as long as they do not contradict the observed data.

TEACHER to TEACHER
Students can use a piece of uncooked spaghetti or other straightedge to model trend lines on scatter plots.

7 Interactive Word Wall, Paraphrasing, Look for a Pattern, Group Presentation In this item, students work in groups to mark a trend line on their scatter plots. Different groups will choose different lines, but the differences should not be great. Each group should come to a consensus about the two points to mark in Item 7a so that they will be able to work together on the following items. Be sure to have students discuss why they positioned the trend line where they did.

ACTIVITY 13 continued

My Notes

MATH TERMS
A **scatter plot** displays the relationship between two sets of numerical data. It can reveal trends in data.

MATH TERMS
A **trend line** is a line drawn on a scatter plot to show the **correlation**, or association, between two sets of data.

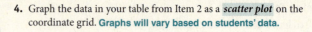
Lesson 13-1
Scatter Plots and Trend Lines

4. Graph the data in your table from Item 2 as a **scatter plot** on the coordinate grid. **Graphs will vary based on students' data.**

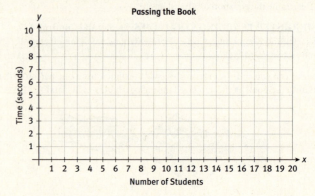

5. Are the data that you collected linear data?
   a. Explain your answer using the scatter plot.
      The data are not linear because all points do not lie on a line.

   b. Explain your answer using the table of data.
      The data are not linear because there is not a constant increase in relay times between 3, 6, and 9 students.

6. Describe how the time to pass the book changes as the number of students increases.
   The time to pass the book increases as the number of students increases.

7. Work as a group to predict the number of seconds it will take to pass the book through the whole class.
   a. Place a **trend line** on the scatter plot in Item 4 in a position that your group feels best models the data. Then, mark two points on the line.
      Answers will vary. Check students' graphs above.

   b. In the spaces provided below, enter the coordinates of the two points identified in Part (a).
      Answers will vary.

      Point 1: (_____, _____)   Point 2: (_____, _____)

   c. Why does your group think that this line gives the best position for modeling the scatter plot data?
      Answers will vary.

Common Core State Standards for Activity 13 (continued)

HSF-IF.C.9	Compare properties of two functions each represented in a different way (algebraically, graphically, numerically in tables, or by verbal descriptions).
HSF-LE.A.1	Distinguish between situations that can be modeled with linear functions and with exponential functions.
HSF-LE.A.1b	Recognize situations in which one quantity changes at a constant rate per unit interval relative to another.
HSF-LE.A.2	Construct linear and exponential functions, including arithmetic and geometric sequences, given a graph, a description of a relationship, or two input-output pairs (include reading these from a table).
HSF-LE.B.5	Interpret the parameters in a linear or exponential function in terms of a context.

194 SpringBoard® Mathematics Algebra 1, Unit 2 • Functions

Lesson 13-1
Scatter Plots and Trend Lines

continued

8. Use the coordinate pairs you recorded in Item 7b to write the equation for your trend line (or linear model) of the scatter plot.
   Answers will vary.

9. Explain what the variables in the equation of your linear model represent.
   x represents the number of students; y represents the time to pass the book in seconds.

10. **Reason abstractly.** Interpret the meaning of the slope in your linear model.
    The slope is the number of additional seconds required to pass the book for each additional student.

11. Use your model to predict how long it would take to pass the book through all the students in your class.
    Students should substitute the number of students in the class for x in the group's trend line equation in Item 8.

    Predicted time to pass the book: _____

12. Using all of the students in your class, find the actual time it takes to pass the book.
    Actual time to pass the book: **Answers will vary.**

13. How do your estimate from Item 1 and your predicted time from Item 11 compare to the actual time that it took to pass the book through the entire class?
    Answers will vary.

14. **Attend to precision.** Suppose that another class took 1 minute and 47 seconds to pass the book through all of the students in the class. Use your linear model to estimate the number of students in the class.
    Solve the equation $mx + b = 107$. The values of m and b will vary depending on the group's trend line.

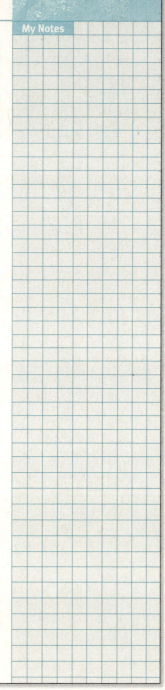

ACTIVITY 13 Continued

Check Your Understanding

Debrief students' answers to these items to make sure they understand concepts related to scatter plots and trend lines. Ask students to explain how they can tell just by looking at their scatter plot from Item 15 whether the slope of an appropriate trend line will be positive or negative.

Answers

15.
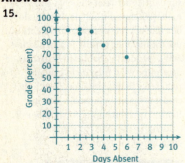

16. The data are not linear. On the scatter plot, the points do not lie on a straight line. In the table, the change in test scores from 0, 1, 2, and 3 days is not a constant amount.
17. As the number of days absent increases, the grades decrease.

ASSESS

Students' answers to lesson practice problems will provide you with a formative assessment of their understanding of the lesson concepts and their ability to apply their learning.

See the Activity Practice for additional problems for this lesson. You may assign the problems here or use them as a culmination for the activity.

LESSON 13-1 PRACTICE
22. Possible answer: (3, 1.2) and (13, 3.6); $y - 1.2 = 0.24(x - 3)$
23. y is the total rainfall in inches and x is the day of the month.
24. The slope is the increase in total rainfall for each additional day of the month. The y-intercept is the total rainfall at the start of the month.

ADAPT

Check students' answers to the Lesson Practice to ensure that they understand how to write and interpret the equation of a trend line. If students have difficulty writing the equation, they may need more practice with writing the equation of a line when given two points on the line. Make sure students understand the general process: (1) Use the points to determine the slope. (2) Write the point-slope form. (3) Convert the point-slope form to slope-intercept form.

18.
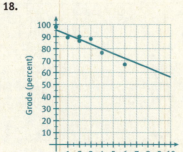
Using the points (0, 95) and (4, 80), the linear equation is $y = -\frac{15}{4}x + 95$.

Lesson 13-1
Scatter Plots and Trend Lines

Check Your Understanding

The table shows the number of days absent and the grades for several students in Ms. Reynoso's Algebra 1 class.

Days Absent	0	3	6	1	2	2	4
Grade (percent)	98	88	69	89	90	86	77

15. Create a scatter plot of the data using Days Absent as the independent variable.
16. Are the data linear? Explain using the scatter plot and the table of data.
17. Based on the data, how do grades change as the number of days absent increases?
18. Draw a trend line on your scatter plot. Identify two points on the trend line and write an equation for the line containing those two points.
19. What is the meaning of the x and y variables in the equation you wrote?
20. Interpret the meaning of the slope and the y-intercept of your trend line.
21. Use your equation to predict the grade of a student who is absent for 5 days.

LESSON 13-1 PRACTICE

Model with mathematics. The scatter plot shows the day of the month and total rainfall for January.

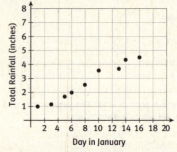

22. Copy the scatter plot and draw a trend line on the scatter plot. Identify two points on the trend line and write a linear equation to model the data containing those two points.
23. Explain the meaning of x and y in your equation.
24. Interpret the meaning of the slope and the y-intercept of your trend line.

19. x is the number of days absent, and y is the grade.
20. The slope is the decrease in the grade for each additional day absent. The y-intercept is the predicted grade if you have 0 absences.
21. Check that students use their equation to find the value of y when $x = 5$.

Lesson 13-2
Linear Regression

Learning Targets:
- Use a linear model to make predictions.
- Use technology to perform a linear regression.

SUGGESTED LEARNING STRATEGIES: Marking the Text, Interactive Word Wall, Look for a Pattern, Think-Pair-Share, Quickwrite

There is a *correlation* between two variables if they share some kind of relationship.

1. Is there a correlation between the variables of your linear model in Item 4 in Lesson 13-1? Explain.
   Yes; there is a relationship between the number of students and the amount of time it takes to pass the book.

MATH TERMS

A scatter plot will show a **positive correlation** if y tends to increase as x increases. Other data may have a **negative correlation**, where y tends to decrease as x increases, or **no correlation**. A correlation is sometimes called an association.

Examples of data with two variables that illustrate a *positive correlation*, a *negative correlation*, and *no correlation* are shown below. The more closely the data resemble a line, the stronger the linear correlation.

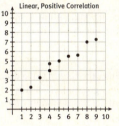
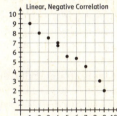
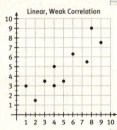
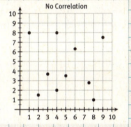

2. Look back at your linear model in Item 4 in Lesson 13-1. Does your linear model represent a positive correlation, a negative correlation, or no correlation? Explain.
   Positive correlation; the line goes up to the right in the graph so the linear model represents a positive correlation (the greater the number of students, the more time it takes to pass the book).

There is *causation* between two variables if a change in one variable causes the other variable to change. For example, doing more exercise causes a greater number of calories to be burned.

3. Does there seem to be causation between the variables of your linear model in Item 4 in Lesson 13-1? Explain.
   Yes; it will take more time for more students to pass the book, and less time for fewer students. So, a change in the number of students causes the time to change.

ACADEMIC VOCABULARY

The idea of **causation** is important in physics. For example, a cause can be represented by a force acting on an object.

TEACHER to TEACHER

If there is a positive or negative correlation between two variables x and y, there are several possible explanations, including the following:
- A change in x causes a change in y.
- A change in y causes a change in x.
- A change in a third variable z causes changes in both x and y, but x and y do not cause each other.
- There is no clear connection between x and y; the correlation is coincidence.

Activity 13 • Equations From Data 197

ACTIVITY 13 Continued

Developing Math Language
In this lesson, students use the terms *positive correlation, negative correlation,* and *no correlation* to describe the associations they see in scatter plots. Ask students to describe the connection between positive correlation and positive slope and between negative correlation and negative slope. The new terms in this lesson also include *line of best fit* and *linear regression*. Be sure to add the terms to the class Word Wall. As students respond to questions or discuss possible solutions to problems, monitor their use of new terms to ensure their understanding.

Check Your Understanding
Debrief students' answers to these items to ensure that they understand the difference between correlation and causation. For Item 5, have students state the direction of the causation. In other words, does a change in the first variable cause a change in the second variable, or vice versa?

Answers
4. **a.** Yes; as shoe size increases, the average price of a movie ticket increases.
   b. No; an increasing shoe size does not cause the price of a movie ticket to increase.
5. Sample answer: the number of beans in a bag and the weight of the bag

6 Shared Reading, Marking the Text, Interactive Word Wall, Think-Pair-Share, Create Representations, Debriefing In this item, students use a graphing calculator to perform a linear regression of their scatterplot data from Lesson 13-1. Monitor students to make sure they understand how enter the data and select the correct regression. Have students determine whether the line of best fit gives a better estimate of the time needed to pass the book through the entire class than did the trend line they wrote in the previous lesson.

ACTIVITY 13 continued

Lesson 13-2
Linear Regression

Correlation does not imply causation. Just because there is a correlation between two variables does not mean that there is causation between them; there may be other factors affecting the situation.

Check Your Understanding

4. Consider the following two variables: your shoe size each year since you were born and the average price of a movie ticket each year since you were born.
   a. Is there a correlation between the variables? Explain.
   b. Is there causation between the variables? Explain.
5. Give an example of two variables for which there is both correlation and causation.

A scatter plot and a **line of best fit**, the most accurate trend line, can be created using a graphing calculator, a spreadsheet program, or other Computer Algebra Systems (CAS).

Linear regression is a method used to find the line of best fit. A line found using linear regression is more accurate than a trend line that has been visually estimated. You can perform linear regression using a graphing calculator.

6. **Use appropriate tools strategically.** Enter the book-passing data you collected in Item 2 in Lesson 13-1 into your graphing calculator. Enter the numbers of students as x-values and the corresponding times to pass the book as y-values.
   Answers below will vary based on students' data.
   a. To find the equation of the line of best fit, use the linear regression feature of your calculator.

   The calculator should return values for a and b. Write these values below.

   $a =$

   $b =$

   b. The value of a is the slope of the line of best fit, and $(0, b)$ is the y-intercept. Round a and b to the nearest hundredth and write the equation of the line of best fit in the form $y = ax + b$. Describe how this equation is different from or similar to your equation in Item 8 in Lesson 13-1.

TEACHER to TEACHER

It can be difficult and tedious to determine a line of best fit by hand. If the class does not have access to graphing calculators, students may be able to use the linear regression function of a spreadsheet (often called LINEST) or an online regression calculator instead. To perform a linear regression on the TI-Nspire calculator, perform these steps:
- From the Home screen, select the Lists & Spreadsheets icon.
- Enter x as the heading of column A and y as the heading of column B. Then enter all x-values in column A and all y-values in column B.
- Highlight both columns and then press [menu] and select Statistics ▶ Stat Calculations ▶ Linear Regression.

198 SpringBoard® Mathematics Algebra 1, Unit 2 • Functions

Lesson 13-2
Linear Regression

Check Your Understanding

7. Enter the following data into your graphing calculator. Make sure that any previous data have been cleared.

   $(6, 1), (9, 0), (12, -3), (3, 3), (0, 5), (-3, 7), (-5, 9), (-7, 13)$

   a. Find the equation of the line of best fit. Round values to the nearest hundredth.
   b. Use the equation of the line of best fit to predict the value of y when $x = 20$.

8. Compare using a graphing calculator to using paper and pencil when plotting data and finding a trend line.

LESSON 13-2 PRACTICE

The owner of a café kept records on the daily high temperature and the number of hot apple ciders sold on that day. Some of the owner's data are shown below.

Daily High Temperature (°F)	32	75	80	48	15
Number of Hot Apple Ciders Sold	51	22	12	40	70

9. Create a scatter plot of the data.
10. Is there a correlation between the variables? If so, what type?
11. Determine the equation of the line of best fit. Round values to the nearest hundredth.
12. What is the slope? What does the slope represent?
13. Identify the y-intercept. What does the y-intercept represent?
14. **Model with mathematics.** Use your model to predict the number of hot apple ciders the café would sell on a day when the high temperature is 92°F. Explain.

ACTIVITY 13 Continued

Check Your Understanding

Debrief students' answers to these items to make sure they understand how to determine a line of best fit and use it to make predictions. Have students discuss whether a prediction made from a line of best fit is guaranteed to be correct.

Answers

7. a. $y = -0.75x + 5.78$
   b. -9.22
8. Sample answer: Using a calculator may give an equation of a trend line that fits the data more accurately.

ASSESS

Students' answers to lesson practice problems will provide you with a formative assessment of their understanding of the lesson concepts and their ability to apply their learning.

See the Activity Practice for additional problems for this lesson. You may assign the problems here or use them as a culmination for the activity.

LESSON 13-2 PRACTICE

9.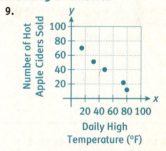

10. Yes; negative correlation
11. $y = -0.82x + 80.23$
12. -0.82; for each increase in the high temperature by 1°F, the number of hot apple ciders sold decreases by about 0.82.
13. 80.23; when the high temperature is 0°F, the café sells about 80 hot apple ciders.
14. 5; Evaluating the linear function for $x = 92$ gives $y = 4.79$, which can be rounded to 5.

ADAPT

Check students' answers to the Lesson Practice to ensure that they understand how to describe correlation in a data set and how to write and interpret the equation of a line of best fit. If students make careless errors when using technology to find the line of best fit, suggest that they graph the line on their hand-drawn scatter plot to confirm that the line is a good model for the data. If not, students should check that they entered the data values correctly in their calculators.

Activity 13 • Equations From Data 199

ACTIVITY 13 Continued

Lesson 13-3

PLAN

Materials
- graphing calculators

Pacing: 1 class period

Chunking the Lesson
#1–2   #3–5   #6
#7   #8
Check Your Understanding
Lesson Practice

TEACH

Bell-Ringer Activity

Have students classify each function as linear or nonlinear and explain their reasoning.

1. $y = x^2 + 3$ [nonlinear]
2. $y = -5x$ [linear]
3. $y = 2^x$ [nonlinear]

Differentiating Instruction

Support students who are learning English by reviewing background information that will help them make sense of the text. To aid student understanding of the problem scenario on this page, explain that a *hit* on a website is a single visit to that website. So, the number of daily hits represents the total number of visitors to the website each day.

1–2 Shared Reading, Look for a Pattern, Quickwrite, Create Representations, Sharing and Responding In these items, students look for patterns in tables of three related data sets. They then graph the data, which may enable them to see other patterns. After students complete Item 1, have them share their observations about the companies' growth. Students should note that a store with a rapid rate of growth can soon overtake a store with a more modest rate of growth.

Lesson 13-3
Quadratic and Exponential Regressions

Learning Targets:
- Use technology to perform quadratic and exponential regressions, and then make predictions.
- Compare and contrast linear, quadratic, and exponential regressions.

SUGGESTED LEARNING STRATEGIES: Look for a Pattern, Create Representations, Quickwrite, Think-Pair-Share, Discussion Groups

Online shopping has experienced tremendous growth since the year 2000. One way to measure the growth is to track the average number of daily hits at the websites of online stores. The tables show the average number of daily hits for three different online stores in various years since the year 2000.

Nile River Retail					
Years Since 2000	0	2	4	6	8
Daily Hits (thousands)	52.1	56.2	60.0	64.1	68.0

eBuy					
Years Since 2000	0	2	4	6	8
Daily Hits (thousands)	1.0	4.9	17.2	37.2	64.9

Spendco					
Years Since 2000	0	2	4	6	8
Daily Hits (thousands)	2.0	6.6	20.9	68.1	220.4

1. **Make sense of problems.** Compare and contrast the growth of the three online stores based on the data in the tables.

   Answers will vary. Nile River Retail started with the greatest number of daily hits in 2000 and its growth was steady, but the other online stores eventually caught up or overtook it. eBuy started with the least number of daily hits, but it experienced rapid growth. Spendco has the fastest growth of the three stores.

Lesson 13-3
Quadratic and Exponential Regressions

2. Plot the data for the three online stores on the graphs below.

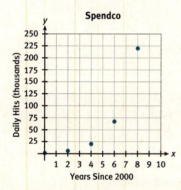

ACTIVITY 13 Continued

3–5 Look for a Pattern, Create Representations, Think-Pair-Share In these items, students identify the graph that best corresponds to a linear model. Then they determine the line of best fit and use it to make a prediction. Monitor group discussions carefully to be sure students have correctly matched Nile River Retail with a linear model before they move on to Items 4 and 5.

TEACHER to TEACHER

When students find regression equations in this lesson, the variable y represents the number of daily hits in thousands. Remind students to consider the units of y when using the regression equations to make predictions. Some students may have difficulty interpreting a value such as 81.99 thousand. Have them discuss how they could write the value as a whole number. One way is as follows: 81.99 thousand = 81.99 × 1000 = 81,990.

Developing Math Language

This lesson introduces the vocabulary terms *quadratic function*, *quadratic regression*, *exponential function*, and *exponential regression*. Students can use their prior knowledge of the terms *linear function* and *linear regression* to help them understand the meaning of the new terms. Have students add the words to their math notebooks, and write the words on your class Word Wall. Be sure students understand that quadratic functions and exponential functions are not the only types of nonlinear functions; there are other types that students will learn about later.

6 Create Representations, Think-Pair-Share In this item, students use technology to perform a quadratic regression and then use the regression equation to make a prediction. You may want to have students graph the quadratic equation on their scatter plot for eBuy. Then have students discuss why a quadratic model appears to be appropriate in this situation.

ACTIVITY 13 continued

My Notes

MATH TERMS

A **quadratic function** is a nonlinear function that can be written in the form $y = ax^2 + bx + c$, where $a \neq 0$. You will study quadratic functions in more detail later in this book.

Lesson 13-3
Quadratic and Exponential Regressions

3. Which online store's growth could best be modeled by a linear function? Explain.
   Nile River Retail; the points in the graph are close to lying on a line.

4. For the online store you identified in Item 3, determine the equation of the line of best fit. Round values to the nearest hundredth.
   y = 1.99x + 52.14

5. Predict the number of daily hits for this online store in 2015.
   81.99 thousand or 81,990

When a line does not appear to be a good fit for a set of data, you may want to model the data using a nonlinear model.

Quadratic regression is a method used to find a **quadratic function** that models a set of data. You can perform quadratic regression using a graphing calculator.

6. Enter the data for eBuy into your graphing calculator. Enter the years since 2000 as the x-values and the corresponding daily hits in thousands as the y-values.
   a. To find the quadratic equation that models the data, use the quadratic regression feature of your calculator.

   The calculator should return values for a, b, and c. Write these values below, rounding to the nearest hundredth.

   $a =$ **0.99**

   $b =$ **0.11**

   $c =$ **0.92**

   b. Write the quadratic equation in the form $y = ax^2 + bx + c$.
      y = 0.99x² + 0.11x + 0.92

   c. Use the quadratic equation to predict the number of daily hits for eBuy in 2015.
      225.32 thousand or 225,320

---

202 SpringBoard® Mathematics **Algebra 1, Unit 2** • Functions

Lesson 13-3
Quadratic and Exponential Regressions

When a set of data shows very rapid growth or decay, an exponential model may be the best choice for modeling the data.

Exponential regression is a method used to find an **exponential function** that models a set of data. You can perform exponential regression using a graphing calculator.

7. Enter the data for Spendco into your graphing calculator. Enter the years since 2000 as the x-values and the corresponding daily hits in thousands as the y-values.

   a. To find the exponential equation that models the data, use the exponential regression feature of your calculator.

   The calculator should return values for a and b. Write these values below, rounding to the nearest hundredth.

   $a =$ **2.01**

   $b =$ **1.80**

   b. Write the exponential equation in the form $y = ab^x$.

   $y = 2.01(1.8)^x$

   c. Use the exponential equation to predict the number of daily hits for Spendco in 2015.

   13,560.75 thousand or 13,560,750

8. **Construct viable arguments.** Based on your predictions for the number of daily hits for each online store in 2015, which type of function has the fastest growth: linear, quadratic, or exponential? Explain.

   Exponential; the exponential function predicts, by far, the greatest number of daily hits in 2015.

MATH TERMS

An **exponential function** is a nonlinear function that can be written in the form $y = ab^x$. You will study exponential functions in more detail later in this book.

ACTIVITY 13 Continued

Check Your Understanding

Debrief students' answers to these items to ensure that they understand concepts related to quadratic and exponential regressions. Have students describe the type of data set they would model with an exponential function rather than a linear or quadratic function.

Answers

9. Sample answer: All three are similar in that you develop an equation to model a set of data. Quadratic and exponential regressions are different from linear regression in that you fit a curve to the set of data rather than a straight line.
10. No; because of the very rapid growth, the model is probably only realistic for a short period of time. At some point in the future, the number of daily hits would level off or begin to decrease.

ASSESS

Students' answers to lesson practice problems will provide you with a formative assessment of their understanding of the lesson concepts and their ability to apply their learning.

See the Activity Practice for additional problems for this lesson. You may assign the problems here or use them as a culmination for the activity.

ADAPT

Check students' answers to the Lesson Practice to ensure that they understand how to write, interpret, and apply linear, quadratic, and exponential models for data sets. If students need more help, give the following information to help them interpret the regression results from their calculators.

Linear model: $y = ax + b$
Quadratic model: $y = ax^2 + bx + c$
Exponential model: $y = ab^x$

Also encourage students to graph the given data along with their function models to confirm that their models are reasonable.

ACTIVITY 13 continued

Lesson 13-3
Quadratic and Exponential Regressions

Check Your Understanding

9. How are quadratic regression and exponential regression similar to and different from linear regression?
10. Do you think the exponential model would be appropriate for predicting the number of daily hits for Spendco in any future year? Explain your reasoning.

LESSON 13-3 PRACTICE

The population of Williston, North Dakota, has grown rapidly over the past decade due to an oil boom. The table gives the population of the town in 2007, 2009, and 2011.

Years Since 2000	7	9	11
Population (thousands)	12.4	13.0	16.0

11. Use your calculator to find the equation of the line of best fit for the data.
12. **Reason quantitatively.** What is the slope of the line? What does it tell you about the population growth of the town?
13. Use your calculator to find a quadratic equation that models the growth of the town.
14. Use your quadratic equation to predict the population of Williston in 2020.
15. Use your calculator to find an exponential equation that models the growth of the town.
16. Use your exponential equation to predict the population of Williston in 2020.
17. According to the exponential model, in what year will the town have a population greater than 40,000 for the first time? (*Hint*: Use the table feature of your calculator.) What assumptions do you make when you use the exponential model to answer this question?

LESSON 13-3 PRACTICE
11. $y = 0.9x + 5.7$
12. 0.9; Each year, the population grows by about 0.9 thousand (or 900).
13. $y = 0.3x^2 - 4.5x + 29.2$
14. 59.2 thousand or 59,200
15. $y = 7.73(1.07)^x$
16. 29.9 thousand or 29,900
17. 2025; This assumes that the population will continue to grow at the same rate for many years into the future.

Equations From Data
Pass the Book

ACTIVITY 13 PRACTICE
Write your answers on notebook paper. Show your work.

Lesson 13-1

The scatter plot shows the relationship between the day of the month and a frozen yogurt stand's daily profit during the month of the July.

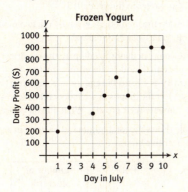

1. Are the data linear? Explain.
2. Draw a trend line on the scatter plot and name two points that your trend line passes through.
3. Write the equation of the trend line you drew in Item 2.
4. What do the variables in your equation represent?
5. What is the slope of the trend line? What does this tell you?
6. Use your trend line to predict the yogurt stand's daily profit on July 20.
7. The owner of a competing frozen yogurt stand finds that her daily profit each day in July is exactly $100 more than that of the stand in the scatter plot. Write the equation of a trend line for the competing stand.

8. The manager of a local history museum experiments with different prices for admission to the museum. For each price, the manager notes the number of visitors who enter the museum on that day. The table shows the data.

Price	$2.75	$3.50	$4.25	$5.75
Number of Daily Visitors	112	88	66	63

Which is a true statement about the data?
A. A trend line on the scatter plot has a positive slope.
B. The y-intercept of the trend line is above the x-axis.
C. The trend line predicts at least 70 visitors when the admission price is $6.25.
D. The trend line fits the data perfectly because the data is linear.

Lesson 13-2

Use your calculator to perform a linear regression for the following data. Use your linear regression for Items 9–12.

$(-6, -3), (-8, -4), (-2, 1), (1, 4), (3, 6), (5, 8), (7, 13)$

9. What is the equation of the line of best fit?
10. What is the value of the slope? What does this tell you about the relationship between x and y?
11. According to your model, what is the value of y when $x = -19$?
12. According to your model, for what value of x is $y = 100$?

ACTIVITY 13 Continued

ACTIVITY PRACTICE
1. No; the points plotted on the scatter plot do not lie on a straight line.
2. Sample answer: (2, 400) and (8, 700)
3. Sample answer: $y = 50x + 300$
4. $x =$ day in July; $y =$ daily profit in dollars
5. Sample answer: 50; For each additional day that goes by, the daily profit increases by about $50.
6. Sample answer: $1300
7. Sample answer: $y = 50x + 400$
8. B
9. $y = 1.07x + 3.57$
10. 1.07; As x increases by 1, y increases by 1.07.
11. -16.76
12. approximately 90.12

ACTIVITY 13 Continued

13. positive correlation
14. D
15. a. Yes; there is a positive correlation since the number of quizzes increases as the height of the plant increases.
    b. No; an increase in the height of the plant does not cause an increase in the number of quizzes, or vice versa.

16.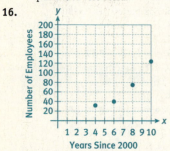

17. Although a linear model could be used, the points do not lie along a straight line. Because of the rapid rate of increase in the number of employees, a quadratic or exponential model might be better.
18. $y = 2.56x^2 - 20.33x + 71.65$
19. approximately 343
20. $y = 11.37(1.26)^x$
21. approximately 364
22. The two predictions are close.
23. Sample answer: The two models give similar predictions at first, but after $x = 20$, the exponential model grows much more quickly than the quadratic model.
24. $y = 12^x$
25. 5,159,780,352
26. Yes; if the equation shows that the slope is positive, then there is a positive correlation; if the equation shows that the slope is negative, then there is a negative correlation.

ADDITIONAL PRACTICE

If students need more practice on the concepts in this activity, see the eBook Teacher Resources for additional practice problems.

ACTIVITY 13 continued

Equations From Data
Pass the Book

13. Look at the scatter plot on the previous page showing the daily profits of a frozen yogurt stand. What type of correlation, if any, does the scatter plot show?

14. Which of the following pairs of variables are likely to show a negative correlation?
    A. the length of a shoe; the size of the shoe
    B. the number of miles on a car's odometer; the age of the car
    C. the weight of a watermelon; the price of the watermelon
    D. the number of minutes you have waited for a bus; the number of minutes remaining until the bus arrives

15. At several times during the school year, Emilio collected data on the height of a plant in the classroom and the total number of quizzes he had taken so far in his science class. The data are shown below.

Height of Plant (cm)	16	19	22	26
Total Number of Quizzes	4	6	7	9

   a. Is there a correlation between the variables? Explain.
   b. Is there causation between the variables? Explain.

Lesson 13-3
The table shows the number of employees at a software company in various years.

Years Since 2000	4	6	8	10
Number of Employees	32	40	75	124

16. Make a scatter plot of the data.
17. Do you think a linear equation would be a good model for the data? Justify your answer.
18. Use your calculator to find a quadratic equation that models the growth of the company.
19. Use the quadratic model to predict the number of employees in the year 2015.
20. Use your calculator to find an exponential equation that models the growth of the company.
21. Use the exponential model to predict the number of employees in the year 2015.
22. How do the predictions given by the two models in Items 19 and 21 compare?
23. Use your calculator to compare the quadratic and exponential models. Enter the equation from Item 19 as Y_1 and the equation from Item 21 as Y_2. View the graphs in a window that allows you to compare their growth. What do you notice?

The table shows the total number of bacteria in a sample over five hours.

Hour	Number of Bacteria
1	12
2	144
3	1728
4	20,736
5	248,832

24. Use your calculator to find an exponential equation that models the bacteria data.
25. If this trend continues, how many bacteria will be growing in the sample after 9 hours?

MATHEMATICAL PRACTICES
Look for and Make Use of Structure

26. Is it possible to tell from the equation of a line of best fit whether there is a positive or negative correlation between two variables? If so, explain how. If not, explain why not.

Linear Models and Slope as Rate of Change
A 10K RUN

Embedded Assessment 3
Use after Activity 13

> **CONNECT TO METRIC MEASUREMENT**
>
> A "10K Run" means that the length of the course for the foot race is 10 kilometers, or 10,000 meters.
>
> $10K = 10 \text{ km} \times \dfrac{1000 \text{ m}}{1 \text{ km}} = 10{,}000 \text{ m}$

Jim was serving as a finish-line judge for the Striders 10K Run. He was interested in finding out how three of his friends were doing out on the course. He was able to get the following data from racing officials.

Runner: J. Matuba

Time (min)	4	5	7	12	20
Distance (m)	1090	1380	2040	3640	6300

Runner: E. Rodriguez

Time (min)	1	6	10	18	25
Distance (m)	500	2000	3280	5510	7700

Runner: T. Donovan

Time (min)	2	4	9	15	20
Distance (m)	620	1250	2900	4690	6250

Answer Items 1–3 below, based on the information Jim received about his three running friends. Use x as the number of minutes elapsed since the race began and y as the number of meters completed.

1. Make a scatter plot showing the data for each runner.
2. Perform a linear regression to find the equation of the line of best fit for each runner. Round values in the equations to the nearest tenth.
3. Explain the order in which the runners will finish the race based on the models you formed using the data.

Answer the following questions for the linear models you formed. Explain your answers.

4. What is the standard form of the linear model for Matuba?
5. What is the domain of the linear model for Rodriguez?
6. What is the slope of the linear model for Donovan? What is its significance in the context of the problem situation?

Common Core State Standards for Embedded Assessment 3

HSF-IF.C.7	Graph functions expressed symbolically and show key features of the graph, by hand in simple cases and using technology for more complicated cases.
HSF-LE.A.2	Construct linear and exponential functions, including arithmetic and geometric sequences, given a graph, a description of a relationship, or two input-output pairs (include reading these from a table).
HSF-LE.B.5	Interpret the parameters in a linear or exponential function in terms of a context.

Embedded Assessment 3

Assessment Focus
- Scatter plots
- Linear regression
- Line of best fit
- Slope and domain
- Comparing data

Materials
- calculator
- graph paper

Answer Key

1.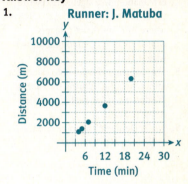
    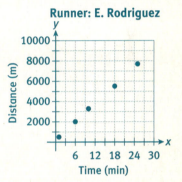
    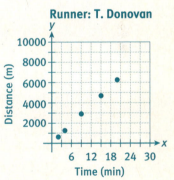

2. Matuba: $y = 326.2x - 241.6$
   Rodriguez: $y = 298.1x + 221.0$
   Donovan: $y = 312.3x + 18.5$

3. The order of finish will be Matuba, Donovan, Rodriguez. To determine this, let $y = 10{,}000$ in each of the linear equations and solve for x to find the finish time. For Matuba, the finish time is 31.4 minutes; for Donovan, the finish time is 32.0 minutes; for Rodriguez, the finish time is 32.8 minutes.

Unit 2 • Functions 207

Embedded Assessment 3

TEACHER to TEACHER

You may wish to read through the scoring guide with students and discuss the differences in the expectations at each level. Check that students understand the terms used.

Answer Key (cont.)

4. $1631x - 5y = 1208$; Sample explanation: Use algebra to rewrite the equation in the form $Ax + By = C$, where A, B and C are integers whose greatest common factor is 1, with $A \geq 0$.

5. The domain is all possible values for minutes elapsed during the race, so the domain is {all real x, such that $0 \leq x \leq 32.8$}.

6. 312.3; Donovan's average rate was 312.3 meters per minute.

Embedded Assessment 3
Use after Activity 13

Linear Models and Slope as Rate of Change
A 10K RUN

Scoring Guide	Exemplary	Proficient	Emerging	Incomplete
	The solution demonstrates the following characteristics:			
Mathematics Knowledge and Thinking (Items 1, 2, 4–6)	• Clear and accurate understanding of scatter plots, linear regression, standard form of a linear model, domain, and slope	• Largely correct understanding of scatter plots, linear regression, standard form of a linear model, domain, and slope	• Partial understanding of scatter plots, linear regression, standard form of a linear model, domain, and slope	• Inaccurate or incomplete understanding of scatter plots, linear regression, standard form of a linear model, domain, and slope
Problem Solving (Item 3)	• Appropriate and efficient strategy that results in a correct answer	• Strategy that may include unnecessary steps but results in a correct answer	• Strategy that results in a partially incorrect answer	• No clear strategy when solving problems
Mathematical Modeling / Representations (Items 1, 2, 4, 6)	• Clear and accurate scatter plot • Fluency in fitting a linear model to real-world data, including how to interpret and draw accurate conclusions from the model	• Largely correct scatter plot • Adequate understanding of how to fit a linear model to real-world data, including how to interpret and draw accurate conclusions from the model	• Partially correct scatter plot • Partial understanding of how to fit a linear model to real-world data, including how to interpret and draw accurate conclusions from the model	• Inaccurate or incomplete scatter plot • Little or no understanding of how to fit a linear model to real-world data, including how to interpret and draw accurate conclusions from the model
Reasoning and Communication (Items 3–6)	• Precise use of appropriate math terms and language to explain the order in which the runners will finish, including justification based on the model • Clear and accurate descriptions of how to find the standard form, identify a reasonable domain, and identify and interpret the slope of a linear model	• Adequate explanation and justification of the order in which the runners will finish • Largely correct description of how to find the standard form, identify a reasonable domain, and identify and interpret the slope of a linear model	• Misleading or confusing explanation and justification of the order in which the runners will finish • Partially correct description of how to find the standard form, identify a reasonable domain, and identify and interpret the slope of a linear model	• Incomplete or inaccurate explanation and justification of the order in which the runners will finish • Incorrect or incomplete description of how to find the standard form, identify a reasonable domain, and identify and interpret the slope of a linear model

Unit 3 Planning the Unit

In this unit, students continue their study of linear concepts by learning about piecewise-defined linear functions, linear inequalities with one and two variables, and systems of linear equations and inequalities.

Vocabulary Development

The key terms for this unit can be found on the Unit Opener page. These terms are divided into Academic Vocabulary and Math Terms. Academic Vocabulary includes terms that have additional meaning outside of math. These terms are listed separately to help students transition from their current understanding of a term to its meaning as a mathematics term. To help students learn new vocabulary:

- Have students discuss meaning and use graphic organizers to record their understanding of new words.
- Remind students to place their graphic organizers in their math notebooks and revisit their notes as their understanding of vocabulary grows.
- As needed, pronounce new words and place pronunciation guides and definitions on the class Word Wall.

Embedded Assessments

Embedded Assessments allow students to do the following:

- Demonstrate their understanding of new concepts.
- Integrate previous and new knowledge by solving real-world problems presented in new settings.

They also provide formative information to help you adjust instruction to meet your students' learning needs.

Prior to beginning instruction, have students unpack the first Embedded Assessment in the unit to identify the skills and knowledge necessary for successful completion of that assessment. Help students create a visual display of the unpacked assessment and post it in your class. As students learn new knowledge and skills, remind them that they will be expected to apply that knowledge to the assessment. After students complete each Embedded Assessment, turn to the next one in the unit and repeat the process of unpacking that assessment with students.

CollegeBoard

AP / College Readiness

Unit 3 continues to develop the algebra and graphing of functions and extends student understanding of the properties and language of functions by:

- Providing contextual situations where piecewise functions, and systems of equations and inequalities, can be applied.
- Giving students opportunities to work with functions in a variety of ways: graphical, numerical, analytical, verbal.
- Allowing students to communicate mathematics and explain solutions.

Unpacking the Embedded Assessments

The following are the key skills and knowledge students will need to know for each assessment.

Embedded Assessment 1

Graphing Inequalities and Piecewise-Defined Functions, *Earnings on a Graph*

- Write, solve, and graph linear inequalities
- Use function notation
- Determine a reasonable domain and range
- Define and graph piecewise-defined functions

Embedded Assessment 2

Systems of Equations and Inequalities, *Tilt the Scales*

- Write, solve, and graph systems of linear equations and inequalities
- Interpret the intersection point of two linear equations in a context
- Represent constraints with equations and inequalities

Planning the Unit continued

Suggested Pacing

The following table provides suggestions for pacing using a 45-minute class period. Space is left for you to write your own pacing guidelines based on your experiences in using the materials.

	45-Minute Period	Your Comments on Pacing
Unit Overview/Getting Ready	1	
Activity 14	4	
Activity 15	3	
Activity 16	2	
Embedded Assessment 1	1	
Activity 17	5	
Activity 18	2	
Embedded Assessment 2	1	
Total 45-Minute Periods	**19**	

Additional Resources

Additional resources that you may find helpful for your instruction include the following, which may be found in the eBook Teacher Resources.

- Unit Practice (additional problems for each activity)
- Getting Ready Practice (additional lessons and practice problems for the prerequisite skills)
- Mini-Lessons (instructional support for concepts related to lesson content)

Extensions of Linear Concepts

3

Unit Overview
In this unit, you will extend your study of linear concepts to the study of piecewise-defined functions and systems of linear equations and inequalities. You will learn to solve systems of equations and inequalities in a variety of ways.

Key Terms
As you study this unit, add these and other terms to your math notebook. Include in your notes your prior knowledge of each word, as well as your experiences in using the word in different mathematical examples. If needed, ask for help in pronouncing new words and add information on pronunciation to your math notebook. It is important that you learn new terms and use them correctly in your class discussions and in your problem solutions.

Math Terms
- piecewise-defined function
- linear inequality
- solutions of a linear inequality
- boundary line
- half-plane
- closed half-plane
- open half-plane
- system of linear equations
- substitution method
- elimination method
- parallel
- coincident
- independent
- dependent
- inconsistent
- consistent
- system of linear inequalities
- solution region

ESSENTIAL QUESTIONS

 Why would you use multiple representations of linear equations and inequalities?

 How are systems of linear equations and systems of linear inequalities useful in analyzing real-world situations?

EMBEDDED ASSESSMENTS
These assessments, following Activities 16 and 18, will give you an opportunity to demonstrate what you have learned about piecewise-defined functions, inequalities, and systems of equations and inequalities.

Embedded Assessment 1:
Graphing Inequalities and Piecewise-Defined Functions   p. 249

Embedded Assessment 2:
Systems of Equations and Inequalities   p. 283

Unit Overview
Ask students to read the unit overview and mark the text to identify key phrases that indicate what they will learn in this unit.

Materials
- colored pencils
- graph paper

Key Terms
Read through the vocabulary list with students. Assess prior knowledge by asking students to identify and give a description of any terms they feel they are already familiar with. As students encounter new terms in this unit, help them to choose an appropriate graphic organizer for their word study. As they complete a graphic organizer, have them place it in their math notebooks and revisit as needed as they gain additional knowledge about each word or concept.

Essential Questions
Read the essential questions with students, and ask them to share possible answers. As students complete the unit, revisit the essential questions to help them revise their initial responses as needed.

Unpacking Embedded Assessments
Prior to beginning the first activity in this unit, turn to Embedded Assessment 1 and have students unpack the assessment by identifying the skills and knowledge they will need to complete the assessment successfully. Guide students through a close reading of the assessment, and use a graphic organizer or other means to capture their identification of the skills and knowledge. Repeat the process for each Embedded Assessment in the unit.

Developing Math Language
As this unit progresses, help students make the transition from general words they may already know to the meanings of those words in mathematics. You may want students to work in pairs or small groups to facilitate discussion and to build confidence and fluency as they internalize new language. Ask students to discuss new mathematics terms as they are introduced, identifying meaning as well as pronunciation and common usage. Remind students to use their math notebooks to record their understanding of new terms and concepts.

As needed, pronounce new terms clearly and monitor students' use of words in their discussions to ensure that they are using terms correctly. Encourage students to practice fluency with new words as they gain greater understanding of mathematical and other terms.

UNIT 3
Getting Ready

Use some or all of these exercises for formative evaluation of students' readiness for Unit 3 topics.

Prerequisite Skills
- Identify linear functions from tables (Item 1) HSF-LE.A.1b
- Write a linear equation from a table (Item 2) HSF-LE.A.2
- Represent linear relationships using tables, equations, and graphs (Item 3) HSF-IF.C.7a, HSF-LE.A.2
- Graph a linear equation in two variables (Item 4) HSF-IF.C.7a
- Graph horizontal lines (Item 5) HSF-IF.C.7a
- Identify solutions of linear inequalities in two variables (Item 6) HSA-REI.D.12
- Graph compound inequalities (Item 7) 6.EE.B.8
- Identify functions with a constant rate of change (Item 8) HSF-LE.A.1b

Answer Key
1. B
2. $y = 2x - 1$
3. table:

m	0	1	2	3	4
A	100	125	150	175	200

algebraic: $A = 25m + 100$ graphical:

4.

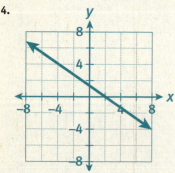

5. a horizontal line 3 units above the x-axis
6. A
7. Both graphs have an open circle at -1 and a closed circle at 3. On the graph of $-1 < x \leq 3$, there is a line segment between -1 and 3. On the graph of $x < -1$ or $x \geq 3$, there is a ray to the left from -1 and to the right from 3.
8. B

UNIT 3
Getting Ready

Write your answers on notebook paper. Show your work.

1. Which of the following tables of values represents linear data?

   A.
x	y
2	3
4	6
6	6
8	3

   B.
x	y
1	1
2	4
3	7
4	10

   C.
x	y
2	3
3	4
4	3
6	4

   D.
x	y
1	3
2	6
3	12
4	24

2. Give an algebraic representation of these data.

x	1	4	7	10
y	1	7	13	19

3. You open a savings account with a deposit of $100. In each month after your initial deposit, you add $25 to your account. Provide a table to display the total amount of money in the account after 1, 2, 3, and 4 months. Give a graphical and an algebraic representation that would allow you to determine the amount of money A in dollars you have deposited after m months.

4. Graph $2x + 3y = 4$.
5. Describe the graph of $y = 3$.
6. Which ordered pair is a solution of $y > x + 5$?
   A. (2, 8) B. (−5, 0)
   C. (1, 6) D. (0, −5)
7. Compare and contrast the graphs of these two compound statements:
   $-1 < x \leq 3$
   $x < -1$ or $x \geq 3$
8. Which of the following represents a function with a constant rate of change?

   A.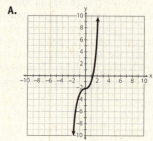

   B.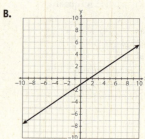

   C. $y = \dfrac{4}{x}$

   D.
x	1	4	16	64
y	5	10	15	20

Getting Ready Practice
For students who may need additional instruction on one or more of the prerequisite skills for this unit, Getting Ready practice pages are available in the Teacher Resources at SpringBoard Digital. These practice pages include worked-out examples as well as multiple opportunities for students to apply concepts learned.

Piecewise-Defined Linear Functions
Breakfast for Bowser
Lesson 14-1 Function Notation and Rate of Change

ACTIVITY 14

Learning Targets:
- Use function notation and interpret statements that use function notation in terms of a context.
- Calculate the rate of change of a linear function presented in multiple representations.

SUGGESTED LEARNING STRATEGIES: Summarizing, Look for a Pattern, Think-Pair-Share, Create Representations, Construct an Argument

Miriam has accepted a job at a veterinarian's office. Her first assignment is to feed the dogs that are housed there. The bag of dog food was already torn open, and she found only part of the label that described the amount of dog food to feed each dog.

Barko Dog Food Feeding Chart

Weight of Dog (pounds)	Daily Amount of Barko (ounces)
3	3
12	12
20	20

Each dog has an information card that gives the dog's name and weight. Using this information, Miriam fed each dog the amount of food she thought was appropriate.

1. How many ounces of dog food per day do you think Miriam gave each dog? Complete the table below.

Dog	Dog's Weight (pounds)	Amount of Dog Food (ounces)
Muffy	4	4
Trixie	14	14
Rags	6	6
Hercules	68	68

2. Based on the dog's weight and the partial label at the top of the page, write a verbal rule that Miriam could use to determine the ounces of dog food per day to feed each dog.

   The number of ounces of food a dog should receive is equal to the weight of the dog in pounds.

Common Core State Standards for Activity 14

HSF-IF.A.2:	Use function notation, evaluate functions from inputs in their domains, and interpret statements that use function notation in terms of a context.
HSF-IF.B.6:	Calculate and interpret the average rate of change of a function (presented symbolically or as a table) over a specified interval. Estimate the rate of change from a graph.
HSF-IF.C.7:	Graph functions expressed symbolically and show key features of the graph, by hand in simple cases and using technology for more complicated cases.
HSF-IF.C.7a:	Graph linear and quadratic functions and show intercepts, maxima and minima.
HSF-IF.C.7b:	Graph square root, cube root, and piecewise-defined functions, including step functions and absolute value functions.
HSF-IF.C.9:	Compare properties of two functions each represented in a different way (algebraically, graphically, numerically in tables, or by verbal descriptions).

ACTIVITY 14
Investigative

Activity Standards Focus
In this activity, students expand their study of functions in general, and linear functions specifically, by interpreting, writing, and graphing piecewise-defined functions.

Lesson 14-1

PLAN

Pacing: 1 class period
Chunking the Lesson
#1–3   #4   #5   #6–8
Check Your Understanding
Lesson Practice

TEACH

Bell-Ringer Activity
Ask students to graph the equation $y = 2x + 4$ over the restricted domain $0 \leq x \leq 5$, and describe the similarities and differences between their graph and the graph of $y = 2x + 4$ with an unrestricted domain.

1–3 Shared Reading, Summarizing, Look for a Pattern, Think-Pair-Share, Create Representations In this set of items, students identify a pattern in a table of data, write a verbal rule that describes the pattern, then write the rule using function notation. Most students should see that the numbers in the first column equal the numbers in the second column. In Item 3, some students may feel uncomfortable writing $A(w) = w$ because it seems to be stating that ounces equals pounds. It is important for students to understand that the equation is describing a relationship in which the number of ounces of food a dog should be fed is a function of the weight of the dog in pounds.

Activity 14 • Piecewise-Defined Linear Functions 211

ACTIVITY 14 Continued

4 Construct an Argument, Sharing and Responding In Item 4, students are presented with a "complete feeding chart" which includes new data that do not fit the pattern students previously derived. Give students an opportunity to suggest a variety of reasons why Miriam's assumption is not correct.

TEACHER to TEACHER

The data for the recommended daily feeding amounts for dogs were based on information from actual dog food bags. For the charts in this activity, the data have been slightly recast for the benefit of easier analysis and understanding by students, and the unit of measure of the amount of food has been changed from cups to ounces.

ACTIVITY 14 continued

My Notes

DISCUSSION GROUP TIPS

As you discuss ideas for the function of the scenario, make notes and listen to what your group members have to contribute. Ask and answer questions to clearly aid comprehension and to ensure understanding of all group members' ideas.

Lesson 14-1
Function Notation and Rate of Change

3. **Model with mathematics.** Let w be a dog's weight in pounds and let $A(w)$ be the amount of food in ounces that the dog should be fed each day. Use the table in Item 1 and function notation to write a function that expresses $A(w)$ in terms of w.
   $A(w) = w$

After several hours, all of the dogs had finished their food except for Hercules. The vet asked, "How much food did you give Hercules?"

After hearing Miriam's answer, the vet told Miriam that she had overfed Hercules and suggested that Miriam check the complete feeding chart posted on the office wall.

Barko Dog Food Feeding Chart

Weight of Dog (pounds)	Daily Amount of Barko (ounces)
3	3
12	12
20	20
50	35
100	60
over 100 pounds	60 ounces plus 1 ounce for each additional 10 pounds of weight

4. Miriam had assumed that each dog should be fed as many ounces of dog food as the dog weighed in pounds. Explain why the data in the new chart indicate that her original assumption was not correct.
   Answers may vary. Each dog, with the exception of Hercules, received the correct amount of food. Hercules received more food than the amount recommended for a dog weighing 100 pounds.

Adult dogs that weigh less than 20 pounds are classified as *small* dogs. Dogs that weigh 20 to 100 pounds are classified as *mid-size* dogs. Finally, dogs that weigh more than 100 pounds are classified as *large* dogs. Miriam had not known that the formulas for feeding small, mid-size, and large dogs are all different.

Lesson 14-1
Function Notation and Rate of Change

5. The vet tells Miriam that the data for feeding mid-size dogs are linear. Verify the vet's statement using the feeding chart.
   Both rates of change, $\left(\frac{35-20}{50-20}\right)$ and $\left(\frac{60-35}{100-50}\right)$, are equal to $\frac{1}{2}$.

6. Determine a rate of change that describes the number of additional ounces of food a mid-size dog should be fed for each pound of weight greater than 20. Explain how you found your answer.
   For weights greater than 20 pounds, the rate of change can be expressed as $\frac{1}{2}$ ounce per pound, or 1 ounce of food for every 2 pounds of weight.

7. Using the rate of change you found in Item 6 and the feeding chart, use function notation to write a function that expresses $A(w)$, the ounces of dog food, in terms of w, the dog's weight in pounds, for mid-size dogs.
   $A(w) = 20 + 0.5(w - 20)$ or $A(w) = 0.5w + 10$

8. How many ounces was Hercules overfed? Justify your response.
   Hercules was fed 68 ounces when he should have been fed $0.5(68) + 10 = 44$ ounces; thus, he was overfed by 24 ounces $(68 - 44 = 24)$.

Check Your Understanding

Use the table for Items 9 and 10.

Grapes (pounds)	Cost (dollars)
0.5	$1.20
1.5	$3.60
2.0	$4.80

9. Determine the rate of change of the data. Express the rate as cost per pound.
10. Do the data in the table represent a linear relationship? Explain.
11. Tom has read the first 40 pages of a book. He plans to read another 12 pages each day. Use function notation to write a function that expresses the number of pages read p after d days.

ACTIVITY 14 Continued

ASSESS

Students' answers to Lesson Practice problems will provide you with a formative assessment of their understanding of the lesson concepts and their ability to apply their learning.

See the Activity Practice for additional problems for this lesson. You may assign the problems here or use them as a culmination for the activity.

LESSON 14-1 PRACTICE
12. 50 miles per gallon
13. $100 per month
14. $9,181.82 per year
15. The total cost increases by $1200 per vehicle.
16. $V(t) = 500 - 20t$

ADAPT

Check student responses to the Lesson Practice to ensure that students are able to determine the rate of change when the data are presented verbally as well as in a table. If students are having difficulty determining rates of change from verbal descriptions, encourage them to create their own tables to represent the information in verbal descriptions such as those in Items 12 and 14. At this point, students should be able to determine a rate of change for a function represented verbally, by data in a table, or by an equation written in function notation.

ACTIVITY 14 continued

My Notes

Lesson 14-1
Function Notation and Rate of Change

LESSON 14-1 PRACTICE

12. Todd had five gallons of gasoline in his motorbike. After driving 100 miles, he has three gallons left. What is Todd's rate of change in miles per gallon?

13. This table shows how the balance in John's savings account has changed over the course of a year.

Month	Balance (dollars)
1	$ 400
3	$ 600
7	$1000
10	$1300
12	$1500

    How much did John save per month during the year?

14. Linda purchased a house for $144,000. Thinking of possibly refinancing after 11 years, she had her house appraised and found that it is now worth $245,000. Find the rate of change of the value of the house in dollars per year.

15. **Make sense of problems.** The cost in dollars of producing x vehicles for a company is given by $C(x) = 1200x + 5500$. Interpret the rate of change of this linear function.

16. A 500-liter tank full of oil is being drained at the constant rate of 20 liters per minute. Use function notation to write a linear function expressing the number of liters in the tank V after t minutes.

Lesson 14-2
Writing Functions and Finding Domain and Range

ACTIVITY 14
continued

Learning Targets:
- Write linear equations in two variables given a table of values, a graph, or a verbal description.
- Determine the domain and range of a linear function, determine their reasonableness, and represent them using inequalities.

SUGGESTED LEARNING STRATEGIES: Identify a Subtask, Think-Pair-Share, Create Representations, Construct an Argument, Interactive Word Wall

1. Rewrite the function that you wrote in Item 7 in Lesson 14-1.
   $A(w) = 20 + 0.5(w - 20)$ or $A(w) = 0.5w + 10$

2. a. Remember that this function is true only for mid-size dogs. Describe the appropriate input values for the function.
   Input values from 20 pounds up to 100 pounds are appropriate.

   b. Use your description of the appropriate input values to express the domain of the function using set notation.
   domain = { w: __$20 \leq w \leq 100$__ }

Miriam knows that she also has several large dogs to feed. When she looks at the chart, she reads the instruction "60 ounces plus 1 ounce for each additional 10 pounds of weight."

3. How much additional dog food should large dogs be fed for each pound of weight greater than 100 pounds?
   Large dogs should receive 0.1 ounce of dog food for each additional pound of weight over 100 pounds.

4. **Critique the reasoning of others.** A 140-pound Great Dane has arrived for a short stay at the kennel. Miriam says that the dog should be fed 64 ounces of food per day. Her friend Chase says that the dog requires 100 ounces of food per day. Who is correct? Justify your response.
   Miriam; because the Great Dane weighs 40 pounds more than the 100 pound minimum for large dogs, it should receive 4 additional ounces of food, or $60 + 4 = 64$ ounces altogether.

My Notes

ACTIVITY 14 Continued

Lesson 14-2

PLAN

Pacing: 1 class period
Chunking the Lesson
#1–2 #3–5 #6–8 #9
Check Your Understanding
Lesson Practice

TEACH

Bell-Ringer Activity
Ask students to recall the functions they wrote in the previous lesson for small and mid-size dogs. Ask them to give appropriate input values for each and to describe ways to determine appropriate output values based on the problem situation.

1–2 Create Representations Some students will find it helpful if you review the fact that w is the weight of a dog in pounds and $A(w)$ is the amount of food in ounces. Be sure students understand that this function is only true for $20 \leq w \leq 100$.

3–5 Shared Reading, Identify a Subtask, Think-Pair-Share, Create Representations Ask a student to reread the paragraph at the bottom of page 212 defining the weights of small, mid-size, and large dogs. Have a copy of the table preceding Item 4 in Lesson 14-1 on display for students to reference throughout the lesson. If students are having difficulty writing the feeding function for large dogs, encourage them to create a table showing the relationships between several weights and amounts of food and look for the pattern.

Developing Math Language
A *piecewise-defined function* is a new type of function. Also, add the words to the class Word Wall. Encourage students to sketch an example of a piecewise-defined graph beside the definition.

Activity 14 • Piecewise-Defined Linear Functions 215

ACTIVITY 14 Continued

6–8 Create Representations, Identify a Subtask, Construct an Argument, Think-Pair-Share Item 6 provides students a way to summarize their findings as related to the "feeding rules." It may be helpful for them to write the words *small*, *mid-size*, and *large* to the right of each line to help them organize the functions and corresponding domains. Before responding to Item 7, students may find it helpful to write the corresponding ranges next to the functions and their domains in Item 6. If groups are having difficulty with Item 7, suggest that they complete Item 8 before finalizing their responses to Item 7.

TEACHER to TEACHER

When denoting domains of the pieces of a piecewise function, the corresponding domains must be mutually disjoint (i.e., not overlapping). This concept is particularly important when the function values of the pieces corresponding to adjacent domains are not equal at a common endpoint of adjacent intervals (i.e., when the piecewise function is not continuous at a common endpoint of adjacent intervals). In such discontinuous cases, if the common endpoint is assigned to both intervals, then the single-valued property of the function is no longer true. However, the function in Miriam's list is a *continuous* piecewise linear function for all values of $w > 0$.

DISCUSSION GROUP TIPS

As you listen to the group discussion, take notes to aid comprehension and to help you describe your own ideas to others in your group. Ask questions to clarify ideas and to gain further understanding of key concepts.

Lesson 14-2
Writing Functions and Finding Domain and Range

5. Write a function that expresses the amount of food, $A(w)$, in ounces that a large dog should be fed, as a function of w, the weight of the dog in pounds. Determine the domain and range of the function.

   $A(w) = 60 + 0.1(w - 100)$ or $A(w) = 0.1w + 50$

   domain: $\{w \mid w > 100\}$; range: $\{A(w) \mid A(w) > 60\}$

Miriam realizes that there are three different algebraic feeding rules to follow because dogs are different sizes. She organizes her feeding rules in a list so that she can quickly refer to them whenever she has to decide how much food to feed a dog.

6. Complete Miriam's list by writing the appropriate function. Indicate the domain by writing the appropriate inequality symbols.

   $A(w) = \underline{\quad w \quad}$, when $0 \underline{\quad < \quad} w \underline{\quad < \quad} 20$

   $A(w) = \underline{\quad 0.5w + 10 \quad}$, when $20 \underline{\quad \leq \quad} w \underline{\quad \leq \quad} 100$

   $A(w) = \underline{\quad 0.1w + 50 \quad}$, when $w \underline{\quad > \quad} 100$

7. The vet has a German Shepard named Max, and Miriam knows that the vet feeds Max 63 ounces of food each day. Miriam also knows that the vet feeds her cocker spaniel Min 32 ounces of food each day. If the two dogs are being correctly fed, what is each dog's weight? Explain your reasoning.

   Since Max receives 63 ounces of dog food, Max must be a large dog; solving the equation $63 = 0.1w + 50$ for w, $w = 130$ pounds. Since Min receives 32 ounces of food, Min must be a mid-size dog; solving the equation $0.5w + 10 = 32$ for w, $w = 44$ pounds.

---

216 SpringBoard® Mathematics **Algebra 1, Unit 3** • Extensions of Linear Concepts

Lesson 14-2
Writing Functions and Finding Domain and Range

My Notes

Miriam decides to make a table that lists the weight of the dogs she will be feeding, in the order that she will feed them. A portion of Miriam's table is shown below.

8. Complete the table using the rules you wrote in Item 6.

Weight of Dog (pounds)	Amount of Dog Food (ounces)
21	**20.5**
70	**45**
16	**16**
120	**62**
15	**15**
56	**38**
27	**23.5**
107	**60.7**
7	**7**
25	**22.5**

The functions in Item 6 can be written as ***piecewise-defined functions.*** Every possible weight w has exactly one feeding amount A assigned to it, but the rule for determining that feeding amount changes for dogs of different sizes. The different feeding rules, along with their domains, are considered to be the pieces of a single piecewise-defined function.

MATH TERMS

A **piecewise-defined function** is a function that is defined differently for different disjoint intervals in its domain.

9. Miriam decides to write a piecewise-defined function to represent the functions she wrote in Item 6.

a. Explain how you know that each equation in Item 6 represents a function.

For each input value, there is exactly one output value. Each equation represents a linear function.

b. Complete the piecewise-defined function for the feeding rules.

$$A(w) = \begin{cases} w, \ \underline{\textbf{when } 0 < w < 20} \\ \underline{\mathbf{0.5}w + \mathbf{10}}, \text{ when } 20 \le w \le 100 \\ 0.1w + 50, \text{ when } \underline{w > 100} \end{cases}$$

c. Explain why the equation from part b is also a function. Use the domain of each rule to help justify your answer.

Each rule represents a function with the same domain as the corresponding part of the piecewise-defined function. Since none of the domains overlap, each input value still has exactly one output value. Therefore the equation represents a function.

Differentiating Instruction

Support students who are struggling to complete the table in Item 6 by suggesting they write the weights in order from least to greatest. Have students draw horizontal lines to separate the data into small, mid-size, and large dog weights.

9 Shared Reading, Create Representations, Interactive Word Wall, Debriefing Ask students to describe the difference between the answers to Item 6 and Item 9b. Students should recognize that the only difference is notation. Some may point out that the notation used for a piecewise-defined function emphasizes the fact that it is a single function. Debrief this lesson by asking students to consider and share real-world examples of situations that could be modeled by piecewise-defined functions, for example, movie ticket prices for children, students, adults, and seniors.

Differentiating Instruction

Extend learning for visual learners by asking them to describe what they think the graph of the piecewise-defined function in Item 9c might look like and how it will differ from a linear function.

ACTIVITY 14 Continued

Check Your Understanding

Debrief students' answers by asking them to describe the steps they took to write a piecewise-defined function from the data they were given. Ensure that they understand why a piecewise-defined function was needed to model the data.

Answers

10. $C(x) = 2x$; domain $= \{x \mid 0 < x < 4\}$; range $= \{C \mid 0 < C < 8\}$
11. $C(x) = 0.5x + 6$; domain $= \{x \mid 4 \leq x \leq 10\}$; range $= \{C \mid 8 \leq C \leq 11\}$
12.

Pounds of Mixed Nuts	Cost
4	$8.00
5	$8.50
8	$10.00

13. $C(x) = \begin{cases} 2x, & \text{when } 0 < x < 4 \\ 0.5x + 6, & \text{when } 4 \leq x \leq 10 \end{cases}$

ASSESS

Students' answers to Lesson Practice problems will provide you with a formative assessment of their understanding of the lesson concepts and their ability to apply their learning.

See the Activity Practice for additional problems for this lesson. You may assign the problems here or use them as a culmination for the activity.

LESSON 14-2 PRACTICE

14. $C(t) = \begin{cases} 40, & \text{when } 0 \leq t \leq 400 \\ 0.5t - 160, & \text{when } t > 400 \end{cases}$

15. $P(h) = \begin{cases} 7h, & \text{when } 0 \leq h \leq 40 \\ 10.5h - 140, & \text{when } h > 40 \end{cases}$

16. $D(t) = \begin{cases} 3t, & \text{when } 0 \leq t \leq 0.75 \\ 5t - 1.5, & \text{when } 0.75 < t \leq 2 \\ 8.5, & \text{when } 2 < t \leq 2.5 \\ 3t + 1, & \text{when } 2.5 < t \leq 4 \end{cases}$

17. $T(i) = \begin{cases} 0, & \text{when } 0 \leq i < 2000 \\ 0.02i - 40, & \text{when } 2000 \leq i \leq 6000 \\ 0.05i - 220, & \text{when } i > 6000 \end{cases}$

ADAPT

Check students' answers to the Lesson Practice to ensure that students can recognize when and why a situation is modeled by a piecewise-defined function. Discuss the parts of each problem that indicated a piecewise function was needed.

Lesson 14-2
Writing Functions and Finding Domain and Range

Check Your Understanding

A wholesale grocery store has the following sale on mixed nuts:

SALE!
Mixed nuts
Less than 4 lbs: $2/lb
4–10 lbs: $6 + $0.50/lb

10. Write a function to represent the cost $C(x)$ of buying less than four pounds of mixed nuts. Identify the domain and range.
11. Write a function to represent the cost $C(x)$ of buying 4 to 10 pounds of mixed nuts. Identify the domain and range.
12. Make a table to show the cost of 4, 5, and 8 pounds of mixed nuts.
13. Use your answers to Items 10 and 11 to write a piecewise-defined function to represent the cost $C(x)$ of x pounds of mixed nuts.

LESSON 14-2 PRACTICE

14. Speed Cell Wireless offers a plan of $40 for the first 400 minutes, and an additional $0.50 for every minute over 400. Let t represent the total talk time in minutes. Write a piecewise-defined function to represent the cost $C(t)$.

15. If Pam works more than 40 hours per week, her hourly wage for every hour over 40 is 1.5 times her normal hourly wage of $7. Write a piecewise-defined function that gives Pam's weekly pay $P(h)$ in terms of the number of hours h that she works.

16. A man walks for 45 minutes at a rate of 3 mi/h, then jogs for 75 minutes at a rate of 5 mi/h, then rests for 30 minutes, and finally walks for 90 minutes at a rate of 3 mi/h. Write a piecewise-defined function expressing the distance $D(t)$ he traveled as a function of time t.

17. **Model with mathematics.** A state income tax law reads as follows:

Annual Income (dollars)	Tax
< $2000	$0 tax
$2000 − $6000	2% of income over $2000
> $6000	$80 plus 5% of income over $6000

Write a piecewise-defined function to represent the income tax law.

218 SpringBoard® Mathematics Algebra 1, Unit 3 • Extensions of Linear Concepts

Lesson 14-3
Evaluating Functions and Graphing Piecewise-Defined Linear Functions

Learning Targets:
- Evaluate a function at specific inputs within the function's domain.
- Graph piecewise-defined functions.

SUGGESTED LEARNING STRATEGIES: Identify a Subtask, Think-Pair-Share, Construct an Argument, Create Representations, Discussion Groups

1. Rewrite the piecewise-defined feeding function that you wrote in Item 9 in Lesson 14-2.

$$A(w) = \begin{cases} w, & \text{when } 0 < w < 20 \\ 0.5w + 10, & \text{when } 20 \leq w \leq 100 \\ 0.1w + 50, & \text{when } w > 100 \end{cases}$$

2. Miriam must feed a dog that weighs 57 pounds.
   a. Which piece of the feeding function should Miriam use? Explain your answer.
   $0.5w + 10$; the given weight, 57 pounds, is between 20 and 100 pounds.

   b. Use your answer to Part (a) to evaluate $A(57)$.
   $A(57) = 38.5$

3. **Make use of structure.** When graphing a piecewise-defined function, it is necessary to graph each piece of the function only for its appropriate interval of the domain. Graph the feeding function on the axes below. When finished, your graph should consist of three line segments.

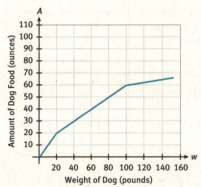

CONNECT TO AP

This is a connection to the concept of continuity in AP Calculus.

TECHNOLOGY TIP

To graph a piecewise-defined function in your calculator, enter the function into Y=, in dot mode, using parentheses to indicate the domain intervals. For example

$$f(x) = \begin{cases} -x^2 - 1, & \text{when } x < 1 \\ x + 2, & \text{when } x \geq 1 \end{cases}$$

would be entered as
$Y_1 = (-x^2 - 1)(x < 1) + (x + 2)(x \geq 1)$.

ACTIVITY 14 Continued

3–5 (continued) Ask students to explain how they can conclude the graph is a function, and ask how they can determine whether or not the function is continuous by looking at the graph. A simple approach is to have students trace the graph with their finger. If they can do so without lifting their finger, the function is continuous.

Differentiating Instruction

Support students who are having difficulty graphing the function in Item 3 by encouraging them to use dashed vertical lines to mark off the three intervals of the domain along the *x*-axis.

Check Your Understanding

Debrief this lesson by asking students how determining the values of $f(1)$ and $f(4)$ helped them graph the function. At this point in the activity, students should be able to graph a piecewise-defined function when given the function written in function notation.

Answers
6. $f(1) = 2$,
   $f(x) = x + 1$, when $-3 < x \leq 1$;
   $f(4) = 8$,
   $f(x) = 2x$, when $1 < x \leq 4$

7.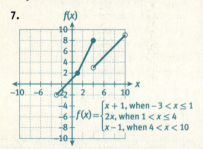

ASSESS

Students' answers to Lesson Practice problems will provide you with a formative assessment of their understanding of the lesson concepts and their ability to apply their learning. See the Activity Practice for additional problems for this lesson. You may assign the problems here or use them as a culmination for the activity.

ADAPT

Check students' answers to the Lesson Practice to ensure they are able to graph a piecewise-defined function. Ask students to describe the steps they should take to graph a piecewise-defined function. To connect to AP, ask students whether the functions they graphed were continuous, and if so, why.

ACTIVITY 14 continued

My Notes

Lesson 14-3
Evaluating Functions and Graphing Piecewise-Defined Linear Functions

4. Why is the graph of a piecewise-defined function more useful than three separate graphs?
   The graph of the piecewise-defined function includes each classification of size, so it is not necessary to first determine which graph to use based on the classification of size. It also combines all of the same information on one graph, so it requires less space.

5. How can you conclude that the graph represents a function?
   No part of a line segment lies above or below part of another line segment. This means each *w*-value is assigned exactly one *A*-value.

Check Your Understanding

Use the piecewise-defined function for Items 6 and 7.

$$f(x) = \begin{cases} x+1, & \text{when } -3 < x \leq 1 \\ 2x, & \text{when } 1 < x \leq 4 \\ x-1, & \text{when } 4 < x < 10 \end{cases}$$

6. Find $f(1)$ and $f(4)$. Explain which piece of the function you used to find each value.

7. Sketch a graph of the function.

LESSON 14-3 PRACTICE

8. **Model with mathematics.** Graph the piecewise-defined function that you wrote in Item 15 in Lesson 14-2 to describe Pam's pay.

For Items 9 and 10, graph each piecewise-defined function.

9. $f(x) = \begin{cases} 2x, \text{when } x > 0 \\ x, \text{when } x \leq 0 \end{cases}$

10. $f(x) = \begin{cases} \frac{1}{2}x + \frac{3}{2}, \text{when } x < 1 \\ -x + 3, \text{when } x \geq 1 \end{cases}$

For Items 11 and 12, consider the piecewise-defined function

$$f(x) = \begin{cases} \frac{1}{2}x - 10, & \text{when } x \leq 3 \\ -x - 1, & \text{when } x > 3 \end{cases}.$$

11. Find $f(-4)$.
12. Find $f(6)$.

LESSON 14-3 PRACTICE
8.

9.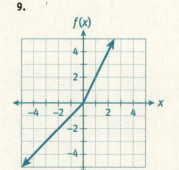

10.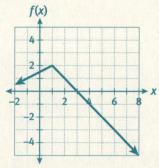

11. -12
12. -7

Lesson 14-4
Comparing Functions

Learning Target:
- Compare the properties of two functions each represented in a different way.

> **SUGGESTED LEARNING STRATEGIES:** Think-Pair-Share, Discussion Groups, Construct an Argument, Create Representations

When Miriam looks at the graph of the feeding function, she notices that for small dogs, each increase of 1 pound in weight causes an increase of 1 ounce of food. She concludes that the rate of change of the feeding rule for small dogs is 1 ounce per pound.

1. **Express regularity in repeated reasoning.** What is the rate of change of the feeding rule for large dogs?

   0.1 ounces per pound for each pound greater than 100

2. Miriam compares the rates of change of the feeding rules for each type of dog. She concludes that the feeding rule for small dogs has the greatest rate of change.
   a. How does the graph in Item 3 in Lesson 14-3 support Miriam's conclusion?

   The answer should reflect recognition of this relationship: the steeper the line, the greater the rate of change.

   b. What parts of the algebraic feeding rules in Item 6 in Lesson 14-2 support Miriam's conclusion?

   The slopes of each equation; small dogs have a rate of 1 oz per pound, medium dogs have a rate of 0.5 oz per pound, and large dogs have a rate of 0.1 oz per pound.

3. Dogs in one size category have a greater rate of change in feeding than other dogs. Explain why you think this might be true.

   Answers may vary. Small dogs are very energetic, and their high-energy temperament suggests that it is reasonable to expect small dogs to receive the greatest increase in dog food per pound increase in their weight.

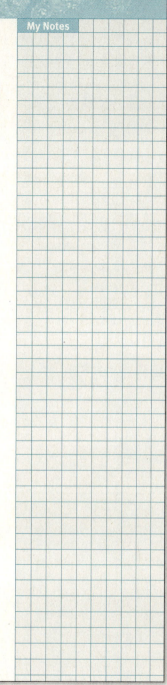

ACTIVITY 14 Continued
Lesson 14-4

PLAN

Pacing: 1 class period
Chunking the Lesson
#1–3   #4–7
Check Your Understanding
Lesson Practice

TEACH

Bell-Ringer Activity
Ask students to describe the relationship between slope and the rate of change of a function. Ask students to determine the slope of each segment in the graph of the piecewise-defined function graphed in Item 3 in Lesson 14-3 [1, 0.5, 0.1].

1–3 Discussion Groups, Construct an Argument
It may be helpful here to have a copy of the graph from Item 3 in Lesson 14-3 and the function from Item 7 in Lesson 14-2 displayed for students to see. While students may use either the graph or the function to respond to Item 1, they must use both the graph and the function to justify their response. It is extremely important for students to make the connection between the slope of the segments of the graph and the indicated slope of the corresponding piece of the function. Student responses to Item 3 will vary, but should focus on reasonableness.

Activity 14 • Piecewise-Defined Linear Functions   221

ACTIVITY 14 Continued

4–7 Think-Pair-Share, Create Representations, Debriefing In these items, students are asked to create a piecewise-defined function given the graph. Similar to the feeding rules, this function represents the amounts of water dogs of different weights should drink. Students make connections between the different notations and compare and contrast this new function with the function describing the feeding rules. As this lesson is debriefed, it is important for students to recognize the three different segments in the graph, be able to determine the slope of each of the segments, and make the connections necessary to write a function with appropriate domains to represent the situation.

Differentiating Instruction

Support students who are having difficulty moving from the visual representation to an algebraic definition of the function by suggesting they use a table to organize ordered pairs from the graph.

Lesson 14-4
Comparing Functions

The graph below shows how many ounces of water a dog should drink daily, based on the dog's weight.

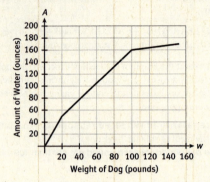

4. How much water should a 20-pound dog consume each day?
   50 ounces

5. How much water should a 100-pound dog consume each day?
   160 ounces

6. Compare the water function with the feeding function. Describe the similarities and differences.
   Both are piecewise-defined functions with three pieces. The domain for each piece is the same. The rates of change for the water function are greater than the rates of change for the feeding function.

7. Write a piecewise-defined function for the graph.

$$A(w) = \begin{cases} 2.5w, & \text{when } 0 < w < 20 \\ 1.5w + 20, & \text{when } 20 \leq w \leq 100 \\ 0.2w + 140, & \text{when } w > 100 \end{cases}$$

Lesson 14-4
Comparing Functions

Check Your Understanding

8. Write a piecewise-defined function for the graph, including the domain for each part.

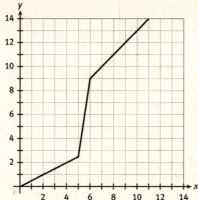

Use the piecewise-defined function below for Items 9 and 10.

$$f(x) = \begin{cases} x-1, & \text{when } 0 < x \leq 5 \\ 1.5x, & \text{when } 5 < x \leq 11 \end{cases}$$

9. Compare the function with the function from Item 8. Describe the similarities and differences.

10. Is the value $x = 5.5$ in the domain of the function you wrote in Item 8? If so, what is the value of the function when $x = 5.5$? Justify your response.

ACTIVITY 14 Continued

ASSESS

Students' answers to Lesson Practice problems will provide you with a formative assessment of their understanding of the lesson concepts and their ability to apply their learning.

See the Activity Practice for additional problems for this lesson. You may assign the problems here or use them as a culmination for the activity.

LESSON 14-4 PRACTICE

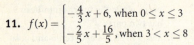

11. $f(x) = \begin{cases} -\frac{4}{3}x + 6, \text{ when } 0 \leq x \leq 3 \\ -\frac{2}{5}x + \frac{16}{5}, \text{ when } 3 < x \leq 8 \end{cases}$

12. Both are piecewise-defined functions with two pieces; pieces are linear. The domain of the function in Item 11 is $\{x \mid 0 \leq x \leq 8\}$, but the domain of the function in Item 12 is the set of real numbers. The range of the function in Item 11 is $0 \leq y \leq 6$ and the range of the function in Item 12 is $\{y \mid y \leq 3\}$.

13.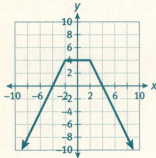

Domain: all real numbers; range: $y \leq 4$

14. Both functions are piecewise defined. The function in Item 13 has three segments and the function in Item 14 has two segments. The domain for both functions is the set of real numbers. The range in Item 13 is $\{y \mid y \leq 4\}$. The range in Item 14 is $\{y \mid y \leq 2\}$.

ADAPT

If students are having difficulty writing piecewise functions from either a graph or a table, encourage them to use the other representation as a transition to function notation. They can create a table using ordered pairs from the graph or a graph using values from a table before trying to move to the algebraic representation.

ACTIVITY 14 continued

My Notes

Lesson 14-4
Comparing Functions

LESSON 14-4 PRACTICE

11. Write a piecewise-defined function for the graph shown.

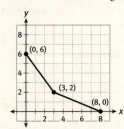

12. **Reason abstractly.** Compare the function in Item 11 to the function $f(x) = \begin{cases} 2x + 1, \text{ when } x < 1 \\ -x + 4, \text{ when } x \geq 1 \end{cases}$. Describe the similarities and differences.

13. Graph the function $f(x) = \begin{cases} 2x + 8, \text{ when } x < -2 \\ 4, \text{ when } -2 \leq x < 2 \\ -2x + 8, \text{ when } x \geq 2 \end{cases}$. State the domain and range.

14. How does the function in Item 13 compare to the function shown in the graph below?

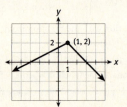

Piecewise-Defined Linear Functions
Breakfast for Bowser

ACTIVITY 14 continued

ACTIVITY 14 PRACTICE
Write your answers on notebook paper.
Show your work.

Lesson 14-1
Use the tables for Items 1–5.

x	y
0	2
1	3
2	4
3	5
4	6
5	7

x	y
6	−5
7	−6
8	−7
9	−8
10	−9
11	−10

1. Find the rate of change for each table.
2. At what x-value does the rate of change switch patterns?
3. Find a function that represents the first table of data, and write the domain for the function.
4. Find a function that represents the second table of data, and write the domain for the function.

Lesson 14-2
5. Write the piecewise-defined function that represents the data in the tables.
6. A store is having a special sale on designer soaps. For every three bars of soap purchased, one is given for free. The bars of soap cost $2 each, and there is a limit of eight bars of soap per customer (including free ones).
   a. Write a piecewise-defined function $C(b)$ that gives the total cost C for b bars of soap.
   b. Write the domain for this function in terms of the context.
7. A museum has the following prices for admission:

   Children under 12: free
   Kids age 12–17: $3
   Adults age 18–64: $8
   Senior citizens 65 and over: $4

   Write a piecewise-defined function that gives the cost $C(n)$ for a museum visitor who is n years old. What is the range of the function?

Lesson 14-3
8. A piecewise-defined function is shown.
$$h(x) = \begin{cases} x+3, & \text{when } x < 3 \\ bx, & \text{when } x \geq 3 \end{cases}$$
   What value of b is needed so that $h(3) = 6$?
   A. 1   B. 2
   C. 3   D. 6

9. Sketch a graph of the function.
$$f(x) = \begin{cases} x, & \text{when } x < 2 \\ -x+4, & \text{when } x \geq 2 \end{cases}$$

10. Ashley participated in a triathlon.
    - She swam for 10 minutes at a rate of 40 meters/min.
    - Then she biked for 40 minutes at a rate of 400 meters/min.
    - Finally, she ran for 25 minutes at a rate of 200 meters/min.
    a. Write a piecewise-defined function expressing the distance $d(t)$ in meters that Ashley traveled as a function of time t in minutes.
    b. Graph your function from Part (a).

Lesson 14-4
11. Refer back to Item 10. Wanda competed in the same triathlon as Ashley. A graph of the distance that Wanda covered as a function of time is shown below.

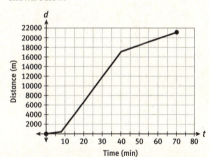

    a. Tell who completed each leg of the triathlon (swimming, biking, and running) faster, Ashley or Wanda. Justify your answers.
    b. Who completed the triathlon first, Ashley or Wanda? Justify your answer.

ACTIVITY 14 Continued

ACTIVITY PRACTICE

1. For values from 0 to 5, the rate of change is 1, and then there is a shift. From 6 to 10, the rate of change is −1.
2. Students may choose either $x = 5$ or $x = 6$. Their choice should be reflected in their answers for Items 3–5.
3. $y = x + 2$; $0 \leq x < 6$, where x is an integer
4. $y = -x + 1$; $6 \leq x \leq 11$, where x is an integer
5. $f(x) = \begin{cases} x+2, & \text{if } 0 \leq x < 6 \\ -x+1, & \text{if } 6 \leq x \leq 10 \end{cases}$
   where x is an integer
6. a. $C(b) = \begin{cases} 2b, & \text{if } 0 \leq b \leq 3 \\ 6, & \text{if } b = 4 \\ 2b-2, & \text{if } 4 < b \leq 6 \\ 10, & \text{if } b = 7 \\ 12, & \text{if } b = 8 \end{cases}$
   where b is an integer
   b. $\{0, 1, 2, 3, 4, 5, 6, 7, 8\}$
7. $C(n) = \begin{cases} 0, & \text{where } n < 12 \\ 3, & \text{where } 12 \leq n \leq 17 \\ 8, & \text{where } 18 \leq n \leq 64 \\ 4, & \text{where } n \geq 65 \end{cases}$
   where n is an integer;
   range = $\{0, 3, 4, 8\}$
8. B
9.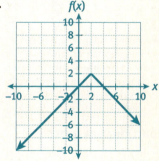

10. a. $d(t) = \begin{cases} 40t, & \text{when } 0 \leq t \leq 10 \\ 400t - 3600, & \text{when } 10 < t \leq 50 \\ 200t + 6{,}400, & \text{when } 50 < t \leq 75 \end{cases}$
    b.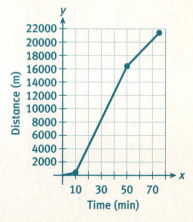

ACTIVITY 14 Continued

11. a. Wanda completed the swimming faster; explanations may vary. It took Ashley 10 minutes to complete the swimming portion, and Wanda's graph shows that she finished swimming before 10 minutes.

Wanda also completed the biking faster; explanations may vary. It took Ashley 40 minutes to complete the biking portion, and Wanda's graph shows that she finished biking in a little over 30 minutes.

Ashley completed the running faster; explanations may vary. It took Ashley 25 minutes to complete the running portion, and Wanda's graph shows that she ran for 30 minutes.

b. Wanda; explanations may vary. It took Ashley $10 + 40 + 25 = 75$ minutes to complete the triathlon, while Wanda's graph shows that she completed the triathlon in 70 minutes.

12. A
13. C
14. B
15. C
16. B
17. B
18. C
19. Answers may vary. In some contexts only positive numbers or whole numbers make sense. The domain and range need not be restricted for temperature values or altitude.

ADDITIONAL PRACTICE

If students need more practice on the concepts in this activity, see the eBook Teacher Resources for additional practice problems.

ACTIVITY 14 continued

Piecewise-Defined Linear Functions
Breakfast for Bowser

Use the graph for Items 12–15.

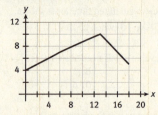

12. What is the domain of the piecewise-defined function?
  A. $0 \leq x \leq 18$
  B. $0 \leq x < 10$
  C. $4 \leq x \leq 10$
  D. $x \neq 6$

13. What is the value of the function for $x = 6$?
  A. 5
  B. 6
  C. 7
  D. 10

14. Which of the following defines the function for the domain $0 \leq x < 6$?
  A. $y = 0.5x$
  B. $y = 0.5x + 4$
  C. $y = 2$
  D. $y = -x + 2$

15. What is the range of the piecewise-defined function?
  A. $0 \leq y \leq 18$
  B. $0 \leq y \leq 10$
  C. $4 \leq y \leq 10$
  D. $y \neq 7$

A function gives Tanya's distance y (in miles) from home x minutes after she leaves her friend's house.

Use the graph of the function for Items 16–18.

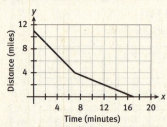

16. What are the domain and range of the function?
  A. $0 \leq x \leq 11, 0 \leq y \leq 17$
  B. $0 \leq x \leq 17, 0 \leq y \leq 11$
  C. $0 \leq x \leq 7, 0 \leq y \leq 4$
  D. $0 \leq x \leq 4, 0 \leq y \leq 7$

17. Which statement is true?
  A. After 7 minutes, Tanya's average speed increased.
  B. After 7 minutes, Tanya's average speed decreased.
  C. After 4 minutes, Tanya's average speed increased.
  D. After 4 minutes, Tanya's average speed decreased.

18. How far does Tanya live from her friend's house?
  A. 4 miles
  B. 7 miles
  C. 11 miles
  D. 17 miles

MATHEMATICAL PRACTICES
Attend to Precision

19. Explain why a restricted domain and a restricted range are sometimes appropriate for a piecewise-defined function. Describe instances when the domain and range need not be restricted.

Comparing Equations
A Tale of a Trucker
Lesson 15-1 Writing Equations from Graphs and Tables

ACTIVITY 15

Learning Targets:
- Write a linear equation given a graph or a table.
- Analyze key features of a function given its graph.

SUGGESTED LEARNING STRATEGIES: Summarizing, Visualization, Look for a Pattern, Create Representations, Discussion Groups, RAFT

Travis Smith and his brother, Roy, are co-owners of a trucking company. One of their regular weekly jobs is to transport fruit grown in Pecos, Texas, to a Dallas, Texas, distributing plant. Travis knows that his customers are concerned about the speed with which the brothers can deliver the produce, because fruit will spoil after a certain length of time.

1. Travis wants to address his customers' concerns with facts and figures. He knows that a typical trip between Pecos and Dallas takes 7.5 hours. He makes the following graph.

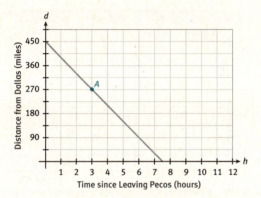

a. What information does the graph provide?
   Answers may vary. Pecos is 450 mi from Dallas, it takes 7.5 hours to drive from Pecos to Dallas, and the graph shows how far the truck is from Dallas after various elapsed times.

b. Locate the point with coordinates of (3, 270) on the graph. Label it point A. Describe the information these coordinates provide.
   See graph above. After 3 hours of travel, Travis is 270 mi from Dallas.

CONNECT TO BUSINESS

Companies that ship fruit from distribution plants to stores around the country use refrigerated trucks to keep the fruit fresher longer. For example, the optimal temperature range for shipping cantaloupes is 36–41°F. What would the temperature range look like on a number line graph?

Common Core State Standards for Activity 15

HSF–IF.B.4:	For a function that models a relationship between two quantities, interpret key features of graphs and tables in terms of the quantities and sketch graphs showing key features given a verbal description of the relationship. *Key features include intercepts; intervals where the function is increasing, decreasing, positive, or negative; relative maximums and minimums; symmetries; end behavior; and periodicity.*
HSF–IF.B.6:	Calculate and interpret the average rate of change of a function (presented symbolically or as a table) over a specified interval. Estimate the rate of change from a graph.*
HSF–IF.7:	Graph functions expressed symbolically and show key features of the graph, by hand in simple cases and using technology for more complicated cases.*
HSF–IF.C.7a:	Graph linear and quadratic functions and show intercepts, maxima and minima.

ACTIVITY 15
Investigative

Activity Standards Focus

In this activity, students will calculate and interpret rates of change in order to compare linear functions. They will write equations and inequalities from graphs and tables and use them to compare and analyze functions and their graphs.

Lesson 15-1

PLAN

Pacing: 1 class period
Chunking the Lesson
#1   #2–3   #4
Check Your Understanding
Lesson Practice

TEACH

Bell-Ringer Activity
Use **Shared Reading** to read the Connect to Business at the beginning of the activity. Give students time to graph the temperature range on a number line graph. Ask students to use the equation $C = \frac{5}{9}(F - 32)$ to convert the temperature range from Fahrenheit to Celsius [2.2–5.0°C]. Students should then graph the Celsius temperatures on a number line graph, and compare and contrast the two graphs they have created.

1 Shared Reading, Summarizing, Visualization, Activating Prior Knowledge Item 1 gives students the opportunity to interpret contextual information presented on a coordinate graph. It is important for students to understand that Travis is leaving Pecos and traveling toward Dallas. Students need to recognize that as Travis gets closer to Dallas, the distance from Dallas decreases, and the time since Travis left Pecos is increasing. By connecting the d-intercept value from the graph to the situation, students should recognize that Pecos is 450 miles from Dallas. In addition, by connecting the h-intercept to the situation, students should recognize that it takes 7.5 hours to drive from Pecos to Dallas.

Activity 15 • Comparing Equations   227

ACTIVITY 15 Continued

1 (continued) Students will activate prior knowledge as they apply the formula $d = r \cdot t$ to determine Travis' average speed. Some students may have difficulty with the concept of average speed. Travis expects to drive at various speeds, including at 65 mi/h, during his trip to Dallas. But, because of traffic slowdowns or construction, his graph shows that he expects to travel a total distance of 450 miles in a total of 7.5 hours.

2–3 Look for a Pattern, Think-Pair-Share, Create Representations, Sharing and Responding Since students have computed the average speed in Item 1d as 60 mi/h, they may carry that value forward. Some students will identify the rate of change from the table. All students should be able to describe patterns from the table. It is important for students to make the connection between the negative rate of change and the fact that the distance from Dallas is decreasing at an average rate of 60 mi/h. Monitor student discussions carefully. This is an excellent opportunity to informally assess student understanding of how the average speed relates to the information in the table, the slope of the line, and the rate of change.

ACTIVITY 15 continued

Lesson 15-1
Writing Equations from Graphs and Tables

c. According to the graph, how many hours will it take Travis to reach Dallas from Pecos? Explain how you determined your answer.
7.5 hours; when Travis arrives in Dallas, his distance from Dallas is 0 mi, which occurs at an elapsed time of 7.5 hours.

d. Interstate 20 is the direct route between Pecos and Dallas. Based upon his graph, at what average speed does Travis expect to travel? Explain how you determined your answer.
60 mi/h. The total distance is 450 mi, and the total time is 7.5 hours. Divide to find the average speed: 450 ÷ 7.5 = 60 mi/h.

2. Complete the table to show Travis's distance from Dallas at each hour.

Hours Since Leaving Pecos, h	Distance from Dallas, d
0	450
1	390
2	330
3	270
4	210
5	150
6	90
7	30

3. Model with mathematics. Use your table and the graph to write an equation that expresses Travis's distance d from Dallas as a function of the number of hours h since he left Pecos.
$d = -60h + 450$ or $d = 450 - 60h$

4. The graph of an equation in two variables is the set of all solutions of the equation plotted in the coordinate plane.
   a. The graph appears to pass through the point (6, 90). Use your equation to verify this.
   $d = 450 - 60h$; substitute 6 for h and evaluate: $d = 450 - 60(6) = 450 - 360 = 90$

WRITING MATH
The graph falls from left to right, and the numbers in the second column of the table decrease. When you write the equation, the coefficient of h should be negative.

Common Core State Standards for Activity 15 *(continued)*

HSF–IF.C.9:	Compare properties of two functions, each represented in a different way (algebraically, graphically, numerically in tables, or by verbal descriptions).
HAS–REI.D.10:	Understand that the graph of an equation in two variables is the set of all its solutions plotted in the coordinate plane, often forming a curve (which could be a line).
HSF–LE.B.5:	Interpret the parameters in a linear or exponential function in terms of a context.

228 SpringBoard® Mathematics **Algebra 1, Unit 3** • Extensions of Linear Concepts

Lesson 15-1
Writing Equations from Graphs and Tables

b. Generate four more ordered-pair solutions to your equation by substituting 1, 5, 8, and 10 for h. Does each point appear on the graph showing Travis's distance from Dallas? Explain.
(1, 390), (5, 150), (8, −30), (10, −150); the two points with negative d-values do not appear on the graph because you cannot have a negative distance.

c. Which of the solution points you generated in part b make sense in this context?
(1, 390) and (5, 150)

d. Explain why only some solution points make sense in this context.
The variables represent number of hours and distance. These do not make sense for negative values.

e. Use what you know about the solution points and the graph to state a reasonable domain and range for the function in the given context.
Domain: $0 \leq h \leq 7.5$; range: $0 \leq d \leq 450$

f. Use the graph to identify the zero of the function.
7.5

Check Your Understanding

The graph shows the amount of water remaining in a tank after a leak occurs. Use the graph for Items 5–7.

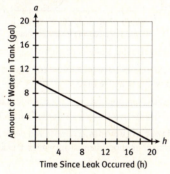

Time Since Leak Occurred (h)

5. Make a table to show the amount of water in the tank at five different times. Use the graph to identify the zero of the function.

6. **Make use of structure.** Use your table and the graph to write an equation that expresses the amount of water in the tank a as a function of the number of hours h since the leak occurred.

7. State a reasonable domain and range for the function in the given context.

ACTIVITY 15 Continued

4 Discussion Groups, Construct an Argument, Debriefing Encourage students to use the graph and the table to describe in words what happens between hours 7 and 8. Students should use the contextual situation to be able to conclude why every input greater than 7.5 is excluded from the reasonable domain of the function and why the least value of both the reasonable domain and range is zero. Debrief this portion of the Activity by allowing students to demonstrate their understanding of the problem situation by explaining what might happen to the graph if Travis was able to increase his average speed or if circumstances required him to decrease his average speed.

Check Your Understanding

Debrief students' answers by asking them to describe how to use their table and the graph to determine the domain and range of the function. These questions also assess whether students can write a linear function from a graph and a table.

Answers

5. Tables will vary.

Time in hours	Amount of water in gallons
0	10
4	8
8	6
16	2
20	0

The zero of the function is 20.

6. $a = -\frac{1}{2}h + 10$

7. domain: $0 \leq h \leq 20$;
range: $0 \leq a \leq 10$

Activity 15 • Comparing Equations

ACTIVITY 15 Continued

12 RAFT Communication is important in all career paths. The memo could be a handwritten note or take the form of an e-mail or text. Ask students how they should write a memo in this professional setting.

ASSESS

Students' answers to lesson practice problems will provide you with a formative assessment of their understanding of the lesson concepts and their ability to apply their learning.

See the Activity Practice for additional problems for this lesson. You may assign the problems here or use them as a culmination for the activity.

LESSON 15-1 PRACTICE

8. -60; Travis and Roy get closer to the distribution plant at a rate of 60 mi/h or they travel away from the farm at a rate of 60 mi/h.
9. -50; the competing shipping company travels at a rate of 50 mi/h.
10. Travis and Roy arrive first.
11. Travis and Roy: $d = -60h + 350$; shipping company: $d = -50h + 350$; Travis and Roy take 5.8 hours and the shipping company takes 7 hours.
12. Answers will vary. Travis and Roy should be the company that is recommended, because they make the trip faster than the other shipping company.

ADAPT

Check students' answers to the Lesson Practice questions to ensure they can find and interpret the rate of change using a table or a graph. Encourage students who are still struggling to transition to the algebraic representations to describe the situation, by verbally identifying the quantity that is changing and how it is changing before they attempt to write the equation. Ask students who are struggling to define a reasonable domain and range by verbally identifying what values make sense (positive, negative, increasing values, decreasing values, zero) before they formally define a reasonable domain or range.

ACTIVITY 15 continued

My Notes

MATH TIP

When writing your answer to Item 12, you can use a RAFT.
- Role—farm worker
- Audience—your supervisor
- Format—a memo
- Topic—which trucking company to use and why

Lesson 15-1
Writing Equations from Graphs and Tables

LESSON 15-1 PRACTICE

A farm in Plainville, Texas, will ship fruit to the distribution plant in Dallas. The farm is 350 miles from the plant.

8. The table shows part of a trip that Travis and Roy made from the farm to the distribution plant.

Distance from Dallas After Leaving Plainville

Time h (hours)	Distance d (miles)
2	230
2.5	200
3	170
3.5	140

Determine and interpret the rate of change of the data in the table.

9. A competing shipping company guarantees the fastest delivery times. The graph shows a recent trip from the farm to the distribution plant.

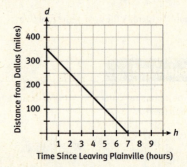

Determine and interpret the rate of change of the data in the graph.

10. Suppose both companies leave the farm at the same time and head to Dallas. Predict who will arrive first.

11. Write equations to represent the trip for each company. Determine how long it takes each company to drive from the farm to the plant in Dallas.

12. **Construct viable arguments.** Suppose you work at the farm. Write a memo to your supervisor about which trucking company to use and why.

Lesson 15-2
Comparing Functions with Inequalities

Learning Targets:
- Graph and analyze functions on the same coordinate plane.
- Write inequalities to represent real-world situations.

> **SUGGESTED LEARNING STRATEGIES:** Think-Pair-Share, Create Representations, Sharing and Responding, Discussion Groups

Travis knows that his first graph represents a model for an ideal travel scenario. In reality, he assumes that his average speed will be lower because he will need to stop to refuel. He also decides that he must account for concerns about road construction and traffic.

1. Experience tells Travis that his average speed will decrease to 45 mi/h. Travis's ideal travel scenario is shown on the grid below. On these same axes, draw the graph of his distance from Dallas, based upon his assumption that he will maintain a 45 mi/h average speed throughout his trip.

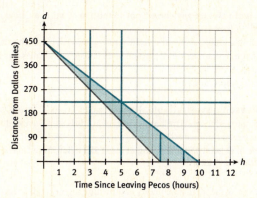

2. Write an equation for the line you drew in Item 1. Interpret the meaning of the constant and the coefficient of h in terms of the context.

   $d = -45h + 450$ or $d = 450 - 45h$; The constant 450 represents Travis's initial distance from Dallas. The coefficient of h is 45, which is Travis's average speed.

ACTIVITY 15 Continued

4–7 Create Representations, Think-Pair-Share, Sharing and Responding In these items, students use the graph to find the possible distances Travis is from Dallas at different times. Students are given various times and are asked to find the range of distances. The segments drawn represent the possible distances that Travis can be from Dallas after traveling 7 hours, 7.5 hours, and 9 hours, if Travis's average speed is between 45 to 60 mi/h. After students share with and respond to other groups following the first chunk of the lesson, students should remember to include the endpoints in each inequality they write for this problem situation. Monitor group discussions carefully to be sure students are making this connection.

ACTIVITY 15 continued

Lesson 15-2
Comparing Functions with Inequalities

My Notes

3. Travis can average anywhere from 45 mi/h to 60 mi/h, as shown on the two previous graphs.
   a. On the graph in Item 1, sketch the vertical line $h = 3$. Highlight the segment of that line that gives all the possible distances from Dallas three hours after Travis leaves Pecos.
   See graph for Item 1.

   b. How far might Travis be from Dallas after 3 hours of travel? Explain in your own words and then write your answer as an inequality.
   The endpoints of the highlighted vertical segment are at $d = 270$ and $d = 315$, so the points of the segment represent all distances from 270 mi to 315 mi. $270 \leq d \leq 315$

The distances you found in Item 3b are the possible distances after three hours of travel. The line segment can be described as the set of ordered pairs (h, d) where $h = 3$ and the value of d is between the values indicated in the inequality you wrote in Item 3b.

4. Travis can average anywhere from 45 mi/h to 60 mi/h, as shown on the graph in Item 1.
   a. On the graph, sketch the vertical line $h = 5$. Highlight the segment of that line that gives all the possible distances from Dallas five hours after Travis leaves Pecos.
   See graph for Item 1.

   b. **Reason quantitatively.** How far might Travis be from Dallas after 5 hours of travel? Explain in your own words, and then write your answer as an inequality.
   He is at least 150 miles, and at most 225 miles, from Dallas; $150 \leq d \leq 225$

   c. Describe what the inequality and the line segment tell Travis about his trip.
   The endpoints of the vertical segment are at $d = 150$ and $d = 225$, so the points of the segment represent all distances from 150 mi to 225 mi; after 5 hours, Travis will be anywhere from 150 to 225 mi from Dallas.

5. Use the graph in Item 1.
   a. Draw the line segment for $h = 7.5$. How far might Travis be from Dallas after 7.5 hours of travel? Write your answer as an inequality.
   See graph for Item 1; $0 \leq d \leq 112.5$.

   b. Describe what the inequality and the line segment tell Travis about his trip.
   The endpoints of the vertical segment are at $d = 0$ and $d = 112.5$, so the points of the segment represent all distances from 0 mi to 112.5 mi; after 7.5 hours, Travis will be anywhere from 0 mi to 112.5 mi away from Dallas.

WRITING MATH

If an inequality includes the greatest and least numbers, write the inequality using $\leq$ or $\geq$. If the greatest and least numbers are not included, write the inequality using $<$ or $>$.

Lesson 15-2
Comparing Functions with Inequalities

6. Use the graph in Item 1.
   a. Draw the line segment for $h = 9$. How far might Travis be from Dallas after 9 hours of travel? Write your answer as an inequality.
   See graph for Item 1; $0 \leq d \leq 45$.

   b. Describe what the inequality and the line segment tell Travis about his trip.
   The endpoints of the vertical segment are at $d = 0$ and $d = 45$, so the points of the segment represent all distances from 0 mi to 45 mi; after 9 hours, Travis will be anywhere from 0 mi to 45 mi away from Dallas.

7. There is a region that could be filled by similar vertical line segments for all values of h that Travis could be on his trip. Shade this region on the graph in Item 1.
   See graph for Item 1.

8. Travis likes to stop for a break after driving half the distance to Dallas.
   a. On the graph in Item 1, sketch the horizontal line that represents half the distance to Dallas. What is the equation of this line? Highlight the segment of the line that gives all possible times that he may stop.
   $d = 225$; see graph for Item 1.

   b. How much time might remain in the trip when Travis stops for a break? Explain how you know.
   Between 3.75 and 5 hours; these are the endpoints of the highlighted horizontal segment. They are also equal to one-half the total time.

9. After Travis drives for 4 hours and 20 minutes at 45 mi/h, he wonders how much farther he has to drive.
   a. Would you prefer to use the graph or the equation to determine the answer? Explain your choice.
   Answers may vary. The equation; the graph can only be used to estimate the value of d after 4 hours and 20 minutes.

   b. How much farther does Travis have to drive? Justify your response.
   255 miles; $d = 450 - 45h = 450 - 45\left(\frac{13}{3}\right) = 255$

ACTIVITY 15 Continued

Check Your Understanding

Debrief students' answers by asking them how they used a graph to answer each question. Discuss ways to solve each problem, if possible, without using a graph and analyze the pros and cons of each method. Students should begin to understand the importance of having a visual representation of data.

Answers

10. Look at the x-axis to find 6 hours; the lower line segment is crossing 90 and the upper line segment is crossing 180 miles from Dallas on the y-axis. Depending on his speed, Travis is between 90 and 180 miles from Dallas after 6 hours of travel.
11. No; he will be between 360 and 330 miles from Dallas if he travels between 45 and 60 mi/h.
12. The graph will start at (0, 450) and end at (9, 0).

ASSESS

Students' answers to lesson practice problems will provide you with a formative assessment of their understanding of the lesson concepts and their ability to apply their learning.

See the Activity Practice for additional problems for this lesson. You may assign the problems here or use them as a culmination for the activity.

LESSON 15-2 PRACTICE

13. 50 mi/h
14.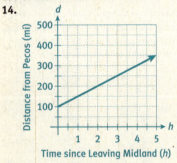
15. $d = 60h + 100$; h is the number of hours he drives.
16. $250 \leq d \leq 280$

ADAPT

Check students' answers to the Lesson Practice to ensure that students can determine the rate of change using a graph and can write the equation of a line using its graph or a verbal description. Encourage students who are still struggling to transition to the algebraic representations to describe the situation verbally, by identifying the quantity that is changing and how it is changing, before attempting to write the equation.

ACTIVITY 15 continued

Lesson 15-2
Comparing Functions with Inequalities

Check Your Understanding

10. Explain how you can use the graph in Item 1 to determine how far Travis might be from Dallas after 6 hours of travel.
11. Assuming Travis can average anywhere from 45 mi/h to 60 mi/h, is it possible for him to be 300 miles from Dallas after 2 hours of travel? Justify your answer.
12. Suppose Travis maintains an average speed of 50 mi/h throughout his trip. Describe how the graph of his distance from Dallas would compare to the graph in Item 1.

LESSON 15-2 PRACTICE

Travis's sister, Amy, lives in Midland, which is 100 miles from Pecos. After visiting his sister, Travis plans to drive his truck to Dallas. The graph shows his planned trip.

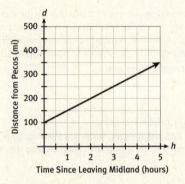

13. At what average speed does Travis expect to travel?
14. Travis hopes he will be able to maintain an average speed of 60 mi/h once he leaves his sister's house. Copy the graph above and draw the graph of Travis's distance from Pecos, based on an average speed of 60 mi/h.
15. **Make sense of problems.** Write an equation for the line you graphed in Item 14. Interpret the meanings of the constant and the coefficient of h in terms of the context.
16. How far might Travis be from Pecos after 3 hours of travel? Write your answer as an inequality.

Lesson 15-3
Writing Equations from Verbal Descriptions

Learning Targets:
- Write a linear equation given a verbal description.
- Graph and analyze functions on the same coordinate plane.

SUGGESTED LEARNING STRATEGIES: Look for a Pattern, Create Representations, Identify a Subtask

1. Suppose that the speed limit on all parts of Interstate 20 has been changed to 70 mi/h. Travis finds that he can now average between 50 mi/h and 70 mi/h on the trip between Pecos and Dallas.

   a. Write an equation that expresses Travis's distance d from Dallas as a function of the hours h since he left Pecos if his average speed is 50 mi/h.

   $d = 450 - 50h$

   b. Write an equation that expresses Travis's distance d from Dallas as a function of the hours h since he left Pecos if his average speed is 70 mi/h.

   $d = 450 - 70h$

   c. **Reason abstractly.** On the grid below, graph the two equations that you found in Parts (a) and (b). Describe how changing the coefficient of h in the equation affects the graph. Explain why this makes sense.

   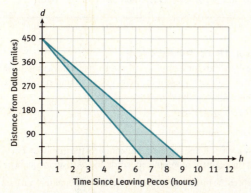

   When the coefficient of h increases, the line becomes steeper. The coefficient is the speed, and a steeper line has a greater rate of change in miles per hour, which is speed.

   d. Shade the region of the graph above for the ordered pairs (h, d) such that h represents all the possible times and d represents all the possible distances from Dallas after h hours of travel.

   See graph above.

ACTIVITY 15 Continued

Check Your Understanding

Debrief this lesson by asking students to explain what information from the verbal description they used to write equations in Items 2 and 3. Ask them also to explain how they graphed the equations.

Answers

2. $d = 10s$
3. $d = 15s$
4.
5. Go to 5 on the *x*-axis and go up until you get to 50 for Sumi's distance and up to 75 for Jerome's distance.

ASSESS

Students' answers to lesson practice problems will provide you with a formative assessment of student understanding of the lesson concepts and their ability to apply their learning.

See the Activity Practice for additional problems for this lesson. You may assign the problems here or use them as a culmination for the activity.

ADAPT

Check students' answers to the Lesson Practice to ensure they know how to graph a linear equation in slope-intercept form and know how to write inequalities to express real-world situations. If necessary, review slope-intercept form of a line and how to use the slope and *y*-intercept to graph a line.

LESSON 15-3 PRACTICE

6. and 7.
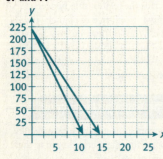

Lesson 15-3
Writing Equations from Verbal Descriptions

e. Describe Travis's possible distances from Dallas after 4 hours of travel.
   Travis will be at least 170 miles, but not more than 250 miles, from Dallas.

f. Write an inequality to represent Travis's possible distances from Dallas after 4 hours.
   $170 \leq d \leq 250$

Check Your Understanding

Sumi and Jerome start riding their bikes at the same time, traveling in the same direction on the same bike path. Sumi rides at a constant speed of 10 ft/s. Jerome rides at a constant speed of 15 ft/s.

2. Write an equation that expresses the distance *d* that Sumi has traveled as a function of the time *s* in seconds since she started riding her bike.
3. Write an equation that expresses the distance *d* that Jerome has traveled as a function of the time *s* in seconds since he started riding his bike.
4. Graph the equations from Items 2 and 3 on the same coordinate plane.
5. Explain how to use your graph to find the distance between Sumi and Jerome after they have ridden their bikes for 5 seconds.

LESSON 15-3 PRACTICE

Graph the equations for Items 6 and 7 in the first quadrant of the same graph. The equations represent the height *y* in centimeters of Ellie's and James's model gliders after *x* seconds.

6. Ellie: $y = -20x + 220$
7. James: $y = -15x + 220$
8. Draw a vertical line segment that connects the graphs at $x = 5$. Describe what the segment represents in words and with an inequality.
9. Interpret the meanings of the coefficients of *x* and the constants in terms of the context.
10. **Construct viable arguments.** Whose glider was in the air longer? Justify your response.
11. Julian's model glider begins at an initial height of 300 centimeters and descends at a rate of 18 cm per second. Write an equation that expresses the height *y* of Julian's glider after *x* seconds.

8.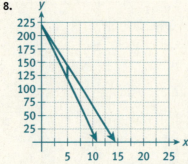
Ellie's glider is 120 cm high, and James' glider is 145 cm high. The segment between the lines represents the distance between the two gliders after 5 seconds; $145 - 120 = 25$ cm. $120 \leq y \leq 145$

9. The coefficients of *x* represent the rate of change of the gliders in cm/sec, and the constant 220 represents the initial height of the gliders in cm.
10. James's glider was in the air longer, because his graph ends at around (15, 0), almost 15 seconds, while Ellie's graph ends at (11, 0), 11 seconds.
11. $y = 300 - 18x$

Comparing Equations
A Tale of a Trucker

ACTIVITY 15
continued

ACTIVITY 15 PRACTICE
Write your answers on notebook paper.
Show your work.

Lesson 15-1

1. The graph shows the number of gallons of water y remaining in a tub x minutes after the tub began draining.

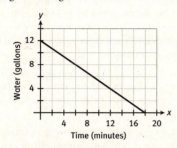

   Which describes something that is being drained at nine times the rate of the tub?
   A. A pool is drained at a rate of nine gallons per minute.
   B. A rain barrel is drained at a rate of six gallons per minute.
   C. A cooler is drained at a rate of 12 gallons per minute.
   D. A pond is drained at a rate of 18 gallons per minute.

2. The equation $d = 30 - 60h$ gives the distance d in miles that a bus is from the station h hours after leaving an arena. How far is the station from the arena?
   A. 0.5 miles   **B.** 30 miles
   C. 60 miles   **D.** 120 miles

3. Claire walks home from a friend's house two miles away. She sketches a graph of her walk, showing her distance from home as a function of time in minutes. Is the slope of the graph positive or negative? Explain.

Use the table for Items 4–9.

Minutes Since Leaving Store, m	Distance in Miles from Home, d
10	1.5
15	1.25
20	1

4. The table represents Holly's walk home from the store. What is Holly's rate in miles per minute?

5. How far is the store from Holly's home? Explain how you know.

6. Write an equation that relates distance d to the number of minutes m.

Lesson 15-2

7. Tim left school on his bike at the same time Holly left the store. The equation $d = 4 - \frac{m}{5}$ gives Tim's distance from Holly's house after m minutes. Sketch a graph of Holly's and Tim's trips on the same coordinate plane.

8. Compare the total time for Tim's trip to the total time for Holly's trip.

9. Part of Tim's trip includes the way Holly will walk. Use your graph to estimate when Tim will run into Holly.

10. Kane researched the cost of a taxi ride in a nearby city. He found conflicting information about the per-mile cost of a ride. The graph below shows his findings.

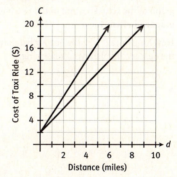

   a. What can you conclude about the cost per mile of a taxi ride?
   b. How much should Kane expect to pay for a five-mile taxi ride? Explain.

ACTIVITY 15 Continued

11. $h = -200t + 1600$
12. domain: $0 \leq x \leq 8$;
    range $0 \leq y \leq 1600$
13.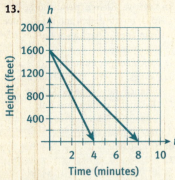
14. The balloon is between 400 ft and 1000 ft; $400 \leq y \leq 1000$.
15. The actual descent is slower than the descent in Item 17 and faster than the descent in Item 15.
16. 6.4 minutes; divide the distance, 1600 ft, by the speed, 250 ft/min.
17. B
18. $d = 62t$
19. $d = 57t$
20.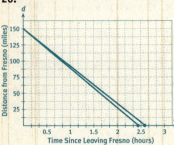
21.
22. Aquarium A is draining more quickly, at 3.5 gal/min, while aquarium B is draining at 2 gal/min; it will take about 15 min to drain.

ADDITIONAL PRACTICE

If students need more practice on the concepts in this activity, see the eBook Teacher Resources for additional practice problems.

ACTIVITY 15 continued

The graph shows the planned descent of a hot-air balloon. Use the graph for Items 11–16.

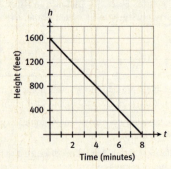

11. Write an equation that expresses the height of the balloon h as a function of the time t since the beginning of the descent.

12. State a reasonable domain and range for the function in Item 11 in this context.

13. The pilot thinks it may be possible to increase the rate of descent to a maximum of 400 ft/min. Copy the above graph and then draw a graph of the descent at 400 ft/min on the same coordinate plane.

14. On your graph from Item 13, draw the vertical line $t = 3$. Use the line to determine all the possible heights of the balloon after 3 minutes. Write your answer as an inequality.

15. The actual descent of the hot-air balloon is given by the equation $h = 1600 - 250t$. Compare the graph of the actual descent to the two lines graphed in Item 13.

16. How long does it take the hot-air balloon to make the descent? Explain your reasoning.

Comparing Equations
A Tale of a Trucker

Lesson 15-3

17. Wendy leaves Sacramento and heads to San Jose. She averages 50 mi/h on the 120-mile trip. Which equation describes Wendy's distance d from San Jose h hours after she leaves Sacramento?
    A. $d = 50h - 120$ B. $d = 120 - 50h$
    C. $d = 120 + 50h$ D. $d = 50h$

18. Fresno is 150 miles from Bakersfield. A driver traveling to Bakersfield leaves Fresno and averages 62 miles per hour. Write an equation for the distance d from Fresno, given the time t in hours since the driver left.

19. A driver traveling from Fresno to Bakersfield averaged 57 miles per hour. Write an equation for the distance d from Fresno, given the time t in hours since the driver left.

20. Sketch a graph of the equations you created in Items 18 and 19 on the same coordinate axis.

21. If a driver traveling between Fresno and Bakersfield knew he could average between 57 and 62 miles per hour, sketch a graph of the region that represents all possible distances from Fresno that the driver could be at time t.

MATHEMATICAL PRACTICES
Make Sense of Problems and Persevere in Solving Them

22. The amount of water w in aquarium A, in gallons, is given by $w = 50 - 3.5m$, where m is the number of minutes since the aquarium started draining. The amount of water w in aquarium B, in gallons, m minutes after the aquarium started draining, is given by the table below. Which aquarium drains more quickly, and how long does it take until it is empty?

Minutes, m	Gallons of Water, w
0	50
5	40

Inequalities in Two Variables
Shared Storage
Lesson 16-1 Writing and Graphing Inequalities in Two Variables

Learning Targets:
- Write linear inequalities in two variables.
- Read and interpret the graph of the solutions of a linear inequality in two variables.

> **SUGGESTED LEARNING STRATEGIES:** Marking the Text, Look for a Pattern, Think-Pair-Share, Create Representations, Sharing and Responding

Axl and Aneeza pay for and share memory storage at a remote Internet site. Their plan allows them upload 250 terabytes (TB) or less of data each month. Let x represent the number of terabytes Axl uploads in a month, and let y represent the number of terabytes Aneeza uploads in a month.

1. Look at the data for four months below. Determine if Axl and Aneeza stayed within the limits of their monthly plan.

Axl's Uploads (x)	Aneeza's Uploads (y)	Within Plan? (Yes/No)
50 TB	120 TB	Yes
135 TB	100 TB	Yes
100 TB	150 TB	Yes
162 TB	128 TB	No

2. Write an inequality that models the plan's restriction on uploading.
   $x + y \leq 250$ or $y \leq 250 - x$

3. **Reason quantitatively.** If Axl uploads 100 terabytes, write an inequality that shows how many terabytes Aneeza can upload.
   $y \leq 150$

4. If Aneeza uploads 75 terabytes, write an inequality that shows how many terabytes Axl can upload.
   $x \leq 175$

Common Core State Standards for Activity 16

HSA-REI.D.12: Graph the solutions to a linear inequality in two variables as a half-plane (excluding the boundary in the case of a strict inequality), and graph the solution set to a system of linear inequalities in two variables as the intersection of the corresponding half-planes.

ACTIVITY 16
Guided

Activity Standards Focus
In this Activity, students will write and graph inequalities in two variables. Students will also read and interpret the graph of the solution set of linear inequalities in two variables within a context.

Lesson 16-1

PLAN

Pacing: 1 class period
Chunking the Lesson
#1–2   #3–4   #5
Check Your Understanding
Lesson Practice

TEACH

Bell-Ringer Activity
Ask students to write examples of ordered pairs that lie in the first quadrant. Then ask students to write two inequalities that could be used to describe the coordinates of the points they have listed. Students can then use repeated reasoning to write similar inequalities for points chosen in the remaining quadrants.

1–2 Shared Reading, Marking the Text, Look for a Pattern, Think-Pair-Share, Create Representations, Sharing and Responding Item 1 is designed to allow students to numerically analyze the situation before attempting to write an inequality to represent it. Students having difficulty writing the inequality should be referred back to the maximum amount of shared storage space and reminded of what the quantities x and y represent in the situation. Having groups share their responses will allow for different (equivalent) forms of the inequality.

3–4 Create Representations These questions can be answered by substituting values for x and y into the inequality from Item 2 or by using the information in the opening paragraph. It is important for students to be aware that because Axl and Aneeza can together upload a total of 250 terabytes, each inequality they write should use $\leq$ or $\geq$ instead of $<$ or $>$.

ACTIVITY 16 Continued

5 Look for a Pattern, Create Representations, Debriefing The purpose of Item 5 is for students to understand what it means for a point to be a solution to a linear inequality. Points which are solutions to linear equations all lie on the same line. Points which are solutions to linear inequalities lie in the same half-plane. Students may use the graph to determine the solutions by determining whether the point lies in the shaded area. Others may add the x- and y-coordinates to determine whether the coordinates satisfy the restrictions of the inequality. As this portion of the activity is debriefed, be sure both methods are validated.

TEACHER to TEACHER

When graphed, the inequality $y \leq 250 - x$ has a solution that extends into all four quadrants. However, in this context, only the first quadrant and its axes are used because negative values for purchasing storage do not make sense. The lines $x = 0$ and $y = 0$ are boundary lines for the solution in this situation.

Developing Math Language

This lesson contains multiple vocabulary terms. Encourage students to mark the text by adding notes explaining the new vocabulary in their own words. Have them draw a graph to further explain the terms *linear inequality, solution to a linear inequality, boundary line, open half-plane* and *closed half-plane*. Have students add the terms to their math notebooks. Add the terms to the class Word Wall and encourage students to use these words in class discussions and written responses to questions.

ACTIVITY 16 continued

My Notes

MATH TIP

Some equations and inequalities are graphed on the coordinate plane, but contain only one variable.
$$x = 1$$
$$y < 3$$
Other equations and inequalities that are graphed on the coordinate plane contain two variables.
$$x + y = 10$$
$$y \geq 7 - x$$

MATH TERMS

A **linear inequality** is an inequality in the form $Ax + By > C$, $Ax + By \geq C$, $Ax + By < C$, or $Ax + By \leq C$, where A, B, and C are constants and A and B are not both zero.
A **solution of a linear inequality** is an ordered pair (x, y) that makes the inequality true.

Lesson 16-1
Writing and Graphing Inequalities in Two Variables

The graph below represents the possible values for the number of terabytes that Axl and Aneeza can upload. An ordered pair (x, y) represents the number of terabytes that Axl and Aneeza upload. For example, the coordinates (100, 75) mean that Axl uploaded 100 TB and Aneeza uploaded 75 TB of data.

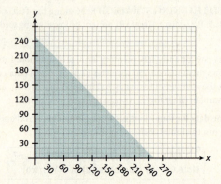

5. Graph each ordered pair on the graph above. Determine if it will allow Axl and Aneeza to remain within their plan, and explain your answers.

   a. (200, 50)
   The graph of (200, 50) is in the shaded region, so it is within the plan.

   b. (150, 150)
   The graph of (150, 150) is not in the shaded region, so it is not within the plan.

   c. (225, 25)
   The graph of (225, 25) is in the shaded region, so it is within the plan.

   d. (20, 120)
   The graph of (20, 120) is in the shaded region, so it is within the plan.

   e. (120, 200)
   The graph of (120, 200) is not in the shaded region, so it is not within the plan.

The graph of Axl and Aneeza's storage plan above is an example of a graph of a **linear inequality** in two variables. All the points in the shaded region are **solutions of the linear inequality**.

Lesson 16-1
Writing and Graphing Inequalities in Two Variables

Check Your Understanding

Axl and Aneeza decide to change to a 350-terabyte plan. Use this information for Items 6–8.

6. How does the graph in Item 5 change?
7. Graph the inequality that represents Axl and Aneeza's new plan. Let x represent the number of terabytes Axl can upload, and let y represent the number of terabytes Aneeza can upload.
8. Which ordered pairs will work with the new plan? Justify your response.
   a. (200, 50)   b. (150, 150)
   c. (240, 180)   d. (0, 300)
   e. (210, 180)   f. (360, −10)
9. Explain how to decide whether an ordered pair is a solution to a linear inequality.

LESSON 16-1 PRACTICE

You are on a treasure-diving ship that is hunting for gold and silver coins. You reel in a wire basket that contains gold and silver coins, among other things. The basket holds no more than 50 pounds of material. Each gold coin weighs about 0.5 ounce, and each silver coin weighs about 0.25 ounce. You want to know the different numbers of each type of coin that could be in the basket.

10. Write an inequality that models the weight in the basket.
11. Give three solutions of the inequality you wrote in Item 10.
12. **Construct viable arguments.** Explain why there cannot be 400 gold coins and 2800 silver coins in the basket.
13. Mitch and Mindy are in charge of purchasing art supplies for their local senior-citizen center. The monthly budget for supplies is $325, so the total they can spend in a month cannot exceed $325. This table displays how much each spent from January through April.

Month	Amount Mitch Spent (x)	Amount Mindy Spent (y)
January	150	155
February	215	110
March	60	252
April	175	150

Write an inequality in two variables that models how much Mitch and Mindy can spend together and stay within the budget.

Answers

6. The intercepts increase from 250 to 350.
7.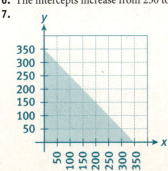
8. (200, 50), (150, 150), (0, 300); These coordinates are nonnegative and have sum ≤ 350.
9. Substitute the values of each coordinate of the point into the inequality. If the result is a true inequality, the point is a solution.

ACTIVITY 16 Continued

Lesson 16-2

PLAN

Pacing: 1 class period
Chunking the Lesson
Example A   Example B
#1–2   #3–4
Check Your Understanding
Lesson Practice

TEACH

Bell-Ringer Activity
Ask students to think of all the places where they know about or have seen boundary lines. These might include football fields, baseball diamonds, tennis courts, property lines and state lines. Ask students to write a working definition of a boundary line based on their examples.

Example A Interactive Word Wall, Vocabulary Organizer, Activating Prior Knowledge, Create Representations, Identify a Subtask Guide students through this first example of how to graph a linear inequality in two variables. Ask students to discuss how graphing a linear inequality is similar to and how it is different from graphing a linear equation. Allow students to make the connection between using solid and open circles when graphing inequalities on the number line and using solid and dashed lines in the coordinate plane.

ACTIVITY 16
continued

My Notes

Lesson 16-2
Graphing Inequalities in Two Variables

Learning Targets:
- Graph on a coordinate plane the solutions of a linear inequality in two variables.
- Interpret the graph of the solutions of a linear inequality in two variables.

SUGGESTED LEARNING STRATEGIES: Vocabulary Organizer, Interactive Word Wall, Create Representations, Identify a Subtask, Construct an Argument

The solutions of a linear inequality in two variables can be represented in the coordinate plane.

Example A
Graph the linear inequality $y \leq 2x + 3$.

Step 1: Graph the corresponding linear equation $y = 2x + 3$. The line you graphed is the **boundary line**.

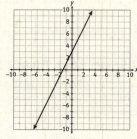

Step 2: Test a point in one of the **half-planes** to see if it is a solution of the inequality. Using $(0, 0)$, $0 \leq 2(0) + 3$ is a true statement. So $(0, 0)$ is a solution.

Step 3: If the point you tested is a solution, shade the half-plane in which it lies. If it is not, shade the other half-plane. $(0, 0)$ is a solution, so shade the half-plane containing the solution point $(0, 0)$.

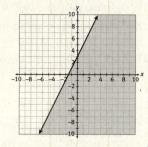

In Example A, the solution set includes the points on the boundary line. The solution set is a **closed half-plane**. In an inequality containing $<$ or $>$ the solution set does not include the points on the boundary line, so the boundary line is dashed, and the solution set is an **open half-plane**.

MATH TERMS

A line that separates the coordinate plane into two regions is a **boundary line**. The two regions are **half-planes**.

MATH TIP

The origin $(0, 0)$ is usually an easy point to test if it is not on the boundary line.

MATH TERMS

A **closed half-plane** includes the points on the boundary line.

An **open half-plane** does not include the points on the boundary line.

Lesson 16-2
Graphing Inequalities in Two Variables

Example B

Graph $x + 2y > 8$.

Step 1: Solve the inequality for y.

$x - x + 2y > 8 - x$ Subtract x from each side.

$2y > -x + 8$

$\dfrac{2y}{2} > \dfrac{-x+8}{2}$ Divide each side by 2.

$y > -\dfrac{1}{2}x + 4$ Simplify.

Step 2: Graph the boundary line, $y = -\dfrac{1}{2}x + 4$. Since the inequality uses the $<$ symbol, the solution set is an open half-plane. Draw a dashed line.

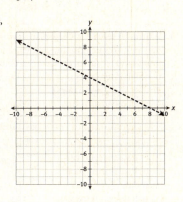

Step 3: Check the test point $(0, 0)$:

$0 > -\dfrac{1}{2}(0) + 4$ is false.

Shade the half-plane that does **not** contain the point $(0, 0)$.

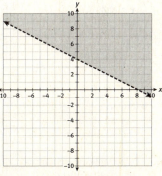

Try These A–B

Graph each inequality.   Check students' graphs.

a. $3x - 4y \geq -12$ b. $y < 3x - 2$
c. $-2y > 6 + x$ d. $y \leq -4x + 5$

MATH TIP

When multiplying or dividing each side of an inequality by a negative number, you must reverse the inequality symbol.

1. When graphing a linear inequality, is it possible for a test point located on the boundary line to determine which half-plane should be shaded? Explain. **No; this will determine only if the line itself is part of the solution set, which it is for $\leq$ and $\geq$ inequalities. To determine which half-plane to shade, the test point must be above or below the boundary line.**

ACTIVITY 16 Continued

Example B Activating Prior Knowledge, Create Representations, Identify a Subtask Example B requires the additional step of solving the inequality for y. Ask students if and why this step is necessary.

Try These A–B

Answers

a.
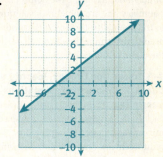

b.
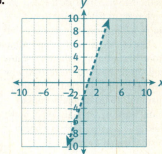

c.
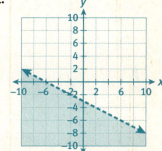

d.
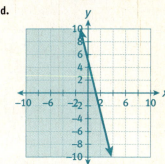

ACTIVITY 16 Continued

1–2 Activating Prior Knowledge, Create Representations, Think-Pair-Share, Construct an Argument These items lead students to understand that a boundary line and the two half-planes it divides are separate entities. The union of the two half-planes and the boundary line makes up the entire coordinate plane. It is important for students to understand that while the points on the boundary line are sometimes solutions to the inequality, these points cannot be used to determine in which half-plane the other solutions lie.

3–4 Discussion Groups, Identify a Subtask, Create Representations, Debriefing These items bring students back to the original problem context and mirror earlier questions, providing reinforcement. Students may determine equations and inequalities using different methods. Some may use the slope and y-intercept as determined from the graph, while some may choose points on the line to determine the slope. Monitor group discussions carefully to be sure students understand the reasons why the graphs are restricted to the first quadrant. To make connections to future learning, a discussion of how the context limits the graph of a linear inequality may be revisited.

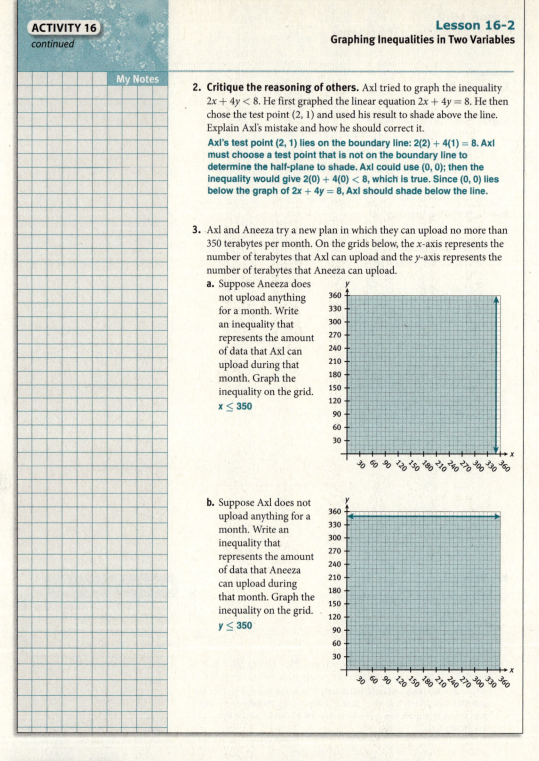

Lesson 16-2
Graphing Inequalities in Two Variables

2. **Critique the reasoning of others.** Axl tried to graph the inequality $2x + 4y < 8$. He first graphed the linear equation $2x + 4y = 8$. He then chose the test point (2, 1) and used his result to shade above the line. Explain Axl's mistake and how he should correct it.

    Axl's test point (2, 1) lies on the boundary line: $2(2) + 4(1) = 8$. Axl must choose a test point that is not on the boundary line to determine the half-plane to shade. Axl could use (0, 0); then the inequality would give $2(0) + 4(0) < 8$, which is true. Since (0, 0) lies below the graph of $2x + 4y = 8$, Axl should shade below the line.

3. Axl and Aneeza try a new plan in which they can upload no more than 350 terabytes per month. On the grids below, the x-axis represents the number of terabytes that Axl can upload and the y-axis represents the number of terabytes that Aneeza can upload.

    a. Suppose Aneeza does not upload anything for a month. Write an inequality that represents the amount of data that Axl can upload during that month. Graph the inequality on the grid.

    $x \leq 350$

    b. Suppose Axl does not upload anything for a month. Write an inequality that represents the amount of data that Aneeza can upload during that month. Graph the inequality on the grid.

    $y \leq 350$

MINI-LESSON: Equations and Inequalities in One Variable

Follow these steps to explore equations and inequalities in one variable.
1. On grid paper, graph $y = 1$, which means for any value of x, $y = 1$.
2. Graph $x = 3$, which means for any value of y, $x = 3$.
3. When the x-coordinate of each point on a line is the same, the line is vertical.
4. When the y-coordinate of each point on a line is the same, the line is horizontal.
5. Which parts of the coordinate plane can be described with inequalities in one variable? Explain your answer. (When the one variable is y, a region above or below a horizontal line is shaded. When the one variable is x, a region to the left or right of a vertical line is shaded.)

Lesson 16-2
Graphing Inequalities in Two Variables

4. The graph below represents another plan that Axl and Aneeza considered. The *x*-axis represents the number of terabytes of data from photos that can be uploaded each month. The *y*-axis represents the number of terabytes of data from text files.

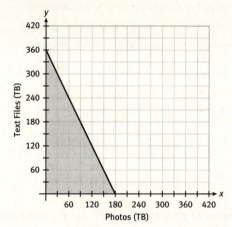

a. Identify the *x*-intercept and *y*-intercept. What do they represent in this context?

The x-intercept is (180, 0); this represents the maximum number of terabytes of photos that can be uploaded during a month in which no text files are uploaded. The y-intercept is (0, 360); this represents the maximum number of terabytes of text files that can be uploaded during a month in which no photos are uploaded.

b. Determine the equation for the boundary line of the graph. Justify your response.

$y = -2x + 360$; the slope of the line is $-\frac{360}{180} = -2$, and the y-intercept is (0, 360).

c. **Model with mathematics.** Write a linear inequality to represent this situation. Then describe the plan in your own words.

$2x + y \leq 360$; the plan allows up to 360 terabytes of uploaded information, but no more than 180 terabytes can be from photos.

ACTIVITY 16 Continued

3–4 (continued) As this portion of the activity is debriefed, discuss how the equation of the boundary line is related to the inequality. Encourage students to describe the meaning of the *x*- and *y*-intercepts of the graph within the provided context.

Differentiating Instruction

Support students who forget to change the direction of the inequality symbol when multiplying or dividing both sides by a negative number. For the following inequalities, have students predict the direction of the inequality symbol when solved for *y* with *y* appearing on the left side of the inequality. Then have them confirm their predictions by solving the inequalities for *y*.

1. $3x - 2y \leq 4$ $[y \geq \frac{3}{2}x - 2]$
2. $-9x - 3y < 8$ $[y > -3x - \frac{8}{3}]$
3. $x - 4y \geq -12$ $[y \leq \frac{1}{4}x + 3]$
4. $5x < 2 - y$ $[y < -5x + 2]$

Extend students' thinking by asking them to evaluate the following statement as always, sometimes or never true and give an explanation for their response.

The solution of a linear inequality includes ordered pairs in each of the four quadrants of a coordinate plane.

Check Your Understanding Answers

5.
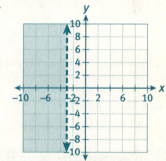

6.
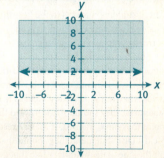

7.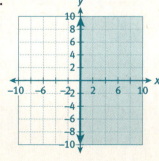

8. $x < 0$

ACTIVITY 16 Continued

Check Your Understanding

Debrief students' answers by asking them how they know which half-plane to shade when graphing an inequality. Have students identify the boundary lines for each equation in Items 5–7. Ask students to share the graphs of their boundary lines with others in their groups before shading. Pay close attention to the shading of the half-planes.

Answers

See previous page.

ASSESS

Students' answers to lesson practice problems will provide you with a formative assessment of their understanding of the lesson concepts and their ability to apply their learning.

See the Activity Practice for additional problems for this lesson. You may assign the problems here or use them as a culmination for the activity.

LESSON 16-2 PRACTICE

9.

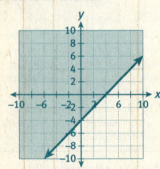

10.

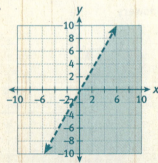

ADAPT

Check students' answers to the Lesson Practice to ensure that students can correctly graph an inequality in two variables on the coordinate plane. If students are struggling to shade the correct half-plane, ask them to plot their test point before they test it. If the test yields a true statement, they should shade toward the point they have graphed. If the test results yield a false statement, they should shade on the side of the boundary line that does not contain the test point. Encourage students to shade lightly so that corrections can easily be made if necessary.

ACTIVITY 16 continued

My Notes

Lesson 16-2
Graphing Inequalities in Two Variables

Check Your Understanding

5. Graph the linear inequality $x < -3$ on the coordinate plane.
6. Graph the linear inequality $y > 2$ on the coordinate plane.
7. Graph the linear inequality $x \geq 0$ on the coordinate plane.
8. Write an inequality whose solutions are all points in the second and third quadrants.

LESSON 16-2 PRACTICE

Graph each inequality on the coordinate plane.

9. $x - y \leq 4$
10. $2x - y > 1$
11. $y \geq 3x + 7$
12. $-x + 6 > y$
13. **Make sense of problems.** Write the inequality whose solutions are shown in the graph.

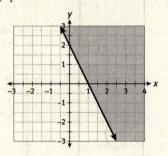

11.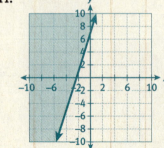

12.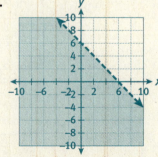

13. $y \geq -2x + 2$

246 SpringBoard® Mathematics Algebra 1, Unit 3 • Extensions of Linear Concepts

Inequalities in Two Variables
Shared Storage

ACTIVITY 16 continued

ACTIVITY 16 PRACTICE
Write your answers on notebook paper.
Show your work.

Lesson 16-1

1. Which ordered pairs are solutions of the inequality $5y - 3x \leq 7$?
   - **A.** (0, 0)
   - **B.** (3, 5)
   - **C.** (−2, −5)
   - **D.** (1, 2.5)
   - **E.** (5, −3)

2. Apple juice costs $2 per bottle, and cranberry juice costs $3 per bottle. Tamiko has at most $18 with which to buy drinks for a club picnic. She lets x represent the number of bottles of apple juice and lets y represent the number of bottles of cranberry juice. Then she graphs the inequality $2x + 3y \leq 18$, as shown below.

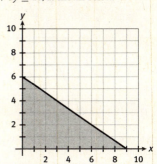

   a. Tamiko states that the graph does not help her decide how many bottles of each type of juice to buy, because there are infinitely many solutions. Do you agree or disagree? Why?
   b. Suppose Tamiko decides to buy two bottles of apple juice. Explain how she can use the graph to determine the possible numbers of bottles of cranberry juice she can buy.

3. Describe a real-world situation that can be represented by the inequality shown in the graph.

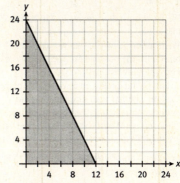

Lesson 16-2

4. Write an inequality for the half-plane. Is the half-plane open or closed?

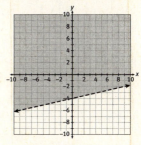

5. Write an inequality for the half-plane. Is the half-plane open or closed?

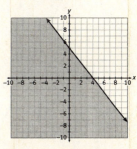

6. Sketch a graph of the inequality $y \geq -\frac{2}{5}x + 2$.

7. Sketch a graph of the inequality $3y > 7x - 15$.

ACTIVITY 16 Continued

ACTIVITY PRACTICE
1. A, C, E
2. a. disagree; Tamiko can count the integer points (x, y) in the solution region.
   b. Tamiko can look at the points $(2, y)$ in the solution region, where y is an integer.
3. Possible answer: Bill wants to build a fence for his garden. He can buy 1-foot and 2-foot sections that lock together. The length of the fence will be at most 24 feet. How many of each size section can Bill buy?
4. $y > \frac{1}{5}x - 4$; open
5. $y \leq -\frac{5}{4}x + 5$; closed
6.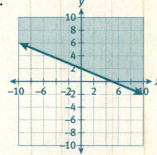
7.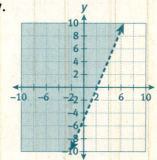

ACTIVITY 16 Continued

8. B
9. a. $b + g \leq 30$
   b.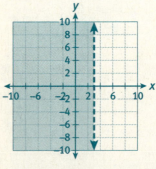
   c. solid line; at *most* includes *exactly* 30 students
   d. (10, 15): There are 10 boys and 15 girls in the class.
10. C
11. a. $3s + 6a \geq 450$
    b. $s + a \leq 120$
    c.

    The intersection (90, 30) represents selling 90 student tickets and 30 adult tickets.
12. When the inequality uses $\leq$ or $\geq$, the boundary line is part of the solution.
13.

    The coordinate plane shows all ordered pairs (x, y) with $x < 3$. The number line shows all real numbers $x < 3$.

ADDITIONAL PRACTICE
If students need more practice on the concepts in this activity, see the eBook Teacher Resources for additional practice problems.

ACTIVITY 16 continued
Inequalities in Two Variables
Shared Storage

8. Which inequality represents all of the points in the first and fourth quadrants?
   A. $x < 0$ B. $x > 0$
   C. $y < 0$ D. $y > 0$

9. There are at most 30 students in Mr. Moreno's history class.
   a. Write an inequality in two variables that represents the possible numbers of boys b and girls g in the class.
   b. Graph the inequality on a coordinate plane.
   c. Explain whether your graph has a solid boundary line or a dashed boundary line.
   d. Choose a point in the shaded region of your graph and explain what the point represents.

10. Which graph represents the solutions of the inequality $2x - y \geq 6$?

    A.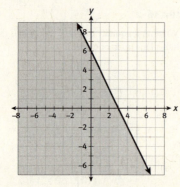

    B.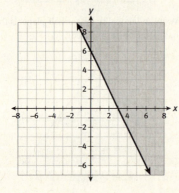

    C.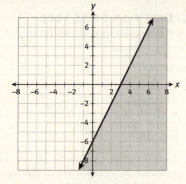

    D.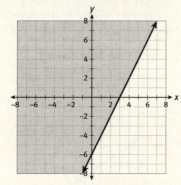

11. Tickets for the school play cost $3 for students and $6 for adults. The drama club hopes to bring in at least $450 in sales. The auditorium has 120 seats. Let a represent the number of adult tickets and s represent the number of student tickets.
    a. Write an inequality in two variables that represents the desired ticket sales.
    b. Write an inequality in two variables that represents the possible numbers of tickets that can be sold.
    c. Sketch both inequalities on the same grid. What does the intersection of the two graphs represent?

12. When is the boundary line of the graph of an inequality in two variables part of the solution?

MATHEMATICAL PRACTICES
Look For and Make Use of Structure

13. Graph the inequality $x < 3$ on a number line and on the coordinate plane. Describe the differences in the graphs.

Graphing Inequalities and Piecewise-Defined Functions
EARNINGS ON A GRAPH

Embedded Assessment 1
Use after Activity 16

1. Steve works at a restaurant. He earns $8.50 per hour.
   a. Write an equation that indicates the amount of money m in dollars that Steve can earn as a function of the hours h that he worked.
   b. Steve works at least 10 hours and not more than 30 hours per week. Describe the reasonable domain and range for your function from Part (a).
   c. The cost of any food that Steve buys while working is deducted from his earnings. Write an inequality that represents the possible amounts of money he can earn after buying food.
   d. Copy the grid below. Graph the inequality you wrote in Part (c).

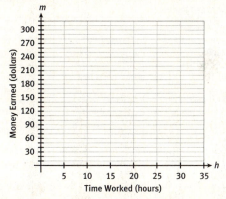

   e. Determine whether the ordered pair (16, 126) is a solution of the inequality you wrote in Part (c). If so, interpret its meaning. If not, explain why not.

Bob has been working at the restaurant longer than Steve. He earns $9 per hour. During some weeks he works more than 40 hours. The hours he works beyond 40 are considered overtime. For overtime pay, Bob earns double time, or $18 per hour.

2. a. Write a function $B(h)$ that will give Bob's pay for working 40 hours or less.
   b. Identify a reasonable domain and range for the function in this context.
3. a. Write a function $B(h)$ that will give Bob's pay for working more than 40 hours.
   b. Identify a reasonable domain and range for the function in this context.
4. Write a piecewise-defined function $B(h)$ that gives Bob's pay for any number of hours h.
5. Graph your function from Item 4.

Embedded Assessment 1

Assessment Focus
- Linear inequalities
- Piecewise functions
- Graphing inequalities
- Graphing piecewise functions

Answer Key
1. a. $m = 8.5h$
   b. domain $= \{h \mid 10 \leq h \leq 30\}$; range $= \{m \mid 85 \leq m \leq 255\}$
   c. $m \leq 8.50h$
   d.

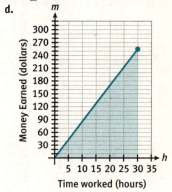

   e. Yes; it means that Steve worked 16 hours and earned $142 but spent $16 on food.
2. a. $B(h) = 9h$
   b. domain $= \{h \mid 0 \leq h \leq 40\}$; range $= \{B(h) \mid 0 \leq B(h) \leq 360\}$
3. a. $B(h) = 18h - 360$ or $B(h) = 18(h - 40) + 360$
   b. domain $= \{h \mid h > 40\}$; range $= \{B(h) \mid B(h) > 360\}$
4. $B(h) = \begin{cases} 9h, \text{ where } h \leq 40 \\ 18(h - 40) + 360, \text{ where } h > 40 \end{cases}$
5.

Common Core State Standards for Embedded Assessment 1

HAS-REI.D.12:	Graph the solutions to a linear inequality in two variables as a half-plane (excluding the boundary in the case of a strict inequality), and graph the solution set to a system of linear inequalities in two variables as the intersection of the corresponding half-planes.
HSF-IF.A.2:	Use function notation, evaluate functions from inputs in their domains, and interpret statements that use function notation in terms of a context.
HSF-IF.C.7:	Graph functions expressed symbolically and show key features of the graph, by hand in simple cases and using technology for more complicated cases.*
HSF-IF.C.7b:	Graph square root, cube root, and piecewise-defined functions, including step functions and absolute value functions.

Embedded Assessment 1

TEACHER to TEACHER

You may wish to read through the scoring guide with students and discuss the differences in the expectations at each level. Check that students understand the terms used.

Unpacking Embedded Assessment 2

Once students have completed this Embedded Assessment, turn to Embedded Assessment 2 and unpack it with them. Use a graphic organizer to help students understand the concepts they will need to know to be successful on Embedded Assessment 2.

Embedded Assessment 1
Use after Activity 16

Graphing Inequalities and Piecewise-Defined Functions
EARNINGS ON A GRAPH

Scoring Guide	Exemplary	Proficient	Emerging	Incomplete
	The solution demonstrates the following characteristics:			
Mathematics Knowledge and Thinking (Items 1b, 2b, 3b)	• Clear understanding and accurate identification of reasonable domain and range	• Adequate understanding and accurate identification of reasonable domain and range	• Partial understanding and partially accurate identification of reasonable domain and range	• No understanding and inaccurate identification of reasonable domain and range
Problem Solving (Item 1e)	• Appropriate and efficient strategy that results in a correct answer	• Strategy that may include unnecessary steps but results in a correct answer	• Strategy that results in some incorrect answers	• No clear strategy when solving problems
Mathematical Modeling / Representations (Items 1a, 1c, 1d, 2a, 3a, 4, 5)	• Effective understanding of how to represent a real-world scenario using equations, inequalities, graphs, and functions	• Little difficulty representing a real-world scenario using equations, inequalities, graphs, and functions	• Partial understanding of how to represent a real-world scenario using equations, inequalities, graphs, and functions	• Little or no understanding of how to represent a real-world scenario using equations, inequalities, graphs, and functions
Reasoning and Communication (Item 1e)	• Precise use of appropriate math terms and language to explain whether an ordered pair is a solution of an inequality • Ease and accuracy describing the relationship between a mathematical result and a real-world scenario	• Adequate explanation of whether an ordered pair is a solution of an inequality • Little difficulty describing the relationship between a mathematical result and a real-world scenario	• Misleading or confusing explanation of whether an ordered pair is a solution of an inequality • Partially correct description of the relationship between a mathematical result and a real-world scenario	• Incomplete or inaccurate explanation of whether an ordered pair is a solution of an inequality • Little or no understanding of how a mathematical result might relate to a real-world scenario

Solving Systems of Linear Equations
A Tale of Two Truckers
Lesson 17-1 The Graphing Method

ACTIVITY 17

Learning Targets:
- Solve a system of linear equations by graphing.
- Interpret the solution of a system of linear equations.

SUGGESTED LEARNING STRATEGIES: Summarizing, Paraphrasing, Marking the Text, Look for a Pattern, Create Representations

Travis Smith and his brother, Roy, are co-owners of a trucking company. The company needs to transport two truckloads of fruit grown in Pecos, Texas, to a distributing plant in Dallas, Texas. If the fruit does not get to Dallas quickly, it will spoil. The farmers offer Travis a bonus if he can get both truckloads to Dallas within 24 hours.

Due to road construction, Travis knows it will take 10 hours to drive from Pecos to Dallas. The return trip to Pecos will take only 7.5 hours. He estimates it will take 1.5 hours to load the fruit onto the truck and 1 hour to unload it.

1. Why is it impossible for Travis to earn the bonus by himself?
   Two trips will require at least
   $1.5 + 10 + 1 + 7.5 + 1.5 + 10 + 1 = 32.5$ hours.
   This length of time is greater than 24 hours.

2. Travis wants to earn the bonus so he asks his brother, Roy, if he will help. With Roy's assistance, can the brothers meet the deadline and earn the bonus? Explain why or why not.
   Yes; explanations will vary.

Roy is in Dallas ready to leave for Pecos. To meet the deadline and earn the bonus, Travis will leave Pecos first and meet Roy somewhere along the interstate to give him a key to the storage area in Pecos.

3. From Pecos to Dallas, Travis averages 45 mi/h. If Dallas is 450 mi from Pecos, write an equation that expresses Travis's distance d in miles from Dallas as a function of the hours h since he left Pecos.
   $d = 450 - 45h$

Common Core State Standards for Activity 17

HSA-REI.D.11: Explain why the x-coordinates of the points where the graphs of the equations $y = f(x)$ and $y = g(x)$ intersect are the solutions of the equation $f(x) = g(x)$; find the solutions approximately, e.g., using technology to graph the functions, make tables of values, or find successive approximations. Include cases where $f(x)$ and/or $g(x)$ are linear, polynomial, rational, absolute value, exponential, and logarithmic functions.

HSA-REI.C.5: Prove that, given a system of two equations in two variables, replacing one equation by the sum of that equation and a multiple of the other produces a system with the same solutions.

HSA-REI.C.6: Solve systems of linear equations exactly and approximately (e.g., with graphs), focusing on pairs of linear equations in two variables.

ACTIVITY 17
Guided

Activity Standards Focus
In this activity, students will use methods including graphing, making a table, substitution, and elimination to solve and classify systems of equations. They will also interpret the solution to a system of equations within a real-world context.

Lesson 17-1

PLAN

Pacing: 1 class period
Chunking the Lesson
#1–2   #3–8   #9–10
Check Your Understanding
Lesson Practice

TEACH

Bell-Ringer Activity
Ask students to graph the lines $y = x - 2$ and $2x - y = 6$ on the same coordinate plane and give the coordinates of the point of intersection [(4, 2)].

1–2 Shared Reading, Marking the Text, Summarizing, Paraphrasing, Graphic Organizer Check students' understanding of the setting and correct use of the information given within the setting. The context should be familiar to them from the earlier *A Tale of A Trucker* activity. The context in this activity, however, involves a second trucker.

TEACHER to TEACHER

In Item 1, some students may not include the loading or the unloading of the fruit in calculating the amount of time needed get two loads to Dallas. Students may find it helpful to use a simple drawing as a graphic organizer before formulating their response, such as a line segment with one endpoint representing Pecos and the other representing Dallas.

3–8 Discussion Groups, Look for a Pattern, Create Representations, Sharing and Responding Make sure students understand they are being asked to express Travis's distance from Dallas in terms of the hours since he left Pecos.

Activity 17 • Solving Systems of Linear Equations **251**

ACTIVITY 17 Continued

Differentiating Instruction

Support students who do not know where to begin when approaching Item 3. Encourage them to make a table showing Travis's distance from Dallas for hours 0, 1, 2, 3, 4, and h. They should see that Travis's distance from Dallas decreases by 45 miles for each hour that he drives. A table will also help students who have difficulty with the −45 mi/h term in the equation. Stress to students that the slope is a rate of change and that Travis's distance from Dallas is decreasing at a rate of 45 mi/h.

3–8 (continued) Students will use the graph to find the solution to the system of equations, so it is important for them to draw as accurate a graph as possible. Encourage students to plot several points on each line and use a straightedge to connect the points.

Monitor group discussions carefully. In Item 5, students may have difficulty writing an expression for the time since Roy left Dallas. Encourage these students to mark the text to highlight important words such as *before* and phrases such as *hours h since Travis left Pecos*. It may also help students to use specific values for h to help identify a pattern before generalizing the expression. It is important that students recognize that Roy is driving *away* from Dallas and increasing his distance from Dallas. They should make the connection between this and the positive slope in the equation representing his distance from Dallas.

In Item 8, students must identify and interpret the point of intersection within the context of the problem. It is important for them to recognize that this is the only coordinate that is a solution to both equations.

Lesson 17-1
The Graphing Method

4. Graph the equation you wrote in Item 3.

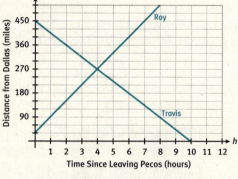

5. Roy leaves Dallas one-half hour before Travis leaves Pecos. In terms of the hours h since Travis left Pecos, write an expression that represents the time since Roy left Dallas.

   $h + \frac{1}{2}$

6. Roy travels 60 mi/h from Dallas to Pecos. Write an equation that expresses Roy's distance d from Dallas as a function of the hours h since Travis left Pecos.

   $d = 60\left(h + \frac{1}{2}\right)$ or $d = 60h + 30$

7. Graph the equation from Item 6 on the grid in Item 4.
   See graph above.

8. Identify the intersection point of the two lines. Describe the information these coordinates provide.
   The coordinates of the point of intersection are (4, 270). This point tell us that four hours after Travis leaves Pecos both he and Roy will meet, and be 270 miles from Dallas.

252 SpringBoard® Mathematics **Algebra 1, Unit 3** • Extensions of Linear Concepts

Lesson 17-1
The Graphing Method

ACTIVITY 17
continued

My Notes

The two equations you wrote in Items 3 and 6 form a *system of linear equations*.

To determine the solution of a system of linear equations, you must identify all the ordered pairs that make both equations true. One method is to graph each equation and determine the intersection point.

9. Graph each system of linear equations. Give each solution as an ordered pair. Check that the point of intersection is a solution of both equations by substituting the solution values into the equations.

   a. $y = 2x - 10$
      $y = -3x + 5$
      $(3, -4)$

      Check:
      $-4 = 2(3) - 10$
      $-4 = 6 - 10$
      $-4 = -4$

      $-4 = -3(3) + 5$
      $-4 = -9 + 5$
      $-4 = -4$

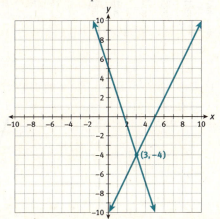

MATH TERMS

Two or more linear equations with the same variables form a **system of linear equations.**

   b. **Reason abstractly.** Edgar has nine coins in his pocket. All of the coins are nickels or dimes and are worth a total of $0.55. The system shown below represents this situation. How many of each type of coin does Edgar have in his pocket?

      $n + d = 9$
      $n + 2d = 11$
      $(7, 2)$

      Check:
      $7 + 2 = 9$
      $9 = 9$

      $7 + 2(2) = 11$
      $7 + 4 = 11$
      $11 = 11$

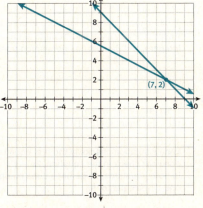

TECHNOLOGY TIP

You can graph each equation on a graphing calculator and use the built-in commands to determine the intersection point.

The solution (7, 2) means that Edgar has 7 nickels and 2 dimes.

ACTIVITY 17 Continued

Developing Math Language
Before students complete Items 9 and 10 there is an opportunity to refine their understanding of the word *solution*. Emphasize the difference between types of solutions. An equation such as $2x + 3 = 5$ typically has one solution. A linear equation such as $x + y = 5$ has infinitely many ordered pair solutions. A new type of solution—the solution to a system of equations—is introduced in this activity.

9–10 Look for a Pattern, Create Representations, Debriefing These questions help students understand the need to check solutions to systems of equations. If students are uncertain of their solutions to 9c remind them to check by substitution. Items 9c and 9d set the stage for needing a more accurate method for solving a system of equations and for the possibility that not all systems have exactly one solution.

MINI-LESSON: Technology Tip

If students need help graphing systems of equations and finding points of intersection on their calculators, a mini-lesson is available to provide practice.

See SpringBoard's *eBook Teacher Resources* for a student page for this mini-lesson.

Activity 17 • Solving Systems of Linear Equations 253

ACTIVITY 17 Continued

9–10 (continued) In Item 9d, students may need some assistance interpreting the meaning of parallel lines in the context of solutions to linear systems. The lines have no intersection point; therefore, the system has no solution. These Items must be debriefed carefully to allow students to clearly understand the limitations of solving by graphing when graphing by hand. The nature of the solutions to a system of linear equations will be explored and refined in subsequent lessons.

Differentiating Instruction

Extend learning for students by having them find the slopes of both equations in Item 9d [both have a slope of -3]. Then have students make a conjecture about the slopes of parallel lines.

Lesson 17-1
The Graphing Method

c. $y = 3x - 5$
   $y = -2x + 4$
   $\left(\dfrac{9}{5}, \dfrac{2}{5}\right)$

Check:

$\dfrac{2}{5} = 3\left(\dfrac{9}{5}\right) - 5 \qquad \dfrac{2}{5} = -2\left(\dfrac{9}{5}\right) + 4$

$\dfrac{2}{5} = \dfrac{27}{5} - 5 \qquad \dfrac{2}{5} = -\dfrac{18}{5} + 4$

$\dfrac{2}{5} = \dfrac{27}{5} - \dfrac{25}{5} \qquad \dfrac{2}{5} = -\dfrac{18}{5} + \dfrac{20}{5}$

$\dfrac{2}{5} = \dfrac{2}{5} \qquad \dfrac{2}{5} = \dfrac{2}{5}$

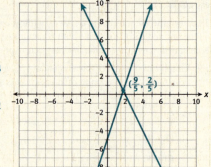

d. $3x + y = 1$
   $6x + 2y = 10$

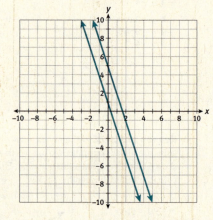

e. What made finding the solutions to parts c and d challenging?
   Answers may vary. It is difficult to estimate solutions that are not integers, and there is no intersection point when the lines are parallel.

10. Lena graphs the system $x - 2y = 3$ and $2x - y = -3$ and determines the solution to be $(1, -1)$. She checks her solution algebraically and decides the solution is correct. Explain her error.
    Lena's solution checks in the first equation, $x - 2y = 3$. However, it does not check in the equation, $2x - y = -3$. The solution must check in both equations, so perhaps Lena made a mistake when graphing one of the equations.

Lesson 17-1
The Graphing Method

Check Your Understanding

11. Solve each system.
    a. $y = -2x + 5$
       $y = \frac{1}{8}x - \frac{7}{2}$
    b. $3x - y = 5$
       $4x - 2y = 4$
    c. $y = -2x + 3$
       $y = x$
    d. $2x + y = 5$
       $4x - 1 = y$

12. Roberto has eight coins that are all dimes or nickels. They are worth $0.50. The system $n + d = 8$ and $n + 2d = 10$ represents this situation. Graph the system to determine how many of each coin Roberto has.

LESSON 17-1 PRACTICE

13. Solve each system.
    a. $y = 2x + 2$
       $y = -2x - 6$
    b. $y = \frac{1}{3}x - 2$
       $y = -x + 2$
    c. $3x + 2y = 6$
       $x - y = -3$
    d. $y = -2$
       $2y = -x - 1$

14. Sandeep ordered peanuts and raisins for his bakery. He ordered a total of eight pounds of these ingredients. Peanuts cost $1 per pound, and raisins cost $2 per pound. He spent a total of $10. The system $p + r = 8$ and $p + 2r = 10$ represents this situation. Graph the system to determine how many pounds of peanuts and raisins Sandeep ordered.

15. **Critique the reasoning of others.**
    Kyla was asked to solve the system of equations below. She made the graph shown and stated that the solution of the system is $(-4, -1)$. Is Kyla correct? Justify your response and identify Kyla's errors, if they exist.
    $x + y = 3$
    $x - 3y = -1$

    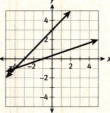

16. Write the system of equations represented by this graph.

    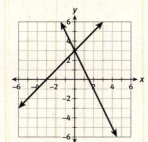

ACTIVITY 17 Continued

Lesson 17-2

PLAN

Pacing: 1 class period
Chunking the Lesson
#1–2   #3–6
Check Your Understanding
#9–11   Example A
Check Your Understanding
Lesson Practice

TEACH

Bell-Ringer Activity
Ask students to solve the following system graphically:
$$\begin{cases} 2x + 4y = 18 \\ x + 27 = 9 \end{cases}$$

Then have students compare their solutions [$(-18, 13.5)$]. This activity should reinforce the need for alternative methods of solving systems of equations.

1–2 Shared Reading, Marking the Text, Think-Pair-Share These items require students to relate the equations to the verbal descriptions in the problem.

Differentiating Instruction

Support students who have difficulty completing the table by suggesting they add an additional column to show the substitution of the variable in the equation.

3–6 Think-Pair-Share, Guess and Check, Debriefing In these items, students solve a system of equations numerically and graphically and compare the solutions determined using each solution method. Students should conclude that if an intersection point is not obvious in a table of values, the graphing method can be used to determine a solution. Students can use technology to facilitate the determination of the solution. Reinforce this idea through debriefing this portion of the lesson.

ACTIVITY 17 continued

Lesson 17-2
Using Tables and the Substitution Method

Learning Targets:
- Solve a system of linear equations using a table or the substitution method.
- Interpret the solution of a system of linear equations.

> **SUGGESTED LEARNING STRATEGIES:** Think-Pair-Share, Note Taking, Marking the Text, Guess and Check, Simplify the Problem

On another trip, Travis is traveling from Pecos to Dallas, and Roy is driving from Dallas to Pecos. They agree to meet for lunch along the way. Each driver averages 60 mi/h but Roy leaves 1.5 hours before his brother does. To determine when and where they will meet, you will solve this system of linear equations.

$$d = 450 - 60h$$
$$d = 60h + 90$$

1. What do the coefficient 60 and the constants 90 and 450 represent in the context of the problem?
   450 = Travis's initial distance from Dallas; 90 = Roy's distance from Dallas 1.5 hours after he left; 60 = the speed of each truck

2. Which equation represents Roy's distance from Dallas? How do you know?
   $d = 60h + 90$. Roy starts in Dallas, and his distance from Dallas is increasing; $60h + 90$ should get larger as h increases.

In addition to graphing, a system of equations can be solved by first making a table of values. Then look for an ordered pair that is common to both equations.

3. Complete each table.

$d = 450 - 60h$

h	d
0	450
1	390
2	330
3	270
4	210

$d = 60h + 90$

h	d
0	90
1	150
2	210
3	270
4	330

4. What ordered pair do the two equations have in common?
   (3, 270)

Lesson 17-2
Using Tables and the Substitution Method

5. **Use appropriate tools strategically.** Graph the equations on a graphing calculator. Identify the solution of the linear system. Describe its meaning in terms of the situation.

   (3, 270); three hours after Travis leaves Pecos, he will meet Roy for lunch. They will each be 270 miles from Dallas when they meet.

6. Is it possible that the intersection point from a graph of a linear system does not show up on a table of values? How could the solution be determined if it is not shown in the table?

   Yes; the table can show only selected values of h and d, so the solution could be skipped. If the solution is not shown in the table, you can refer back to the graph.

TECHNOLOGY TIP
You can use the table feature of a graphing calculator to quickly generate a table of values.

Check Your Understanding

7. Solve each system.
   a. $y = 2x + 6$
      $y = -3x + 16$
   b. $x + y = 8$
      $3x + 2y = 14$
   c. $y = 100 - 2x$
      $y = 20 + 6x$
   d. $2x + y = 6$
      $2x + 3y = 8$

8. What challenges did you encounter when solving these systems of linear equations?

Sometimes it is difficult to solve a system of equations by graphing or by using tables of values, and another solution method is necessary.

On another trip, Roy leaves from Pecos one hour before his brother and averages 55 mi/h. Travis leaves from Dallas during rush hour so he averages only 45 mi/h.

This system of linear equations represents each brother's distance d from Dallas, h hours after Roy leaves Pecos.

$$\text{Roy:} \quad d = 450 - 55h$$
$$\text{Travis:} \quad d = 45(h - 1)$$

Solve the system by finding when Travis's distance from Dallas is the same as Roy's distance.

$$\text{Travis's distance} = \text{Roy's distance}$$
$$45(h - 1) = 450 - 55h$$

9. Solve the equation for h and show your work.

   $45(h - 1) = 450 - 55h$
   $45h - 45 = 450 - 55h$
   $100h - 45 = 450$
   $100h = 495$
   $h = 4.95$

ACTIVITY 17 Continued

9–11 (continued) Students are asked to find only the value of h in Item 9. However, it should be emphasized as students share out that the complete solution to any system of equations includes both values of the ordered pair. One way to solve for d is to substitute the value of h into either equation and simplify. Another way is to solve by graphing using a calculator, as in Item 11.

TEACHER to TEACHER

In these items, students are taught to solve a system algebraically by solving each equation for one variable and then setting the two equations equal to each other. This method can be justified by the Transitive Property of Equality: if a, b, and c are real numbers and if $a = b$ and $b = c$, then $a = c$.

Example A Note Taking In Example A, students are introduced to the substitution method for solving systems of equations. Students need to understand that substitution is used to create a single equation in one variable. This equation can be solved using properties of equality and then using substitution to determine the value of the second variable.

TEACHER to TEACHER

The solution method used in Item 9 was actually an example of the substitution method. In Item 9, students were told to set the two expressions equal to each other, because both expressions in the real-world context equaled d. Another way for students to think about the method is to say that the expression from the second equation is *substituted* into the first equation. Since the second equation is $d = 60h + 90$, we substituted the expression $60h + 90$ into the first equation for d. The resulting equation could then be solved for the single variable h.

Lesson 17-2
Using Tables and the Substitution Method

10. Reason quantitatively. What does the answer to Item 9 represent? Is your answer reasonable?

The elapsed time when the brothers are the same distance from Dallas; almost 5 hours after Roy leaves Pecos, the two brothers pass each other. Yes, the answer is reasonable because $4.95 \approx 5$, and 5 hours at 50 miles per hour is 250 miles, which is about halfway between Dallas and Pecos.

11. Use a graphing calculator to graph the system above.
   a. What is the intersection point of the graphs? How would you use this point to answer Item 9?
   (4.95, 177.75); the first coordinate of the point is the solution to the equation.

   b. The second coordinate of the intersection point represents the distance that the brothers are from Dallas. How could you confirm this using the equations?
   When the h-value of 4.95 is substituted into either of the original equations, $d = 177.75$. This is the second coordinate of the intersection point.

Another method for solving systems of equations is the **substitution method**, in which one equation is solved for one of the variables. Then the expression for that variable is substituted into the other equation.

Example A

For a Valentine's Day dance, tickets for couples cost $12 and tickets for individuals cost $8. Suppose 250 students attended the dance, and $1580 was collected from ticket sales. How many of each type of ticket was sold?

Step 1: Let $x =$ number of couples, and $y =$ number of individuals.

Step 2: Write an equation to represent the number of people attending.
$2x + y = 250$ The number of attendees is 250.

Write another equation to represent the money collected.
$12x + 8y = 1580$ The total ticket sales is $1580.

Step 3: Use substitution to solve this system.
$2x + y = 250$ Solve the first equation for y.
$\mathbf{y = 250 - 2x}$
$12x + 8(\mathbf{250 - 2x}) = 1580$ Substitute for y in the second equation.
$12x + 2000 - 16x = 1580$ Solve for x.
$-4x = -420$
$x = 105$

---

258 SpringBoard® Mathematics Algebra 1, Unit 3 • Extensions of Linear Concepts

Lesson 17-2
Using Tables and the Substitution Method

Step 4: Substitute the value of x into one of the original equations to find y.

$$2x + y = 250$$
$$2(105) + y = 250 \quad \text{Substitute 105 for } x.$$
$$210 + y = 250$$
$$y = 40$$

Solution: For the dance, 105 couples' tickets and 40 individual tickets were sold.

Try These A
Solve each system using substitution.

a. $x + 2y = 8$ and $3x - 4y = 4$
   (4, 2)

b. $5x - 2y = 0$ and $3x + y = -1$
   $\left(-\frac{2}{11}, -\frac{5}{11}\right)$

c. Patty and Toby live 345 miles apart. They decide to drive to meet one another. Patty leaves at noon, traveling at an average rate of 45 mi/h, and Toby leaves at 3:00 P.M., traveling at an average speed of 60 mi/h. At what time will they meet?
   They will meet at 5:00 P.M.

12. Write the system of equations represented by these tables of values. Then use any method to solve the system.

x	−2	0	3	4
y	−4	0	6	8

x	−1	1	3	5
y	7	5	3	1

   $2x - y = 0$ and $x + y = 6$; (2, 4)

Check Your Understanding

Solve the systems by any method. Explain why you chose the method you did.

13. $x - 5y = 2$ and $x + y = 8$

14. $2y = x$ and $x - 7y = 10$

15. Theo buys a slice of pizza and a bottle of water for $3. Ralph buys 3 slices of pizza and 2 bottles of water for $8. How much does a slice of pizza cost?

ACTIVITY 17 Continued

ASSESS

Students' answers to lesson practice problems will provide you with a formative assessment of their understanding of the lesson concepts and their ability to apply their learning.

See the Activity Practice for additional problems for this lesson. You may assign the problems here or use them as a culmination for the activity.

LESSON 17-2 PRACTICE

16. **a.** No; the same point would need to appear in each table.
    b. The solution has an x-coordinate between 3 and 4. The first function is increasing and the second function is decreasing, and the y-values of the first function become greater than the y-values of the second function between the x-values of 3 and 4. This means the graphs intersect between an x-value of 3 and 4.
    c. Algebraically; this ensures an exact answer.
17. **a.** $y = 3x + 2.75$
    b. $y = 2.5x + 4.25$
    c. (3, 11.75); Answers may vary; sample answer: I used a graphing calculator, because each equation is solved for y.
    d. Three hours of climbing costs the same ($11.75) at each gym.
    e. Climb the Walls; the costs were the same for three hours, and an additional hour costs $0.50 less than at Rock-and-Roll.
18. $2x - y = -2$ and $x - 2y = -4$; (0, 2)

ADAPT

Check students' answers to the Lesson Practice to ensure that students can write a system of equations from a problem statement and then solve it. Remind students who are struggling to decide which method to use to solve a system of equations that there is not any one correct way to solve. Students who are more comfortable with technology may be more comfortable using the graphing method, while students who are uncomfortable with graphing may prefer the substitution method. Encourage students to identify clues in the equations, such as a variable with one as a coefficient, that signal opportunities when one method might be preferable to another. Students will get additional practice writing systems in the Activity Practice.

ACTIVITY 17
continued

Lesson 17-2
Using Tables and the Substitution Method

LESSON 17-2 PRACTICE

16. Delano was asked to solve the system $y = 2x - 4.5$ and $y = -x + 6$. He made the tables shown below.

$y = 2x - 4.5$

x	y
0	−4.5
1	−2.5
2	−0.5
3	1.5
4	3.5
5	5.5

$y = -x + 6$

x	y
0	6
1	5
2	4
3	3
4	2
5	1

   a. Are any of the ordered pairs in the tables solutions of the system? Why or why not?
   b. What can you conclude about the solution of the system? Explain your reasoning.
   c. Would you choose to solve the system graphically or algebraically? Justify your choice.

17. A rock-climbing gym called Rock-and-Roll charges $2.75 to rent shoes, and $3 per hour to climb. A competing gym, Climb the Walls, charges $4.25 to rent shoes, and $2.50 per hour to climb.
    a. Write an equation that gives the cost y of renting shoes and climbing for x hours at Rock-and-Roll.
    b. Write an equation that gives the cost y of renting shoes and climbing for x hours at Climb the Walls.
    c. Solve the system of the two equations you wrote in parts a and b using any method. Why did you choose the solution method you did?
    d. What does your solution to the system represent in the context of the problem?
    e. Construct viable arguments. Suppose you plan to rent shoes and go rock climbing for 4 hours. Which gym offers a better deal in this case? Justify your response.

18. Write the system of equations represented by these tables of values. Then use any method to solve the system.

x	−2	−1	1	2
y	−2	0	4	6

x	−2	2	4	6
y	1	3	4	5

260 SpringBoard® Mathematics **Algebra 1, Unit 3** • Extensions of Linear Concepts

Lesson 17-3
The Elimination Method

ACTIVITY 17 *continued*

My Notes

Learning Targets:

- Use the elimination method to solve a system of linear equations.
- Write a system of linear equations to model a situation.

SUGGESTED LEARNING STRATEGIES: Note Taking, Discussion Groups, Critique Reasoning, Vocabulary Organizer, Marking the Text

Elimination is another algebraic method that may be used to solve a system of equations. Two equations can be combined to yield a third equation that is also true. The ***elimination method*** creates like terms that add to zero.

Example A

Solve the system using the elimination method: $4x - 5y = 30$
$3x + 4y = 7$

Step 1: To solve this system of equations by elimination, decide to eliminate the y variable.

Original system	Multiply the first equation by 4. Multiply the second equation by 5.	Add the two equations to eliminate y.
$4x - 5y = 30$ →	$4(4x - 5y) = 4(30)$ →	$16x - 20y = 120$
$3x + 4y = 7$ →	$5(3x + 4y) = 5(7)$ →	$15x + 20y = 35$
		$31x = 155$
	Solve for x.	$x = 5$

Step 2: Find y by substituting the value of x into one of the original equations.

$4x - 5y = 30$
$4(5) - 5y = 30$ Substitute 5 for x.
$20 - 5y = 30$
$-5y = 10$
$y = -2$

Step 3: Check $(5, -2)$ in the second equation $3x + 4y = 7$.

$3x + 4y = 7$
$3(5) + 4(-2) \ ? \ 7$
$15 - 8 \ ? \ 7$
$7 = 7$ check

Solution: The solution is $(5, -2)$.

Try These A

Solve each system using elimination.

a. $3x - 2y = -21$
$2x + 5y = 5$
(−5, 3)

b. $7x + 5y = 9$
$4x - 3y = 11$
(2, −1)

MATH TERMS

The **elimination method,** also called the linear combination method, for solving a system of two linear equations involves *eliminating* one variable. To eliminate one variable, multiply each equation in the system by an appropriate number so that the terms for one of the variables will combine to zero when the equations are added. Then substitute the value of the known variable to find the value of the unknown variable. The ordered pair is the *solution* of the system.

Lesson 17-3

PLAN

Pacing: 1 class period

Chunking the Lesson

Example A #1
Example B #2
Check Your Understanding
Lesson Practice

TEACH

Bell-Ringer Activity

Ask students to make a flowchart to describe the process of solving a system of equations using the substitution method Have students share and compare their flowcharts.

Example A Shared Reading, Note Taking, Discussion Groups, Critique Reasoning Use the **Shared Reading** strategy to read the verbal description of the elimination method before beginning work on Example A. This example illustrates a case when both equations must be multiplied by appropriate numbers to create a zeroing out of one of the variables, x or y. Be sure students note that while either variable can be eliminated when using this method, it may be easier to focus on one variable or the other, depending on the coefficients in the system.

Debrief the Try These items by assigning different groups the responsibility of reporting their solution on whiteboards to the class. In this way, students can compare and contrast their answers with those of other groups.

Developing Math Language

This activity presents many methods for solving systems of equations. After students have completed this lesson, have them create a **Vocabulary Organizer** showing the methods for solving systems of equations. Students should include graphing, making a table, substitution, and elimination. Have them discuss and include on their organizers suggestions for when a particular method may be a good choice for solving a system. Debrief the Try These items by assigning different groups the responsibility of reporting their solution on whiteboards to the class. In this way, students can compare and contrast their answers with those of other groups.

ACTIVITY 17 Continued

1 Close Reading, Create Representations Ask student groups not only to share their system of equations, but also explain how they used elimination to solve the system. Have students explain why their answers are reasonable.

Example B Shared Reading, Marking the Text, Note Taking Problems like this—sometimes referred to as "mixture problems"—provide a context for solving systems of linear equations with coefficients and solutions that are decimals. In this case, students find the amount of each solution that should be mixed to produce a 5% solution.

TEACHER to TEACHER

Allow time for students to think as they read Example B to themselves and give them an opportunity talk about the solution. Then debrief the class on the solution. When you discuss the example, be clear about which meaning of the word *solution* you are using.

Differentiating Instruction

Support students may need clarification about the two different uses of the word *solution*—one is used to describe the liquids and the other refers to the answer. While context clues can be used to determine meaning, as the example is discussed, be careful to be clear as to which meaning of *solution* is being used.

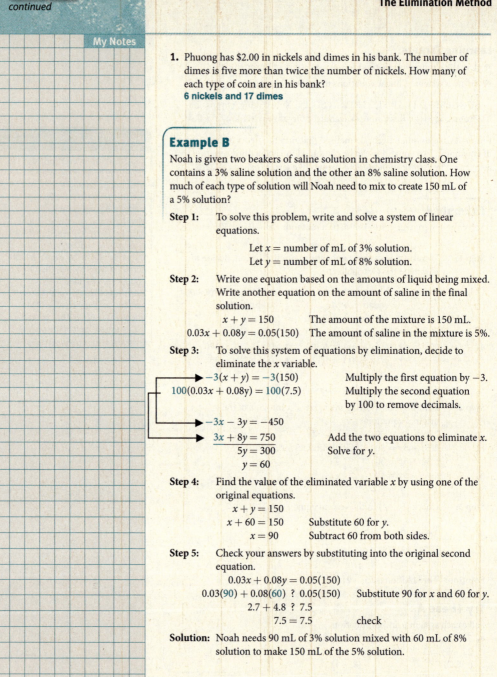

Lesson 17-3
The Elimination Method

1. Phuong has $2.00 in nickels and dimes in his bank. The number of dimes is five more than twice the number of nickels. How many of each type of coin are in his bank?
   6 nickels and 17 dimes

Example B

Noah is given two beakers of saline solution in chemistry class. One contains a 3% saline solution and the other an 8% saline solution. How much of each type of solution will Noah need to mix to create 150 mL of a 5% solution?

Step 1: To solve this problem, write and solve a system of linear equations.

Let $x =$ number of mL of 3% solution.
Let $y =$ number of mL of 8% solution.

Step 2: Write one equation based on the amounts of liquid being mixed. Write another equation on the amount of saline in the final solution.

$x + y = 150$ The amount of the mixture is 150 mL.
$0.03x + 0.08y = 0.05(150)$ The amount of saline in the mixture is 5%.

Step 3: To solve this system of equations by elimination, decide to eliminate the x variable.

$-3(x + y) = -3(150)$ Multiply the first equation by -3.
$100(0.03x + 0.08y) = 100(7.5)$ Multiply the second equation by 100 to remove decimals.

$-3x - 3y = -450$
$3x + 8y = 750$ Add the two equations to eliminate x.
$5y = 300$ Solve for y.
$y = 60$

Step 4: Find the value of the eliminated variable x by using one of the original equations.

$x + y = 150$
$x + 60 = 150$ Substitute 60 for y.
$x = 90$ Subtract 60 from both sides.

Step 5: Check your answers by substituting into the original second equation.

$0.03x + 0.08y = 0.05(150)$
$0.03(90) + 0.08(60) \;?\; 0.05(150)$ Substitute 90 for x and 60 for y.
$2.7 + 4.8 \;?\; 7.5$
$7.5 = 7.5$ check

Solution: Noah needs 90 mL of 3% solution mixed with 60 mL of 8% solution to make 150 mL of the 5% solution.

Lesson 17-3
The Elimination Method

Try These B
Solve each system using elimination.
a. $7x + 5y = -1$
   $4x - y = -16$ **(−3, 4)**

b. **Make sense of problems.** Mary had $25,000 to invest. She invested part of that amount at 3% interest and part at 5% interest for one year. The amount of interest she earned for both investments was $1100. How much was invested at each rate?
$7500 at 3% and $17,500 at 5%

2. Sylvia wants to mix 100 pounds of Breakfast Blend coffee that will sell for $25 per pound. She is using two types of coffee to create the mixture. Kona coffee sells for $51 per pound and Columbian coffee sells for $11 per pound. How many pounds of each type of coffee should Sylvia use?
35 lbs of Kona and 65 lbs of Columbian

MATH TIP

Recall the formula for simple interest,
$$i = prt,$$
where
i = interest earned
p = principal amount
r = rate
t = time in years

Check Your Understanding

3. Fay wants to solve the system $2x - 3y = 5$ and $3x + 2y = -5$ using elimination. She multiplies the second equation by 1.5 to get $4.5x + 3y = -7.5$.
   a. Do you think Fay's approach is correct? Explain why or why not.
   b. Describe how Fay could have multiplied to avoid decimal coefficients.

LESSON 17-3 PRACTICE

For each situation, write and solve a system of equations.

4. **Attend to precision.** A pharmacist has a 10% alcohol solution and a 25% alcohol solution. How many milliliters of each solution will she need to mix together in order to have 200 mL of a 20% alcohol solution?

5. Alyssa invested a total of $1500 in two accounts. One account paid 2% annual interest, and the other account paid 4% annual interest. After one year, Alyssa earned a total of $44 interest. How much did she invest in each account?

6. Kendall is Jamal's older brother. The sum of their ages is 39. The difference of their ages is 9. How old are Kendall and Jamal?

7. Yolanda wants to solve the system shown below.
   $3x - 4y = 5$
   $2x + 3y = -2$
   She decides to use the elimination method to eliminate the x variable. Describe how she can do this.

Answers
3. a. Yes; possible explanation: she can now use elimination because the y-terms will add to zero.
   b. She could have multiplied the first equation by 3, and the second equation by −2.

ACTIVITY 17 Continued

Lesson 17-4

PLAN

Pacing: 1 class period
Chunking the Lesson
#1–3   #4–6
Check Your Understanding
Lesson Practice

TEACH

Bell-Ringer Activity
Ask students to use two pencils to model all the ways two lines in a plane can be related. Have students discuss their answers. Watch for students who describe *skew* lines. You may need to define skew lines and lead students to understand that they do not lie in the same plane.

> **TEACHER to TEACHER**
>
> Until now, most of the systems of equations students have examined have had one and only one solution. However, in a plane, two lines may result in three different types of solutions. Students examine graphs of these types of systems and their equations.
> - Two lines may intersect at one point and have exactly one solution.
> - Two lines may be parallel, and the system has no solution.
> - Two lines may coincide so that they have all points in common, and the system has an infinite number of solutions.

1–3 Close Reading, Marking the Text, Create Representations These items guide students to the understanding that a system of equations yielding two parallel lines will have no solution. While students should intuitively understand that if both men are traveling at the same speed and leaving from the same place, that Travis can never catch up to Roy; it is important to examine the graph to connect the algebraic and graphical representations of the system of equations. If the solution(s) to a system of equations is (are) identified as the point(s) where the graphs intersect and the graphs do not intersect, there can be no solution to the problem.

ACTIVITY 17 continued

Lesson 17-4
Systems Without a Unique Solution

My Notes

Learning Targets:
- Explain when a system of linear equations has no solution.
- Explain when a system of linear equations has infinitely many solutions.

SUGGESTED LEARNING STRATEGIES: Create Representations, Marking the Text, Close Reading, Think-Pair-Share

The graph of a system of linear inequalities does not always result in a unique intersection point. **Parallel** lines have graphs that do not intersect. **Coincident** lines have graphs that intersect infinitely many times.

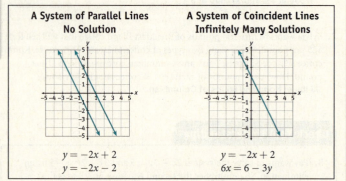

A System of Parallel Lines — No Solution
$y = -2x + 2$
$y = -2x - 2$

A System of Coincident Lines — Infinitely Many Solutions
$y = -2x + 2$
$6x = 6 - 3y$

> **MATH TERMS**
>
> Objects that are coincident lie in the same place. **Coincident** lines occupy the same location in the plane and pass through the same set of ordered pairs.

On a particular trip, Roy leaves from Pecos driving 55 mi/h. After 20 minutes, Travis realizes Roy forgot two cases of peaches. In 4 minutes, Travis loads the cases into his truck and heads out after Roy, traveling 55 mi/h.

This system of equations represents each brother's distance d from Dallas h hours after Roy leaves Pecos.

Roy: $d = 450 - 55h$ Travis: $d = 450 - 55(h - 0.4)$

1. **Model with mathematics.** Graph the system of equations. Describe the graph.

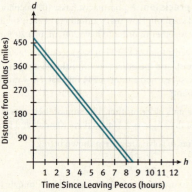

The lines do not appear to intersect.

Lesson 17-4
Systems Without a Unique Solution

2. What information about Travis and Roy does the graph provide?
   Travis will never catch up to Roy. He will not be able to give him the cases of peaches until he also arrives in Dallas, more than 8.5 hours after Roy left Pecos.

3. How many solutions exist to the system of equations? How is this shown in the graph?
   None; the lines are parallel (they do not intersect).

The system below represents a return trip from Dallas to Pecos, where d represents the distance from Pecos h hours after Roy leaves Dallas.

   Travis: $d = 450 - 60h$
   Roy: $d = -10(6h - 45)$

4. **Model with mathematics.** Graph the system of equations. Describe the graph.

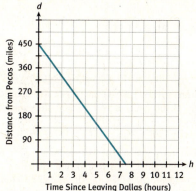

   There is only one graph, so the lines are coincident lines. The equations of the lines are equivalent.

5. What information about Travis and Roy does the graph provide?
   Travis and Roy are always the same distance from Pecos. They are making the trip from Dallas to Pecos at the same time.

6. How many solutions exist to the system of equations? How is this shown in the graph?
   Infinitely many; the lines are coincident.

ACTIVITY 17 Continued

Check Your Understanding

Debrief these items by asking students to describe the solutions of systems of equations represented by intersecting, parallel, or coincident lines.

Answers

7. a. (10, 80)
   b. $\left(-\frac{6}{11}, \frac{35}{11}\right)$
8. 13 hours after she leaves
9. after 10 years
10. a. (2.5, 0)
    b. no solution

ASSESS

Students' answers to lesson practice problems will provide you with a formative assessment of their understanding of the lesson concepts and their ability to apply their learning.

See the Activity Practice for additional problems for this lesson. You may assign the problems here or use them as a culmination for the activity.

LESSON 17-4 PRACTICE

11. $40h + 75 = c$, $60h + 35 = c$; 2 hours cost $155 at each company.
12. $40h + 35 = c$, $60h + 35 = c$; Drains-R-Us is always more expensive (the cost is the same for 0 hours).
13. $40h + 75 = c$, $40h + 35 = c$; Speedy Plumber is always $40 more for any number of hours.

ADAPT

Check students' answers to the Lesson Practice to ensure that they understand the meaning of no solution or infinitely many solutions to systems that represent real-world problems. Help students who are having difficulty understanding systems with infinitely many solutions by comparing them to statements that are *always* true. Systems of equations with one solution are like statements that are *sometimes* true. Systems having no solution can be compared to statements that are *never* true. Students will get additional practice in interpreting solutions in a real-world context in the Activity Practice.

ACTIVITY 17 continued

My Notes

Lesson 17-4
Systems Without a Unique Solution

Check Your Understanding

7. Solve each system.
   a. $y = 50 + 3x$
      $y = 100 - 2x$
   b. $x + 3y = 9$
      $-3x + 2y = 8$

8. Tom leaves for Los Angeles averaging 65 mi/h. Michelle leaves for Los Angeles one hour later than Tom from the same location. She travels the same route averaging 70 mi/h. When will she pass Tom?

9. Juan bought a house for $200,000 and each year its value increases by $10,000. Tia bought a house for $350,000 and its value is decreasing annually by $5000. When will the two homes be worth the same amount of money?

10. Solve each system by graphing. If the lines are parallel, write *no solution*. If the lines are coincident, write *infinitely many solutions*.
    a. $4x + 2y = 10$
       $y = -2x + 5$
    b. $y - 2x = 4$
       $y = 5 + 2x$

LESSON 17-4 PRACTICE

Speedy Plumber charges $75 for a house call and $40 per hour for work done during the visit. Drains-R-Us charges $35 for a house call and $60 per hour for work done during the visit. Use this information for Items 11–13.

11. **Reason quantitatively**. Write and solve a system of equations to determine how many hours of work result in the same total cost for a house call from either company. What is the cost in this case? How many hours must each company work to charge the same amount?

12. Eliza has a coupon for $40 off the fee for a house call from Speedy Plumber. How does this change your answer to Item 11? Is the total cost for the two companies ever the same? If so, after how many hours?

13. Tyrone has a coupon that lowers the hourly rate for Drains-R-Us to $40 per hour. How does this change your answer to Item 11? Is the total cost for the two companies ever the same? If so, after how many hours?

Lesson 17-5
Classifying Systems of Equations

Learning Targets:
- Determine the number of solutions of a system of equations.
- Classify a system of linear equations as independent or dependent and as consistent or inconsistent.

> **SUGGESTED LEARNING STRATEGIES:** Summarizing, Paraphrasing, Sharing and Responding, Interactive Word Wall, Predict and Confirm

When a system of two linear equations in two variables is solved, three possible relationships can occur.
- Two distinct lines that intersect produce one ordered pair as the solution.
- Two distinct parallel lines that do not intersect produce no solutions.
- Two lines that are coincident produce the same solution set—an infinite set of ordered pairs that satisfy both equations.

The three systems in the chart represent each of the possible relationships described above.

Relationship of Lines	Sketch	Number of Solutions
two intersecting lines		one solution
two parallel lines		no solution
two coincident lines		infinitely many solutions

1. Use the system $\begin{matrix} 4x - 2y = 21 \\ y - 2x = 10 \end{matrix}$ to answer parts a–d below.

    a. **Make use of structure.** Write each equation in the system in *slope-intercept form*. Compare the slopes and y-intercepts.
    $y = 2x - 10.5$ and $y = 2x + 10$; the slopes are the same, but the y-intercepts are different.

    b. Make a conjecture about the graph and the solution of this system.
    The lines are parallel, because they have the same slope but different y-intercepts. The system has no solutions.

MATH TERMS

The **slope-intercept form** of a linear equation is $y = mx + b$; m is the slope of the line and $(0, b)$ is the y-intercept.

ACTIVITY 17 Continued
Lesson 17-5

PLAN
Pacing: 1 class period
Chunking the Lesson
#1–2   #3–6   #7
Check Your Understanding
Lesson Practice

TEACH

Bell-Ringer Activity
Ask students to write each equation in slope intercept form:
$2x + 4y = 12 \; [y = -\frac{1}{2}x + 3]$
$x - y = 8 \; [y = x - 8]$
$-2x = 7 - 3y \; [y = \frac{2}{3}x + \frac{7}{3}]$

1–2 Think-Pair-Share, Predict and Confirm, Sharing and Responding In these items, students learn how to identify whether systems of equations have one solution, no solution, or infinitely many solutions by examining the slopes and y-intercepts of the lines. Students verify or revise their conjectures by graphing the systems and using substitution or elimination to solve the systems.

Activity 17 • Solving Systems of Linear Equations **267**

ACTIVITY 17 Continued

3–6 Discussion Groups, Construct an Argument, Sharing and Responding, Interactive Word Wall

In contrast to the last chunk of the lesson, these items require students to justify, using slopes and *y*-intercepts, why a system of equations will have only one solution and then determine the actual solution using the method of their choice. Allow students to share different solution methods and reasons for choosing those methods.

Developing Math Language

This activity contains many new vocabulary words and terms that sound and look very similar. It is important for students to be able to differentiate among consistent and inconsistent and dependent and independent systems. Encourage students to use a sketch or a diagram to help them remember the meanings of the words and terms. Add the words to your class Word Wall and encourage students to use these words to describe systems of linear equations.

Differentiating Instruction

Support students who need help writing equations in slope-intercept form to work closely with others in their groups. The objective of this lesson is to classify systems, and a student's inability to quickly write an equation in slope-intercept form may cause unnecessary frustration.

ACTIVITY 17 continued

My Notes

Lesson 17-5
Classifying Systems of Equations

c. Check the conjecture from part b by graphing the system on a graphing calculator. Revise the conjecture if necessary.

d. Solve the system using either the substitution method or the elimination method. Describe the result.
   Answers may vary based on the solution method chosen, but students should arrive at a false equation, such as $0 = 41$.

2. Use the system $\begin{array}{l} 2x = -y + 1 \\ 6x + 3y = 3 \end{array}$ to answer parts a–d below.

   a. Write each equation in the system in slope-intercept form. Compare the slopes and *y*-intercepts.
      $y = -2x + 1$ and $y = -2x + 1$; the slopes and *y*-intercepts are the same.

   b. Make a conjecture about the graph and the solution of this system.
      The lines are coincident because they have the same slope and *y*-intercept. The system has infinitely many solutions.

   c. Check the conjecture from part b by graphing the system on a graphing calculator. Revise the conjecture if necessary.

   d. Solve the system using either the substitution method or the elimination method. Describe the result.
      Answers may vary based on the solution method chosen, but students should arrive at a true equation, such as $0 = 0$.

3. Write each equation in the system $\begin{array}{l} 2x - y = 6 \\ x = 6 - y \end{array}$ in slope-intercept form.
   $y = 2x - 6$ and $y = -x + 6$

4. **Construct viable arguments.** Without graphing or solving, describe the solution of the system in Item 3. Justify your response.
   According to the slope-intercept forms, the lines have different slopes. Therefore the lines intersect and the system has one solution.

5. Verify your answer to Item 4 by solving the system using any method. Revise your answer to Item 4 if necessary.
   The solution is (4, 2).

Systems of linear equations are classified by the relationships of their lines.
- Systems that produce two distinct lines when graphed are **independent**. Systems that produce coincident lines are **dependent**.
- Systems that have no solution are **inconsistent**. Systems that have at least one solution are **consistent**.

6. Classify the systems in Items 1, 2, and 3.
   The system in Item 1 is independent and inconsistent. The system in Item 2 is dependent and consistent. The system in Item 3 is independent and consistent.

268 SpringBoard® Mathematics **Algebra 1, Unit 3** • Extensions of Linear Concepts

Lesson 17-5
Classifying Systems of Equations

7. For each system below, complete the table with the information requested.

The Nature of Solutions to a System of Two Linear Equations		
Equations in Standard Form:		
$2x + y = 2$ $6x + 3y = 6$	$2x + y = 2$ $x + y = 3$	$2x + y = 2$ $4x + 2y = -4$
Graph Each System:		
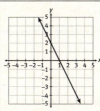	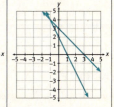	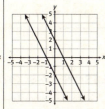
Write the Number of Solutions:		
Infinitely many	One solution $(-1, 4)$	none
Write the Relationship of the Lines:		
2 Coincident Lines	2 Intersecting Lines	2 Parallel Lines
Solve Algebraically:		
Multiply the 1st equation by (-3) to get $-6x - 3y = -6$. Add the result to the 2nd equation to get $0 = 0$. A statement that is TRUE for any choice of x.	Subtract the 2nd equation from the 1st equation $2x + y = 2$ $-x - y = -3$ to get $x = -1$. Substitute -1 for x in the 1st equation to get $2(-1) + y = 2 \rightarrow y = 4$ Solution $(-1, 4)$	Multiply the 1st equation by (-2) to get $-4x - 2y = -4$. Add the result to the 2nd equation to get $0 = -8$. A statement that is FALSE for any choice of x.
Write the Equations in Slope-Intercept Form:		
$y = -2x + 2$ $y = -2x + 2$	$y = -2x + 2$ $y = -x + 3$	$y = -2x + 2$ $y = -2x - 2$
Compare the Slopes and y-intercepts:		
Same slopes, same y-intercepts	Different slopes, different y-intercepts	Same slopes, different y-intercepts
Classify the System:		
Dependent and consistent	Independent and consistent	Independent and inconsistent

ACTIVITY 17 Continued

Check Your Understanding

Debrief students' answers by asking them how having the equations in slope-intercept form can help them classify a system of linear equations.

Answers

8. infinitely many; coincident lines; dependent and consistent
9. no solution; parallel lines; independent and inconsistent
10. infinitely many; coincident lines; dependent and consistent
11. one solution; intersecting lines; independent and consistent

ASSESS

Students' answers to lesson practice problems will provide you with a formative assessment of their understanding of the lesson concepts and their ability to apply their learning.

See the Activity Practice for additional problems for this lesson. You may assign the problems here or use them as a culmination for the activity.

LESSON 17-5 PRACTICE

12. $\left(\frac{8}{3}, 0\right)$; independent and consistent
13. $(4, 7)$; answers will vary.
14. The second equation in each system is the same. The first equation in the second system is two times the first equation in the first system.
15. no solution; independent and inconsistent
16. Disagree; the lines have the same slope and different y-intercepts, so the system is independent and inconsistent.

ADAPT

Check students' answers to the Lesson Practice to ensure that they understand what happens when solving an inconsistent or consistent, dependent system using the methods of substitution or elimination. Encourage students who are having difficulty managing the amount of information and number of new vocabulary terms to use the chart in Item 7 extensively as they complete practice problems. Students will get additional practice in classifying systems in the Activity Practice.

ACTIVITY 17 continued

Lesson 17-5
Classifying Systems of Equations

Check Your Understanding

For each system below:
a. Tell how many solutions the system has.
b. Describe the graph.
c. Classify the system.

8. $2x - 2y = 6$
   $y - x = -3$

9. $y = 1.5x + 5$
   $3x - 2y = 10$

10. $y = \frac{2}{3}x + 1$
    $4x - 6y = -6$

11. $3x + 4y = 1$
    $2x - 5y = 16$

LESSON 17-5 PRACTICE

12. Solve the system $3x + 4y = 8$ and $y = \frac{3}{4}x - 2$ using any method. Classify the system.

13. Approximate the point of intersection for the system of linear equations graphed below. Verify algebraically using substitution or elimination that the selected point is a solution for the system.

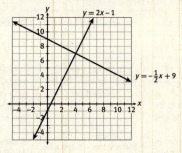

14. Monica claims that the system $3x + y = 16$ and $2x + 2y = 12$ has the same solutions as the system $6x + 2y = 32$ and $2x + 2y = 12$. Explain how you can tell whether Monica is correct without solving the systems.

15. Find the solution of the system $y = -\frac{2}{5}x + 1$ and $2x + 5y = 3$ by any method. Classify the system.

16. **Critique the reasoning of others.** Kristen graphed the system $y = 3x + 5$ and $10y = 30x + 51$ on her graphing calculator. She saw a single line on the screen and concluded that the system was dependent and consistent. Do you agree or disagree? Explain.

Solving Systems of Linear Equations
A Tale of Two Truckers

ACTIVITY 17
continued

ACTIVITY 17 PRACTICE
Write your answers on notebook paper. Show your work.

Lesson 17-1

1. Solve each system of linear equations.
   a. $y = 3x - 4$
      $y = \frac{2}{5}x + 9$

   b. $x + y = 7$
      $x - 3y = -1$

2. Which ordered pair is a solution to the system shown at right? $y = \frac{2}{3}x + 3$
   $y = -3x + 14$

   A. $(-3, 1)$ B. $(5, -1)$
   C. $(3, 5)$ D. $(5, 3)$

3. Which system's solution is represented by the graphs shown below?

   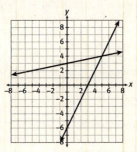

   A. $y = 2x + 3$
      $y = \frac{1}{5}x - 6$

   B. $y = 2x - 6$
      $y = \frac{1}{5}x + 3$

   C. $y = 5x - 6$
      $y = \frac{1}{2}x + 3$

   D. $y = 5x + 3$
      $y = \frac{1}{2}x - 6$

Lesson 17-2

4. A bushel of apples currently costs $10 and the price is increasing by $0.50 per week. A bushel of pears currently costs $15 and the price is decreasing by $0.25 per week. Which system of linear equations could be used to determine when the two fruits will cost the same amount per bushel?

   A. $y = 0.5x + 10$
      $y = -0.25x + 15$

   B. $y = 0.25x + 15$
      $y = -0.5x + 10$

   C. $y = 10x + 0.5$
      $y = 15x - 0.25$

   D. $y = 5x + 3$
      $y = -\frac{1}{2}x - 6$

5. Ray starts walking to school at a rate of 2 mi/h. Ten minutes later, his sister runs after him with his lunch, averaging 6 mi/h.
   a. Write a system of linear equations to represent this situation.
   b. Solve the system to determine how much time it took Ray's sister to catch up to him.

6. Colleen is in charge of ordering office supplies for her company. Last month she ordered ink cartridges and toner cartridges for the office printers. The cost of a toner cartridge is $19.50 more than the cost of an ink cartridge. Colleen ordered 11 ink cartridges and 4 toner cartridges, and the total cost was $460.50. Write and solve a system of equations to find the cost of each cartridge.

ACTIVITY 17 Continued

ACTIVITY PRACTICE

1. a. $(5, 11)$
   b. $(5, 2)$
2. A
3. B
4. A
5. a. $d = \frac{1}{3} + 2h$
      $d = 6h$
   b. 5 minutes
6. $\begin{cases} t = i + 19.50 \\ 11i + 4t = 460.50 \end{cases}$; the cost of an ink cartridge is $25.50 and the cost of a toner cartridge is $45.

ACTIVITY 17 Continued

7. **a.** $\begin{cases} c = 22.50h + 45 \\ c = 35h + 30 \end{cases}$

   b. (1.2, 72); a trail ride that lasts 1.2 hours will cost $72 at either ranch.

   c. Rocking Horse Ranch; a three-hour trail ride at Rocking Horse Ranch will cost $112.50. A three-hour trail ride at Saddlecreek Ranch will cost $150, which is more money than Janelle has.

8. **a.** $\begin{cases} n + d = 9 \\ 5n + 10d = 65 \end{cases}$

   b. 5 nickels, 4 dimes

9. apple $12, azalea $15

10. $\begin{cases} s + f = 17 \\ 2s + f = 28 \end{cases}$; 11 2-point shots and 6 free throws

11. **a.**

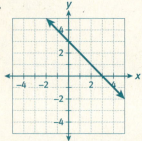

infinitely many solutions on the line $x + y = 3$

b.

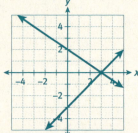

one solution at intersection (3, 0)

c.

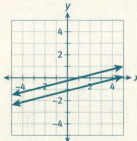

no solution

12. A
13. A

ADDITIONAL PRACTICE

If students need more practice on the concepts in this activity, see the eBook Teacher Resources for additional practice problems.

ACTIVITY 17 continued

7. At Rocking Horse Ranch, admission is $45 and trail rides are $22.50 per hour. At Saddlecreek Ranch, admission is $30 and trail rides are $35 per hour.
   a. Write a system of equations that shows the total cost c for a trail ride that lasts h hours at each ranch.
   b. Solve your system from part a. Interpret the meaning of the solution in the context of the problem.
   c. Janelle has $125, and she wants to go on a three-hour trail ride. Which ranch should Janelle choose? Justify your answer.

Lesson 17-3

8. Lawrence has 10 coins in his pocket. One coin is a quarter, and the others are all nickels or dimes. The coins are worth 90 cents.
   a. Write a system that represents this situation.
   b. Solve the system to determine the number of dimes and nickels in Lawrence's pocket.

9. Pedro placed an order with an online nursery for 6 apple trees and 5 azaleas and the order came to $147. The next order for 3 apple trees and 4 azaleas came to $96. What was the unit cost for each apple tree and for each azalea?

10. Jeremiah scored 28 points in yesterday's basketball game. He made a total of 17 baskets. Some of the baskets were field goals (worth two points) and the rest were free throws (worth one point). Write and solve a system of equations to find the number of field goals and the number of free throws that Jeremiah made.

Lesson 17-4

11. Graph each system of linear equations and describe the solutions.
    a. $x + y = 3$ and $2x + 2y = 6$
    b. $2x + 3y = 6$ and $-x + y = -3$
    c. $x - 4y = 1$ and $2x - 9 = 8y$

12. Ming graphs a system of linear equations on his calculator. When he looks at the result, he sees only one line. Assuming he graphed the equations correctly, what could this mean?
    A. The system has infinitely many solutions.
    B. The system has no solution.
    C. The system has exactly one solution.
    D. Every possible ordered pair (x, y) is a solution.

14. Infinitely many solutions; dependent and consistent
15. Sample: $2y = 6x + 2$
16. **a.** $y = 7 + 1.5x$, $y = 3 + 3.5x$; 2 weeks
    b. $y = 3 + 3.5x$, $y = 5 + 3.5x$
    c. part a: independent and consistent; part b: independent and inconsistent
    d. The lines are parallel so there is no solution.
17. Solve each equation for y. If the equations are identical, there are infinitely many solutions. If the

---

Solving Systems of Linear Equations
A Tale of Two Truckers

Lesson 17-5

13. Which is the best way to classify the system $y = 1.5x - 2$ and $3y = 4.5x - 7$?
    A. Independent and inconsistent
    B. Independent and consistent
    C. Dependent and inconsistent
    D. Dependent and consistent

14. Solve the system $3x + 4y = 8$ and $y = -\frac{3}{4}x + 2$ and classify the system.

15. The equation $3x - y = -1$ is part of a system of linear equations that is dependent and consistent. Write an equation that could be the other equation in the system.

16. Three friends decide to increase their exercise programs at the same time. Carolyn walks 7 miles per week and decides to increase the number of miles she walks by 1.5 miles per week. Eduardo walks 3 miles per week and decides to increase the number of miles he walks by 3.5 miles per week. Kendra walks 5 miles per week and decides to increase the number of miles she walks by 3.5 miles per week.
    a. Write and solve a system of equations to determine how many weeks it will be until Carolyn and Eduardo are walking the same distance each week.
    b. Write a system of equations you could use to determine how many weeks it will be until Eduardo and Kendra are walking the same distance each week.
    c. Classify the systems you wrote in parts a and b.
    d. Describe what would happen if you tried to solve the system you wrote in part b using the graphing method

MATHEMATICAL PRACTICES
Look for and Make Use of Structure

17. Explain how you can determine whether a system of two linear equations has a unique solution by examining the equations.

slopes are the same but the y-intercepts are different, there is no solution. Otherwise, there is a unique solution.

272 SpringBoard® Mathematics **Algebra 1, Unit 3** • Extensions of Linear Concepts

Solving Systems of Linear Inequalities
Which Region Is It?
Lesson 18-1 Representing the Solution of a System of Inequalities

ACTIVITY 18

Learning Targets:
- Determine whether an ordered pair is a solution of a system of linear inequalities.
- Graph the solutions of a system of linear inequalities.

SUGGESTED LEARNING STRATEGIES: Create Representations, Look for a Pattern, Discussion Groups, Quickwrite, Graphic Organizer

1. Graph each inequality on the number lines and grids.

Inequality	Graph all x	Graph all (x, y)
$x < 2$		
$x \geq -3$		
$x < 2$ and $x \geq -3$		

2. Compare and contrast the graphs you made in the third row of the table. In your explanation, compare the graphs to those in the first two rows and use the following words: dimension, half-line, half-plane, open, closed, and intersection. **Answers will vary.**
Contrast should include the fact that the number line graphs are in one dimension while the coordinate grids are in two dimensions.
Comparisons should include the fact that the third row in one dimension is the intersection (overlap) of two half-lines, while the third row in two dimensions is the intersection of two half-planes.
Contrast should include the fact that the first row graphs are an open half-line and an open plane, while the second row includes a closed half-line and a closed half-plane. In the third row, the intersections include boundaries that are inclusive and exclusive as part of the intersection.

Common Core State Standards for Activity 18

HSA-CED.A.3 Represent constraints by equations or inequalities, and by systems of equations and/or inequalities, and interpret solutions as viable or nonviable options in a modeling context.

HAS-REI.D.12 Graph the solutions to a linear inequality in two variables as a half-plane (excluding the boundary in the case of a strict inequality), and graph the solution set to a system of linear inequalities in two variables as the intersection of the corresponding half-planes.

ACTIVITY 18
Guided

Activity Standards Focus
In Activity 18, students solve systems of linear inequalities by graphing. They also interpret solutions of the systems as viable or nonviable within the context of a problem. Throughout this activity, it will be important for students to pay attention to which side of a boundary line should be shaded and whether the line should be solid or dashed.

Lesson 18-1

PLAN

Pacing: 1 class period
Chunking the Lesson
#1–3 #4–5
Example A #6
Check Your Understanding
Lesson Practice

TEACH

Bell-Ringer Activity
Write the following system of equations on the board.
$$\begin{cases} y = 2x - 1 \\ y = 3x - 3 \end{cases}$$

Ask students to solve the system by graphing [(2, 3)]. Then have students discuss how they determined the solution from their graphs.

1–3 Activating Prior Knowledge, Create Representations, Look for a Pattern, Think-Pair-Share, Quickwrite In these items, students graph simple and compound inequalities in one variable, both on number lines and on coordinate planes, in order to make connections between one- and two-dimensional representations of solution sets. They also compare the graphs of the simple inequalities to the graphs of the related compound inequalities.

TEACHER to TEACHER

In one dimension, the solution to the simple inequalities is a half-line, while the solution to the compound inequality is the intersection of the two half-lines. In two dimensions, the solution to the simple inequalities is a half-plane, while the solution to the compound inequality is the intersection of the two half-planes.

Activity 18 • Solving Systems of Linear Inequalities **273**

Lesson 18-1
Representing the Solution of a System of Inequalities

3. On the coordinate grid, graph the solutions common to the inequalities $y \leq 4$ and $y > 1$.

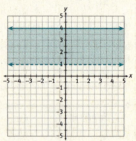

Solving a *system of linear inequalities* means finding all solutions that are common to all inequalities in the system.

4. **Reason quantitatively.** For the inequalities below, complete the table showing whether each ordered pair is a solution of the system of inequalities by deciding whether the ordered pair is a solution of **both** inequalities. Justify your responses.

$$x + y > 2$$
$$2x - y \geq -5$$

Ordered Pair	Is it a Solution?	Why or Why Not?
(−2, −3)	No	Satisfies the second inequality, but not the first
(3, 2)	Yes	Satisfies both inequalities
(3, −1)	No	Satisfies the second inequality, but not the first (because it is not inclusive)
(0, 5)	Yes	Satisfies both inequalities (The second is satisfied because it is inclusive.)

5. **Make use of structure.** Find two more solutions of the system in Item 4.

Answers may vary; students will likely use a "guess and check" strategy.

ACTIVITY 18 Continued

1–3 (continued) If students have difficulty graphing the common solutions in Item 3, suggest that they first graph the solutions of each simple inequality on separate coordinate planes, as they did in Item 1.

Differentiating Instruction

Some students may find it difficult to graph an inequality in one variable, such as $x < 2$, in two dimensions.

Support these students by suggesting that they first graph the related linear equation, $x = 2$, on a coordinate plane. Ask students to choose any two points on the line and explain why these points are solutions of $x = 2$. Then have students discuss how they could show all of the points on a coordinate plane that have an x-coordinate less than 2. To do so, they will have to shade all of the points to the left of the line $x = 2$. Remind students that whether they are graphing on a number line or in a coordinate plane, the solution to an inequality must show all values of the variable for which the inequality is a true statement.

4–5 Graphic Organizer, Guess and Check, Discussion Groups, Debriefing In this set of items, students use substitution to determine whether ordered pairs satisfy both of two given inequalities. Students may have difficulties with the last two ordered pairs in the table because these points lie on the boundary lines of the inequalities. Remind students to pay attention to whether the inequality signs include "equal to." Students may use a variety of methods to find two additional ordered pairs that satisfy both inequalities, but most will probably use guess and check. Debriefing will help students realize that there are many (in fact, infinitely many) ordered pairs that are solutions of both inequalities and that the set of these ordered pairs is difficult to describe numerically. As students realize this, the need for a graphic representation of the solutions of a system of inequalities becomes more obvious.

MATH TERMS

A **system of linear inequalities** consists of two or more linear inequalities with the same variables.

Lesson 18-1
Representing the Solution of a System of Inequalities

Since a system of inequalities has infinitely many solutions, you can represent all solutions using a graph. To solve a system of inequalities, graph each inequality on the same coordinate grid by shading a half-plane. The region that is the intersection of the two shaded half-planes, called the *solution region*, represents all solutions of the system.

Example A
Solve the system of inequalities.
$x + y \geq 1$
$x - 3y > 3$

Step 1: First, graph $x + y \geq 1$.

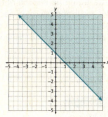

Step 2: Next, graph $x - 3y > 3$.

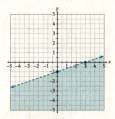

Step 3: Identify the solution region.

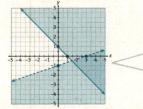

The solution region is the double shaded region.

Solution: The solution region is the double shaded region shown in Step 3.

MATH TERMS
A **solution region** is the part of the coordinate plane in which the ordered pairs are solutions to all inequalities in a system.

MATH TIP
To graph an inequality, it may be helpful to solve the inequality for y first.

MATH TIP
When finding the solution region for a system of inequalities, it may be helpful to shade each half-plane with a different pattern or color.

ACTIVITY 18 Continued

6 Discussion Groups, Create Representations, Look for a Pattern, Debriefing In this item, students make connections between the solutions they found for the pair of inequalities in Item 4 and the graph of this system they made in Try These part a. As this lesson is debriefed, be sure to focus on student understanding of the fact that solutions of a system of linear inequalities lie either in the double-shaded region or on the solid boundary of that region. Be sure students are able to explain why points on a solid boundary of a solution region are solutions while points on a dashed boundary are not solutions.

Differentiating Instruction

Support visual learners by suggesting they shade each inequality in a given system using a different color. The colors overlap to show the solution region for the system of inequalities.

Check Your Understanding

Debrief students' answers to these items to make sure they understand how to solve systems of inequalities and identify solutions from the solution region. For Item 9, have students explain how they decided whether to shade above or below the boundary lines when graphing the inequalities.

Answers

7. a. Sample answer: $(-4, 2)$
   b. Sample answer: $(3, 2)$
   c. Yes; the boundary line of each inequality is included in its solution set and the boundary lines intersect at $(0, 3)$.

8.

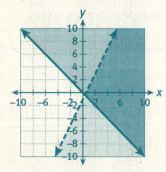

a.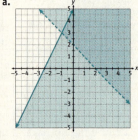

b.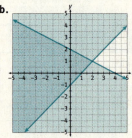

c.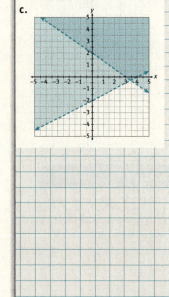

Lesson 18-1
Representing the Solution of a System of Inequalities

Try These A
Solve each system of inequalities using the coordinate grids in the *My Notes* section.

a. $x + y > 2$
$2x - y \geq -5$

b. $y \geq x - 1$
$y \leq -\frac{1}{2}x + 2$

c. $2x + 3y > 6$
$x - 2y < 4$

6. The system in Try These Part (a) is the same as the system in Item 4. Plot the points listed in Items 4 and 5 on the graph you made for Try These A part a. Where do the points that are solutions lie? Where do points that are not solutions lie? Give three additional points that are solutions.

After plotting the points, students should notice that the points representing solutions are in the double shaded region or on the solid boundary line of that region, and points that do not represent solutions lie outside the double shaded region or on the dashed boundary line of that region. Given solutions will vary.

Check Your Understanding

7. The graph below shows the solution of a system of inequalities.

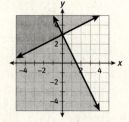

   a. Name an ordered pair that is a solution of the system.
   b. Name an ordered pair that is **not** a solution of the system.
   c. Is the ordered pair $(0, 3)$ a solution of the system? Explain how you know.

Solve each system of inequalities.

8. $y < 2x - 1$
$y \geq -x$

9. $3x + y < 3$
$x - y > 1$

10. $y \geq \frac{2}{3}x + 4$
$x + 2y < 6$

9.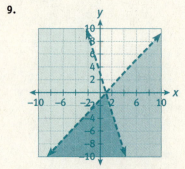

10.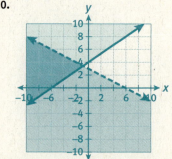

Lesson 18-1
Representing the Solution of a System of Inequalities

ACTIVITY 18 continued

LESSON 18-1 PRACTICE

Solve each system of inequalities.

11. $y \geq x - 3$
    $y \leq \frac{1}{2}x + 1$

12. $3x + 3y > 1$
    $2y < 11$

13. $y < \frac{1}{3}x$
    $x + y \geq -2$

14. Name three ordered pairs that are solutions of the system of inequalities shown below.

    $y < 4x + 4$
    $x - y > 3$

15. Write a system of inequalities whose solution is shown by the overlapping regions in the graph below.

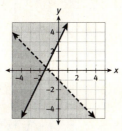

16. **Critique the reasoning of others.** A student was asked to solve the system $y \geq \frac{1}{2}x + 4$ and $x + 4y \leq -20$. She made the graph shown below. She noticed that the two shaded regions do not overlap, so she concluded that the system of inequalities has no solution. Do you agree or disagree? Justify your reasoning.

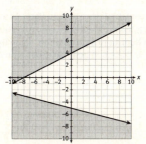

13.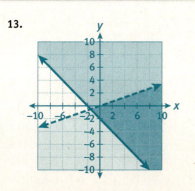

16. I disagree. If the student used a different scale on the graph, she would see that the shaded regions do overlap, as shown below. So, the system of inequalities does have solutions.

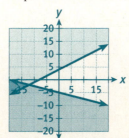

14. Sample answer: $(3, -4), (-1, -7), (0, -5)$
15. $y \geq 2x + 3$
    $y < -x - 1$

ACTIVITY 18 Continued

ASSESS

Students' answers to lesson practice problems will provide you with a formative assessment of their understanding of the lesson concepts and their ability to apply their learning.

See the Activity Practice for additional problems for this lesson. You may assign the problems here or use them as a culmination for the activity.

LESSON 18-1 PRACTICE

11.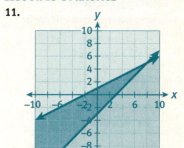

12.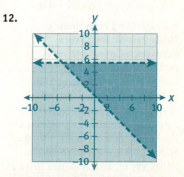

ADAPT

Check students' answers to the Lesson Practice to ensure that they understand concepts related to systems of linear inequalities. Some students may have difficulty with these items if they have not yet mastered the skill of graphing a linear inequality in two variables. Review the following steps with these students:

1. Solve the inequality for y. Reverse the inequality sign if you multiply or divide both sides by a negative number.

2. Graph the boundary line. Use a dashed line for > or < and a solid line for $\geq$ or $\leq$.

3. Choose a test point. Substitute the test point into the given inequality to decide which half-plane should be shaded to represent all of the solutions to the linear inequality.

Activity 18 • Solving Systems of Linear Inequalities 277

ACTIVITY 18 Continued

Lesson 18-2

PLAN

Pacing: 1 class period
Chunking the Lesson
#1–2 #3
Check Your Understanding
Lesson Practice

TEACH

Bell-Ringer Activity
Write the following equations on the board.
a. $y = 3x + 4$
b. $y + 3x = 5$
c. $y - 2 = 3(x - 5)$
d. $y = \frac{1}{3}x$

Ask students to determine which of the lines are parallel without graphing the lines and to explain how they know [a and c]. If needed, remind students that parallel lines have the same slope.

1–2 Activating Prior Knowledge, Think-Pair-Share, Create Representations, Look for a Pattern These items allow students to explore systems of linear inequalities with parallel boundary lines. They identify two of the possibilities for this type of system: no solution or solutions that lie between the parallel lines. It is important for students to be able to justify why the system in Item 1a has no solution.

Differentiating Instruction

Extend learning for students by having them explore a third possibility for a system of inequalities with parallel boundary lines. Ask students to solve the system $\begin{cases} 3x + y < 3 \\ y + 2 \leq -3x \end{cases}$ by graphing. In this case, the solution region is identical to the solution set of one of the inequalities in the system, $y + 2 \leq -3x$. Have students compare and contrast this system with the one in Item 1b.

ACTIVITY 18 continued

Lesson 18-2
Interpreting the Solution of a System of Inequalities

Learning Targets:
- Identify solutions to systems of linear inequalities when the solution region is determined by parallel lines.
- Interpret solutions of systems of linear inequalities.

SUGGESTED LEARNING STRATEGIES: Create Representations, Look for a Pattern, Discussion Groups, Close Reading, Marking the Text

As you share your ideas, be sure to use mathematical terms and academic vocabulary precisely. Make notes to help you remember the meaning of new words and how they are used to describe mathematical concepts.

1. Determine the solutions to the systems of inequalities by graphing.
   a. $y > \frac{1}{2}x + 2$
      $x - 2y > 8$

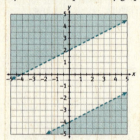

   b. $3x + y < 3$
      $y + 2 \geq -3x$

2. Compare and contrast the systems in Items 1a and 1b.
   Both systems in Item 1 have boundary lines that are parallel. The system in Item 1a has no solution, since the half-planes do not intersect. The solutions in Item 1b lie between the parallel lines.

Lesson 18-2
Interpreting the Solution of a System of Inequalities

3. Ray plays on the basketball team. Last week, he scored all of his points from free throws (worth one point each) and field goals (worth two points each). He has forgotten how many points he scored, but he remembers some facts from the game. Review with your group the background information that is given as you solve the items below.

   a. **Reason abstractly.** Let f represent the number of free throws and g represent the number of field goals. Write an inequality for each of the facts below.

      Ray scored fewer than 20 points.
      $f + 2g < 20$

      Ray made fewer than six free throws.
      $f < 6$

      At most, Ray made twice as many free throws as he made field goals.
      $f \leq 2g$

   b. Graph the solutions to the system represented by your inequalities from part a.

   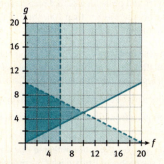

   c. **Reason quantitatively.** In the solution region you graphed in part b not every point makes sense in this context. Give two solutions that make sense in the context of the problem and one that does not. Explain your reasoning.
   Possible answers: (4, 6) and (5, 4); (3.5, 5). Explanations may vary; Ray can only score whole numbers of free throws and field goals, so points with fraction or decimal coordinates do not make sense.

ACTIVITY 18 Continued

3 Close Reading, Marking the Text, Discussion Groups, Create Representations, Debriefing

Students write and solve a system of inequalities representing a real-world problem situation by graphing and interpreting possible solutions. Monitor group discussions carefully to be sure students have written the inequalities correctly before moving on to graph the system. As groups discuss which types of solutions within the solution region make sense in this situation and which do not make sense, they should come to realize that Ray could only score a whole number of free throws and a whole number of field goals. So, only ordered pairs in the solution region with whole-number coordinates make sense in this situation. Debrief this portion of the activity by asking students to discuss the characteristics of a system of inequalities having no solution. Are there instances when every point in a solution region may not be a solution to the given problem?

TEACHER to TEACHER

In mathematics, a constraint is a restriction on the value of a variable or the value of an expression involving one or more variables. Constraints can be expressed as statements of equality or inequality. For example, the inequalities that students write in Item 3 represent constraints on the values of f and g. Solutions that meet all of the constraints of a real-world situation are said to be *viable* if they make sense in the situation and *nonviable* if they do not make sense. Thus, in Item 3, the solution (4, 6) is viable because it is possible for Ray to score 4 free throws and 6 field goals. By contrast, the solution (3.5, 5) is nonviable because it is not possible for Ray to score 3.5 free throws.

ACTIVITY 18 Continued

Check Your Understanding

Debrief students' answers to these items to ensure that they can solve systems of inequalities and write a system of inequalities to represent a situation. Ask students to write and graph a system of two linear inequalities that has no solution.

Answers

4. $(4, -4)$
5.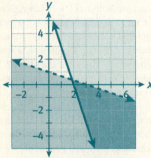

$(4, -4)$ lies in the region where the solutions of the inequalities overlap.

6.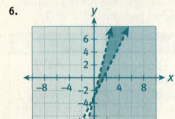

7. $x < 0$
   $y < 0$

ASSESS

Students' answers to lesson practice problems will provide you with a formative assessment of their understanding of the lesson concepts and their ability to apply their learning.

See the Activity Practice for additional problems for this lesson. You may assign the problems here or use them as a culmination for the activity.

ADAPT

Check students' answers to the Lesson Practice to make sure that they understand how to write a system of linear inequalities to model a real-world situation and can identify viable solutions of the system. If students have difficulty recognizing viable solutions, remind them to consider what the variables in the inequalities represent and what types of numbers make sense for those quantities (only integers? only positive numbers? only whole numbers?).

LESSON 18-2 PRACTICE

8. $y < 100 - x$
   $y < x$

9.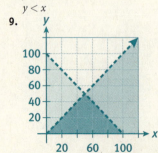

Lesson 18-2
Interpreting the Solution of a System of Inequalities

Check Your Understanding

4. Use guess and check to identify an ordered pair that is a solution to the system of inequalities.
$$3x + y \geq 6$$
$$x + 3y < 3$$

5. Solve the system in Item 4 by graphing. Confirm that the ordered pair you wrote in Item 4 is a solution. Explain.

6. Solve the system of linear inequalities.
$$y > 2x - 3$$
$$y < 4x - 3$$

7. Write a system of inequalities whose solution region is all the points in the third quadrant.

LESSON 18-2 PRACTICE

Catherine bought apples and pears for a school picnic. The total number of pieces of fruit that she bought was less than 100. She bought more apples than pears. Use this information for Items 8–11.

8. **Model with mathematics.** Write a system of inequalities that describes the situation. Let x represent the number of apples Catherine bought and let y represent the number of pears Catherine bought.

9. Graph the solutions of the system you wrote in Item 8.

10. Give two ordered pairs from the solution region that make sense in the context of the problem.

11. Give an ordered pair from the solution region that does not make sense in the context of the problem. Explain why the ordered pair does not make sense.

12. Describe the solution region of the system of the inequalities $-x + 2y \leq 6$ and $2y \geq x + 6$.

10. Sample answer: (6, 2) and (60, 30)
11. Sample answer: (40, 10.3); Catherine can only buy a whole number of pears.
12. The solution set is the line $y = \frac{1}{2}x + 3$.

Solving Systems of Linear Inequalities
Which Region Is It?

ACTIVITY 18 continued

ACTIVITY 18 PRACTICE
Write your answers on notebook paper. Show your work.

Lesson 18-1

1. Determine which of the following ordered pairs are solutions to each of the given systems of inequalities.

   $\{(5, 3), (-2, 1), (1, 2), (2, -3), (3, 5), (-2, 3), (2, 0)\}$

   a. $y < -x + 3$
      $y \geq x - 2$

   b. $2x - y \leq 0$
      $y > -\frac{1}{2}x + 1$

   c. $3x - y > 6$
      $y \leq 3$

2. Graph each of the systems of inequalities in Item 1. Graph the points that you chose as solutions from Item 1 on the same coordinate grid to verify that they are solutions.

3. Identify four ordered pairs that are solutions to the following system of equations. Explain how you chose your points.

   $4x + 3y \geq -12$
   $2x - y < 4$

4. Solve the following systems of inequalities.
   a. $2x + 3y \geq 15$
      $5x - y \geq 3$
   b. $x - 4y \geq 4$
      $4y - x > 8$

5. Tickets for the school play cost $3 for students and $6 for adults. The drama club hopes to bring in at least $450 in sales, and the auditorium has 120 seats. Let a represent the number of adult tickets, and let s represent the number of student tickets.
   a. Write a system of inequalities representing this situation.
   b. Show the solutions to the system of inequalities by graphing.
   c. If the show sells out, what is the greatest number of student tickets that could be sold to get the desired amount of sales?

6. Which system of inequalities represents all of the points in the second quadrant?
   A. $x > 0, y < 0$
   B. $x > 0, y > 0$
   C. $x < 0, y < 0$
   D. $x < 0, y > 0$

7. Which graph represents the system of inequalities shown?

   $2x - y > 3$
   $x + y > 4$

   A.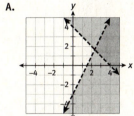

   B.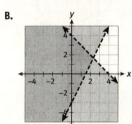

   C.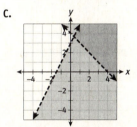

   D.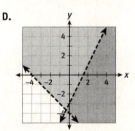

ACTIVITY 18 Continued
ACTIVITY PRACTICE

1. a. $(-2, 1), (-2, 3), (2, 0)$
   b. $(1, 2), (-2, 3)$
   c. $(5, 3), (2, -3)$

2. a.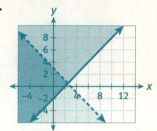

   b.

   c.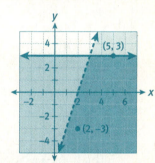

3. Sample answers: $(1, 0), (0, 0), (2, 5), (-3, 0)$; Explanations will vary.

4. a.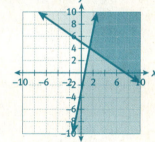

   b.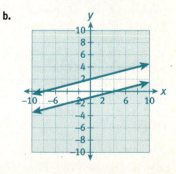

5. a. $6a + 3s \geq 450$
      $a + s \leq 120$

   b.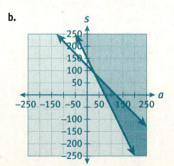

   c. 90 student tickets

6. D
7. A

ACTIVITY 18 Continued

8. B
9. D
10. a.

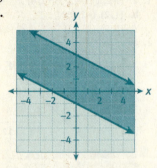

b. The solution region lies between the parallel boundary lines and includes both boundary lines.
c. Sample answer: (0, 1), (−2, 2), (3, 0)
d. There would be no solutions. The shading for each inequality would be outside the parallel boundary lines, so there would be no overlap in the shaded regions.

11. a. $2p + 3h \leq 90$
$p + h \leq 35$;
p represents paperbacks, and h represents hardcovers.

b.

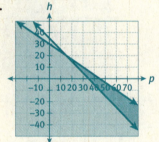

c. Answers will vary. (28, 4) means 28 paperbacks and 4 hardcovers or 32 books that cost $68. (4, 25) means 4 paperbacks and 25 hardcovers or 29 books that cost $83.
d. He bought 15 paperbacks and 20 hardcovers. The point (15, 20) is where the lines $2p + 3h = 90$ and $p + h = 35$ intersect. $15 + 20 = 35$ and $2(15) + 3(20) = 90$.

ADDITIONAL PRACTICE

If students need more practice on the concepts in this activity, see the eBook Teacher Resources for additional practice problems.

ACTIVITY 18 continued

8. Which system of inequalities has the solution region shown in the graph below?

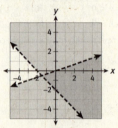

A. $3y < x$
   $x + y > 2$
B. $3y < x$
   $x + y > -2$
C. $3y > x$
   $x + y > 2$
D. $3y > x$
   $x + y > -2$

9. The graph shows the solution of a system of inequalities. Which statement is true?

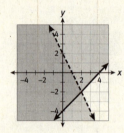

A. (2, −2) is a solution.
B. The origin is not a solution.
C. Any ordered pair with a negative x-coordinate is a solution.
D. (2, −4) is not a solution.

12. a. Answers will vary; $y > 2x + 1$
$y < 2x - 4$
b. Yes; if a system of linear inequalities has coincident boundary lines and at least one of the inequalities is represented by an open half-plane (dashed boundary line), the system will have no solutions. An example is $y > x + 3$
$y - 3 \leq x$.

Solving Systems of Linear Inequalities
Which Region Is It?

Lesson 18-2

10. Consider the system of inequalities shown below.
$$x + 2y \leq 6$$
$$x + 2y \geq -2$$

a. Graph the solution of the system of inequalities.
b. Describe the solution region.
c. Name three ordered pairs that are solutions of the system.
d. How would the solution be different if the inequality signs were reversed? That is, what is the solution of the system $x + 2y \geq 6$ and $x + 2y \leq -2$? Explain why your answer makes sense.

11. Connor bought used books from a Web site. The paperbacks cost $2 each, and the hardcovers cost $3 each. He spent no more than $90 on the books, and he bought no more than 35 books.
a. Write a system of inequalities to represent the situation. State what the variables represent.
b. Graph the solution of the system of inequalities you wrote in part a.
c. Name two ordered pairs that are solutions. Interpret the meaning of each ordered pair in the context of the problem.
d. Suppose you know that Connor spent exactly $90 and that he bought exactly 35 books. What can you conclude in this case? Why?

12. a. Write a system of linear inequalities that has no solutions.
b. Is it possible for a system of linear inequalities with no solutions to have nonparallel boundary lines? If so, explain why and give an example. If not, explain why not. (Hint: Consider coincident boundary lines.)

MATHEMATICAL PRACTICES
Attend to Precision

13. Explain why graphing is the preferred method of representing the solutions of a system of linear inequalities.

---

13. Since a system of inequalities may have an infinite number of solutions, graphing is preferred because it represents all of the possible solutions to a system of inequalities.

Systems of Equations and Inequalities
TILT THE SCALES

Embedded Assessment 2
Use after Activity 18

1. Rajesh and his brother Mohib are each mailing a birthday gift to a friend. Rajesh's package weighs three more pounds than twice the weight of Mohib's package. The combined weight of both packages is 15 pounds.
   a. Write a system to represent this situation. Define each variable that you use.
   b. Solve the system using substitution or elimination to determine the weight of each package. Justify the reasonableness of your solution.
   c. Rajesh and Mohib each graph the system that represents this situation. Who is correct? Explain why.

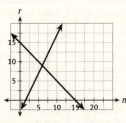
Rajesh's Graph

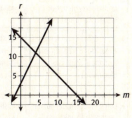
Mohib's Graph

2. Rajesh and Mohib will mail rectangular packages that meet the weight and height requirements of their delivery service. Their packages are described in the table.

	Height	Length	Width
Rajesh's package	7 in.	18 in.	15 in.
Mohib's package	8 in.	19 in.	13 in.

   a. The delivery service also requires that both the length and width of the packages be 20 inches or less and that the length plus the width be no more than 32 inches. Write a system of inequalities to represent this situation.
   b. Graph the system from Part (a) to show all of the possible dimensions of length and width that the packages could have.
   c. Is either package unacceptable? Justify your answer using the table or the graph.

Embedded Assessment 2

Assessment Focus
- Systems of linear equations
- Systems of linear inequalities

Materials
- Colored pencils

Answer Key

1. a. Let x represent the weight of Mohib's package. Let y represent the weight of Rajesh's package.
   $$y = 3 + 2x$$
   $$y = 15 - x$$
   b. $x = 4$; $y = 11$; The combined weight is 15 pounds. Also, 11 is 3 more than 2 times 4.
   c. Mohib's graph is correct. The point of intersection is (4, 11).

   The lines on Rajesh's graph intersect at (6, 9); 9 is not 3 more than twice 6.

2. a. $L \leq 20$
      $W \leq 20$
      $L + W \leq 32$
   b.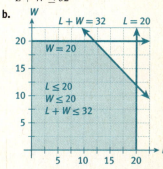
   c. Rajesh's package is unacceptable because $L + W = 33$ inches; the point (18, 15) is not in the shaded region.

Common Core State Standards for Embedded Assessment 2

HSA-CED.A.3: Represent constraints by equations or inequalities, and by systems of equations and/or inequalities, and interpret solutions a viable or nonviable options in a modeling context. *For example, represent inequalities describing nutritional and cost constraints on combinations of different foods.*

HSA-REI.C.6: Solve systems of linear equations exactly and approximately (e.g., with graphs), focusing on pairs of linear equations in two variables.

Embedded Assessment 2

Teacher to Teacher

You may wish to read through the scoring guide with students and discuss the differences in the expectations at each level. Check that students understand the terms used.

Embedded Assessment 2
Use after Activity 18

Systems of Equations and Inequalities
TILT THE SCALES

Scoring Guide	Exemplary	Proficient	Emerging	Incomplete
	The solution demonstrates the following characteristics:			
Mathematics Knowledge and Thinking (Items 1b, 2b)	• Effective understanding of and accuracy in solving systems of equations and inequalities	• Adequate understanding of and accuracy in solving systems of equations and inequalities	• Partial understanding of and some difficulty solving systems of equations and inequalities	• Incomplete understanding of and significant difficulty solving systems of equations and inequalities
Problem Solving (Items 1b, 2c)	• Appropriate and efficient strategy that results in a correct answer	• Strategy that may include unnecessary steps but results in a correct answer	• Strategy that results in some incorrect answers	• No clear strategy when solving problems
Mathematical Modeling / Representations (Items 1a, 2a)	• Fluency in representing real-world scenarios using systems of equations and inequalities	• Little difficulty representing a real-world scenario using systems of equations and inequalities	• Partial understanding of how to represent real-world scenarios using systems of equations and inequalities	• Little or no understanding of how to represent real-world scenarios using systems of equations and inequalities
Reasoning and Communication (Items 1c, 2c)	• Precise use of appropriate math terms and language to identify and explain an error • Ease and accuracy describing the relationship between a table or a graph and a real-world scenario	• Correct identification of an error with an adequate explanation • Little difficulty describing the relationship between a table or a graph and a real-world scenario	• Misleading or confusing explanation of an error • Partially correct description of the relationship between a table or a graph and a real-world scenario	• Inaccurate identification of an error with an incomplete or inaccurate explanation • Little or no understanding of how a table or a graph might relate to a real-world scenario

Common Core State Standards for Embedded Assessment 2 (cont.)

HSA-REI.D.11: Explain why the x-coordinates of the points where the graphs of the equations $y = f(x)$ and $y = g(x)$ intersect are the solutions of the equation $f(x) = g(x)$; find the solutions approximately, e.g., using technology to graph the functions, make tables of values, or find successive approximations. Include cases where $f(x)$ and/or $g(x)$ are linear, polynomial, rational, absolute value, exponential, and logarithmic functions.*

HSA-REI.D.12: Graph the solutions to a linear inequality in two variables as a half-plane (excluding the boundary in the case of a strict inequality), and graph the solution set to a system of linear inequalities in two variables as the intersection of the corresponding half-planes.

Unit 4 Planning the Unit

In prior units students have generally studied linear relationships. Now students focus on exponent rules and functions, and extends into operations with radical and polynomial functions and operations. Rational expressions are also introduced.

Vocabulary Development

The key terms for this unit can be found on the Unit Opener page. These terms are divided into Academic Vocabulary and Math Terms. Academic Vocabulary includes terms that have additional meaning outside of math. These terms are listed separately to help students transition from their current understanding of a term to its meaning as a mathematics term. To help students learn new vocabulary:

- Have students discuss meaning and use graphic organizers to record their understanding of new words.
- Remind students to place their graphic organizers in their math notebooks and revisit their notes as their understanding of vocabulary grows.
- As needed, pronounce new words and place pronunciation guides and definitions on the class Word Wall.

Embedded Assessments

Embedded Assessments allow students to do the following:

- Demonstrate their understanding of new concepts.
- Integrate previous and new knowledge by solving real-world problems presented in new settings.

They also provide formative information to help you adjust instruction to meet your students' learning needs.

Prior to beginning instruction, have students unpack the first Embedded Assessment in the unit to identify the skills and knowledge necessary for successful completion of that assessment. Help students create a visual display of the unpacked assessment and post it in your class. As students learn new knowledge and skills, remind them that they will be expected to apply that knowledge to the assessment. After students complete each Embedded Assessment, turn to the next one in the unit and repeat the process of unpacking that assessment with students.

CollegeBoard

AP / College Readiness

Unit 4 expands on students' understanding of the concept of rate of change, as well as the properties, language, and algebra of some nonlinear functions by:

- Giving students the opportunity to look further at exponential and polynomial functions graphically, numerically, algebraically, and verbally, both in and out of contextual situations.
- Introducing geometric sequences and the formulas for calculating their terms.

Unpacking the Embedded Assessments

The following are the key skills and knowledge students will need to know for each assessment.

Embedded Assessment 1

Exponents, Radicals, and Geometric Sequences *Taking Stock*

- Rational and irrational numbers
- Exponential expressions
- Radical expressions
- Geometric sequences

Embedded Assessment 2

Exponential Functions, *Family Bonds*

- Exponential functions
- Compound interest

Embedded Assessment 3

Polynomial Operations, *Measuring Up*

- Adding polynomials
- Multiplying polynomials

Planning the Unit continued

Embedded Assessment 4

Factoring and Simplifying Rational Expressions, *Rock Star Demands*
- Factoring trinomials
- Dividing polynomials
- Rational expressions

Suggested Pacing

The following table provides suggestions for pacing using a 45-minute class period. Space is left for you to write your own pacing guidelines based on your experiences in using the materials.

	45-Minute Period	Your Comments on Pacing
Unit Overview/Getting Ready	1	
Activity 19	3	
Activity 20	3	
Activity 21	2	
Embedded Assessment 1	1	
Activity 22	3	
Activity 23	2	
Embedded Assessment 2	1	
Activity 24	3	
Activity 25	3	
Embedded Assessment 3	1	
Activity 26	2	
Activity 27	2	
Activity 28	4	
Embedded Assessment 4	1	
Total 45-Minute Periods	**32**	

Additional Resources

Additional resources that you may find helpful for your instruction include the following, which may be found in the eBook Teacher Resources.

- Unit Practice (additional problems for each activity)
- Getting Ready Practice (additional lessons and practice problems for the prerequisite skills)
- Mini-Lessons (instructional support for concepts related to lesson content)

Exponents, Radicals, and Polynomials

Unit Overview
In this unit you will explore multiplicative patterns and representations of nonlinear data. Exponential growth and decay will be the basis for studying exponential functions. You will investigate the properties of powers and radical expressions. You will also perform operations with radical and rational expressions.

Key Terms
As you study this unit, add these and other terms to your math notebook. Include in your notes your prior knowledge of each word, as well as your experiences in using the word in different mathematical examples. If needed, ask for help in pronouncing new words and add information on pronunciation to your math notebook. It is important that you learn new terms and use them correctly in your class discussions and in your problem solutions.

Math Terms
- radical expression
- principal square root
- negative square root
- cube root
- rationalize
- tree diagram
- geometric sequence
- common ratio
- arithmetic sequence
- recursive formula
- exponential growth
- exponential function
- exponential decay
- compound interest
- exponential regression
- term
- polynomial
- coefficient
- constant term
- degree of a term
- degree of a polynomial
- standard form of a polynomial
- descending order
- leading coefficient
- monomial
- binomial
- trinomial
- like terms
- difference of two squares
- square of a binomial
- greatest common factor of a polynomial
- perfect square trinomial
- rational expression

ESSENTIAL QUESTIONS

 How do multiplicative and exponential patterns model the physical world?

 How are adding and multiplying polynomial expressions different from each other?

EMBEDDED ASSESSMENTS

This unit has four embedded assessments, following Activities 21, 23, 25, and 28. They will give you an opportunity to demonstrate what you have learned.

Embedded Assessment 1:
Exponents, Radicals, and Geometric Sequences   p. 323

Embedded Assessment 2:
Exponential Functions   p. 353

Embedded Assessment 3:
Polynomial Operations   p. 383

Embedded Assessment 4:
Factoring and Simplifying Rational Expressions   p. 419

285

Unit Overview
Ask students to read the unit overview and mark the text to identify key phrases that indicate what they will learn in this unit.

Materials
- Algebra tiles
- Graphing calculators

Key Terms
As students encounter new terms in this unit, help them to choose an appropriate graphic organizer for their word study. As they complete a graphic organizer, have them place it in their math notebooks and revisit as needed as they gain additional knowledge about each word or concept.

Essential Questions
Read the essential questions with students and ask them to share possible answers. As students complete the unit, revisit the essential questions to help them adjust their initial answers as needed.

Unpacking Embedded Assessments
Prior to beginning the first activity in this unit, turn to Embedded Assessment 1 and have students unpack the assessment by identifying the skills and knowledge they will need to complete the assessments successfully. Guide students through a close reading of the assignment, and use a graphic organizer or other means to capture their identification of the skills and knowledge. Repeat the process for each Embedded Assessment in the unit.

Developing Math Language
As this unit progresses, help students make the transition from general words they may already know to the meanings of those words in mathematics. You may want students to work in pairs or small groups to facilitate discussion and to build confidence and fluency as they internalize new language. Ask students to discuss new mathematics terms as they are introduced, identifying meaning as well as pronunciation and common usage. Remind students to use their math notebooks to record their understanding of new terms and concepts.

As needed, pronounce new terms clearly and monitor students' use of words in their discussions to ensure that they are using terms correctly. Encourage students to practice fluency with new words as they gain greater understanding of mathematical and other terms.

285

UNIT 4
Getting Ready

Use some or all of these exercises for formative evaluation of students' readiness for Unit 4 topics.

Prerequisite Skills
- Factors and greatest common factors (Items 1, 2) 6.NS.B.4, 4.OA.B.4
- Exponential expressions (Items 3, 4) 6.EE.A.1, 6.EE.A.2c, 6.EE.A.2b
- Distributive property (Item 5) 3.OA.B.5
- Linear functions (Item 6) 8.F.B.4
- Graph linear functions (Item 7) HSF-IF.C.7a
- Ratios (Item 8) 6.RP.A.1
- Recognize rational and irrational numbers (Item 9) 8.NS.A.1
- Fraction operations (Item 10) 5.NF.A.1, 5.NF.B.4, 6.NS.A.1

Answer Key
1. 18
2. 1, 2, 3, 5, 6, 9, 10, 15, 18, 30, 45, 90
3. D
4. The coefficient is 4. The base is x. The exponent is 5.
5. Sample explanation: One way: First subtract $90 - 3$. Then multiply the difference by 15. Another way: Multiply 15 by 90. Multiply 15 by 3. Then subtract the second product from the first.
6.

x	2	4	6	8	10
y	3	5	7	9	11

7.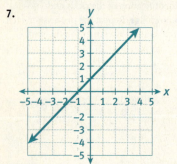

8. a. $\frac{75}{10}$
   b. $\frac{341}{436}$
   c. $\frac{12}{19}$
   d. $\frac{19}{31}$
9. a. rational
   b. rational
   c. rational
   d. irrational
10. a. $\frac{7}{8}$
    b. $\frac{1}{12}$
    c. $\frac{3}{5}$
    d. $\frac{5}{6}$

UNIT 4
Getting Ready

Write your answers on notebook paper. Show your work.

1. Find the greatest common factor of 36 and 54.
2. List all the factors of 90.
3. Which of the following is equivalent to $39 \cdot 26 + 39 \cdot 13$?
   A. 13^9 B. $13^4 \cdot 14$
   C. $13^2 \cdot 3^2 \cdot 2$ D. $13^2 \cdot 3^2$
4. Identify the coefficient, base, and exponent of $4x^5$.
5. Explain two ways to evaluate $15(90 - 3)$.
6. Complete the following table to create a linear relationship.

x	2	4	6	8	10
y	3	5			

7. Graph the function described in the table in Item 6.
8. Use ratios to model the following:
   a. 7.5
   b. Caleb receives 341 of the 436 votes cast for class president.
      Students in Mr. Bulluck's Class

Girls	Boys
12	19

   c. girls to boys
   d. boys to total class members
9. Tell whether each number is rational or irrational.
   a. $\sqrt{25}$ b. $\frac{4}{3}$
   c. 2.16 d. π
10. Calculate.
    a. $\frac{1}{2} + \frac{3}{8}$ b. $\frac{5}{12} - \frac{1}{3}$
    c. $\frac{3}{2} \cdot \frac{2}{5}$ d. $\frac{5}{8} \div \frac{3}{4}$

Getting Ready Practice
For students who may need additional instruction on one or more of the prerequisite skills for this unit, Getting Ready practice pages are available in the Teacher Resources at SpringBoard Digital. These practice pages include worked-out examples as well as multiple opportunities for students to apply concepts learned.

Exponent Rules
Icebergs and Exponents
Lesson 19-1 Basic Exponent Properties

Learning Targets:
- Develop basic exponent properties.
- Simplify expressions involving exponents.

SUGGESTED LEARNING STRATEGIES: Create Representations, Predict and Confirm, Look for a Pattern, Think-Pair-Share, Discussion Groups, Sharing and Responding

An *iceberg* is a large piece of freshwater ice that has broken off from a glacier or ice shelf and is floating in open seawater. Icebergs are classified by size. The smallest sized iceberg is called a "growler."

A growler was found floating in the ocean just off the shore of Greenland. Its volume above water was approximately 27 cubic meters.

CONNECT TO GEOLOGY

Because ice is not as dense as seawater, about one-tenth of the volume of an iceberg is visible above water. It is difficult to tell what an iceberg looks like underwater simply by looking at the visible part. Growlers got their name because the sound they make when they are melting sounds like a growling animal.

GROUP DISCUSSION TIPS

Work with your peers to set rules for:
- discussions and decision-making
- clear goals and deadlines
- individual roles as needed

MATH TERMS

The expression 3^4 is a **power**. The **base** is 3 and the **exponent** is 4. The term **power** may also refer to the **exponent**.

1. **Reason quantitatively.** Two icebergs float near this growler. One iceberg's volume is 3^4 times greater than the growler. The second iceberg's volume is 2^8 times greater than the growler. Which iceberg has the larger volume? Explain.

   The iceberg with a volume that is 2^8 times greater than the growler is larger since $3^4 = 81$ and $2^8 = 256$, and $256 > 81$.

2. What is the meaning of 3^4 and 2^8? Why do you think *exponents* are used when writing numbers?

   $3^4 = 3 \cdot 3 \cdot 3 \cdot 3$ and $2^8 = 2 \cdot 2 \cdot 2 \cdot 2 \cdot 2 \cdot 2 \cdot 2 \cdot 2$. Answers may vary. Exponents make it easy to represent products of repeated factors.

3. Suppose the original growler's volume under the water is 9 times the volume above. How much of its ice is below the surface?

   243 cubic meters

4. Write your solution to Item 3 using powers. Complete the equation below. Write the missing terms as a *power* of 3.

   volume above water $\cdot\, 3^2 =$ volume below the surface

   $\boxed{3^3} \cdot 3^2 = \boxed{3^5}$

5. Look at the equation you completed for Item 4. What relationship do you notice between the exponents on the left side of the equation and the exponent on the right?

   The exponent on the right side of the equation is the sum of the exponents on the left.

Common Core State Standards for Activity 19

HSN-RN.A.1	Explain how the definition of the meaning of rational exponents follows from extending the properties of integer exponents to those values, allowing for a notation for radicals in terms of rational exponents.
HSN-RN.A.2	Rewrite expressions involving radicals and rational exponents using the properties of exponents.
HSA-SSE.B.3c	Use the properties of exponents to transform expressions for exponential functions.

ACTIVITY 19
Guided

Activity Standards Focus

In Activity 19, students investigate and apply properties of exponents to simplify numeric and algebraic expressions. Students will learn that the properties of whole-number exponents also apply to integer exponents as well as rational exponents.

Lesson 19-1

PLAN

Pacing: 1 class period

Chunking the Lesson

#1 #2–5 #6–8
#9–11 #12–14 Examples A, B
Check Your Understanding
Lesson Practice

TEACH

Bell-Ringer Activity

Write the expression $x + x + x$ on the board. Then ask students to simplify the expression [$3x$]. Next, write $x \cdot x \cdot x$ on the board, and ask students to write an equivalent expression [x^3]. Have students discuss why $x \cdot x \cdot x$ is not equal to $3x$.

1 Shared Reading, Marking the Text, Activating Prior Knowledge, Think-Pair-Share Students will use the analysis of icebergs to develop the ideas of exponent rules and negative exponents.

2–5 Activating Prior Knowledge, Create Representations, Look for a Pattern, Discussion Groups, Predict and Confirm These questions are designed as ease of entry questions. They serve to activate students' prior knowledge of exponents. Monitor group discussions carefully to ensure students are expressing the quantities requested in Item 4 as powers of 3. Some students may already know or be able to predict the Product of Powers Property. They will verify their predictions in the subsequent items.

Activity 19 • Exponent Rules 287

ACTIVITY 19 Continued

6–8 Predict and Confirm, Look for a Pattern, Create Representations, Think-Pair-Share, Sharing and Responding Throughout this activity, students complete tables to illustrate patterns with exponents. After each table, students use the patterns they observe to generalize their findings in symbolic forms for the exponent rules. In Item 7, students should be given the opportunity to share their thoughts on what the missing exponent should be, but as groups share out and respond to the thoughts of other groups, be sure the property is clarified as $a^m \cdot a^n = a^{m+n}$. It is important that students also be able to verbalize the rule: *To multiply two powers with the same base, add the exponents and keep the base the same.* Item 8 will likely be students' first exposure to rational exponents. Rational exponents will be explored in depth in the next activity. At this point, students should understand the properties of exponents also apply to any type of exponent, including rational exponents. Allow time for students to add the Product of Powers Property to their Properties of Exponents graphic organizer.

9–11 Guess and Check, Look for a Pattern, Discussion Groups, Predict and Confirm, Sharing and Responding It is expected that students use a calculator for Items 9 and 10. It is understood that some teachers prefer to have students compute by hand to improve their numeracy skills. However, the intent of this lesson is for students to explore exponents, and computing by hand may distract from this purpose and slow student progress on the lesson. Students may use guess and check to rewrite quantities in Item 9. Monitor group discussions carefully to ensure students are expressing the quantities requested in Item 9 as powers of 9. Some students may already know or be able to predict the Quotient of Powers Property. They will use the items that follow to verify their predictions.

ACTIVITY 19 continued

Lesson 19-1
Basic Exponent Properties

6. Use the table below to help verify the pattern you noticed in Item 5. First write each product in the table in expanded form. Then express the product as a single power of the given base. The first one has been done for you.

Original Product	Expanded Form	Single Power
$2^2 \cdot 2^4$	$2 \cdot 2 \cdot 2 \cdot 2 \cdot 2 \cdot 2$	2^6
$5^3 \cdot 5^2$	$5 \cdot 5 \cdot 5 \cdot 5 \cdot 5$	5^5
$x^4 \cdot x^7$	$x \cdot x \cdot x \cdot x \cdot x \cdot x \cdot x \cdot x \cdot x \cdot x \cdot x$	x^{11}
$a^6 \cdot a^2$	$a \cdot a \cdot a \cdot a \cdot a \cdot a \cdot a \cdot a$	a^8

7. **Express regularity in repeated reasoning.** Based on the pattern you observed in the table in Item 6, write the missing exponent in the box below to complete the **Product of Powers Property** for exponents.

$$a^m \cdot a^n = a^{\boxed{m+n}}$$

8. Use the Product of Powers Property to write $x^{\frac{3}{4}} \cdot x^{\frac{5}{4}}$ as a single power.
$x^{\frac{8}{4}}$, or x^2

CONNECT TO SCIENCE

The formula for density is
$$D = \frac{M}{V}$$
where D is density, M is mass, and V is volume.

9. The density of an iceberg is determined by dividing its mass by its volume. Suppose a growler had a mass of 59,049 kg and a volume of 81 cubic meters. Compute the density of the iceberg.
729 kg/m³

10. Write your solution to Item 9 using powers of 9.

$$\frac{\text{Mass}}{\text{Volume}} = \text{Density} \qquad \frac{9^5}{9^2} = 9^3$$

11. What pattern do you notice in the equation you completed for Item 10?
The exponent on the right side of the equation is the difference between the exponents on the left.

Developing Math Language

This lesson contains the vocabulary terms *power*, *base* and *exponent*. These terms will likely be review for many students. However, be sure to distinguish the base of a power from other mathematical definitions of *base*, such as the base of a triangle or the base of a pyramid. It is also important for students to understand that the combination of a base and its corresponding exponent is referred to as a *power*. Students will be introduced to the *Product of Powers Property* and the *Quotient of Powers Property*. Encourage students to create a graphic organizer in their math notebooks where they can list the properties of exponents they encounter in this activity. For each property, they can give a verbal description, the symbolic form, and a numeric example. A sample graphic organizer for the activity is shown at the end of this lesson on page 290. Information for the first property has been filled in as an example.

Lesson 19-1
Basic Exponent Properties

12. Use the table to help verify the patterns you noticed in Item 11. First write each quotient in the table below in expanded form. Then express the quotient as a single power of the given base. The first one has been done for you.

Original Quotient	Expanded Form	Single Power
$\dfrac{2^5}{2^2}$	$\dfrac{2 \cdot 2 \cdot 2 \cdot 2 \cdot 2}{2 \cdot 2} = \dfrac{\cancel{2} \cdot \cancel{2} \cdot 2 \cdot 2 \cdot 2}{\cancel{2} \cdot \cancel{2}}$	2^3
$\dfrac{5^8}{5^6}$	$\dfrac{5 \cdot 5 \cdot 5 \cdot 5 \cdot 5 \cdot 5 \cdot 5 \cdot 5}{5 \cdot 5 \cdot 5 \cdot 5 \cdot 5 \cdot 5} = \dfrac{\cancel{5} \cdot \cancel{5} \cdot \cancel{5} \cdot \cancel{5} \cdot \cancel{5} \cdot \cancel{5} \cdot 5 \cdot 5}{\cancel{5} \cdot \cancel{5} \cdot \cancel{5} \cdot \cancel{5} \cdot \cancel{5} \cdot \cancel{5}}$	5^2
$\dfrac{a^3}{a^1}$	$\dfrac{a \cdot a \cdot a}{a} = \dfrac{\cancel{a} \cdot a \cdot a}{\cancel{a}}$	a^2
$\dfrac{x^7}{x^3}$	$\dfrac{x \cdot x \cdot x \cdot x \cdot x \cdot x \cdot x}{x \cdot x \cdot x} = \dfrac{\cancel{x} \cdot \cancel{x} \cdot \cancel{x} \cdot x \cdot x \cdot x \cdot x}{\cancel{x} \cdot \cancel{x} \cdot \cancel{x}}$	x^4

13. Based on the pattern you observed in Item 12, write the missing exponent in the box below to complete the **Quotient of Powers Property** for exponents.

$$\dfrac{a^m}{a^n} = a^{\boxed{m-n}}, \text{ where } a \neq 0$$

14. Use the Quotient of Powers Property to write $\dfrac{a^{\frac{11}{3}}}{a^{\frac{2}{3}}}$ as a single power.

$a^{\frac{9}{3}}$, or a^3

The product and quotient properties of exponents can be used to simplify expressions.

Example A
Simplify: $2x^5 \cdot 5x^4$

Step 1: Group powers with the same base.
$$2x^5 \cdot 5x^4 = 2 \cdot 5 \cdot x^5 \cdot x^4$$

Step 2: Product of Powers Property $= 10x^{5+4}$

Step 3: Simplify the exponent. $= 10x^9$

Solution: $2x^5 \cdot 5x^4 = 10x^9$

ACTIVITY 1 Continued

Example B Close Reading, Note Taking In this example, students are exposed to the application of the Quotient of Powers Property to simplify an algebraic expression. In Step 1, point out that the variable x in the denominator can be written as x^1. So, the factor $\frac{x^5}{x}$ is equivalent to $\frac{x^5}{x^1}$. The Try These items for the two examples provide a valuable tool for determining student understanding of exponent properties when applied to more complex expressions. These items should be used as a formative assessment and can aid instruction by having groups share their answers and processes on whiteboards.

Check Your Understanding

Debrief students' answers to these items to make sure they understand how to apply the Product and Quotient of Powers Properties. For Items 15 and 16, have students list a property or explanation for each step of their work.

Answers
15. $15y^3z^3$
16. $3fg$
17. 3^2 kg/m³

ASSESS

Students' answers to lesson practice problems will provide you with a formative assessment of their understanding of the lesson concepts and their ability to apply their learning.

See the Activity Practice for additional problems for this lesson. You may assign the problems here or use them as a culmination for the activity.

LESSON 19-1 PRACTICE

18. They both have the same value of 256.
19. x^6 grams/cm³
20. y^{14} grams
21. $(3x)^2$ or $9x^2$
22. Sample answer: First, use the Product of Powers Property to multiply the first two powers: $8^7 \cdot 8^3 \cdot 8^2 = 8^{7+3} \cdot 8^2 = 8^{10} \cdot 8^2$. Then, apply the property again: $8^{10} \cdot 8^2 = 8^{12}$.

ADAPT

Check students' answers to the Lesson Practice to ensure that they understand concepts related to multiplying and dividing powers with the same base. If students have difficulty remembering or applying the properties of exponents, suggest that they first write each expression in expanded form before attempting to simplify it.

ACTIVITY 19 continued

My Notes

Lesson 19-1
Basic Exponent Properties

Example B
Simplify: $\dfrac{2x^5 y^4}{xy^2}$

Step 1: Group powers with the same base. $\quad \dfrac{2x^5 y^4}{xy^2} = 2 \cdot \dfrac{x^5}{x} \cdot \dfrac{y^4}{y^2}$

Step 2: Quotient of Powers Property $\quad = 2x^{5-1} \cdot y^{4-2}$

Step 3: Simplify the exponents. $\quad = 2x^4 y^2$

Solution: $\dfrac{2x^5 y^4}{xy^2} = 2x^4 y^2$

Try These A–B
Simplify each expression.

a. $(4xy^4)(-2x^2y^5)$
   $-8x^3y^9$

b. $\dfrac{2a^2 b^5 c}{4ab^2 c}$
   $\dfrac{ab^3}{2}$

c. $\dfrac{6y^3}{18x} \cdot 2xy$
   $\dfrac{2y^4}{3}$

Check Your Understanding

15. Simplify $3yz^2 \cdot 5y^2z$.

16. Simplify $\dfrac{21f^2 g^{\frac{7}{4}}}{7fg^{\frac{3}{4}}}$.

17. A growler has a mass of 243 kg and a volume of 27 cubic meters. Compute the density of the iceberg by completing the following. Write your answer using powers of 3. $\dfrac{3^5}{3^3} =$

LESSON 19-1 PRACTICE

18. Which expression has the greater value? Explain your reasoning.
    a. $2^3 \cdot 2^5$
    b. $\dfrac{4^7}{4^3}$

19. The mass of an object is x^{15} grams. Its volume is x^9 cm³. What is the object's density?

20. The density of an object is y^{10} grams/cm³. Its volume is y^4 cm³. What is the object's mass?

21. Simplify the expression $\dfrac{(3x)^{\frac{1}{3}} \cdot (3x)^{\frac{7}{3}}}{(3x)^{\frac{2}{3}}}$.

22. **Make sense of problems.** Tanika asks Toby to multiply the expression $8^7 \cdot 8^3 \cdot 8^2$. Toby says he doesn't know how to do it, because he believes the Product of Powers Property works with only two exponential terms, and this problem has three terms. Explain how Toby could use the Product of Powers Property with three exponential terms.

MATH TIP
Use a graphic organizer to record the properties of exponents you learn in this activity.

Property of Exponents	Verbal Description	Symbolic Form	Numerical Example
Product of Powers	To multiply two powers with the same base, add the exponents and keep the base the same.	$a^m \cdot a^n = a^{m+n}$	$2^4 \cdot 2^2 = 2^6$
Quotient of Powers			
Negative Power			
Zero Power			
Power of a Power			
Power of a Product			
Power of a Quotient			

As students work through the Activity, they can continue to complete this graphic organizer.

290 SpringBoard® Mathematics Algebra 1, Unit 4 • Exponents, Radicals, and Polynomials

Lesson 19-2
Negative and Zero Powers

Learning Targets:
- Understand what is meant by negative and zero powers.
- Simplify expressions involving exponents.

SUGGESTED LEARNING STRATEGIES: Look for a Pattern, Discussion Groups, Sharing and Responding, Think-Pair-Share, Close Reading, Note Taking

1. **Attend to precision.** Write each quotient in expanded form and simplify it. Then apply the Quotient of Powers Property. The first one has been done for you.

Original Quotient	Expanded Form	Single Power
$\dfrac{2^5}{2^8}$	$\dfrac{2 \cdot 2 \cdot 2 \cdot 2 \cdot 2}{2 \cdot 2 \cdot 2 \cdot 2 \cdot 2 \cdot 2 \cdot 2 \cdot 2} = \dfrac{\cancel{2} \cdot \cancel{2} \cdot \cancel{2} \cdot \cancel{2} \cdot \cancel{2}}{\cancel{2} \cdot \cancel{2} \cdot \cancel{2} \cdot \cancel{2} \cdot \cancel{2} \cdot 2 \cdot 2 \cdot 2} = \dfrac{1}{2^3}$	$2^{5-8} = 2^{-3}$
$\dfrac{5^3}{5^6}$	$\dfrac{5 \cdot 5 \cdot 5}{5 \cdot 5 \cdot 5 \cdot 5 \cdot 5 \cdot 5} = \dfrac{\cancel{5} \cdot \cancel{5} \cdot \cancel{5}}{\cancel{5} \cdot \cancel{5} \cdot \cancel{5} \cdot 5 \cdot 5 \cdot 5} = \dfrac{1}{5^3}$	$5^{3-6} = 5^{-3}$
$\dfrac{a^3}{a^8}$	$\dfrac{a \cdot a \cdot a}{a \cdot a \cdot a \cdot a \cdot a \cdot a \cdot a \cdot a} = \dfrac{\cancel{a} \cdot \cancel{a} \cdot \cancel{a}}{\cancel{a} \cdot \cancel{a} \cdot \cancel{a} \cdot a \cdot a \cdot a \cdot a \cdot a} = \dfrac{1}{a^5}$	$a^{3-8} = a^{-5}$
$\dfrac{x^4}{x^{10}}$	$\dfrac{x \cdot x \cdot x \cdot x}{x \cdot x \cdot x \cdot x \cdot x \cdot x \cdot x \cdot x \cdot x \cdot x} = \dfrac{\cancel{x} \cdot \cancel{x} \cdot \cancel{x} \cdot \cancel{x}}{\cancel{x} \cdot \cancel{x} \cdot \cancel{x} \cdot \cancel{x} \cdot x \cdot x \cdot x \cdot x \cdot x \cdot x} = \dfrac{1}{x^6}$	$x^{4-10} = x^{-6}$

2. Based on the pattern you observed in Item 1, write the missing exponent in the box below to complete the **Negative Power Property** for exponents.

$$\dfrac{1}{a^n} = a^{\boxed{-n}}, \text{ where } a \neq 0$$

3. Write each quotient in expanded form and simplify it. Then apply the quotient property of exponents. The first one has been done for you.

Original Quotient	Expanded Form	Single Power
$\dfrac{2^4}{2^4}$	$\dfrac{2 \cdot 2 \cdot 2 \cdot 2}{2 \cdot 2 \cdot 2 \cdot 2} = \dfrac{\cancel{2} \cdot \cancel{2} \cdot \cancel{2} \cdot \cancel{2}}{\cancel{2} \cdot \cancel{2} \cdot \cancel{2} \cdot \cancel{2}} = 1$	$2^{4-4} = 2^0$
$\dfrac{5^6}{5^6}$	$\dfrac{5 \cdot 5 \cdot 5 \cdot 5 \cdot 5 \cdot 5}{5 \cdot 5 \cdot 5 \cdot 5 \cdot 5 \cdot 5} = \dfrac{\cancel{5} \cdot \cancel{5} \cdot \cancel{5} \cdot \cancel{5} \cdot \cancel{5} \cdot \cancel{5}}{\cancel{5} \cdot \cancel{5} \cdot \cancel{5} \cdot \cancel{5} \cdot \cancel{5} \cdot \cancel{5}} = 1$	$5^{6-6} = 5^0$
$\dfrac{a^3}{a^3}$	$\dfrac{a \cdot a \cdot a}{a \cdot a \cdot a} = \dfrac{\cancel{a} \cdot \cancel{a} \cdot \cancel{a}}{\cancel{a} \cdot \cancel{a} \cdot \cancel{a}} = 1$	$a^{3-3} = a^0$

CONNECT TO AP

In calculus, an expression containing a negative exponent is often preferable to one written as a quotient. For example, $\dfrac{1}{x^3}$ is written x^{-3}.

CONNECT TO AP

In calculus, it is easier to take the derivative of some functions if the function rule is rewritten without variables in any denominators. For example, it is easier to find the derivative of $y = \dfrac{6}{x}$ if you first rewrite the function as $y = 6x^{-1}$ so that you can use the power shortcut rule.

Activity 19 • Exponent Rules 291

ACTIVITY 19 Continued

Developing Math Language
This lesson includes the vocabulary terms *Negative Power Property* and *Zero Power Property*. Students can add these new terms to the graphic organizer they made in Lesson 19-1 for properties of exponents. As students respond to questions or discuss how to simplify expressions in this lesson, monitor their use of the new terms to ensure their understanding and ability to use language correctly and precisely.

5 Activating Prior Knowledge, Think-Pair-Share, Sharing and Responding
Students apply properties of exponents, including the Negative and Zero Power Properties, to simplify numerical expressions. The expressions in this item focus on typical student problem areas. To solidify student understanding, use questioning strategies and activate prior knowledge by referring back to the properties students learned in Lesson 19-1.

Example A Close Reading, Note Taking
In this example, students are exposed to the application of the Negative and Zero Power Properties to simplify an algebraic expression. Have students identify the property of exponents that is applied in each step of the example. The Try These items provide a valuable tool for formative assessment and can aid instruction by having groups share their answers and processes.

Differentiating Instruction
Support learning for students who are not yet ready to rewrite the terms using the Commutative Property by using positive exponents and expanded form to simplify the expression in Example A.

$5x^{-2}yz^0 \cdot \dfrac{3x^4}{y^4} =$

$5 \cdot \dfrac{1}{x^2} \cdot y \cdot z^0 \cdot 3 \cdot x^4 \cdot \dfrac{1}{y^4} =$

$5 \cdot 3 \cdot \dfrac{x^4}{x^2} \cdot \dfrac{y}{y^4} \cdot z^0 =$

$15 \cdot \dfrac{\cancel{x} \cdot \cancel{x} \cdot x \cdot x}{\cancel{x} \cdot \cancel{x}} \cdot \dfrac{\cancel{y}}{\cancel{y} \cdot y \cdot y \cdot y} \cdot 1 =$

$15 \cdot x^2 \cdot \dfrac{1}{y^3} = \dfrac{15x^2}{y^3}$

ACTIVITY 19 continued

My Notes

Lesson 19-2
Negative and Zero Powers

4. Based on the pattern you observed in Item 3, fill in the box below to complete the **Zero Power Property** of exponents.

$a^0 = \boxed{1}$, where $a \neq 0$

5. Use the properties of exponents to evaluate the following expressions.

a. 2^{-3}
$\dfrac{1}{8}$

b. $\dfrac{10^2}{10^{-2}}$
10^4

c. $3^{-2} \cdot 5^0$
$\dfrac{1}{9}$

d. $(-3.75)^0$
1

When evaluating and simplifying expressions, you can apply the properties of exponents and then write the answer without negative or zero powers.

Example A
Simplify $5x^{-2}yz^0 \cdot \dfrac{3x^4}{y^4}$ and write without negative powers.

Step 1: Commutative Property $5x^{-2}yz^0 \cdot \dfrac{3x^4}{y^4}$

$= 5 \cdot 3 \cdot x^{-2} \cdot x^4 \cdot y^1 \cdot y^{-4} \cdot z^0$

Step 2: Apply the exponent rules.

$= 5 \cdot 3 \cdot x^{-2+4} \cdot y^{1-4} \cdot z^0$

Step 3: Simplify the exponents.

$= 15 \cdot x^2 \cdot y^{-3} \cdot 1$

Step 4: Write without negative exponents.

$= \dfrac{15x^2}{y^3}$

Solution: $5x^{-2}yz^0 \cdot \dfrac{3x^4}{y^4} = \dfrac{15x^2}{y^3}$

Try These A
Simplify and write without negative powers.

a. $2a^2b^{-3} \cdot 5ab$
$\dfrac{10a^3}{b^2}$

b. $\dfrac{10x^2y^{-4}}{5x^{-3}y^{-1}}$
$\dfrac{2x^5}{y^3}$

c. $(-3xy^{-5})^0$
1

---

292 SpringBoard® Mathematics **Algebra 1, Unit 4** • Exponents, Radicals, and Polynomials

Lesson 19-2
Negative and Zero Powers

Check Your Understanding

Simplify each expression. Write your answer without negative exponents.

6. $(z)^{-3}$
7. $12(xyz)^0$
8. $\dfrac{6^{-4}}{6^{-2}}$
9. $2^3 \cdot 2^{-6}$
10. $\dfrac{4x^{-2}}{x^3}$
11. $\dfrac{-5}{(ab)^0}$

LESSON 19-2 PRACTICE

12. For what value of v is $a^v = 1$, if $a \neq 0$?
13. For what value of w is $b^{-w} = \dfrac{1}{b^9}$, if $b \neq 0$?
14. For what value of y is $\dfrac{3^3}{3^y} = \dfrac{1}{9}$?
15. For what value of z is $5^8 \cdot 5^z = 1$?
16. Determine the values of n and m that would make the equation $7^n \cdot 7^m = 1$ a true statement. Assume that $n \neq m$.
17. For what value of x is $\dfrac{3^x \cdot 2^2}{3^4} = \dfrac{4}{3}$?
18. **Reason abstractly.** What is the value of $2^0 \cdot 3^0 \cdot 4^0 \cdot 5^0$? What is the value of any multiplication problem in which all of the factors are raised to a power of 0? Explain.

ACTIVITY 19 Continued

Check Your Understanding

Debrief students' answers to these items to ensure that they understand how to apply properties of exponents to simplify numerical and algebraic expressions. Ask students to explain why $\dfrac{x}{x^{-2}}$ is equivalent to x^3 for $x \neq 0$.

Answers

6. $\dfrac{1}{z^3}$
7. 12
8. $\dfrac{1}{36}$
9. $\dfrac{1}{8}$
10. $\dfrac{4}{x^5}$
11. -5

ASSESS

Students' answers to lesson practice problems will provide you with a formative assessment of their understanding of the lesson concepts and their ability to apply their learning.

See the Activity Practice for additional problems for this lesson. You may assign the problems here or use them as a culmination for the activity.

LESSON 19-2 PRACTICE

12. $v = 0$
13. $w = 9$
14. $y = 5$
15. $z = -8$
16. Possible answer: $m = 2, n = -2$
17. $x = 3$
18. 1; 1; Each factor is equal to 1, so the product will be 1.

ADAPT

Check students' answers to the Lesson Practice to make sure that they understand concepts related to the Negative Power and Zero Power Properties. Reinforce students' understanding of these properties by using patterns, such as the one introduced in the Bell-Ringer Activity for this lesson.

Power of 2	Expanded Form	Value
2^3	$2 \cdot 2 \cdot 2$	8
2^2	$2 \cdot 2$	4
2^1	2	2
2^0	1	1
2^{-1}	$\dfrac{1}{2}$	$\dfrac{1}{2}$
2^{-2}	$\dfrac{1}{2 \cdot 2}$	$\dfrac{1}{4}$
2^{-3}	$\dfrac{1}{2 \cdot 2 \cdot 2}$	$\dfrac{1}{8}$

Activity 19 • Exponent Rules

ACTIVITY 19 Continued

Lesson 19-3

PLAN

Pacing: 1 class period
Chunking the Lesson
#1–3   #4–6   #7–8
Example A   Example B
Check Your Understanding
Lesson Practice

TEACH

Bell-Ringer Activity
Ask students to review the properties of exponents they have learned so far by asking them simplify the following expressions.

1. $\dfrac{x^5}{x^2}$, for $x \neq 0$ $[x^3]$
2. n^0, for $n \neq 0$ $[1]$
3. 3^{-2} $\left[\dfrac{1}{9}\right]$
4. $a^2 \cdot a^6$ $[a^8]$

TEACHER to TEACHER

The repetition of learning strategies in this activity can help to speed the flow of the lesson, but be careful that a small number of students who can see patterns quickly are not driving the lesson. Students who are still struggling with the earlier properties may need some extra assistance. Using small-group questioning strategies and the expanded form of expressions will help them to see the patterns and solidify their understanding of the properties.

1–3 Look for a Pattern, Create Representations, Think-Pair-Share
Students develop the Power of a Power Property using strategies they employed earlier in this activity. As in previous items, students should be given the opportunity to share their thoughts on the patterns they notice. Sharing and Responding should clarify the rule as $(a^m)^n = a^{mn}$. It is important that students also be able to verbalize the rule: *To determine the power of a power, multiply the exponents.*

4–6 Look for a Pattern, Create Representations, Think-Pair-Share, Sharing and Responding Students develop the Power of a Product Property. Have students discuss the patterns they notice in the table. Sharing and Responding should clarify the rule as $(ab)^m = a^m \cdot b^m$. It is important that students also be able to verbalize the rule: *A product raised to a power is equal to the product of each factor raised to the power.*

ACTIVITY 19 continued

Lesson 19-3
Additional Properties of Exponents

Learning Targets:
- Develop the Power of a Power, Power of a Product, and the Power of a Quotient Properties.
- Simplify expressions involving exponents.

SUGGESTED LEARNING STRATEGIES: Note Taking, Look for a Pattern, Create Representations, Think-Pair-Share, Sharing and Responding, Close Reading

1. Write each expression in expanded form. Then write the expression using a single exponent with the given base. The first one has been done for you.

Original Expression	Expanded Form	Single Power
$(2^2)^4$	$2^2 \cdot 2^2 \cdot 2^2 \cdot 2^2 = 2 \cdot 2 \cdot 2 \cdot 2 \cdot 2 \cdot 2 \cdot 2 \cdot 2$	2^8
$(5^5)^3$	$5^5 \cdot 5^5 \cdot 5^5 =$ $5 \cdot 5 \cdot 5 \cdot 5 \cdot 5 \cdot 5 \cdot 5 \cdot 5 \cdot 5 \cdot 5 \cdot 5 \cdot 5 \cdot 5 \cdot 5 \cdot 5$	5^{15}
$(x^3)^4$	$x^3 \cdot x^3 \cdot x^3 \cdot x^3 =$ $x \cdot x \cdot x \cdot x \cdot x \cdot x \cdot x \cdot x \cdot x \cdot x \cdot x \cdot x$	x^{12}

2. Based on the pattern you observed in Item 1, write the missing exponent in the box below to complete the **Power of a Power Property** for exponents.
$$(a^m)^n = a^{\boxed{m \cdot n}}$$

3. Use the Power of a Power Property to write $\left(x^{\frac{6}{5}}\right)^{25}$ as a single power.
x^{30}

4. Write each expression in expanded form and group like terms. Then write the expression as a product of powers. The first one has been done for you.

Original Expression	Expanded Form	Product of Powers
$(2x)^4$	$2x \cdot 2x \cdot 2x \cdot 2x =$ $2 \cdot 2 \cdot 2 \cdot 2 \cdot x \cdot x \cdot x \cdot x$	$2^4 x^4$
$(-4a)^3$	$-4a \cdot -4a \cdot -4a =$ $-4 \cdot -4 \cdot -4 \cdot a \cdot a \cdot a$	$(-4)^3 a^3$
$(x^3 y^2)^4$	$x^3 y^2 \cdot x^3 y^2 \cdot x^3 y^2 \cdot x^3 y^2 =$ $x^3 \cdot x^3 \cdot x^3 \cdot x^3 \cdot y^2 \cdot y^2 \cdot y^2 \cdot y^2$	$x^{12} y^8$

Lesson 19-3
Additional Properties of Exponents

5. Based on the pattern you observed in Item 4, write the missing exponents in the boxes below to complete the **Power of a Product Property** for exponents.
$$(ab)^m = a^{\boxed{m}} \cdot b^{\boxed{m}}$$

6. Use the Power of a Product Property to write $\left(c^{\frac{1}{2}} d^{\frac{1}{4}}\right)^8$ as a product of powers.
   $c^4 d^2$

7. **Make use of structure.** Use the patterns you have seen. Predict and write the missing exponents in the boxes below to complete the **Power of a Quotient Property** for exponents.
$$\left(\frac{a}{b}\right)^m = \frac{a^{\boxed{m}}}{b^{\boxed{m}}}, \text{ where } b \neq 0$$

8. Use the Power of a Quotient Property to write $\left(\frac{x^3}{y^6}\right)^{\frac{1}{3}}$ as a quotient of powers.
   $\dfrac{x}{y^2}$

You can apply these power properties and the exponent rules you have already learned to simplify expressions.

MATH TIP
Create an organized summary of the properties used to simplify and evaluate expressions with exponents.

Example A
Simplify $(2x^2y^5)^3 (3x^2)^{-2}$ and write without negative powers.

Step 1: Power of a Power Property
$$(2x^2y^5)^3 (3x^2)^{-2} = 2^3 x^{2 \cdot 3} y^{5 \cdot 3} \cdot 3^{-2} \cdot x^{2 \cdot -2}$$

Step 2: Simplify the exponents and the numerical terms.
$$= 8 \cdot x^6 y^{15} \cdot \frac{1}{3^2} \cdot x^{-4}$$

Step 3: Commutative Property
$$= 8 \cdot \frac{1}{9} x^6 \cdot x^{-4} y^{15}$$

Step 4: Product of Powers Property
$$= \frac{8}{9} x^{6-4} y^{15}$$

Step 5: Simplify the exponents.
$$= \frac{8}{9} x^2 y^{15}$$

Solution: $(2x^2y^5)^3 (3x^2)^{-2} = \dfrac{8}{9} x^2 y^{15}$

ACTIVITY 19 Continued

Example B Close Reading, Note Taking Students are exposed to the application of the Power of a Quotient and Power of a Power Properties to simplify an algebraic expression. The Try These items can be used for formative assessment to identify students who need more help and support. Suggest that struggling students first rewrite the expressions using only positive exponents and then write powers in expanded form as needed.

Check Your Understanding

Debrief students' answers to these items to ensure that they understand how to simplify expressions by using properties of exponents. Have students give a property or explanation for each step of their work on these items.

Answers

9. $\dfrac{16x^6}{y^2}$
10. $\dfrac{125x^3}{y^6}$
11. $\dfrac{-24a^7c^8}{b^2}$
12. $\dfrac{1}{3gh^2}$
13. $\dfrac{a^3}{8b^9}$
14. 1

ASSESS

Students' answers to lesson practice problems will provide you with a formative assessment of their understanding of the lesson concepts and their ability to apply their learning.

See the Activity Practice for additional problems for this lesson. You may assign the problems here or use them as a culmination for the activity.

ADAPT

Check students' answers to the Lesson Practice to make sure that they understand and can apply the properties of exponents. To reinforce the application of the properties of exponents for all students, but especially for those who may still be having difficulty, ask groups to do a jigsaw-type presentation. Each group can be assigned a property to become "experts" on. Students then move to new groups to share information about their properties and give examples.

ACTIVITY 19 continued

My Notes

Lesson 19-3
Additional Properties of Exponents

Example B

Simplify $\left(\dfrac{x^2 y^{-3}}{z}\right)^2$.

Step 1: Power of a Quotient Property $\qquad \left(\dfrac{x^2 y^{-3}}{z}\right)^2 = \dfrac{x^{2\cdot 2} y^{-3\cdot 2}}{z^2}$

Step 2: Simplify the exponents. $\qquad\qquad\qquad = \dfrac{x^4 y^{-6}}{z^2}$

Step 3: Negative Power Property $\qquad\qquad\qquad = \dfrac{x^4}{y^6 z^2}$

Solution: $\left(\dfrac{x^2 y^{-3}}{z}\right)^2 = \dfrac{x^4}{y^6 z^2}$

Try These A–B

Simplify and write without negative powers.

a. $(2x^2 y)^3(-3xy^3)^2$
$72x^8 y^9$

b. $-2ab(5b^2 c)^3$
$-250ab^7 c^3$

c. $\left(\dfrac{4x}{y^3}\right)^{-2}$
$\dfrac{y^6}{16x^2}$

d. $\left(\dfrac{5x}{y}\right)^2 \left(\dfrac{y^3}{10x^2}\right)$
$\dfrac{5y}{2}$

e. $(3xy^{-2})^2 (2x^3 yz)(6yz^2)^{-1}$
$\dfrac{3x^5}{y^4 z}$

Check Your Understanding

Simplify each expression. Write your answer without negative exponents.

9. $(4x^3 y^{-1})^2$
10. $\left(\dfrac{5x}{y^2}\right)^3$
11. $(-2a^2 b^{-2} c)^3 (3ab^4 c^5)(xyz)^0$
12. $(4fg^3)^{-2}(-4fg^3 h)^2 (3gh^4)^{-1}$
13. $\left(\dfrac{2ab}{a^2 b^{-2}}\right)^{-3}$
14. $\left[(-7nm^2)^{-3}\right]^0$

LESSON 19-3 PRACTICE

Simplify.

15. a. $\left(\dfrac{2}{3}\right)^2$ b. $\left(\dfrac{2}{3}\right)^{-2}$
16. a. $(3x)^3$ b. $(3x)^{-3}$
17. a. $(2^5)^4$ b. $(2^5)^{-4}$
18. **Model with mathematics.** The formula for the area of a square is $A = s^2$, where s is the side length. A square garden has a side length of $x^4 y$. What is the area of the garden?

LESSON 19-3 PRACTICE

15. a. $\dfrac{4}{9}$
    b. $\dfrac{9}{4}$
16. a. $27x^3$
    b. $\dfrac{1}{27x^3}$
17. a. 2^{20} or $1{,}048{,}576$
    b. $\dfrac{1}{2^{20}}$ or $\dfrac{1}{1{,}048{,}576}$
18. $x^8 y^2$

Exponent Rules
Icebergs and Exponents

ACTIVITY 19
continued

ACTIVITY 19 PRACTICE
Write your answers on notebook paper.
Show your work.

Lesson 19-1

For Items 1–5, evaluate the expression. Write your answer without negative powers.

1. $x^8 \cdot x^7$

2. $\dfrac{6a^{10}b^9}{3ab^3}$

3. $(6a^2b)(-3ab^3)$

4. $\dfrac{7x^2y^5}{14xy^4}$

5. $\dfrac{2xy^2}{x^5y^3} \cdot \dfrac{5xy^3}{-30y^{-2}}$

6. The volume of an iceberg that is below the water line is 2^5 cubic meters. The volume that is above the water line is 2^2 cubic meters. How many times greater is the volume below the water line than above it?
   A. $2^{2.5}$
   B. 2^3
   C. 2^7
   D. 2^{10}

7. A megabyte is equal to 2^{20} bytes, and a gigabyte is equal to 2^{30} bytes. How many times larger is a gigabyte than a megabyte?

8. A jackpot is worth 10^5 dollars. The contestant who wins the jackpot has the opportunity to put it all on the line with the single spin of a prize wheel. If the contestant spins the number 7 on the wheel, she will win 10^2 times more money. How many dollars will the contestant win if she risks her prize money and spins a 7?

The number of earthquakes of a given magnitude that are likely to occur in any given year is represented by the formula $10^{(8-M)}$, where M is the magnitude. Use this formula for Items 9 and 10.

9. How many earthquakes of magnitude 8 are likely to occur next year?

10. If an earthquake of magnitude 10 occurred last year, how many years will it be before another one of that magnitude is likely to occur?

Lesson 19-2

11. Which of the following expressions is **not** equal to 1?
    A. $x^3 \cdot x^{-3}$
    B. 1001^0
    C. $\dfrac{a^2b}{ba^2}$
    D. $\dfrac{y^2}{y^{-2}}$

12. Which of the following expressions is equal to $\dfrac{y}{x^2}$?
    A. $x^{-2}y^3 \cdot y^{-2}$
    B. $xy^2 \cdot x^{-3}y^{-2}$
    C. $\dfrac{y^2x}{yx^{-3}}$
    D. $\dfrac{x^2y}{y^{-2}}$

Determine whether each statement is always, sometimes, or never true.

13. For $a \neq 0$, the value of a^{-1} is positive.

14. If n is an integer, then $3^n \cdot 3^{-n}$ equals 1.

15. If $6^p > 0$, then $p > 0$.

16. 4^{-x} equals $\dfrac{1}{4^x}$.

17. If m is an integer, then the value of 2^m is negative.

ACTIVITY 19 Continued

ACTIVITY PRACTICE
1. x^{15}
2. $2a^9b^6$
3. $-18a^3b^4$
4. $\dfrac{xy}{2}$
5. $\dfrac{-y^4}{3x^3}$
6. B
7. 2^{10}, or 1024 times larger
8. 10^7 dollars ($10,000,000)
9. 1
10. 100
11. D
12. A
13. sometimes
14. always
15. sometimes
16. always
17. never

ACTIVITY 19 Continued

18. $a = 2$
19. $b = -4$
20. undefined
21. 1
22. undefined
23. undefined
24. $a^4 b^2$
25. x^6
26. $64c^9 d^3$ units^3
27. x^9; $(x^9)^3 = x^{9 \cdot 3} = x^{27}$
28. $\dfrac{625 x^8}{y^4}$
29. $\dfrac{c^5}{d^{10}}$
30. $x^{11} y^8 z$
31. $\dfrac{1}{m^7}$
32. $\dfrac{3}{8}$
33. B
34. No; the first expression is equal to $a^7 b^7$ while the second is equal to $a^{12} b^{12}$.

ADDITIONAL PRACTICE

If students need more practice on the concepts in this activity, see the eBook Teacher Resources for additional practice problems.

ACTIVITY 19 continued

Exponent Rules
Icebergs and Exponents

18. For what value of a is $w^{a-2} = 1$, if $w \neq 0$?
19. For what value of b is $p^{b-1} = \dfrac{1}{p^5}$, if $p \neq 0$?

For each of the following, give the value of the expression or state that the expression is undefined.

20. x^0 when $x = 0$
21. 2^{-a} when $a = 0$
22. $\dfrac{1}{x^p}$ when $x = 0$ and $p > 0$
23. $0^n \cdot 0^{-n}$ when n is an integer

Lesson 19-3

24. The area of a square is given by the formula $A = s^2$, where s is the length of the side. What is the area of the square shown?

The volume of a cube is given by the formula $V = s^3$, where s is the length of the side. Use this formula for Items 25–27.

25. What is the volume of the cube shown?

26. What is the volume of the cube shown?

27. The volume of a cube is x^{27} cubic inches. What expression represents the length of one side of the cube? Justify your reasoning.

Simplify each expression. Write your answer without negative exponents.

28. $(-5x^2 y^{-1})^4$
29. $\left(\dfrac{c^2 d^{-2}}{c} \right)^5$
30. $(x^2 y^2 z^{-1})^3 (xyz^4)(x^3 y)$
31. $(m^2 n^{-5})^0 m^{-7}$
32. $\left(\dfrac{2x^{-2}}{3} \right) \left(\dfrac{3x}{4} \right)^2$
33. Which of the following is a true statement about the expression $a^4 \left(\dfrac{1}{a} \right)^2$, given that $a \neq 0$?
    A. The expression is always equal to 1.
    B. The value of the expression is positive.
    C. If a is negative, then the value of the expression is also negative.
    D. The expression cannot be simplified any further.

MATHEMATICAL PRACTICES
Construct Viable Arguments and Critique the Reasoning of Others

34. Alana says that $(ab)^3 \cdot (ab)^4$ is the same as $[(ab)^3]^4$. Is Alana correct? Justify your response.

Operations with Radicals

Go Fly a Kite
Lesson 20-1 Radical Expressions

ACTIVITY 20

Learning Targets:
- Write and simplify radical expressions.
- Understand what is meant by a rational exponent.

> **SUGGESTED LEARNING STRATEGIES:** Create Representations, Close Reading, Discussion Groups, Sharing and Responding, Note Taking, Think-Pair-Share

The frame of a box kite has four "legs" of equal length and four pairs of crossbars, all of equal length. The legs of the kite form a square base. The crossbars are attached to the legs so that each crossbar is positioned as a diagonal of the square base.

1. a. Label the legs of the kite pictured to the right. How many legs are in a kite? How many crossbars?

   4 legs; 8 crossbars

   b. Label the points on the top view where the ends of the crossbars are attached to the legs *A*, *B*, *C*, and *D*. Begin at the bottom left and go clockwise.

   c. Use one color to show the sides of the square and another color to show crossbar *AC*. What two figures are formed by two sides of the square and one diagonal?

   two right triangles

Members of the Windy Hill Science Club are building kites to explore aerodynamic forces. Club members will provide paper, plastic, or lightweight cloth for the covering of their kite. The club will provide the balsa wood for the frames.

2. **Model with mathematics.** The science club advisor has created the chart below to help determine how much balsa wood he needs to buy.
   a. For each kite, calculate the exact length of one crossbar that will be needed to stabilize the kite. Use your drawing from Item 1c as a guide for the rectangular base of these box kites.

Kite	Dimensions of Base (in feet)	Exact Length of One Crossbar (in feet)	Kite	Dimensions of Base (in feet)	Exact Length of One Crossbar (in feet)
A	1 by 1	$\sqrt{2}$	D	1 by 2	$\sqrt{5}$
B	2 by 2	$\sqrt{8}$	E	2 by 4	$\sqrt{20}$
C	3 by 3	$\sqrt{18}$	F	3 by 6	$\sqrt{45}$

   b. How much wood would you recommend buying for the crossbars of Kite A? Explain your reasoning.

   $8\sqrt{2}$; there are 8 crossbars in each kite and the length of each crossbar is $\sqrt{2}$.

MATH TIP

Pythagorean Theorem

$$a^2 + b^2 = c^2$$

MATH TIP

If you take the square root of a number that is not a perfect square, the result is a decimal number that does not terminate or repeat and is called an **irrational number**. The exact value of an irrational number must be written using a radical sign.

Common Core State Standards for Activity 20

HSN-RN.A.2	Rewrite expressions involving radicals and rational exponents using the properties of exponents.
HSA-SSE.A.2	Use the structure of an expression to identify ways to rewrite it.

ACTIVITY 20 Directed

Activity Standards Focus

In Activity 20, students simplify and perform operations with radical expressions. They also investigate the meaning of rational exponents and learn to write powers with rational exponents in radical form.

Lesson 20-1

PLAN

Pacing: 1 class period

Chunking the Lesson
- #1–2 — Example A
- #3 — Example B
- Check Your Understanding
- #6–7 — Example C
- Check Your Understanding
- Lesson Practice

TEACH

Bell-Ringer Activity
Ask students to solve this problem: *A square rug has an area of 36 square feet. What is the length of a side of the rug?* [6 feet] Have students discuss how they determined their answers.

1–2 Marking the Text, Close Reading, Visualization, Discussion Groups, Sharing and Responding Item 1 allows all students access to the situation and provides a break from the introductory text. It is important that students understand the construction of the kite and that each crossbar forms the hypotenuse of a right triangle. Labeling the diagram and using different colors helps students to visualize this construction. Students also need to understand the difference in the exact value of a square root and an approximate decimal value. Ask students to share ideas about why the exact value of an irrational number cannot be written as a decimal number. Ask students to explain why the exact lengths of the crossbars in Item 2 must be written using radical signs.

Differentiating Instruction

Support students who need a review of the Pythagorean Theorem by showing a quick example on the board. For instance, show students how to find the length of the hypotenuse of a right triangle with legs having lengths of 3 units and 5 units. [$\sqrt{34}$ units]

ACTIVITY 20 Continued

Developing Math Language
This lesson includes several new vocabulary terms: *irrational number, radical expression, principal square root, negative square root* and *cube root*. Encourage students to mark the text by adding notes explaining the new vocabulary in their own words. Be sure to add the words to the class Word Wall. You may wish to point out that the prefix *ir-* means "not." Therefore, an *irrational* number is a number that is not rational; in other words, it cannot be expressed as the ratio of two integers. Some students may be able to use their prior knowledge of the definitions of *square root* and the *square* of a number to determine the meaning of the new term *cube root*.

TEACHER to TEACHER
A positive number *a* has two distinct square roots, one positive and one negative. The positive, or principal, square root is denoted $\sqrt{a}$, and the negative square root is denoted $-\sqrt{a}$. A negative number has no real square roots. In later courses, students will learn that the square roots of negative numbers are imaginary.

Example A Note Taking, Think-Pair-Share
Students are exposed to one way to simplify square roots when the radicand is not a perfect square. You may wish to point out that the square root of a product can be rewritten as the product of the square roots of the factors. Thus, in part a of the example, $\sqrt{25 \cdot 3}$ can be rewritten as $\sqrt{25} \cdot \sqrt{3}$. After students complete the Try These problems, have them discuss how they could check that they have simplified the expressions correctly. One way is to use a calculator to evaluate both the original expression and the simplified form to check that they have the same value.

Differentiating Instruction

Support students who have trouble identifying perfect-square factors of radicands by having them make factor trees for the radicands. Students can look for both perfect square factors and repeated factors in their factor trees.

ACTIVITY 20 continued

My Notes

MATH TIP
When there is no root index given, it is assumed to be 2 and is called a *square root*.
$\sqrt{36} = \sqrt[2]{36}$

READING MATH
$a\sqrt{b}$ is read "*a* times the square root of *b*." Example A Part (c) is read "7 times the square root of 12."

Lesson 20-1
Radical Expressions

Each amount of wood in the table in Item 2 is a **radical expression**.

Radical Expression
An expression of the form $\sqrt[n]{a}$, where *a* is the radicand, $\sqrt{\ }$ is the radical symbol, and *n* is the root index.
$\sqrt[n]{a} = b$, if $b^n = a$.   *b* is the *n*th root of *a*.

Finding the square root of a number or expression is the inverse operation of squaring a number or expression.

$\sqrt{25} = 5$, because $(5)(5) = 25$
$\sqrt{81} = 9$, because $(9)(9) = 81$
$\sqrt{x^2} = x$, because $(x)(x) = x^2, x \geq 0$

Notice also that $(-5)(-5) = (-5)^2 = 25$. The **principal square root** of a number is the positive square root value. The expression $\sqrt{25}$ simplifies to 5, the principal square root. The **negative square root** is the negative root value, so $-\sqrt{25}$ simplifies to -5.

To simplify square roots in which the radicand is not a perfect square:

Step 1: Write the radicand as a product of numbers, one of which is a perfect square.

Step 2: Find the square root of the perfect square.

Example A
Simplify each expression.
a. $\sqrt{75} = \sqrt{25 \cdot 3} = 5\sqrt{3}$
b. $\sqrt{72} = \sqrt{36 \cdot 2} = 6\sqrt{2}$
   $\sqrt{72} = \sqrt{9 \cdot 4 \cdot 2} = (3 \cdot 2)\sqrt{2} = 6\sqrt{2}$
c. $7\sqrt{12} = 7\sqrt{4 \cdot 3} = 7(2\sqrt{3}) = 14\sqrt{3}$
d. $\sqrt{c^3} = \sqrt{c^2 \cdot c} = c\sqrt{c}, c \geq 0$

Try These A
Simplify each expression.
a. $\sqrt{18}$ b. $5\sqrt{48}$ c. $\sqrt{126}$
   $3\sqrt{2}$ $20\sqrt{3}$ $3\sqrt{14}$

d. $\sqrt{24y^2}$ e. $\sqrt{45b^3}$
   $2y\sqrt{6}$ $3b\sqrt{5b}$

MINI-LESSON: Using Prime Factorization to Simplify Square Roots

If students need more practice with simplifying square roots, a mini-lesson is available. In this mini-lesson, students write the prime factorization of the radicand and then use the prime factorization to identify perfect-square factors.

See SpringBoard's eBook Teacher Resources for a student page for this mini-lesson.

Lesson 20-1
Radical Expressions

3. Copy the lengths of the crossbars from the chart in Item 1. Then express the lengths of the crossbars in simplified form.

Kite	Dimensions of Base (feet)	Exact Length of One Crossbar (feet)	Simplified Form of Length of Crossbar
A	1 by 1	$\sqrt{2}$	$\sqrt{2}$
B	2 by 2	$\sqrt{8}$	$2\sqrt{2}$
C	3 by 3	$\sqrt{18}$	$3\sqrt{2}$
D	1 by 2	$\sqrt{5}$	$\sqrt{5}$
E	2 by 4	$\sqrt{20}$	$2\sqrt{5}$
F	3 by 6	$\sqrt{45}$	$3\sqrt{5}$

The process of finding roots can be expanded to **cube roots**. Finding the cube root of a number or an expression is the inverse operation of cubing that number or expression.

$$\sqrt[3]{125} = 5, \text{ because } (5)(5)(5) = 125$$
$$\sqrt[3]{y^3} = y, \text{ because } (y)(y)(y) = y^3$$

To simplify cube roots in which the radicand is not a perfect cube, follow the same two-step process that you used for square roots.

Step 1: Write the radicand as a product of numbers, one of which is a perfect cube.

Step 2: Find the cube root of the perfect cube.

Example B
Simplify each expression.

a. $\sqrt[3]{16} = \sqrt[3]{8 \cdot 2} = \sqrt[3]{8} \cdot \sqrt[3]{2} = 2\sqrt[3]{2}$

b. $3\sqrt[3]{128} = 3\sqrt[3]{64 \cdot 2} = 3\sqrt[3]{64} \cdot \sqrt[3]{2} = 3(4\sqrt[3]{2}) = 12\sqrt[3]{2}$

c. $\sqrt[3]{x^5} = \sqrt[3]{x^3 \cdot x^2} = x\sqrt[3]{x^2}$

Try These B
Simplify each expression.

a. $\sqrt[3]{24}$ b. $\sqrt[3]{54z^3}$ c. $\sqrt[3]{40b^4}$

   $2\sqrt[3]{3}$ $3z\sqrt[3]{2}$ $2b\sqrt[3]{5b}$

MATH TERMS

The root index n can be any integer greater than or equal to 2. A **cube root** has $n = 3$. The cube root of 8 is $\sqrt[3]{8} = 2$ because $2 \cdot 2 \cdot 2 = 8$.

ACTIVITY 20 Continued

3 Create Representations, Think-Pair-Share Students simplify the radical expressions that represent the lengths of the crossbars. After students complete the table, ask them to describe any patterns they notice in the simplified forms. Students may notice that the lengths for kites A, B, and C have a common factor of $\sqrt{2}$ and the lengths for kites D, E, and F have a common factor of $\sqrt{5}$.

TEACHER to TEACHER

A positive number has one real cube root. For example, $\sqrt[3]{8} = 2$ because $2 \cdot 2 \cdot 2 = 8$. A negative number also has one real cube root. For example, $\sqrt[3]{-8} = -2$ because $-2 \cdot (-2) \cdot (-2) = -8$. The real cube root of a positive number is positive, and the real cube root of a negative number is negative.

Example B Note Taking, Think-Pair-Share In this example, students are exposed to one way to simplify cube roots when the radicand is not a perfect cube. Ask students to compare and contrast the process of simplifying square roots with the process of simplifying cube roots. As students work to complete the Try These problems, encourage those who are having difficulty to use factor trees.

Differentiating Instruction

Support students who have trouble recognizing perfect cubes by writing the first 10 perfect cubes on the board.

$1^3 = 1$ $2^3 = 8$ $3^3 = 27$
$4^3 = 64$ $5^3 = 125$ $6^3 = 216$
$7^3 = 343$ $8^3 = 512$ $9^3 = 729$
$10^3 = 1000$

Activity 20 • Operations with Radicals

ACTIVITY 20 Continued

Check Your Understanding
Debrief students' answers to these items to ensure they understand how to simplify square roots and cube roots. Debrief by asking students to explain why $\sqrt[3]{-1000}$ is equal to -10.

Answers
4. $\sqrt{13}$ feet
5. a. $2\sqrt{31}$
   b. $5d^2\sqrt{5}$
   c. $5\sqrt[3]{2}$
   d. $3m^2\sqrt[3]{3m}$

6–7 Activating Prior Knowledge, Look for a Pattern, Predict and Confirm, Discussion Groups, Sharing and Responding Students use their knowledge of radicals and the Product of Powers Property to develop a definition of the fractional exponent $\frac{1}{n}$, where n is a positive integer. Students should notice that within each row, the radical form and the fractional exponent form simplify to the same value. Therefore, the radical form and the fractional exponent form are equivalent. For Item 7, allow students to share their thoughts on how to write the radical expression, but as groups share and respond to each other, be sure to clarify that $a^{\frac{1}{n}} = \sqrt[n]{a}$, where $a \geq 0$.

Example C Close Reading, Note Taking, Think-Pair-Share In this example, students are exposed to writing a power with a fractional exponent as a radical expression. Have students compare and contrast the two methods shown. Also ask students to explain how the Power of a Power Property is applied in this example.

Lesson 20-1
Radical Expressions

Check Your Understanding

4. A kite has a base with dimensions of 2 feet by 3 feet. What is the length of one crossbar that will be needed to stabilize the kite?

5. Simplify.
   a. $\sqrt{124}$
   b. $\sqrt{125d^4}$
   c. $\sqrt[3]{250}$
   d. $\sqrt[3]{81m^7}$

Another way to write radical expressions is with fractional exponents.

6. **Make use of structure.** Use the definition of a radical and the properties of exponents to simplify the expressions of each row of the table. The first row has been done for you.

Radical Form	Simplified Form	Fractional Exponent Form	Simplified Form
$\sqrt{16} \cdot \sqrt{16}$	$4 \cdot 4 = 16$	$16^{\frac{1}{2}} \cdot 16^{\frac{1}{2}}$	$16^{\frac{1}{2}+\frac{1}{2}} = 16^1 = 16$
$\sqrt[3]{8} \cdot \sqrt[3]{8} \cdot \sqrt[3]{8}$	$2 \cdot 2 \cdot 2 = 8$	$8^{\frac{1}{3}} \cdot 8^{\frac{1}{3}} \cdot 8^{\frac{1}{3}}$	$8^{\frac{1}{3}+\frac{1}{3}+\frac{1}{3}} = 8^1 = 8$
$\sqrt[4]{81} \cdot \sqrt[4]{81} \cdot \sqrt[4]{81} \cdot \sqrt[4]{81}$	$3 \cdot 3 \cdot 3 \cdot 3 = 81$	$81^{\frac{1}{4}} \cdot 81^{\frac{1}{4}} \cdot 81^{\frac{1}{4}} \cdot 81^{\frac{1}{4}}$	$81^{\frac{1}{4}+\frac{1}{4}+\frac{1}{4}+\frac{1}{4}} = 81^1 = 81$
$\sqrt{a} \cdot \sqrt{a}$	a	$a^{\frac{1}{2}} \cdot a^{\frac{1}{2}}$	a

7. Identify and describe any patterns in the table. Write $a^{\frac{1}{n}}$ as a radical expression.

$a^{\frac{1}{n}} = \sqrt[n]{a}$

The general rule for fractional exponents when the numerator is not 1 is
$$a^{\frac{m}{n}} = \sqrt[n]{a^m} = \left(\sqrt[n]{a}\right)^m.$$

Example C

Write $6^{\frac{2}{3}}$ as a radical expression.

Method 1: $6^{\frac{2}{3}} = 6^{2 \cdot \frac{1}{3}} = (6^2)^{\frac{1}{3}} = \sqrt[3]{6^2}$

Method 2: $6^{\frac{2}{3}} = 6^{\frac{1}{3} \cdot 2} = \left(6^{\frac{1}{3}}\right)^2 = \left(\sqrt[3]{6}\right)^2$

Try These C

Write each of the following as a radical expression.

a. $13^{\frac{1}{4}}$
   $\sqrt[4]{13}$

b. $7^{\frac{3}{5}}$
   $\sqrt[5]{7^3}$ or $(\sqrt[5]{7})^3$

c. $x^{\frac{3}{2}}$
   $\sqrt{x^3}$ or $(\sqrt{x})^3$

Lesson 20-1
Radical Expressions

Check Your Understanding

8. a. What is the value of $16^{\frac{1}{4}}$?
   b. What is the value of $16^{\frac{3}{4}}$?
9. For each radical expression, write an equivalent expression with a fractional exponent.
   a. $\sqrt{15}$ b. $\sqrt[3]{21}$

LESSON 20-1 PRACTICE

10. A square has an area of 72 square inches. What is its side length s? Give the exact answer using simplified radicals.

11. A cube has a volume of 216 cubic centimeters. What is its edge length s? Give the exact answer using simplified radicals.

12. A square has an area of $12x^2$ square feet. What is the length of its sides?
13. A cube has a volume of $128y^3$ cubic millimeters. What is its edge length?
14. A kite has a square base with dimensions of 4 feet by 4 feet. What is the length of one of the diagonal crossbars that will be needed to stabilize the kite?
15. For each radical expression, write an equivalent expression with a fractional exponent.
    a. $\sqrt{6}$ b. $\sqrt{10}$ c. $\sqrt[3]{5}$ d. $\sqrt[3]{18}$
16. **Reason abstractly.** Devise a plan for simplifying the fourth root of a number that is not a perfect fourth power. Explain to a friend how to use your plan to simplify the fourth root. Be sure to include examples.

ACTIVITY 20 Continued

Lesson 20-2

PLAN

Pacing: 1 class period
Chunking the Lesson
Example A   #1–2
Check Your Understanding
Lesson Practice

TEACH

Bell-Ringer Activity
Write the following expression on the board: $3x + 7x$. Ask students why the terms of this expression are like terms. [*The terms have the same variable raised to the same power.*] Have students explain how they can apply the Distributive Property to combine the like terms. [*Rewrite the expression as $(3 + 7)x$. Then add within the parentheses to get $10x$.*]

Example A Close Reading, Note Taking, Think-Pair-Share In this example, students are introduced to the addition and subtraction of radicals by applying the Addition Property of Radicals. Ask students to describe how adding and subtracting radicals is similar to combining like terms. Ask students how to identify each sum or difference in the example as rational or irrational.

Differentiating Instruction

Support students who are visual learners by suggesting that they use colored pencils or highlighters to mark radicals with the same index and radicand. They can use a different color for each set of like radicals.

Developing Math Language

This lesson includes the vocabulary terms *closed* and *Addition Property of Radicals*. Students will already be familiar with everyday meanings of the word *closed*, but the mathematical definition will likely be new to them. Explain the meaning in words that are accessible for students. As possible, provide concrete examples to help students gain understanding. It may be helpful to present an example of a set of numbers that is *not* closed under an operation. For instance, the set of whole numbers is not closed under subtraction. In other words, the difference of two whole numbers is not always a whole number, as in $5 - 8 = -3$. Encourage students to make notes about new terms and their understanding of what the terms mean.

ACTIVITY 20 continued

My Notes

MATH TIP
The rational numbers are **closed** under addition and subtraction. This means that the sum or difference of two rational numbers is rational.

Lesson 20-2
Adding and Subtracting Radical Expressions

Learning Targets:
- Add radical expressions.
- Subtract radical expressions.

SUGGESTED LEARNING STRATEGIES: Discussion Groups, Close Reading, Note Taking, Think-Pair-Share, Identify a Subtask

The Windy Hill Science Club advisor wants to find the total length of the balsa wood needed to make the frames for the kites. To do so, he will need to add radicals.

Addition Property of Radicals
$a\sqrt{b} \pm c\sqrt{b} = (a \pm c)\sqrt{b}$, where $b \geq 0$.

To add or subtract radicals, the index and radicand must be the same.

Example A
Add or subtract each expression and simplify. State whether the sum or difference is rational or irrational.

a. $3\sqrt{5} + 7\sqrt{5}$
  $= (3+7)\sqrt{5}$ ←Add or subtract→
  $= 10\sqrt{5}$ the coefficients.
  irrational

b. $10\sqrt[3]{3} - 4\sqrt[3]{3}$
  $= (10-4)\sqrt[3]{3}$
  $= 6\sqrt[3]{3}$
  irrational

c. $2\sqrt{5} + 8\sqrt{3} + 6\sqrt{5} - 3\sqrt{3}$

Step 1: Group terms with like radicands. $2\sqrt{5} + 6\sqrt{5} + 8\sqrt{3} - 3\sqrt{3}$
Step 2: Add or subtract the coefficients. $= (2+6)\sqrt{5} + (8-3)\sqrt{3}$
  $= 8\sqrt{5} + 5\sqrt{3}$

Solution: $2\sqrt{5} + 8\sqrt{3} + 6\sqrt{5} - 3\sqrt{3} = 8\sqrt{5} + 5\sqrt{3}$; irrational

Try These A
Add or subtract each expression and simplify. State whether the sum or difference is rational or irrational.

a. $2\sqrt{7} + 3\sqrt{7} + \frac{2}{3}$

  $5\sqrt{7} + \frac{2}{3}$; irrational

b. $5\sqrt{6} + 2\sqrt{5} - \sqrt{6} + 7\sqrt{5}$

  $4\sqrt{6} + 9\sqrt{5}$; irrational

c. $2 + 2\sqrt{2} + \sqrt{8} + 3\sqrt{2}$

  $2 + 7\sqrt{2}$; irrational

---

304 SpringBoard® Mathematics **Algebra 1, Unit 4** • Exponents, Radicals, and Polynomials

Lesson 20-2
Adding and Subtracting Radical Expressions

1. The club advisor also needs to know how much wood to buy for the legs of the kites. Each kite will be 3 feet tall.
   a. Complete the table below.

Kite	Dimensions of Base (feet)	Length of One Crossbar (feet)	Length of One Leg (feet)	Wood Needed for Legs (feet)	Wood Needed for Crossbars (feet)
A	1 by 1	$\sqrt{2}$	3	12	$8\sqrt{2}$
B	2 by 2	$2\sqrt{2}$	3	12	$16\sqrt{2}$
C	3 by 3	$3\sqrt{2}$	3	12	$24\sqrt{2}$
D	1 by 2	$\sqrt{5}$	3	12	$8\sqrt{5}$
E	2 by 4	$2\sqrt{5}$	3	12	$16\sqrt{5}$
F	3 by 6	$3\sqrt{5}$	3	12	$24\sqrt{5}$

   b. **Reason quantitatively.** How much balsa wood should the club advisor buy if the club is going to build the six kites described above? Is the result rational or irrational?

   $(72 + 48\sqrt{2} + 48\sqrt{5})$ feet; irrational

   c. Explain how you reached your conclusion.
   Answers may vary. I added the wood needed for the legs, 72 feet, and the wood needed for the crossbars, $(48\sqrt{2} + 48\sqrt{5})$ feet. The total is $(72 + 48\sqrt{2} + 48\sqrt{5})$ feet. 72 is rational, but $48\sqrt{2}$ and $48\sqrt{5}$ are irrational, so their sum is irrational.

2. **Use appropriate tools strategically.** Approximately how much balsa wood, in decimal notation, will the club advisor need to buy?
   a. Use your calculator to approximate the amount of balsa wood, and then decide on a reasonable way to round.
   ~247.21 feet

   b. Explain why the club advisor would need this approximation rather than the exact answer expressed as a radical.
   Answers will vary. Shopping for wood requires the decimal approximation.

ACTIVITY 20 Continued

Check Your Understanding
Debrief students' answers to these items to ensure that they understand how to add and subtract like radicals. Ask students to explain why the expression $\sqrt{2} + \sqrt[3]{2}$ cannot be simplified further. [*The radicals have the same radicand, but the indexes are different, so the radicals cannot be added.*]

Answers
3. a. $16\sqrt{6}$; irrational
   b. 18; rational
4. a. $9\sqrt{3}$; irrational
   b. $9\sqrt{2}$; irrational
5. a. $\sqrt[3]{2}$; irrational
   b. $3\sqrt[3]{3} + 16$; irrational
6. a. 24; rational
   b. $8\sqrt{10}$; irrational
7. irrational; Examples will vary. 1 is rational, and $\sqrt{2}$ is irrational. The sum is $1 + \sqrt{2}$, which is irrational.

ASSESS

Students' answers to lesson practice problems will provide you with a formative assessment of their understanding of the lesson concepts and their ability to apply their learning.

See the Activity Practice for additional problems for this lesson. You may assign the problems here or use them as a culmination for the activity.

LESSON 20-2 PRACTICE
8. $20\sqrt{2}$ units
9. $16\sqrt{2}$ units
10. $\sqrt{2}$ units longer
11. $4\sqrt{2}$ units greater
12. $4\sqrt{29}$ units

ADAPT

Check students' answers to the Lesson Practice to ensure they understand how to apply the Addition Property of Radicals to solve problems. If students have not yet mastered addition and subtraction of radicals, work with them to develop a series of steps, such as the following, for these types of problems.

1. Simplify each radical in the expression.
2. Identify radicals with the same index and radicand—the like radicals.
3. Add or subtract terms with like radicals by adding or subtracting the coefficients and keeping the radical the same.

ACTIVITY 20 continued

My Notes

Lesson 20-2
Adding and Subtracting Radical Expressions

Check Your Understanding

Perform the indicated operations. Be sure to completely simplify your answer. State whether each sum or difference is rational or irrational.

3. a. $7\sqrt{6} + 9\sqrt{6}$
   b. $12\sqrt{4} - 5\sqrt{4} + 2\sqrt{4}$
4. a. $8\sqrt{3} - \sqrt{12} + 3\sqrt{3}$
   b. $\sqrt{18} + \sqrt{8} + \sqrt{32}$
5. a. $\sqrt[3]{16} - \sqrt[3]{2}$
   b. $\sqrt[3]{3} + \sqrt[3]{24} + 16$
6. a. $3\sqrt{9} + 5\sqrt{9}$
   b. $3\sqrt{10} + 5\sqrt{10}$

7. Is the sum of a rational number and an irrational number rational or irrational? Support your response with an example.

LESSON 20-2 PRACTICE

Use the figures of a rectangle and kite for Items 8–12.

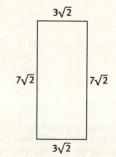
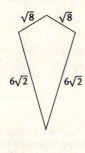

8. Determine the perimeter of the rectangle.
9. Determine the perimeter of the kite.
10. How much longer is the long side of the rectangle than the longer side of the kite?
11. How much greater is the perimeter of the rectangle than the perimeter of the kite?
12. **Make sense of problems.** How much wood would be required to insert diagonal crossbars in the rectangle?

Lesson 20-3
Multiplying and Dividing Radical Expressions

ACTIVITY 20 continued

Learning Targets:
- Multiply and divide radical expressions.
- Rationalize the denominator of a radical expression.

SUGGESTED LEARNING STRATEGIES: Think-Pair-Share, Predict and Confirm, Discussion Groups, Close Reading, Marking the Text, Note Taking

1. **a.** Complete the table below and simplify the radical expressions in the third and fifth columns.

a	b	$\sqrt{a} \cdot \sqrt{b}$	ab	$\sqrt{ab}$
4	9	$\sqrt{4} \cdot \sqrt{9} = 2 \cdot 3 = 6$	36	$\sqrt{36} = 6$
100	25	$\sqrt{100} \cdot \sqrt{25} = 10 \cdot 5 = 50$	2500	$\sqrt{2500} = 50$
9	16	$\sqrt{9} \cdot \sqrt{16} = 3 \cdot 4 = 12$	144	$\sqrt{144} = 12$

b. Express regularity in repeated reasoning. Use the patterns you observe in the table above to write an equation that relates $\sqrt{a}$, $\sqrt{b}$, and $\sqrt{ab}$.
$\sqrt{a} \cdot \sqrt{b} = \sqrt{ab}$

c. All the values of *a* and *b* in part a are perfect squares. In the table below, choose some values for *a* and *b* that are *not* perfect squares and use a calculator to show that the equation you wrote in Part (b) is true for those numbers as well.
Answers may vary.

a	b	$\sqrt{a} \cdot \sqrt{b}$	ab	$\sqrt{ab}$
3	5	$\sqrt{3} \cdot \sqrt{5} \approx 1.73 \cdot 2.24 = 3.88$	15	$\sqrt{15} \approx 3.87$
7	8	$\sqrt{7} \cdot \sqrt{8} \approx 2.65 \cdot 2.83 = 7.50$	56	$\sqrt{56} \approx 7.48$

d. Simplify the products in Columns A and B below.

A	Simplified Form	B	Simplified Form
$(2\sqrt{4})(\sqrt{9})$	12	$2\sqrt{4 \cdot 9}$	12
$(3\sqrt{4})(5\sqrt{16})$	120	$(3 \cdot 5)\sqrt{4 \cdot 16}$	120
$(2\sqrt{7})(3\sqrt{14})$	$42\sqrt{2}$	$(2 \cdot 3)\sqrt{7 \cdot 14}$	$42\sqrt{2}$

e. Which products in the table in Part (d) are rational and which are irrational?
The products in the first two rows are rational; the products in the last row are irrational.

TECHNOLOGY TIP
Approximate values of square roots that are not perfect squares can be found using a calculator.

TEACHER to TEACHER
Item 1c asks students to use a calculator to approximate the square roots of numbers that are not perfect squares. If your class does not have access to calculators, you can ask students to estimate the square roots to the nearest whole number. Students should find that their estimates for $\sqrt{a} \cdot \sqrt{b}$ are close to their estimates for $\sqrt{ab}$.

ACTIVITY 20 Continued
Lesson 20-3

PLAN
Pacing: 1 class period
Chunking the Lesson
#1   Example A
Example B   Example C
Check Your Understanding
Lesson Practice

TEACH

Bell-Ringer Activity
Write the following expressions on the board, and ask students to simplify them. Tell students to assume that all variables are nonnegative.
1. $4\sqrt{40}$ $[8\sqrt{10}]$
2. $3\sqrt{27}$ $[9\sqrt{3}]$
3. $2\sqrt{8a^3}$ $[4a\sqrt{2a}]$
4. $\sqrt{60x^6}$ $[2x^3\sqrt{15}]$

1 Look for a Pattern, Predict and Confirm, Create Representations, Discussion Groups, Debriefing
Students use patterns to develop the Multiplication Property of Radicals. They start by using perfect squares to discover that the product of two square roots is equal to the square root of the product of the radicands. Monitor student discussions carefully to verify that students have written a correct equation to describe this pattern. Students then use calculators to verify that the pattern holds for numbers that are not perfect squares. It is important for students to understand that they are working with approximations and that answers may be close but not exact. A discussion of rounding might be appropriate if students find that their approximations for $\sqrt{a} \cdot \sqrt{b}$ do not exactly equal their approximations for $\sqrt{ab}$. Finally, students extend their conjectures to include products of radical expressions with coefficients. Again, monitor group discussions carefully to ensure students are applying the rule from Item 1b correctly and remembering to completely simplify products. Debrief this portion of the lesson by focusing on the rules in Items 1b and 1f. Allow groups an opportunity, perhaps using whiteboards, to share their rules from Item 1f so that students have an opportunity to hear the rule described in different ways.

Activity 20 • Operations with Radicals   307

ACTIVITY 20 Continued

Developing Math Language
This lesson includes the vocabulary terms *Multiplication Property of Radicals* and *Division Property of Radicals*. Suggest that students make a table or other graphic organizer in their math notebooks where they can list each property and give a verbal description, the symbolic form, and a numerical example. Students can also include the Addition Property of Radicals from the previous lesson.

TEACHER to TEACHER
The Multiplication and Division Properties of Radicals are presented to students in symbolic language using square roots. Note that these properties also apply for higher indices as long as the radical expressions being multiplied or divided have the same index. Students may ask why the values of b and d are restricted to nonnegative values or positive values. This could lead to a class discussion around an example such as finding $\sqrt{-4}$.

Example A Close Reading, Marking the Text, Note Taking, Identify a Subtask, Discussion Groups In this example, students are introduced to the multiplication of radical expressions. Allow time for students to discuss each step of parts a, b and c of the example and explain why each step is performed. Monitor discussions to ensure that students are using appropriate vocabulary with specific mathematical details to communicate their understanding of math concepts. Use the Try These items as a formative assessment opportunity to monitor how well students understand the Multiplication Property of Radicals.

CONNECT TO AP
Later in this book, students will be introduced to the set of complex numbers, which includes both the real numbers and the imaginary numbers. Within the complex numbers, $\sqrt{-1}$ is defined as the imaginary unit i. Students will learn that equations that have no real solutions may have complex solutions with imaginary parts.

ACTIVITY 20 continued

Lesson 20-3
Multiplying and Dividing Radical Expressions

My Notes

MATH TIP
The rational numbers are **closed** under multiplication. This means that the product of two rational numbers is rational. Since the coefficients a and c are rational, their product will also be rational.

MATH TIP

coefficient, index, radicand ($5\sqrt[3]{2}$)

CONNECT TO AP
Later in this course, you will study another system of numbers, called the complex numbers. In the complex number system, $\sqrt{-1}$ is defined as the imaginary number i.

---

f. Write a verbal rule that explains how to multiply radical expressions.
Answers may vary. First, multiply the coefficients. Then, multiply the radicands. Simplify the result.

Multiplication Property of Radicals

$$\left(a\sqrt{b}\right)\left(c\sqrt{d}\right) = ac\sqrt{bd},$$
where $b \geq 0, d \geq 0$.

To multiply radical expressions, the index must be the same. Find the product of the coefficients and the product of the radicands. Simplify the radical expression.

Example A
Multiply each expression and simplify.

a. $\left(3\sqrt{6}\right)\left(4\sqrt{5}\right) = (3 \cdot 4)\left(\sqrt{6 \cdot 5}\right) = 12\sqrt{30}$

b. $\left(2\sqrt{10}\right)\left(3\sqrt{6}\right)$
$= (2 \cdot 3)\sqrt{10 \cdot 6}$ Step 1: Multiply.
$= 6\sqrt{60}$
$= 6\left(\sqrt{4 \cdot 15}\right)$ Step 2: Simplify.
$= (6 \cdot 2)\sqrt{15}$
$= 12\sqrt{15}$

c. $\left(2x\sqrt{6x}\right)\left(5\sqrt{3x^2}\right)$
$= 10x\sqrt{6x \cdot 3x^2}$
$= 10x\left(\sqrt{18x^3}\right)$
$= 10x\left(\sqrt{9x^2 \cdot 2x}\right)$
$= (10x)(3x)\left(\sqrt{2x}\right)$
$= 30x^2\sqrt{2x}$

Try These A
Multiply each expression and simplify.

a. $\left(2\sqrt{10}\right)\left(5\sqrt{3}\right)$
$10\sqrt{30}$

b. $\left(3\sqrt{8}\right)\left(2\sqrt{6}\right)$
$24\sqrt{3}$

c. $\left(4\sqrt{12}\right)\left(5\sqrt{18}\right)$
$120\sqrt{6}$

d. $\left(3\sqrt{5a}\right)\left(2a\sqrt{15a^2}\right)$
$30a^2\sqrt{3a}$

Division Property of Radicals

$$\frac{a\sqrt{b}}{c\sqrt{d}} = \frac{a}{c}\sqrt{\frac{b}{d}}$$
where $b \geq 0, d \geq 0$.

To divide radical expressions, the index must be the same. Find the quotient of the coefficients and the quotient of the radicands. Simplify the expression.

Lesson 20-3
Multiplying and Dividing Radical Expressions

ACTIVITY 20 continued

Example B
Divide each expression and simplify.

a. $\dfrac{\sqrt{6}}{\sqrt{2}} = \sqrt{\dfrac{6}{2}} = \sqrt{3}$

b. $\dfrac{2\sqrt{10}}{3\sqrt{2}} = \dfrac{2}{3}\sqrt{\dfrac{10}{2}} = \dfrac{2}{3}\sqrt{5}$

c. $\dfrac{8\sqrt{24x^2}}{2\sqrt{3}} = \dfrac{8}{2}\sqrt{\dfrac{24x^2}{3}} = 4\sqrt{8x^2} = 4\sqrt{4 \cdot 2 \cdot x^2}$
$= 4(2x\sqrt{2}) = 8x\sqrt{2}$

Try These B
Divide each expression and simplify.

a. $\dfrac{4\sqrt{42}}{5\sqrt{6}}$

$\dfrac{4}{5}\sqrt{7}$

b. $\dfrac{10\sqrt{54}}{2\sqrt{2}}$

$15\sqrt{3}$

c. $\dfrac{12\sqrt{75}}{3\sqrt{3}}$

20

d. $\dfrac{16\sqrt[3]{8x^{11}}}{8\sqrt[3]{x^2}}$

$4x^3$

A radical expression in simplified form does not have a radical in the denominator. Most frequently, the denominator is **rationalized**. You **rationalize the denominator** by simplifying the expression to get a perfect square under the radicand in the denominator.

$$\dfrac{\sqrt{a}}{\sqrt{b}} \cdot 1 = \left(\dfrac{\sqrt{a}}{\sqrt{b}}\right)\left(\dfrac{\sqrt{b}}{\sqrt{b}}\right) = \dfrac{\sqrt{ab}}{\sqrt{b^2}} = \dfrac{\sqrt{ab}}{b}$$

Example C
Rationalize the denominator of $\dfrac{\sqrt{5}}{\sqrt{3}}$.

Step 1: Multiply the numerator and denominator by $\sqrt{3}$.

$\dfrac{\sqrt{5}}{\sqrt{3}} = \dfrac{\sqrt{5}}{\sqrt{3}} \cdot \dfrac{\sqrt{3}}{\sqrt{3}} = \dfrac{\sqrt{15}}{\sqrt{9}}$

Step 2: Simplify.

$= \dfrac{\sqrt{15}}{3}$

Solution: $\dfrac{\sqrt{5}}{\sqrt{3}} = \dfrac{\sqrt{15}}{3}$

Try These C
Rationalize the denominator in each expression.

a. $\dfrac{\sqrt{11}}{\sqrt{6}}$

$\dfrac{\sqrt{66}}{6}$

b. $\dfrac{2\sqrt{7}}{\sqrt{5}}$

$\dfrac{2\sqrt{35}}{5}$

c. $\dfrac{3\sqrt{5}}{\sqrt{8}}$

$\dfrac{3\sqrt{10}}{4}$

MATH TERMS

Rationalize means to make rational. You can **rationalize the denominator** without changing the value of the expression by multiplying the fraction by an appropriate form of 1.

CONNECT TO AP

In calculus, both numerators and denominators are rationalized. The procedure for rationalizing a numerator is similar to that for rationalizing a denominator.

ACTIVITY 20 Continued

Check Your Understanding

Debrief students' answers to these items to ensure that they understand how to multiply and divide radical expressions. Have students give a property or explanation for each step of their work for Items 2–7.

Answers

2. $8\sqrt{21}$; irrational
3. $2 + 6\sqrt{3}$; irrational
4. $\sqrt{15}$; irrational
5. $\dfrac{\sqrt{10}}{4}$; irrational
6. $240y\sqrt{2}$
7. $42x\sqrt{x}$
8. irrational; Examples may vary. 3 is rational, and $\sqrt{2}$ is irrational. Their product is $3\sqrt{2}$, which is irrational.

ASSESS

Students' answers to lesson practice problems will provide you with a formative assessment of their understanding of the lesson concepts and their ability to apply their learning.

See the Activity Practice for additional problems for this lesson. You may assign the problems here or use them as a culmination for the activity.

LESSON 20-3 PRACTICE

9. $\dfrac{\sqrt{30}}{10}$; irrational
10. 1; rational
11. $\dfrac{2\sqrt{21}}{3}$; irrational
12. $\dfrac{2\sqrt{10}}{3}$; irrational
13. $40m^2\sqrt[3]{m}$
14. $4x^4$
15. Possible answer: There must not be any radicals in a denominator. A radicand does not contain any fractions. A radicand does not have any factors that are perfect nth powers, where n is the index of the radical. There are no terms with like radicals.

ADAPT

Check students' answers to the Lesson Practice to make sure that they understand and can apply the Multiplication and Division Properties of Radicals. Reinforce students' understanding of these properties by having them complete the following activity in pairs. Each student in the pair writes a radical expression in the form $a\sqrt{b}$, such as $3\sqrt{12}$. Then, pairs work together to both multiply and divide their expressions.

310 SpringBoard® Mathematics **Algebra 1, Unit 4** • Exponents, Radicals, and Polynomials

---

ACTIVITY 20 continued

My Notes

Lesson 20-3
Multiplying and Dividing Radical Expressions

Check Your Understanding

Express each expression in simplest radical form. State whether each result in Items 2–5 is rational or irrational.

2. $(4\sqrt{7})(2\sqrt{3})$
3. $\sqrt{2}(\sqrt{2} + 3\sqrt{6})$
4. $\dfrac{\sqrt{75}}{\sqrt{5}}$
5. $\sqrt{\dfrac{5}{8}}$
6. $(3\sqrt{32y})(4\sqrt{25y})$
7. $\dfrac{6\sqrt{98x^4}}{\sqrt{2x}}$
8. Is the product of a nonzero rational number and an irrational number rational or irrational? Support your response with an example.

LESSON 20-3 PRACTICE

Express each expression in simplest radical form. State whether each result in Items 9–12 is rational or irrational.

9. $\left(\sqrt{\dfrac{1}{2}}\right)\left(\sqrt{\dfrac{3}{5}}\right)$
10. $\sqrt{27} \cdot \sqrt{\dfrac{1}{27}}$
11. $\dfrac{2\sqrt{7}}{\sqrt{3}}$
12. $\dfrac{4\sqrt{5}}{3\sqrt{2}}$
13. $\left(4\sqrt[3]{4m^2}\right)\left(5m\sqrt[3]{2m^2}\right)$
14. $\dfrac{2\sqrt{52x^9}}{\sqrt{13x}}$
15. **Attend to precision.** What conditions must be satisfied for a radical expression to be in simplified form?

Operations with Radicals
Go Fly a Kite

ACTIVITY 20 PRACTICE
Write your answers on notebook paper.
Show your work.

Lesson 20-1

Write each expression in simplest radical form.

1. $\sqrt{40}$
2. $\sqrt{128}$
3. $\sqrt{162}$

Use the Pythagorean Theorem and the triangle below for Items 4 and 5. Recall that the Pythagorean Theorem states that for all right triangles, $a^2 + b^2 = c^2$.

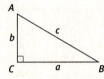

4. In the right triangle, if $a = 3$ and $b = 6$, what is the value of c?
   A. $3\sqrt{5}$ B. 9
   C. $9\sqrt{5}$ D. 45

5. In the right triangle, if $a = 12$ and $b = 15$, what is the value of c?

Simplify each expression.

6. $\sqrt{4m^7}$
7. $3\sqrt[4]{16n^8}$
8. $\sqrt[3]{16x^4}$

Write each of the following as a radical expression.

9. $15^{\frac{2}{5}}$
10. $(2p)^{\frac{1}{3}}$
11. $16x^{\frac{3}{4}}$

12. Which of the following expressions is **not** equivalent to $(8x)^{\frac{2}{3}}$?
    A. $4\sqrt[3]{x^2}$ B. $\sqrt[3]{8x^2}$
    C. $4x^{\frac{2}{3}}$ D. $8^{\frac{2}{3}}x^{\frac{2}{3}}$

Lesson 20-2

Write each expression in simplest radical form. State whether each result is rational or irrational.

13. $4\sqrt{27} + 6\sqrt{12}$
14. $8\sqrt{6} + 2\sqrt{12} + 5\sqrt{3} - \sqrt{54}$
15. $3\sqrt{36} - 5\sqrt{16} + 4$
16. Which of the following is the difference of $9\sqrt{20}$ and $2\sqrt{5}$?
    A. $7\sqrt{15}$ B. $16\sqrt{5}$
    C. $9\sqrt{5}$ D. $7\sqrt{5}$

The figure below is composed of a rectangle and a right triangle. Use the figure for Items 17–19.

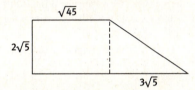

17. Determine the perimeter of the rectangle.
18. Determine the perimeter of the triangle.
19. Determine the perimeter of the composite figure.
20. A student was asked to completely simplify the expression $3\sqrt{3} + \sqrt{12} + 2\sqrt{3}$. The student wrote $5\sqrt{3} + \sqrt{12}$. Do you agree with the student's answer? Explain.

ACTIVITY 20 Continued
ACTIVITY PRACTICE

1. $2\sqrt{10}$
2. $8\sqrt{2}$
3. $9\sqrt{2}$
4. A
5. $3\sqrt{41}$
6. $2m^3\sqrt{m}$
7. $6n^2$
8. $2x\sqrt[3]{2x}$
9. $\sqrt[5]{15^2}$ or $(\sqrt[5]{15})^2$
10. $\sqrt[3]{2p}$
11. $16\sqrt[4]{x^3}$ or $16(\sqrt[4]{x})^3$
12. B
13. $24\sqrt{3}$; irrational
14. $5\sqrt{6} + 9\sqrt{3}$; irrational
15. 2; rational
16. B
17. $10\sqrt{5}$ units
18. $(5\sqrt{5} + \sqrt{65})$ units
19. $(11\sqrt{5} + \sqrt{65})$ units
20. No; the term $\sqrt{12}$ can be written as $2\sqrt{3}$, so the correct answer is $7\sqrt{3}$.

ACTIVITY 20 Continued

21. 6 cubic units; rational
22. $60\sqrt{15}$ cubic units; irrational
23. $\dfrac{4\sqrt{15}}{3}$ units
24. $\dfrac{2\sqrt{15}}{5}$ units
25. $6x^3\sqrt{2}$
26. $4.8p^3$
27. $56m^3$
28. 12
29. $3\sqrt[3]{x}$
30. xy^2
31. D
32. **a.** Answers may vary. Use the calculator to evaluate both expressions for several values of x; the two results should be the same.
    b. yes
33. $\dfrac{\pi\sqrt{2}}{4}$ seconds
34. $\dfrac{\pi}{2}$ seconds
35. Possible answer: It will tick more often, because it will take the pendulum less time to swing from side to side.
36. ≈ 3.24 feet
37. No; he forgot to square the radical. The correct area is 20π square feet.

ADDITIONAL PRACTICE

If students need more practice on the concepts in this activity, see the eBook Teacher Resources for additional practice problems.

ACTIVITY 20 continued

Operations with Radicals
Go Fly a Kite

Lesson 20-3

The figure shows a rectangular prism. The volume of the rectangular prism is the product of the length, width, and height. Use the figure for Items 21–24.

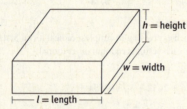

21. If $l = \sqrt{3}$, $w = \sqrt{2}$, and $h = \sqrt{6}$, what is the volume of the rectangular prism? Is the volume rational or irrational?
22. If $l = 3\sqrt{3}$, $w = 2\sqrt{2}$, and $h = 5\sqrt{10}$, what is the volume of the rectangular prism? Is the volume rational or irrational?
23. If the volume of the rectangular prism is 20, the length is $\sqrt{3}$, and the width is $\sqrt{5}$, what is the height?
24. If the volume of the rectangular prism is $24\sqrt{3}$, the height is $2\sqrt{2}$, and the width is $3\sqrt{10}$, what is the length?

Write each expression in simplest form.

25. $\left(2\sqrt{2x^2}\right)\left(3x\sqrt{x^2}\right)$
26. $\left(6p\sqrt{p^3}\right)\left(0.2\sqrt{16p}\right)$
27. $\left(4\sqrt[3]{8m}\right)\left(7m\sqrt[3]{m^5}\right)$
28. $\dfrac{3\sqrt{32}}{\sqrt{2}}$
29. $\dfrac{\sqrt[3]{81x}}{\sqrt[3]{3}}$
30. $\dfrac{\sqrt{x^2y^5}}{\sqrt{y}}$

31. Which of the following expressions cannot be simplified any further?
    A. $\sqrt{\dfrac{5}{2}}$ B. $\dfrac{\sqrt{5}}{\sqrt{2}}$
    C. $\sqrt{52}$ D. $5\sqrt{2}$

32. Elena was asked to simplify the expression $\left(2\sqrt{12x}\right)\left(4x\sqrt{3}\right)$. Her answer was $48x\sqrt{x}$.
    a. Explain how Elena can use her calculator to check whether her answer is reasonable.
    b. Is Elena's answer correct? If not, explain Elena's mistake and give the correct answer.

The time, T, in seconds, it takes the pendulum of a clock to swing from one side to the other side is given by the formula $T = \pi\sqrt{\dfrac{l}{32}}$, where l is the length of the pendulum, in feet. The clock ticks each time the pendulum is at the extreme left or right point.

Use this information for Items 33–36.

33. If the pendulum is 4 feet long, how long does it take the pendulum to swing from left to right? Give an exact value in terms of π.
34. If the pendulum is 8 feet long, how long does it take the pendulum to swing from left to right? Give an exact value in terms of π.
35. If the pendulum is shortened, will the clock tick more or less often? Explain how you arrived at your conclusion.
36. Approximately what length of the pendulum will result in its swinging from one side to the other every second?

MATHEMATICAL PRACTICES
Construct Viable Arguments and Critique the Reasoning of Others

37. Amil knows that the formula for the area of a circle is $A = \pi r^2$. He says that the area of a circle with a radius of $2\sqrt{5}$ feet is $4\sqrt{5}\pi$ square feet. Is he correct? If not, describe his error.

Geometric Sequences

Go Viral!
Lesson 21-1 Identifying Geometric Sequences

ACTIVITY 21

Learning Targets:
- Identify geometric sequences and the common ratio in a geometric sequence.
- Distinguish between arithmetic and geometric sequences.

> **SUGGESTED LEARNING STRATEGIES:** Visualization, Look for a Pattern, Create Representations, Think-Pair-Share, Sharing and Responding

For her Electronic Communications class, Keisha has been tasked with investigating the effects of social media. She decides to post a video in cyberspace to see if she can make it go viral.

To get things started, Keisha e-mails the video link to three of her friends. In the message, she asks each of the recipients to forward the link to three of his or her friends. Whenever a recipient forwards the link, Keisha asks him or her to attach the following message: *After watching, please forward this video link to three of your friends who have not yet received it.*

One way to visually represent this situation is with a tree diagram. A **tree diagram** shows all the possible outcomes of an event.

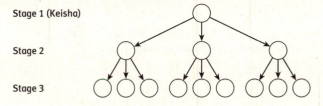

Stage 1 (Keisha)
Stage 2
Stage 3

1. Use the tree diagram to help you complete the table. (Assume that everyone who receives the video link watches the video.)

Stage	Number of People Who Watch the Video
1	1
2	3
3	9
4	27

2. **Express regularity in repeated reasoning.** Describe any patterns you notice in the table.

   Answers may vary. Each subsequent stage is 1 more than the previous stage. Each subsequent number of people is 3 times the previous number.

Common Core State Standards for Activity 21

HSF-IF.A.3	Recognize that sequences are functions, sometimes defined recursively, whose domain is a subset of the integers.
HSF-BF.A.2	Write arithmetic and geometric sequences both recursively and with an explicit formula, use them to model situations, and translate between the two forms.*
HSF-LE.A.1	Distinguish between situations that can be modeled with linear functions and with exponential functions.
HSF-LE.A.1.c	Recognize situations in which a quantity grows or decays by a constant percent rate per unit interval relative to another.

Activity 21 • Geometric Sequences 313

ACTIVITY 21 Continued

Developing Math Language
Students should understand that a *common ratio* is a constant ratio between consecutive terms of a sequence. Help students understand that the common ratio is the value that each term in a geometric series is multiplied by to generate the next term. Equivalently, the quotient of any term and the preceding term is equal to the common ratio. Be sure to add these terms to the class Word Wall.

3–5 Create Representations, Look for a Pattern, Think-Pair-Share, Sharing and Responding Students use a graph of the relationship between the stage number and the number of people who watch the video to determine that the relationship is a nonlinear function. If students do not mention the rate of change as a way for determining whether the relationship is linear, encourage them to discuss whether the rate of change shown by the relationship is constant and how this can be justified. If needed, remind students that the domain of a function is the set of possible input values.

TEACHER to TEACHER

Some students may wonder why a geometric sequence is called *geometric* when it involves a pattern of numbers and not shapes. Explain that each positive term (after the first term) in a geometric sequence is the *geometric* mean of the two terms adjacent to it. The geometric mean of two numbers a and b is equal to $\sqrt{ab}$. Similarly, each term (after the first term) in an arithmetic sequence is equal to the *arithmetic* mean of the two terms adjacent to it. The arithmetic mean of two numbers a and b is equal to $\frac{a+b}{2}$.

6–7 Close Reading, Marking the Text, Look for a Pattern, Discussion Group, Debriefing Some students may have difficulty determining common ratios when the numbers in the sequence are decreasing. Monitor group discussions carefully and redirect students, if necessary. Remind groups that are having difficulty that division can also be expressed as multiplication by a fraction. Debrief this portion of the lesson by asking students to verbalize the difference between an arithmetic and a geometric sequence. Ask students to each create one arithmetic and one geometric sequence to share with their group.

ACTIVITY 21 continued

My Notes

MATH TERMS
A **function** is a relation in which every input is paired with exactly one output.

Lesson 21-1
Identifying Geometric Sequences

3. Use the table of values to graph the viral video situation.

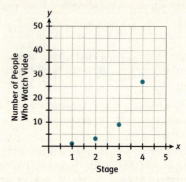

4. Is the relationship a linear relationship? Justify your response.
   No; it is not linear, because the graph is not a line; there is no constant rate of change.

5. Is the graph the graph of a *function*? If so, what is the domain?
   Yes; the domain is the natural numbers, {1, 2, 3, . . .}

The number of people who have received the video link at each stage form a *geometric sequence*. A **geometric sequence** is a sequence of values in which a nonzero constant ratio exists between consecutive terms. The constant ratio is called the common ratio and is typically denoted by the letter r. The common ratio is the value that each term is multiplied by to get the next term.

6. **a.** Write the numbers of people who have received the video as a sequence.
   1, 3, 9, 27, . . .

   b. Reason quantitatively. Identify the common ratio. Justify your response.
   3; each term is multiplied by 3 to get the next term.

---

314 SpringBoard® Mathematics **Algebra 1, Unit 4** • Exponents, Radicals, and Polynomials

Lesson 21-1
Identifying Geometric Sequences

7. Identify each sequence as arithmetic, geometric, or neither. If it is arithmetic, state the common difference. If it is geometric, state the common ratio.
   a. 5, 8, 11, 14, …
      arithmetic; common difference = 3
   b. 18, 6, 2, $\frac{2}{3}$, …
      geometric; common ratio = $\frac{1}{3}$
   c. 1, 4, 9, 16, …
      neither
   d. −1, 4, −16, 64, …
      geometric; common ratio = −4
   e. 16, −8, 4, −2, …
      geometric; common ratio = −$\frac{1}{2}$

Check Your Understanding

Identify each sequence as arithmetic, geometric, or neither. If it is arithmetic, state the common difference. If it is geometric, state the common ratio.

8. 10, 8, 6, 4, …   9. 2, $\frac{1}{2}$, $\frac{1}{8}$, $\frac{1}{32}$, …   10. 9, −3, 1, −$\frac{1}{3}$, …

LESSON 21-1 PRACTICE

A cell divides in two every day. The tree diagram shows the first few stages of this process. Use the tree diagram for Items 11–13.

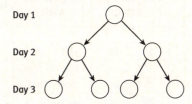

11. Make a table of values to represent the scenario shown in the tree diagram.
12. Does the tree diagram represent a geometric sequence? If so, what is the common ratio?
13. If the diagram were extended to a sixth day, how many circles would there be on Day 6?
14. **Reason abstractly.** Can a geometric sequence ever have a term equal to 0? Explain.

MATH TERMS

An **arithmetic sequence** is a sequence in which the difference between consecutive terms is constant. This difference is called the **common difference**.

ACTIVITY 21 Continued

Lesson 21-2

PLAN

Materials:
- graphing calculators

Pacing: 1 class period

Chunking the Lesson
#1–3
Check Your Understanding
#7–8   #9–11   #12
#13–14   #15
Check Your Understanding
#20
Check Your Understanding
Lesson Practice

TEACH

Bell-Ringer Activity
Give students the sequence 12, 27, 42, 57, 72, ... This is the same sequence that was identified as arithmetic at the beginning of the previous lesson. Ask students to write an explicit formula and a recursive formula for the sequence.
$[a_n = 12 + (n-1)15;$
$\begin{cases} a_1 = 12 \\ a_n = a_{n-1} + 15 \end{cases}]$

TEACHER to TEACHER

Remind students that a recursive formula tells how to find the nth term of a sequence based on the previous term. In sequence notation, the nth term is a_n and the term before a_n is a_{n-1}. In function notation, the nth term of a sequence is $f(n)$, and the term before $f(n)$ is $f(n-1)$.

1–3 Look for a Pattern, Create Representations, Discussion Groups, Debriefing In these items, students use patterns to develop recursive formulas for a geometric sequence. Monitor group discussions carefully to be sure students have a clear understanding of Item 1b before moving on to complete the table. As you debrief this portion of the lesson, allow students to share the formulas they have written and to describe their meanings. Ask students to describe the purpose of a recursive formula and when it might be appropriate to use a recursive formula. Encourage students to relate their responses back to the viral video scenario.

ACTIVITY 21 continued

Lesson 21-2
Formulas for Geometric Sequences

My Notes

Learning Targets:
- Write a recursive formula for a geometric sequence.
- Write an explicit formula for a geometric sequence.
- Use a formula to find a given term of a geometric sequence.

SUGGESTED LEARNING STRATEGIES: Create Representations, Look for a Pattern, Discussion Groups, Think-Pair-Share, Construct an Argument, Sharing and Responding

Remember that the numbers in a sequence are called *terms*, and you can use sequence notation a_n or function notation $f(n)$ to refer to the nth term.

1. a. Use the above notation to rewrite the first four terms of the viral video sequence. Also write the common ratio.
   $a_1 = f(1) = $ **1**
   $a_2 = f(2) = $ **3**
   $a_3 = f(3) = $ **9**
   $a_4 = f(4) = $ **27**
   $r = $ **3**

   b. What is the value of the term following $a_4 = f(4)$? Write an expression to represent this term using a_4 and the common ratio.
   $81 = 3a_4$

Just as with arithmetic sequences, you can use a **recursive formula** to represent a geometric sequence.

MATH TERMS

A **recursive formula** is a formula that gives any term as a function of preceding terms.

2. Complete the table below for the viral video sequence.

Term	Sequence Representation Using Common Ratio	Function Representation Using Common Ratio	Numerical Value (number of people who have seen the video)
$a_1 = f(1)$	1	1	1
$a_2 = f(2)$	$3(1) = 3a_1$	$3(1) = 3f(1)$	3
$a_3 = f(3)$	$3(3) = 3a_2$	$3(3) = 3f(2)$	9
$a_4 = f(4)$	$3(9) = 3a_3$	$3(9) = 3f(3)$	27
$a_5 = f(5)$	$3(27) = 3a_4$	$3(27) = 3f(4)$	81
...	...	...	...
$a_n = f(n)$	$3a_{n-1}$	$3f(n-1)$	—

3. The recursive formulas for the viral video sequence are partially given below. Complete the formulas by writing the expressions for a_n and $f(n)$.

$\begin{cases} a_1 = 1 \\ a_n = \mathbf{3a_{n-1}} \end{cases}$ $\begin{cases} f(1) = 1 \\ f(n) = \mathbf{3f(n-1)} \end{cases}$

316 SpringBoard® Mathematics Algebra 1, Unit 4 • Exponents, Radicals, and Polynomials

Lesson 21-2
Formulas for Geometric Sequences

Check Your Understanding

Write a recursive formula for each geometric sequence. Include the recursive formula in function notation.

4. 64, 16, 4, 1, ...
5. 64, −16, 4, −1, ...
6. −1, 1, −1, 1, ...

7. Use the recursive formula to find a_6, a_7, and a_8 for the viral video sequence. Explain your results.
   243, 729, 2187; multiply 81 by 3 to get 243, multiply by 3 again to get 729, and multiply by 3 again to get 2187.

8. Why might it be difficult to find the 100th term of the viral video sequence using the recursive formula?
   You would have to find all 100 terms of the sequence by repeatedly multiplying by 3, which would be time-consuming.

As with arithmetic sequences, geometric sequences can be represented with explicit formulas. The terms in a geometric sequence can be written as the product of the first term and a power of the common ratio.

9. For the viral video sequence, identify a_1 and r. Then fill in the missing exponents and blanks.

   $a_1 = \underline{1}$ $r = \underline{3}$
   $a_2 = 1 \cdot 3^{\boxed{1}} = 3$
   $a_3 = 1 \cdot 3^{\boxed{2}} = 9$
   $a_4 = 1 \cdot 3^{\boxed{3}} = \underline{27}$
   $a_5 = 1 \cdot 3^{\boxed{4}} = \underline{81}$
   $a_6 = 1 \cdot 3^{\boxed{5}} = \underline{243}$
   $a_{10} = 1 \cdot 3^{\boxed{9}} = \underline{19{,}683}$

10. **Express regularity in repeated reasoning.** Describe any patterns you observe in your responses to Item 9. Then use a_1, r, and n to write a formula for the nth term of any geometric sequence.
    Answers may vary; $a_n = a_1 r^{n-1}$

11. Write the explicit formula for the viral video sequence. Use the formula to determine the 12th term of the sequence. What does the 12th term represent?
    $a_n = 1(3)^{n-1}$; $a_{12} = 1 \times 3^{11} = 177{,}147$; it represents the number of people who will watch the video at Stage 12.

ACTIVITY 21 Continued

12 Activating Prior Knowledge, Create Representations, Think-Pair-Share
Students use function notation to represent the explicit formula for a geometric sequence and identify characteristics of this function. Students should compare the graph of the function in part (d) to the graph they made of the viral video situation in Item 3 of the previous lesson. Ask students how the graphs help to confirm that they wrote the explicit formula for the viral video sequence correctly.

Teacher to Teacher

Item 12d asks students to use a graphing calculator to graph a function. If your class does not have access to graphing calculators, students can create the graph using pencil and paper.

13–14 Look for a Pattern, Create Representations, Think-Pair-Share
Students write explicit formulas for geometric sequences and apply them to find given terms of the sequences. Monitor student discussions carefully to ensure students are taking time to check formulas using terms from the given sequences. Students who find their initial formulas are not correct can work with group members to revise their formulas.

Differentiating Instruction

Extend learning for students by challenging them to write an explicit formula for the sequence 1, 4, 9, 16, 25, . . . , which is neither arithmetic nor geometric. [$a_n = n^2$] Then ask them to find the 20th term of the sequence. [$a_{20} = 400$]

ACTIVITY 21 continued

Lesson 21-2
Formulas for Geometric Sequences

12. The explicit formula for a geometric sequence can be thought of as a function.
   a. What is the input? What is the output?
      Input: n, the term number; output: a_n, the value of the term

   b. State the domain of the function.
      Positive integers

   c. Rewrite the explicit formula for the viral video sequence using function notation.
      $f(n) = 1(3)^{n-1}$

   d. **Use appropriate tools strategically.** Use a graphing calculator to graph your function from Part (c). Is the function linear or nonlinear? Justify your response.
      Nonlinear; the graph is not a line.

13. Consider the geometric sequence 5, 10, 20, 40, . . .
   a. Write the explicit formula for the sequence.
      $a_n = 5(2)^{n-1}$

   b. How can you check that your formula is correct?
      Answers may vary. Check that the formula gives the correct value for one of the terms. For example, $a_3 = 5(2)^{3-1} = 5(2)^2 = 20$, which is correct.

   c. Determine the 16th term in the sequence.
      163,840

   d. Use function notation to write the explicit formula for the sequence.
      $f(n) = 5(2)^{n-1}$

   e. What is the value of $f(10)$? What does it represent?
      $f(10) = 2560$; **it is the 10th term of the sequence.**

14. a. Write the explicit formula for the geometric sequence 32, 16, 8, 4, . . .
      $a_n = 32\left(\frac{1}{2}\right)^{n-1}$

   b. Determine the 9th term in the sequence.
      0.125, or $\frac{1}{8}$

318 SpringBoard® Mathematics **Algebra 1, Unit 4** • Exponents, Radicals, and Polynomials

Lesson 21-2
Formulas for Geometric Sequences

15. The explicit formula for a geometric sequence is $a_n = 6 \cdot 3^{n-1}$. State the recursive formula for the sequence. Include the recursive formula in function notation.

$\begin{cases} a_1 = 6 \\ a_n = 3a_{n-1} \end{cases}$ $\quad\quad$ $\begin{cases} f(1) = 6 \\ f(n) = 3f(n-1) \end{cases}$

Check Your Understanding

16. How can you use the recursive formula for a geometric sequence to write the explicit formula?

Write the explicit formula for each geometric sequence. Then determine the 6th term of each sequence.

17. 1, 5, 25, ...

18. 48, −24, 12, ...

19. $\begin{cases} a_1 = -81 \\ a_n = -\frac{1}{3}a_{n-1} \end{cases}$

20. **Make sense of problems.** Revisit the viral video scenario at the beginning of the activity. How many stages will it take until 1 million new people receive the link to the viral video? Explain how you found your answer.

    14; students will likely use a guess and check method, substituting various values for *n* into the explicit formula, until a_n is 1 million or greater.

Check Your Understanding

21. Write a recursive formula for the geometric sequence whose explicit formula is $a_n = 1 \cdot (-2)^{n-1}$. Include the recursive formula in function notation.

22. Write an explicit formula for the sequence $\begin{cases} a_1 = 3 \\ a_n = 2a_{n-1} \end{cases}$.

Check Your Understanding

Ask students to share their answers to these questions to ensure that they understand how to convert between the recursive and the explicit formulas of geometric sequences. You may also want to present students with this scenario: *The terms of a geometric sequence are decreasing, and all of the terms are positive.* Ask students what they can conclude about the common ratio of this sequence. [*It is a positive fraction less than 1.*]

Answers

21. $\begin{cases} a_1 = 1 \\ a_n = -2a_{n-1} \end{cases}$;

    $\begin{cases} f(1) = 1 \\ f(n) = -2f(n-1) \end{cases}$

22. $a_n = 3(2)^{n-1}$

ACTIVITY 21 Continued

15 Create Representations, Work Backward, Discussion Groups, Debriefing In this item, students use the explicit formula of a geometric sequence to write the recursive formula. As this portion of the lesson is debriefed, allow groups to share the methods they used. Some students may use the explicit formula to write the first few terms of the sequence and then use these terms to write the recursive formula. Other students may identify the first term and the common ratio directly from the explicit formula and use these to write the recursive formula without finding any other terms of the sequence. Many will use much the same logic to write the explicit formula from the recursive one. No matter which the method they choose, the focus should be on the identification of a_1 and r.

Check Your Understanding

Debrief students' answers to these items to ensure that they understand how to write and apply explicit formulas of geometric sequences. You may also want to have students compare and contrast the explicit formula for a geometric sequence with the explicit formula for an arithmetic sequence.

Answers

16. From the recursive formula, you could identify the values of a_1 and r. Then you could substitute these values into the explicit formula.

17. $a_n = 1(5)^{n-1}$; 3125

18. $a_n = 48\left(-\frac{1}{2}\right)^{n-1}$; $-\frac{3}{2}$

19. $a_n = -81\left(-\frac{1}{3}\right)^{n-1}$; $\frac{1}{3}$

20 Guess and Check, Sharing and Responding Students are being asked to apply a geometric sequence to solve a real-world problem. Most students will likely use a guess and check strategy and the explicit formula to determine the least value of *n* for which a_n is 1 million or greater. However, other students may prefer to apply the recursive formula repeatedly. If possible, allow students to use calculators as they work on this item, since basic operations are not the focus of this problem. Allow students to share reasoning, comparing and contrasting the differences between the use of the recursive and explicit formulas.

Activity 21 • Geometric Sequences 319

ACTIVITY 21 Continued

ASSESS

Students' answers to lesson practice problems will provide you with a formative assessment of their understanding of the lesson concepts and their ability to apply their learning. See the Activity Practice for additional problems for this lesson. You may assign the problems here or use them as a culmination for the activity.

LESSON 21-2 PRACTICE

23. $\frac{1}{2}, \frac{1}{4}, \frac{1}{8}, \frac{1}{16}, \ldots$

24. $a_n = \frac{1}{2}\left(\frac{1}{2}\right)^{n-1} = \left(\frac{1}{2}\right)^n$

25. $\frac{1}{1024}$; the area of the smallest region formed when the square has been divided as described 10 times

26. $\begin{cases} f(1) = 5 \\ f(n) = -2f(n-1) \end{cases}$

ADAPT

Check students' answers to the Lesson Practice questions to make sure that they understand how to write and apply both recursive and explicit formulas for geometric sequences. As in Lesson 21-1, pair students who need more practice with geometric sequences for the following activity: Ask each student create a geometric sequence and write its first 5 terms. Students then trade sequences and write recursive and explicit formulas for their partner's sequence. Partners can check each other's work and discuss any disagreements.

ACTIVITY 21 continued

My Notes

Lesson 21-2
Formulas for Geometric Sequences

LESSON 21-2 PRACTICE

The diagram below shows a square repeatedly divided in half. The entire square has an area of 1 square unit. The number in each region is the area of the region. Use the diagram for Items 23–25.

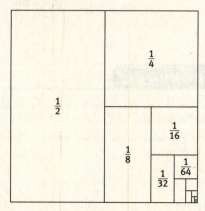

23. Write a geometric sequence to describe the areas of successive regions.
24. Write an explicit formula for the geometric sequence that you wrote in Item 23.
25. **Model with mathematics.** What is the 10th term of the sequence? What does it represent?
26. The explicit formula for a geometric sequence is $f(n) = 5(-2)^{n-1}$. Give the recursive formula for the sequence.

Geometric Sequences
Go Viral!

ACTIVITY 21
continued

ACTIVITY 21 PRACTICE
Write your answers on notebook paper.
Show your work.

Lesson 21-1

In Items 1 and 2, assume that the first term of a sequence is -3.

1. Write the first four terms of the sequence if it is an arithmetic sequence with common difference $-\frac{1}{3}$.
2. Write the first four terms of the sequence if it is a geometric sequence with common ratio $-\frac{1}{3}$.

The tree diagram below shows the number of possible outcomes when tossing a coin a number of times. For example, if you toss a coin once (Stage 1), there are two possible outcomes: heads (H) and tails (T). If you toss a coin twice (Stage 2), there are four possible outcomes for the two tosses: HH, HT, TH, and TT.

Use the tree diagram for Items 3–5.

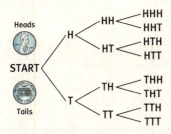

3. How many possible outcomes are there when you toss a coin 4 times?
4. Identify the common ratio of the sequence represented by the tree diagram.
5. How many possible outcomes are there when you toss a coin 23 times? Express your answer using exponents.

For Items 6–10, identify each sequence as arithmetic, geometric, or neither. If it is arithmetic, state the common difference. If it is geometric, state the common ratio.

6. 17, 25, 33, 41, ...
7. 1, 3, 6, 10, 15, ...
8. $-27, -9, -3, -1, ...$
9. 0.1, 0.5, 0.9, 1.3, ...
10. $\frac{1}{2}, \frac{1}{3}, \frac{1}{4}, \frac{1}{5}, ...$

11. A geometric sequence begins with the value 1 and has a common ratio of -2. Identify the eighth term in the sequence.
    A. -128 B. 128
    C. -256 D. 256

12. A geometric sequence begins with the value 2 and has a common ratio of $\frac{1}{2}$. Identify the fifth term in the sequence.
    A. 4 B. $\frac{1}{4}$
    C. $\frac{1}{8}$ D. $\frac{1}{16}$

13. Which of the following is a *false* statement about the sequence 2, 4, 8, 16, 32, ... ?
    A. The common ratio of the sequence is 2.
    B. The tenth term of the sequence is 2^{10}.
    C. Every term of the sequence is even.
    D. The number 216 appears in the sequence.

14. Give an example of a geometric sequence with a common ratio of 0.2. Write at least the first four terms of the sequence.

ACTIVITY 21 Continued

ACTIVITY PRACTICE

1. $-3, -\frac{10}{3}, -\frac{11}{3}, -4$
2. $-3, 1, -\frac{1}{3}, \frac{1}{9}$
3. 16
4. 2
5. 2^{23}
6. arithmetic; common difference $= 8$
7. neither
8. geometric; common ratio $= \frac{1}{3}$
9. arithmetic; common difference $= 0.4$
10. neither
11. A
12. C
13. D
14. Answers may vary. 1, 0.2, 0.04, 0.008, ...

Activity 21 • Geometric Sequences 321

ACTIVITY 21 Continued

15. $\begin{cases} a_1 = 7 \\ a_n = 3a_{n-1} \end{cases}$; $\begin{cases} f(1) = 7 \\ f(n) = 3f(n-1) \end{cases}$

16. $\begin{cases} a_1 = 100 \\ a_n = 0.1a_{n-1} \end{cases}$; $\begin{cases} f(1) = 100 \\ f(n) = 0.1f(n-1) \end{cases}$

17. $\begin{cases} a_1 = -10 \\ a = -2a_{n-1} \end{cases}$; $\begin{cases} f(1) = -10 \\ f(n) = -2f(n-1) \end{cases}$

18. $a_n = 4(4)^{n-1}$; 65,536

19. $a_n = \frac{1}{2}\left(\frac{1}{3}\right)^{n-1}$; $\frac{1}{4374}$

20. $a_n = 5(-3)^{n-1}$; $-10,935$

21. $a_n = 100(2)^{n-1}$

22. $51,200

23. 15 rounds

24. a

25. $a_n = 80(0.75)^{n-1}$

26. ≈ 14.24 cm

27. 9 swings

28. $a_n = 80(1.5)^{n-1}$

29. Graphs may vary.

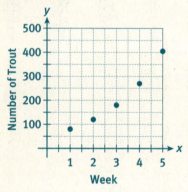

30. 405
31. Week 6
32. Agree; If the sequence is a constant sequence, such as 4, 4, 4, ..., then it is both arithmetic, because it has a common difference of 0, and geometric, because it has a common ratio of 1.

ADDITIONAL PRACTICE

If students need more practice on the concepts in this activity, see the eBook Teacher Resources for additional practice problems.

ACTIVITY 21 continued

Lesson 21-2

Write a recursive formula for each geometric sequence. Include the recursive formula in sequence notation.

15. 7, 21, 63, 189, ...
16. 100, 10, 1, 0.1, ...
17. −10, 20, −40, 80, ...

Write the explicit formula for each geometric sequence. Then determine the 8th term of each sequence.

18. 4, 16, 64, 256, ...
19. $\frac{1}{2}, \frac{1}{6}, \frac{1}{18}, \ldots$
20. $\begin{cases} a_1 = 5 \\ a_n = -3a_{n-1} \end{cases}$

A contestant on a game show wins $100 for answering a question correctly in Round 1. In each subsequent round, the contestant's winnings are doubled if she gives a correct answer. If the contestant gives an incorrect answer, she loses everything. Use this information for Items 21–23.

21. Write an explicit formula that gives the contestant's winnings in round *n*, assuming she answers all questions correctly.

22. How much does the contestant win in Round 10, assuming she answers all questions correctly?

23. How many rounds does a contestant need to play in order to answer a question worth at least $1,000,000?

24. A geometric sequence is given by the recursive formula $\begin{cases} f(1) = -6 \\ f(n) = -\frac{1}{2}f(n-1) \end{cases}$. Which of the following is a term in the sequence?

A. $\frac{3}{4}$ B. $-\frac{3}{4}$
C. $\frac{3}{2}$ D. -3

Geometric Sequences
Go Viral!

Each time a pendulum swings, the distance it travels decreases, as shown in the figure.

Pendulum Swing

The table shows how far the pendulum travels with each swing. Use this table for Items 25–27.

Swing Number	Distance Traveled (cm)
1	80
2	60
3	45
4	33.75

25. Write the explicit formula for the pendulum situation.
26. How far will the pendulum travel on the seventh swing?
27. How many swings will it take for the pendulum to travel less than 10 cm?

The game commission observes the fish population in a stream and notices that the number of trout increases by a factor of 1.5 every week. The commission initially observed 80 trout in the stream. Use this information for Items 28–31.

28. Write the explicit formula for the trout situation.
29. Make a graph of the population growth.
30. If this pattern continues, how many trout will be in the stream on the fifth week?
31. If this pattern continues, on what week will the trout population exceed 500?

MATHEMATICAL PRACTICES
Construct Viable Arguments and Critique the Reasoning of Others

32. Samir says that it is possible for a sequence to be both an arithmetic sequence and a geometric sequence. Do you agree or disagree? Explain.

Exponents, Radicals, and Geometric Sequences
TAKING STOCK

Embedded Assessment 1
Use after Activity 21

Stocking a lake is the process of adding fish to the lake. Once the fish have been added to the lake, their population growth depends upon many factors, such as the species of the fish, the number of predators in the lake, the quality of the water, and the lake's food supply.

Sapphire Lake was stocked with five species of fish several years ago. Michelle, a new employee at the parks and recreation commission, wants to analyze the population growth of the fish. She is able to find only the information shown below.

Species	Population Model	Meaning of Variables
A	$(3x^2y)^2$	x = average length of species in centimeters
B	$6(xy)^3$	
C	$20\sqrt{xy}$	y = year of project (Year 1 is the year in which fish are first added to the lake.)
D	$4\sqrt{x^2y}$	

Species E

Year	Population (hundreds)
1	4
2	6
3	9
4	13.5

1. Michelle wants to know the ratio of the population of species A to species B.
   a. Write the ratio as a fraction in simplified form without negative exponents.
   b. Write the ratio using negative exponents.

2. Next, Michelle analyzes the population of species C and D.
   a. Write the ratio of the population of species C to species D as a fraction in simplified form.
   b. Michelle needs to know the total population of species C and D in Year 1. She learns that the average length of both species is 8 centimeters. Write a simplified expression for the approximate total population of species C and D in Year 1.
   c. Is the expression you wrote in Part (b) rational or irrational? Explain your reasoning.

3. Michelle assumes that the population of species E continues to grow as shown in the table.
   a. Write an explicit formula for the sequence.
   b. Write a recursive formula for the sequence. Include the recursive formula in function notation.
   c. According to the model, what was the approximate population of species E in year 7?

Embedded Assessment 1

Assessment Focus
- Properties of exponents
- Integer exponents
- Simplifying expressions involving exponents
- Simplifying radical expressions
- Performing operations with radical expressions
- Distinguishing rational and irrational numbers
- Identifying geometric sequences
- Recursive and explicit formulas for geometric sequences
- Finding a given term of a geometric sequence

Answer Key

1. a. $\dfrac{3x}{2y}$

   b. $\dfrac{3}{2}xy^{-1}$ or $3x(2y)^{-1}$

2. a. $32 + 40\sqrt{2}$
   b. irrational; The sum of a rational number (32) and an irrational number ($40\sqrt{2}$) is irrational.
   c. $\dfrac{5\sqrt{x}}{x}$

3. a. $a_n = 4(2)^{n-1}$

   b. $\begin{cases} a_1 = 4 \\ a_n = 2a_{n-1} \end{cases}$ $\begin{cases} f(1) = 4 \\ f(n) = 2f(n-1) \end{cases}$

   c. 25,600

Common Core State Standards for Embedded Assessment 1

HSN-RN.A.2	Rewrite expressions involving radicals and rational exponents using the properties of exponents.
HSA-SSE.A.2	Use the structure of an expression to identify ways to rewrite it.
HSA-SSE.B.3c	Use the properties of exponents to transform expressions for exponential functions.
HSF-IF.A.3	Recognize that sequences are functions, sometimes defined recursively, whose domain is a subset of the integers.
HSF-BF.A.2	Write arithmetic and geometric sequences both recursively and with an explicit formula, use them to model situations, and translate between the two forms.*

Embedded Assessment 1

TEACHER to TEACHER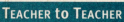

You may wish to read through the scoring guide with students and discuss the differences in the expectations at each level. Check that students understand the terms used.

Unpacking Embedded Assessment 2

Once students have completed this Embedded Assessment, turn to Embedded Assessment 2 and unpack it with them. Use a graphic organizer to help students understand the concepts they will need to know to be successful on Embedded Assessment 2.

Embedded Assessment 1
Use after Activity 21

Exponents, Radicals, and Geometric Sequences
TAKING STOCK

Scoring Guide	Exemplary	Proficient	Emerging	Incomplete
	The solution demonstrates the following characteristics:			
Mathematics Knowledge and Thinking (Items 1a, 1b, 2a, 2b, 3c)	• Clear and accurate understanding of how to write and simplify exponential and radical expressions, including expressions with negative exponents • Fluency in determining a specified term of a geometric sequence	• Adequate understanding of how to write and simplify exponential and radical expressions, including expressions with negative exponents • Correct identification of a specified term of a geometric sequence	• Partial understanding of how to write and simplify exponential and radical expressions, including expressions with negative exponents • Partial understanding of and some difficulty determining a specified term of a geometric sequence	• Little or no understanding of how to write and simplify exponential and radical expressions, including expressions with negative exponents • Incomplete understanding of and significant difficulty determining a specified term of a geometric sequence
Problem Solving (Item 3)	• Appropriate and efficient strategy that results in a correct answer	• Strategy that may include unnecessary steps but results in a correct answer	• Strategy that results in some incorrect answers	• No clear strategy when solving problems
Mathematical Modeling/ Representations (Items 3a, 3b)	• Effective understanding of how to describe a real-world data set using explicit and recursive formulas	• Little difficulty describing a real-world data set using explicit and recursive formulas	• Partial understanding of how to describe a real-world data set using explicit and recursive formulas	• Inaccurate or incomplete understanding of how to describe a real-world data set using explicit and recursive formulas
Reasoning and Communication (Items 2c, 3c)	• Precise use of appropriate math terms and language to explain whether an expression is rational or irrational • Ease and accuracy describing the relationship between a sequence and a real-world scenario	• Correct identification of an expression as rational or irrational with an adequate explanation • Little difficulty describing the relationship between a sequence and a real-world scenario	• Misleading or confusing explanation of whether an expression is rational or irrational • Partially correct description of the relationship between a sequence and a real-world scenario	• Incomplete or inaccurate explanation of whether an expression is rational or irrational • Little or no understanding of how a sequence might relate to a real-world scenario

Common Core State Standards for Embedded Assessment 1 (cont.)

HSF-LE.A.1	Distinguish between situations that can be modeled with linear functions and with exponential functions.
HSF-LE.A.1c	Recognize situations in which a quantity grows or decays by a constant percent rate per unit interval relative to another.

324 SpringBoard® Mathematics **Algebra 1**

Exponential Functions
Protecting Your Investment
Lesson 22-1 Exponential Functions and Exponential Growth

ACTIVITY 22

Learning Targets:
- Understand the definition of an exponential function.
- Graph and analyze exponential growth functions.

SUGGESTED LEARNING STRATEGIES: Marking the Text, Create Representations, Look for a Pattern, Interactive Word Wall, Predict and Confirm, Think-Pair-Share

The National Association of Realtors estimates that, on average, the price of a house doubles every ten years. Tony's grandparents bought a house in 1960 for $10,000. Assume that the trend identified by the National Association of Realtors applies to Tony's grandparents' house.

1. What was the value of Tony's grandparents' house in 1970 and in 1980?
   1970: $20,000   **1980: $40,000**

2. Compute the difference in value from 1960 to 1970.
   $10,000

3. Compute the ratio of the 1970 value to the 1960 value.
   $\frac{\$20{,}000}{\$10{,}000} = 2$

4. Complete the table of values for the years 1960 to 2010.

		House Value		
Year	Decades Since 1960	Value of House	Difference Between Values of Consecutive Decades	Ratio of Values of Consecutive Decades
1960	0	$10,000	—	—
1970	1	$20,000	$10,000	2
1980	2	$40,000	$20,000	2
1990	3	$80,000	$40,000	2
2000	4	$160,000	$80,000	2
2010	5	$320,000	$160,000	2

5. What patterns do you recognize in the table?
   Answers may vary. Differences of values from decade to decade are different; differences of values increase; ratio of values from decade to decade is constant.

MATH TIP

The ratio of the quantity a to the quantity b is evaluated by dividing a by b (ratio of a to $b = \frac{a}{b}$).

Common Core State Standards for Activity 22

HSA-CED.A.1: Create equations and inequalities in one variable and use them to solve problems. *Include equations arising from linear and quadratic functions, and simple rational and exponential functions.*

HSA-CED.A.2: Create equations in two or more variables to represent relationships between quantities; graph equations on coordinate axes with labels and scales.

HSF-IF.B.4: For a function that models a relationship between two quantities, interpret key features of graphs and tables in terms of the quantities, and sketch graphs showing key features given a verbal description of the relationship. Key features include: intercepts; intervals where the function is increasing, decreasing, positive, or negative; relative maximums and minimums; symmetries; end behavior; and periodicity.

ACTIVITY 22
Investigative

Activity Standards Focus
In this activity, students explore exponential functions, their application, and their graphs. They learn how different values of the constant factor and the exponent affect the shape of the graph.

Lesson 22-1

PLAN

Pacing: 1 class period
Chunking the Lesson
#1–3   #4–5   #6–8
#9–10   #11–13   #14–17
Check Your Understanding
Lesson Practice

TEACH

Bell-Ringer Activity
Tell students that their new boss offers them two pay schedules:
Plan 1: $300 the first day, $303 the second day, $306 the third day, adding $3 each day.
Plan 2: $0.03 the first day, $0.09 the second day, $0.27 the third day, and so on, tripling the rate each day.
Ask students to choose which plan they believe is better. They should include a table or graph to support their conclusions.

1–3 Marking the Text, Think-Pair-Share The doubling pattern is introduced, and these questions provide an opportunity to ensure that the context is understood. Have students highlight "price of a house doubles every ten years" and discuss its meaning.

4–5 Create Representations, Look for a Pattern, Sharing and Responding Students use the information from Items 1–3 to complete the table. Be sure that students compute the ratio correctly. While patterns identified by students for Item 5 may vary, as groups share their thoughts, encourage connections to sequences and linear versus nonlinear relationships.

Activity 22 • Exponential Functions 325

ACTIVITY 22 Continued

6–8 Discussion Groups, Look for a Pattern, Create Representations
Students should identify the sequence of house values as a geometric sequence with a common ratio of 2. This should be reinforced as they graph in Item 7. Be sure that students graph the data from the appropriate columns in the table, columns 2 and 3. Students must address the noncollinear points on the graph and the nonconstant differences in the table when identifying the data as nonlinear.

9–10 Look for a Pattern, Discussion Groups, Construct an Argument, Predict and Confirm Students should have little difficulty extending the pattern to find the value of the house in 2020. In Item 10, students may assume linearity and find the average of 2000 and 2010, an incorrect response. It is not important to correct incorrect responses here as long as students can provide some justification for their responses. Students will revisit this response in later items as they develop the idea of exponential growth.

Developing Math Language
This lesson contains the terms *exponential function* and *exponential growth*. Have students add the terms to their math notebooks. Encourage students to include a sample scenario with drawings to clarify the definitions. Add the terms to the Word Wall.

> **Differentiating Instruction**
>
> **Support** students having difficulty understanding exponential functions by having the students choose two integers, *a* and *b*, that are greater than 1 and less than or equal to 6. Have them use a table to show a sequence by writing *a* for the first term, *ab* for the second, and then multiplying each successive term by *b* again to get the next term. Have students explain their sequences and how they developed them to another student. Then have them explain why the sequence they just wrote fits the definition.

ACTIVITY 22 continued

Lesson 22-1
Exponential Functions and Exponential Growth

6. Write the house values as a sequence. Identify the sequence as arithmetic or geometric and justify your answer.
   $10,000, $20,000, $40,000, $80,000, $160,000, $320,000, ...; this is a geometric sequence because it has a common ratio of 2.

7. Using the data from the table, graph the ordered pairs (decades since 1960, house value) on the coordinate grid below.

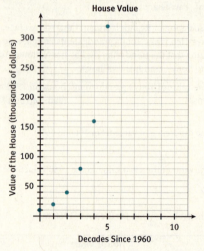

> **DISCUSSION GROUP TIPS**
>
> As you discuss ideas about the housing scenario, make notes and listen to what your group members have to contribute. Ask and answer questions to clearly aid comprehension and to ensure understanding of all group members' ideas.

8. The data comparing the number of decades since 1960 and value of the house are not linear. Explain why using the table and the graph.
   In the table, the rate of change is not constant. Graphically, the points do not form a line.

9. **Make use of structure.** Using the information that you have regarding the house value, predict the value of the house in the year 2020. Explain how you made your prediction.
   $640,000. Following the pattern, the house value will double from 2010 to 2020.

10. Tony would like to know what the value of the house was in 2005. Using the same data, predict the house value in 2005. Explain how you made your prediction.
    Answers may vary. The value of the house in 2005 was about $240,000. Since house value in 2000 = $160,000 and house value in 2010 = $320,000, a linear relationship would give Value in 2005 = $\frac{\$320{,}000 + 160{,}000}{2}$, or $240,000.

The increase in house value for Tony's grandparents' house is an example of *exponential growth*. Exponential growth can be modeled using an *exponential function*.

> **Exponential Function**
> A function of the form $f(x) = a \cdot b^x$, where x is the domain value, $f(x)$ is the range value, $a \neq 0$, $b > 0$, and $b \neq 1$.

Common Core State Standards for Activity 22 (continued)

HSF-IF.C.7:	Graph functions expressed symbolically and show key features of the graph, by hand in simple cases and using technology for more complicated cases.
HSF-BF.A.1b:	Combine standard function types using arithmetic operations.
HSF-LE.A.3:	Observe using graphs and tables that a quantity increasing exponentially eventually exceeds a quantity increasing linearly, quadratically, or (more generally) as a polynomial function.
HSF-LE.B.5:	Interpret the parameters in a linear or exponential function in terms of a context.

Lesson 22-1
Exponential Functions and Exponential Growth

In exponential growth, a quantity is multiplied by a constant factor greater than 1 during each time period.

11. The value of Tony's grandparents' house is growing exponentially because it is multiplied by a constant factor for each decade. What is this constant factor?
    2

A function that can be used to model the house value is $h(t) = 10{,}000 \cdot (2)^t$. Use this function for Items 12–17.

12. Identify the meaning of $h(t)$ and t. What are the reasonable domain and range?
    $h(t)$ is the value of the house; t is the number of decades since 1960. The reasonable domain and range are nonnegative real numbers.

13. Describe how your answer to Item 11 is related to the function $h(t) = 10{,}000 \cdot (2)^t$.
    It is the number that is raised to a power, or b.

14. Complete the table of values for t and $h(t)$. Then graph the function $h(t)$ on the grid below.
    Tables will vary.

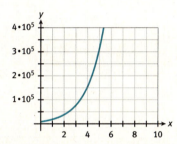

15. What was the value of the house in 1960? Describe how this value is related to the function $h(t) = 10{,}000 \cdot (2)^t$ and to the graph.
    $10{,}000$; in the function, it is the number multiplied by the power, or a. On the graph, it is the y-intercept.

16. Calculate the value of the house in the year 2020. How does the value compare with your prediction in Item 9?
    $640{,}000. This value should agree with the response to Item 9.

17. Calculate the value of the house in the year 2005. How does the value compare with your prediction in Item 10?
    $226{,}274.17. This value will probably not match responses to Item 10 exactly, but should be close.

TECHNOLOGY TIP
Graph $h(t)$ on a graphing calculator. Find the y-coordinate when x is about 4.5. The value should be close to your calculated value in Item 17.

ACTIVITY 22 Continued

Check Your Understanding
As you debrief students' answers, look for the reasons for any errors and correct the misconceptions. Make sure students can identify the common factor and the exponent and explain the role of each.

Answers

18.

x	g(x)
0	1
1	3
2	9
3	27
4	81

19.

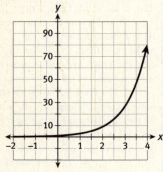

20. 3
21. $56,568.54
22. 1985

ASSESS

Students' answers to lesson practice problems will provide you with a formative assessment of their understanding of the lesson concepts and their ability to apply their learning.

See the Activity Practice for additional problems for this lesson. You may assign the problems here or use them as a culmination for the activity.

ADAPT

Check students' answers to ensure that they understand exponential growth. If students struggle to determine the constant growth factor, make connections to the identification of the common ratio of a geometric sequence using the terms of the sequence.

ACTIVITY 22 continued

My Notes

Lesson 22-1
Exponential Functions and Exponential Growth

Check Your Understanding

18. Copy and complete the table for the exponential function $g(x) = 3^x$.
19. Use your table to make a graph of $g(x)$.
20. Identify the constant factor for this exponential function.

Isaac evaluates the function modeling Tony's grandparents' house value, $h(t) = 10,000 \cdot (2)^t$, at $t = 2.5$. The variable t represents the number of decades since 1960.

21. What is $h(2.5)$?
22. For which year is Isaac estimating the house's value?

x	g(x)
0	
1	
2	
3	
4	

LESSON 22-1 PRACTICE

The value of houses in different locations can grow at different rates. The table below shows the value of Maddie's house from 1960 until 2010. Use the table for Items 23–25.

Year	Decades Since 1960	Value of House
1960	0	$10,000
1970	1	$15,000
1980	2	$22,500
1990	3	$33,750
2000	4	$50,625
2010	5	$75,938

23. Create a graph showing the value of Maddie's house from 1960 until 2010.
24. Explain how you know that the value of Maddie's house is growing exponentially.
25. What was the approximate value of Maddie's house in 1995?

The function $f(t) = 20,000 \cdot (1.2)^t$ can be used to find the value of Eduardo's house between 1970 and 2010, where the initial value of the function is the value of Eduardo's house in 1970.

26. **Model with mathematics.** Describe what the domain and range of the function mean in the context of Eduardo's house value.
27. What was the value of Eduardo's house in 1970?
28. Approximately how much was the house worth in 2000?

LESSON 22-1 PRACTICE

23.

24. The house value is multiplied by 1.5 each decade.

25. $41,335.14
26. The reasonable domain is $t \geq 0$, since t is the decades since 1970. The range is the value of the house, and the starting value is $20,000, so the reasonable range is $f(t) \geq 20,000$.
27. $20,000
28. $34,560

328 SpringBoard® Mathematics Algebra 1, Unit 4 • Exponents, Radicals, and Polynomials

Lesson 22-2
Exponential Decay

ACTIVITY 22
continued

Learning Targets:
- Describe characteristics of exponential decay functions.
- Graph and analyze exponential decay functions.

SUGGESTED LEARNING STRATEGIES: Look for a Pattern, Create Representations, Predict and Confirm, Discussion Groups, Visualization

Radon, a naturally occurring radioactive gas, was identified as a health hazard in some homes in the mid 1980s. Since radon is colorless and odorless, it is important to be aware of the concentration of the gas. Radon has a *half-life* of approximately four days.

Tony's grandparents' house was discovered to have a radon concentration of 400 pCi/L. Renee, a chemist, isolated and eliminated the source of the gas. She then wanted to know the quantity of radon in the house in the days following so that she could determine when the house would be safe.

1. **Make sense of problems.** What is the amount of the radon in the house four days after the source was eliminated? Explain your reasoning.
   The radon in the house four days after the source was eliminated was 200 pCi/L. Before the source was eliminated, the amount of radon was 400 pCi/L. Since radon has a half life of approximately four days, half of the gas is eliminated every four days.

2. Compute the difference in the amount of radon from Day 0 to Day 4.
   200 pCi/L

3. Determine the ratio of the amount of radon on Day 4 to the amount of radon on Day 0.
   $\frac{1}{2}$

CONNECT TO SCIENCE

All radioactive elements have a *half-life*. A half-life is the amount of time in which a radioactive element decays to half of its original quantity.

CONNECT TO SCIENCE

The US Environmental Protection Agency (EPA) recommends that the level of radon be below 4 pCi/L (picoCuries per liter) in any home. The EPA recommends that all homes be tested for radon.

ACTIVITY 22 Continued
Lesson 22-2

PLAN

Pacing: 1 class period
Chunking the Lesson
#1–3   #4–7   #8–9
#10–14   #15–17
Check Your Understanding
Lesson Practice

TEACH

Bell-Ringer Activity
Have students suppose that they are stranded in the desert with only 8,000 milliliters of water. They use exactly $\frac{1}{2}$ of the water each day. Have them use a table to find out how many days it will take to use all of the water.

1–3 Shared Reading, Marking the Text, Create Representations, Think-Pair-Share It is important that students understand the context of the problem before they consider the mathematics involved. The introduction to radon is preparation for the exponential decay part of the activity. The unit of pCi/L will likely be new to students and teachers alike. To respond to Item 1, students need to understand the concept of half-life and that in this situation, four days is equivalent to one half-life. Students are being asked to determine one difference and one ratio before completing the table in Item 4. Be certain students are computing the ratio in the correct order, i.e., Day 4 to Day 0 rather than Day 0 to Day 4.

Activity 22 • Exponential Functions   329

Lesson 22-2
Exponential Decay

4. Complete the table for the radon concentration.

		Radon Concentration		
Half-Lives	Days After Radon Source Was Eliminated	Concentration of Radon in pCi/L	Difference Between Concentration of Consecutive Half-Lives	Ratio of Concentrations of Consecutive Half-Lives
0	0	400	—	—
1	4	200	200	$\frac{1}{2}$
2	8	100	100	$\frac{1}{2}$
3	12	50	50	$\frac{1}{2}$

5. **Express regularity in repeated reasoning.** What patterns do you recognize in the table?

   Answers may vary. Differences of values from half-life to half-life are different; differences of values decrease; ratio of values from half-life to half-life is constant.

6. Graph the data in the table as ordered pairs in the form (half-lives, concentration).

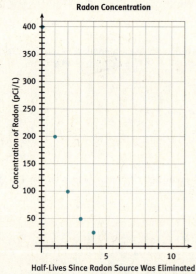

7. The data that compares the number of half-lives and the concentration of radon are not linear. Explain why using the table of values and the graph.

   In the table, the rate of change is not constant. Graphically, the points do not form a line.

Lesson 22-2
Exponential Decay

8. Renee needs to know the concentration of radon in the house after 20 days. How many radon half-lives are in 20 days? What is the concentration after 20 days?
   Five half-lives, 12.5 pCi/L

9. How many radon half-lives are in 22 days? Predict the concentration after 22 days.
   5.5 half-lives. Answers may vary; about 9.375 pCi/L based on a linear assumption.

The decrease in radon concentration in Tony's grandparents' house is an example of *exponential decay*. Exponential decay can be modeled using an exponential function.

In exponential decay, a quantity is multiplied by a constant factor that is greater than 0 but less than 1 during each time period.

10. The concentration of radon is multiplied by a constant factor for each half-life. What is this constant factor?
    $\frac{1}{2}$

A function that can be used to model the radon concentration is $r(t) = 400 \cdot \left(\frac{1}{2}\right)^t$. Use the function for Items 11–17.

11. Identify the meaning of $r(t)$ and t. What are the reasonable domain and range?
    $r(t)$ is the radon concentration in pCi/L; t is the number of half-lives. The reasonable domain and range are nonnegative real numbers.

12. Describe how your answer to Item 10 is related to the function $r(t) = 400 \cdot \left(\frac{1}{2}\right)^t$.
    It is the number that is raised to a power, or b.

13. Graph the function $r(t)$.

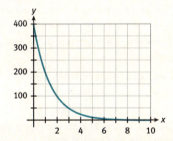

14. Describe how the original concentration of radon is related to the function $r(t) = 400 \cdot \left(\frac{1}{2}\right)^t$ and to the graph.
    400 pCi/L; in the function, it is the number multiplied by the power, or a. On the graph, it is the y-intercept.

ACTIVITY 22 Continued

15–17 Predict and Confirm, Discussion Groups, Debriefing

Students are provided the opportunity to use the function to make predictions and to then compare those predictions to earlier ones. Attention to computational detail, especially the order of operations, is important when students make their predictions. Item 16 will be the first time students are able to answer the 5.5 half-life question accurately. Allow for the comparison of the predicted and actual 5.5 half-life values as this portion of the activity is debriefed. Students may use a graphing calculator to verify that the function will never intersect the *x*-axis to support the conclusion that the concentration of radon will never be zero.

Check Your Understanding

Debrief the students' answers to make sure they understand the parts of an exponential function and how to use it to generate values. This basic understanding is important as students continue to graph exponential functions and use them to solve problems.

Answers
18.

x	g(x)
0	1
1	$\frac{1}{4}$
2	$\frac{1}{16}$
3	$\frac{1}{64}$

19. $\frac{1}{4}$

ASSESS

Students' answers to lesson practice problems will provide you with a formative assessment of their understanding of the lesson concepts and their ability to apply their learning.

See the Activity Practice for additional problems for this lesson. You may assign the problems here or use them as a culmination for the activity.

ADAPT

If students have difficulty distinguishing between exponential growth and decay, encourage the use of graphing to make connections to the growth factor. Students may use graphing calculators to graph the function. Visual learners often make connections to graphical representations before connecting with algebraic representations.

ACTIVITY 22 continued

CONNECT TO AP
In calculus, you will discover what happens as functions approach 0.

Lesson 22-2
Exponential Decay

15. Use the function to identify the concentration of radon after 20 days. How does the concentration compare with your prediction in Item 8?
    $r(5) = 400\left(\frac{1}{2}\right)^5 = 12.5$. This value should agree with the response to Item 8.

16. Use the function to calculate the concentration of radon after 22 days. How does the concentration compare with your prediction in Item 9?
    $r(5.5) = 400\left(\frac{1}{2}\right)^{5.5} \approx 8.839$. This value will probably not match responses to Item 9 exactly, but should be close.

17. **Construct viable arguments.** Will the concentration of radon ever be 0? Explain your reasoning.
    No; as *t* increases in value, *r(t)* decreases in value but will never be 0. There is no exponent for which the exponential function will be 0.

Check Your Understanding

18. Copy and complete the table for the exponential function $g(x) = \left(\frac{1}{4}\right)^x$.
19. Identify the constant factor for this exponential function.

x	g(x)
0	
1	
2	
3	

LESSON 22-2 PRACTICE

20. The amount of medication in a patient's bloodstream decreases exponentially from the time the medication is administered. For a particular medication, a function that gives the amount of medication in a patient's bloodstream *t* hours after taking a 100 mg dose is $A(t) = 100\left(\frac{7}{10}\right)^t$. Use this function to find the amount of medication remaining after 2 hours.

21. Make a table of values and graph each function.
    a. $h(x) = 2^x$
    b. $l(x) = 3^x$
    c. $m(x) = \left(\frac{1}{2}\right)^x$
    d. $p(x) = \left(\frac{1}{3}\right)^x$

22. Which of the functions in Item 21 represent exponential growth? Explain using your table of values and graph.

23. Which of the functions in Item 21 represent exponential decay? Explain using your table of values and graph.

24. **Reason abstractly.** How can you identify which of the functions represent growth or decay by looking at the function?

LESSON 22-2 PRACTICE
20. 49 mg
21. a. Students' tables may vary.

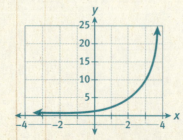

22. 21a and 21b represent growth, since both of the functions have range values that are increasing as their domain values increase (evidenced both in tables and graphs).

23. 21c and 21d represent decay, since both of the functions have range values that are decreasing as their domain values increase.

24. The value of the base indicates if the function represents growth or decay. If the base is greater than 1, the function represents growth. If the base is greater than 0 but less than 1, the function represents decay.

Lesson 22-3
Graphs of Exponential Functions

Learning Targets:
- Describe key features of graphs of exponential functions.
- Compare graphs of exponential and linear functions.

SUGGESTED LEARNING STRATEGIES: Predict and Confirm, Discussion Groups, Create Representations, Look for a Pattern, Sharing and Responding, Summarizing

Recall that an exponential function is a function of the form $f(x) = ab^x$, where $a \neq 0$, $b > 0$, and $b \neq 1$.

1. Use a graphing calculator to graph each function. Sketch each graph on the coordinate grid provided.

$a > 0$	$a < 0$
a. $y = 3 \cdot 2^x$ 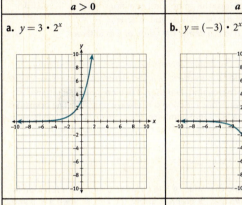	b. $y = (-3) \cdot 2^x$ 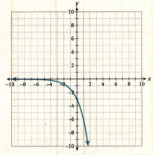
c. $y = 3 \cdot (0.5^x)$ 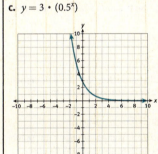	d. $y = -2 \cdot (0.5^x)$ 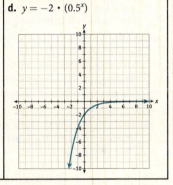

2. Compare and contrast the graphs and equations in Items 1a and 1b above.
   a. How are the equations similar and different?
   The value of *a* is different; the value of *b* is the same.

LESSON 22-2 PRACTICE (cont.)

21. b. Students' tables may vary.
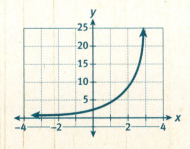

c. Students' tables may vary.
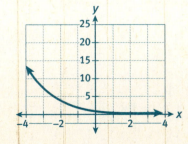

d. Students' tables may vary.
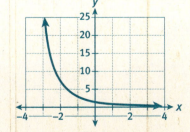

Activity 22 • Exponential Functions 333

ACTIVITY 22 Continued

4 Summarizing, Group Discussion, Debriefing It is important here that students recognize that the sign on *a* in the equation of the function determines whether the graph lies above or below the *x*-axis. In addition, this sign also determines whether the graph is increasing or decreasing. These rules apply whether the function represents exponential decay or exponential growth. Encourage students to draw sketches to illustrate each case discussed.

Differentiating Instruction

Support students' development of descriptive vocabulary. Sketch the curve $y = 2^x$ on the board. Trace the curve as you say, "The curve is going up from left to right." Have students suggest other words that mean *going up*. List them on the board, introducing other terms students do not mention. Describe the curve using each word, such as *The curve is ascending from left to right.* and *The curve is increasing from left to right.* Then point out the asymptote $y = 0$. Help students describe that the curve gets very close to the *x*-axis but does not touch it. Once students can describe the curve, sketch $y = -2^x$ on the board. Again, work through the vocabulary needed to describe the curve.

ACTIVITY 22 continued

Lesson 22-3
Graphs of Exponential Functions

b. Use words like *increasing, decreasing, positive, negative, domain,* and *range* to describe the similarities and differences in the graphs.

Function a is increasing. Function b is decreasing. Function a is always positive, while function b is always negative. The domain of both functions is all real numbers. The range of function a is all real numbers greater than 0 and the range of function b is all real numbers less than 0.

c. What connections can be made between the graphs and their equations?

When $b > 1$ and a is positive, the function is always positive, and its graph is above the *x*-axis. When $b > 1$ and a is negative, the function is always negative, and its graph is below the *x*-axis.

3. Compare and contrast the graphs and equations in Items 1c and 1d.
a. How are the equations similar and different?

The value of *b* is the same; the value of *a* is different.

b. Use words like *increasing, decreasing, positive, negative, domain,* and *range* to describe the similarities and differences between the graphs.

Function c is decreasing. Function d is increasing. Function c is always positive, while function d is always negative. The domain of both functions is all real numbers. The range of function c is all real numbers greater than 0, and the range of function d is all real numbers less than 0.

c. What connections can be made between the graphs and their equations?

When $a > 0$ and the value of b is between 0 and 1, the function is always positive and its graph is above the *x*-axis. When $a < 0$ and the value of b is between 0 and 1, the function is always negative and its graph is below the *x*-axis.

4. Describe the effects of the values of *a* and *b* on the graph of the exponential function $f(x) = ab^x$.
a. Describe the graph of an exponential function when $a > 0$.

The graph is above the *x*-axis and approaches, but does not reach, the *x*-axis. The graph may ascend or descend from left to right.

b. Describe the graph of an exponential function when $a < 0$.

The graph is below the *x*-axis and approaches, but does not reach, the *x*-axis. The graph may ascend or descend from left to right.

c. Describe the graph of an exponential function when $b > 1$.

The graph approaches, but does not reach, the negative *x*-axis. The graph may ascend or descend from left to right.

d. Describe the graph of an exponential function when $0 < b < 1$.

The graph approaches, but does not reach, the positive *x*-axis. The graph may ascend or descend from left to right.

Lesson 22-3
Graphs of Exponential Functions

Check Your Understanding

5. Describe the values of a and b for which the exponential function $f(x) = ab^x$ is always positive.
6. Describe the values of a and b for which the exponential function $f(x) = ab^x$ is increasing.

7. Let $f(x) = 2x$ and $g(x) = 2^x$. Complete the tables below for each function.

x	0	1	2	3	4	5
f(x)	0	2	4	6	8	10

x	0	1	2	3	4	5
g(x)	1	2	4	8	16	32

8. Graph $f(x)$ and $g(x)$ below.

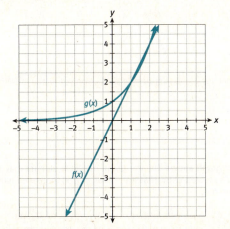

9. Examine the graphs of $f(x)$ and $g(x)$. Compare the values of each function from $x = -2$ through $x = 2$. Which function is greater on this interval?
   Students should note that the greater function changes on this interval. From $x = -2$ through $x < 1$, $g(x)$ is greater. For $1 < x < 2$, $f(x)$ is greater. At $x = 1$ and $x = 2$, the graphs are equal.

10. Examine the values of $f(x)$ and $g(x)$ for $x > 2$.
    a. Which function is greater on this interval?
       $g(x)$ is greater.

ACTIVITY 22 Continued

Check Your Understanding
As you debrief students' answers, encourage them to supply reasons why their equations satisfy the conditions stated. Encourage students to visualize or sketch graphs as they discuss the equations.

Answers
5. $f(x) = ab^x$ is positive when $a > 0$ (regardless of the value of b).
6. $f(x) = ab^x$ is increasing when $a > 0$ and $b > 1$, or when $a < 0$ and $0 < b < 1$.

7–13 Predict and Confirm, Discussion Groups Monitor group discussions carefully. If students have difficulty visualizing the intervals, have them identify where the two functions are equal. Then encourage them to examine the two graphs to the left of $x = 1$, between $x = 1$ and $x = 2$, and to the right of $x = 2$, describing what they see.

Activity 22 • Exponential Functions

ACTIVITY 22 Continued

7–13 (continued) Students may also need help understanding what "increase faster" means in terms of a graph. Suggest words such as *steep* and *flat* to help them describe the concept. Note that they may need to answer the questions by saying that one graph increases more quickly in one interval, but the other graph increases faster in another interval. Students may also use the concept of slope to explain the increase in the linear function.

14–18 Predict and Confirm, Discussion Groups, Debriefing It is important for students to understand that while a linear function may yield greater values on some interval than an exponential growth function, the values of the exponential growth function will eventually overtake the values of the linear function. While this may sometimes be difficult to show using a table, a graph will show it, given an appropriate window and scale.

Lesson 22-3
Graphs of Exponential Functions

b. Do you think this will continue to be true as x continues to increase? Explain your reasoning.

Answers will vary. Yes, because g(x) will continue to curve upward more steeply than f(x).

11. To take a closer look at the graphs of $f(x)$ and $g(x)$ for larger values of x, regraph the two functions below. Note the new scale.

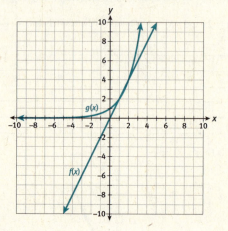

12. Does the new graph support the prediction you made in Item 10b?

Answers will vary.

13. Which function increases faster, $f(x)$ or $g(x)$? Explain your reasoning using the graph and the tables.

g(x); the graph shows that the graph of g(x) is steeper than the graph of f(x). The tables show that the values of g(x) increase faster than the values of f(x).

Alex believes that for the linear function $f(x) = 50x$ and the exponential function $g(x) = 2^x$, the value of $f(x)$ is always greater than the value of $g(x)$.

Glenda believes that for a linear function $f(x)$ to always be greater than an exponential function $g(x)$, the graph of $f(x)$ must be very steep while the graph of $g(x)$ must be very flat. She proposes graphing $f(x) = 50x$ and $g(x) = 1.1^x$ to test her conjecture.

14. a. Test Alex's conjecture by graphing $f(x) = 50x$ and $g(x) = 2^x$ on your graphing calculator. Do you agree or disagree with Alex's conjecture? Explain your reasoning.

Answers may vary. Depending on students' viewing window, it will likely appear that f(x) is always greater than g(x).

---

336 SpringBoard® Mathematics **Algebra 1, Unit 4** • Exponents, Radicals, and Polynomials

Lesson 22-3
Graphs of Exponential Functions

b. Use appropriate tools strategically. Now adjust your viewing window to match the coordinate plane below. Sketch the graphs of $f(x)$ and $g(x)$.

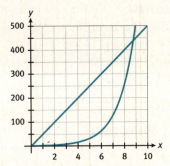

15. Should Alex revise his conjecture? Use the graph in Item 14b to explain.

Yes; the graph shows that somewhere between $x = 8$ and $x = 9$, $g(x)$ overtakes $f(x)$ and begins to increase faster than $f(x)$, disproving Alex's conjecture.

16. a. Test Glenda's conjecture by graphing $f(x) = 50x$ and $g(x) = 1.1^x$ on your graphing calculator. Do you agree with Glenda's conjecture?

Answers may vary. Depending on students' viewing window, it will likely appear that $f(x)$ is always greater/steeper than $g(x)$.

b. Now adjust your viewing window to match the coordinate plane below. Sketch the graphs of $f(x)$ and $g(x)$.

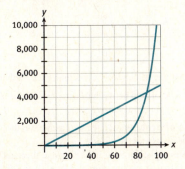

17. Should Glenda revise her conjecture? Use the graph in Item 16b to support your response.

Yes; the graph shows that somewhere around $x = 90$, $g(x)$ overtakes $f(x)$ and begins to increase faster than $f(x)$, disproving Glenda's conjecture.

18. Attend to precision. Is an exponential function always greater than a linear function? Explain your reasoning.

No; as shown in the graphs above, there are intervals for which a linear function is greater. However, an exponential function will always eventually overtake a linear function.

ACTIVITY 22 Continued

Check Your Understanding

As you debrief students' answers, ask them explain how they visualized the functions in order to predict which increases more gradually. How did the defined window on the graphing calculator assist in making decisions?

Answers

19. $b(x)$, because it is exponential and $a(x)$ is linear.
20. Check students' graphs.

ASSESS

Students' answers to lesson practice problems will provide you with a formative assessment of their understanding of the lesson concepts and their ability to apply their learning.

See the Activity Practice for additional problems for this lesson. You may assign the problems here or use them as a culmination for the activity.

LESSON 22-3 PRACTICE

21. Isaac needs to zoom out or graph the functions on a larger coordinate plane to see that $f(x)$ will eventually be greater than $g(x)$.
22. The graphs intersect near the origin, and again at approximately (2.8, 25.5). Only between these two points is $f(x) < g(x)$. Check students' graphs.
23. Answers may vary. Check students' graphs. If the club wants to continue its recruitment for 14 months or fewer, they should choose Julia's proposal. If they want to continue for longer than 14 months, they should choose Jorge's proposal.

ADAPT

Check students' answers to make sure they understand the relative behaviors of linear and exponential functions and the implication of those behaviors for a real-world situation. Reinforce the importance of examining the functions at different intervals before drawing conclusions.

ACTIVITY PRACTICE

1.

Months after January (m)	Earnings $e(m)$
0	$175,000
1	$192,500
2	$211,750
3	$232,925
4	$256,217.50

ACTIVITY 22
continued

My Notes

Lesson 22-3
Graphs of Exponential Functions

Check Your Understanding

Use the functions $a(x) = 25x$ and $b(x) = 5 \cdot 3^x$ for Items 19 and 20.

19. Without graphing, tell which function increases more quickly. Explain your reasoning.
20. Use your graphing calculator to justify your answer to Item 19. Sketch the graphs from your calculator, and be sure to label your viewing window.

LESSON 22-3 PRACTICE

Isaac graphs $f(x) = 0.5 \cdot 4^x$ and $g(x) = 9x$. His graphs are shown below.

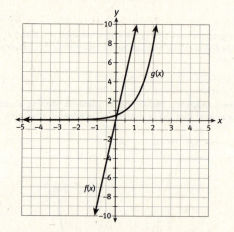

21. Isaac states that $f(x)$ will always be less than $g(x)$. Explain Isaac's error.
22. Describe the relationship between the graphs of $f(x)$ and $g(x)$. Make a new graph to support your answer.
23. **Make sense of problems.** The math club has only 10 members and wants to increase its membership.
    - Julia proposes a goal of recruiting 2 new members each month. If the club meets this goal, the function $y = 2x + 10$ will give the total number of members y after x months.
    - Jorge proposes a goal to increase membership by 10% each month. If the club meets this goal, the function $y = 10 \cdot 1.1^x$ will give the total number of members y after x months.

    Club members want to choose the goal that will cause the membership to grow more quickly. Assume that the club will meet the recruitment goal that they choose. Which proposal should they choose? Use a graph to support your answer.

2.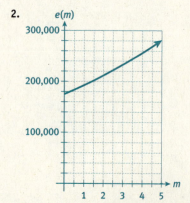

3. about $412,640.85
4. Yes; the number of bacteria is multiplied by a constant factor of 2.5 each minute.
5. B

6.

Years after purchase (t)	Value of car $v(t)$
0	$25,000
1	$21,250
2	$18,062.50
3	$15,353.13
4	$13,050.16

338 SpringBoard® Mathematics Algebra 1, Unit 4 • Exponents, Radicals, and Polynomials

Exponential Functions
Protecting Your Investment

ACTIVITY 22 PRACTICE
Write your answers on notebook paper.
Show your work.

Lesson 22-1
In January of this year, a clothing store earned $175,000. Since then, earnings have increased by 10% each month. A function that models the store's earnings after m months is $e(m) = 175,000 \cdot (1.1)^m$. Use this information for Items 1–3.

1. Copy and complete the table.

Months After January (m)	Earnings $e(m)$
0	$175,000
1	
2	
3	
4	

2. Make a graph of the function.
3. Predict the store's earnings after 9 months.
4. A scientist studying a bacteria population recorded the data in the table below.

Time (min)	0	1	2	3
Number of Bacteria	8	20	50	125

Is the number of bacteria growing exponentially? Justify your response.

5. The function $f(x) = 3 \cdot b^x$ is an exponential growth function. Which statement about the value of b is true?
   A. Because $f(x)$ is an exponential growth function, b must be positive.
   B. Because $f(x)$ is an exponential growth function, b must be greater than 1.
   C. Because $f(x)$ is an exponential growth function, b must be between 0 and 1.
   D. The function represents exponential growth because $3 > 1$, so b can have any value.

Lesson 22-2
A new car *depreciates*, or loses value, each year after it is purchased. A general rule is that a car loses 15% of its value each year.

Christopher bought a new car for $25,000. A function that models the value of Christopher's car after t years is $v(t) = 25,000 \cdot (0.85)^t$. Use this information for Items 6–8.

6. Copy and complete the table.

Years After Purchase (t)	Value of Car $v(t)$
0	$25,000
1	
2	
3	
4	

7. Make a graph of the function.
8. Predict the value of Christopher's car after 10 years.

For Items 9–12, graph each function and tell whether it represents exponential growth, exponential decay, or neither.

9. $y = (2.5)^x$
10. $y = -0.75x$
11. $y = 3(1.5)^x$
12. $y = 80(0.25)^x$

For Items 13–15, tell whether each function represents exponential growth, exponential decay, or neither. Justify your responses.

13.
x	0	1	2	3
y	2	60	118	176

14.
x	0	1	2	3
y	25	5	1	0.2

ACTIVITY 22 Continued
ACTIVITY PRACTICE

7.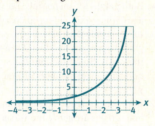

8. $4921.86
9. exponential growth

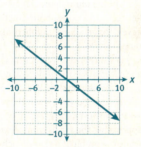

10. neither

11. exponential growth

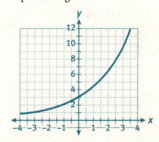

12. exponential decay

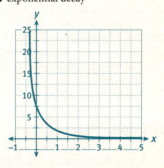

13. neither; Explanations may vary. There is a constant rate of change of 58, so this function is linear.
14. Exponential decay; there is a constant factor of 0.2, which is greater than 0 but less than 1.
15. Exponential growth; there is a constant factor of 9, which is greater than 1.
16. 50,000
17. Sample answer: $f(x) = 3(0.75)^x$; the value of b, 0.75, is greater than 0 but less than 1.

For Items 18–21, check students' graphs. Descriptions may vary.
18. $a = -4$; $b = 2$; Because $a < 0$, the graph is below the x-axis; because $b > 1$, the function approaches but does not reach the negative x-axis.
19. $a = -1$; $b = 2.5$; Because $a < 0$, the graph is below the x-axis; because $b > 1$, the function approaches but does not reach the negative x-axis.
20. $a = 1.5$; $b = 2$; Because $a > 0$, the graph is above the x-axis; because $b > 1$, the function approaches but does not reach the negative x-axis.
21. $a = 0.5$; $b = 0.2$; Because $a > 0$, the graph is above the x-axis; because $0 < b < 1$, the function approaches but does not reach the positive x-axis.

ACTIVITY 22 Continued

For Items 22–29, check students' graphs. Descriptions may vary.

22. Both graphs are increasing and above the x-axis. $f(x)$ increases more quickly than $a(x)$. Both graphs approach but do not reach the x-axis. Both functions have domain all real numbers and range $y > 0$.
23. Both graphs are increasing and above the x-axis. $a(x)$ increases much more quickly than $f(x)$. Both graphs approach but do not reach the x-axis. Both functions have domain all real numbers and range $y > 0$.
24. Both graphs are increasing and above the x-axis, but $f(x)$ increases more quickly than $a(x)$. Both graphs approach but do not reach the x-axis. Both functions have domain all real numbers and range $y > 0$.
25. Both functions are increasing and both graphs are above the x-axis. $f(x)$ increases more quickly than $a(x)$. Both graphs approach but do not reach the x-axis. Both functions have domain all real numbers and range $y > 0$.
26. $f(x)$ is always negative while $a(x)$ is always positive. $f(x)$ is decreasing, and $a(x)$ is increasing. Both graphs approach but do not reach the x-axis; however, the graph of $a(x)$ approaches the x-axis from above while the graph of $f(x)$ approaches the x-axis from below. Both functions have domain all real numbers; $a(x)$ has range $y > 0$ and $f(x)$ has range $y < 0$.
27. $f(x)$ is always less than $a(x)$. Both functions are increasing and both graphs approach but do not reach the x-axis. The graph of $a(x)$ approaches the negative x-axis from above, while the graph of $f(x)$ approaches the positive x-axis from below. Both functions have domain all real numbers. $a(x)$ has range $y > 0$, and $f(x)$ has range $y < 0$.
28. $f(x)$ is always less than $a(x)$. Both functions are increasing and always positive, and both graphs approach but do not reach the x-axis. Both functions have domain all real numbers and range $y > 0$.

ADDITIONAL PRACTICE
If students need more practice on the concepts in this activity, see the eBook Teacher Resources for additional practice problems.

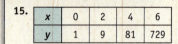

ACTIVITY 22 continued

15.

x	0	2	4	6
y	1	9	81	729

16. A wildlife biologist is studying an endangered species of salamander in a particular region. She finds the following data.

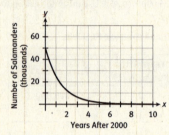

What was the initial number of salamanders in 2000?

17. Write a function that represents exponential decay. Explain how you know that your function represents exponential decay.

Lesson 22-3
Graph each of the following functions. Identify the values of a and b, and describe how these values affect the graphs.

18. $y = -4(2)^x$
19. $y = -1(2.5)^x$
20. $y = 1.5(2)^x$
21. $y = 0.5(0.2)^x$

For Items 22–29, use a graphing calculator to graph each function. Compare each function to $a(x) = 2^x$, graphed below. Describe the similarities and differences between the graphs.

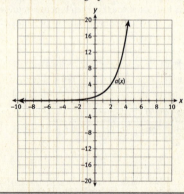

29. $f(x)$ is decreasing, and $a(x)$ is increasing. Both functions are always positive and both graphs approach but do not reach the x-axis. The graphs intersect at (0, 1) and appear to be reflections of each other across the y-axis. Both functions have domain all real numbers and range $y > 0$.

Exponential Functions
Protecting Your Investment

22. $f(x) = 0.5 \cdot 5^x$
23. $f(x) = 2 \cdot (1.1)^x$
24. $f(x) = 12^x$
25. $f(x) = 0.25 \cdot 4^x$
26. $f(x) = -3 \cdot 6^x$
27. $f(x) = -1 \cdot (0.3)^x$
28. $f(x) = 0.1 \cdot 2^x$
29. $f(x) = (0.5)^x$
30. Which function increases the fastest?
    A. $y = 104x$
    B. $y = -2 \cdot 15^x$
    C. $y = 12^x$
    D. $y = -220x$
31. Examine the graphs of $f(x) = 3^x$ and $g(x) = 5x$, shown below.

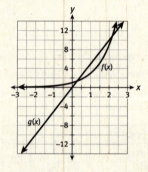

a. Estimate the values of x for which $f(x)$ is greater than $g(x)$.
b. Estimate the values of x for which $g(x)$ is greater than $f(x)$.
c. As the values of x decrease, the graph of $f(x)$ gets closer and closer to 0, or the x-axis. Will the graph ever intersect the x-axis? Explain.

MATHEMATICAL PRACTICES
Reason Abstractly and Quantitatively

32. Why can't the value of a in an exponential function be 0? Why can't the value of b be equal to 1?

---

30. C
31. a. $f(x)$ is greater than $g(x)$ when x is less than about 0.2 and when x is greater than about 2.1.
    b. $g(x)$ is greater than $f(x)$ when x is between about 0.2 and about 2.1.
    c. No; 3^x will always be greater than 0 for any value of x, so the graph will not intersect the x-axis.

32. An exponential function with $a = 0$ would equal 0 for all values of x. The graph would be the horizontal line $y = 0$ (the x-axis), which is a linear function, not exponential. An exponential function with $b = 1$ would equal a for all values of x. The graph would be the horizontal line $y = a$, which is also linear, not exponential.

Modeling with Exponential Functions

Growing, Growing, Gone
Lesson 23-1 Compound Interest

ACTIVITY 23

Learning Target:
- Create an exponential function to model compound interest.

> **SUGGESTED LEARNING STRATEGIES:** Create Representations, Look for a Pattern, Predict and Confirm, Discussion Groups, Think-Pair-Share, Critique Reasoning

Madison received $10,000 in gift money when she graduated from college. She deposits the money into an account that pays 5% *compound interest* annually.

1. To find the total amount of money in her account after the first year, Madison must add the interest earned in the first year to the initial amount deposited.
    a. Calculate the earned interest for the first year by multiplying the amount of Madison's deposit by the interest rate of 5%.
    $500

    b. Including interest, how much money did Madison have in her account at the end of the first year?
    $10,500

2. Madison wants to record the amount of money she will have in her account at the end of each year. Complete the table. Round amounts to the nearest cent.

Year	Account Balance
0	$10,000.00
1	$10,500.00
2	$11,025.00
3	**$11,576.25**
4	**$12,155.06**
5	**$12,762.81**
6	**$13,400.95**
7	**$14,071.00**
8	**$14,774.55**
9	**$15,513.28**
10	**$16,288.94**

> **MATH TERMS**
>
> **Compound interest** is interest, or money paid by a bank to an account holder, that is earned on both initial account funds, or *principal*, and previously earned interest.

Common Core State Standards for Activity 23

- **HSA-SSE.B.3:** Choose and produce an equivalent form of an expression to reveal and explain properties of the quantity represented by the expression.
- **HSA-SSE.B.3c:** Use the properties of exponents to transform expressions for exponential functions.
- **HSA-CED.A.1:** Create equations and inequalities in one variable and use them to solve problems.

ACTIVITY 23
Investigative

Activity Standards Focus
In this activity, students use exponential functions to model compound interest and population growth.

Lesson 23-1

PLAN

Chunking the Lesson
#1–4   #5–8   #9–13
#14–16   #17–19
#20–21   #22–23
Check Your Understanding
Lesson Practice

TEACH

Bell-Ringer Activity
Write the formula for simple interest, $I = prt$, on the board. Review the meaning of each variable. Then ask students to determine the values of the missing variables for examples such as these:
a. $p = \$200$, $r = 4.25\%$, $t = 18$ months
b. $I = \$35$, $r = 3.5\%$, $t = 4$ years
c. $I = \$3.75$, $p = \$600$, $t = 6$ months
d. $I = \$250$, $p = \$2500$, $r = 2.5\%$
[$I = \$12.75$; $p = \$250$, $i = 1.25\%$, $t = 4$ years]

1–4 Shared Reading, Create Representations, Look for a Pattern, Discussion Groups, Debriefing It is important for students to realize through repeated calculations that there is a constant growth factor in the account balances. Monitor group discussions carefully and encourage students who are having difficulty determining the constant growth factor to write the account balances as a sequence and determine the constant ratio. Although the amount of computation may seem excessive, students need it to understand an important feature of compound interest: the amount of interest increases each year, because the money in the account, on which the interest is calculated, increases each year.

Developing Math Language
Ask students to compute the amount of simple interest earned by $10,000 invested at 5% for 10 years and the total of the investment plus interest. ($5,000; $15,000) Have students compare the total amount with the amount in the table for the 10th year. Ask students to give reasons for the differences in the amounts. As you guide the discussion of the idea that the amount of the principal keeps increasing, model the term *compound interest*.

ACTIVITY 23 Continued

1–4 (continued) As you debrief this portion of the activity, help students connect the growth factor, 1.05, to the process of finding interest and total amount. You may wish to demonstrate the concept algebraically.

principal = x
interest = $0.05x$
total amount = $x + 0.05x$
$= 1x + 0.05x = 1.05x$

Students should understand that multiplying the principal by 1.05 is the same as multiplying it by 0.05 and then adding the result to the principal.

5–8 Look for a Pattern, Create Representations, Discussion Groups It is important for students to realize why the exponent in each expression is equal to the year number. Items 8b and 8c give students an opportunity to verify their expressions using data from the initial table although small discrepancies may exist in the results due to rounding errors.

ACTIVITY 23 continued

Lesson 23-1
Compound Interest

The amount of money in the account increases by a constant growth factor each year.

3. Identify the constant growth factor to the nearest hundredth. **1.05**

4. How is the interest rate on Madison's account related to the constant growth factor in Item 3? **5% written as a decimal is 0.05; this is the decimal portion of the constant factor.**

5. Instead of calculating the amount of money in the account after each year, write an expression for each amount of money using $10,000 and repeated multiplication of the constant factor. Then rewrite each expression using exponents.

Year	Account Balance
0	$10,000.00
1	$10,000 · 1.05 = $10,000 · 1.05^1
2	($10,000 · 1.05) · 1.05 = $10,000 · 1.05^2
3	($10,000 · 1.05 · 1.05) · 1.05 = $10,000 · 1.05^3
4	($10,000 · 1.05 · 1.05 · 1.05) · 1.05 = $10,000 · 1.05^4
5	($10,000 · 1.05 · 1.05 · 1.05 · 1.05) · 1.05 = $10,000 · 1.05^5
6	($10,000 · 1.05 · 1.05 · 1.05 · 1.05 · 1.05) · 1.05 = $10,000 · 1.05^6

6. Describe the relationship between the year number and the exponential expression.
   The power of 1.05 is equal to the year number.

7. Write an expression to represent the amount of money in the account at the end of Year 8.
   $10,000 · 1.05^8

8. Let t equal the number of years.
   a. **Express regularity in repeated reasoning.** Write an expression to represent the amount of money in the account after t years.
      $10,000 · 1.05^t

   b. Evaluate the expression for $t = 6$ to confirm that the expression is correct.
      $10,000 · 1.05^6 = $13,400.96

   c. Evaluate the expression for $t = 10$.   **$16,288.95**

9. Write your expression as a function $m(t)$, where $m(t)$ is the total amount of money in Madison's account after t years.   **$m(t) = 10,000 · 1.05^t$**

Lesson 23-1
Compound Interest

10. Use the data from the table in Item 2 to graph the function.

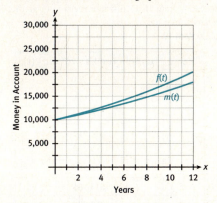

11. Describe the function as linear or non-linear. Justify your response.
Non-linear. Justifications will vary. The graph is a curve. The table does not exhibit a constant rate of change.

12. Identify the reasonable domain and range. Explain your reasoning.
The domain is the years, and the range is the account balance. The reasonable domain is $t \geq 0$ because number of years cannot be negative, and the reasonable range is $m(t) \geq \$10,000$, because the account balance when $x = 0$ is $10,000.

13. Madison's future plans include purchasing a home. She estimates that she will need at least $20,000 for a down payment. Determine the year in which Madison will have enough funds in her account for the down payment.
the 15th year

At the same time that Madison opens her account, her friend Frank deposits $10,000 in an account with an annual compound interest rate of 6%.

14. Write a new function to represent the total funds in Frank's account, $f(t)$, after t years.
The interest rate changes the value raised to an exponent. The value raised to the exponent is 1 + interest rate as a decimal = 1.06; $f(t) = \$10,000 \cdot (1.06)^t$.

15. Predict how the graph of Frank's bank account balance will differ from the graph of Madison's account balance.
Answers may vary. The higher interest rate will cause the account balance to increase at a faster rate. This will be shown in a graph as a steeper curve, and the graph of $f(t) = \$10,000 \cdot (1.06)^t$ will always be above the graph of $m(t) = \$10,000 \cdot (1.05)^t$.

ACTIVITY 23 Continued

17–19 Discussion Groups, Critique Reasoning, Construct an Argument, Sharing and Responding As students compare various savings scenarios, they should realize that three factors influence the total amount of money in an account: the initial amount invested, the interest rate, and the time the money is in the account. As students share responses, it is important for them to conclude that over a long period of time, the interest rate makes the greatest impact on the account balance regardless of the initial deposit.

Differentiating Instruction

Extend students' understanding of compound interest by introducing the Rule of 72. Say that to estimate how long it will take an investment to double, divide 72 by the percent of interest earned. For example, they can predict that at a 5% interest rate, money will double in about 14.4 years, because 72 ÷ 5 = 14.4. Note that the estimate of 15 years from the graphing calculator is very close to 14.4. Have students use the Rule of 72 to estimate how long it will take $10,000 to double at various interest rates, such as 2%, 4%, and 8%. Have students write and evaluate functions to see if their estimates do result in a total of $20,000.

Lesson 23-1
Compound Interest

16. Create a table of values for $f(t)$, rounding to the nearest dollar. Then graph $f(t)$ on the grid in Item 10. Confirm or revise your prediction in Item 15 using the table and graph.

See graph in Item 10.

Year, t	Funds in Frank's Account, $f(t)$
0	$10,000
1	$10,600
2	$11,236
3	$11,910
4	$12,625
5	$13,382
6	$14,185
7	$15,036
8	$15,938
9	$16,895
10	$17,908

At the same time that Madison and Frank open their accounts, another friend, Kasey, opens a savings account in a different bank. Kasey deposits $12,000 at an annual compound interest rate of 4%.

17. How does Kasey's situation change the function? Write a new function $k(t)$ to represent Kasey's account balance at any year t.

The initial amount is represented by the number that the power is multiplied by, so that the value will change to $12,000. The interest rate changes the value raised to an exponent to 1.04; $k(t) = \$12{,}000 \cdot (1.04)^t$

18. Critique the reasoning of others. Kasey believes that since she started her account with more money than Madison or Frank, she will always have more money in her account than either of them, even though her interest rate is lower. Is Kasey correct? Justify your response using a table, graph, or both.

Kasey is not correct. Both a table and a graph show that Madison will have more money than Kasey around the 20th year, and Frank will have more money than Kasey around the 10th year.

Lesson 23-1
Compound Interest

19. Over a long period of time, does the initial deposit or the interest rate have a greater effect on the amount of money in an account that has interest compounded yearly? Explain your reasoning.

The interest rate. Explanations may vary. The greater the interest rate, the greater the number raised to a power. The resulting numbers increase exponentially, while the initial deposit never increases.

Most savings institutions offer compounding intervals other than annual compounding. For example, a bank that offers *quarterly compounding* computes interest on an account every quarter; that is, at the end of every 3 months. Instead of computing the interest once each year, interest is computed four times each year. If a bank advertises that it is offering 8% interest compounded quarterly, 8% is not the actual growth factor. Instead, the bank will use $\frac{8\%}{4} = 2\%$ to determine the quarterly growth factor.

20. What is the quarterly interest rate for an account with an annual interest rate of 5%, compounded quarterly?

$\frac{5\%}{4} = 1.25\%$

21. Suppose that Madison invested her $10,000 in the account described in Item 20.

a. In the table below, determine Madison's account balance after the specified times since her initial deposit.

Time Since Initial Deposit	Number of Times Interest Has Been Compounded	Account Balance
3 months	1	$10,125
6 months	2	$10,251.56
9 months	3	$10,379.71
1 year	4	$10,509.45
4 years	16	$12,198.90
t years	$4t$	$10,000(1.0125)^{4t}$

b. Write a function $A(t)$ to represent the balance in Madison's account after t years.

$A(t) = 10,000(1.0125)^{4t}$

c. Calculate the balance in Madison's account after 20 years.

$27,014.85

CONNECT TO FINANCE

Interest can be compounded semiannually (every 6 months), quarterly (every 3 months), monthly, and daily.

ACTIVITY 23 Continued

Check Your Understanding

As you debrief these items, ask students to explain the steps they used to rewrite Frank's function to calculate interest compounded monthly.

Answers

24. $y = 10{,}000 \cdot (1.005)^{12t}$
25. Tables may vary.

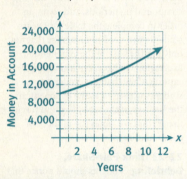

ASSESS

Students' answers to lesson practice problems will provide you with a formative assessment of their understanding of the lesson concepts and their ability to apply their learning.

See the Activity Practice for additional problems for this lesson. You may assign the problems here or use them as a culmination for the activity.

LESSON 23-1 PRACTICE

26.

Years	Money in Account
0	$5000
1	$5200
2	$5408
3	$5624.32
4	$5849.29
5	$6083.26
6	$6326.60
7	$6579.66
8	$6842.85

27. $y = \$5000 \cdot (1.04)^t$; since t is the time in years after the account was opened, $t \geq 0$ is the reasonable domain. The initial account balance is $5000, and the balance is growing as t increases, so the reasonable range is $y \geq \$5000$.

ADAPT

If students are writing appropriate functions, but having difficulty accurately computing account balances, be sure that they are following the order of operations. The base of the exponential expression must be raised to the appropriate exponent before multiplying by the initial deposit.

ACTIVITY 23 continued

My Notes

MATH TIP

For a given annual interest rate, properties of exponents can be used to approximate equivalent semi-annual, quarterly, monthly, and daily interest rates. For example, the function $f(t) = 5000(1.03)^t$ is used to approximate the balance in an account with an initial deposit of $5000 and an annual interest rate of 3%. $f(t)$ can be rewritten as

$$f(t) = 5000\left(1.03^{\frac{1}{12}}\right)^{12t}$$ and is equivalent to the function $g(t) = 5000(1.0025)^{12t}$, which reveals that the approximate equivalent monthly interest rate is 0.25%.

$$f(t) = 5000\left(1.03^{\frac{1}{4}}\right)^{4t}$$ is equivalent to $h(t) = 5000(1.0074)^{4t}$, which reveals that the approximate equivalent quarterly interest rate is 0.74%.

Lesson 23-1 Compound Interest

22. For the compounding periods given below, write a function to represent the balance in Madison's account after t years. Then calculate the balance in the account after 20 years. She is investing $10,000 at a rate of 5% annual compound interest.

   a. Yearly: $A(t) = 10{,}000(1.05)^t$; $26,532.98
   b. Quarterly: $A(t) = 10{,}000(1.0125)^{4t}$; $27,014.85
   c. Monthly: $A(t) = 10{,}000(1.00417)^{12t}$; $27,126.40
   d. Daily (assume there are 365 days in a year): $A(t) = 10{,}000(1.000137)^{365t}$; $27,180.96

23. What is the effect of the compounding period on the amount of money in the account after 20 years as the number of times the interest is compounded each year increases?

   Observations may vary. As the compounding period shortens and the number of times interest is compounded each year increases, the amount of money increases. But shortening the compounding period decreases the rate at which the amount of money in the account increases.

Check Your Understanding

24. Write a function that gives the amount of money in Frank's account after t years when 6% annual interest is compounded monthly.
25. Create a table and a graph for the function in Item 24. Be sure to label the units on the x-axis correctly.

LESSON 23-1 PRACTICE

Model with mathematics. Nick deposits $5000 into an account with a 4% annual interest rate, compounded annually.

26. Create a table showing the amount of money in Nick's account after 0–8 years.
27. Write a function that gives the amount of money in Nick's account after t years. Identify the reasonable domain and range.
28. Create a graph of your function.
29. Explain how Nick's account balance would be different if he deposited his money into an account that pays 2% annual interest, compounded annually. Graph this situation on the same coordinate plane that you used in Item 28. Describe the similarities and differences between the graphs.

28.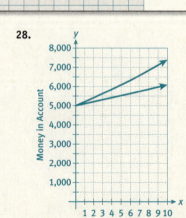

29. The account with a 2% interest rate will grow at a slower rate, so its graph is less steep. The account with a 2% interest rate will always have less money than the account with a 4% interest rate, so the graph for the account with 2% interest is below the graph for the account with 4% interest. (See left for graph.)

346 SpringBoard® Mathematics Algebra 1, Unit 4 • Exponents, Radicals, and Polynomials

Lesson 23-2
Population Growth

Learning Targets:
- Create an exponential function to fit population data.
- Interpret values in an exponential function.

SUGGESTED LEARNING STRATEGIES: Create Representations, Look for a Pattern, Predict and Confirm, Think Aloud, Sharing and Responding, Construct Arguments

The population of Nevada since 1950 is shown in the table in the *My Notes* section.

1. Graph the data from the table.

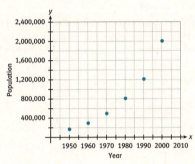

Year	Resident Population
1950	160,083
1960	285,278
1970	488,738
1980	800,508
1990	1,201,833
2000	1,998,257

2. Use the table and the graph to explain why the data are not linear.
   The table shows that there is not a constant rate of change. The graph shows that the points do not lie on a line.

3. **a.** Complete the table by finding the approximate ratio between the populations in each decade.

Decades Since 1950	Resident Population	Ratio
0	160,083	--
1	285,278	$\frac{285{,}278}{160{,}083} \approx 1.782$
2	488,738	$\frac{488{,}738}{285{,}278} \approx$ **1.713**
3	800,508	$\frac{800{,}508}{488{,}738} \approx$ **1.638**
4	1,201,833	$\frac{1{,}201{,}833}{800{,}508} \approx$ **1.501**
5	1,998,257	$\frac{1{,}998{,}257}{1{,}201{,}833} \approx$ **1.663**

b. Explain how the table shows that the data are not exponential.
   The table shows that there is not a constant factor.

The data are not exactly exponential, but the shape of the graph resembles an exponential curve. Also, the table in Item 3a shows a near-constant factor. These suggest that the data are approximately exponential. Use **exponential regression** to find an exponential function that models the data.

MATH TERMS

Exponential regression is a method used to find an exponential function that models a set of data.

Activity 23 • Modeling with Exponential Functions 347

ACTIVITY 23 Continued

4 Discussion Groups, Create Representations If students are not familiar with the regression function on the calculator, demonstrate how to use it. Monitor students discussions carefully to be sure students can explain the meanings of *a* and *b* within the context of the problem situation.

CONNECT TO TECHNOLOGY

Some students have a difficult time with the more complex functions on a graphing calculator. To encourage independence when using calculators, allow students to create a list of the steps necessary to create an exponential regression. Students may add sketches, arrows, words in another language, or other cues that will help them find exponential regression efficiently.

Students using a Casio calculator can perform an exponential regression by selecting the CALC function menu in the STAT mode.

For additional technology resources, visit SpringBoard online.

5–8 Think Aloud, Sharing and Responding Students make connections between the different representations of the function and the context of the problem. It is important for students to realize that there is no value of *x* for which the function value is 0. In the context, there has not been a time in which the population of Nevada was 0.

ACTIVITY 23 continued

My Notes

TECHNOLOGY TIP

On a Texas Instruments (TI) calculator, perform exponential regression using the ExpReg function.

Lesson 23-2
Population Growth

4. Use a graphing calculator to determine the exponential regression equation to model the relationship between the decades since 1950 and the population.

   a. The calculator returns two values, *a* and *b*. Write these values below. Round *a* to the nearest whole number and *b* to the nearest thousandth, if necessary.

   $a = 170{,}377$ $b = 1.645$

   b. The general form of an exponential function is $y = ab^x$. Use this general form and the values of *a* and *b* from Part (a) to write an exponential function that models Nevada's population growth.

   $y = 170{,}377 \cdot (1.645)^x$

   c. Use a graphing calculator to graph the data points and the function from Part (b). Sketch the graph and the data points below. Is the exponential function a good approximation of the data? Explain.

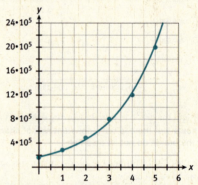

   Yes; the graph is very close to the data points.

5. **Reason abstractly.** What does the value of *b* tell you about Nevada's population growth?

   The approximate growth rate of the population each decade is 64.5%.

6. Interpret the value of *a* in terms of Nevada's population. How is this value related to the graph?

   The value of *a* is the approximate population at year 0, or in 1950, when the data set begins. On the graph, this value corresponds to the *y*-intercept.

7. What do the domain values represent?

   the number of decades since 1950

8. What would the *x*-intercept represent in terms of Nevada's population? Does the graph have an *x*-intercept? Explain.

   The *x*-intercept represents the year in which the population of Nevada is 0. At a negative value of *x*, or in some year prior to 1950, it is possible that the population of Nevada was 0. However, the graph does not have an *x*-intercept because there is no value of *x* for which the function is equal to 0.

348 SpringBoard® Mathematics **Algebra 1, Unit 4** • Exponents, Radicals, and Polynomials

Lesson 23-2
Population Growth

9. **Make sense of problems.** Describe how to estimate the population of Nevada in 1995 using each of the following:
   a. the function identified in Item 4b
   Evaluate at $x = 4.5$.

   b. the graph of the function
   Identify the point on the curve whose x-coordinate is 4.5 and estimate its y-coordinate.

   c. a table
   Examine a table of function values and determine the corresponding y-value when $x = 4.5$.

10. Estimate the population in 1995. Which method did you use, and why?
    The population in 1995 was about 1,600,129. Answers vary, but using the function or table likely gives a more accurate estimate.

11. a. Estimate Nevada's population in 2010.
    about 3,376,031

    b. **Construct viable arguments.** Which estimate do you think is likely to be more accurate, your estimate of the population in 1995 or in 2010? Explain.
    1995; the estimate for 2010 assumes that the population continues to grow at the same rate, and we do not know that this is true.

The function for the growth rate of Nevada's population estimates the growth per decade. You can use this rate to estimate the growth per year, or the annual growth rate.

12. Let n be the number of years since 1950. Write an equation that gives the number of years n in x decades. Solve your equation for x.
    $n = 10x$; $x = \frac{n}{10}$

13. Rewrite the function that models Nevada's population from Item 4b. Then write the function again, but replace x with the equivalent expression for x from Item 12.
    $y = 170,377 \cdot (1.645)^x$; $y = 170,377 \cdot (1.645)^{\frac{n}{10}}$

14. Simplify to write the function in the form $y = ab^n$.

    $$y = 170,377 \cdot (1.645)^{\frac{n}{10}} = 170,377 \cdot \left(1.645^{\frac{1}{10}}\right)^n$$

    $$\approx 170,377 \cdot (1.051)^n$$

TECHNOLOGY TIP

Use the table function on a graphing calculator to determine the value of the function when $x = 4.5$.

ACTIVITY 23 Continued

17–18 Think-Pair-Share,

Debriefing Students should now use the number of years since 1950 for the exponent rather than for the number of decades since 1950. They should realize that the 2013 estimate should be larger than the 2010 estimate if the population is, in fact, growing. However, comparisons should show reasonable growth. As this lesson is debriefed, ask students to make comparisons between compound interest and population growth.

Check Your Understanding

Check student work to ensure students are able to graph both functions and to explain why the two functions are similar but not exactly the same.

Answers

19.

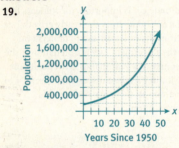

20. Answers may vary. The *x*-values of the domain are larger, because 5 decades of population growth is 50 years. The overall shape of the graph is the same.

ASSESS

Students' answers to lesson practice problems will provide you with a formative assessment of their understanding of the lesson concepts and their ability to apply their learning.

See the Activity Practice for additional problems for this lesson. You may assign the problems here or use them as a culmination for the activity.

ADAPT

If students are having difficulty using technology to determine exponential regression equations, have them "teach" the steps to another student, using their list of steps if necessary.

LESSON 23-2 PRACTICE

21. $y = 7{,}732{,}255 \cdot (1.219)^x$
22.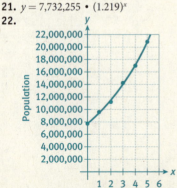

23. It appears to be a good fit, because it is close to the data points.
24. Domain: decades since 1950; range: population; *y*-intercept: population in 1950; *x*-intercept: year (the number of decades from 1950) with a population of 0 (however the exponential function does not have an *x*-intercept).

Lesson 23-2
Population Growth

15. What is the approximate annual growth rate of Nevada's population? How do you know?

    The decimal portion of the value of *b* in $y = ab^x$ represents the approximate growth rate. Nevada's population grew approximately 5.1% per year.

16. To find the approximate population of Nevada in 2013, what value should you use for *n*? Explain.

    63; the number of years from 1950 until 2013 is 63.

17. Use the function from Item 14 to find the approximate population of Nevada for the year 2013.

    $y = 170{,}377(1.051)^{63} \approx 3{,}911{,}849$

18. Compare the approximate population for 2013 that you found in Item 17 to the approximate population you found for 2010 in Item 11. Does your estimate for 2013 seem reasonable? Why or why not?

    Yes; the population is increasing, so the 2013 estimate should be greater than the 2010 estimate, and $3{,}911{,}849 > 3{,}376{,}031$.

Check Your Understanding

19. Create a graph showing the annual growth of Nevada's population.
20. Describe the similarities and differences between the graph in Item 19 and the previous graph of Nevada's population from Item 4c.

LESSON 23-2 PRACTICE

The population of Texas from 1950 to 2000 is shown in the table below.

Year	Resident Population
1950	7,711,194
1960	9,579,677
1970	11,198,655
1980	14,225,513
1990	16,986,510
2000	20,851,820

21. **Use appropriate tools strategically.** Use a graphing calculator to find a function that models Texas's population growth.
22. Create a graph showing the actual population from the table and the approximate population from the function in Item 21.
23. Is the function a good fit for the data? Why or why not?
24. Describe the meanings of the domain, range, *y*-intercept, and *x*-intercept in the context of Texas's population growth.

Modeling with Exponential Functions
Growing, Growing, Gone

ACTIVITY 23 PRACTICE
Write your answers on notebook paper.
Show your work.

Lesson 23-1

1. Four friends deposited money into savings accounts. The amount of money in each account is given by the functions below.

   Marisol: $m(t) = 100 \cdot (1.01)^t$
   Iris: $i(t) = 200 \cdot (1.04)^t$
   Brenda: $b(t) = 300 \cdot (1.05)^t$
   José: $j(t) = 400 \cdot (1.03)^t$

   Which statement is correct?
   A. José has the greatest interest rate.
   B. Brenda has the greatest initial deposit.
   C. The person with the least initial deposit also has the least interest rate.
   D. The person with the greatest initial deposit also has the greatest interest rate.

Darius makes an initial deposit into a bank account, and then earns interest on his account. He records the amount of money in his account each year in the table below. Use this table for Items 2–5.

Year	Amount
0	$4000.00
1	$4120.00
2	$4243.60
3	$4370.91
4	$4502.04

2. Make a graph showing the amount of money in Darius's account each year.
3. Identify the constant factor. Round to the nearest hundredth.
4. Identify the reasonable domain and range. Explain your answers.
5. What is the annual interest rate? How do you know?

The amount of money y in Jesse's checking account t years after the account was opened is given by the function $j(t) = 15{,}000 \cdot (1.02)^t$. Use this information for Items 6–10.

6. What was the initial amount of money deposited in Jesse's account?
7. What is the annual interest rate?
8. Create a graph of the amount of money in Jesse's checking account.
9. Interpret the meaning of the y-intercept in the context of Jesse's account.
10. Find the amount of money in the account after 4 years.

The two graphs on the coordinate grid below represent the amounts of money in two different savings accounts. Graph a represents the amount of money in Allison's account, and graph b represents the amount of money in Boris's account. Use the graph for Items 11–13.

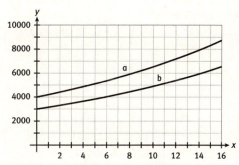

11. Whose account had a higher initial deposit? Use the graph to justify your answer.
12. What was the amount of Allison's initial deposit?
13. Identify the reasonable domain and range for each function, and explain your answers.

ACTIVITY 23 Continued
ACTIVITY PRACTICE

1. C
2.

(graph: line from (0, 4000) rising to about (5, 4500); x-axis "Years" 1–5; y-axis "Amount ($)" 1,000–5,000)

3. 1.03
4. The reasonable domain is $x \geq 0$, because x represents a number of years, which cannot be negative. The reasonable range is $y \geq 4000$, because the initial amount is $4000 and it increases.
5. 3%; The decimal portion of the constant factor, 1.03, represents the interest rate.
6. $15,000
7. 2%
8.

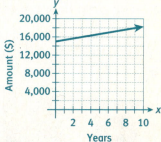

9. The y-intercept is the initial account balance, or the amount initially deposited in the account.
10. $16,236.48
11. Allison's; Allison's graph is above Boris's graph at $x = 0$.
12. $4000
13. For both functions, the reasonable domain is $x \geq 0$, because x represents a number of years, which cannot be negative. For Allison's function, the reasonable range is $y \geq \$4000$ because her account started with $4000, and the amount increases. For Boris's function, the reasonable range is $y \geq \$3000$ because his account started with $3000, and the amount increases.

ACTIVITY 23 Continued

14. $1259.71
15. $y = \$1000 \cdot (1.08)^t$
16. $y = \$1000 \cdot (1.0067)^{12t}$
17. 0.67%
18. $1079.56
19. The account balances will be approximately equal after 5 years, with a balance of $1469.33. The function for the monthly compounding gives a slightly smaller amount, but that is due to the rounding of the interest rate.
20. B
21. Yes; explanations may vary. Students may graph the data to see that the points are close to lying on an exponential curve, or they may notice that each y-value is about twice the preceding y-value, so there is an almost constant factor of 2.
22. No; explanations may vary. Students may graph the data to see that the points are almost linear, or they may notice that each rate of change is about $-\frac{4}{3}$, so the rate of change is close to constant.
23. $y = 41.78 \cdot (1.013)^x$
24. The reasonable range is $x \geq 0$, because x represents a number of months, which cannot be negative. The reasonable range is $y \geq 0$, because y represents a circumference, which cannot be negative.

25.
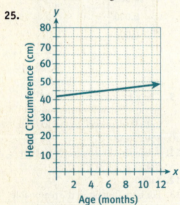

ADDITIONAL PRACTICE

If students need more practice on the concepts in this activity, see the eBook Teacher Resources for additional practice problems.

ACTIVITY 23 continued

Maria's bank offers two types of savings accounts. The first has an annual interest rate of 8% compounded annually. The second also has an annual interest rate of 8%, but it is compounded monthly. She is going to open an account by depositing $1000. Use this information for Items 14–19.

14. If Maria chooses the first account, determine the amount of money she will have in the account after 3 years.
15. Write a function that gives the amount of money in the first account after t years.
16. Write a function that gives the amount of money in the second account after t years.
17. What is the *monthly* interest rate for the second account?
18. If Maria chooses the second account, determine the amount of money she will have in the account after 1 year.
19. After 10 years, which account will have the higher balance?

Lesson 23-2

20. Which function is the best model for the data in the table?

x	y
0	19
1	44.5
2	112
3	282
4	704

A. $y = 172x + 18$ B. $y = 44x^2$
C. $y = 44x$ D. $y = 172 \cdot 18^x$

For Items 21 and 22, tell whether an exponential function would be a good model for each data set. Explain your answers.

21.
x	0	1	2	3	4
y	33	58	120	247	506

26. Approximately 1.3% each month; the decimal portion of the number raised to a power is the growth rate.
27. Nathan's head circumference in cm at birth, or age 0 months
28. The function is only an approximation of the context, such as population. A population can reasonably have been 0 at some point in time, but an exponential function will never equal 0. It is important to recognize that the function only estimates the behavior.

Modeling with Exponential Functions
Growing, Growing, Gone

22.
x	0	3	6	9	4
y	16.5	12	8.5	4.6	0

The head circumference of an infant is measured and recorded to track the infant's growth and development. Nathan's head circumferences from age 3 months through 12 months are recorded in the table below. Use the table for Items 23–27.

Age (months)	Head Circumference (cm)
3	43
4	44
5	44.7
6	45.2
7	45.8
8	46.3
9	47.1
10	47.6
11	48.0
12	48.3

23. Use a graphing calculator to find an exponential function to approximate Nathan's head circumference.
24. Identify the reasonable domain and range for the function in Item 23. Explain your answers.
25. Create a graph showing Nathan's head circumference.
26. Determine the growth rate, and explain how you found your answer.
27. Interpret the meaning of the y-intercept in the context of Nathan's head circumference.

MATHEMATICAL PRACTICES
Attend to Precision

28. Explain why the x-intercept can have a meaning in the context of a situation, such as population growth, but cannot be shown on the graph.

Exponential Functions
FAMILY BONDS

Embedded Assessment 2
Use after Activity 23

Mr. Davis has just become a grandfather! He wants to invest money for his new granddaughter's college education.

Mr. Davis has done some research on savings bonds. He has learned that you buy a savings bond from the government or from a bank. After one year, you can cash in your bond and get back the money you paid for it. However, if you wait at least five years, you will get back your money plus interest.

Mr. Davis has also learned that he can buy paper bonds or electronic bonds. While there are many similarities and differences between the bonds, Mr. Davis has summarized the most important information below.

Paper Bond	Electronic Bond
Current rate of interest: 1.8% annual, but the interest rate may change over the life of the bond.	Current rate of interest: 1.4% annual; this rate will not change.
For both bonds, the interest is compounded semiannually (every 6 months, or twice per year) for 30 years or until the bond is cashed in, whichever comes first.	

Mr. Davis decides to buy a $5000 bond that he will give to his granddaughter on her 18th birthday. He will use the current interest rates to decide which bond he will purchase.

1. Using the current interest rate, a function that gives the value of a $5000 paper bond after t years is $p(t) = 5000 \cdot (1.009)^{2t}$.
    a. How is the interest rate of 1.8% related to the function?
    b. Why is the exponent $2t$ instead of t?
    c. Use the function to determine the value of the bond in 18 years. Round your answer to the nearest cent.

2. a. Write a function $e(t)$ that gives the value of a $5000 electronic bond after t years.
    b. Use your function to determine the value of the bond in 18 years. Round your answer to the nearest cent.

3. Identify the reasonable domain and range for each function. Explain your answers.

4. Use a graphing calculator to graph both functions on the same coordinate plane. Sketch the graphs and label each function.

5. Explain to Mr. Davis which bond you think he should purchase and why.

6. Mr. Davis's accountant has more information about electronic bonds. She tells Mr. Davis that if you keep an electronic bond for 20 years, the value becomes double what you paid for it. Would this change your advice to Mr. Davis? Explain.

Common Core State Standards for Embedded Assessment 2

HSA-SSE.B.3	Choose and produce an equivalent form of an expression to reveal and explain properties of the quantity represented by the expression.
HSA-SSE.B.3c	Use the properties of exponents to transform expressions for exponential functions.
HSA-CED.A.2	Create equations in two or more variables to represent relationships between quantities; graph equations on coordinate axes with labels and scales.
HSF-IF.B.4	For a function that models a relationship between two quantities, interpret key features of graphs and tables in terms of the quantities, and sketch graphs showing key features given a verbal description of the relationship. Key features include: intercepts; intervals where the function is increasing, decreasing, positive, or negative; relative maximums and minimums; symmetries; end behavior; and periodicity.

Embedded Assessment 2

Assessment Focus
- Exponential functions
- Compound interest

Materials
- graphing calculators

Answer Key

1. a. The interest rate is twice the decimal portion of the number that is raised to a power.
   b. The interest is compounded every six months, or every one-half year. Therefore, if n represents years, $2n$ represents the number of times per year the interest is compounded.
   c. $6,903.22

2. a. $e(t) = 5000(1.007)^{2t}$
   b. $6,427.36

3. The reasonable domain and range are the same for each function. The reasonable domains are $n \geq 0$ and $t \geq 0$, because both n and t represent a number of years, which cannot be negative. The reasonable ranges are $p(n) \geq 5000$ and $e(t) \geq 5000$, because the value of the savings bond starts out at $5,000 and increases.

4.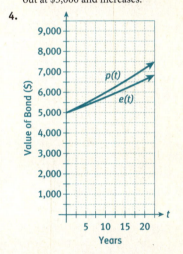

5. Answers may vary. The graph shows that the value of the paper bond increases faster than the graph of the electronic bond. Students may advise Mr. Davis to buy the paper bond because it will have a value of $6,003.24 after 18 years while the electronic bond will have a value of only $6,427.36. However, students may also cite the variable interest rate of the paper bond; the interest rate on the paper bond may go down. Students may advise Mr. Davis to purchase the electronic bond for this reason.

Embedded Assessment 2

6. Answers may vary. He should choose the electronic bond, even though his granddaughter would have to wait two more years for the money. At 20 years, the value of the electronic bond becomes $10,000. The value of the paper bond at 20 years is $7,155.12. It could be worth waiting two more years for this much of an increase in value.

> **TEACHER to TEACHER**
>
> You may wish to read through the scoring guide with students and discuss the differences in the expectations at each level. Check that students understand the terms used.

Unpacking Embedded Assessment 3

Once students have completed this Embedded Assessment, turn to Embedded Assessment 3 and unpack it with them. Use a graphic organizer to help students understand the concepts they will need to know to be successful on Embedded Assessment 3.

Embedded Assessment 2
Use after Activity 23

Exponential Functions
FAMILY BONDS

Scoring Guide	Exemplary	Proficient	Emerging	Incomplete
	The solution demonstrates the following characteristics:			
Mathematics Knowledge and Thinking (Items 1c, 2b)	• Effective understanding of and accuracy in evaluating an exponential function	• Largely correct understanding of and accuracy in evaluating an exponential function	• Partial understanding of and some difficulty in evaluating an exponential function	• Incomplete understanding of and significant difficulty in evaluating an exponential function
Problem Solving (Items 5, 6)	• Appropriate and efficient strategy that results in a correct answer	• Strategy that may include unnecessary steps but results in a correct answer	• Strategy that results in some incorrect answers	• No clear strategy when solving problems
Mathematical Modeling/ Representations (Item 1a, 1b, 2a, 3, 4)	• Fluency in representing a real-world scenario using an exponential function, including identification of a reasonable domain and range • Clear and accurate graphs of exponential functions	• Adequate understanding of how to represent a real-world scenario using an exponential function, including identification of a reasonable domain and range • Little difficulty graphing exponential functions	• Partial understanding of how to represent a real-world scenario using an exponential function, including identification of a reasonable domain and range • Partially accurate graphs of exponential functions	• Inaccurate or incomplete understanding of how to represent a real-world scenario using an exponential function, including identification of a reasonable domain and range • Inaccurate or incomplete graphs of exponential functions
Reasoning and Communication (Items 5, 6)	• Precise use of appropriate math terms and language to make and justify a recommendation	• Recommendation with an adequate justification	• Recommendation with a misleading or confusing justification	• No recommendation or a recommendation with an inaccurate or incomplete justification

Common Core State Standards for Embedded Assessment 2 (cont.)

HSF-IF.C.7	Graph functions expressed symbolically and show key features of the graph, by hand in simple cases and using technology for more complicated cases.
HSF-LE.B.5	Interpret the parameters in a linear or exponential function in terms of a context.

Adding and Subtracting Polynomials
Polynomials in the Sun
Lesson 24-1 Polynomial Terminology

ACTIVITY 24

Learning Targets:
- Identify parts of a polynomial.
- Identify the degree of a polynomial.

SUGGESTED LEARNING STRATEGIES: Create Representations, Vocabulary Organizer, Interactive Word Wall, Think-Pair-Share, Close Reading

A solar panel is a device that collects and converts solar energy into electricity or heat. The solar panel consists of interconnected solar cells. The panels can have differing numbers of solar cells and can come in square or rectangular shapes.

1. How many solar cells are in the panel below?
   16

CONNECT TO SCIENCE

Solar panels, also known as photovoltaic panels, are made of semiconductor materials. A panel has both positive and negative layers of semiconductor material. When sunlight hits the semiconductor, electrons travel across the intersection of the two different layers of materials, creating an electric current.

2. **Reason abstractly.** If a solar panel has four rows as the picture does, but can be extended to have an unknown number of columns, x, write an expression to give the number of solar cells that could be in the panel.
   $4x$

3. Write an expression that would give the total number of cells in the panel for a solar panel having x rows and x columns.
   x^2

4. If there were 5 panels like those found in Item 3, write an expression to represent the total number of solar cells.
   $5x^2$

All the answers in Items 1–4 are called *terms*. A **term** is a number, variable, or the product of a number and/or variable(s).

5. Write an expression to represent the sum of your answers from Items 1, 2, and 4.
   $5x^2 + 4x + 16$

ACTIVITY 24 Guided

Activity Standards Focus
In Activity 24, students classify and identify components of polynomials. They also add and subtract polynomials.

Lesson 24-1

PLAN
Pacing: 1 class period
Chunking the Lesson
#1–5   #6–7
Check Your Understanding
#10–11   #12–13
Check Your Understanding
Lesson Practice

TEACH

Bell-Ringer Activity
A town is organizing a children's parade in which participants may walk or ride any type of nonmotorized vehicle. Parade participants will line up in the parade according to the number of wheels on their vehicles. Students should consider bicycles, tricycles, wagons, unicycles and pedestrians and describe the order in which the participants are lined up.

1–5 Shared Reading, Create Representations, Vocabulary Organizer, Interactive Word Wall
Students develop monomials in the context of solar panels. While the term *monomial* is not introduced to students until after Item 10, this series of items is designed to help students understand the concept of a *term*. Some students will have trouble visualizing an x by x solar panel. A sketch may help students visualize it.

Common Core State Standards for Activity 24

HSA-SSE.1:	Interpret expressions that represent a quantity in terms of its context.
HAS-SSE.A.1.a:	Interpret parts of an expression, such as terms, factors, and coefficients.
HAS-SSE.A.1.b:	Interpret complicated parts of expressions by viewing one or more of their parts as a single entity.
HSA-APR.A.1:	Understand that polynomials form a system analogous to the integers, namely, they are closed under the operations of addition, subtraction, and multiplication; add, subtract, and multiply polynomials.

ACTIVITY 24 Continued

6–7 Think-Pair-Share, Debriefing
These problems are designed as a formative assessment to determine if students have a general understanding of the vocabulary just introduced. Note that a polynomial can have only one term. Students will later learn that *monomial* is the specific word for a polynomial with one term. Debrief this portion of the lesson by asking students to give examples of polynomials with two, three, four or another specific number of terms.

Check Your Understanding
Ask students to give more examples and nonexamples of polynomials and to identify terms, coefficients and constant terms in their examples. Discuss subtraction within a polynomial. Because a polynomial with more than one term is defined as a *sum*, subtraction within a polynomial indicates the addition of a negative term. For example, the polynomial $2x^2 - x - 5$ is equivalent to $2x^2 + (-x) + (-5)$, and so the terms are $2x^2$, $-x$, and -5. This will be important in upcoming lessons and activities in which students will add, subtract and multiply polynomials.

Answers
8. a. No; the exponent of x is a negative number.
   b. Yes; it is the sum of terms with whole-number powers.
   c. Yes; it is a constant term.
   d. No; the exponent of x is a fraction.
9. For the polynomial in 8b, the terms are $6x$ and $4x^2$. The coefficients are 6 and 4. There is no constant term.

   For the polynomial in 8c, the only term is 15, and it is a constant term. There are no coefficients.

10–11 Close Reading, Discussion Groups, Vocabulary Organizer, Interactive Word Wall
These items will help students clarify the meaning of the degree of a term. As students read and apply the definition, emphasize that if a term has more than one variable, the degree is the sum of the exponents of all the variables. Students should understand that x can be written as x^1 and that any constant c can be written as cx^0.

ACTIVITY 24 continued

My Notes

MATH TERMS
A **coefficient** is the numeric factor of a term.

A **constant term** is a term that contains only a number, such as the answer to Item 1. The constant term of a polynomial is a term of degree zero.

DISCUSSION GROUP TIPS
As needed, refer to the Glossary to review definitions and pronunciations of key terms. Incorporate your understanding into group discussion to confirm your knowledge and use of key mathematical language.

Lesson 24-1
Polynomial Terminology

Expressions like the answer to Item 5 are called polynomials. A *polynomial* is a single term or the sum of two or more terms with *whole-number powers*.

6. List the terms of the polynomial you wrote in Item 5.
   $5x^2$, $4x$, 16

7. What are the **coefficients** of the polynomial in Item 5? What is the **constant term**?
   5 and 4 are the coefficients, and 16 is the constant term.

Check Your Understanding

8. Tell whether each expression is a polynomial. Explain your reasoning.
   a. $3x^{-2} - 5$ b. $6x + 4x^2$ c. 15 d. $2 + x^{\frac{1}{2}}$

9. For the expressions in Item 8 that are polynomials, identify the terms, coefficients, and constant terms.

The **degree of a term** is the sum of the exponents of the variables contained in the term.

10. Identify the degree and coefficient of each term in the polynomial $4x^5 + 12x^3 + x^2 - x + 5$.

Term	Degree	Coefficient
$4x^5$	5	4
$12x^3$	3	12
x^2	2	1
$-x$	1	-1

11. **Make use of structure.** For the polynomial $2x^3y - 6x^2y^2 + 9xy - 13y^5 + 5x + 15$, list each term and identify its degree and coefficient. Identify the constant term.
    $2x^3y$: the degree is 4, the coefficient is 2.
    $-6x^2y^2$: the degree is 4, the coefficient is -6.
    $9xy$: the degree is 2, the coefficient is 9.
    $-13y^5$: the degree is 5, the coefficient is -13.
    $5x$: the degree is 1, the coefficient is 5.
    15 is a constant term.

Developing Math Language
The vocabulary introduced here (*term*, *polynomial*, *coefficient*, and *constant term*) is used throughout the activity and the unit. Because the concepts and vocabulary in this lesson are important, model the terminology in context throughout this activity. The classroom Word Wall and a vocabulary organizer to which students add vocabulary as they move through the lesson will be valuable tools to help students make connections between vocabulary words and develop their understanding of polynomials.

Lesson 24-1
Polynomial Terminology

The *degree of a polynomial* is the greatest degree of any term in the polynomial.

12. Identify the degree and constant term of each polynomial.

Polynomial	Degree of Polynomial	Constant Term
$2x^2 + 3x + 7$	2	7
$-5y^3 + 4y^2 - 8y - 3$	3	-3
$36 + 12x + x^2$	2	36

The *standard form of a polynomial* is a polynomial whose terms are written in *descending order* of degree. The *leading coefficient* is the coefficient of a polynomial's leading term when the polynomial is written in standard form.

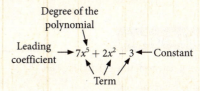

A polynomial can be classified by the number of terms it has when it is in simplest form.

Name	Number of Terms n	Examples
monomial	1	8 or $-2x$ or $3x^2$
binomial	2	$3x + 2$ or $4x^2 - 7x$
trinomial	3	$-x^2 - 3x + 9$
polynomial	$n > 3$	$9x^4 - x^3 - 3x^2 + 7x - 2$

MATH TERMS

Descending order of degree means that the term that has the highest degree is written first, the term with the next highest degree is written next, and so on.

READING MATH

The prefixes mono (one), bi (two), tri (three), and poly (many) appear in many math terms such as bisect (cut in half), triangle (three-sided figure), and polygon (many-sided figure).

ACTIVITY 24 Continued

10–11 (continued) It is important to note that a polynomial may be "missing" terms. For example, the polynomial in Item 10 starts with a term whose degree is 5, but there is no 4th degree term. That means the coefficient of that term is 0. The polynomial would be $4x^5 + 0x^4 + 12x^3 + x^2 - x + 5$, but we do not write the terms with a coefficient of 0. Students often confuse a degree of 0 with an exponent of 0.

12–13 Close Reading, Think-Pair-Share, Interactive Word Wall, Vocabulary Organizer Debriefing These items allow students an opportunity to use the vocabulary they are learning. Monitor student discussions carefully to be sure that as students determine the degree of a polynomial, they can articulate the difference between finding the degree of a term and the degree of a polynomial.

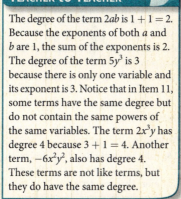

TEACHER to TEACHER

The degree of the term $2ab$ is $1 + 1 = 2$. Because the exponents of both a and b are 1, the sum of the exponents is 2. The degree of the term $5y^3$ is 3 because there is only one variable and its exponent is 3. Notice that in Item 11, some terms have the same degree but do not contain the same powers of the same variables. The term $2x^3y$ has degree 4 because $3 + 1 = 4$. Another term, $-6x^2y^2$, also has degree 4. These terms are not like terms, but they do have the same degree.

Activity 24 • Adding and Subtracting Polynomials

ACTIVITY 24 Continued

12–13 (continued) The table in Item 13 will allow for further formative assessment of students' understanding of the vocabulary presented in the unit. Debrief this portion of the lesson by allowing students to build and share their own polynomials given a set of characteristics. For example, a binomial with degree 4 and constant term 2. A sample student response could be $3x^2y^2 - 2$.

Developing Math Language
In this section of the activity, students are introduced to numerous vocabulary terms. Be sure that students, especially those who struggle with language, understand all of the vocabulary related to polynomials. You may wish to have students create posters, booklets, or other ways of demonstrating that they understand polynomial vocabulary.

Check Your Understanding
Students' answers to these items should reflect a basic understanding of polynomials. As you debrief, ask students to give examples of monomials, binomials and trinomials. Students can then practice writing their examples in standard form, identifying the terms, determining the degree of each term and of the polynomial, and identifying the leading coefficient.

Answers
14. False. All binomials are polynomials, but some polynomials may have more than two terms.
15. Answers may vary. Identify the term whose variable has the greatest exponent.

ASSESS
Students' answers to lesson practice problems will provide you with a formative assessment of their understanding of the lesson concepts and their ability to apply their learning.

See the Activity Practice for additional problems for this lesson. You may assign the problems here or use them as a culmination for the activity.

ADAPT
For students who have difficulty finding the degree of a term, suggest that they rewrite variable terms that have no apparent exponent as terms whose exponent is 1. For example, they can rewrite $4x$ as $4x^1$. This is often particularly helpful for terms that have more than one variable; rewriting $2x^2y$ as $2x^2y^1$ helps students more easily see that the degree is $2 + 1 = 3$.

ACTIVITY 24 continued

My Notes

Lesson 24-1
Polynomial Terminology

13. Fill in the missing information in the table below.

Polynomial	Number of Terms	Name	Leading Coefficient	Constant Term	Degree
$3x^2 - 5x$	2	Binomial	3	0	2
$-2x^2 + 13x + 6$	3	Trinomial	-2	6	2
$15x^2$	1	Monomial	15	0	2
$5p^3 + 2p^2 - p - 7$	4	Polynomial	5	-7	3
$a^2 - 25$	2	Binomial	1	-25	2
$0.23x^3 + 0.54x^2 - 0.58x + 0.0218$	4	Polynomial	0.23	0.0218	3
$-9.8t^2 - 20t + 150$	3	Trinomial	-9.8	150	2

Check Your Understanding

14. Is the following statement true or false? Explain.
"All polynomials are binomials."

15. Describe your first step for writing $3x - 5x^2 + 7$ in standard form.

LESSON 24-1 PRACTICE

For Items 16–20, use the polynomial $4x^3 + 3x^2 - 9x + 7$.

16. Name the coefficients of the terms in the polynomial that have variables.
17. List the terms, and give the degree of each term.
18. What is the degree of the polynomial?
19. Identify the leading coefficient of the polynomial.
20. Identify the constant term of the polynomial.

Write each polynomial in standard form.

21. $9 + 8x^2 + 2x^3$ **22.** $y^2 + 1 + 4y^3 - 2x$

23. Construct viable arguments. Is the expression $5x^2 + \sqrt{2}x$ a polynomial? Justify your response.

LESSON 24-1 PRACTICE
16. 4, 3, -9
17. $4x^3$, 3; $3x^2$, 2; $-9x$, 1; 7, 0
18. 3
19. 4
20. 7
21. $2x^3 + 8x^2 + 9$
22. $4y^3 + y^2 - 2x + 1$
23. Yes. Only the variables are restricted to having whole-number exponents, which would also mean they could not be part of a square root. But a coefficient does not have this restriction. So, $\sqrt{2}x$ can be a term of a polynomial.

Lesson 24-2
Adding Polynomials

Learning Targets:
- Use algebra tiles to add polynomials.
- Add polynomials algebraically.

SUGGESTED LEARNING STRATEGIES: Discussion Groups, Use Manipulatives, Create Representations, Close Reading, Note Taking

Notice that in the solar panels at the right, there are 4^2 or 16 cells. Each column has 4 cells.

1. If a square solar panel with an unknown number of cells along the edge can be represented by x^2, how many cells would be in one column of the panel?
   x

A square solar panel with x rows and x columns can be represented by the algebra tile:

A column of x cells can be represented by using the tile ▯, and a single solar cell can be represented by +1.

Suppose there were 3 square solar panels that each had x columns and x rows, 2 columns with x cells, and 3 single solar cells. You can represent $3x^2 + 2x + 3$ using algebra tiles.

2. Represent $2x^2 - 3x + 2$ using algebra tiles. Draw a picture of the representation below.
   Students' pictures should show two x^2 tiles, three $-x$ tiles, and two $+1$ tiles.

MATH TIP

The additive inverse of the x^2, x, and 1 algebra tiles can be represented with another color, or the flip side of the tile.

Lesson 24-2
Adding Polynomials

Adding polynomials using algebra tiles can be done by:
- modeling each polynomial
- identifying and removing zero pairs
- writing the new polynomial

Example A
Add $(3x^2 - 3x - 5) + (2x^2 + 5x + 3)$ using algebra tiles.

Step 1: Model the polynomials.
$3x^2 - 3x - 5$ $2x^2 + 5x + 3$

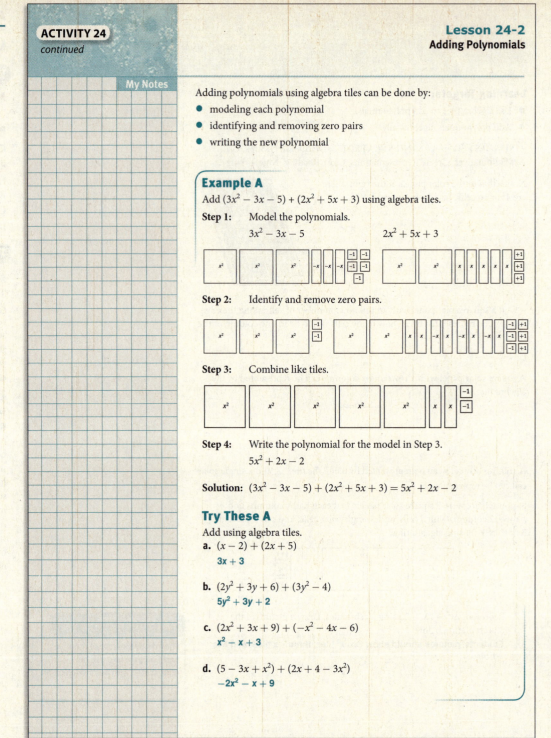

Step 2: Identify and remove zero pairs.

Step 3: Combine like tiles.

Step 4: Write the polynomial for the model in Step 3.
$5x^2 + 2x - 2$

Solution: $(3x^2 - 3x - 5) + (2x^2 + 5x + 3) = 5x^2 + 2x - 2$

Try These A
Add using algebra tiles.
a. $(x - 2) + (2x + 5)$
   $3x + 3$

b. $(2y^2 + 3y + 6) + (3y^2 - 4)$
   $5y^2 + 3y + 2$

c. $(2x^2 + 3x + 9) + (-x^2 - 4x - 6)$
   $x^2 - x + 3$

d. $(5 - 3x + x^2) + (2x + 4 - 3x^2)$
   $-2x^2 - x + 9$

Lesson 24-2
Adding Polynomials

ACTIVITY 24
continued

3. **Use appropriate tools strategically.** Can you use algebra tiles to add $(4x^4 + 3x^2 + 15) + (x^4 + 10x^3 - 4x^2 + 22x - 23)$? If so, model the polynomials and add. If not, explain why.
   Answers may vary. No, there are no tiles to represent x^4 or x^3.

Like terms in an expression are terms that have the same variable and exponent for that variable. All constants are like terms.

4. State whether the terms are like or unlike terms. Explain.
   Answers may vary.

   a. $2x$; $2x^3$
   Unlike terms; the exponents of the variables are different.

   b. 5; $5x$
   Unlike terms; one term is a constant. The other term has a variable.

   c. $-3y$; $3y$
   Like terms; both terms have the same variable and exponent.

   d. x^2y; xy^2
   Unlike terms; the exponents for the same variables are different.

   e. 14; -0.6
   Like terms; both terms are constants.

5. **Attend to precision.** Using vocabulary from this activity, describe a method that could be used to add polynomials without using algebra tiles.
   Answers may vary. One possible method would be to add the coefficients of like terms.

ACTIVITY 24 Continued

3 Sharing and Responding Algebra tiles are powerful models for polynomials. But because there are at most two dimensions, they can only be used to model polynomials with degree 2 or less. Be sure students understand the limitations of using algebra tiles.

4–5 Interactive Word Wall, Vocabulary Organizer, Discussion Groups, Debriefing It is important for students to understand that like terms must have the same variables with like exponents. Monitor student discussions carefully to ensure students are making this connection. Debrief this portion of the lesson by asking students to describe how like terms are represented with algebra tiles.

Developing Math Language
The concept of *like terms* is important when students add and subtract polynomials. Students should understand that while the variable(s) and exponent(s) of like terms must be the same, the coefficients can be different. Add *like terms* to the Word Wall and have students write it in their math notebooks along with examples and nonexamples.

TEACHER to TEACHER

Some students mistakenly add the exponents as well as the coefficients when adding like terms. Emphasize that when adding two like terms, the exponent does not change. This is usually easy for students to see when using algebra tiles. If students make this mistake later on, refer back to the algebra tile models.

ACTIVITY 24 Continued

Example B Close Reading, Note Taking To help students maintain the proper alignment in the vertical method, model the use of $0x^3$, $0x^2$, $0x$ or 0 when necessary for "missing" terms. When using the horizontal method, remind students that they can remove parentheses and reorder the terms because of the Associative and Commutative Properties of Addition. When students are rearranging terms, make sure they view a subtraction sign as part of a term and keep it with the term in its new position. Use the Try These exercises to assess students' understanding and determine the need for extra support.

Differentiating Instruction

Support students who have difficulty identifying like terms to combine. For example, ask students to circle all terms of degree 3, draw squares around all terms of degree 2, and draw triangles around all terms of degree 1. Then students can combine those terms that are in the same geometric shape.

6–7 Think-Pair-Share, Construct and Argument, Debriefing Students are asked to reflect on the definition of a polynomial and the process of adding polynomials. Students also extend their understanding of closure to the set of polynomials. As this portion of the lesson is debriefed, be sure students understand that when adding polynomials, only the coefficients and constants change and that the sum of any two polynomials will always be a polynomial.

ACTIVITY 24 continued

My Notes

MATH TIP
The Commutative and Associative Properties of Addition allow you to re-order and group like terms.

MATH TIP
Polynomials are *closed* under addition. A set is closed under addition if the sum of any two elements in the set is also an element of the set.

Lesson 24-2
Adding Polynomials

Use the properties of real numbers to add polynomials algebraically.

Example B
Add $(3x^3 + 2x^2 - 5x + 7) + (4x^3 + 2x - 3)$ horizontally and vertically. Write your answer in standard form.

Horizontally

Step 1: Identify like terms. $(3x^3 + 2x^2 - 5x + 7) + (4x^3 + 2x - 3)$

Step 2: Group like terms. $(3x^3 + 4x^3) + (2x^2) + (-5x + 2x) + (7 - 3)$

Step 3: Add the coefficients of like terms. $\quad 7x^3 + 2x^2 - 3x + 4$

Solution: $(3x^3 + 2x^2 - 5x + 7) + (4x^3 + 2x - 3) = 7x^3 + 2x^2 - 3x + 4$

Vertically

Step 1: Vertically align like terms. $\quad\quad 3x^3 + 2x^2 - 5x + 7$
Step 2: Add the coefficients of like terms. $\quad\quad \underline{+4x^3 \quad\quad\;\; + 2x - 3}$
$\quad\quad\quad\quad\quad\quad\quad\quad\quad\quad\quad\quad 7x^3 + 2x^2 - 3x + 4$

Solution: $(3x^3 + 2x^2 - 5x + 7) + (4x^2 + 2x - 3) = 7x^3 + 2x^2 - 3x + 4$

Try These B
Add. Write your answers in standard form.

a. $(4x^2 + 3) + (x^2 - 3x + 5)$ b. $(10y^2 + 8y + 6) + (17y^2 - 11)$
$\quad 5x^2 - 3x + 8$ $27y^2 + 8y - 5$

c. $(9x^2 + 15x + 21) + (-13x^2 - 11x - 26)$
$\quad -4x^2 + 4x - 5$

6. Are the answers to Try These B polynomials? Justify your response.
   Yes; the expressions are sums of terms with whole-number powers.

7. Explain why the sum of two polynomials will always be a polynomial.
   Answers may vary. The terms in a polynomial must have whole-number powers. When you add two polynomials, you are adding like terms with whole-number powers. Adding terms does not affect their exponents, so the sum will also have whole-number powers. Therefore, the sum is a polynomial.

Lesson 24-2
Adding Polynomials

Check Your Understanding

Write a polynomial for each expression represented by algebra tiles.

8.

9.

10. Add the expressions you wrote for Items 8 and 9.
11. What property or properties justify Steps 1 and 2 below?

$$(2x^2 + x + 1) + (x^2 + 2x - 1)$$

Step 1: $(2x^2 + x^2) + (x + 2x) + (1 - 1)$

Step 2: $(2 - 1)x^2 + (1 + 2)x + (1 - 1)$

Step 3: $x^2 + 3x$

LESSON 24-2 PRACTICE

Add. Write your answers in standard form.

12. $(3x^2 + x + 5) + (2x^2 + x - 5)$
13. $(-4x^2 + 2x - 1) + (x^2 - x + 9)$
14. $(7x^2 - 2x + 3) + (3x^2 + 2x + 7)$
15. $(-x^2 + 5x + 2) + (-3x^2 + x - 9)$

Write the perimeter of each figure as a polynomial in standard form.

16.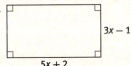

17.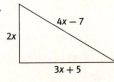

18. **Critique the reasoning of others.** A student added the expressions $x^4 + 5x^2 - 2x + 1$ and $2x^4 + x^3 + 2x - 7$. Identify and correct the student's error.

$$\begin{array}{r} x^4 + 5x^2 - 2x + 1 \\ 2x^4 + x^3 + 2x - 7 \\ \hline 3x^4 + 6x^2 - 6 \end{array}$$

ACTIVITY 24 Continued

Lesson 24-3

PLAN

Pacing: 1 class period
Chunking the Lesson
Example A   #1–2
Check Your Understanding
Lesson Practice

TEACH

Bell-Ringer Activity
Write $2 + 3$, $2 + (-3)$, $2 - 3$, and $2 - (-3)$ on the board. After students simplify each expression, have them explain how $2 + 3$ and $2 - (-3)$ are related and how $2 + (-3)$ and $2 - 3$ are related.

Example A Note Taking, Close Reading Make the connection back to the process of adding horizontally and vertically. Show students that the process is the same with one change: when you clear the parentheses, you represent the polynomial that is being subtracted as an additive inverse, and then you add. Have students brainstorm ways they can show this step that help them perform it accurately.

Developing Math Language
Ask students to describe the mathematical meaning of *opposite* in their own words. Review the concept of additive inverse and relate it to subtracting polynomials. Have students write and define the word *opposite* in their math notebooks, and add the word to the Word Wall.

TEACHER to TEACHER

Students can use addition to check their work. If the difference is correct, the sum of the difference and the subtrahend will equal the minuend.

ACTIVITY 24 continued

Lesson 24-3
Subtracting Polynomials

My Notes

Learning Target:
- Subtract polynomials algebraically.

SUGGESTED LEARNING STRATEGIES: Note Taking, Close Reading, Think-Pair-Share

To subtract a polynomial you add its *opposite*, or subtract each of its terms.

MATH TERMS

The **opposite** of a number or a polynomial is its additive inverse.

Example A

Subtract $(2x^3 + 8x^2 + x + 10) - (5x^2 - 4x + 6)$ horizontally and vertically. Write the answer in standard form.

Horizontally

Step 1: Distribute the negative. $(2x^3 + 8x^2 + x + 10) - (5x^2 - 4x + 6)$

Step 2: Identify like terms. $2x^3 + 8x^2 + x + 10 - 5x^2 + 4x - 6$

Step 3: Group like terms. $2x^3 + (8x^2 - 5x^2) + (x + 4x) + (10 - 6)$

Step 4: Combine coefficients of like terms. $2x^3 + 3x^2 + 5x + 4$

Solution: $(2x^3 + 8x^2 + x + 10) - (5x^2 - 4x + 6) = 2x^3 + 3x^2 + 5x + 4$

Vertically

Step 1: Vertically align like terms.
$$2x^3 + 8x^2 + x + 10$$
$$-(5x^2 - 4x + 6)$$

Step 2: Distribute the negative.
$$2x^3 + 8x^2 + x + 10$$
$$-5x^2 + 4x - 6$$

Step 3: Combine coefficients of like terms. $2x^3 + 3x^2 + 5x + 4$

Solution: $(2x^3 + 8x^2 + x + 10) - (5x^2 - 4x + 6) = 2x^3 + 3x^2 + 5x + 4$

MINI-LESSON: Subtracting Polynomials Using Algebra Tiles

Students may need a concrete representation of polynomial subtraction before attempting to subtract algebraically. If so, a mini-lesson on subtracting polynomials using algebra tiles is available.

See SpringBoard's eBook Teacher Resources for a student page for this mini-lesson.

Lesson 24-3
Subtracting Polynomials

Try These A
Subtract. Write your answers in standard form.

a. $(5x - 5) - (x + 7)$
   $4x - 12$

b. $(2x^2 + 3x + 2) - (-5x^2 - 2x - 9)$
   $7x^2 + 5x + 11$

c. $(y^2 + 3y + 8) - (4y^2 - 9)$
   $-3y^2 + 3y + 17$

d. $(12 + 5x + 14x^2) - (8x + 15 - 7x^2)$
   $21x^2 - 3x - 3$

1. Are the answers to Try These A polynomials? Justify your response.
   Yes; the expressions are sums of terms with whole-number powers.

2. Explain why the difference of two polynomials will always be a polynomial.
   Answers may vary. The terms in a polynomial must have whole-number powers. When you subtract two polynomials, you are subtracting terms with whole-number powers. Subtracting terms does not affect their exponents, so the difference will also have whole-number powers. Therefore, the difference is a polynomial.

MATH TIP

Polynomials are *closed* under subtraction. A set is closed under subtraction if the difference of any two elements in the set is also an element of the set.

ACTIVITY 24 Continued

Check Your Understanding

As you debrief these items, ask students to describe the steps they use to identify and combine like terms.

Answers
3. $(x^2 + 2x + 3) + (-4x^2 + x - 5)$
4. $(5y^2 + y - 2) + (y^2 + 3y - 4)$
5. Gil did not distribute the negative sign to each term in the second polynomial.

ASSESS

Students' answers to lesson practice problems will provide you with a formative assessment of their understanding of the lesson concepts and their ability to apply their learning.

See the Activity Practice for additional problems for this lesson. You may assign the problems here or use them as a culmination for the activity.

LESSON 24-3 PRACTICE
6. $-5x^2 + 7x + 5$
7. $x - 1$
8. $-5x^2 + 8x - 10$
9. $x^2 - 2x$
10. $2y^4 + y^2 + 2y - 3$
11. Answers may vary.
    $(8x + 5) - (2x + 2)$
12. $x^2 + 2x + 1$

ADAPT

If there are students who continue to have difficulty identifying and combining like terms algebraically, continue to allow them to use algebra tiles and/or drawings for as long as they need a concrete representation of the process.

Lesson 24-3
Subtracting Polynomials

Check Your Understanding

Rewrite each difference as addition of the opposite, or additive inverse, of the second polynomial.

3. $(x^2 + 2x + 3) - (4x^2 - x + 5)$
4. $(5y^2 + y - 2) - (-y^2 - 3y + 4)$
5. **Critique the reasoning of others.** Gil used the vertical method to subtract $(3x^2 - 5x + 2) - (x^2 + 2x + 4)$ as shown below. Identify Gil's error.

$$\begin{array}{r} 3x^2 - 5x + 2 \\ -x^2 + 2x + 4 \\ \hline 2x^2 - 3x + 6 \end{array}$$

LESSON 24-3 PRACTICE

Subtract. Write your answers in standard form.

6. $(2x^2 + 4x + 1) - (7x^2 - 3x - 4)$
7. $(x^2 + 3x - 9) - (x^2 + 2x - 8)$
8. $(9x^2 + x - 12) - (14x^2 - 7x - 2)$
9. $(x^2 + 3x - 6) - (5x - 6)$
10. $(y^4 + y^2 + 2y) - (-y^4 + 3)$
11. Write two polynomials whose difference is $6x + 3$.
12. **Model with mathematics.** A rectangular piece of paper has area $4x^2 + 3x + 2$. A square is cut from the rectangle and the remainder of the rectangle is discarded. The area of the discarded paper is $3x^2 + x + 1$. What is the area of the square?

Adding and Subtracting Polynomials
Polynomials in the Sun

ACTIVITY 24
continued

ACTIVITY 24 PRACTICE
Write your answers on notebook paper.
Show your work.

Lesson 24-1

For Items 1–5, use the polynomial
$5x^4 - 2x^2 + 8x - 3$.

1. Identify the coefficients of the variable terms of the polynomial.
2. List the terms, and give the degree of each term.
3. State the degree of the polynomial.
4. Identify the leading coefficient of the polynomial.
5. Identify the constant term of the polynomial.
6. Consider the expressions $3x^2 + 2x - 7$ and $3x^2 + 2x - 7x^0$. Are the expressions equivalent? Explain.

Write each polynomial in standard form.

7. $5x^2 - 2x^3 - 10$
8. $11 + y^2 - 8y$
9. $y^2 - 12 - y^3 + y^4$
10. $9x + 7 - 5x^3$

Lesson 24-2

Add. Write your answers in standard form.

11. $(4x + 9) + (3x - 5)$
12. $(2x^2 + 3x - 1) + (x^2 - 5x + 2)$
13. $(x^3 + 5x^2 + 3) + (2x^3 - 5x^2)$
14. $(7y^3 - 2y^2 + 5) + (4y^3 - 3y)$

15. Which expression represents the perimeter of the trapezoid?

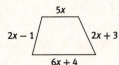

A. $11x + 4$
B. $4x + 2$
C. $15x + 6$
D. $13x + 8$

16. The length of each side of a square is $4y + 5$. Draw and label the square, and write an expression to represent its perimeter.

Lesson 24-3

17. Which expression is equivalent to $10x - (7x - 1)$?
A. $3x - 1$
B. $3x + 1$
C. $17x - 1$
D. $17x + 1$

Subtract. Write your answers in standard form.

18. $(5x - 4) - (3x + 2)$
19. $(3x^2 - 2x + 7) - (2x^2 + 2x - 7)$
20. $(8y^2 - 3y + 6) - (-2y^2 - 3)$
21. $(x^2 - 5x) - (4x - 6)$

ACTIVITY 24 Continued

ACTIVITY PRACTICE
1. $5, -2, 8$
2. $5x^4, 4; -2x^2, 2; 8x, 1; -3, 0$
3. 4
4. 5
5. -3
6. Yes; since $x^0 = 1$, the second expression can be simplified to $3x^2 + 2x - 7 \cdot 1$, or $3x^2 + 2x - 7$.
7. $-2x^3 + 5x^2 - 10$
8. $y^2 - 8y + 11$
9. $y^4 - y^3 + y^2 - 12$
10. $-5x^3 + 9x + 7$
11. $7x + 4$
12. $3x^2 - 2x + 1$
13. $3x^3 + 3$
14. $11y^3 - 2y^2 - 3y + 5$
15. C
16. $16y + 20$
17. B
18. $2x - 6$
19. $x^2 - 4x + 14$
20. $10y^2 - 3y + 9$
21. $x^2 - 9x + 6$

ACTIVITY 24 Continued

22. $5y^2 + 10y + 5$
23. $x^2 + 7$
24. $-4x^3 - 4x^2 + 12$
25. $12x^3 + 3x^2 - 8x + 3$
26. $-7y^3 - 5y^2 - y - 4$
27. $-x + 10y$
28. $6x^2 + 4xy + 6y^2$
29. $6x^2 - 8x + 3$
30. $36x^2 + 12x + 44$
31. **a.** Yes; the sum of two integers is an integer, and the difference of two integers is an integer.
    b. Answers will vary. $5 - 8 = -3$ shows that the whole numbers are not closed under subtraction, because the difference of these two whole numbers is not a whole number.

ADDITIONAL PRACTICE

If students need more practice on the concepts in this activity, see the eBook Teacher Resources for additional practice problems.

ACTIVITY 24 continued

Adding and Subtracting Polynomials
Polynomials in the Sun

Determine the sum or difference. Write your answers in standard form.

22. $(5y^2 + 3y + 7) + (7y - 2)$
23. $(3x^2 + x + 9) - (2x^2 + x + 2)$
24. $(x^3 + 3x^2 + 12) - (5x^3 + 7x^2)$
25. $(8x^3 - 5x + 7) + (4x^3 + 3x^2 - 3x - 4)$
26. $(-4y^2 - 2y + 1) - (7y^3 + y^2 - y + 5)$
27. $(3x + 7y) + (-4x + 3y)$
28. $(5x^2 + 8xy + y^2) - (-x^2 + 4xy - 5y^2)$
29. A playground has a sidewalk border around a play area.

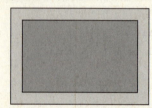

The total area of the playground, the larger rectangle, is $16x^2 - 5x + 2$. The area of the play area, the smaller rectangle, is $10x^2 + 3x - 1$. Write an expression to represent the area of the sidewalk.

30. To make a box, four corners of a rectangular piece of cardboard are cut out and the box is folded and taped.

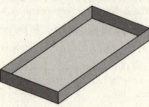

The area of the cardboard, after the corners are cut out, is $28x^2 + 12x + 32$. The area of each cut-out corner is $2x^2 + 3$. Write an expression to represent the area of the original piece of cardboard.

MATHEMATICAL PRACTICES
Reason Abstractly and Quantitatively

31. The set of polynomials is closed under the operations of addition and subtraction. This means that when you add or subtract two polynomials, the result is also a polynomial.
    a. Are the integers closed under addition and subtraction? In other words, when you add or subtract two integers, is the result always an integer? Justify your response.
    b. Give a counterexample to show that the whole numbers are **not** closed under subtraction.

Multiplying Polynomials
Tri-Com Computers
Lesson 25-1 Multiplying Binomials

ACTIVITY 25

Learning Targets:
- Use a graphic organizer to multiply expressions.
- Use the Distributive Property to multiply expressions.

SUGGESTED LEARNING STRATEGIES: Think-Pair-Share, Look for a Pattern, Discussion Groups, Create Representations, Graphic Organizer

Tri-Com Computers is a company that sets up local area networks in offices. In setting up a network, consultants need to consider not only where to place computers, but also where to place peripheral equipment, such as printers.

Tri-Com typically sets up local area networks of computers and printers in square or rectangular offices. Printers are placed in each corner of the room. The primary printer A serves the first 25 computers and the other three printers, B, C, and D, are assigned to other regions in the room. Below is an example.

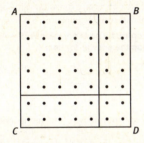

1. If each dot represents a computer, how many computers in this room will be assigned to each of the printers?
    Printer A serves 25 computers.
    Printer B serves 10 computers.
    Printer C serves 10 computers.
    Printer D serves 4 computers.

2. What is the total number of computers in the room? Describe two ways to find the total.
    49 computers. Answers may vary. You can add the number of computers assigned to each printer, 25 + 10 + 10 + 4 = 49, or you can multiply the number of rows in the diagram times the number of columns, 7 • 7 = 49.

Common Core State Standards for Activity 25

HSA-SSE.A.1	Interpret expressions that represent a quantity in terms of its context.
HSA-SSE.A.1a	Interpret parts of an expression, such as terms, factors, and coefficients.
HSA-SSE.A.1b	Interpret complicated parts of expressions by viewing one or more of their parts as a single entity.
HSA-APR.A.1	Understand that polynomials form a system analogous to the integers, namely, they are closed under the operations of addition, subtraction, and multiplication; add, subtract, and multiply polynomials.

ACTIVITY 25
Investigative

Activity Standards Focus
In Activity 25, students explore multiplying polynomials. They also examine the patterns exhibited by the special products of the difference of two squares and the square of a binomial.

Lesson 25-1

PLAN

Pacing: 1 class period
Chunking the Lesson
#1–2 #3–4 #5–6 #7–8
#9   #10   #11–12 #13–15
#16–17 #18–19

Check Your Understanding
Lesson Practice

TEACH

Bell-Ringer Activity
Ask students to respond to these items:
a. Show that 4(10 − 6) = (4 • 10) + (4 • 6).
b. Write an explanation of the distributive property. Include diagrams or examples.

1–2 Shared Reading, Think-Pair-Share Be sure students explain the scenario in their own words before attempting to answer the questions. The questions are designed to familiarize students with the situation and help them visualize how the computer networks are configured.

ACTIVITY 25 Continued

3–4 Visualization, Look for a Pattern, Think-Pair-Share These items are designed to encourage students to begin thinking about the regions of the square as they are related to the area of the square in order to make connections to the multiplication of binomials. Students may recognize that (5 + 4)(5 + 4) represents the total number of computers. This is because the number of computers along the length of the wall can be multiplied by the number of computers along the width of the wall. They should also determine that the total number of computers can be determined by multiplying to find the number of computers in each region, and then adding.

5–6 Discussion Groups, Create Representations This may be the first time students are using the distributive property outside of the context of a monomial times a binomial. Monitor group discussions carefully to be sure students are making the connections to the multiplication of binomials.

TEACHER to TEACHER

When students find the number of computers using the two methods, they are exploring the use of the distributive property.
(5 + 3)(5 + 3) =
5(5 + 3) + 3(5 + 3) =
(5 · 5) + (5 · 3) + (3 · 5) + (3 · 3) =
25 + 15 + 15 + 9 = 64
Students should refer to the distributive property as they work through this activity.

ACTIVITY 25 continued

**Lesson 25-1
Multiplying Binomials**

Another example of an office in which Tri-Com installed a network had 9 computers along each wall. The computers are aligned in an array with the number of computers in each region determined by the number of computers along the wall.

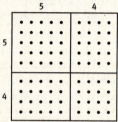

3. A technician claimed that since 9 = 5 + 4, the number of computers in the office could be written as an expression using only the numbers 5 and 4. Is the technician correct? Explain.

   The technician is correct. The expression is (5 + 4)(5 + 4), or the expression given as the answer for Item 4.

4. Show another way to determine the total number of computers in the office.

   5 · 5 + 5 · 4 + 4 · 5 + 4 · 4

5. Rewrite the expression (5 + 4)(5 + 4) using the Distributive Property.
   (5 + 4)(5 + 4) = (5)(5 + 4) + (4)(5 + 4)
   = (5)(5) + (5)(4) + (4)(5) + (4)(4)

6. **Make sense of problems.** Explain why (5 + 4)(5 + 4) could be used to determine the total number of computers.

   The four products in the sum above represent the numbers of computers in the four areas of the room.

Lesson 25-1
Multiplying Binomials

7. The office to the right has 8^2 computers. Fill in the number of computers in each section if it is split into a $(5 + 3)^2$ configuration.

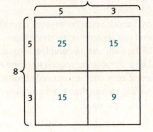

8. What is the total number of computers? Describe two ways to find the total.
   64 computers. Answers may vary. One way to find the total is to multiply the number of computers along the length of the wall times the number of computers along the width of the wall, $(5 + 3)(5 + 3) = 25 + 15 + 15 + 9 = 64$. Another way is to multiply the number of rows in the diagram by the number of columns, $8 \cdot 8 = 64$.

9. For each possible office configuration below, draw a diagram like the one next to Item 7. Label the number of computers on the edge of each section and determine the total number of computers in the room by adding the number of computers in each section.

   a. $(2 + 3)^2$

   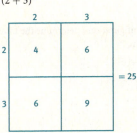
   $= 25$

   b. $(4 + 1)^2$

   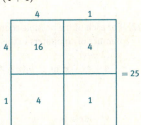
   $= 25$

   c. $(3 + 7)^2$

   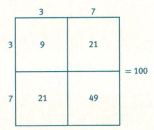
   $= 100$

ACTIVITY 25 Continued

10 Quickwrite, Self Revision/Peer Revision It is important for students to understand from the diagram and the patterns they have observed that $5^2 + x^2$ is not a correct representation of the diagram. Encourage students who are having difficulty to again write the products representing the number of computers in each region of the graphic organizer.

11–12 Look for a Pattern, Graphic Organizer, Discussion Groups, Create Representations There are different possibilities for students' responses at this point. All should be encouraged. Some possible responses could be $(5 + x)^2$ or $5^2 + 5x + 5x + x^2$. Students may also approach these problems differently. Some will continue to use the graphic organizer to make sure they include all four partial products. Others will be able to apply the patterns they have observed or use the distributive property to find the products. As students share their strategies in their groups, encourage them to make connections among the various processes used.

> **Differentiating Instruction**
>
> Algebra tiles may be used as a tool to help kinesthetic learners visualize the multiplication more easily.

ACTIVITY 25 continued

Lesson 25-1 Multiplying Binomials

Tri-Com has a minimum requirement of 25 computers per installation arranged in a 5 by 5 array. Some rooms are larger than others and can accommodate more than 5 computers along each wall to complete a square array. Use a variable expression to represent the total number of computers needed for any office having x more than the 5 computer minimum along each wall.

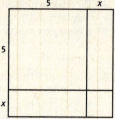

10. One technician said that $5^2 + x^2$ would be the correct way to represent the total number of computers in the office space. Use the diagram to explain how the statement is incorrect.

    Answers may vary. The expression $5^2 + x^2$ only represents the number of computers in the square spaces in the office. The computers in the rectangular spaces are not taken into account.

11. **Model with mathematics.** Write an expression for the sum of the number of computers in each region in Item 10.

    $25 + 10x + x^2$ or $x^2 + 10x + 25$

12. For each of the possible room configurations, determine the total number of computers in the room.
    a. $(2 + x)^2$
       $4 + 4x + x^2$ or $x^2 + 4x + 4$

    b. $(x + 3)^2$
       $x^2 + 6x + 9$

    c. $(x + 6)^2$
       $x^2 + 12x + 36$

Lesson 25-1
Multiplying Binomials

The graphic organizer below can be used to help arrange the multiplications of the Distributive Property. It does not need to be related to the number of computers in an office. For example, this graphic organizer shows $5 \cdot 7 = (3 + 2)(4 + 3)$.

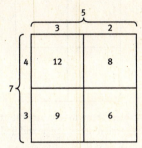

13. Draw a graphic organizer to represent the expression $(5 + 2)(2 + 3)$. Label each inner rectangle and find the sum.

	5	2
2	10	4
3	15	6

= 35

14. Draw a graphic organizer to represent the expression $(6 - 3)(4 - 2)$. Label each inner rectangle and find the sum.

	6	−3
4	24	−12
−2	−12	6

= 6

15. Multiply the binomials in Item 14 using the Distributive Property. What do you notice?

    Answers may vary. The product is 6. Multiplying binomials using the Distributive Property results in the same answer as multiplying using the graphic organizer.

Lesson 25-1
Multiplying Binomials

You can use the same graphic organizer to multiply binomials that contain variables. The following diagram represents $(x - 2)(x - 3)$.

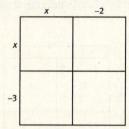

16. Use the graphic organizer above to represent the expression $(x - 2)(x - 3)$. Label each inner rectangle and find the sum.

$x^2 + (-2x) + (-3x) + 6 = x^2 - 5x + 6$

17. Multiply the binomials in Item 16 using the Distributive Property. What do you notice?

    $(x - 2)(x - 3) = (x \cdot x) + (x \cdot (-3)) + ((-2) \cdot x) + ((-2) \cdot (-3)) = x^2 - 3x - 2x + 6 = x^2 - 5x + 6$. The product is the same as the one in Item 16.

18. Determine the product of the binomials.

    a. $(x - 7)(x - 5)$
       $x^2 - 12x + 35$

    b. $(x - 7)(x + 5)$
       $x^2 - 2x - 35$

    c. $(x + 7)(x + 5)$
       $x^2 + 12x + 35$

    d. $(x + 7)(x - 5)$
       $x^2 + 2x - 35$

    e. $(4x + 1)(x + 3)$
       $4x^2 + 13x + 3$

    f. $(2x - 1)(3x + 2)$
       $6x^2 + x - 2$

19. **Reason abstractly.** Examine the products in Item 18. How can you predict the sign of the last term?

    If the signs of the last two terms of the factors are the same, the last term will be positive. If the signs of the last two terms of the factors are different, the last term will be negative.

Lesson 25-1
Multiplying Binomials

Check Your Understanding

20. Use a graphic organizer to calculate $(6 + 2)^2$. Explain why the product is not $6^2 + 2^2$.

Determine the product of the binomials using a graphic organizer or by using the Distributive Property.

21. $(x + 7)(x + 2)$
22. $(x + 7)(3x - 2)$
23. Compare the use of the graphic organizer and the use of the Distributive Property to find the product of two binomials.

LESSON 25-1 PRACTICE

Determine each product.

24. $(2 + 1)(3 + 5)$
25. $(2 + 3)(2 + 7)$
26. $(x + 9)(x + 3)$
27. $(x + 5)(x + 1)$
28. $(x - 3)(x + 4)$
29. $(x + 1)(x - 5)$
30. $(x + 3)(x - 3)$
31. $(x + 3)(x + 3)$
32. $(2x - 3)(x - 1)$
33. $(x + 7)(3x - 5)$
34. $(4x + 3)(2x + 1)$
35. $(6x - 2)(5x + 1)$

36. **Critique the reasoning of others.** A student determined the product $(x - 2)(x - 4)$. Identify and correct the student's error.

$$(x - 2)(x - 4)$$
$$x(x - 4) - 2(x - 4)$$
$$x^2 - 4x - 2x - 8$$
$$x^2 - 6x - 8$$

ACTIVITY 25 Continued

Lesson 25-2

PLAN

Pacing: 1 class period
Chunking the Lesson
#1–3   #4–6
Check Your Understanding
Lesson Practice

TEACH

Bell-Ringer Activity
Write $1^2 = 1$ and $2^2 = 4$ on the board. Ask students to continue the list, identifying as many perfect squares as time allows.

1–3 Look for a Pattern, Think-Pair-Share Students should notice that the factors in each problem are the sum and the difference of the same two terms. As they find the products, they should discover the pattern that $(a + b)(a - b) = a^2 - b^2$. As students share the patterns they have identified, ask them to explain why the product has only two terms.

Developing Math Language
Break down the phrase *difference of two squares* to help students remember its meaning. Once students define both *difference* and *squares*, have them relate the two definitions to a *difference of two squares*. Add the term to the classroom Word Wall and have students add it to their notebooks.

Differentiating Instruction

Extend students' learning by having them find products of numbers that can be written as $(a + b)(a - b)$ using the difference of two squares. Demonstrate how to multiply $38 \cdot 42$ as $(40 - 2)(40 + 2) = 1600 - 4 = 1596$. Have students use the method to find the product of $21 \cdot 19$, $27 \cdot 33$, and $195 \cdot 205$. Discuss why it would not be an efficient method for finding the product of $51 \cdot 48$.

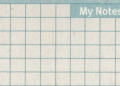

Lesson 25-2
Special Products of Binomials

Learning Targets:
- Multiply binomials.
- Find special products of binomials.

SUGGESTED LEARNING STRATEGIES: Think-Pair-Share, Look for a Pattern

1. Determine each product.

   a. $(x + 1)(x - 1)$
      $x^2 - 1$

   b. $(x + 4)(x - 4)$
      $x^2 - 16$

   c. $(x - 7)(x + 7)$
      $x^2 - 49$

   d. $(2x - 3)(2x + 3)$
      $4x^2 - 9$

2. Describe any patterns in the binomials and products in Item 1.
   Answers may vary. Each product has the form (variable term + number)(variable term − number). When this product is multiplied, the linear terms add to 0 and the answer is in the form (variable term)^2 − (number)^2.

3. **Express regularity in repeated reasoning.** The product of binomials of the form $(a + b)(a - b)$, has a special pattern called a **difference of two squares**. Use the patterns you found in Items 1 and 2 to explain how to find the product $(a + b)(a - b)$.
   Answers may vary. $(a + b)(a - b)$ equals the square of the first term minus the square of the second term: $a^2 - b^2$.

MATH TERMS

A binomial of the form $a^2 - b^2$ is known as the **difference of two squares**.

DISCUSSION GROUP TIP

As you read and define new terms, discuss their meanings with other group members and make connections to prior learning.

376 SpringBoard® Mathematics **Algebra 1, Unit 4** • Exponents, Radicals, and Polynomials

Lesson 25-2
Special Products of Binomials

4. Determine each product.

   a. $(x + 1)^2$
   $x^2 + 2x + 1$

   b. $(4 + y)^2$
   $16 + 8y + y^2$ or $y^2 + 8y + 16$

   c. $(x + 7)^2$
   $x^2 + 14x + 49$

   d. $(2y + 3)^2$
   $4y^2 + 12y + 9$

   e. $(x - 5)^2$
   $x^2 - 10x + 25$

   f. $(4 - x)^2$
   $16 - 8x + x^2$ or $x^2 - 8x + 16$

   g. $(y - 7)^2$
   $y^2 - 14y + 49$

   h. $(2x - 3)^2$
   $4x^2 - 12x + 9$

5. Describe any patterns in the binomials and products in Item 4.

   Answers may vary. Each product has the form (variable term + number)^2 or (variable term − number)^2. When this product is multiplied, the answer is in the form (variable term)^2 + 2 • variable term • number + (number)^2 or (variable term)^2 − 2 • variable term • number + (number)^2.

6. **Reason abstractly.** The **square of a binomial**, $(a + b)^2$ or $(a - b)^2$, also has a special pattern. Use the pattern you found in Items 4 and 5 to explain how to determine the square of any binomial.

   To find the square of a binomial, square the first term, add or subtract twice the second term, and add the square of the second term. The square of the binomial $(a + b)^2$ is $a^2 + 2ab + b^2$ and $(a - b)^2$ is $a^2 - 2ab + b^2$.

MATH TERMS

A binomial of the form $(a + b)^2$ or $(a - b)^2$ is known as the **square of a binomial.**

ACTIVITY 25 Continued

Check Your Understanding
When debriefing this activity, ask why it is efficient to apply the patterns. Then discuss whether using a graphic organizer or the distributive property would also result in the correct answer.

Answers
7. $p^2 - k^2$
8. $p^2 + 2pk + k^2$
9. Since the signs are different, the only formula to consider is the difference of squares. Since the last terms in the two factors are not the same, that formula does not apply.

ASSESS

Students' answers to Lesson Practice problems will provide you with a formative assessment of their understanding of the lesson concepts and their ability to apply their learning.

See the Activity Practice for additional problems for this lesson. You may assign the problems here or use them as a culmination for the activity.

LESSON 25-2 PRACTICE
10. $x^2 - 16$
11. $x^2 + 8x + 16$
12. $y^2 - 100$
13. $y^2 - 20y + 100$
14. $4x^2 - 12x + 9$
15. $4x^2 - 9$
16. $x^2 + 12x + 27$
17. $4y^2 - 4y + 1$
18. Sample answer: The first product will have three terms because it will be in the form $a^2 - 2ab + b^2$. The second product will only have two terms because it will be in the form $a^2 - b^2$.

ADAPT

If students continue to struggle to use the distributive property effectively to multiply binomials, encourage the use of the graphic organizer to support their work.

Lesson 25-2
Special Products of Binomials

Check Your Understanding

7. Use the difference of two squares pattern to find the product $(p + k)(p - k)$.
8. Use the square of a binomial pattern to determine $(p + k)^2$.
9. Can you use a special products pattern to determine $(x + 1)(x - 2)$? Explain your reasoning.

LESSON 25-2 PRACTICE
Determine each product.
10. $(x - 4)(x + 4)$
11. $(x + 4)^2$
12. $(y + 10)(y - 10)$
13. $(y - 10)^2$
14. $(2x - 3)^2$
15. $(2x - 3)(2x + 3)$
16. $(5x + 1)^2$
17. $(2y - 1)(2y - 1)$
18. **Construct viable arguments.** Explain why the products $(x - 3)^2$ and $(x + 3)(x - 3)$ have a different number of terms.

Lesson 25-3
Multiplying Polynomials

Learning Targets:
- Use a graphic organizer to multiply polynomials.
- Use the Distributive Property to multiply polynomials.

SUGGESTED LEARNING STRATEGIES: Graphic Organizer, Create Representations, Think-Pair-Share, Look for a Pattern

A graphic organizer can be used to multiply polynomials that have more than two terms, such as a binomial times a trinomial. The graphic organizer at right can be used to multiply $(x + 2)(x^2 + 2x + 3)$.

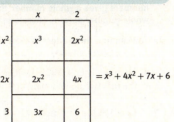

$= x^3 + 4x^2 + 7x + 6$

1. Draw a graphic organizer in the space provided in the *My Notes* section to represent $(x - 3)(x^2 + 5x + 6)$. Label each inner rectangle and find the sum.
   $x^3 + 2x^2 - 9x - 18$

2. How many boxes would you need to represent the multiplication of $(x^3 + 5x^2 + 3x - 3)(x^4 - 6x^3 - 7x^2 + 5x + 6)$ using the graphic organizer?
   20 boxes

   a. Explain how you determined your answer. **I drew the frame of a graphic organizer and counted the number of boxes.**

   b. **Use appropriate tools strategically.** Would you use the graphic organizer for other multiplications with this many terms? Explain your reasoning. **Answers may vary, but students should realize that the table will eventually become unwieldy.**

The Distributive Property can be used to multiply any polynomial by another. Multiply each term in the first polynomial by each term in the second polynomial.

$(x - 3)(5x^2 - 2x + 1) = 5x^3 - 17x^2 + 7x - 3$

3. Determine each product.

   a. $x(x + 5)$
      $x^2 + 5x$

   b. $(x - 3)(x + 6)$
      $x^2 + 3x - 18$

   c. $(x + 7)(3x^2 - x - 1)$
      $3x^3 + 20x^2 - 8x - 7$

   d. $(3x - 7)(4x^2 + 4x - 3)$
      $12x^3 - 16x^2 - 37x + 21$

4. How can you predict the number of terms the product will have before you combine like terms?
   The number of terms in the product before combining like terms will be the product of the number of the terms in each polynomial.

My Notes

	x	-3
x^2	x^3	$-3x^2$
$5x$	$5x^2$	$-15x$
6	$6x$	-18

ACTIVITY 25 Continued

7–8 Think-Pair-Share, Debriefing
Students can multiply any of the two binomials first and then multiply the resulting trinomial by the third binomial. To help students focus on the two they begin with, introduce the use of square brackets. For example, they can envision $(x - 2)(x - 6)(x + 2)$ as either $[(x - 2)(x - 6)](x + 2)$ or $(x - 2)[(x - 6)(x + 2)]$. Debrief this portion of the lesson by asking students to identify any patterns they have identified as they have multiplied polynomials.

Check Your Understanding
Each of these answers results in a general rule for multiplying polynomials. As you debrief the answers, discuss how the patterns can be used to multiply similar polynomials with constant terms. At the same time, be sure students know that there are only a few patterns—two of which they saw in the previous lesson—that it is really useful for them to memorize.

Answers
9. $ab + bc$
10. $a^2 + ac + ab + bc$
11. $a^3 + ab + ac + a^2b + b^2 + bc$
12. $a^2b + ab^2 + b^2c + a^2c + ac^2 + bc^2 + 2abc$

ASSESS

Students' answers to Lesson Practice problems will provide you with a formative assessment of their understanding of the lesson concepts and their ability to apply their learning.

See the Activity Practice for additional problems for this lesson. You may assign the problems here or use them as a culmination for the activity.

LESSON 25-3 PRACTICE
13. $x^2 + 7x$
14. $2x^2 - 5x$
15. $y^2 + 9y + 18$
16. $y^2 - 3y - 18$
17. $2x^3 - 5x^2 + x$
18. $2x^3 - 7x^2 + 6x - 1$
19. $10x^3 - 35x^2 - 2x + 7$
20. $10x^3 - 41x^2 + 19x + 7$
21. $x^3 - 7x - 6$
22. $4x^3 + 4x^2 - 11x - 6$

ADAPT

If students have difficulty with this lesson because they still do not multiply and add monomials and signed numbers fluently, provide support for these skills.

ACTIVITY 25 continued

My Notes

MATH TIP
Polynomials are *closed* under multiplication. A set is closed under multiplication if the product of any two elements in the set is also an element of the set.

23. The degree of a polynomial is the largest degree of any term in the polynomial. So 2 and 4 must be the largest degrees of some term in each of the original polynomials. Since $x^2 \cdot x^4 = x^6$, the degree of the new polynomial will be 6.

Lesson 25-3
Multiplying Polynomials

5. Are all of the answers to Item 3 polynomials? Justify your response.
   Yes; the expressions are sums of terms with whole-number powers.

6. Explain why the product of two polynomials will always be a polynomial.
   Answers may vary. The terms in a polynomial must have whole-number powers. When you multiply two polynomials, you add the exponents of terms with the same base. Because adding two whole numbers gives a whole number, the terms in the product will have whole-number powers. Therefore, the product is a polynomial.

7. You can find the product of more than two polynomials, such as $(x + 3)(2x + 1)(3x - 2)$.

   a. To multiply $(x + 3)(2x + 1)(3x - 2)$, first determine the product of the first two polynomials, $(x + 3)(2x + 1)$.

   $$(x + 3)(2x + 1) = 2x^2 + 7x + 3$$

   b. Multiply your answer to Part (a) by the third polynomial, $(3x - 2)$.
   $(2x^2 + 7x + 3)(3x - 2) = 6x^3 + 17x^2 - 5x - 6$

8. Determine each product.

   a. $(x - 2)(x + 1)(2x + 2)$
      $2x^3 - 6x - 4$
   b. $(x + 3)(3x + 1)(2x - 1)$
      $6x^3 + 17x^2 - 4x - 3$
   c. $(x - 1)(3x - 2)(x + 4)$
      $3x^3 + 7x^2 - 18x + 8$
   d. $(2x - 4)(4x + 1)(3x + 3)$
      $24x^3 - 18x^2 - 54x - 12$

Check Your Understanding

Determine each product.
9. $a(b + c)$
10. $(a + b)(a + c)$
11. $(a + b)(a^2 + b + c)$
12. $(a + b)(a + c)(b + c)$

LESSON 25-3 PRACTICE

Determine each product.
13. $x(x + 7)$
14. $x(2x - 5)$
15. $(y + 3)(y + 6)$
16. $(y + 3)(y - 6)$
17. $x(2x^2 - 5x + 1)$
18. $(x - 1)(2x^2 - 5x + 1)$
19. $(2x - 7)(5x^2 - 1)$
20. $(2x - 7)(5x^2 - 3x - 1)$
21. $(x + 2)(x - 3)(x + 1)$
22. $(x + 2)(2x - 3)(2x + 1)$

23. **Attend to precision.** A binomial of degree 2 and variable x and a trinomial of degree 4 and variable x are multiplied. What will be the degree of the product? Explain your reasoning.

Multiplying Polynomials
Tri-Com Computers

ACTIVITY 25 continued

ACTIVITY 25 PRACTICE
Write your answers on notebook paper.
Show your work.

Lesson 25-1
Determine each product.

1. $(10-3)(10-8)$
2. $(x-3)(x-8)$
3. $(y-7)(y+2)$
4. $(x+5)(x-9)$
5. $(2y-6)(3y-8)$
6. $(4x+3)(x-11)$
7. Which expression represents the area of the rectangle?

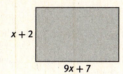

A. $10x+9$
B. $9x^2+14$
C. $20x+18$
D. $9x^2+25x+14$

Lesson 25-2
Determine each product.

8. $(x-7)(x+7)$
9. $(y+6)(y-6)$
10. $(2x-5)(2x+5)$
11. $(3y+1)(3y-1)$
12. $(x-11)^2$
13. $(x+8)^2$
14. $(2y-3)^2$
15. $(3y-2)^2$
16. Which expression represents the area of the square?

A. $20y^2+28$
B. $20y-28$
C. $25y^2-70y+49$
D. $25y^2-49$

ACTIVITY 25 Continued

ACTIVITY PRACTICE
1. 14
2. $x^2-11x+24$
3. $y^2-5y-14$
4. $x^2-4x-45$
5. $6y^2-34y+48$
6. $4x^2-41x-33$
7. D
8. x^2-49
9. y^2-36
10. $4x^2-25$
11. $9y^2-1$
12. $x^2-22x+121$
13. $x^2+16x+64$
14. $4y^2-12y+9$
15. $9y^2-12y+4$
16. C

ACTIVITY 25 Continued

17. $x^3 - 7x$
18. $2x^3 - 6x^2 + 4x$
19. $4x^3 + x^2 - 9x + 10$
20. $4y^3 - 15y^2 - 23y - 10$
21. $25x^2 - 90x + 81$
22. $9x^2 - 16$
23. $2y^3 + 7y^2 - 7y - 5$
24. B
25. $7x^3 + 6x^2 - 15x + 2$
26. $8x^3 + 50x^2 + 47x - 15$
27. $y^3 + 3y^2 + 3y + 1$
28. Sample answer: First find the product of the first two polynomials. Multiply this product by the third polynomial, and then multiply this product by the fourth polynomial.
29. $x^2 - 1$, $x^3 - 1$, $x^4 - 1$, $x^5 - 1$. Sample answer: Multiplying the second factor by the x in $(x - 1)$ gives positive middle terms with coefficients of 1. Multiplying the second factor by the -1 in $(x - 1)$ gives negative middle terms with coefficients of -1. All of the terms with the exception of the first term and last term result in zero pairs.

ADDITIONAL PRACTICE

If students need more practice on the concepts in this activity, see the eBook Teacher Resources for additional practice problems.

ACTIVITY 25 continued

Multiplying Polynomials
Tri-Com Computers

Lesson 25-3
Determine each product.

17. $x(x^2 - 7)$
18. $2x(x^2 - 3x + 2)$
19. $(x + 2)(4x^2 - 7x + 5)$
20. $(y - 5)(4y^2 + 5y + 2)$
21. $(5x - 9)^2$
22. $(3x - 4)(3x + 4)$
23. $(2y + 1)(y^2 + 3y - 5)$
24. Which expression represents the area of the triangle? Use the formula $A = \frac{1}{2}bh$.

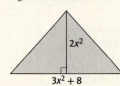

A. $3x^4 + x^2$
B. $3x^4 + 8x^2$
C. $6x^2 + 8$
D. $6x^4 + 16x$

Determine each product.

25. $(x - 1)(7x - 1)(x + 2)$
26. $(x + 5)(4x - 1)(2x + 3)$
27. $(y + 1)^3$
28. Devise a plan for finding the product of four polynomials.

MATHEMATICAL PRACTICES
Look for and Make Use of Structure

29. Determine each product and describe any patterns you observe.

$(x - 1)(x + 1)$
$(x - 1)(x^2 + x + 1)$
$(x - 1)(x^3 + x^2 + x + 1)$

From the patterns you see, predict the product of $(x - 1)(x^4 + x^3 + x^2 + x + 1)$. Describe the pattern that helps you know the answer without needing to multiply.

Polynomial Operations
MEASURING UP

Embedded Assessment 3
Use after Activity 25

Employees at Ship-It-Quik must perform computations involving volume and surface area. As part of the job application, potential employees must take a test that involves surface area, volume, and algebraic skills.

1. The surface area of a figure is the total area of all faces. The areas of the faces of a rectangular prism are shown. The surface area of this prism is $18x^2 + 12x + 22$. Complete the first part of the job application by finding the area of the missing face.

The formula for the volume of a rectangular prism is $V = lwh$, where l, w, and h are length, width, and height, respectively. The formula for the surface area is $SA = 2lw + 2wh + 2lh$.

2. Complete the second part of the job application by verifying whether or not the following computations are correct. Explain your reasoning by showing your work.

   Volume:
   $2x^2 + 15x + 36$

   Surface Area:
   $10x^2 + 90x + 72$

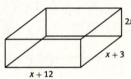

3. Complete the final part of the job application by writing an expression for the volume of a cylinder with radius $3x^2y$ and height $2xy$. Use the formula $V = \pi r^2 h$ where r is the radius and h is the height. Simplify your answer as much as possible.

Embedded Assessment 3

Assessment Focus
- Adding polynomials
- Multiplying polynomials

Answer Key

1. The missing side has area $4x^2 - x + 3$. The surface area is $2(x^2 + 2x + 1) + 2(4x^2 + 5x + 7) + 2(\text{missing area}) = 18x^2 + 12x + 22$. First, multiply $2(x^2 + 2x + 1)$ and $2(4x^2 + 5x + 7)$ by using the Distributive Property. Then, add the two resulting polynomials and subtract the sum from $18x^2 + 12x + 22$ to get $8x^2 - 2x + 6 = 2(\text{missing area})$. Therefore, the missing area is $4x^2 - x + 3$.

2. Explanations may vary. The volume is calculated incorrectly. The surface area is correct.
   $V = 2x(x + 3)(x + 12)$
   $ = 2x(x^2 + 15x + 36)$
   $ = 2x^3 + 30x^2 + 72x$
   $SA = 2(2x)(x + 3) + 2(2x)(x + 12) + 2(x + 3)(x + 12)$
   $ = 4x^2 + 12x + 4x^2 + 48x + 2x^2 + 30x + 72$
   $ = 10x^2 + 90x + 72$

3. The volume is $18\pi x^5 y^3$.
   $V = \pi r^2 h$
   $V = \pi(3x^2y)^2(2xy)$
   $V = \pi(3x^2y)(3x^2y)(2xy)$
   $V = 18\pi x^5 y^3$

Common Core State Standards for Embedded Assessment 3

HSA-APR.A.1: Understand that polynomials form a system analogous to the integers, namely, they are closed under the operations of addition, subtraction, and multiplication; add, subtract, and multiply polynomials.

Unit 4 • Exponents, Radicals, and Polynomials   **383**

Embedded Assessment 3

Teacher to Teacher

You may wish to read through the scoring guide with students and discuss the differences in the expectations at each level. Check that students understand the terms used.

Unpacking Embedded Assessment 4

Once students have completed this Embedded Assessment, turn to the Embedded Assessment 4 and unpack it with them. Use a graphic organizer to help students understand the concepts they will need to know to be successful on Embedded Assessment 4.

Embedded Assessment 3
Use after Activity 25

Polynomial Operations
MEASURING UP

Scoring Guide	Exemplary	Proficient	Emerging	Incomplete
	The solution demonstrates the following characteristics:			
Mathematics Knowledge and Thinking (Items 1–3)	• Effective understanding of and accuracy in adding, subtracting, and multiplying polynomials	• Addition, subtraction, and multiplication of polynomials that are usually correct	• Difficulty adding, subtracting, and multiplying polynomials	• Inaccurate addition, subtraction, and multiplication of polynomials
Problem Solving (Item 1)	• Appropriate and efficient strategy that results in a correct answer	• Strategy that may include unnecessary steps but results in a correct answer	• Strategy that results in some incorrect answers	• No clear strategy when solving problems
Mathematical Modeling/ Representations (Items 1–3)	• Clear and accurate understanding of geometric formulas	• Functional understanding of geometric formulas that results in correct answers	• Partial understanding of geometric formulas that results in some incorrect answers	• Little or no understanding of geometric formulas
Reasoning and Communication (Item 2)	• Precise use of appropriate math terms and language to justify each step in verifying an answer	• Adequate justification of each step to verify an answer	• Misleading or confusing justification of the steps to verify an answer	• Incomplete or inaccurate justification of the steps to verify an answer

Factoring

Factors of Construction
Lesson 26-1 Factoring by Greatest Common Factor (GCF)

ACTIVITY 26

Learning Targets:
- Identify the GCF of the terms in a polynomial.
- Factor the GCF from a polynomial.

SUGGESTED LEARNING STRATEGIES: Look for a Pattern, Think-Pair-Share, Discussion Groups, Note Taking

Factor Steele Buildings is a company that manufactures prefabricated metal buildings that are customizable. All the buildings come in square or rectangular designs. Most office buildings have an entrance area or great room, large offices, and cubicles. The diagram below shows the front face of one of their designs. The distance c represents space available for large offices, and p represents the space available for the great room.

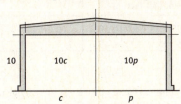

1. To determine how much material is needed to cover the front wall of the building, represent the total area as a product of a **monomial** and a **binomial**.
   $10(c + p)$

2. Represent the same area from Item 1 as a sum of two monomials.
   $10c + 10p$

3. **Make use of structure.** What property can be used to show that the two quantities in Items 1 and 2 are equal?
   The Distributive Property

4. Factor Steele Buildings inputs the length of the large office space c into an expression that gives the area of an entire space: $6c^2 + 12c - 9$. Determine the **greatest common factor (GCF)** of the terms in this polynomial. Explain your choice.
   3; 3 is the greatest monomial that divides into each term of the polynomial without a remainder.

MATH TERMS

A **monomial** is a number, a variable, or a product of numbers and variables with whole-number exponents. For example, 4, $-9x$, and $5xy^2$ are all monomials. A **binomial** is a sum or difference of two monomials.

MATH TERMS

The **greatest common factor (GCF)** of the terms in a polynomial is the greatest monomial that divides into each term without a remainder.

Common Core State Standards for Activity 26

HSA-SSE.A.1	Interpret expressions that represent a quantity in terms of its context.*
HSA-SSE.A.1.a	Interpret parts of an expression, such as terms, factors, and coefficients.
HSA-SSE.A.1.b	Interpret complicated expressions by viewing one or more of their parts as a single entity.
HSA-SSE.A.2	Use the structure of an expression to identify ways to rewrite it.

ACTIVITY 26 Guided

Activity Standards Focus
In Activity 26, students are introduced to the factoring of polynomials. They begin by identifying the GCF of the terms of a polynomial and then using the GCF to factor. Students also use patterns to factor perfect square trinomials and differences of two squares. Throughout this activity, be sure to emphasize that the factored form of a polynomial is equivalent to the original form.

Lesson 26-1

PLAN

Pacing: 1 class period
Chunking the Lesson
#1–3   #4   Example A
Check Your Understanding
Lesson Practice

TEACH

Bell-Ringer Activity
Ask students to find the greatest common factor of each pair of numbers. Then have them explain how they found each GCF.
1. 20 and 64   [4]
2. -75 and 175   [25]
3. -18 and 72   [18]

1–3 Shared Reading, Activating Prior Knowledge, Look for a Pattern, Think-Pair-Share These first items are designed to ease students into the activity and prepare them for the factoring that will be done later in the lesson. In Item 1, students are representing the total area by multiplying the height of the front face by the base of the front face. In Item 2, students are representing the total area by adding the area of the front face of the large offices and the area of the front face of the great room.

4 Discussion Groups, Look for a Pattern, Think-Pair-Share, Debriefing Students determine the GCF of the terms of a polynomial. Be sure students can identify the terms of the polynomial. [$6c^2$, $12c$, and -9] As students share out, be sure that the discussion includes the reason why c is not a common factor of the terms. [c is not a factor of -9.] Some students may need more practice finding the GCF of two or more monomials. If so, use the mini-lesson on the next page.

Activity 26 • Factoring   385

ACTIVITY 26 Continued

Developing Math Language
This lesson includes the vocabulary terms *monomial*, *binomial* and *factor*. Students are likely familiar with the use of *factor* as a noun (e.g., 3 and x are factors of $3x$), but they may not have used it as a verb. Explain that to factor a polynomial means to write the polynomial as a product. For example, you can factor the polynomial $3x + 3$ by using the Distributive Property to write it as the product $3(x + 1)$.

Example A Note Taking In this example, students are introduced to factoring a polynomial using the GCF of its terms. Ask students to explain how they know that $2x$ is the GCF of the polynomial $6x^3 + 2x^2 - 8x$. Then have students discuss how they can check that a polynomial has been factored correctly. One way is by using the Distributive Property to multiply the factors and check that the product is equal to the original polynomial.

Differentiating Instruction

Support students who need extra help with the Try These items by giving them time to work in small groups. Provide assistance as needed.

Extend learning for students who understand the process by giving them polynomials with multiple variables to factor, such as $12x^3y + 24x^2y^2 - 15xy$.

Check Your Understanding
Debrief students' answers to these items to be sure they understand how to factor the GCF from a polynomial. Then ask students to explain how to identify the GCF of two or more monomials.

Answers
5. $7x$
6. $9(4x + 1)$
7. $6x(x^3 + 2x - 3)$
8. $25n^3(5n^3 + 10n^2 + 1)$
9. $3x(x^2 + 3x + 2)$
10. $\frac{1}{3}y^2(2y^2 + y - 4)$
11. $4xy(xy + 3y - 2x - 1)$

ACTIVITY 26 continued

MATH TERMS
A **factor** is any of the numbers or symbols that when multiplied together form a product. For example, 2 and x are factors of $2x$ because 2 and x are multiplied to get $2x$. *Factor* can be used as a noun or a verb.

Lesson 26-1
Factoring by Greatest Common Factor (GCF)

To *factor* a number or expression means to write the number or expression as a product of its *factors*.

Example A

To Factor a Monomial (the GCF) from a Polynomial	
Steps to Factoring	Example
• Determine the GCF of all terms in the polynomial.	$6x^3 + 2x^2 - 8x$ GCF = $2x$
• Write each term as the product of the GCF and another factor.	$2x(3x^2) + 2x(x) + 2x(-4)$
• Use the Distributive Property to factor out the GCF.	$2x(3x^2 + x - 4)$

Try These A
Find the greatest common factor of the terms in each polynomial. Then write each polynomial with the GCF factored out.

a. $36y - 24$
   12; $12(3y - 2)$

b. $4x^5 - 6x^3 + 10x^2$
   $2x^2$; $2x^2(2x^3 - 3x + 5)$

c. $15t^2 + 10t - 5$
   5; $5(3t^2 + 2t - 1)$

Check Your Understanding

5. Identify the GCF of the terms in the polynomial $21x^3 + 14x^2 + 35x$.

Factor a monomial (the GCF) from each polynomial.

6. $36x + 9$
7. $6x^4 + 12x^2 - 18x$
8. $125n^6 + 250n^5 + 25n^3$
9. $3x^3 + 9x^2 + 6x$
10. $\frac{2}{3}y^4 + \frac{1}{3}y^3 - \frac{4}{3}y^2$
11. $4x^2y^2 + 12xy^2 - 8x^2y - 4xy$

MINI-LESSON: Greatest Common Factor of Monomials

If students need more practice with finding the GCF of two or more monomials, a mini-lesson is available.

See SpringBoard's eBook Teacher Resources for a student page for this mini-lesson.

Lesson 26-1
Factoring by Greatest Common Factor (GCF)

LESSON 26-1 PRACTICE

Use the cylinder for Items 12–15.

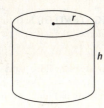

The surface area of a cylinder is given by the formula $SA = 2\pi r^2 + 2\pi rh$, where r is the radius and h is the height.

12. Factor a monomial (the GCF) from the formula.
13. Suppose the radius of the cylinder is y and the height is $y + 2$. Rewrite the formula in this case, using multiplication and exponent rules as needed to simplify the expression.
14. Factor the expression from Item 13 completely.
15. **Construct viable arguments.** Answer each of the following questions and justify your responses.
    a. If the radius of a cylinder doubles, what happens to the GCF of its surface area?
    b. What happens to the GCF of the cylinder's surface area if its radius is squared?

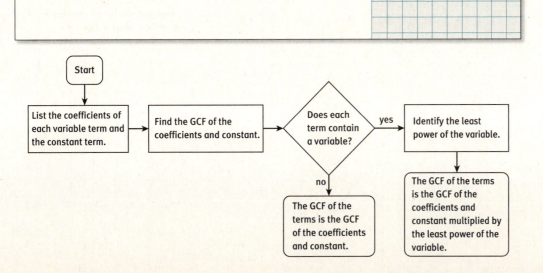

ACTIVITY 26 Continued
Lesson 26-2

PLAN

Materials
- algebra tiles (optional)

Pacing: 1 class period
Chunking the Lesson
#1–3   #4   #5–7
Check Your Understanding
#11–12   #13–14   #15
Check Your Understanding
Lesson Practice

TEACH

Bell-Ringer Activity
Have students find the product of each pair of binomials.
1. $(x + 3)(x + 3)$ $[x^2 + 6x + 9]$
2. $(x - 6)(x - 6)$ $[x^2 - 12x + 36]$
3. $(x + 8)(x - 8)$ $[x^2 - 64]$

1–3 Activating Prior Knowledge, Create Representations, Discussion Groups Students use area models to relate polynomials to their factored forms. Monitor student discussions carefully to be sure students can interpret the floor plan at the top of the page. If needed, prompt discussion by asking groups to give the area, in square units, of the great room $[x \cdot x = x^2]$, a large office $[x \cdot 1 = x]$, and a cubicle $[1 \cdot 1 = 1]$. In Item 2, it is important for students to understand that the area of the entire office can be written as a product of two binomials by multiplying an expression for its length by an expression for its width.

4 Look for a Pattern, Sharing and Responding This item presents yet another opportunity for students to identify multiplication patterns. Allow students to share the patterns they observe in Item 4c. There are numerous patterns that can be noted, so it will be helpful for students to hear a variety of answers.

Differentiating Instruction

Support kinesthetic learners by allowing them to model the office spaces by using algebra tiles. They can use an x^2 tile to represent a great room, an x tile to represent a large office, and a $+1$ tile to represent a cubicle.

ACTIVITY 26
continued

Lesson 26-2
Factoring Special Products

Learning Targets:
- Factor a perfect square trinomial.
- Factor a difference of two squares.

SUGGESTED LEARNING STRATEGIES: Create Representations, Discussion Groups, Look for a Pattern, Sharing and Responding, Think-Pair-Share

Factor Steele Buildings can create many floor plans with different size spaces. In the diagram below the great room has a length and width of x units, and each cubicle has a length and width of 1 unit. Use the diagram below for Items 1–3.

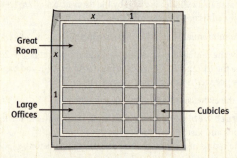

1. **Model with mathematics.** Represent the area of the entire office above as a sum of the areas of all the rooms.
   $x^2 + 6x + 9$

2. Write the area of the entire office as a product of two binomials.
   $(x + 3)(x + 3)$

3. What property can you use to show how the answers to Items 1 and 2 are related? Show this relationship.
   The Distributive Property;
   $(x + 3)(x + 3) = x^2 + 3x + 3x + 9 = x^2 + 6x + 9$

4. For each of the following floor plans, write the area of the office as a sum of the areas of all the rooms and as a product of binomials.
   a.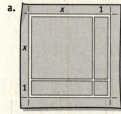

   $x^2 + 2x + 1; (x + 1)(x + 1)$

   b.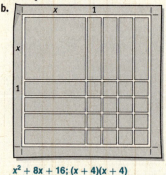

   $x^2 + 8x + 16; (x + 4)(x + 4)$

Lesson 26-2
Factoring Special Products

c. What patterns do you observe?
   Answers may vary. When the area is written as a sum of the areas of all the rooms, the constant term is the square of the constant terms in the binomials and the coefficient of x is the sum of the constant terms in the binomials.

5. Complete the following table. The first row has been done for you.

Polynomial	1st Factor	2nd Factor	First Term in Each Factor	Second Term in Each Factor
$x^2 + 6x + 9$	$(x + 3)$	$(x + 3)$	x	3
$x^2 - 6x + 9$	$(x - 3)$	$(x - 3)$	x	-3
$x^2 + 8x + 16$	$(x + 4)$	$(x + 4)$	x	4
$x^2 - 8x + 16$	$(x - 4)$	$(x - 4)$	x	-4
$x^2 + 10x + 25$	$(x + 5)$	$(x + 5)$	x	5
$x^2 - 10x + 25$	$(x - 5)$	$(x - 5)$	x	-5

6. **Express regularity in repeated reasoning.** Describe any patterns that you observe in the table from Item 5.
   Answers may vary. The constant term of the polynomial is the square of the second term in each factor and the coefficient of x in the polynomial is twice the second term in each factor.

7. Explain how to factor polynomials of the form $a^2 + 2ab + b^2$ and $a^2 - 2ab + b^2$.
   The polynomial $a^2 + 2ab + b^2$ factors into $(a + b)^2$, and $a^2 - 2ab + b^2$ factors into $(a - b)^2$.

Polynomials of the form $a^2 + 2ab + b^2$ and $a^2 - 2ab + b^2$ are called **perfect square trinomials**.

Check Your Understanding

Factor each perfect square trinomial.

8. $x^2 - 14x + 49$
9. $m^2 + 20m + 100$
10. $y^2 - 16y + 64$

11. Complete the table by finding the polynomial product of each pair of binomial factors. The first row has been done for you.

1st Factor	2nd Factor	Polynomial
$(x + 3)$	$(x - 3)$	$x^2 - 9$
$(x + 4)$	$(x - 4)$	$x^2 - 16$
$(x - 5)$	$(x + 5)$	$x^2 - 25$
$(9 - x)$	$(9 + x)$	$81 - x^2$
$(2x - 7)$	$(2x + 7)$	$4x^2 - 49$
$(6x - 2y)$	$(6x + 2y)$	$36x^2 - 4y^2$

ACTIVITY 26 Continued

13–14 Look for a Pattern, Create Representations, Think-Pair-Share These items lead students to apply the patterns they have observed to factor the difference of two squares. Allow students to share any additional patterns they notice as they complete these items. Be sure to ask students how they can check that they have factored the polynomials correctly.

15 Look for a Pattern, Create Representations, Discussion Groups, Debriefing Students write a general rule for factoring a difference of two squares. Debrief this item to make sure all students have written the rule correctly: $a^2 - b^2 = (a + b)(a - b)$.

Check Your Understanding

Debrief students' answers to these items to ensure that they understand how to factor differences of two squares. Ask students to explain why the polynomial $x^2 + 36$ is not a difference of two squares.

Answers
16. $(x + 11)(x - 11)$
17. $(4m + 9)(4m - 9)$
18. $(3 + 5p)(3 - 5p)$

ASSESS

Students' answers to lesson practice problems will provide you with a formative assessment of their understanding of the lesson concepts and their ability to apply their learning. See the Activity Practice for additional problems for this lesson. You may assign the problems here or use them as a culmination for the activity.

ADAPT

Check students' answers to the Lesson Practice questions to make sure that they understand concepts related to factoring perfect square trinomials and differences of squares. If students have not yet mastered these skills, it may help them to use algebra tiles to model some of the polynomials from this lesson. This will allow students to see the binomials that are multiplied as factors to yield the polynomials. For example, the perfect square trinomial $x^2 - 6x + 9$ can be modeled as follows to show that its factored form is $(x - 3)(x - 3)$.

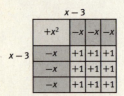

ACTIVITY 26 continued

My Notes

Lesson 26-2
Factoring Special Products

12. Describe any patterns you observe in the table from Item 11.
    Answers may vary. The polynomial is always a difference. The first term of the polynomial is the square of the first term of the factors; the second term of the polynomial is the square of the second term of the factors.

13. a. One factor of $36 - y^2$ is $6 + y$. What is the other factor?
       $6 - y$
    b. One factor of $p^2 - 144$ is $p - 12$. What is the other factor?
       $p + 12$
    c. Describe any patterns you observe.
       The two factors have the form $a + b$ and $a - b$.

14. Factor each of the following.
    a. $49 - x^2$ b. $n^2 - 9$ c. $64w^2 - 25$
       $(7 + x)(7 - x)$ $(n + 3)(n - 3)$ $(8w + 5)(8w - 5)$
    d. Describe any patterns you observe.
       The two factors have the form $a + b$ and $a - b$, where a is the square root of the first term of the polynomial and b is the square root of the second term of the polynomial.

15. Explain how to factor a polynomial of the form $a^2 - b^2$.
    The polynomial factors into $(a + b)(a - b)$ or $(a - b)(a + b)$.

A polynomial of the form $a^2 - b^2$ is referred to as the **difference of two squares**.

Check Your Understanding

Factor each difference of two squares.
16. $x^2 - 121$
17. $16m^2 - 81$
18. $9 - 25p^2$

LESSON 26-2 PRACTICE

Identify each polynomial as a perfect square trinomial, a difference of two squares, or neither. Then factor the polynomial if it is a perfect square trinomial or a difference of two squares.

19. $z^2 + 6z + 12$ 20. $4x^2 - 121$ 21. $y^2 - 8y + 16$
22. $y^2 - 8y - 16$ 23. $n^2 + 25$ 24. $169 - 9x^2$

25. What factor would you need to multiply by $(4c + 7)$ to get $16c^2 - 49$?

26. What factor would you need to multiply by $(3d + 1)$ to get $9d^2 + 6d + 1$?

Factor completely. (*Hint*: First look for a GCF.)
27. $2x^2 + 8x + 8$ 28. $3y^2 - 75$ 29. $12x^2 - 12x + 3$

30. **Use appropriate tools strategically.** Explain how you can use your calculator to check that you have factored a polynomial correctly.

LESSON 26-2 PRACTICE
19. neither
20. difference of two squares; $(2x + 11)(2x - 11)$
21. perfect square trinomial; $(y - 4)^2$
22. neither
23. neither
24. difference of two squares; $(13 + 3x)(13 - 3x)$
25. $(4c - 7)$
26. $(3d + 1)$
27. $2(x + 2)^2$
28. $3(x + 5)(x - 5)$
29. $3(2x - 1)^2$
30. Sample answer: Enter the polynomial in the calculator as y_1 and enter the factored form as y_2. Check that the graphs are the same or that a table gives the same values for both functions.

390 SpringBoard® Mathematics Algebra 1, Unit 4 • Exponents, Radicals, and Polynomials

Factoring
Factors of Construction

ACTIVITY 26
continued

ACTIVITY 26 PRACTICE
Write your answers on notebook paper.
Show your work.

Lesson 26-1

1. What is the greatest common factor of the terms in the polynomial $24x^8 + 6x^5 + 9x^2$?
   A. 3
   B. $3x^2$
   C. $6x$
   D. $6x^2$

Factor a monomial (the GCF) from each polynomial.

2. $15x^4 + 20x^3 + 35x$
3. $12m^3 - 8m^2 + 16m + 8$
4. $32y^2 + 48y - 16$
5. $x^5 + x^4 + 3x^3 + 3x^2$

6. Which of these polynomials cannot be factored by factoring out the GCF?
   A. $7x^2 + 14x + 21$
   B. $49x^3 + 21x^2 + x$
   C. $x^2 + 14x + 7$
   D. $35x^3 + 28x^2 + 7x$

7. The figure shows the dimensions of a garden plot in the shape of a trapezoid. Write and simplify a polynomial for the perimeter of the plot. Then factor the polynomial completely.

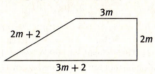

8. The area of the rectangle shown below is $6x^2 + 9x$ square feet. The width of the rectangle is given in the figure. What is the length of the rectangle? Justify your answer.

9. Marcus saw the factorization shown below in his textbook, but part of the factorization was covered by a drop of ink. What expression was covered by the drop of ink?

   $-24x^5 - 16x^3 = -8x^3(\text{\textbullet} + 2)$

10. Write a polynomial with four terms that has a GCF of $4x^2$.

Lesson 26-2
Identify each polynomial as a perfect square trinomial, a difference of two squares, or neither. Then factor the polynomial if it is a perfect square trinomial or a difference of two squares.

11. $9x^2 - 121$
12. $m^2 - 16m + 64$
13. $y^2 + 12y - 36$
14. $16z^2 + 25$
15. $25 - 144p^2$
16. $x^2 + 50x + 625$

Factor completely.

17. $2x^2 - 32$
18. $32 - 8p$
19. $3x^3 + 12x^2 + 12x$
20. $4y^3 - 32y^2 + 64y$
21. $5x^4 - 125x^2$

22. What factor would you need to multiply by $(4x - 1)$ to get $16x^2 - 8x + 1$?
    A. $4x - 1$
    B. $4x + 1$
    C. $4x^2$
    D. $4x$

Use the rectangle for Items 23–25.

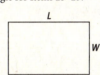

23. The area of a rectangle is $64b^2 - 4$ and $W = 8b - 2$. What is L?

24. The area of another rectangle is $144c^2 - 4$ and $L = 12c + 2$. What is W?

ACTIVITY 26 Continued

ACTIVITY PRACTICE
1. B
2. $5x(3x^3 + 4x^2 + 7)$
3. $4(3m^3 - 2m^2 + 4m + 2)$
4. $16(2y^2 + 3y - 1)$
5. $x^2(x^3 + x^2 + 3x + 3)$
6. C
7. $10m + 4 = 2(5m + 2)$
8. $2x + 3$ ft; The product of the length and width is the area, and $3x(2x + 3) = 6x^2 + 9x$.
9. $3x^2$
10. Sample answer: $8x^5 + 4x^4 + 12x^3 + 4x^2$
11. difference of two squares; $(3x + 11)(3x - 11)$
12. perfect square trinomial; $(m - 8)^2$
13. neither
14. neither
15. difference of two squares; $(5 + 12p)(5 - 12p)$
16. perfect square trinomial; $(x + 25)^2$
17. $2(x + 4)(x - 4)$
18. $8(4 - p)$
19. $3x(x + 2)^2$
20. $4y(y - 4)^2$
21. $5x^2(x + 5)(x - 5)$
22. A
23. $8b + 2$
24. $12c - 2$

Activity 26 • Factoring 391

ACTIVITY 26 Continued

25. a. $2x - 1$
    b. It is a square because the length and width are the same.
26. C
27. C
28. C
29. Sample answer: $9x^2 + 6x + 1$
30. $b = 10$ or $b = -10$; the factored form of the polynomial could be $(x + 5)^2$ or $(x - 5)^2$.
31. Although both students have factored correctly, neither student has factored the polynomial completely. Sasha could have factored a GCF of 3 from each factor of $3x + 3$. Pedro could have factored the expression $x^2 - 1$ as a difference of squares. In either case, the complete factorization is $9(x + 1)(x - 1)$.
32. C
33. B
34. $(x^2 + 9)(x + 3)(x - 3)$
35. $(y^4 + 25)(y^2 + 5)(y^2 - 5)$
36. Yes, when $b = 0$; $(a + b)(a - b)$ is equal to the difference of squares $a^2 - b^2$; if $a^2 - b^2 = a^2 + b^2$, then $-b^2 = b^2$ and $b = 0$.

ADDITIONAL PRACTICE

If students need more practice on the concepts in this activity, see the eBook Teacher Resources for additional practice problems.

ACTIVITY 26 continued

Factoring
Factors of Construction

25. Suppose the area of a rectangle is $4x^2 - 4x + 1$ and $L = 2x - 1$.
    a. What is W?
    b. What must be true about the rectangle in this case? Explain.

26. The area of a square window is given by the expression $m^2 - 16m + 64$. Which expression represents the length of one side of the window?
    A. $m - 4$
    B. $m + 4$
    C. $m - 8$
    D. $m + 8$

27. What value of k makes the polynomial $x^2 + 6x + k$ a perfect square trinomial?
    A. 3
    B. 6
    C. 9
    D. 36

28. Consider the following values of c in the polynomial $36x^2 + c$.
    I. $c = -25$
    II. $c = 25$
    III. $c = -36$

    Which value or values of c make it possible to factor the polynomial?
    A. I only
    B. I and II only
    C. I and III only
    D. I, II, and III

29. Write a perfect square trinomial that includes the term $9x^2$.

30. The polynomial $x^2 + bx + 25$ is a perfect square trinomial. What is the value of b? Is there more than one possibility? Explain.

31. Sasha and Pedro were asked to factor the polynomial $9x^2 - 9$ completely and explain their process. Their work is shown below. Has either student factored the polynomial completely? Explain. If not, give the complete factorization.

Sasha's Work
$9x^2 - 9 = (3x + 3)(3x - 3)$ I used the fact that $9x^2 - 9$ is a difference of two squares.

Pedro's Work
$9x^2 - 9 = 9(x^2 - 1)$ I factored out the GCF.

32. Which of the following polynomials has $m - 4$ as a factor?
    A. $m^2 - 4$
    B. $m^2 + 16$
    C. $m^2 - 8m + 16$
    D. $m^2 - 8m - 16$

33. Given that $x^2 + \square + 100$ is a perfect square trinomial, which of these could be the missing term?
    A. $10x$
    B. $20x$
    C. $50x$
    D. $100x$

34. Factor $x^4 - 81$ completely. (*Hint*: Use the fact that $x^4 = (x^2)^2$ to factor $x^4 - 81$ as a difference of two squares. Then consider whether any of the resulting factors can be factored again.)

35. Use the method in Item 34 to factor $y^8 - 625$ completely.

MATHEMATICAL PRACTICES
Reason Abstractly and Quantitatively

36. Could a product in the form $(a + b)(a - b)$ ever be equal to $a^2 + b^2$? Justify your answer.

Factoring Trinomials
Deconstructing Floor Plans
Lesson 27-1 Factoring $x^2 + bx + c$

Learning Targets:
- Use algebra tiles to factor trinomials of the form $x^2 + bx + c$.
- Factor trinomials of the form $x^2 + bx + c$.

> **SUGGESTED LEARNING STRATEGIES:** Marking the Text, Create Representations, Think-Pair-Share, Look for a Pattern, Discussion Groups

Recall that Factor Steele Buildings can create many floor plans with different-size spaces. Custom Showrooms has asked Factor Steele Buildings for a floor plan with one great room, five large offices, and six cubicles. Each great room has a length and width equal to x units, each large office has a width of x units and a length of 1 unit, and each cubicle has a length and width of 1 unit.

Factor Steele Buildings proposes the rectangular floor plan shown below.

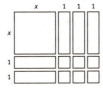

1. Represent the area of the entire office as a sum of the areas of all the rooms.
   $x^2 + 5x + 6$

2. Write the area of the entire office as a product of two binomials by multiplying the length of the entire office by the width of the entire office.
   $(x + 3)(x + 2)$

3. **Make use of structure.** Multiply the binomials in Item 2 to check that their product is the expression you wrote in Item 1. Justify your steps and name any properties you use to multiply the binomials.
   $(x + 3)(x + 2)$
   $= x^2 + 2x + 3x + 6$ Distributive Property
   $= x^2 + 5x + 6$ Simplify.

Common Core State Standards for Activity 27

HSA-SSE.A.1   Interpret expressions that represent a quantity in terms of its context.*

HSA-SSE.A.1.a   Interpret parts of an expression, such as terms, factors, and coefficients.

HSA-SSE.A.1.b   Interpret complicated expressions by viewing one or more of their parts as a single entity.

HSA-SSE.A.2   Use the structure of an expression to identify ways to rewrite it.

ACTIVITY 27
Directed

Activity Standards Focus
In Activity 27, students continue factoring polynomials. They learn to factor trinomials of the form $x^2 + bx + c$, both with and without models. They also learn to factor trinomials of the form $ax^2 + bx + c$ by using a guess and check method. Emphasize that students can always check that they have factored a polynomial correctly by multiplying the factors.

Lesson 27-1

PLAN
Materials
- algebra tiles (optional)

Pacing: 1 class period

Chunking the Lesson
#1–3   #4–6   Example A
Check Your Understanding
Lesson Practice

TEACH

Bell-Ringer Activity
Briefly review what students learned in the last activity by having them factor each of these polynomials and identify the method they used to factor each one.
1. $2x^2 + 6x$
   [$2x(x + 3)$; GCF]
2. $x^2 - 10x + 25$
   [$(x - 5)^2$; perfect square trinomial]
3. $x^2 - 49$
   [$(x - 7)(x + 7)$; difference of squares]

1–3 Shared Reading, Marking the Text, Activating Prior Knowledge, Create Representations, Think-Pair-Share Students use an area model to relate a trinomial of the form $x^2 + bx + c$ to its factored form. After students complete the items, ask them to identify the factored form of $x^2 + 5x + 6$. It is important for students to understand that the expressions $x^2 + 5x + 6$ and $(x + 3)(x + 2)$ are equivalent.

Activity 27 • Factoring Trinomials

ACTIVITY 27 Continued

4–6 Look for a Pattern, Discussion Groups, Sharing and Responding

Students look for patterns between the values of b and c in trinomials of the form $x^2 + bx + c$ and the constant terms of the factored form of the trinomials. As students share and respond to other students, be sure they have recognized that numerical terms in the binomial factors have a sum of b (the coefficient of bx) and a product of c (the constant term in the polynomial). Ask students how they could use the patterns they have identified to determine the product of $(x + 2)$ and $(x + 6)$ without using the Distributive Property.

Differentiating Instruction

The items on this page require students to read and comprehend text that is heavy with mathematical vocabulary. For example, Item 4b includes the phrase *the coefficient of the trinomial's middle term*. This phrase requires students to first recognize that the trinomial is the expression $x^2 - 2x - 15$, then that its middle term is $-2x$, and finally that the coefficient of the middle term is -2.

Support students who are struggling with the mathematical terms by asking key questions about the wording of the items and the meaning of difficult phrases. Questions might include:

- What is the trinomial in this problem?
- How do you know that this expression is a trinomial?
- What is a term of a polynomial?
- What are the terms of the trinomial $x^2 - 2x - 15$?
- What is the middle term?

ACTIVITY 27 continued

My Notes

Lesson 27-1
Factoring $x^2 + bx + c$

Items 1 through 3 show how to use algebra tiles to factor a trinomial. However, drawing tiles to factor a trinomial can become time-consuming. Analyzing patterns and using graphic organizers can help factor a trinomial of the form $x^2 + bx + c$ without using tiles.

4. Consider the binomials $(x - 5)$ and $(x + 3)$.
   a. Determine their product.
   $x^2 - 2x - 15$

   b. How is the coefficient of the trinomial's middle term related to the constant terms of the binomials?
   $-2 = -5 + 3$

   c. How is the constant term of the trinomial related to the constant terms of the binomials?
   $-15 = (-5)(3)$

5. Consider the binomials $(x + 6)$ and $(x + 1)$.
   a. Determine their product.
   $x^2 + 7x + 6$

   b. How is the coefficient of the trinomial's middle term related to the constant terms of the binomials?
   $7 = 6 + 1$

   c. How is the constant term of the trinomial related to the constant terms of the binomials?
   $6 = (6)(1)$

6. **Express regularity in repeated reasoning.** Use the patterns you observed in Items 4 and 5 to analyze a trinomial of the form $x^2 + bx + c$. Describe how the numbers in the binomial factors are related to the constant term c, and to b, the coefficient of x.
   The product of the numbers in the binomial factors is c and their sum is b.

Lesson 27-1
Factoring $x^2 + bx + c$

Example A

Factor $x^2 + 12x + 32$.

Step 1: Create a graph organizer as shown. Place the first term in the upper left region. Place the last term in the lower right region.

x^2	
	32

Step 2: Identify the factors of c that add to b. Use a table to help you test factors.

Factors of 32		Sum of the Factors		
32	1	32 + 1	=	33
16	2	16 + 2	=	18
8	4	8 + 4	=	12 ✓

Step 3: Fill in the missing factors and products in the graphic organizer.

	x	8
x	x^2	8x
4	4x	32

Step 4: Write the original trinomial as the product of two binomials.

$$x^2 + 12x + 32 = (x + 4)(x + 8)$$

ACTIVITY 27 Continued

Example A Note Taking, Graphic Organizer In this example, students are introduced to factoring a trinomial of the form $x^2 + bx + c$ using a graphic organizer. A table is used to test pairs of factors in an organized way. In Step 2, be sure students can explain why the goal is to find factors of 32 that add to 12. In Step 3, ask students to explain how the graphic organizer shows the original trinomial and how it also shows the factored form of the trinomial.

TEACHER to TEACHER

The graphic organizer used in this example is an area model. Explain that a label inside a box represents the area of that box; a label outside a box represents the length or width of that box. For example, the upper right box has a length of x, a width of 8, and an area of $8x$. There are two ways to find the total area of the model. One is by adding the individual areas of the boxes, which results in the original trinomial, $x^2 + 12x + 32$. The other way is to multiply the total length by the total width, which results in the factored form of the trinomial, $(x + 4)(x + 8)$. Thus, the model demonstrates that $x^2 + 12x + 32 = (x + 4)(x + 8)$.

ACTIVITY 27 Continued

TEACHER to TEACHER

As students work on the Try These problems, encourage them to make tables in the My Notes column to help them test factors. As they set up each table, ask them what number they need to list factors for and what sum those factors should have. Be sure students recognize that in Item (a) the sum of the factors of 8 needs to be −6, not 6. Have them discuss what the signs of the factors of 8 must be for the sum of the factors to be negative.

Lesson 27-1
Factoring $x^2 + bx + c$

Try These A

a. Fill in the missing sections of the graphic organizer for the trinomial $x^2 - 6x + 8$. Express the trinomial as a product of two binomials.

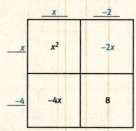

$x^2 - 6x + 8 = (x - 4)(x - 2)$

b. Make a graphic organizer like the one above for the trinomial $x^2 + 14x + 45$. Express the trinomial as a product of two binomials.

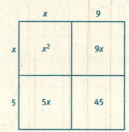

$x^2 + 14x + 45 = (x + 5)(x + 9)$

c. Factor $x^2 + 6x - 27$.
$x^2 + 6x - 27 = (x + 9)(x - 3)$

d. Factor $x^2 + 10x + 1$.
cannot be factored

MATH TIP

If there are no factors of *c* that add to *b*, the trinomial cannot be factored. A polynomial that cannot be factored is called *unfactorable* or a *prime polynomial*.

Lesson 27-1
Factoring $x^2 + bx + c$

Check Your Understanding

Factor each trinomial. Then multiply your factors to check your work.

7. $x^2 + 15x + 56$
8. $x^2 + 22x + 120$
9. $x^2 + 6x - 27$
10. $x^2 - 14x + 48$
11. $x^2 - x + 1$

LESSON 27-1 PRACTICE

Factor each trinomial.

12. $x^2 + 8x + 15$
13. $x^2 - 5x - 14$
14. $x^2 - 5x + 3$
15. $x^2 - 16x + 48$
16. $24 + 10x + x^2$
17. Custom Showrooms has expanded and now wants Factor Steele Buildings to create a floor plan with one great room, 15 large offices, and 50 cubicles.
    a. Write the area of the new floor plan as a trinomial.
    b. Factor the trinomial.
    c. Multiply the binomials in Part (b) to check your work.
18. **Reason abstractly.** Suppose $x^2 + bx + c$ is a factorable trinomial in which c is a positive prime number.
    a. Write an expression to represent the value of b.
    b. Write $x^2 + bx + c$ as the product of two factors using only c as an unknown constant.

MATH TIP

A prime number has only itself and 1 as factors. For example, the numbers 3 and 11 are prime numbers.

ACTIVITY 27 Continued

Lesson 27-2

PLAN

Pacing: 1 class period
Chunking the Lesson
#1 Examples A, B
Example C Example D
Check Your Understanding
Lesson Practice

TEACH

Bell-Ringer Activity
Ask students to factor the following trinomials, and ask them to discuss how they determined the constants of the binomials in the factored forms.
1. $x^2 + 11x + 30$ $[(x + 5)(x + 6)]$
2. $x^2 + 4x - 32$ $[(x - 4)(x + 8)]$
3. $x^2 - 10x + 21$ $[(x - 7)(x - 3)]$

1 Activating Prior Knowledge, Think-Pair-Share This item is designed to introduce students to trinomials of the form $ax^2 + bx + c$. By noting differences between the trinomial given in this item and the trinomials they factored in Lesson 27-1, students may recognize the need to adjust their factoring methods.

Example A Note Taking, Work Backward, Guess and Check In this example, students are introduced to using a guess and check method to factor a trinomial of the form $ax^2 + bx + c$. Be sure to ask questions to aid student understanding of each step of the example. For instance, ask students how they know that the variable terms of the factored form must be $2x$ and x. Also have students discuss how they can determine that both constants of the factored form will be positive. Be sure to emphasize that in the factored form, the coefficients of the variable terms will be factors of a, and the constant terms will be factors of c.

Example B Note Taking, Work Backward, Guess and Check, Look for a Pattern, Discussion Groups In this example, students are introduced to factoring a trinomial of the form $ax^2 + bx + c$, for which c is negative rather than positive. Again, be sure to ask questions to aid student understanding of each step of the example. For instance, ask students how they know that the constants of the factored form will have different signs. Note that students need not continue to check binomial factor pairs after the correct product is found. However, discuss all of the factor pairs listed in this example so that students can see how to confirm their work.

398 SpringBoard® Mathematics **Algebra 1, Unit 4** • Exponents, Radicals, and Polynomials

---

ACTIVITY 27 continued

My Notes

MATH TIP
The factors of c will both have the same sign if $c > 0$. If $b < 0$, both factors will be negative. If $b > 0$, both factors will be positive.

Lesson 27-2
Factoring $ax^2 + bx + c$

Learning Targets:
- Factor trinomials of the form $ax^2 + bx + c$ when the GCF is 1.
- Factor trinomials of the form $ax^2 + bx + c$ when the GCF is not 1.

SUGGESTED LEARNING STRATEGIES: Think-Pair-Share, Note Taking, Guess and Check, Look for a Pattern, Work Backward

Custom Showrooms now wants Factor Steele Buildings to create a floor plan with more than one great room. Instead, Custom Showrooms wants two great rooms, seven large offices, and six cubicles.

The trinomial $2x^2 + 7x + 6$ can be factored to determine the length and width of the entire office space.

1. **Attend to precision.** How is the trinomial $2x^2 + 7x + 6$ different from the trinomials you factored in Lesson 27-1?

   The x^2 term has a coefficient other than 1.

Example A
Factor $2x^2 + 7x + 6$ using a guess and check method.

Possible Binomial Factors	Reasoning
$(2x \quad)(x \quad)$	$a = 2$ can be factored as $2 \cdot 1$.
$(2x + \quad)(x + \quad)$	$c = 6$, so both factors have the same sign. $b = 7$, so both factors are positive. 6 can be factored as $1 \cdot 6$, $6 \cdot 1$, $2 \cdot 3$, or $3 \cdot 2$.
$(2x + 1)(x + 6)$	Product: $2x^2 + 13x + 6$, incorrect
$(2x + 6)(x + 1)$	Product: $2x^2 + 8x + 6$, incorrect
$(2x + 2)(x + 3)$	Product: $2x^2 + 8x + 6$, incorrect
$(2x + 3)(x + 2)$	Product: $2x^2 + 7x + 6$, correct factors

Example B
Factor $3x^2 + 8x - 11$ using a guess and check method.

Possible Binomial Factors	Reasoning
$(3x \quad)(x \quad)$	$a = 3$ can be factored as $3 \cdot 1$.
$(3x + \quad)(x - \quad)$ or $(3x - \quad)(x + \quad)$	$c = -11$, so the factors have different signs. 11 can be factored as $11 \cdot 1$ or $1 \cdot 11$.
$(3x + 11)(x - 1)$	Product: $3x^2 + 8x - 11$, correct factors
$(3x - 11)(x + 1)$	Product: $3x^2 - 8x - 11$, incorrect
$(3x + 1)(x - 11)$	Product: $3x^2 - 32x - 11$, incorrect
$(3x - 1)(x + 11)$	Product: $3x^2 + 32x - 11$, incorrect

Lesson 27-2
Factoring $ax^2 + bx + c$

Try These A–B

Factor the trinomials.

a. $3x^2 + 5x + 2$
   $(3x + 2)(x + 1)$

b. $2x^2 + 5x - 18$
   $(2x + 9)(x - 2)$

c. $2x^2 + 6x - 7$
   cannot be factored

Example C

Factor $4x^2 - 4x - 15$ using a guess and check method.

Possible Binomial Factors	Reasoning
$(4x\ \ )(x\ \ )$ or $(2x\ \ )(2x\ \ )$	$a = 4$ can be factored as $4 \cdot 1$ or $2 \cdot 2$.
$(4x -\ \ )(x +\ \ )$ or $(4x +\ \ )(x -\ \ )$ or $(2x -\ \ )(2x +\ \ )$ or $(2x +\ \ )(2x -\ \ )$	$c = -15$, so the factors have different signs. 15 can be factored as $1 \cdot 15$, $15 \cdot 1$, $3 \cdot 5$, or $5 \cdot 3$.
$(4x - 1)(x + 15)$	Product: $4x^2 + 59x - 15$, incorrect
$(4x + 1)(x - 15)$	Product: $4x^2 - 59x - 15$, incorrect
$(4x - 15)(x + 1)$	Product: $4x^2 - 11x - 15$, incorrect
$(4x + 15)(x - 1)$	Product: $4x^2 + 11x - 15$, incorrect
$(4x - 3)(x + 5)$	Product: $4x^2 + 17x - 15$, incorrect
$(4x + 3)(x - 5)$	Product: $4x^2 - 17x - 15$, incorrect
$(4x - 5)(x + 3)$	Product: $4x^2 + 7x - 15$, incorrect
$(4x + 5)(x - 3)$	Product: $4x^2 - 7x - 15$, incorrect
$(2x - 1)(2x + 15)$	Product: $4x^2 + 28x - 15$, incorrect
$(2x + 1)(2x - 15)$	Product: $4x^2 - 28x - 15$, incorrect
$(2x - 3)(2x + 5)$	Product: $4x^2 + 4x - 15$, incorrect
$(2x + 3)(2x - 5)$	Product: $4x^2 - 4x - 15$, correct factors

Try These C

Factor the trinomials.

a. $6x^2 - 11x - 2$
   $(6x + 1)(x - 2)$

b. $6x^2 - 13x - 4$
   cannot be factored

c. $4x^2 - 20x + 21$
   $(2x - 3)(2x - 7)$

ACTIVITY 27 Continued

Developing Math Language

The values of a and c in the trinomial from Example B are prime numbers. Ask students how this fact helps to simplify the factoring process.

Example C Note Taking, Work Backward, Guess and Check

In this example, students factor a trinomial of the form $ax^2 + bx + c$ for which the value of a is not prime. Thus, the guess and check method must be applied not only to the constant terms of the binomials, but also to the variable terms. Ask students to discuss why there are two different possibilities for the variable terms in the binomial factors in this example, while there was only one possibility for the variable terms in Examples A and B. When students work on the Try These problems, have them compare their answers to those of others in their group.

Differentiating Instruction

Extend learning for students in Example C by having them use logical reasoning to determine which constant in the binomial factors will have the greater absolute value. Because the constants have different signs and the value of b is negative, the negative constant must have a greater absolute value than the positive constant. This fact can help students eliminate some of the factor pairs so that they do not have to check as many. For instance, the factor pair -1 and 15 and the factor pair -3 and 5 can be eliminated in Example C because they would result in a positive value of b.

ACTIVITY 27 Continued

Example D Note Taking, Work Backward, Guess and Check

Students factor a trinomial of the form $ax^2 + bx + c$ in which the terms have a common factor. Students are shown how to factor by using the GCF of the terms, followed by the guess and check method. In Step 1, ask students to explain how they know that the GCF of the terms is 3. In Step 2, have students discuss how they know that both constants of the binomial factors will be negative.

Check Your Understanding

Debrief students' answers to these items to ensure that they understand how to factor trinomials of the form $ax^2 + bx + c$. Have students explain the steps they used to factor the trinomials in each of these items.

Answers

2. $(5x - 4)(x + 1)$
3. $7(7x - 4)(x - 2)$
4. $3x(3x + 2)(x - 5)$

ASSESS

Students' answers to lesson practice problems will provide you with a formative assessment of their understanding of the lesson concepts and their ability to apply their learning.

See the Activity Practice for additional problems for this lesson. You may assign the problems here or use them as a culmination for the activity.

ADAPT

Check students' answers to the Lesson Practice questions to make sure that they understand concepts related to writing and factoring trinomials of the form $ax^2 + bx + c$. If students are struggling to factor trinomials, encourage them to develop a set of steps, such as the following, for factoring trinomials.

- Factor out the GCF if needed.
- List factor pairs of a.
- Use the factor pairs of a to determine possible combinations for the variable terms of the binomials.
- Use the values of b and c to determine the signs of the constant terms in the binomials.
- List factor pairs of c.
- Check factor pairs of c in the binomials until you get the correct product.

LESSON 27-2 PRACTICE

5. $(3x + 2)(x + 7)$
6. $(5x + 3)(2x + 5)$
7. $4(2x + 5)(x + 9)$
8. $2(3x + 2)(2x + 5)$
9. If a and c are prime, their only factors are 1 and themselves. Therefore, the factorization is either $(ax + c)(x + 1)$ or $(ax + 1)(x + c)$. Therefore, $b = a + c$ or $b = 1 + ac$.

ACTIVITY 27 continued

Lesson 27-2
Factoring $ax^2 + bx + c$

Example D

Factor $9x^2 - 24x + 12$.

Step 1: The coefficients 9, −24, and 12 are all divisible by 3. Factor out the GCF.
$$9x^2 - 24x + 12 = 3(3x^2 - 8x + 4)$$

Step 2: Factor $3x^2 - 8x + 4$ using a guess and check method.

Possible Binomial Factors	Reasoning
$(3x\quad)(x\quad)$	$a = 3$ can be factored as $3 \cdot 1$.
$(3x - \quad)(x - \quad)$	$c = 4$, so the factors have the same sign. $b = -8$, so both factors are negative. 4 can be factored as $4 \cdot 1$, $1 \cdot 4$, or $2 \cdot 2$.
$(3x - 4)(x - 1)$	Product: $3x^2 - 7x + 4$, incorrect
$(3x - 1)(x - 4)$	Product: $3x^2 - 13x + 4$, incorrect
$(3x - 2)(x - 2)$	Product: $3x^2 - 8x + 4$, correct factors

Solution: Write the complete factorization, including the GCF from Step 1:
$3(3x - 2)(x - 2)$

Check: Multiply to check your answer.
$3(3x - 2)(x - 2)$
$= 3(3x^2 - 6x - 2x + 4) = 3(3x^2 - 8x + 4) = 9x^2 - 24x + 12$

Try These D

Factor the trinomials completely. Check your work by multiplying the factors.

a. $10x^2 + 19x + 6$ b. $8x^2 + 20x - 28$ c. $8x^3 - 14x^2 + 6x$
   $(5x + 2)(2x + 3)$ $4(2x + 7)(x - 1)$ $2x(4x - 3)(x - 1)$

Check Your Understanding

Factor each trinomial completely. Check your work by multiplying the factors.

2. $5x^2 + x - 4$ 3. $49x^2 - 126x + 56$ 4. $9x^3 - 39x^2 - 30x$

LESSON 27-2 PRACTICE

Model with mathematics. Factor Steele Buildings has received several floor plan requests. For Items 5–8, factor each floor plan scenario completely to help Factor Steele Buildings determine the space's dimensions.

5. 3 great rooms, 23 large offices, 14 cubicles
6. 10 great rooms, 31 large offices, 15 cubicles
7. 8 great rooms, 92 large offices, 180 cubicles
8. 12 great rooms, 38 large offices, 20 cubicles
9. Suppose $ax^2 + bx + c$ is a factorable trinomial in which both a and c are positive prime numbers. Write an expression to represent the value of b.

Factoring Trinomials
Deconstructing Floor Plans

ACTIVITY 27
continued

ACTIVITY 27 PRACTICE
Write your answers on notebook paper.
Show your work.

Lesson 27-1
Factor each trinomial.

1. $x^2 + 11x + 30$
2. $x^2 + 22x + 121$
3. $x^2 + x - 30$
4. $x^2 - 7x - 18$
5. $x^2 - 169$
6. $x^2 + 9x - 36$

Mrs. Harbrook can choose from two rectangular pool sizes. The pool manufacturer provides her with the area of the pool, but she needs to find the dimensions in order to determine if the pool will fit in her yard. Use the rectangle for Items 7 and 8.

7. If the area of the pool is $x^2 - 17x + 72$, what are possible expressions to represent the length L and the width W?

8. **a.** If the area of the pool is $x^2 + 24x + 144$, what are possible expressions to represent the length L and the width W?
   b. What do these dimensions tell you about the shape of the pool?

The area of a parallelogram is given by the formula $A = bh$, where b is the base and h is the height. Use this information for Items 9 and 10.

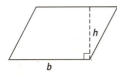

9. If the area of the parallelogram is $x^2 + x - 42$, what are possible expressions to represent the base b and the height h?

10. If the area of the parallelogram is $x^2 + 4x - 117$, what are possible expressions to represent the base b and the height h?

11. Which of the following trinomials cannot be factored?
    A. $x^2 + 3x + 2$ **B.** $x^2 + 3x - 2$
    C. $x^2 - 3x + 2$ **D.** $x^2 + 2x - 3$

12. Which of the following binomials is a factor of the trinomial $y^2 - y - 20$?
    A. $y - 4$ **B.** $y + 4$
    C. $y - 10$ **D.** $y + 10$

For Items 13–15, consider the trinomial $x^2 + 2x + c$. Determine whether each statement is always, sometimes, or never true.

13. If c is a prime number, then the trinomial cannot be factored.

14. If c is an even number, then the GCF of the terms in the trinomial is 2.

15. If $c < 0$, then the trinomial can be factored.

16. Write a trinomial that can be factored such that one of the binomial factors is $x - 5$. Explain how you found the trinomial.

Lesson 27-2
Factor each trinomial completely.

17. $3x^2 + 8x - 11$
18. $5x^2 - 7x + 2$
19. $2x^2 - 9x - 5$
20. $3x^2 + 17x - 28$
21. $7x^2 + 9x + 2$
22. $6x^2 - 11x - 7$
23. $12x^2 - 11x + 2$
24. $8x^2 + 16x + 6$

ACTIVITY 27 Continued

ACTIVITY PRACTICE
1. $(x + 5)(x + 6)$
2. $(x + 11)^2$
3. $(x + 6)(x - 5)$
4. $(x - 9)(x + 2)$
5. $(x + 13)(x - 13)$
6. $(x + 12)(x - 3)$
7. $x - 8$ and $x - 9$
8. **a.** $x + 12$ for both the length and width
   b. The pool is square.
9. $x - 6$ and $x + 7$
10. $x + 13$ and $x - 9$
11. B
12. B
13. always
14. never
15. sometimes
16. Sample answer: $x^2 - 3x - 10$; I found the trinomial by multiplying $(x - 5)$ by $(x + 2)$.
17. $(3x + 11)(x - 1)$
18. $(5x - 2)(x - 1)$
19. $(2x + 1)(x - 5)$
20. $(3x - 4)(x + 7)$
21. $(7x + 2)(x + 1)$
22. $(3x - 7)(2x + 1)$
23. $(3x - 2)(4x - 1)$
24. $2(2x + 1)(2x + 3)$

Activity 27 • Factoring Trinomials **401**

ACTIVITY 27 Continued

25. D
26. A
27. $3x$, $2x - 3$, and $x + 2$
28. 5, $2x - 3$, and $x - 4$
29. 2, $3x + 1$, and $2x + 3$
30. B
31. D
32. No; she should have started by factoring out the GCF, which is 2.
33. C
34. $(px - q)^2$
35. Sample answer: $2x^2 + 10x + 3$; Based on the values of a and c, the only possible factorizations are $(2x \pm 1)(x \pm 3)$ and $(2x \pm 3)(x \pm 1)$, but none of these produce the correct value of b.
36. A
37. No, he is not correct. One factor will involve subtraction and the other addition. If both involved subtraction, then the constant term of the trinomial would be $+8$.

ADDITIONAL PRACTICE

If students need more practice on the concepts in this activity, see the eBook Teacher Resources for additional practice problems.

ACTIVITY 27
continued

Factoring Trinomials
Deconstructing Floor Plans

25. Which of the following is **not** a factor of the trinomial $24x^3 - 6x^2 - 9x$?
    A. $3x$ B. $4x - 3$
    C. $2x + 1$ D. $2x - 1$

26. Which binomial is a factor of $4x^2 + 12x + 5$?
    A. $2x + 5$ B. $2x - 5$
    C. $4x + 1$ D. $4x - 1$

The volume of a rectangular prism is found using the formula $V = lwh$, where l is the length, w is the width, and h is the height. Use the rectangular prism for Items 27–29.

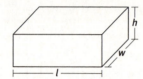

27. If the volume of a rectangular prism is $6x^3 + 3x^2 - 18x$, what are possible expressions to represent the length, width, and height?

28. If the volume of a rectangular prism is $10x^2 - 55x + 60$, what are possible expressions to represent the length, width, and height?

29. If the volume of a rectangular prism is $12x^2 + 22x + 6$, what are possible expressions to represent the length, width, and height?

30. For which value of k is it possible to factor the trinomial $2x^2 + 3x + k$?
    A. -1 B. 1
    C. 2 D. 3

31. Which of the following trinomials has the binomial $x + 1$ as a factor?
    A. $2x^2 - x - 1$
    B. $2x^2 - 3x + 1$
    C. $3x^2 - 5x + 2$
    D. $3x^2 + x - 2$

32. Mayumi was asked to completely factor the trinomial $4x^2 + 10x + 4$. Her work is shown below. Is her solution correct? Justify your response.

> 4 can be factored as 4 · 1 or 2 · 2.
> Try $(2x + \quad)(2x + \quad)$.
> $(2x + 2)(2x + 2) = 4x^2 + 8x + 4$; incorrect
> $(2x + 4)(2x + 1) = 4x^2 + 10x + 4$; correct!
> The factorization is $(2x + 4)(2x + 1)$.

33. Given that the trinomial $5x^2 + bx + 10$ can be factored, which of the following statements must be true?
    A. The value of b must be positive.
    B. The value of b must be negative.
    C. The value of b cannot be 3.
    D. The value of b cannot be -27.

34. What is the factorization of the trinomial $p^2x^2 - 2pqx + q^2$?

35. Write a trinomial of the form $ax^2 + bx + c$ (with $a \neq 1$) that cannot be factored into binomial factors. Explain how you know the trinomial cannot be factored.

36. The area of a rectangular carpet is $6x^2 - 11x + 4$ square yards. The length of the carpet is $3x - 4$ yards. Which of the following is the width?
    A. $2x - 1$ yards B. $2x + 1$ yards
    C. $3x - 1$ yards D. $3x + 1$ yards

MATHEMATICAL PRACTICES
Construct Viable Arguments and Critique the Reasoning of Others

37. Guillaume is asked to factor a trinomial of the form $x^2 + bx - 8$. He says that because the constant term is negative, both binomial factors of the trinomial will involve subtraction. Is he correct? Explain.

Simplifying Rational Expressions
Totally Rational
Lesson 28-1 Simplifying Rational Expressions

Learning Targets:
- Simplify a rational expression by dividing a polynomial by a monomial.
- Simplify a rational expression by dividing out common factors.

> **SUGGESTED LEARNING STRATEGIES:** Think-Pair-Share, Note Taking, Identify a Subtask

A field trips costs $800 for the charter bus plus $10 per student for x students. The cost per student is represented by the expression $\frac{10x + 800}{x}$.

The cost-per-student expression is a rational expression. A **rational expression** is a ratio of two polynomials.

Like fractions, rational expressions can be simplified and combined using the operations of addition, subtraction, multiplication, and division.

When a rational expression has a polynomial in the numerator and a monomial in the denominator, it may be possible to simplify the expression by dividing each term of the polynomial by the monomial.

Example A
Simplify by dividing: $\dfrac{12x^5 + 6x^4 - 9x^3}{3x^2}$

Step 1: Rewrite the rational expression to indicate each term of the numerator divided by the denominator.

$$\frac{12x^5}{3x^2} + \frac{6x^4}{3x^2} - \frac{9x^3}{3x^2}$$

Step 2: Divide. Use the Quotient of Powers Property.

$$\frac{12x^5}{3x^2} + \frac{6x^4}{3x^2} - \frac{9x^3}{3x^2}$$
$$4x^{5-2} + 2x^{4-2} - 3x^{3-2}$$
$$4x^3 + 2x^2 - 3x^1$$

Solution: $4x^3 + 2x^2 - 3x$

Try These A
Simplify by dividing.

a. $\dfrac{5y^4 - 10y^3 - 5y^2}{5y^2}$

$y^2 - 2y - 1$

b. $\dfrac{32n^6 - 24n^4 + 16n^2}{-8n^2}$

$-4n^4 + 3n^2 - 2$

Common Core State Standards for Activity 28

HSA-APR.D.6 Rewrite simple rational expressions in different forms; write $\dfrac{a(x)}{b(x)}$ in the form $q(x) + \dfrac{r(x)}{b(x)}$, where $a(x)$, $b(x)$, $q(x)$, and $r(x)$ are polynomials with the degree of $r(x)$ less than the degree of $b(x)$, using inspection, long division, or, for the more complicated examples, a computer algebra system.

HSA-APR.D.7 (+) Understand that rational expressions form a system analogous to the rational numbers, closed under addition, subtraction, multiplication, and division by a nonzero rational expression; add, subtract, multiply, and divide rational expressions.

ACTIVITY 28 Continued

TEACHER to TEACHER

Students sometimes try to divide across the operation of addition. A numerical example can help students understand why this is not correct. Ask them to identify the error in the following work:
$\frac{3+2}{3} = \frac{\cancel{3}+2}{\cancel{3}} = \frac{1+2}{1} = \frac{3}{1}$. The error is that each term of the numerator must be divided by the denominator, not just the first term: $\frac{3+2}{3} = \frac{\cancel{3}}{\cancel{3}} + \frac{2}{3} = 1\frac{2}{3}$. Ask students whether it would be valid to write $\frac{x+4}{x}$ as $\frac{x+4}{\cancel{x}}$.

Example B Note Taking, Identify a Subtask In this example, students are introduced to simplifying the ratio of two monomials by dividing out common factors. In this method, powers of the variable are written in expanded form. Using the expanded form may be easier for students who have difficulty remembering and applying the properties of exponents.

Example C Note Taking, Activating Prior Knowledge, Identify a Subtask, Think-Pair Share In this example, students are introduced to simplifying a rational expression in which the numerator and denominator have binomial factors. As in Example B, the numerator and denominator are factored, and then common factors are divided out. Ask students to discuss the steps they would use to factor the numerator $2x^2 - 8$ and the steps they would use to factor the denominator $x^2 - 2x - 8$. Have students compare solutions with others in their groups.

ACTIVITY 28 continued

My Notes

Lesson 28-1
Simplifying Rational Expressions

To simplify a rational expression, first factor the numerator and denominator. Remember that factors can be monomials, binomials, or even polynomials. Then, divide out the common factors.

Example B
Simplify $\frac{12x^2}{6x^3}$.

Step 1: Factor the numerator and denominator.
$$\frac{2 \cdot 6 \cdot x \cdot x}{6 \cdot x \cdot x \cdot x}$$

Step 2: Divide out the common factors.
$$\frac{2 \cdot \cancel{6} \cdot \cancel{x} \cdot \cancel{x}}{\cancel{6} \cdot x \cdot \cancel{x} \cdot \cancel{x}}$$

Solution: $\frac{2}{x}$

> **MATH TIP**
> If a, b, and c are polynomials, and b and c do not equal 0, then $\frac{ac}{bc} = \frac{a}{b}$, because $\frac{c}{c} = 1$.

Example C
Simplify $\frac{2x^2 - 8}{x^2 - 2x - 8}$.

Step 1: Factor the numerator and denominator.
$$\frac{2(x+2)(x-2)}{(x+2)(x-4)}$$

Step 2: Divide out the common factors.
$$\frac{2\cancel{(x+2)}(x-2)}{\cancel{(x+2)}(x-4)}$$

Solution: $\frac{2(x-2)}{x-4}$

> **MATH TIP**
> The graph of $y = \frac{2}{x}$ will never cross the x-axis since x cannot equal 0.
>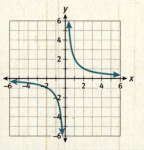

Try These B–C
Simplify each rational expression.

a. $\frac{6x^4 y}{15xy^3}$
   $\frac{2x^3}{5y^2}$

b. $\frac{x^2 + 3x - 4}{x^2 - 16}$
   $\frac{x-1}{x-4}$

c. $\frac{15x^2 - 3x}{25x^2 - 1}$
   $\frac{3x}{5x+1}$

The value of the denominator in a rational expression cannot be zero because division by zero is undefined.
- In Example B, x cannot equal 0 because $6 \cdot (0)^3 = 0$.
- To find the excluded values of x in Example C, first factor the denominator. This shows that $x \neq -2$ because that would make the factor $x + 2 = 0$. Also, $x \neq 4$ because that would make the factor $x - 4 = 0$. Therefore, in Example C, x cannot equal -2 or 4.

Lesson 28-1
Simplifying Rational Expressions

Example D

Divide $\frac{1-x}{x-1}$. Simplify your answer if possible.

Step 1: Factor the numerator. $\quad -1(x-1)$

Step 2: Divide out the common factor. $\quad \frac{-1\cancel{(x-1)}}{\cancel{x-1}}$

Solution: -1

Try These D

Divide. Simplify your answer if possible. Identify any excluded values of the variable.

a. $\frac{x-5}{5-x}$ $-1; x \neq 5$
b. $\frac{3x-3}{1-x}$ $-3; x \neq 1$

Check Your Understanding

Attend to precision. Describe the steps you would take to simplify each rational expression. Identify any excluded values of the variable.

1. $\frac{x^2-36}{6-x}$
2. $\frac{x^2-10x+24}{4x-16}$

MATH TIP

The graph of the rational function $f(x) = \frac{1-x}{x-1}$ looks like:

It looks like the graph of $y = -1$.

LESSON 28-1 PRACTICE

Simplify by dividing.

3. $\frac{16x^5 - 8x^3 + 4x^2}{4x^2}$
4. $\frac{15x^6 - 20x^4}{-5x^3}$
5. $\frac{24x^6 + 18x^5 - 15x^3 + 12x^2}{3x^2}$

Simplify.

6. $\frac{3x^2yz}{12xyz^3}$
7. $\frac{25x^4y^3z^4}{-5x^5y^2z^3}$
8. $\frac{x^2-2x+1}{x^2+3x-4}$
9. $\frac{2x^2}{4x^3-16x}$
10. $\frac{x+1}{-4-4x}$
11. $\frac{x^2+6x+9}{x^2-9}$

12. **Model with mathematics.** The four algebra classes at Sanchez School are going on a field trip to a museum. Each class contains s students. The museum charges $8 per student for admission. There is also a flat fee of $200 for the buses.
    a. Write an expression for the total cost of the buses and the museum admission fees for all four classes.
    b. Write a rational expression for the cost per student. Simplify the expression as much as possible.
    c. Use the expression you wrote in Part (b) to find the cost per student if each class has 20 students.

LESSON 28-1 PRACTICE

3. $4x^3 - 2x + 1$
4. $-3x^3 + 4x$
5. $8x^4 + 6x^3 - 5x + 4$
6. $\frac{x}{4z^2}$
7. $\frac{-5yz}{x}$
8. $\frac{x-1}{x+4}$
9. $\frac{x}{2(x-2)(x+2)}$
10. $-\frac{1}{4}$
11. $\frac{x+3}{x-3}$
12. a. $200 + 32s$
    b. $\frac{50+8s}{s}$
    c. $10.50

ACTIVITY 28 Continued

Lesson 28-2

PLAN

Chunking the Lesson
Example A   Example B
Example C   Example D   #1
Check Your Understanding
Lesson Practice

TEACH

Bell-Ringer Activity
Ask students to simplify each rational expression by dividing.

1. $\dfrac{3x^2 + 4x}{x}$ $[3x + 4]$
2. $\dfrac{4x^6 - 6x^5 + 2x^2}{2x^2}$ $[2x^4 - 3x^3 + 1]$

Example A Activating Prior Knowledge, Summarizing In this example, students review the steps of long division. Reminding students of this process helps to prepare them for using similar steps when they divide polynomials later in the lesson.

> **TEACHER to TEACHER**
>
> You may want to have students create a KWL chart for this lesson. They can fill in what they already know about dividing whole numbers and what they want to know about dividing polynomials. As students work through the examples, they can add to their charts by summarizing what they have learned.

ACTIVITY 28 continued

Lesson 28-2
Dividing Polynomials

Learning Targets:
- Divide a polynomial of degree one or two by a polynomial of degree one or two.
- Express the remainder of polynomial division as a rational expression.

SUGGESTED LEARNING STRATEGIES: Think-Pair-Share, Identify a Subtask, Close Reading, Note Taking, Discussion Groups

Division of polynomials is similar to long division of real numbers.

Example A

Divide $\dfrac{525}{25}$ using long division.

Step 1: Divide 52 by 25.

$$\begin{array}{r} 2 \\ 25{\overline{\smash{)}525}} \\ -\,50 \\ \hline 2 \end{array}$$

Step 2: Bring down 5.

$$\begin{array}{r} 2 \\ 25{\overline{\smash{)}525}} \\ -\,50\downarrow \\ \hline 25 \end{array}$$

Step 3: Divide 25 by 25.

$$\begin{array}{r} 21 \\ 25{\overline{\smash{)}525}} \\ -\,50\downarrow \\ \hline 25 \\ -\,25 \\ \hline 0 \end{array}$$

Solution: The quotient is 21.

Lesson 28-2
Dividing Polynomials

Division with polynomials can be done in the same way as long division with whole numbers.

Example B
Simplify using long division: $\dfrac{12x^5 + 6x^4 - 9x^3}{3x^2}$

Step 1: Divide $12x^5$ by $3x^2$.

$$\begin{array}{r} 4x^3 \\ 3x^2 \overline{\smash{)}12x^5 + 6x^4 - 9x^3} \\ \underline{-12x^5} \\ 0 \end{array}$$

Step 2: Bring down $6x^4$.

$$\begin{array}{r} 4x^3 \\ 3x^2 \overline{\smash{)}12x^5 + 6x^4 - 9x^3} \\ \underline{-12x^5}\downarrow \\ 6x^4 \end{array}$$

Step 3: Divide $6x^4$ by $3x^2$.

$$\begin{array}{r} 4x^3 + 2x^2 \\ 3x^2 \overline{\smash{)}12x^5 + 6x^4 - 9x^3} \\ \underline{-12x^5} \\ 6x^4 \\ \underline{-6x^4} \\ 0 \end{array}$$

Step 4: Bring down $-9x^3$.

$$\begin{array}{r} 4x^3 + 2x^2 \\ 3x^2 \overline{\smash{)}12x^5 + 6x^4 - 9x^3} \\ \underline{-12x^5} \\ 6x^4 \\ \underline{-6x^4} \\ -9x^3 \end{array}$$

Step 5: Divide $-9x^3$ by $3x^2$.

$$\begin{array}{r} 4x^3 + 2x^2 - 3x \\ 3x^2 \overline{\smash{)}12x^5 + 6x^4 - 9x^3} \\ \underline{-12x^5} \\ 6x^4 \\ \underline{-6x^4} \\ -9x^3 \\ \underline{-(-9x^3)} \\ 0 \end{array}$$

Solution: The quotient is $4x^3 + 2x^2 - 3x$.

ACTIVITY 28 Continued

Example C Close Reading, Note Taking, Identify a Subtask, Think-Pair-Share In this example, students are introduced to the use of long division to divide a polynomial by a binomial. In Step 1, ask students why it is important that x in the quotient is written directly above x^2 in the dividend. Make sure students understand that after you write x in the quotient, you must multiply it by *both* terms of the divisor: $x(x + 7) = x^2 + 7x$. Thus, you need to subtract $x^2 + 7x$ from $x^2 + 9x$ in the dividend. In Step 2, ask students what is meant by distributing the negative and why distributing the negative is necessary when subtracting $x^2 + 7x$. Draw students' attention to the Math Tip, which shows how to check the quotient. As students complete the Try These items, monitor their discussions carefully to ensure that they are not skipping terms or dropping signs as they multiply and subtract.

Differentiating Instruction

Support students who have trouble keeping terms aligned during long division of polynomials by suggesting that they solve these problems on grid paper. Writing one term or operation symbol in each grid square may help them to avoid skipping a term in the dividend or bringing down the same term twice.

ACTIVITY 28 continued

My Notes

MATH TIP

You can check the quotient in a division problem by using multiplication. Multiply the quotient, $x + 2$, by the divisor, $x + 7$. If you have divided correctly, the product will be the dividend, $x^2 + 9x + 14$.

$(x + 2)(x + 7) =$
$x^2 + 7x + 2x + 14 =$
$x^2 + 9x + 14$

Lesson 28-2
Dividing Polynomials

Example C
Simplify using long division: $\dfrac{x^2 + 9x + 14}{x + 7}$.

Step 1: Divide x^2 by x.

Step 2: Distribute the negative and subtract $x^2 + 7x$ from $x^2 + 9x$.

Step 3: Bring down the next term, 14.

Step 4: Divide $2x$ by x.

Step 5: Distribute the negative and subtract $2x + 14$ from $2x + 14$.

Solution: The quotient is $x + 2$.

Try These A–B–C
Simplify using long division.

a. $\dfrac{24x^5 - 8x^4 + 12x^3 - 4x^2}{4x^2}$

  $8x^3 - 2x^2 + 3x - 1$

b. $\dfrac{x^2 - x - 12}{x - 4}$

  $x + 3$

---

408 SpringBoard® Mathematics **Algebra 1, Unit 4** • Exponents, Radicals, and Polynomials

Lesson 28-2
Dividing Polynomials

Sometimes there are remainders when dividing integers. In a similar way, sometimes there are remainders when dividing polynomials.

Example D
Simplify using long division: $\dfrac{2x^3 - 6x + 15}{x + 1}$.

Step 1: Divide. Add the term $0x^2$ to the dividend as a placeholder.

$$\begin{array}{r} 2x^2 - 2x - 4 \\ x+1 \overline{\smash{\big)} 2x^3 + 0x^2 - 6x + 15} \\ -\underline{(2x^3 + 2x^2)} \\ -2x^2 - 6x \\ -\underline{(-2x^2 - 2x)} \\ -4x + 15 \\ -\underline{(-4x - 4)} \\ 19 \end{array}$$

Step 2: Write the remainder as $\dfrac{19}{x+1}$.

$$\begin{array}{r} 2x^2 - 2x - 4 + \dfrac{19}{x+1} \\ x+1 \overline{\smash{\big)} 2x^3 + 0x^2 - 6x + 15} \\ -\underline{(2x^3 + 2x^2)} \\ -2x^2 - 6x \\ -\underline{(-2x^2 - 2x)} \\ -4x + 15 \\ -\underline{(-4x - 4)} \\ 19 \end{array}$$

Solution: The quotient is $2x^2 - 2x - 4 + \dfrac{19}{x+1}$.

Try These D
Simplify using long division.

a. $(3x^2 + 6x + 1) \div (3x)$

$x + 2 + \dfrac{1}{3x}$

b. $(6x^2 + 5x - 20) \div (3x + 4)$

$2x - 1 + \dfrac{-16}{3x+4}$

c. $(3x^3 + 5x - 10) \div (x - 4)$

$3x^2 + 12x + 53 + \dfrac{202}{x-4}$

MATH TIP

When dividing with integers, the remainder is often written as a fraction whose denominator is the divisor.

$$\begin{array}{r} 76\tfrac{1}{2} \\ 2\overline{\smash{\big)}153} \\ \underline{-14} \\ 13 \\ \underline{-12} \\ 1 \end{array}$$

ACTIVITY 28 Continued

1 Identify a Subtask, Discussion Groups, Debriefing In this item, students use long division to divide polynomials when the degree of both the dividend and the divisor is 2. As you debrief this item, ask students to explain how they know when to stop dividing. Ask them to describe any similarities and differences between this solution and the solutions to the Try These items they just completed.

Check Your Understanding

Debrief students' answers to these items to ensure that they understand the relationship between long division for polynomials and long division for whole numbers. You may also want to ask students whether the quotient of two polynomials is always a polynomial and to tell how the quotient in Item 4 supports their answer.

ASSESS

Students' answers to lesson practice problems will provide you with a formative assessment of their understanding of the lesson concepts and their ability to apply their learning.

See the Activity Practice for additional problems for this lesson. You may assign the problems here or use them as a culmination for the activity.

ADAPT

Check students' answers to the Lesson Practice questions to make sure that they understand how to divide polynomials using long division. Pair the students who may need more practice with polynomial division. Ask students to write two binomials and multiply them, without showing their partner. For example, a student might write $(x + 5)(2x - 3) = 2x^2 + 7x - 15$. Students can then create a long division problem for their partners by using the product as the dividend and one of the binomial factors as the divisor. For instance, a student's problem might read, "What is $2x - 3 \overline{\smash{)}2x^2 + 7x - 15}$?" After students solve each other's division problems, they can check each other's work.

Lesson 28-2
Dividing Polynomials

1. **Make sense of problems.** Consider the following polynomial division problem: $\dfrac{3x^2 - 8x + 15}{x^2 + 3x - 4}$.

   a. How does this division problem differ from those in the examples?
   The degree of the polynomial in the denominator matches the degree of the polynomial in the numerator.

   b. Use long division to perform the division.

   c. Write the remainder as a rational expression.
   $\dfrac{-17x + 27}{x^2 + 3x - 4}$

   d. What is the quotient?
   $3 + \dfrac{-17x + 27}{x^2 + 3x - 4}$

Check Your Understanding

2. **Make use of structure.** Describe how dividing polynomials using long division is similar to dividing whole numbers using long division.

3. Explain how to check a division problem involving whole numbers that has a remainder.

4. Explain how to check a division problem involving polynomials that has a remainder. To demonstrate, use
$(3x^2 + x - 2) \div (x^2 + 2x + 3) = 3 + \dfrac{-5x - 11}{x^2 + 2x + 3}$.

LESSON 28-2 PRACTICE
Simplify using long division.

5. $\dfrac{4x^2 + 6x}{2x}$

6. $\dfrac{3x^4 - 9x^3 + 6x^2}{3x^2}$

7. $\dfrac{12x^5 + 24x^4 - 16x^3 - 12x^2}{-4x^2}$

8. $\dfrac{3x^2 - 6x - 24}{3x - 6}$

9. $\dfrac{5x^2 - 21x + 4}{5x - 1}$

10. $\dfrac{12x^2 - 15}{x + 5}$

11. $\dfrac{3x^2 + 6x - 9}{x + 1}$

12. $\dfrac{25x^2 + 20x - 15}{5x^2 + 5x + 5}$

13. **Reason abstractly.** The area of a rectangular swimming pool is $2x^2 + 11x + 4$ square feet. The width of the pool is $x - 2$ feet.

    a. Write a rational expression that represents the length of the pool. Simplify the expression using long division.

    b. What are the length, width, and area of the pool when $x = 19$?

Answers

2. Sample answer: Dividing polynomials is like dividing whole numbers using long division because the procedure uses the same steps. As with whole numbers, the remainder can be written as a fraction.

3. If the remainder is written as a fraction, then multiply the whole-number part of the quotient by the divisor and multiply the remainder by the divisor. The sum should be the dividend.

4. Multiply $\left(3 + \dfrac{-5x - 11}{x^2 + 2x + 3}\right)$ by $(x^2 + 2x + 3)$ to show that this product equals $3x^2 + x - 2$.

LESSON 28-2 PRACTICE

5. $2x + 3$

6. $x^2 - 3x + 2$

7. $-3x^3 - 6x^2 + 4x + 3$

8. $x - \dfrac{8}{x - 2}$

9. $x - 4$

10. $12x - 60 + \dfrac{285}{x + 5}$

11. $3x + 3 - \dfrac{12}{x + 1}$

12. $5 - \dfrac{x + 8}{x^2 + x + 1}$

13. a. $\dfrac{2x^2 + 11x + 4}{x - 2}$; $2x + 15 + \dfrac{34}{x - 2}$

    b. The length is 55 ft, the width is 17 ft, and the area is 935 ft².

Lesson 28-3
Multiplying and Dividing Rational Expressions

Learning Targets:
- Multiply rational expressions.
- Divide rational expressions.

SUGGESTED LEARNING STRATEGIES: Note Taking, Close Reading

To multiply rational expressions, first factor the numerator and denominator of each expression. Next, divide out any common factors. Then simplify, if possible.

Example A
Multiply $\dfrac{2x-4}{x^2-1} \cdot \dfrac{3x+3}{x^2-2x}$. Simplify your answer if possible.

Step 1: Factor the numerators and denominators.
$$\dfrac{2(x-2)}{(x+1)(x-1)} \cdot \dfrac{3(x+1)}{x(x-2)}$$

Step 2: Divide out common factors.
$$\dfrac{2(\cancel{x-2}) \cdot 3(\cancel{x+1})}{(\cancel{x+1})(x-1)(x)(\cancel{x-2})}$$

Solution: $\dfrac{6}{x(x-1)}$

Try These A
Multiply. Simplify your answer.

a. $\dfrac{y^2+5y+6}{y+2} \cdot \dfrac{y}{2y+6}$ $\left[\dfrac{y}{2}\right]$

b. $\dfrac{2x+2}{x^2-16} \cdot \dfrac{x^2-5x+4}{4x^2-4}$ $\left[\dfrac{1}{2(x+4)}\right]$

To divide rational expressions, use the same process as dividing fractions. Write the division as multiplication of the reciprocal. Then simplify.

Example B
Divide: $\dfrac{x^2-5x+6}{x^2-9} \div \dfrac{2x-4}{x^2+2x-3}$. Simplify your answer.

Step 1: Rewrite the division as multiplication by the reciprocal.
$$\dfrac{x^2-5x+6}{x^2-9} \cdot \dfrac{x^2+2x-3}{2x-4}$$

Step 2: Factor the numerators and the denominators.
$$\dfrac{(x-2)(x-3)}{(x+3)(x-3)} \cdot \dfrac{(x+3)(x-1)}{2(x-2)}$$

Step 3: Divide out common factors.
$$\dfrac{(\cancel{x-2})(\cancel{x-3})(\cancel{x+3})(x-1)}{(\cancel{x+3})(\cancel{x-3})(2)(\cancel{x-2})}$$

Solution: $\dfrac{x-1}{2}$

MATH TIP

When dividing fractions, write the division as multiplication by the reciprocal.
$$\dfrac{a}{b} \div \dfrac{c}{d} = \dfrac{a}{b} \cdot \dfrac{d}{c} = \dfrac{ad}{bc}$$
If a, b, c, and d have any common factors, you can divide them out before you multiply.
$$\dfrac{4}{15} \div \dfrac{8}{3} = \dfrac{4}{15} \cdot \dfrac{3}{8}$$
$$= \dfrac{\cancel{4}}{\cancel{3} \cdot 5} \cdot \dfrac{\cancel{3}}{2 \cdot \cancel{4}} = \dfrac{1}{10}$$

MINI-LESSON: Dividing out Common Factors

A common misconception of students when they multiply rational expressions is that they can only "cross cancel." As a result, they may neglect to divide out factors that are common to the numerator and denominator of the same rational expression. A mini-lesson is available to address this misconception.

See SpringBoard's eBook Teacher Resources for a student page for this mini-lesson.

ACTIVITY 28 Continued

Check Your Understanding

Debrief students' answers to these items to ensure that they understand concepts related to multiplying and dividing rational expressions. You may also want to ask students how the student described in Item 1 could have checked his or her answer.

Answers
1. The incorrect rational expression is inverted. The student should have multiplied by the reciprocal of $\frac{a+3}{a-3}$. The correct quotient is $\frac{(a-3)^2}{3a}$.
2. $x+3$

ASSESS

Students' answers to lesson practice problems will provide you with a formative assessment of their understanding of the lesson concepts and their ability to apply their learning.

See the Activity Practice for additional problems for this lesson. You may assign the problems here or use them as a culmination for the activity.

LESSON 28-3 PRACTICE

3. $\dfrac{x+1}{(x-2)(x-6)}$ 4. $\dfrac{x^2}{2(x-1)}$
5. $3(x-y)$ 6. $b+1$
7. $\dfrac{(x-5)(x-1)}{2(x-2)}$ 8. $\dfrac{1}{x}$
9. $\dfrac{2(x+5)}{(x+1)}$ 10. $\dfrac{m+4}{m+1}$
11. a. $\dfrac{x^2+2x-15}{x+1}$
    b. $\dfrac{x^2+2x-15}{x+1} \cdot \dfrac{x^2+2x+1}{x+5}$
    $= (x-3)(x+1)$

ADAPT

Check students' answers to the Lesson Practice questions to make sure that they understand how to multiply and divide rational expressions. If students have not yet mastered these skills, they may need more practice with factoring binomials and trinomials. Remind students of the methods they have learned for factoring.

- Factor out the GCF of all terms.
- Factor perfect square trinomials: $a^2 + 2ab + b^2 = (a+b)^2$ and $a^2 - 2ab + b^2 = (a-b)^2$.
- Factor differences of two squares: $a^2 - b^2 = (a+b)(a-b)$.
- Factor $x^2 + bx + c$ by finding factors of c that have a sum of b.
- Factor $ax^2 + bx + c$ by using factors of a, factors of c, and guess and check.

412 SpringBoard® Mathematics Algebra 1, Unit 4 • Exponents, Radicals, and Polynomials

---

ACTIVITY 28 continued

Lesson 28-3
Multiplying and Dividing Rational Expressions

Try These B
Divide. Simplify your answer.

a. $\dfrac{w^2 - 2w - 3}{w^2 - 6w + 9} \div \dfrac{5}{w-3}$ b. $\dfrac{3xy}{3x^2 - 12} \div \dfrac{xy+y}{x^2+3x+2}$

$\dfrac{w+1}{5}$ $\dfrac{x}{x-2}$

Check Your Understanding

1. **Critique the reasoning of others.** A student was asked to divide the rational expressions shown below. Examine the student's solution, and then identify and correct the error.

$$\dfrac{a^2-9}{3a} \div \dfrac{a+3}{a-3} = \dfrac{3a}{a^2-9} \cdot \dfrac{a+3}{a-3}$$
$$= \dfrac{3a}{(a+3)(a-3)} \cdot \dfrac{(a+3)}{a-3} = \dfrac{3a}{(a-3)^2}$$

2. What is the quotient when $\dfrac{2x+6}{x+5}$ is divided by $\dfrac{2}{x+5}$?

LESSON 28-3 PRACTICE
Multiply or divide.

3. $\dfrac{x^2 - 5x - 6}{x^2 - 4} \cdot \dfrac{x+2}{x^2 - 12x + 36}$ 4. $\dfrac{x^3}{x^2-1} \cdot \dfrac{2x+2}{4x}$

5. $\dfrac{x^2 - y^2}{12} \cdot \dfrac{36}{x+y}$ 6. $(b^2 + 12b + 11) \cdot \dfrac{b+9}{b^2 + 20b + 99}$

7. $\dfrac{x^2 + 4x + 4}{2x+4} \div \dfrac{x^2-4}{x^2-6x+5}$ 8. $\dfrac{1}{x-1} \div \dfrac{x}{x-1}$

9. $\dfrac{2x+4}{x^2+11x+18} \div \dfrac{x+1}{x^2+14x+45}$ 10. $\dfrac{m^2+m-6}{m^2+8m+15} \div \dfrac{m^2-m-2}{m^2+9m+20}$

11. **Make sense of problems.** The figure shows a rectangular prism. The area of the rectangular face $ABCD$ is $x^2 + 2x - 15$.

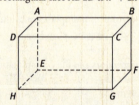

a. The length of edge $\overline{DC}$ is $x+1$. Write a rational expression that represents the length of edge $\overline{BC}$.

b. The length of edge $\overline{BF}$ is $\dfrac{x^2+2x+1}{x+5}$. Write and simplify a product to find the area of face $BFGC$.

© 2014 College Board. All rights reserved.

Lesson 28-4
Adding and Subtracting Rational Expressions

ACTIVITY 28
continued

My Notes

Learning Targets:
- Identify the least common multiple (LCM) of algebraic expressions.
- Add and subtract rational expressions.

SUGGESTED LEARNING STRATEGIES: Note Taking, Close Reading, Sharing and Responding, Identify a Subtask

To add or subtract rational expressions with the same denominator, add or subtract the numerators and then simplify if possible.

Example A
Simplify $\frac{10}{x} - \frac{5}{x}$.

Step 1: Subtract the numerators. $\quad \frac{10}{x} - \frac{5}{x} = \frac{10-5}{x}$

Solution: $\frac{5}{x}$

Example B
Simplify $\frac{2x}{x+1} + \frac{2}{x+1}$.

Step 1: Add the numerators. $\quad \frac{2x}{x+1} + \frac{2}{x+1} = \frac{2x+2}{x+1}$

Step 2: Factor. $\quad = \frac{2(x+1)}{x+1}$

Step 3: Divide out common factors. $\quad = \frac{2(\cancel{x+1})}{\cancel{x+1}}$

Solution: 2

Try These A–B
Add or subtract. Simplify your answer.

a. $\frac{3}{x^2} - \frac{x}{x^2}$ $\quad \frac{3-x}{x^2}$

b. $\frac{2}{x+3} - \frac{6}{x+3} + \frac{x}{x+3}$ $\quad \frac{x-4}{x+3}$

c. $\frac{x}{x^2-x} + \frac{4x}{x^2-x}$ $\quad \frac{5}{x-1}$

To add or subtract rational expressions with unlike denominators, first identify a common denominator. The **least common multiple** (LCM) of the denominators is used for the common denominator.

One way to determine the LCM is to factor each expression. The LCM is the product of each factor common to the expressions as well as any non-common factors.

MATH TERMS

The **least common multiple** is the smallest multiple that two or more numbers or expressions have in common.

The numbers 10 and 25 have many common multiples. The number 50 is the least common multiple.

ACTIVITY 28 Continued

Developing Math Language
This lesson includes the vocabulary term *least common multiple* (LCM). Have students add the term to their math notebooks. In previous courses, students learned to identify the least common multiple of two whole numbers by identifying factors of the numbers. Students can determine the least common multiple of two algebraic expressions by using a similar process. As students respond to questions or discuss possible solutions to problems, monitor their use of the new term to ensure their understanding and ability to use language correctly and precisely.

Example C Activating Prior Knowledge, Note Taking, Identify a Subtask, Sharing and Responding
In this example, students determine the LCM of two polynomials. The purpose of this example is to prepare students for adding and subtracting rational expressions with unlike denominators. Ask students why the LCM needs to include the common factors as well as the factors that are not in common.

Example D Activating Prior Knowledge, Close Reading, Note Taking In this example, students are introduced to the subtraction of rational expressions with unlike denominators. In Step 1, ask students why the LCM of the denominators is $x(x-2)$. Before Step 2, use the example in the Math Tip to reinforce the idea that multiplication by any form of 1 changes the way an expression looks, but not the value of the expression. In Step 2, have students discuss the purpose of multiplying the first rational expression by $\frac{(x-2)}{(x-2)}$. Also ask students to explain why the rational expressions must be rewritten with a common denominator before you can subtract them.

ACTIVITY 28 continued

My Notes

Lesson 28-4
Adding and Subtracting Rational Expressions

Example C
Determine the LCM of $x^2 - 4$ and $2x + 4$.
Step 1: Factor each expression.
$$x^2 - 4 = (x+2)(x-2)$$
$$2x + 4 = 2(x+2)$$
Step 2: Identify the factors.
Common Factor: $(x + 2)$
Factors Not in Common: 2 and $(x - 2)$
Step 3: The LCM is the product of all of the factors in Step 2.
Solution: The LCM is $2(x+2)(x-2)$.

Try These C
a. Determine the LCM of $2x + 2$ and $x^2 + x$. Use the steps below.
Factor each expression: **$2(x + 1)$ and $x(x + 1)$**

Common Factor(s): **$(x + 1)$**

Factors Not in Common: **2 and x**

LCM: **$2x(x + 1)$**

b. Determine the LCM of $x^2 - 2x - 15$ and $3x + 9$.
$3(x + 3)(x - 5)$

Now add and subtract rational expressions with different denominators. First, determine the LCM of the denominators. Next, write each fraction with the LCM as the denominator. Then, add or subtract. Simplify if possible.

Example D
Subtract $\frac{2}{x} - \frac{3}{x^2 - 2x}$. Simplify your answer if possible.
Step 1: Determine the LCM.
Factor the denominators: x and $x(x - 2)$
The LCM is $x(x - 2)$.
Step 2: Multiply the numerator and denominator of the first term by $(x - 2)$. The denominator of the second term is the LCM.
$$\frac{2}{x} \cdot \frac{(x-2)}{(x-2)} - \frac{3}{x(x-2)}$$
Step 3: Use the Distributive Property in the numerator.
$$\frac{2x - 4}{x(x-2)} - \frac{3}{x(x-2)}$$
Step 4: Subtract the numerators. $\frac{2x - 7}{x(x-2)}$
Solution: $\frac{2x - 7}{x(x-2)}$

> **MATH TIP**
> Multiplying a fraction by a form of 1 gives an equivalent fraction.
> $\frac{1}{3} = \frac{1 \cdot 2}{3 \cdot 2} = \frac{2}{6}$, because $\frac{2}{2} = 1$.
> The same is true for rational expressions. Multiplying by $\frac{(x-2)}{(x-2)}$ gives an equivalent expression because $\frac{(x-2)}{(x-2)} = 1$ when $x \ne 2$.

MINI-LESSON: Least Common Multiple

Students who are struggling with Example C may need to review the process of finding the LCM of two whole numbers by using prime factorization. A mini-lesson is available to provide practice.

See SpringBoard's eBook Teacher Resources for a student page for this mini-lesson.

Lesson 28-4
Adding and Subtracting Rational Expressions

Example E

Add $\dfrac{-4}{5-p} + \dfrac{3}{p-5}$. Simplify your answer if possible.

Step 1: Determine a common denominator. $\quad p-5$

Step 2: Multiply the numerator and denominator of the first term by -1.

$$\dfrac{-4}{5-p} \cdot \dfrac{-1}{-1} + \dfrac{3}{p-5}$$

Step 3: Multiply. $\quad \dfrac{4}{p-5} + \dfrac{3}{p-5}$

Step 4: Add. $\quad \dfrac{7}{p-5}$

Solution: $\dfrac{7}{p-5}$

MATH TIP

If you multiply $(5-p)$ by -1, the product is $p-5$.

Try These D–E

a. Add $\dfrac{1}{x^2-1} + \dfrac{2}{x+1}$. Use the steps below.

Factor each denominator: $(x+1)(x-1)$ and $x+1$

Common Factors: $x+1$

Factors Not in Common: $x-1$

LCM: $(x+1)(x-1)$

Factor the denominator of the first term. Multiply the numerator and denominator of the second term by _____ $x-1$ _____ :

$$\dfrac{1}{(x+1)(x-1)} + \dfrac{2}{x+1} \cdot \dfrac{x-1}{x-1}$$

Add the numerators: $\dfrac{1+2(x-1)}{(x+1)(x-1)}$

Use the Distributive Property: $\dfrac{1+2x-2}{(x+1)(x-1)}$

Combine like terms: $\dfrac{2x-1}{(x+1)(x-1)}$

Solution: $\dfrac{2x-1}{(x+1)(x-1)}$

Add or subtract. Simplify your answer.

b. $\dfrac{3}{x+1} - \dfrac{x}{x-1}$

$\dfrac{-x^2+2x-3}{(x+1)(x-1)}$

c. $\dfrac{2}{x} - \dfrac{3}{x^2-3x}$

$\dfrac{2x-9}{x(x-3)}$

d. $\dfrac{2}{x^2-4} + \dfrac{x}{x^2+4x+4}$

$\dfrac{x^2+4}{(x+2)^2(x-2)}$

ACTIVITY 28 Continued

Check Your Understanding

Debrief students' answers to these items to make sure that they understand concepts related to adding and subtracting rational expressions. Ask students to explain how to find the least common multiple of two polynomials.

Answers

1. **a.** Sample answer: $\frac{1}{2}$ and $\frac{1}{8}$
   b. Sample answer: $\frac{1}{x+1}$ and $\frac{1}{(x+1)^2}$
   c. Sample answer:
   $\frac{1}{x+1} + \frac{1}{(x+1)^2} = \frac{x+2}{(x+1)^2}$

2. Sample answer: Factor the denominators, use the factors to find a common denominator, rewrite the expressions with the common denominator, and then add or subtract.

ASSESS

Students' answers to lesson practice problems will provide you with a formative assessment of their understanding of the lesson concepts and their ability to apply their learning.

See the Activity Practice for additional problems for this lesson. You may assign the problems here or use them as a culmination for the activity.

LESSON 28-4 PRACTICE

3. $2(x-2)(x+2)$
4. $2(x-4)$
5. $(x+3)(x-3)$
6. $(x+6)(x+1)(x+7)$
7. $(x-10)(x+3)^2$
8. $\frac{x^2+x-2}{(x+1)(x+3)}$

ADAPT

Use the Lesson Practice to verify that students can find the least common multiple of polynomials and add and subtract rational expressions. If students have not yet mastered these skills, they may need more practice with finding the LCM. You may wish to show students how to find the LCM of two polynomials by writing the factors of the polynomials in columns, with a different factor in each column. This method is shown below for the polynomials $2x^2$ and $x^2 + 2x$.

Factored forms:
$2x^2 = 2(x)(x)$ and $x^2 + 2x = x(x+2)$

$2x^2$	2	x	x
$x^2 + 2x$		x	x+2
LCM	2	x	x	x+2

The LCM is $2x^2(x+2)$.

ACTIVITY 28
continued

My Notes

Lesson 28-4
Adding and Subtracting Rational Expressions

Check Your Understanding

1. **Make use of structure.** Sometimes the denominator of one fraction or one rational expression works as a common denominator for all fractions or rational expressions in a set.
   a. Write two fractions (rational numbers) in which the denominator of one of the fractions is a common denominator.
   b. Write two rational expressions in which the denominator of one of the expressions is a common denominator.
   c. Show how to add the two rational expressions you wrote in Part (b).
2. List the steps you usually use to add or subtract rational expressions with unlike denominators.

LESSON 28-4 PRACTICE

Determine the least common multiple of each set of expressions.

3. $2x + 4$ and $x^2 - 4$
4. $2x - 8$ and $x - 4$
5. $x - 3$ and $x + 3$
6. $x + 6$, $x + 7$, and $x^2 + 7x + 6$
7. $x + 3$, $x^2 + 6x + 9$, and $x^2 - 7x - 30$

Perform the indicated operation.

8. $\frac{x}{x+1} - \frac{2}{x+3}$
9. $\frac{2}{3x-3} - \frac{x}{x^2-1}$
10. $\frac{x}{x+5} - \frac{2}{x+3}$
11. $\frac{3}{x-3} - \frac{x}{x+4}$
12. $\frac{x}{3x-2} + \frac{2}{x-5}$
13. $\frac{x-2}{x^2+4x+4} + \frac{x-2}{x+2}$

14. **Model with mathematics.** In the past week, Emilio jogged for a total of 7 miles and biked for a total of 7 miles. He biked at a rate that was twice as fast as his jogging rate.
   a. Suppose Emilio jogs at a rate of r miles per hour. Write an expression that represents the amount of time he jogged last week and an expression that represents the amount of time he biked last week. (Hint: distance = rate × time, so time = $\frac{\text{distance}}{\text{rate}}$.)
   b. Write and simplify an expression for the total amount of time Emilio jogged and biked last week.
   c. Emilio jogged at a rate of 5 miles per hour. What was the total amount of time Emilio jogged and biked last week?

---

9. $\frac{2-x}{2(x-1)(x+1)}$
10. $\frac{x^2+x-10}{(x+5)(x+3)}$
11. $\frac{-x^2+6x+12}{(x-3)(x+4)}$
12. $\frac{x^2+x-4}{(3x-2)(x-5)}$
13. $\frac{x^2+x-6}{(x+2)^2}$
14. **a.** $\frac{7}{r}$; $\frac{7}{2r}$
    b. $\frac{7}{r} + \frac{7}{2r} = \frac{21}{2r}$
    c. 2.1 hours (or 2 hours and 6 minutes)

416 SpringBoard® Mathematics Algebra 1, Unit 4 • Exponents, Radicals, and Polynomials

Simplifying Rational Expressions
Totally Rational

ACTIVITY 28 PRACTICE
Write your answers on notebook paper.
Show your work.

Lesson 28-1

1. Allison correctly simplified the rational expression shown below by dividing.
$$\frac{35x^7 + 15x^5 - 10x^3}{5x^3}$$
   Which of these is a term in the resulting expression?
   A. $3x^8$ B. $3x^4$
   C. $-2x$ D. -2

For Items 2–5, simplify each expression.

2. $\frac{56x^2 y}{70x^3 y}$

3. $\frac{28x^2}{49xy}$

4. $\frac{x^2 - 25}{5x + 25}$

5. $\frac{x+5}{x^2 + x - 20}$

6. Which of the following expressions is equivalent to a negative integer?
   A. $\frac{5y+5}{5y-5}$ B. $\frac{6y-6}{3-3y}$
   C. $\frac{2-2y}{4y-4}$ D. $\frac{8y-8}{4y-4}$

7. A rental car costs $24 plus $3 per mile.
   a. Write an expression that represents the total cost of the rental if you drive the car m miles.
   b. Write and simplify an expression that represents the cost per day if you keep the car for 3 days.
   c. What is the cost per day if you drive 50 miles?

8. The expression $\frac{x^2 + 8x + c}{x+4}$ can be simplified to $x+4$. What is the value of c?
   A. -16 B. 0
   C. 16   D. 64

Lesson 28-2
For Items 9–14, determine each quotient by using long division.

9. $(3x^2 + 6x + 2) \div 3x$

10. $(3x^2 - 7x - 6) \div (3x + 2)$

11. $\frac{2x^2 - 7x - 16}{2x+3}$

12. $\frac{x^2 - 19x + 9}{x-4}$

13. $\frac{4x^2 + 17x - 1}{4x+1}$

14. $\frac{5x^3 + x - 2}{x-1}$

15. The area A and length ℓ of a rectangle are shown below. Write a rational expression that represents the width w of the rectangle. Then simplify the expression using long division.

   w | $A = (x^2 + 4x + 9)$ cm²
   $\ell = (x+4)$ cm

16. Greg was asked to simplify each expression below using long division. For which expression should he have a remainder?
   A. $\frac{6x^2 + 9x + 3}{3x}$ B. $\frac{8x^4 + 12x^3 + 16x^2}{4x^2}$
   C. $\frac{15x^2 + 5x + 25}{5}$ D. $\frac{6x^4 + 12x^3 + 6x^2}{6x}$

17. A student performed the long division shown below. Is the student's work correct? Justify your response.

   $$x^2 + 2x - 5 \overline{\smash{\big)}\, 4x^2 - 6x + 11}$$
   quotient: 4, remainder line: $4x^2 + 8x - 20$, then $2x - 9$

   The quotient is $4 + \frac{2x-9}{x^2 + 2x - 5}$.

ACTIVITY 28 Continued

ACTIVITY PRACTICE
1. D
2. $\frac{4}{5x}$
3. $\frac{4x}{7y}$
4. $\frac{x-5}{5}$
5. $\frac{1}{x-4}$
6. B
7. a. $24 + 3m$
   b. $8 + m$
   c. $58
8. C
9. $x + 2 + \frac{2}{3x}$
10. $x - 3$
11. $x - 5 + \frac{-1}{2x+3}$
12. $x - 15 + \frac{-51}{x-4}$
13. $x + 4 + \frac{-5}{4x+1}$
14. $5x^2 + 5x + 6 + \frac{4}{x-1}$
15. $\frac{x^2 + 4x + 9}{x+4} = \left(x + \frac{9}{x+4}\right)$ cm
16. A
17. It is not correct; the student should have subtracted $4x^2 + 8x - 20$ to get $-14x + 31$. The quotient is $4 + \frac{-14x + 31}{x^2 + 2x - 5}$.

ACTIVITY 28 Continued

18. $\dfrac{4x}{3(x+5)}$
19. $\dfrac{3x}{x-3}$
20. $\dfrac{x+4}{x+2}$
21. $\dfrac{2}{x+5}$
22. 1
23. always
24. sometimes
25. always
26. The student should have factored $x^2 - 6x + 9$ as $(x-3)^2$. The correct result is $\dfrac{(x-3)(x+3)}{5x}$.
27. B
28. $(x-5)(x+5)$
29. $y^2(y+3)$
30. $(x+2)(x+3)(x+4)$
31. $(x-2)^3$
32. B
33. $\dfrac{7}{x}$
34. x
35. 1
36. $\dfrac{1-5x}{x-4}$
37. $\dfrac{3x+5}{3(x-2)}$
38. $\dfrac{25}{x-3}$
39. a. $\dfrac{1}{r}$
    b. $\dfrac{1}{2r}$
    c. $\dfrac{1}{r} + \dfrac{1}{2r} = \dfrac{3}{2r}$
    d. 3 minutes

ADDITIONAL PRACTICE

If students need more practice on the concepts in this activity, see the eBook Teacher Resources for additional practice problems.

ACTIVITY 28
continued

Simplifying Rational Expressions
Totally Rational

Lesson 28-3

Multiply or divide. Simplify your answer if possible.

18. $\dfrac{x+4}{3x} \cdot \dfrac{4x^2}{x^2+9x+20}$

19. $\dfrac{3x+9}{x} \cdot \dfrac{x^2}{x^2-9}$

20. $\dfrac{x^2-x-6}{x^2-9} \cdot \dfrac{x^2+7x+12}{x^2+4x+4}$

21. $\dfrac{x^2-25}{x^2-10x+25} \div \dfrac{x^2+10x+25}{2x-10}$

22. $\dfrac{n^2-4n-5}{n^2+2n+1} \div \dfrac{n^2-6n+5}{n^2-1}$

In the expression $\dfrac{1}{(x+5)^2} \div \dfrac{k}{(x+5)^2}$, k is a real number with $k \neq 0$. For Items 23–25, determine whether each statement is always, sometimes, or never true.

23. The expression may be simplified so that the variable x does not appear.

24. The value of the expression is a real number less than 1.

25. When $k > 0$, the value of the expression is also greater than 0.

26. A student was asked to divide the rational expressions shown below. Examine the student's solution, then identify and correct the error.

$$\dfrac{x^2-6x+9}{5x} \div \dfrac{x-3}{x+3} = \dfrac{(x+3)(x-3)}{5x} \cdot \dfrac{x+3}{x-3}$$
$$= \dfrac{(x+3)\cancel{(x-3)}}{5x} \cdot \dfrac{x+3}{\cancel{x-3}}$$
$$= \dfrac{(x+3)^2}{5x}$$

27. Which expression is equivalent to $\dfrac{3x+3}{x^2} \cdot \dfrac{x^2-x}{x^2-1}$?
    A. 3
    B. $\dfrac{3}{x}$
    C. $3-x$
    D. $\dfrac{3x+3}{x}$

Lesson 28-4

For Items 28–31, determine the least common multiple of each set of expressions.

28. $x^2 - 25$ and $x+5$
29. $y+3$, y, and y^2
30. $x^2 + 5x + 6$ and $x^2 + 7x + 12$
31. $x^2 - 4x + 4$, $x-2$, and $(x-2)^3$

32. Which pair of expressions has a least common multiple that is the product of the expressions?
    A. $x+7$ and $x^2+14x+49$
    B. $x+7$ and $x-7$
    C. $x-3$ and x^2-9
    D. $x-3$ and $(x-3)^2$

Add or subtract. Express in simplest form.

33. $\dfrac{4}{x} + \dfrac{3}{x}$ 34. $\dfrac{x}{2} + \dfrac{x}{2}$

35. $\dfrac{x}{x+1} + \dfrac{1}{x+1}$ 36. $\dfrac{x}{x^2-4x} - \dfrac{5x}{x-4}$

37. $\dfrac{3x+2}{3x-6} + \dfrac{x+2}{x^2-4}$ 38. $\dfrac{-18}{3-x} + \dfrac{7}{x-3}$

MATHEMATICAL PRACTICES
Reason Abstractly and Quantitatively

39. Justine lives one mile from the grocery store. While she was driving to the store, there was a lot of traffic. On her way home, there was no traffic at all, and her average rate (speed) was twice the average rate of her trip to the store.

    a. Let r represent Justine's average rate on her way to the store. Write an expression for the time it took her to get to the store. (*Hint:* distance = rate × time, so time = $\dfrac{\text{distance}}{\text{rate}}$.)

    b. Write an expression for the time it took Justine to drive home.

    c. Write and simplify an expression for the total time of the round trip to and from the store.

    d. If Justine drove at 30 miles per hour to the store, what was the total time for the round trip? Write your answer in minutes.

Factoring and Simplifying Rational Expressions
ROCK STAR DEMANDS

Embedded Assessment 4
Use after Activity 28

Rockstar Platforms sets up outdoor stages for rock concerts. The musician Fuchsia requires a square stage at all of her concerts. Rockstar Platforms lays down a stage with an area of $x^2 + 8x + 16$ square feet for her.

1. **a.** Draw a diagram to represent the area of the stage.
   b. Write expressions to represent the side lengths of Fuchsia's stage.

Fuchsia looks at the stage and says, "That's too small!" So Rockstar Platforms goes back to the drawing board and designs another square stage with an area of $4x^2 + 20x + 25$ square feet.

2. What are the side lengths of Fuchsia's new stage?

The company calls Fuchsia back out to look at the stage. "I guess it will do," she says, "but I would have preferred a stage with an area of $5x^2 + 12x + 4$ square feet." Rockstar Platforms' foreman explains, "But then your demand for a square stage would not be met."

3. Explain why Fuchsia's demand for a square stage would not be met if the stage had an area of $5x^2 + 12x + 4$ square feet.

"Leave it like it is," concedes Fuchsia. "Now, I need you to put up a video screen." Fuchsia wants a large rectangular video screen set up behind her so her fans can see her from far away. Rockstar Platforms' foreman has a plan for a video screen with an area of $2x^2 + 11x + 14$. "Boss, that's not going to work," his assistant cautions. "It's going to be too long for the stage." "It'll work," the foreman insists.

4. Who is correct, the foreman or his assistant? Justify your response.

After finishing Fuchsia's stage, Rockstar Platforms is contracted to set up a stage for Mich.i.el. Mich.i.el wants a rectangular stage with an area of $2x^2 + 7x + 5$ square feet, but he makes one very specific request. In order to fit all of his backup dancers, the length of the stage must be $x + 4$ feet.

5. How wide will Rockstar Platforms have to make the stage to meet Mich.i.el's request?

Two nights later, Mich.i.el and Fuchsia run into each other at a party and start arguing over who had the bigger stage.

6. How many times larger was Fuchsia's stage than Mich.i.el's stage? Write an expression in simplest form that represents the ratio of the area of Fuchsia's stage to the area of Mich.i.el's stage.

Embedded Assessment 4

Assessment Focus
- Factoring perfect square trinomials
- Factoring trinomials of the form $ax^2 + bx + c$
- Dividing polynomials
- Expressing the remainder of polynomial division as a rational expression
- Dividing rational expressions
- Simplifying rational expressions

Answer Key
1. **a.** Possible sketch:

   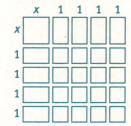

   b. $(x + 4)$ ft by $(x + 4)$ ft
2. $(2x + 5)$ ft by $(2x + 5)$ ft
3. The stage would no longer be square. It would measure $(5x + 2)$ ft by $(x + 2)$ ft.
4. The assistant is correct. The dimensions of the video screen will be $(2x + 7)$ ft by $(x + 2)$ ft, and the length of the video screen, $(2x + 7)$ ft, is greater than the length of the stage, $(2x + 5)$ ft.
5. $\left(2x - 1 + \dfrac{9}{x + 4}\right)$ ft
6. $\dfrac{2x + 5}{x + 1}$

Common Core State Standards for Embedded Assessment 4

HSA-SSE.A.1	Interpret expressions that represent a quantity in terms of its context.*
HSA-SSE.A.1a	Interpret parts of an expression, such as terms, factors, and coefficients.
HSA-SSE.A.1b	Interpret complicated parts of expressions by viewing one or more of their parts as a single entity.
HSA-SSE.A.2	Use the structure of an expression to identify ways to rewrite it.
HSA-APR.D.6	Rewrite simple rational expressions in different forms; write $\dfrac{a(x)}{b(x)}$ in the form $q(x) + \dfrac{r(x)}{b(x)}$, where $a(x)$, $b(x)$, $q(x)$, and $r(x)$ are polynomials with the degree of $r(x)$ less than the degree of $b(x)$, using inspection, long division, or, for the more complicated examples, a computer algebra system.

Unit 4 • Exponents, Radicals, and Polynomials

Embedded Assessment 4

Teacher to Teacher

You may wish to read through the scoring guide with students and discuss the differences in the expectations at each level. Check that students understand the terms used.

Embedded Assessment 4
Use after Activity 28

Factoring and Simplifying Rational Expressions
ROCK STAR DEMANDS

Scoring Guide	Exemplary	Proficient	Emerging	Incomplete
	The solution demonstrates the following characteristics:			
Mathematics Knowledge and Thinking (Items 1b, 2, 5, 6)	• Clear understanding of and accuracy in factoring polynomials and simplifying rational expressions	• Adequate understanding of and accuracy in factoring polynomials and simplifying rational expressions	• Partial understanding of factoring polynomials and simplifying rational expressions that results in some incorrect answers	• Little or no understanding of factoring polynomials and simplifying rational expressions
Problem Solving (Items 1b, 2, 5, 6)	• Appropriate and efficient strategy that results in a correct answer	• Strategy that may include unnecessary steps but results in a correct answer	• Strategy that results in some incorrect answers	• No clear strategy when solving problems
Mathematical Modeling/ Representations (Item 1a)	• Clear and accurate diagram of the stage	• Correct diagram of the stage	• Partially accurate diagram of the stage	• Incomplete or inaccurate diagram of the stage
Reasoning and Communication (Items 3, 4)	• Ease and accuracy describing the relationship between a mathematical expression and a real-world constraint • Precise use of appropriate math terms and language to identify and explain whether the foreman or his assistant is correct	• Little difficulty describing the relationship between a mathematical expression and a real-world constraint • Correct identification of whether the foreman or his assistant is correct with an adequate explanation	• Partially correct description of the relationship between a mathematical expression and a real-world constraint • Misleading or confusing explanation of whether the foreman or his assistant is correct	• Little or no understanding of how a mathematical expression might relate to a real-world constraint • Incomplete or inaccurate explanation of whether the foreman or his assistant is correct

Common Core State Standards for Embedded Assessment 4 (cont.)

HSA-APR.D.7 (+) Understand that rational expressions form a system analogous to the rational numbers, closed under addition, subtraction, multiplication, and division by a nonzero rational expression; add, subtract, multiply, and divide rational expressions.

Unit 5 Planning the Unit

In this unit, students will use a variety of methods to solve quadratic equations, as well as systems of two equations that contain linear and quadratic or exponential functions. They will apply this to modeling real-world situations.

Vocabulary Development

The key terms for this unit can be found on the Unit Opener page. These terms are divided into Academic Vocabulary and Math Terms. Academic Vocabulary includes terms that have additional meaning outside of math. These terms are listed separately to help students transition from their current understanding of a term to its meaning as a mathematics term. To help students learn new vocabulary:

- Have students discuss meaning and use graphic organizers to record their understanding of new words.
- Remind students to place their graphic organizers in their math notebooks and revisit their notes as their understanding of vocabulary grows.
- As needed, pronounce new words and place pronunciation guides and definitions on the class Word Wall.

Embedded Assessments

Embedded Assessments allow students to do the following:

- Demonstrate their understanding of new concepts.
- Integrate previous and new knowledge by solving real-world problems presented in new settings.

They also provide formative information to help you adjust instruction to meet your students' learning needs.

Prior to beginning instruction, have students unpack the first Embedded Assessment in the unit to identify the skills and knowledge necessary for successful completion of that assessment. Help students create a visual display of the unpacked assessment and post it in your class. As students learn new knowledge and skills, remind them that they will be expected to apply that knowledge to the assessment. After students complete each Embedded Assessment, turn to the next one in the unit and repeat the process of unpacking that assessment with students.

CollegeBoard

AP / College Readiness

This unit helps prepare students for Advanced Placement courses by

- Modeling motion using quadratic relationships.
- Making connections between multiple ways to represent mathematical information: numerically, graphically, verbally and algebraically.
- Increasing student ability to solve a wide-variety of equations and to choose the most appropriate solution method when needed.

Unpacking the Embedded Assessments

The following are the key skills and knowledge students will need to know for each assessment.

Embedded Assessment 1

Graphing Quadratic Functions, *Parabolic Paths*

- Writing quadratic functions
- Analyzing quadratic functions
- Graphing quadratic functions
- Transforming quadratic functions

Embedded Assessment 2

Solving Quadratic Equations, *Egg Drop*

- Solving quadratic equations
- Writing the equation of a quadratic function to fit data
- Using a quadratic model to solve problems
- Interpreting solutions of a quadratic equation

Embedded Assessment 3

Solving Systems of Equations, *Sports Collector*

- Graphing linear, quadratic, and exponential functions
- Identifying the domain of a function
- Identifying the function with the greatest maximum value
- Solving systems of equations

Planning the Unit continued

Suggested Pacing

The following table provides suggestions for pacing using a 45-minute class period. Space is left for you to write your own pacing guidelines based on your experiences in using the materials.

	45-Minute Period	Your Comments on Pacing
Unit Overview/Getting Ready	1	
Activity 29	2	
Activity 30	3	
Embedded Assessment 1	1	
Activity 31	3	
Activity 32	5	
Activity 33	2	
Embedded Assessment 2	1	
Activity 34	3	
Activity 35	2	
Embedded Assessment 3	1	
Total 45-Minute Periods	**24**	

Additional Resources

Additional resources that you may find helpful for your instruction include the following, which may be found in the eBook Teacher Resources.

- Unit Practice (additional problems for each activity)
- Getting Ready Practice (additional lessons and practice problems for the prerequisite skills)
- Mini-Lessons (instructional support for concepts related to lesson content)

Quadratic Functions

Unit Overview
In this unit you will study a variety of ways to solve quadratic functions and systems of equations and apply your learning to analyzing real world problems.

Key Terms
As you study this unit, add these and other terms to your math notebook. Include in your notes your prior knowledge of each word, as well as your experiences in using the word in different mathematical examples. If needed, ask for help in pronouncing new words and add information on pronunciation to your math notebook. It is important that you learn new terms and use them correctly in your class discussions and in your problem solutions.

Math Terms
- quadratic function
- standard form of a quadratic function
- parabola
- vertex of a parabola
- maximum
- minimum
- parent function
- axis of symmetry
- translation
- vertical stretch
- vertical shrink
- transformation
- reflection
- factored form
- zeros of a function
- roots
- completing the square
- discriminant
- imaginary numbers
- imaginary unit
- complex numbers
- piecewise-defined function
- nonlinear system of equations

ESSENTIAL QUESTIONS

 How are quadratic functions used to model, analyze, and interpret mathematical relationships?

 Why is it advantageous to know a variety of ways to solve and graph quadratic functions?

EMBEDDED ASSESSMENTS
This unit has three embedded assessments, following Activities 30, 33, and 35. They will allow you to demonstrate your understanding of graphing, identifying, and modeling quadratic functions, solving quadratic equations and solving nonlinear systems of equations.

Embedded Assessment 1:
Graphing Quadratic Functions — p. 453

Embedded Assessment 2:
Solving Quadratic Equations — p. 493

Embedded Assessment 3:
Solving Systems of Equations — p. 519

UNIT 5
Getting Ready

Use some or all of these exercises for formative evaluation of students' readiness for Unit 5 topics.

Prerequisite Skills

- Operations on polynomials (Item 1) HSA-APR.A.1
- Factoring polynomials (Item 2) HSA-SSE.A.2
- Evaluating functions (Item 3) HSF-IF.A.2
- Solving equations (Item 4) 7.EE.B.4a
- Solving inequalities (Item 5) 7.EE.B.4b
- Graphing linear functions (Item 6) HSF-IF.C.7
- Interpreting graphs of linear functions (Items 7–8) HSF-IF.B.4

Answer Key

1. **a.** $3x^2 - x - 10$
   b. $2y^3 + 10y^2 - 12y$
2. **a.** $2x(x + 7)$
   b. $3(x - 5)(x + 5)$
   c. $(x + 5)(x + 2)$
3. **a.** $f(4) = 7$
   b. $f(-2) = 11$
4. $x = 6$
5. $x > 12$
6. Sample answer: Make a table of values, plot the ordered pairs, and draw a line through the points.
7. 300 calories are burned when walking 1.5 hours.
8. running: 3600 calories; walking: 800 calories

UNIT 5
Getting Ready

Write your answers on notebook paper. Show your work.

1. Determine each product.
   a. $(x - 2)(3x + 5)$
   b. $2y(y + 6)(y - 1)$

2. Factor each polynomial.
   a. $2x^2 + 14x$
   b. $3x^2 - 75$
   c. $x^2 + 7x + 10$

3. If $f(x) = 3x - 5$, find each value.
   a. $f(4)$
   b. $f(-2)$

4. Solve the equation. $4x - 5 = 19$

5. Solve the inequality.
   $\frac{1}{3}x + 9 > 13$

6. Explain how to graph $2x + y = 4$.

The following graph compares calories burned when running and walking at constant rates of 10 mi/h and 2 mi/h, respectively.

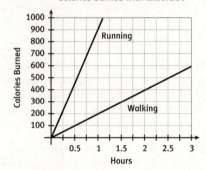

7. What does the ordered pair (1.5, 300) represent on this graph?

8. How many calories would be burned after four hours when running and after four hours when walking?

Getting Ready Practice

For students who may need additional instruction on one or more of the prerequisite skills for this unit, Getting Ready practice pages are available in the Teacher Resources at SpringBoard Digital. These practice pages include worked-out examples as well as multiple opportunities for students to apply concepts learned.

Introduction to Quadratic Functions

Touchlines
Lesson 29-1 Modeling with a Quadratic Function

ACTIVITY 29

Learning Targets:
- Model a real-world situation with a quadratic function.
- Identify quadratic functions.
- Write a quadratic function in standard form.

SUGGESTED LEARNING STRATEGIES: Create Representations, Interactive Word Wall, Marking the Text, Look for a Pattern, Discussion Groups

Coach Wentworth coaches girls' soccer and teaches algebra. Soccer season is starting, and she needs to mark the field by chalking the touchlines and goal lines for the soccer field. Coach Wentworth can mark 320 yards for the total length of all the touchlines and goal lines combined. She would like to mark the field with the largest possible area.

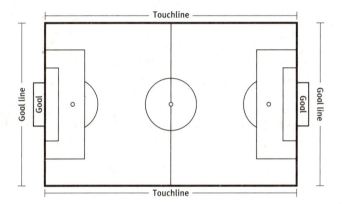

FIFA regulations require that all soccer fields be rectangular in shape.

1. How is the perimeter of a rectangle determined? How is the area of a rectangle determined?
   The perimeter of a rectangle is found by adding the lengths of the 4 sides. The area is found by multiplying the length times the width.

CONNECT TO SPORTS
FIFA stands for *Fédération Internationale de Football Association* (International Federation of Association Football) and is the international governing body of soccer.

Common Core State Standards for Activity 29

HSF-IF.B.4	For a function that models a relationship between two quantities, interpret key features of graphs and tables in terms of the quantities, and sketch graphs showing key features given a verbal description of the relationship. *Key features include: intercepts; intervals where the function is increasing, decreasing, positive, or negative; relative maximums and minimums; symmetries; end behavior; and periodicity.*
HSF-IF.C.7	Graph functions expressed symbolically and show key features of the graph, by hand in simple cases and using technology for more complicated cases.*
HSF-IF.C.7a	Graph linear and quadratic functions and show intercepts, maxima, and minima.
HSF-BF.A.1	Write a function that describes a relationship between two quantities.*
HSF-BF.A.1a	Determine an explicit expression, a recursive process, or steps for calculation from a context.

ACTIVITY 29 Continued

Differentiating Instruction

Support students who have a difficult time understanding that rectangles with a fixed perimeter can have different areas. You can use a piece of string to make the connections for visual and kinesthetic learners. Tie the string in a loop—a fixed perimeter—and have students make different rectangles using the loop. If this is done on top of graph paper, students can then draw the representative rectangles and determine the areas of the rectangles. This should help students visualize that while the perimeter remains constant, the area can change.

1–2 (continued) Verify that area computations in the table are correct. It is essential that students understand that the sum of the length and the width is 160 yards. Twice this length, 320 yards, is the perimeter. The numerical answers in the first eight rows of the table may be easier for students than the last row. Students may struggle with the abstraction of including a variable, so careful attention is needed to this response.

3–5 Look for a Pattern, Create Representations, Discussion Groups
The patterns students identify will vary, but should focus on the fact that while the length increases, the areas increase, reaching a maximum, and then decrease. Additionally, students should notice the symmetry in the area values about the value $l = 80$. Some students may insist that the length of a rectangle is greater than its width. Take care to mention that this is not true, referring students to the table to see that in the first four rows, the width is greater than the length. At this point in the lesson, the graph that represents the data should consist of only the points represented by the entries in the table. Some students may wish to "connect" the points into a continuous graph.

Lesson 29-1
Modeling with a Quadratic Function

2. Complete the table below for rectangles with the given side lengths. The first row has been completed for you.

Length (yards)	Width (yards)	Perimeter (yards)	Area (square yards)
10	150	320	1500
20	**140**	320	**2800**
40	**120**	320	**4800**
60	**100**	320	**6000**
80	**80**	320	**6400**
100	**60**	320	**6000**
120	**40**	320	**4800**
140	**20**	320	**2800**
150	**10**	320	**1500**
l	**$160 - l$**	320	**$l(160 - l)$**

3. **Express regularity in repeated reasoning.** Describe any patterns you observe in the table above.
   Answers may vary.
   - **As length increases, width decreases.**
   - **As length increases, area increases, reaches a maximum, and then decreases.**
   - **The greatest area is when the length and width are 80 yards.**

4. Is a 70-yd by 90-yd rectangle the same as a 90-yd by 70-yd rectangle? Explain your reasoning.
   Answers may vary. The two rectangles are congruent. They are the same rectangle, just in different orientations.

5. Graph the data from the table in Item 2 as ordered pairs.

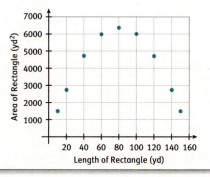

Lesson 29-1
Modeling with a Quadratic Function

My Notes

6. Use the table and the graph to explain why the data in Items 2 and 5 are not linear.
   Answers may vary. The data in the table are not linear because there is no constant rate of change. The data on the graph are not linear because the points do not lie on a line.

7. Describe any patterns you see in the graph above.
   Answers may vary. As lengths increase, areas increase, reach a maximum, and then decrease. There is symmetry about the line $l = 80$.

8. What appears to be the largest area from the data in Items 2 and 5?
   6400 square yards appears to be the largest area, occurring when the length is 80 yards.

9. Write a function $A(l)$ that represents the area of a rectangle whose length is l and whose perimeter is 320.
   $A(l) = l(160 - l)$

The function $A(l)$ is called a **quadratic function** because the greatest degree of any term is 2 (an x^2 term). The **standard form of a quadratic function** is $y = ax^2 + bx + c$ or $f(x) = ax^2 + bx + c$, where a, b, and c are real numbers and $a \neq 0$.

10. Write the function $A(l)$ in standard form. What are the values of a, b, and c?
    $A(l) = -l^2 + 160l$; $a = -1$, $b = 160$, $c = 0$

MINI-LESSON: Identifying Quadratic Functions

If students need additional work with quadratic equations, a mini-lesson is available to provide more practice.

See SpringBoard's eBook Teacher Resources for a student page for this mini-lesson.

ACTIVITY 29 Continued

Check Your Understanding
Debrief students' answers by asking them to describe the graph of a quadratic function. Have them describe how to identify a quadratic function and how write a quadratic function in standard form.

Answers
11. Check students' tables.

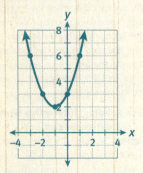

12. $l(2 + l)$
13. Sally thought that any function with an x^2 term is quadratic; however, no other terms can have a degree higher than 2. The x^3 term means the function is not quadratic.
14. $f(x) = x^2 - 6x + 9$

ASSESS
Students' answers to Lesson Practice problems will provide you with a formative assessment of their understanding of the lesson concepts and their ability to apply their learning.

See the Activity Practice for additional problems for this lesson. You may assign the problems here or use them as a culmination for the activity.

LESSON 29-1 PRACTICE
15. Check students' graphs.
- $y = 3x$ is a linear function whose graph is a straight line and whose rate of change is constant.
- $y = 3x^2$ is a quadratic function whose graph is parabolic (curved, nonlinear) and whose rate of change is not constant.

ADAPT
Check students' answers to the Lesson Practice to ensure that they can identify and graph a quadratic function. Use Items 16 and 17 to compare and contrast the equations of linear and quadratic functions. See the Activity Practice for additional problems for this lesson. You may assign the problems here or use them as a culmination for the Activity.

ACTIVITY 29 continued

My Notes

Lesson 29-1
Modeling with a Quadratic Function

Check Your Understanding

11. For the function $f(x) = x^2 + 2x + 3$, create a table of values for $x = -3, -2, -1, 0, 1$. Then sketch a graph of the quadratic function on grid paper.

12. Barry needs to find the area of a rectangular room with a width that is 2 feet longer than the length. Write an expression for the area of the rectangle in terms of the length.

13. **Critique the reasoning of others.** Sally states that the equation $g(x) = x^3 + 10x^2 - 3x$ represents a quadratic function. Explain why Sally is incorrect.

14. Write the quadratic function $f(x) = (3 - x)^2$ in standard form.

LESSON 29-1 PRACTICE

15. Create tables to graph $y = 3x$ and $y = 3x^2$ on grid paper. Explain the differences between the graphs.

16. Pierre uses the function $r(t) = t + 2$ to model his rate r in mi/h t minutes after leaving school. Complete the table and use the data points to graph the function.

t	r(t)
0	
1	
2	
3	
4	

17. Pierre uses the function $d(t) = t(t + 2)$ to model his distance d in miles from home. Add another column to your table in Item 16 to represent $d(t)$. Sketch a graph of $d(t)$ on the same coordinate plane as the graph of $r(t)$.

18. What types of functions are represented in Items 16 and 17?

19. Write each quadratic function in standard form.
    a. $g(x) = x(x - 2) + 4$
    b. $f(t) = 3 - 2t^2 + t$

20. **Attend to precision.** Determine whether each function is a quadratic function. Justify your responses.
    a. $S(r) = 2\pi r^2 + 20\pi r$
    b. $f(a) = \dfrac{a^2 + 4a - 3}{2}$
    c. $f(x) = 4x^2 - 3x + 2$
    d. $g(x) = 3x^{-2} + 2x - 1$
    e. $f(x) = 4^2 x - 3$
    f. $h(x) = \dfrac{4}{x^2} - 3x + 2$

16–17.

t	r(t)	d(t)
0	2	0
1	3	3
2	4	8
3	5	15
4	6	24

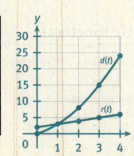

18. $r(t)$ is linear, and $d(t)$ is quadratic.
19. a. $g(x) = x^2 - 2x + 4$
    b. $f(t) = -2t^2 + t + 3$
20. a. Yes; the term with the greatest degree has degree 2.
    b. Yes; the term with the greatest degree has degree 2.
    c. Yes; the term with the greatest degree has degree 2.
    d. No; the term with the greatest degree has degree 1.
    e. No; the term with the greatest degree has degree 1.
    f. No; the variable is in the denominator of a term.

Lesson 29-2
Graphing and Analyzing a Quadratic Function

Learning Targets:
- Graph a quadratic function.
- Interpret key features of the graph of a quadratic function.

SUGGESTED LEARNING STRATEGIES: Interactive Word Wall, Create Representations, Construct an Argument, Marking the Text, Discussion Groups

1. Use a graphing calculator to graph $A(l)$ from Item 9 in Lesson 29-1. Sketch the graph on the grid below.

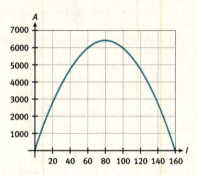

The graph of a quadratic function is a curve called a **parabola**. A parabola has a point at which a maximum or minimum value of the function occurs. That point is called the **vertex of a parabola**. The y-value of the vertex is the **maximum** or **minimum** of the function.

2. Identify the vertex of the graph of $A(l)$ in Item 1. Does the vertex represent a maximum or a minimum of the function?
   The vertex of the graph of $A(l)$ is (80, 6400). The vertex represents a maximum of the function.

3. Examine the graph of $A(l)$. For what values of the length is the area increasing? For what values of the length is the area decreasing?
   The area is increasing when the length is between 0 and 80 yards.
   The area is decreasing when the length is between 80 and 160 yards.

TECHNOLOGY TIP
Be sure that the RANGE on your calculator's graph matches the range shown in the grid.

CONNECT TO AP
AP Calculus students find these same maximum or minimum values of functions in *optimization* problems.

ACTIVITY 29 Continued

3–6 Discussion Groups, Construct an Argument, Sharing and Responding Students may use the graph or the table from Lesson 29-1 to respond to these items. Monitor student discussions carefully to be sure domain and range values make sense within the context of the problem situation. Since FIFA regulations indicate that the touchline must have a length greater than that of the goal line, the 80 yard by 80 yard rectangle would not be an appropriate choice. Students should realize that a soccer field can never be a square because of this regulation.

Lesson 29-2
Graphing and Analyzing a Quadratic Function

4. Describe the point where the area changes from increasing to decreasing.
   The vertex, or maximum, is where the area changes from increasing to decreasing.

5. Use the table, the graph, and/or the function to determine the reasonable domain and range of the function $A(l)$. Describe each using words and an inequality.
   Domain: $0 < l < 160$; the length can be any value greater than 0 and less than 160.
   Range: $0 < A(l) \leq 6400$; the area can be any value greater than 0 and less than or equal to 6400.

FIFA regulations state that the length of the touchline of a soccer field must be greater than the length of the goal line.

6. **Reason abstractly.** Can Coach Wentworth use the rectangle that represents the largest area of $A(l)$ for her soccer field? Explain why or why not.
   Answers may vary. The rectangle that represents the largest area cannot be used for the soccer field since it is an 80 yard by 80 yard rectangle, or a square. It does not meet the FIFA regulation that states that the length of the touchline be greater than the length of the goal line.

Lesson 29-2
Graphing and Analyzing a Quadratic Function

FIFA regulations also state that the length of the touchlines of a soccer field must be at least 100 yds, but no more than 130 yds. The goal lines must be at least 50 yds, but no more than 100 yds.

7. **Construct viable arguments.** Determine the dimensions of the FIFA regulation soccer field with the largest area and a 320-yd perimeter. Support your reasoning with multiple representations.
   Answers may vary. The maximum area of the FIFA regulation soccer field is 6000 square yards, with a touchline length of 100 yards and a goal line length of 60 yards. This is supported using the data table where the length is 100 and the width is 60, the graph at the point (100, 6000), and the equation, $A(l) = -l^2 + 160l$; $-(100)^2 + 160(100) = -10{,}000 + 16{,}000 = 6000$.

8. Consider the quadratic function $f(x) = x^2 - 2x - 3$.
   a. Write the function in factored form by factoring the polynomial $x^2 - 2x - 3$.
   $f(x) = (x - 3)(x + 1)$

   b. To find the x-intercepts of $f(x)$, use the factored form of $f(x)$ and solve the equation $f(x) = 0$.
   $f(x) = (x - 3)(x + 1) = 0$; $x = 3$ or $x = -1$.
   The x-intercepts are (3, 0) and (−1, 0).

   c. A parabola is symmetric over the vertical line that contains the vertex. How do you think the x-coordinate of the vertex relates to the x-coordinates of the x-intercepts? Use the symmetry of a parabola to support your answer.
   Answers may vary. The x-coordinate of the vertex is halfway between the x-coordinates of the intercepts. This is because points on each side of the parabola are the same horizontal distance from the vertex.

   d. Write the vertex of the quadratic function.
   (1, −4)

ACTIVITY 29 Continued

7 Think-Pair-Share, Construct an Argument, Debriefing The additional FIFA regulations indicate that the length of the rectangle must be at least 100 yards. The 320-yard perimeter constraint in the original problem restricts the length to a maximum of 110 yards. As this lesson is debriefed, focus on the reasonable domain and range. Why can't the soccer field be 80 yards by 80 yards? Is it acceptable for soccer fields to be different sizes? Why or why not?

8 Guess and Check, Create Representations Have students graph the function to confirm their answers to Item 8. It is important for students to understand that they can find the x-intercepts and vertex of a quadratic function graphically or algebraically.

Differentiating Instruction

Extend learning for students who are easily able to complete Item 8 by asking them to find the y-intercept of the function. Have them explain why there is only one y-intercept. Ask students whether every quadratic function will have two x-intercepts and one y-intercept. Encourage them to sketch different graphs on the coordinate plane to find the answer.

To **support** students' comprehension of their reading, you may want to add a section of your Word Wall on basic sentence structure in English writing (simple sentence, compound sentence, complex sentence, transition words, etc.). Review these structures with students prior to the assignment, and provide an opportunity to clarify any questions about language structure.

ACTIVITY 29 Continued

Check Your Understanding

Debrief students' answers by asking them how to graph a quadratic equation. Ask students to describe how to use a graph to determine the domain, range, and maximum or minimum of a quadratic function.

Answers

9.

x	f(x)
−4	−3
−3	0
−2	1
−1	0
0	−3

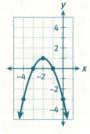

10. The function has a maximum of 1 when $x = -2$.

11. a.

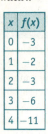

x	f(x)
0	−3
1	−2
2	−3
3	−6
4	−11

b.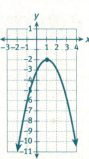

c. Domain: $-\infty < x < \infty$; range: $y \leq -2$; decreasing for $x > 1$

ASSESS

Students' answers to Lesson Practice problems will provide you with a formative assessment of their understanding of the lesson concepts and their ability to apply their learning.

See the Activity Practice for additional problems for this lesson. You may assign the problems here or use them as a culmination for the activity.

ADAPT

Check students' answers to the Lesson Practice to ensure that students can identify all the properties of a quadratic function. Students who are still having difficulty with all the new vocabulary can draw examples of a parabola opening up and a parabola opening down. They can label the vertex as a maximum or minimum and make connections to the equation of the parabola.

ACTIVITY 29 continued

My Notes

Lesson 29-2
Graphing and Analyzing a Quadratic Function

Check Your Understanding

9. Complete the table for the quadratic function $f(x) = -x^2 - 4x - 3$. Then graph the function.

x	f(x)
−4	
−3	
−2	
−1	
0	

10. Identify the maximum or minimum value of the quadratic function in Item 9.

11. Consider the quadratic function $y = -x^2 + 2x - 3$.
   a. Create a table of values for the function for domain values 0, 1, 2, 3, and 4.
   b. Sketch a graph of the function. Identify and label the vertex. What is the maximum value of the function?
   c. Use inequalities to write the domain, range, and values of x for which y is decreasing.

LESSON 29-2 PRACTICE

12. Write the quadratic function $g(x) = x(x - 2) + 4$ in standard form.

For Items 13–16, use the quadratic function $f(x) = x^2 + 6x + 5$.

13. Create a table of values and graph $f(x)$.

14. Use your graph to identify the maximum or minimum value of $f(x)$.

15. Write the domain and range of $f(x)$ using inequalities.

16. Determine the values of x for which $f(x)$ is increasing.

17. **Make use of structure.** Sketch a graph of a quadratic function with a maximum. Now sketch another graph of a quadratic function with a minimum. Explain the difference between the increasing and decreasing behavior of the two functions.

LESSON 29-2 PRACTICE

12. $g(x) = x^2 - 2x + 4$
13. Tables may vary.

x	f(x)
−4	−3
−3	−4
−2	−3
−1	0
0	5
1	12
2	21

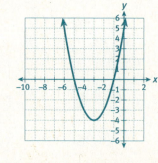

14. The function has a minimum of −4 when $x = -3$.
15. Domain: $-\infty < x < \infty$; range: $y \geq -4$
16. The function is increasing for all values $x > -3$.
17. Graphs may vary. Quadratic functions that have a maximum value at the vertex change from increasing to decreasing (from left to right). Quadratic functions that have a minimum value at the vertex change from decreasing to increasing (from left to right).

430 SpringBoard® Mathematics Algebra 1, Unit 5 • Quadratic Functions

Introduction to Quadratic Functions
Touchlines

ACTIVITY 29
continued

ACTIVITY 29 PRACTICE
Write your answers on notebook paper.
Show your work.

Lesson 29-1

1. The base of a rectangular window frame must be 1 foot longer than the height. Which of the following is an equation for the area of the window in terms of the height?
   A. $A(h) = h + 1$
   B. $A(h) = (h + 1)h$
   C. $A(h) = h^2 - 1$
   D. $A(h) = h^2 + h + 1$

2. Which of the following is an equation for the area of an isosceles right triangle, in terms of the base?

   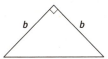

   A. $A(b) = b^2$
   B. $A(b) = \frac{1}{2}b^2$
   C. $A(b) = \frac{\sqrt{b^2}}{2}$
   D. $A(b) = \frac{1}{2}b$

3. Ben is creating a triangle that has a base that is twice the length of the height.
   a. Write an expression for the base of the triangle in terms of the height.
   b. Write a function for the area, $A(h)$, of the triangle in terms of the height.
   c. Complete the table and then graph the function.

h	A(h)
1	
2	
3	
4	
5	

Use the following information for Items 4–8.

Jenna is in charge of designing the screen for a new smart phone. The design specs call for a screen that has an outside perimeter of 12 inches.

4. Complete the table for the screen measurements.

Width, w	Length, l	Area, A(w)
1	5	5
2		
3		
4		
5		
w		

5. Write a function $A(w)$ for the area of the screen in terms of the width.

6. Graph the function $A(w)$. Label each axis.

7. Determine the domain and range of the function.

8. Determine the maximum area of the screen that Jenna can design. What are the dimensions of this screen?

9. Samantha's teacher writes the function $f(x) = 2x^3 - 2x(3 - x + x^2) + 3$.
   a. Barry tells Samantha that the function cannot be quadratic because it contains the term $2x^3$. What should Barry do to the function before making this assumption?
   b. Is the function a quadratic function? Explain.

ACTIVITY 29 Continued

ACTIVITY PRACTICE
1. B
2. B
3. a. $b = 2h$
   b. $A(h) = \frac{1}{2}(2h)h = h^2$
   c.
h	A(h)
1	1
2	4
3	9
4	16
5	25

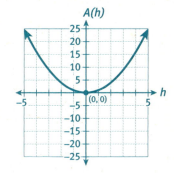

4.
Width, w	Length, l	Area, A(w)
1	5	5
2	4	8
3	3	9
4	2	8
5	1	6
w	6 − w	w(6 − w)

5. $A(w) = w(6 - w)$

6.

(graph with points (1,5), (2,8), (3,9), (4,8), (5,5))

7. Domain: $0 < w < 6$; The width of the screen can be greater than 0 inches and less than 6 inches.

   Range: $0 < A(w) < 9$; The area of the screen can be greater than 0 square inches and less than 9 square inches.

8. The maximum area is 9 square inches and the dimensions are 3 inches by 3 inches.

9. a. Barry should first simplify the function, to see if any terms cancel out or the degree of the terms changes.
   b. Yes; the simplified function is of the form $ax^2 + bx + c$.

ACTIVITY 29 Continued

10. a. no
 b. yes
 c. no
11. The data in the first table are linear because there is a constant rate of change. The data in the second table are not linear because the rate of change is not constant.
12. a. $y = x^2 + 3x - 5$
 b. $y = -5x^2 + 6$
 c. $y = \frac{3}{4}x^2 - 0.5x - \pi$
13. a.

x	y
−3	−4
−2	−5
−1	−4
0	−1
1	4

The minimum value is −5.

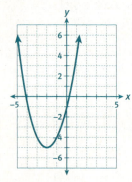

b.

x	y
2	−1
3	2
4	3
5	2
6	−1

The maximum value is 3.

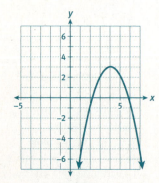

ADDITIONAL PRACTICE

If students need more practice on the concepts in this activity, see the eBook Teacher Resources for additional practice problems.

ACTIVITY 29 continued

Introduction to Quadratic Functions
Touchlines

10. Identify whether each function is quadratic.
 a. $y = 2x - 3^2$
 b. $y = 3x^2 - 2x$
 c. $y = 2 - \frac{3}{x^2} + x$

11. State whether the data in each table are linear. Explain why or why not.

x	y
0	5
1	2
2	−1
3	−4
4	−7

x	y
0	5
1	2
2	1
3	2
4	5

Lesson 29-2

12. Write each quadratic function in standard form.
 a. $y = 3x - 5 + x^2$
 b. $y = 6 - 5x^2$
 c. $y = -0.5x + \frac{3}{4}x^2 - \pi$

13. For each function, complete the table of values. Graph the function and identify the maximum or minimum of the function.
 a. $y = x^2 + 4x - 1$

x	y
−3	
−2	
−1	
0	
1	

c.

x	y
0	−1
−1	0
−2	3
−3	8
−4	15

The minimum value is −1.

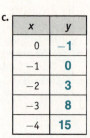

b. $y = -x^2 + 8x - 13$

x	y
2	
3	
4	
5	
6	

c. $y = x^2 - 1$

x	y
0	
−1	
−2	
−3	
−4	

14. Use your graphing calculator to graph several functions $y = ax^2 + b$ using positive and negative values of a and b. Do the signs of a and b appear to affect whether the function has a minimum or maximum value? Make a conjecture.

MATHEMATICAL PRACTICES
Construct Viable Arguments and Critique the Reasoning of Others

15. As part of her math homework, Kylie is graphing a quadratic function. After plotting several points, she notices that the dependent values are increasing as the independent values increase. She reasons that the function will eventually reach a maximum value and then begin to decrease. Is Kylie correct? Why or why not?

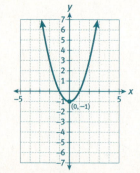

14. It appears that negative a yields a maximum value, and positive a yields a minimum value.
15. Kylie is not necessarily correct. It is possible that her data are increasing because the dependent values are all greater than that of the *minimum* value. In other words, she may be looking at data to the right of the vertex. If this is the case, the function will never turn and decrease.

Graphing Quadratic Functions
Transformers
Lesson 30-1 Translations of the Quadratic Parent Function

ACTIVITY 30

Learning Targets:
- Graph translations of the quadratic parent function.
- Identify and distinguish among transformations.

SUGGESTED LEARNING STRATEGIES: Look for a Pattern, Create Representations, Sharing and Responding, Quickwrite, Think-Pair-Share

The function $y = x^2$ or $f(x) = x^2$ is the quadratic *parent function*.

1. Complete the table for $y = x^2$. Then graph the function.

x	y
−3	9
−2	4
−1	1
0	0
1	1
2	4
3	9

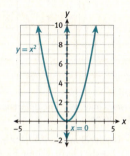

MATH TERMS

A **parent function** is the most basic function of a particular category or type. For example, the linear parent function is $y = x$ or $f(x) = x$.

2. Use words such as *vertex, maximum, minimum, increasing,* and *decreasing* to describe the graph of $y = x^2$.
   Answers may vary.
   - The graph is a parabola with vertex (0, 0).
   - Minimum value of the function is 0 at $x = 0$.
   - Function decreases from negative infinity to 0.
   - Function increases from 0 to infinity.

3. The line that passes through the vertex of a parabola and divides the parabola into two symmetrical parts is the parabola's *axis of symmetry*.
   a. Draw the axis of symmetry for the graph of $y = x^2$ as a dashed line on the graph in Item 1.
      See graph above.

   b. Write the equation for the axis of symmetry you drew in Item 1.
      $x = 0$

ACTIVITY 30 Guided

Activity Standards Focus
In Activity 30, students develop fluency in understanding function behavior by graphing, identifying, and distinguishing transformations of the parent quadratic function $y = x^2$.

Lesson 30-1

PLAN

Pacing: 1 class period
Chunking the Lesson
#1–3   #4–5   #6–9
Check Your Understanding
#13–16
Check Your Understanding
Lesson Practice

TEACH

Bell-Ringer Activity
Sketch a graph of a quadratic function on a coordinate plane. Ask students to translate the graph 5 spaces up. Then ask them to translate the new graph to the left 2 spaces. Ask students to describe which features change when a graph is translated (position) and which do not (shape, size, orientation).

1–3 Create Representations, Quickwrite, Interactive Word Wall, Sharing and Responding By graphing and describing the parent quadratic function, students will review the behavior of quadratics and the stage will be set for transformations. Be certain that students graph a smooth curve as opposed to "connecting the dots." Monitor students' writing to ensure that they are using vocabulary correctly and including adequate details. As students share descriptions of the graph, encourage the connection between the equation of the axis of symmetry and the x-coordinate of the vertex of the parabola.

Common Core State Standards for Activity 30

HSF-IF.B.4:	For a function that models a relationship between two quantities, interpret key features of graphs and tables in terms of the quantities, and sketch graphs showing key features given a verbal description of the relationship.
HSF-IF.B.5:	Relate the domain of a function to its graph and, where applicable, to the quantitative relationship it describes.
HSF-IF.C.7:	Graph functions expressed symbolically and show key features of the graph, by hand in simple cases and using technology for more complicated cases.
HSF-IF.C.7.a:	Graph linear and quadratic functions and show intercepts, maxima, and minima.

Activity 30 • Graphing Quadratic Functions   433

ACTIVITY 30 Continued

4–5 Look for a Pattern, Activating Prior Knowledge Students complete the table to recognize that a constant rate of change does not exist. Therefore, the function is not linear. It is worth noting that the differences between range values in this chart are "first differences." By inspecting the "second differences," the differences of the first differences, students will recognize an important numeric quality of quadratic functions: Second differences of quadratic functions are constant.

Developing Math Language

This lesson contains multiple vocabulary terms. Encourage students to mark the text by adding notes explaining the new vocabulary in their own words. The terms *parent function*, *axis of symmetry*, and *translation* are important for students to develop fluency in their understanding. Have students add the words to their math notebooks. As needed, pronounce new terms clearly and monitor students' pronunciation of terms in their class discussions. Use the class Word Wall to keep new terms in front of students. Include pronunciation guides as needed. Encourage students to review the Word Wall regularly and to monitor their own understanding and use of new terms in their group discussions.

ACTIVITY 30 continued

My Notes

Lesson 30-1
Translations of the Quadratic Parent Function

4. Complete the second and third columns of the table below. (Leave the fourth column blank for now.) Use your results to explain why the function $y = x^2$ is not linear.

x	$y = x^2$	Difference Between Consecutive y-Values ("First Differences")	"Second Differences"
−3	9	----	----
−2	4	−5	----
−1	1	−3	2
0	0	−1	2
1	1	1	2
2	4	3	2
3	9	5	2

Answers may vary. The differences between consecutive y-values are not constant.

5. The fourth column in the table in Item 4 is titled "Second Differences." Complete the fourth column by finding the change in consecutive values in the third column. What do you notice about the values?
Answers may vary. The differences in the fourth column are all equal to 2. The second differences are constant.

6. **a.** Complete the table for $f(x) = x^2$ and $g(x) = x^2 + 3$. Then graph each function on the same coordinate grid.

x	$f(x) = x^2$	$g(x) = x^2 + 3$
−3	9	12
−2	4	7
−1	1	4
0	0	3
1	1	4
2	4	7
3	9	12

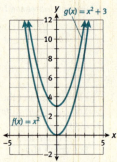

b. Identify the vertex, domain, range, and axis of symmetry for each function.

	$f(x) = x^2$	$g(x) = x^2 + 3$
Vertex	(0, 0)	(0, 3)
Domain	all real numbers	all real numbers
Range	$f(x) \geq 0$	$g(x) \geq 3$
Axis of symmetry	$x = 0$	$x = 0$

Common Core State Standards for Activity 30 *(continued)*

HSF-IF.C.9: Compare properties of two functions each represented in a different way (algebraically, graphically, numerically in tables, or by verbal descriptions).

HSF-BF.B.3: Identify the effect on the graph of replacing $f(x)$ by $f(x) + k$, $k f(x)$, $f(kx)$, and $f(x + k)$ for specific values of k (both positive and negative); find the value of k given the graphs. Experiment with cases and illustrate an explanation of the effects on the graph using technology.

Lesson 30-1
Translations of the Quadratic Parent Function

7. a. Complete the table for $f(x) = x^2$ and $h(x) = x^2 - 4$. Then graph each function on the same coordinate grid.

x	$f(x) = x^2$	$h(x) = x^2 - 4$
-3	9	5
-2	4	0
-1	1	-3
0	0	-4
1	1	-3
2	4	0
3	9	5

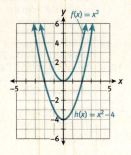

b. Identify the vertex, domain, range, and axis of symmetry for each function.

	$f(x) = x^2$	$h(x) = x^2 - 4$
Vertex	(0, 0)	(0, -4)
Domain	all real numbers	all real numbers
Range	$f(x) \geq 0$	$h(x) \geq -4$
Axis of symmetry	$x = 0$	$x = 0$

8. Compare the functions $g(x)$ and $h(x)$ in Items 6 and 7 to the parent function $f(x) = x^2$. Describe any patterns you notice in the following.

a. equations
Answers may vary. The equations for *g(x)* and *h(x)* changed from that of the parent function by adding or subtracting a constant.

b. tables
Answers may vary. The *y*-values for *g(x)* are 3 more than the *y*-values for *f(x)*. The *y*-values for *h(x)* are 4 less than the *y*-values for *f(x)*. The change in the *y*-values is determined by the constant added to or subtracted from the parent function.

c. graphs
Answers may vary. For *g(x)*, 3 was added to the parent function, and the graph moved up 3 units. For *h(x)*, 4 was subtracted from the parent function, and the graph moved down 4 units. In both cases, the shape of the graph is maintained.

ACTIVITY 30 Continued

6–9 (continued) As this portion of the lesson is debriefed, it is important that the effect of k on vertical translation be generalized not only in terms of direction (positive = translate up and negative = translate down), but also in terms of magnitude (the larger the absolute value of k, the greater the distance translated). It is worth noting that if the original function has a point (x, y) on the graph, then the point $(x, y + k)$ will be on the translated graph.

Check Your Understanding

Debrief students' answers to these items to ensure they understand how to graph vertical translations of quadratic functions. This part of the lesson prepares students to move on to graphing horizontal translations.

Answers

10. The graph will be translated 2 units down.

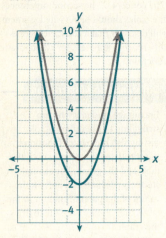

11. The graph will be translated 4 units up.

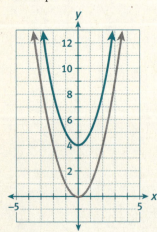

12. a. $k = 1$
    b. $k = -6$

Lesson 30-1
Translations of the Quadratic Parent Function

d. **vertices**
Answers may vary. For g(x), 3 was added to the parent function, and the y-coordinate of the vertex increased by 3. For h(x), 4 was subtracted from the parent function, and the y-coordinate of the vertex decreased by 4.

e. **domain and range**
Answers may vary. The domain is the same for all three functions, f(x), g(x), and h(x). For g(x) and h(x), the least value of the range increased or decreased by the amount that was added to or subtracted from the parent function.

f. **axes of symmetry**
Answers may vary. The axis of symmetry for all three functions is the same.

The changes to the parent function in Items 6 and 7 are examples of vertical translations. A *translation* of a graph is a change that shifts the graph horizontally, vertically, or both. A translation does not change the shape of the graph.

9. **Make use of structure.** How does the value of k in the equation $g(x) = x^2 + k$ change the graph of the parent function $f(x) = x^2$?
The graph moves k units up if $k > 0$ or $|k|$ units down if $k < 0$. The shape of the graph remains unchanged.

Check Your Understanding

For Items 10 and 11, predict the translations of the graph of $f(x) = x^2$ for each function. Confirm your predictions by graphing each equation on the same coordinate grid as the graph of the parent function.

10. $g(x) = x^2 - 2$ 11. $h(x) = x^2 + 4$

12. The graphs of two functions are shown below, along with the graph of the parent function. Determine the value of c for each function.
    a. $g(x) = x^2 + c$ b. $h(x) = x^2 + c$

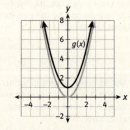 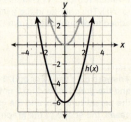

Lesson 30-1
Translations of the Quadratic Parent Function

13. a. Complete the table for $f(x) = x^2$ and $k(x) = (x + 3)^2$. Then graph both functions on the same coordinate grid.

x	$f(x) = x^2$	$k(x) = (x + 3)^2$
−3	9	0
−2	4	1
−1	1	4
0	0	9
1	1	16
2	4	25
3	9	36

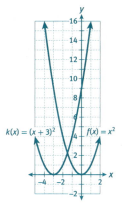

b. Identify the vertex, domain, range, and axis of symmetry for each function.

	$f(x) = x^2$	$k(x) = (x + 3)^2$
Vertex	(0, 0)	(−3, 0)
Domain	all real numbers	all real numbers
Range	$f(x) \geq 0$	$k(x) \geq 0$
Axis of symmetry	$x = 0$	$x = -3$

14. a. Complete the table for $f(x) = x^2$ and $p(x) = (x - 4)^2$. Then graph both functions on the same coordinate grid.

x	$f(x) = x^2$	$p(x) = (x - 4)^2$
−1	1	25
0	0	16
1	1	9
2	4	4
3	9	1
4	16	0
5	25	1

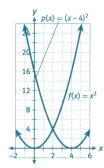

ACTIVITY 30 Continued

13–16 Look for a Pattern, Create Representations, Think-Pair-Share, Debriefing By completing the table for both the parent function and the translated functions, students will be able to visualize the translations. The comparison of the symbolic, numerical, and graphic representations is essential for students to make the appropriate connections between the parent function and the translations. Encourage students to focus on the relationship between the translations and patterns of change in the vertex, domain, range, and axis of symmetry.

Activity 30 • Graphing Quadratic Functions 437

ACTIVITY 30 Continued

13–16 (continued) As this portion of the lesson is debriefed, the effect of h on horizontal translation should be generalized not only in terms of direction (positive = translate left and negative = translate right), but also in terms of magnitude (the larger the absolute value of h, the greater the distance translated). Students should recognize that the relationship between the sign of h and the direction of the shift seems counterintuitive: Although it would seem a positive h would indicate a shift to the right, the graph shifts to the left instead. Similarly, a negative h shifts right, in a positive direction. It is worth noting that if the original function has a point (x, y) on the graph, then the point $(x + h, y)$ will be on the translated graph.

Check Your Understanding

Debrief students' answers to these items to ensure that they understand how to graph horizontal translations of quadratic functions. This lesson prepares students to move from translations to other transformations such as stretching and shrinking the quadratic parent function.

Answers

17. The graph will be translated 2 units to the right.

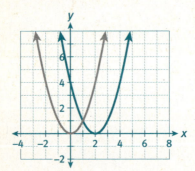

18. The graph will be translated 1 unit to the left.

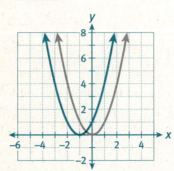

19. a. $h = -5;\ k(x) = x^2 - 10x + 25$
b. $h = 6;\ p(x) = x^2 + 12x + 36$

Lesson 30-1
Translations of the Quadratic Parent Function

b. Identify the vertex, domain, range, and axis of symmetry for each function.

	$f(x) = x^2$	$p(x) = (x - 4)^2$
Vertex	(0, 0)	(4, 0)
Domain	all real numbers	all real numbers
Range	$f(x) \geq 0$	$p(x) \geq 0$
Axis of symmetry	$x = 0$	$x = 4$

15. Compare the functions $k(x)$ and $p(x)$ in Items 13 and 14 to the parent function $f(x) = x^2$. Describe any patterns you notice in the following.

a. equations
Answers may vary. For $k(x)$ and $p(x)$, the equation changed from that of the parent function by adding or subtracting a constant inside the parentheses.

b. graphs
Answers may vary. For $k(x)$, 3 was added inside the parentheses, and the graph moved left 3 units. For $p(x)$, 4 was subtracted inside the parentheses, and the graph moved right 4 units. In both cases, the shape of the graph is maintained.

c. vertices
Answers may vary. For $k(x)$, 3 was added inside the parentheses, and the x-coordinate of the vertex decreased by 3. For $p(x)$, 4 was subtracted inside the parentheses, and the x-coordinate of the vertex increased by 4.

d. domain and range
The domain and range are the same for all three functions.

e. axes of symmetry
Answers may vary. For $k(x)$, 3 was added inside the parentheses, and the axis of symmetry moved 3 units to the left. For $p(x)$, 4 was subtracted inside the parentheses, and the axis of symmetry moved 4 units to the right.

Lesson 30-1
Translations of the Quadratic Parent Function

The changes to the parent function in Items 13 and 14 are examples of horizontal translations.

16. Express regularity in repeated reasoning. How does the value of h in the equation $k(x) = (x + h)^2$ change the graph of the parent function $k(x) = x^2$?
Answers may vary. The graph moves horizontally h units to the left if $h > 0$ and $|h|$ units to the right if $h < 0$. The shape of the graph remains unchanged.

Check Your Understanding

For Items 17 and 18, predict the translations of the graph of $f(x) = x^2$ for each equation. Confirm your predictions by graphing each equation on the same coordinate grid as the graph of the parent function.

17. $k(x) = (x - 2)^2$ **18.** $p(x) = (x + 1)^2$

19. The graphs of two functions are shown below, along with the graph of the parent function. Determine the value of h for each function. Then write each function in standard form.

a. $k(x) = (x + h)^2$ **b.** $p(x) = (x + h)^2$

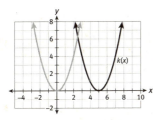 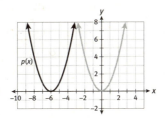

LESSON 30-1 PRACTICE

20. The graph of $g(x)$ is a vertical translation 2 units up from the graph of $f(x) = x^2$. Which of the following equations describes $g(x)$?
 A. $g(x) = x^2 + 2$ **B.** $g(x) = x^2 - 2$
 C. $g(x) = (x + 2)^2$ **D.** $g(x) = (x - 2)^2$

21. For each function, use translations to graph the function on the same coordinate grid as the parent function $f(x) = x^2$. Then identify the vertex, domain, range, and axis of symmetry.
 a. $g(x) = x^2 + 1$ **b.** $h(x) = x^2 - 1$
 c. $k(x) = (x + 1)^2$ **d.** $n(x) = (x - 1)^2$

22. Attend to precision. Write an ordered pair that represents the vertex of the graph of each quadratic function.
 a. $f(x) = x^2 + c$ **b.** $f(x) = (x - k)^2$

21. c.

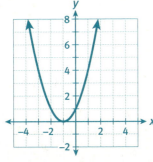

Vertex: $(-1, 0)$; domain: all real numbers; range: $y \geq 0$; axis of symmetry: $x = -1$

d.

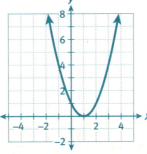

Vertex: $(1, 0)$; domain: all real numbers; range: $y \geq 0$; axis of symmetry: $x = 1$

22. a. $(0, c)$
b. $(k, 0)$

ACTIVITY 30 Continued

ASSESS

Students' answers to lesson practice problems will provide you with a formative assessment of their understanding of the lesson concepts and their ability to apply their learning.

See the Activity Practice for additional problems for this lesson. You may assign the problems here or use them as a culmination for the activity.

LESSON 30-1 PRACTICE
20. A
21. a.

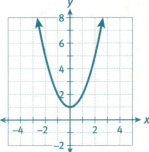

Vertex: $(0, 1)$; domain: all real numbers; range: $y \geq 1$; axis of symmetry: $x = 0$

b.

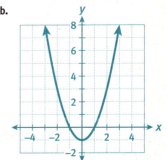

Vertex: $(0, -1)$; domain: all real numbers; range: $y \geq -1$; axis of symmetry: $x = 0$

ADAPT

Check students' answers to the Lesson Practice to ensure that they understand vertical and horizontal translations of the quadratic parent function and are ready to transition to stretching and shrinking it. Students having difficulty should continue to be encouraged to use different colors when graphing. In addition, students may benefit from drawing arrows above the numbers in the equation indicating the way in which the parent graph will "slide" based on the number.

Activity 30 • Graphing Quadratic Functions 439

ACTIVITY 30 Continued

Lesson 30-2

PLAN

Pacing: 1 class period
Chunking the Lesson
#1–3   #4–5
Check Your Understanding
Lesson Practice

TEACH

Bell-Ringer Activity
Display the graph of the quadratic parent function on a coordinate plane. Then, sketch a "skinnier" version of the graph inside it. Have students predict how the relationship between x- and y-values will change as the graph of the parent function moves closer to the y-axis.

1–3 Create Representations, Think-Pair-Share, Look for a Pattern
Carefully monitor student discussions as students graph these transformations. Order of operations may prove to be problematic for some students who want to multiply by the coefficient prior to squaring the x-value. Again, the comparison of the symbolic, numerical, and graphic representations is essential for students to make the appropriate connections between the parent function and the translations. Encourage students to examine the relationships between the second and third columns of the tables.

Developing Math Language
This lesson contains multiple vocabulary terms. Encourage students to mark the text by adding notes explaining the new vocabulary in their own words. The terms *vertical stretch*, *vertical shrink*, and *transformation* are important for students to develop fluency in their understanding. Have students add the words to their math notebooks. Students can make notes about new terms and how to use them to describe precise mathematical concepts and processes. Add these terms to the classroom Word Wall.

ACTIVITY 30 continued

Lesson 30-2
Stretching and Shrinking the Quadratic Parent Function

Learning Targets:
- Graph vertical stretches and shrinks of the quadratic parent function.
- Identify and distinguish among transformations.

SUGGESTED LEARNING STRATEGIES: Create Representations, Look for a Pattern, Sharing and Responding, Think-Pair-Share, Discussion Groups

1. Complete the table for $f(x) = x^2$ and $g(x) = 2x^2$.

x	$f(x) = x^2$	$g(x) = 2x^2$
-2	4	8
-1	1	2
0	0	0
1	1	2
2	4	8

a. The graph of $f(x) = x^2$ is shown below. Graph $g(x) = 2x^2$ on the same coordinate grid.

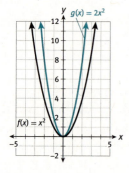

b. Identify the vertex, domain, range, and axis of symmetry for each function.

	$f(x) = x^2$	$g(x) = 2x^2$
Vertex	(0, 0)	(0, 0)
Domain	all real numbers	all real numbers
Range	$f(x) \geq 0$	$g(x) \geq 0$
Axis of symmetry	$x = 0$	$x = 0$

440 SpringBoard® Mathematics Algebra 1, Unit 5 • Quadratic Functions

Lesson 30-2
Stretching and Shrinking the Quadratic Parent Function

2. Complete the table for $f(x) = x^2$ and $h(x) = \frac{1}{2}x^2$.

x	$f(x) = x^2$	$h(x) = \frac{1}{2}x^2$
−2	4	2
−1	1	$\frac{1}{2}$
0	0	0
1	1	$\frac{1}{2}$
2	4	2

a. The graph of $f(x) = x^2$ is shown below. Graph $h(x) = \frac{1}{2}x^2$ on the same coordinate grid.

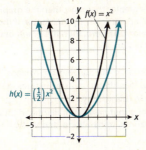

b. Identify the vertex, domain, range, and axis of symmetry for each function.

	$f(x) = x^2$	$h(x) = \frac{1}{2}x^2$
Vertex	(0, 0)	(0, 0)
Domain	all real numbers	all real numbers
Range	$f(x) \geq 0$	$h(x) \geq 0$
Axis of symmetry	$x = 0$	$x = 0$

3. Compare the functions $g(x)$ and $h(x)$ in Items 1 and 2 to the parent function $f(x) = x^2$. Describe any patterns you notice in the following.
   a. equations
   Answers may vary. The equations for both $g(x)$ and $h(x)$ have the form of a positive constant multiplied by the parent function.

ACTIVITY 30 Continued

4–6 Discussion Groups, Debriefing
Generalizing the effect of a on the graph of the equation may pose a challenge to some students. Debriefing should focus students on the change in shape (narrower, wider) of the graph while the vertex remains in the same location. It is worth noting that if the original function has a point (x, y) on the graph, then the point (x, ay) will be on the transformed graph. Students are introduced to "transformation," a general term for different types of changes in a graph of a parent function. Encourage students to draw sketches to illustrate each of the transformations they have seen so far.

Check Your Understanding
Debrief students' answers to these items to ensure that they understand how to graph a stretch and a shrink of the graph of the quadratic parent function. This lesson prepares students to move on to graphing multiple transformations of the quadratic parent function including reflections.

Answers
7. The graph will be a vertical stretch of the parent graph by a factor of 3.

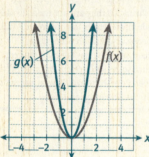

8. The graph will be a vertical shrink of the parent graph by a factor of $\frac{1}{4}$.

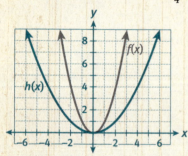

9. $g(x) = 5x^2$
10. $y = \frac{1}{5}x^2$

Lesson 30-2
Stretching and Shrinking the Quadratic Parent Function

b. tables
Answers may vary. For $g(x)$, the y-values are twice the y-values for $f(x)$. For $h(x)$, the y-values are $\frac{1}{2}$ the y-values for $f(x)$.

c. graphs
Answers may vary. The graph of $g(x)$ is narrower than the graph of $f(x)$. The graph of $h(x)$ is wider than the graph of $f(x)$.

d. vertex, domain and range, and axes of symmetry
Answers may vary. The vertex, domain and range, and axis of symmetry are the same for all three functions.

The change to the parent function in Item 1 is a **vertical stretch** by a factor of 2 and the change in Item 2 is a **vertical shrink** by a factor of $\frac{1}{2}$. A vertical stretch or shrink changes the shape of the graph.

4. **Reason quantitatively.** For $a > 0$, how does the value of a in the equation $g(x) = ax^2$ change the graph of the parent function $f(x) = x^2$?
   If the value of a is greater than one, then the graph of the parent function is stretched vertically by that factor. If the value of a is less than one but greater than zero, then the graph of the parent function is shrunk by that factor.

5. A change in the position, size, or shape of a parent graph is a **transformation**. Identify the transformations that have been introduced so far in this activity.
   The transformations that have been identified thus far are vertical translations, horizontal translations, and vertical stretches and shrinks.

Vertical stretching and shrinking also applies to linear functions. Below is the graph of the parent linear function $f(x) = x$.

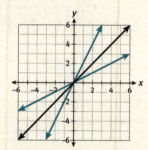

6. Graph the functions $g(x) = 2x$ and $h(x) = \frac{1}{2}x$ on the coordinate grid above. How do these graphs compare with the parent linear function?
   The graph of $g(x)$ is a vertical stretch, making the line steeper; the graph of $h(x)$ is a vertical shrink, making the line less steep.

MATH TIP
Create an organized summary of the single transformations of a quadratic function. Include a graph of the parent function, graph of the transformation, the equation, and a verbal description of the transformation.

Lesson 30-2
Stretching and Shrinking the Quadratic Parent Function

Check Your Understanding

For Items 7 and 8, predict the change from the graph of $f(x) = x^2$ for each function. Confirm your predictions by graphing each function on the same coordinate grid as the graph of the parent function.

7. $g(x) = 3x^2$
8. $h(x) = \frac{1}{4}x^2$

Write the equation for each transformation of the graph of $f(x) = x^2$.

9. a vertical stretch by a factor of 5
10. a vertical shrink by a factor of $\frac{1}{5}$

LESSON 30-2 PRACTICE

11. The graph of $g(x)$ is a vertical stretch of the graph of $f(x) = x^2$ by a factor of 7. Which of the following equations describes $g(x)$?
    A. $g(x) = x^2 + 7$
    B. $g(x) = 7x^2$
    C. $g(x) = (x + 7)^2$
    D. $g(x) = \frac{1}{7}x^2$

12. For each function, use transformations to graph the function on the same coordinate grid as the parent function $f(x) = x^2$.
    a. $g(x) = 4x^2$
    b. $h(x) = \frac{1}{4}x^2$
    c. $k(x) = 0.5x^2$
    d. $p(x) = \frac{3}{4}x^2$

13. **Reason abstractly.** Chen begins with a quadratic data set, which contains the vertex (0, 0). He multiplies every value in the range by 3 to create a new data set. Which of the following statements are true?
    A. The function that represents the new data set must be $y = 3x^2$.
    B. The graphs of the original data set and the new data set will have the same vertex.
    C. The graph of the new data set will be a vertical stretch of the graph of the original data set.
    D. The graph of the new data set will be translated up 3 units from the graph of the original data set.

14. How are the graphs of linear functions $f(x) = x$, $k(x) = \frac{3}{4}x$, and $t(x) = \frac{4}{3}x$ the same? How are they different?

12. d.
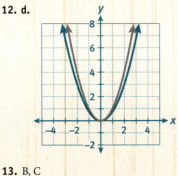

14. Their graphs are all lines that pass through the origin. Since $\frac{3}{4} < 1$, the graph of $k(x)$ is not as steep as the graph of $f(x)$. Since $\frac{4}{3} > 1$, the graph of $t(x)$ is steeper than the graph of $f(x)$.

13. B, C

ACTIVITY 30 Continued

ASSESS

Students' answers to Lesson Practice problems will provide you with a formative assessment of their understanding of the lesson concepts and their ability to apply their learning.

See the Activity Practice for additional problems for this lesson. You may assign the problems here or use them as a culmination for the activity.

LESSON 30-2 PRACTICE
11. B
12. a.
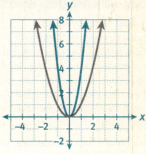

b.
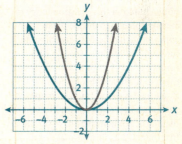

c.
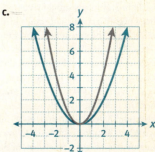

ADAPT

Check students' answers to the Lesson Practice to ensure that they understand shrinks and stretches of the quadratic parent function and are ready to transition to multiple transformations including reflections. Students having difficulty should continue to be encouraged to use different colors when graphing. In addition, students may benefit from drawing arrows or making notes above the constants and coefficients in the equation indicating the way in which the parent graph will change based on the constant or coefficient indicated.

Activity 30 • Graphing Quadratic Functions 443

ACTIVITY 30 Continued

Lesson 30-3

PLAN

Pacing: 1 class period
Chunking the Lesson
#1–3   #4–7   #8–9
#10   #11–12
Check Your Understanding
Example A   #14–15
Check Your Understanding
Lesson Practice

TEACH

Bell-Ringer Activity
Ask each student to sketch the graph of the quadratic parent function. Then, ask students to flip the sketch upside down. Have students discuss how this graph would be related to the original if the coordinate plane were still in its original orientation.

1–3 Create Representations, Look for a Pattern, Interactive Word Wall, Sharing and Responding The table is not included in this question to try to enable students to decide on an appropriate method for graphing the function. If students struggle with finding points for the graph, encourage them to make a table of values for the function. Once the graph is completed, students will recognize the reflection over the x-axis. Students may have vocabulary that differs (e.g., mirror image or flip). For Item 3, students should recognize that a negative coefficient results in a reflection of the parent graph over the x-axis. As students share responses, encourage the use of this new vocabulary. It is worth noting that if the original function has a point (x, y) on the graph, then the point $(x, -y)$ will be on the reflection. Extend your students' thinking by asking them to evaluate the following statement as always, sometimes, or never true and provide explanations for their response: If $y = ax^2 + c$ and $a > 0$, then the graph of the parabola will have a vertex at the maximum value of the range (never).

ACTIVITY 30
continued

Lesson 30-3
Multiple Transformations of the Quadratic Parent Function

My Notes

Learning Targets:
- Graph reflections of the quadratic parent function.
- Identify and distinguish among transformations.
- Compare functions represented in different ways.

SUGGESTED LEARNING STRATEGIES: Look for a Pattern, Create Representations, Predict and Confirm, Sharing and Responding, Think-Pair-Share

1. The quadratic parent function $f(x) = x^2$ is graphed below. Graph $g(x) = -x^2$ on the same coordinate grid.

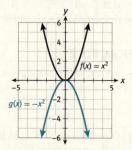

2. Compare $g(x)$ and its graph to the parent function $f(x) = x^2$.
   Answers may vary. The equation of $g(x)$ differs from the parent function in that it is multiplied by -1. The graph of $g(x)$ is the graph of $f(x)$ flipped over the x-axis.

The change to the parent function in Item 1 is a **reflection** over the x-axis. A *reflection* of a graph is the mirror image of the graph over a line. A reflection preserves the shape of the graph.

3. How does the sign of a in the function $g(x) = ax^2$ affect the graph?
   Answers may vary. If a is positive, then the parabola will open upward. If a is negative, then the parabola will open downward.

A graph may represent more than one transformation of the graph of the parent function. The order in which multiple transformations are performed is determined by the order of operations as indicated in the equation.

---

444 SpringBoard® Mathematics Algebra 1, Unit 5 • Quadratic Functions

Lesson 30-3
Multiple Transformations of the Quadratic Parent Function

4. How is the graph of $f(x) = x^2$ transformed to produce the graph of $g(x) = 2x^2 - 5$?

   Answers may vary. Stretch the graph of $f(x) = x^2$ by a factor of 2 and then translate the graph 5 units down.

5. Use the transformations you described in Item 4 to graph the function $g(x) = 2x^2 - 5$.

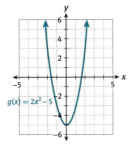

6. How would you transform of the graph of $f(x) = x^2$ to produce the graph of the function $h(x) = -\frac{1}{2}x^2 + 4$?

   Answers may vary. There will be a reflection over the x-axis, a vertical shrink by a factor of one-half, and a vertical translation 4 units up.

7. Use the transformations you described in Item 6 to graph the function $h(x) = -\frac{1}{2}x^2 + 4$. Identify the vertex and axis of symmetry of the parabola.

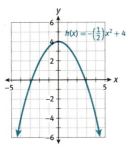

   vertex: (0, 4); axis of symmetry: $x = 0$

Lesson 30-3
Multiple Transformations of the Quadratic Parent Function

8. How would you transform of the graph of $f(x) = x^2$ to produce the graph of the function $k(x) = -\frac{1}{2}(x+4)^2$?

   Answers may vary. There will be a reflection over the x-axis, a vertical shrink by a factor of one-half, and a horizontal translation of 4 units left.

9. Use the transformations you described in Item 8 to graph the function $k(x) = -\frac{1}{2}(x+4)^2$. Identify the vertex and axis of symmetry of the parabola.

   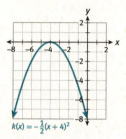

   $k(x) = -\frac{1}{2}(x+4)^2$

   vertex: (0, −4): axis of symmetry: x = 0

10. **Reason abstractly.** The graph of $g(x) = a(x-h)^2 + k$ represents multiple transformations of the graph of the parent function, $f(x) = x^2$. Describe how each value transforms the graph of $g(x)$ from the graph of $f(x)$.

    a. k

       A vertical translation of |k| units, up if k > 0 and down if k < 0

    b. h

       A horizontal translation of |h| units, right if h > 0 and left if h < 0

    c. |a|

       A vertical shrink for |a| < 1, and a vertical stretch for |a| > 1

    d. the sign of a

       If a is negative, a reflection over the x-axis

Lesson 30-3
Multiple Transformations of the Quadratic Parent Function

11. Examine the function $r(x) = 2(x - 4)^2 + 3$.
    a. Describe the transformations from the graph of $f(x) = x^2$ to the graph of $r(x) = 2(x - 4)^2 + 3$.
    A vertical stretch by a factor of 2, a horizontal translation 4 units to the right, and a vertical translation 3 units up
    b. Use the transformations you described in Part (a) to graph the function.

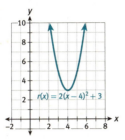

    c. Identify the vertex and axis of symmetry of the parabola.
    Vertex: (4, 3); axis of symmetry: $x = 4$

12. The vertex of the graph of $f(x) = x^2$ is $(0, 0)$.
    a. Describe how transformations can be used to determine the vertex of the graph of $g(x) = a(x - h)^2 + k$ without graphing.
    This represents a horizontal shift of $|h|$ units and a vertical shift of $|k|$ units. So the vertex will move from (0, 0) to $(0 + h, 0 + k) = (h, k)$.
    b. Identify the vertex of the graph of $y = 3(x - 4)^2 + 7$.
    $h = 4$ and $k = 7$, so the vertex is (4, 7).

Check Your Understanding

13. Without graphing, determine the vertex of the graph of each function.
    a. $y = x^2 + 6$
    b. $y = 3x^2 - 8$
    c. $y = (x - 4)^2$
    d. $y = (x - 2)^2 + 1$
    e. $y = 2(x + 2)^2 - 5$

ACTIVITY 30 Continued

Example A Note Taking Students are introduced to the process of writing the equation of a quadratic function given the graph of the function. This example transitions students from the "knowns" that can be determined from the graph of the function and written in the form $f(x) = a(x − h)^2 + k$ to the "unknowns" when its equation is in standard form $y = ax^2 + bx + c$.

TEACHER TO TEACHER

Some students may have difficulty with the algebra in Example A. There may be a great temptation to simply skim over the steps and equations, with the notion that the general point of the exercise can be gleaned without following each step in detail. Encourage students not to take this course of action. Suggest that they take the time to review this example, multiple times if necessary, in order to thoroughly understand what is being done, as representing functions in different forms will be an important skill as they advance in algebra.

The words students need to understand to follow the example will likely be on your Word Wall already. Remind students to refer to the Word Wall to refresh their understanding of the meanings. If there are words that are not on the Word Wall, ask students to choose words to add.

ACTIVITY 30 continued

My Notes

Lesson 30-3
Multiple Transformations of the Quadratic Parent Function

Example A

For the quadratic function shown in the graph, write the equation in standard form.

Step 1: Identify the vertex.
$$(h, k) = (−1, 1)$$

Step 2: Choose another point on the graph.
$$(x, y) = (0, 3)$$

Step 3: Substitute the known values.
$$y = a(x − h)^2 + k$$
$$3 = a(0 − (−1))^2 + 1$$

Step 4: Solve for a.
$$3 = a(1)^2 + 1$$
$$3 = a + 1$$
$$2 = a$$

Step 5: Write the equation of the function by substituting the values of a, h, and k.
$$y = a(x − h)^2 + k$$
$$y = 2(x + 1)^2 + 1$$

Step 6: Multiply and combine like terms to write the equation in standard form.
$$y = 2(x + 1)^2 + 1$$
$$y = 2(x^2 + 2x + 1) + 1$$
$$y = 2x^2 + 4x + 2 + 1$$
$$y = 2x^2 + 4x + 3$$

Solution: $y = 2x^2 + 4x + 3$

Try These A

For each quadratic function, write its equation in standard form.

a.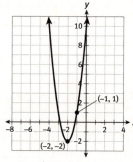

$y = 3x^2 + 12x + 10$

b.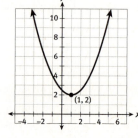

$y = \frac{1}{2}x^2 − x + \frac{5}{2}$

---

448 SpringBoard® Mathematics **Algebra 1, Unit 5** • Quadratic Functions

Lesson 30-3
Multiple Transformations of the Quadratic Parent Function

The vertex of a parabola represents the *minimum* value of the quadratic function if the parabola opens upward. The vertex of a parabola represents the *maximum* value of the quadratic function if the parabola opens downward.

14. **Make sense of problems.** The equation and graph below represent two different quadratic functions.

    Function 1: $y = x^2 - 3$

    Function 2:

    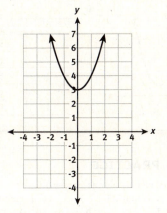

    Which function has the greater minimum value? Justify your response.
    Function 2; the graph shows that the vertex is (0, 3), so function 2 has a minimum value of 3. For function 1, the equation shows that its graph has vertex (0, −3), so function 1 has a minimum value of −3.

15. The verbal description and the table below represent two different quadratic functions.

    Function 1: The graph of function 1 is the graph of the parent function stretched vertically by a factor of 4 and translated 1 unit up.

    Function 2:

x	y
−5	4
−4	1
−3	0
−2	1
−1	4

    Which function has the greater minimum value? Justify your response.
    Function 1; from the description, the graph of function 1 has vertex (0, 1), so the minimum value of function 1 is 1. The table shows that the value of function 2 at $x = -3$ is 0, so the minimum value of function 2 cannot be greater than 0.

DISCUSSION GROUP TIPS

If you do not understand something in your group discussions, ask for help or raise your hand for help. Describe your questions as clearly as possible, using synonyms or other words when you do not know the precise words to use.

ACTIVITY 30 Continued

Check Your Understanding

Debrief students' answers to this item to ensure that they understand how to compare functions in different forms.

Answers

16. C

ASSESS

Students' answers to Lesson Practice problems will provide you with a formative assessment of their understanding of the lesson concepts and their ability to apply their learning.

See the Activity Practice for additional problems for this lesson. You may assign the problems here or use them as a culmination for the activity.

LESSON 30-3 PRACTICE

17. **a.** Shrink vertically by a factor of $\frac{1}{2}$ and translate up 3 units.

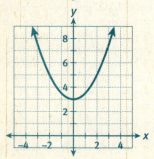

b. Reflect over the x-axis and translate up 4 units.

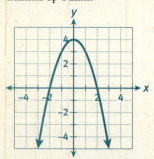

ADAPT

Check students' answers to the Lesson Practice to ensure that they understand multiple transformations of the quadratic parent function and are confident comparing functions represented in different ways. Again, for students still having difficulty applying multiple transformations, encourage the use of multiple colors and allow them to apply one translation at a time. For example, draw the parent function in one color, shrinks/stretches in a second color, reflections in a third color, and translations left or right in a fourth color. Finally, draw translations up or down in a fifth color.

18. **a.** $y = (x - 2)^2 + 9$
    b. $y = 5x^2 - 12$
19. **a.** $y = -3x^2 + 6$
    b. $y = \frac{1}{4}x^2 - 3$
20. Phillip is not necessarily correct. The vertex of Kumara's graph may be a minimum value. If that is the case, then the maximum of Kumara's graph is infinity and is therefore greater than the maximum of Phillip's data.

Lesson 30-3
Multiple Transformations of the Quadratic Parent Function

Check Your Understanding

16. Which of the following represents a quadratic function whose minimum value is less than the minimum value of the function $y = 3x^2 + 4$?

    A. $y = 2(x + 1)^2 + 6$
    B. The function whose graph is translated right 1 unit and up 4 units from the graph of the parent function
    C.
    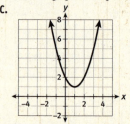
    D.

x	y
−3	9
−2	6
−1	5
0	6
1	9

LESSON 30-3 PRACTICE

17. For each function, identify how its graph has been transformed from the graph of the parent function $f(x) = x^2$. Then graph each function.
    a. $g(x) = \frac{1}{2}x^2 + 3$
    b. $h(x) = -x^2 + 4$

18. Write the equation of the function whose graph has been transformed from the graph of the parent function as described.
    a. Translated up 9 units and right 2 units
    b. Stretched vertically by a factor of 5 and translated down 12 units

19. Write an equation for each function.
    a.
    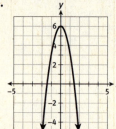
    b.
    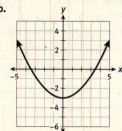

20. **Critique the reasoning of others.** Kumara graphs a quadratic function, and the vertex of the graph is (2, 3). Phillip states that his function $y = -x^2 + 7$ has a greater maximum value than Kumara's. Is Phillip correct? Justify your response.

Graphing Quadratic Functions
Transformers

ACTIVITY 30 PRACTICE
Write your answers on notebook paper. Show your work.

Lesson 30-1

For each quadratic function, describe the transformation of its graph from the graph of the parent function $f(x) = x^2$.

1. $g(x) = x^2 + 7$
2. $y = x^2 - \frac{3}{5}$
3. $h(x) = (x + 4)^2$
4. $y = \left(x - \frac{1}{2}\right)^2$
5. The graph of which of the following has a vertex at the point (0, 8)?
   A. $y = x^2 + 8$
   B. $y = x^2 - 8$
   C. $y = (x + 8)^2$
   D. $y = (x - 8)^2$
6. Identify the range of the function $y = x^2 - 3$.
7. Determine the equation of the axis of symmetry of the graph of $y = x^2 - 3$.
8. Identify the range of the function $y = (x - 11)^2$.
9. Determine the equation of the axis of symmetry of the graph of $y = (x - 11)^2$.

10. Write the equation of each quadratic function.
    a.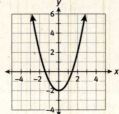
    b. The quadratic function whose graph is translated 10 units to the left from the graph of the parent function $f(x) = x^2$
    c.

x	y
−2	4.5
−1	1.5
0	0.5
1	1.5
2	4.5

ACTIVITY 30 Continued
ACTIVITY PRACTICE
1. Translate up 7 units.
2. Translate down $\frac{3}{5}$ unit.
3. Translate left 4 units.
4. Translate right $\frac{1}{2}$ unit.
5. A
6. $y \geq -3$
7. $x = 0$
8. $y \geq 0$
9. $x = 11$
10. a. $y = x^2 - 2$
    b. $y = (x + 10)^2$
    c. $y = x^2 + \frac{1}{2}$

ACTIVITY 30 Continued

11. B
12. a. vertical stretch by a factor of 12
  b. vertical shrink by a factor of $\frac{2}{5}$
13. a. $y = 6x^2$
  b. $y = \frac{1}{2}x^2$
14. $y = \frac{1}{3}x^2$
15. a. vertical stretch by a factor of 2 and translation up 1 unit

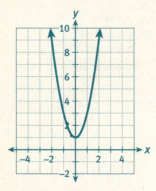

  b. translation left 3 units and reflection across the x-axis

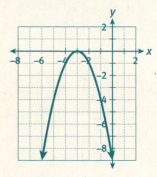

  c. translation left 1 unit and up 4 units

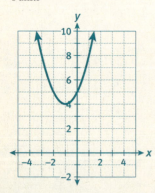

ADDITIONAL PRACTICE
If students need more practice on the concepts in this activity, see the eBook Teacher Resources for additional practice problems.

ACTIVITY 30 continued

Lesson 30-2

11. The graph of which function shares a vertex with the graph of $y = 5x^2 + 3$?
  A. $y = 3x^2 + 5$
  B. $y = x^2 + 3$
  C. $y = (x + 3)^2$
  D. $y = (x - 5)^2$

12. Identify the transformations from the graph of $f(x) = x^2$ to the graph of each function.
  a. $y = 12x^2$
  b. $y = \frac{2}{5}x^2$

13. Write the equation of the function whose graph has been transformed from the graph of the parent function as described.
  a. a vertical stretch by a factor of 6
  b. a vertical shrink by a factor of $\frac{1}{2}$

14. Write the equation of the quadratic function shown in the graph.

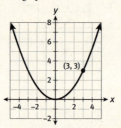

  d. vertical shrink by a factor of $\frac{1}{2}$, reflection across the x-axis, and translation 1 unit to the right and 1 unit up

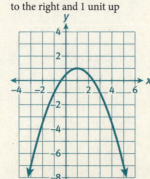

Graphing Quadratic Functions
Transformers

Lesson 30-3

15. Identify the transformations from the graph of the parent function $f(x) = x^2$ to the graph of each function. Then graph the function.
  a. $y = 2x^2 + 1$
  b. $y = -(x + 3)^2$
  c. $y = (x + 1)^2 + 4$
  d. $y = -\frac{1}{2}(x - 1)^2 + 1$

16. Write the equation of the quadratic function whose graph has the given vertex and passes through the given point.
  a. vertex: $(0, -1)$; point $(1, 0)$
  b. vertex: $(2, 0)$; point $(0, 4)$
  c. vertex: $(-1, 5)$; point $(2, -4)$

MATHEMATICAL PRACTICES
Construct Viable Arguments and Critique the Reasoning of Others

17. Explain why the domains of the functions $y = x^2$ and $y = x^2 + c$, for any real number c, are the same. Then write a statement about the ranges of the functions $y = x^2$ and $y = (x + c)^2$.

16. a. $y = x^2 - 1$
  b. $y = (x - 2)^2$
  c. $y = -(x + 1)^2 + 5$
17. The value of c will determine the vertical translation of the graph. However, the domain of all quadratic functions is all real numbers. The ranges of $y = x^2$ and $y = (x + c)^2$ are each $y \geq 0$, because a horizontal translation does not affect the y-values covered by the graph.

Graphing Quadratic Functions
PARABOLIC PATHS

Embedded Assessment 1
Use after Activity 30

In 1680, Isaac Newton, scientist, astronomer, and mathematician, used a comet visible from Earth to prove that some comets follow a parabolic path through space as they travel around the sun. This and other discoveries like it help scientists to predict past and future positions of comets.

1. Assume the path of a comet is given by the function $y = -x^2 + 4$.
   a. Graph the path of the comet. Explain how you graphed it.
   b. Identify the vertex of the function.
   c. Identify the maximum or minimum value of the function.
   d. Identify the domain and range.
   e. Write the equation for the axis of symmetry.

2. Identify the table that represents a parabolic comet path. Explain and justify your choice.

   a.
x	y
−2	−1
−1	−4
0	−5
1	−4
2	−1

   b.
x	y
−2	4
−1	−3
0	−5
1	−3
2	4

3. The graph at right shows a portion of the path of a comet represented by a function in the form $y = a(x-h)^2 + k$. Determine the values of a, h, and k and write the equation of the function.

4. The equation and graph below represent two different quadratic functions for the parabolic paths of comets.

   Function 1:   Function 2:
   $y = -3x^2 + 4$

   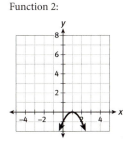

   Identify the maximum value of each function. Which function has the greater maximum value?

Embedded Assessment 1

Assessment Focus
- Writing quadratic functions
- Analyzing quadratic functions
- Graphing quadratic functions
- Transforming quadratic functions

Answer Key

1. a.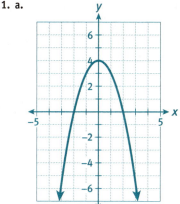

   Possible explanation: I made a table of values, plotted the points, and drew a parabola that opens downward through the points.
   b. The vertex is (0, 4).
   c. The maximum value of the function is 4.
   d. Domain: all reals; range: $y \leq 4$
   e. Axis of symmetry: $x = 0$

2. Table A represents a quadratic function and therefore represents a parabolic path.
   Sample explanation: The second differences of the range values are constant.

3. $a = 2, c = -3, y = 2x^2 - 3$

4. Function 1 maximum value: 4; Function 2 maximum value: 0. Function 1 has a greater maximum value.

Common Core State Standards for Embedded Assessment 1

HSF-IF.B.4:	For a function that models a relationship between two quantities, interpret key features of graphs and tables in terms of the quantities, and sketch graphs showing key features given a verbal description of the relationship. *Key features include: intercepts; intervals where the function is increasing, decreasing, positive, or negative; relative maximums and minimums; symmetries; end behavior; and periodicity.* *
HSF-IF.B.5:	Relate the domain of a function to its graph and, where applicable, to the quantitative relationship it describes.*
HSF-IF.C.7:	Graph functions expressed symbolically and show key features of the graph, by hand in simple cases and using technology for more complicated cases.*
HSF-IF.C.7.a:	Graph linear and quadratic functions and show intercepts, maxima, and minima.

Embedded Assessment 1

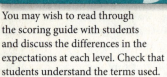

TEACHER to TEACHER

You may wish to read through the scoring guide with students and discuss the differences in the expectations at each level. Check that students understand the terms used.

Unpacking Embedded Assessment 2

Once students have completed this Embedded Assessment, turn to Embedded Assessment 2 and unpack it with them. Use a graphic organizer to help students understand the concepts they will need to know to be successful on Embedded Assessment 2.

Embedded Assessment 1
Use after Activity 30

Graphing Quadratic Functions
PARABOLIC PATHS

Scoring Guide	Exemplary	Proficient	Emerging	Incomplete
	The solution demonstrates the following characteristics:			
Mathematics Knowledge and Thinking (Items 2, 3)	• Clear and accurate understanding of how to determine whether a table of values represents a quadratic function • Effective understanding of quadratic functions as transformations of $y = x^2$	• Largely correct understanding of how to determine whether a table of values represents a quadratic function • Adequate understanding of quadratic functions as transformations of $y = x^2$	• Difficulty determining whether a table of values represents a quadratic function • Partial understanding of quadratic functions as transformations of $y = x^2$	• Inaccurate or incomplete understanding of how to determine whether a table of values represents a quadratic function • Little or no understanding of quadratic functions as transformations of $y = x^2$
Problem Solving (Items 2, 4)	• Appropriate and efficient strategy that results in a correct answer	• Strategy that may include unnecessary steps but results in a correct answer	• Strategy that results in a partially incorrect answer	• No clear strategy when solving problems
Mathematical Modeling / Representations (Items 1–4)	• Fluency in using tables, equations, and graphs to represent quadratic functions • Effective understanding of how to graph a quadratic function and identify key features from the graph	• Little difficulty using tables, equations, and graphs to represent quadratic functions • Largely correct understanding of how to graph a quadratic function and identify key features from the graph	• Some difficulty using tables, equations, and/or graphs to represent quadratic functions • Partial understanding of how to graph a quadratic function and/or identify key features from the graph	• Significant difficulty using tables, equations, and/or graphs to represent quadratic functions • Inaccurate or incomplete understanding of how to graph a quadratic function and/or identify key features from the graph
Reasoning and Communication (Items 1a, 2)	• Precise use of appropriate math terms and language to explain how to graph a quadratic function and whether a table of values represents a quadratic function	• Adequate explanation of how to graph a quadratic function and whether a table of values represents a quadratic function	• Misleading or confusing explanation of how to graph a quadratic function and/or whether a table of values represents a quadratic function	• Incomplete or inaccurate explanation of how to graph a quadratic function and/or whether a table of values represents a quadratic function

Common Core State Standards for Embedded Assessment 1 (cont.)

HSF-IF.C.9:	Compare properties of two functions each represented in a different way (algebraically, graphically, numerically in tables, or by verbal descriptions).
HSF-BF.A.1:	Write a function that describes a relationship between two quantities.*
HSF-BF.A.1.a:	Determine an explicit expression, a recursive process, or steps for calculation from a context.
HSF-BF.B.3:	Identify the effect on the graph of replacing $f(x)$ by $f(x) + k$, $k f(x)$, $f(kx)$, and $f(x + k)$ for specific values of k (both positive and negative); find the value of k given the graphs. Experiment with cases and illustrate an explanation of the effects on the graph using technology. *Include recognizing even and odd functions from their graphs and algebraic expressions for them.*

Solving Quadratic Equations by Graphing and Factoring ACTIVITY 31

Trebuchet Trials

Lesson 31-1 Solving by Graphing or Factoring

Learning Targets:
- Use a graph to solve a quadratic equation.
- Use factoring to solve a quadratic equation.
- Describe the connection between the zeros of a quadratic function and the *x*-intercepts of the function's graph.

> **SUGGESTED LEARNING STRATEGIES:** Visualization, Summarizing, Paraphrasing, Think-Pair-Share, Quickwrite

Carter, Alisha, and Joseph are building a *trebuchet* for an engineering competition. A trebuchet is a medieval siege weapon that uses gravity to launch an object through the air. When the counterweight at one end of the throwing arm drops, the other end rises and a projectile is launched through the air. The path the projectile takes through the air is modeled by a parabola.

To win the competition, the team must build their trebuchet according to the competition specifications to launch a small projectile as far as possible. After conducting experiments that varied the projectile's mass and launch angle, the team discovered that the ball they were launching followed the path given by the quadratic equation $y = -\frac{1}{8}x^2 + 2x$.

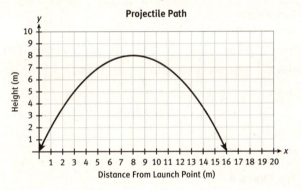

CONNECT TO PHYSICS
The throwing arm of a trebuchet is an example of a class I lever.

1. **Make sense of problems.** How far does the ball land from the launching point?
   16 m

2. What is the maximum height of the ball?
   8 m

3. What are the *x*-coordinates of the points where the ball is on the ground?
   The *x*-coordinates are 0 and 16.

Common Core State Standards for Activity 31

HSA-SSE.A.1:	Interpret expressions that represent a quantity in terms of a context.
HSA-SSE.A.1a:	Interpret parts of an expression, such as terms, factors, and coefficients.
HSA-SSE.A.1b:	Interpret complicated expressions by viewing one or more of their parts as a single entity.
HSA-SSE.B.3:	Choose and produce an equivalent form of an expression to reveal and explain properties of the quantity represented by the expression.
HSA-SSE.B.3a:	Factor a quadratic expression to reveal the zeros of the function it defines.

ACTIVITY 31
Guided

Activity Standards Focus
In this activity, students make connections between the solutions or roots of a quadratic equation and the *x*-intercepts or zeros of the related function. They determine both the equation of the axis of symmetry and the coordinates of the vertex of a parabola by calculating $\frac{-b}{2a}$. This information is used to graph the parabola.

Lesson 31-1

PLAN

Pacing: 1 class period

Chunking the Lesson
#1–3   #4–5   Example A
#6   #7–11   #12
Check Your Understanding
Lesson Practice

TEACH

Bell-Ringer Activity
Ask students to quickly calculate these products without using a calculator:
489 • 197 • 0   8895 • 0 • 345
0 • 50,895 • 756   4843 • 1896 • 0

Ask students to explain why this was an easy task.

> **Differentiating Instruction**
>
> **Support** students' comprehension of the problem scenario in the introduction. Review the pronunciations and meanings of words that students may not have encountered in the past. Evaluate students' understanding of their reading by asking them to summarize what a trebuchet is and what it does before they use mathematics to describe the path of a projectile.

1–3 Shared Reading, Summarizing, Paraphrasing, Visualization, Think-Pair-Share Ask students to interpret the graph of the projectile's path. This model is simplified because it shows the projectile launching from a height of 0 m. In reality, it would launch from a point above the ground. Item 1 in Lesson 31-3 will refine this model.

Activity 31 • Solving Quadratic Equations by Graphing and Factoring 455

ACTIVITY 31 Continued

4–5 Think-Pair-Share Item 4 begins to make the connection between solutions of a quadratic equation and the *x*-intercepts of the related function. Item 5 introduces the idea that solving an equation written as a sum with multiple *x* terms is not efficient or effective, thus motivating the need to rewrite the sum as a product.

Example A Close Reading, Note Taking Students must understand the Zero Product Property in order to make sense of why setting an equation equal to 0 and then factoring leads to the solutions. Be sure to debrief each of the Try These questions, especially the last one. Students who do not rewrite the equation so that it is equal to 0 before factoring do not understand how to use the Zero Product Property. Once students have used factoring to find solutions, ask them to justify their answers using substitution.

Developing Math Language
Review the idea of *factored form* by writing $30 = 2 \cdot 3 \cdot 5$ and explain that this is an example of writing 30 in factored form. Read the definition of the *factored form* of a polynomial and provide examples to help students gain understanding. Write the phrase on the Word Wall as students add it to their math notebooks.

TEACHER to TEACHER

Emphasize the word *zero* in the Zero Product Property. Help students understand that this property applies only when the product is equal to 0. Students often try to solve equations like $x(x + 3) = 5$ by setting the variable factors equal to the factors of 5. Also point out that the Zero Product Property can apply to an equation with any number of factors as long as the product is equal to 0.

ACTIVITY 31
continued

Lesson 31-1
Solving by Graphing or Factoring

My Notes

To determine how far the ball lands from the launching point, you can solve the equation $-\frac{1}{8}x^2 + 2x = 0$, because the height *y* equals 0.

4. Verify that $x = 0$ and $x = 16$ are solutions to this equation.

$-\frac{1}{8}(0)^2 + 2 \cdot 0 = 0$ and $-\frac{1}{8}(16)^2 + 2 \cdot 16 = 0$
$0 + 0 = 0$ $\qquad\qquad -32 + 32 = 0$

5. Without the graph, could you have determined these solutions? Explain.
   Answers may vary. It is easy to see that 0 is a solution, but it would be difficult to guess and check that 16 is also a solution.

The *factored form* of a polynomial equation provides an effective way to determine the values of *x* that make the equation equal 0.

MATH TERMS

A polynomial is in **factored form** when it is expressed as the product of one or more polynomials.

Zero Product Property
If $ab = 0$, then either $a = 0$ or $b = 0$.

Example A
Solve $-\frac{1}{8}x^2 + 2x = 0$ by factoring.
Step 1: Factor. $\qquad\qquad x\left(-\frac{1}{8}x + 2\right) = 0$
Step 2: Apply the Zero Product Property. $x = 0$ or $-\frac{1}{8}x + 2 = 0$
Step 3: Solve each equation for *x*. $\qquad x = 0$ or $-\frac{1}{8}x = -2$
$\qquad\qquad\qquad (-8)\left(-\frac{1}{8}\right)x = -2(-8)$
$\qquad\qquad\qquad x = 16$
Solution: $x = 0$ or $x = 16$

Try These A
Solve each quadratic equation by factoring.
a. $x^2 - 5x - 14 = 0$
   $x = 7$ or $x = -2$
b. $3x^2 - 6x = 0$
   $x = 0$ or $x = 2$
c. $x^2 + 3x = 18$
   $x = -6$ or $x = 3$

Common Core State Standards for Activity 31 *(continued)*

HSA-REI.B.4:	Solve quadratic equations in one variable.
HSA-REI.B.4b:	Solve quadratic equations by inspection (e.g., for $x^2 = 49$), taking square roots, completing the square, the quadratic formula and factoring, as appropriate to the initial form of the equation. Recognize when the quadratic formula gives complex solutions and write them as $a \pm bi$ for real numbers *a* and *b*.
HSF-BF.A.1:	Write a function that describes a relationship between two quantities.

Lesson 31-1
Solving by Graphing or Factoring

6. How do the solutions to the projectile path equation $-\frac{1}{8}x^2 + 2x = 0$ in Example A relate to the equation's graph?
   The solutions to the equation are the x-coordinates of the x-intercepts.

The graph of the function $y = x^2 - 6x + 5$ is shown below.

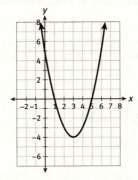

7. Identify the x-intercepts of the graph.
   The x-intercepts of the graph are (1, 0) and (5, 0).

8. What is the x-coordinate of the vertex?
   The x-coordinate of the vertex is 3.

9. Describe the x-coordinate of the vertex with respect to the two x-intercepts.
   The x-coordinate of the vertex is the average of the x-coordinates of the x-intercepts.

10. Solve the related quadratic equation $x^2 - 6x + 5 = 0$ by factoring.
    x = 1 or x = 5

11. How do the solutions you found in Item 10 relate to the x-intercepts of the above graph?
    The solutions are the x-coordinates of the x-intercepts.

ACTIVITY 31 Continued

6 Think-Pair-Share Monitor discussion to be sure that students name these specifically as the x-coordinates of the x-intercepts. After students see that the solutions of the given equation are the x-coordinates of the x-intercepts of its graph, discuss whether or not the solutions of all quadratic equations are the x-coordinates of the x-intercepts of the graphs of the related functions.

7–11 Think-Pair-Share, Look for a Pattern, Sharing and Responding These items make the connection that the real solutions of a quadratic equation and the x-coordinates of the x-intercepts of the related function are the same. They also establish that the x-coordinate of the vertex is halfway between the x-intercepts, due to the symmetry of a parabola. As students share out and respond to other groups, be sure these points are brought out.

Item 10 provides formative assessment information. Do students use what they learned earlier in the lesson to verify the conclusions they are reaching?

Differentiating Instruction

Extend students' learning by asking them to name the y-values of the points where the graph intersects the x-axis. Have them relate their answers to the ordered pairs that are solutions of the function's equation.

Activity 31 • Solving Quadratic Equations by Graphing and Factoring

ACTIVITY 31 Continued

12 Quickwrite, Debriefing Continue to clarify the connection between the graph of the function and the solutions, or real roots, of the related equation. You might suggest that students think of the related equation as $f(0)$. It is important to use the term *real roots* here because complex roots of a polynomial equation are not related to the x-intercepts.

Developing Math Language

Students should understand that the *zeros of the function* are the x-coordinates of the points at which the graph of the function $f(x) = ax^2 + bx + c$ crosses the x-axis. The *roots* of the equation are the solutions of the related equation $0 = ax^2 + bx + c$. Have students record the terms *x-intercept*, *real roots of an equation*, and *zeros of a function* in their math notebooks. As students respond to questions or discuss possible solutions to problems, monitor their use of these terms to ensure correct and precise usage.

Check Your Understanding

As you debrief this lesson, have students describe the steps they followed to answer each Check Your Understanding item. Be sure they understand the distinction between the solutions or roots in Items 13 and 14 and the zeros in Items 15 and 16.

Answers
13. $x = -9$ or $x = 9$
14. $x = 0$ or $x = 4$
15. -1 and -7
16. 1 and 2

ASSESS

Students' answers to Lesson Practice problems will provide you with a formative assessment of their understanding of the lesson concepts and their ability to apply their learning.

See the Activity Practice for additional problems for this lesson. You may assign the problems here or use them as a culmination for the activity.

ADAPT

In addition to solving quadratic equations, students must understand what the solutions mean. In Item 21, if students have difficulty finding an equation when given two roots, encourage them to reverse the process they use to find the roots of a quadratic equation. In the final problem, if students cannot relate the roots of the related equation to the context of the problem, have them graph the function and answer the questions visually.

ACTIVITY 31 continued

My Notes

MATH TIP

The relationship described in Item 12b is between the *real* roots of a quadratic equation and the zeros of the related quadratic function. Later in this unit you will see quadratic equations that have non-real roots. The same relationship does not exist between non-real roots and zeros of the related function.

Lesson 31-1
Solving by Graphing or Factoring

The x-coordinates of the x-intercepts of a quadratic function $y = ax^2 + bx + c$ are the **zeros of the function**. The solutions of a quadratic equation $ax^2 + bx + c = 0$ are the **roots** of the equation.

12. The quadratic function $y = ax^2 + bx + c$ is related to the equation $ax^2 + bx + c = 0$ by letting y equal zero.

   a. Why do you think the x-coordinates of the x-intercepts are called the zeros of the function?
   Answers may vary. These are x-values for which $y = 0$.

   b. Describe the relationship between the real *roots* of a quadratic equation and the *zeros* of the related quadratic function.
   Answers may vary. The real roots of a quadratic equation are equal to the zeros of the related quadratic function.

Check Your Understanding

Solve by factoring.

13. $x^2 - 81 = 0$
14. $\frac{1}{2}x^2 - 2x = 0$

Identify the zeros of the quadratic function.

15. $y = x^2 + 8x + 7$
16. $y = x^2 - 3x + 2$

LESSON 31-1 PRACTICE

17. **Make use of structure.** Use the graph of the quadratic function $y = 2x^2 + 6x$ shown to determine the roots of the quadratic equation $0 = 2x^2 + 6x$.

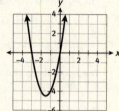

Solve by factoring.

18. $x^2 - 2x + 1 = 0$
19. $2x^2 - 7x - 4 = 0$
20. $x^2 + 5x = 0$

21. Write a quadratic equation whose roots are 3 and -6.

22. A whale jumps vertically from a pool at Ocean World. The function $y = -16x^2 + 32x$ models the height of a whale in feet above the surface of the water after x seconds.
   a. What is the maximum height of the whale above the surface of the water?
   b. How long is the whale out of the water? Justify your answer.

23. **Construct viable arguments.** Is it possible for two different quadratic functions to share the same zeros? Use a graph to justify your response.

LESSON 31-1 PRACTICE
17. 0 and -3
18. $x = 1$
19. $x = 4$ or $x = -\frac{1}{2}$
20. $x = 0$ or $x = -5$
21. Answers may vary. $0 = x^2 + 3x - 18$
22. a. 16 feet
    b. 2 seconds; $y = 0$ when $x = 0$ (the beginning of the jump) and $x = 2$ (the end of the jump)
23. Yes; it is possible for two different quadratic functions that are vertical stretches or shrinks of one another to share the same roots. For example: $y = x^2 - 1$ and $y = 2x^2 - 2$ share the roots -1 and 1. The graphs of both functions intersect the x-axis when $x = -1$ and when $x = 1$.

Lesson 31-2
The Axis of Symmetry and the Vertex

ACTIVITY 31
continued

Learning Targets:
- Identify the axis of symmetry of the graph of a quadratic function.
- Identify the vertex of the graph of a quadratic function.

SUGGESTED LEARNING STRATEGIES: Close Reading, Marking the Text, Think-Pair-Share, Predict and Confirm, Discussion Groups

The *axis of symmetry* of the parabola determined by the function $y = ax^2 + bx + c$ is the vertical line that passes through the vertex.

The equation for the axis of symmetry is $x = -\frac{b}{2a}$.

The vertex is on the axis of symmetry. Therefore, the x-coordinate of the vertex is $-\frac{b}{2a}$.

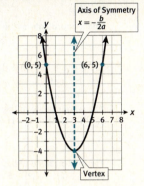

MATH TIP

Substitute the x-coordinate of the vertex into the quadratic equation to find the y-coordinate of the vertex.

Each point on a quadratic graph will have a mirror image point with the same y-coordinate that is equidistant from the axis of symmetry. For example, the point (0, 5) is reflected over the axis of symmetry to the point (6, 5) on the graph.

1. Give the coordinates of two additional points that are reflections over the axis of symmetry in the graph above.
   Answers will vary. (1, 0) and (5, 0)

2. For each function, identify the zeros graphically. Confirm your answer by setting the function equal to 0 and solving by factoring.

   a. $y = x^2 + 2x - 8$

   The zeros are −4 and 2.

   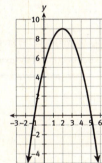

   b. $y = -x^2 + 4x + 5$

   The zeros are −1 and 5.

   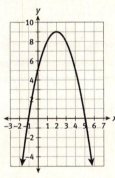

ACTIVITY 31 Continued

3–5 Discussion Groups, Debriefing
Students must incorporate many different concepts as they answer these questions. The teacher can provide support as they work by asking, "What do you already know? How does it relate to what you are trying to find?" As these questions are used to debrief the lesson, ask students to describe any connections between graphs of quadratic functions and the equations of the related quadratic functions and their solutions.

Differentiating Instruction
Support students who have difficulty with the vocabulary by having them make a graphic organizer that clarifies the terms used in this lesson. The poster should show a parabola with the vertex, axis of symmetry, and x-intercepts clearly labeled. On the side, students should list information about each feature. For example, for the vertex, they could include that it is a maximum or minimum and the formula for finding its coordinates algebraically. For the x-intercepts, students could say that they are also called zeros of the function and that they are the real roots or solutions of the related equation.

Check Your Understanding
As you debrief this lesson, use students' answers to these items to determine which students understand how to determine features of a parabola algebraically. Be sure students identify both the x- and the y-coordinates of the vertex when asked to give the vertex of a quadratic function.

Answers
6. $(0.5, 6.25)$
7. $x = 0.5$

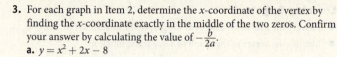

Lesson 31-2
The Axis of Symmetry and the Vertex

3. For each graph in Item 2, determine the x-coordinate of the vertex by finding the x-coordinate exactly in the middle of the two zeros. Confirm your answer by calculating the value of $-\frac{b}{2a}$.
   a. $y = x^2 + 2x - 8$
      The x-coordinate of the vertex is -1. $-\frac{2}{2(1)} = -1$

   b. $y = -x^2 + 4x + 5$
      The x-coordinate of the vertex is 2. $-\frac{4}{2(-1)} = 2$

4. For each graph in Item 2, determine the y-coordinate of the vertex from the graph. Confirm your answer by evaluating the function at $x = -\frac{b}{2a}$.
   a. $y = x^2 + 2x - 8$
      The y-coordinate of the vertex is -9.
      $y(-1) = (-1)^2 + 2(-1) - 8 = 1 - 10 = -9$

   b. $y = -x^2 + 4x + 5$
      The y-coordinate of the vertex is 9.
      $y(2) = -(2)^2 + 4(2) + 5 = -4 + 13 = 9$

5. **Reason quantitatively.** For each graph in Item 2, determine the maximum or minimum value of the function.
   a. $y = x^2 + 2x - 8$
      The minimum value is -9.

   b. $y = -x^2 + 4x + 5$
      The maximum value is 9.

Check Your Understanding

For the function $y = -x^2 + x + 6$, use the value of $-\frac{b}{2a}$ to respond to the following.

6. Identify the vertex of the graph.
7. Write the equation of the axis of symmetry.

Lesson 31-2
The Axis of Symmetry and the Vertex

LESSON 31-2 PRACTICE

Use the value of $-\frac{b}{2a}$ to determine the vertex and to write the equation for the axis of symmetry for the graph of each of the following quadratic functions.

8. $y = x^2 - 10x$
9. $y = x^2 - 4x - 32$
10. $y = x^2 + x - 12$
11. $y = 6x - x^2$
12. Describe the graph of a quadratic function that has its vertex and a zero at the same point.
13. **Model with mathematics.** An architect is designing a tunnel and is considering using the function $y = -0.12x^2 + 2.4x$ to determine the shape of the tunnel's entrance, as shown in the figure. In this model, y is the height of the entrance in feet and x is the distance in feet from one end of the entrance.

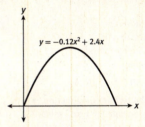

   a. How wide is the tunnel's entrance at its base?
   b. What is the vertex? What does it represent?
   c. Could a truck that is 14 feet tall pass through the tunnel? Explain.

ACTIVITY 31 Continued

Lesson 31-3

PLAN

Pacing: 1 class period
Chunking the Lesson
Example A   #1–2
Check Your Understanding
Lesson Practice

TEACH

Bell-Ringer Activity
Ask students to work in pairs. One student should explain to the other how to find the zeros of the function $y = x^2 - 2x + 5$. Then have them switch roles and have the second student tell the first how to find the vertex and the axis of symmetry. Repeat the exercise with $y = x^2 + 9x + 18$.

Example A Close Reading, Note Taking The focus of this example is graphing a quadratic function by finding the vertex and the zeros (x-intercepts). After discussing each step, have students write it in their math notebooks. Students should be able to summarize the process before attempting the items in Try These A.

For the first item in Try These A, if the chosen points do not appear to be on the graph, ask questions to help students identify the error. Reinforce the idea that if the point is on the graph, substituting its coordinates into the equation must create a true statement, and vice versa.

ACTIVITY 31 continued

Lesson 31-3
Graphing a Quadratic Function

Learning Targets:
- Use the axis of symmetry, the vertex, and the zeros to graph a quadratic function.
- Interpret the graph of a quadratic function.

SUGGESTED LEARNING STRATEGIES: Create Representations, Close Reading, Note Taking, Identify a Subtask

If a quadratic function can be written in factored form, you can graph it by finding the vertex and the zeros.

Example A
Graph the quadratic function $y = x^2 - x - 12$.

Step 1: Determine the axis of symmetry using $x = -\dfrac{b}{2a}$.

The axis of symmetry is $x = 0.5$.

$y = x^2 - x - 12$
$a = 1, b = -1$
$x = -\dfrac{(-1)}{2(1)} = \dfrac{1}{2} = 0.5$

Step 2: Determine the vertex.
The x-coordinate is 0.5.
Substitute 0.5 for x to find the y-coordinate.

The vertex is $(0.5, -12.25)$.

$y = (0.5)^2 - (0.5) - 12$
$= 0.25 - 0.5 - 12$
$= -12.25$

Step 3: Determine the zeros of the function.
Set the function equal to 0 and solve.

The zeros are -3 and 4.

$x^2 - x - 12 = 0$
$(x - 4)(x + 3) = 0$
$x - 4 = 0$ or $x + 3 = 0$
$x = 4$ or $x = -3$

Step 4: Graph the function.

Plot the vertex $(0.5, -12.25)$ and the points where the function crosses the x-axis $(-3, 0)$ and $(4, 0)$. Connect the points with a smooth parabolic curve.

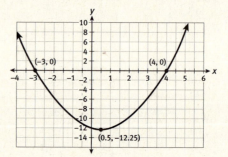

Lesson 31-3
Graphing a Quadratic Function

Try These A

a. Check the graph in Example A by plotting two more points on the graph. First choose an *x*-value, and then find the *y*-value by evaluating the function. Plot this point. Then plot the reflection of the point over the axis of symmetry to get another point. Verify that both points are on the graph.
Check students' work.

Graph the quadratic functions by finding the vertex and the zeros. Check your graphs. **See below.**

b. $y = x^2 - 2x - 8$
c. $y = 4x - x^2$
d. $y = x^2 + 4x - 5$

Joseph, Carter, and Alisha tested a new trebuchet designed to launch the projectile even further. They also refined their model to reflect a more accurate launch height of 1 m. The new projectile path is given by the function $y = -\frac{1}{19}(x^2 - 18x - 19)$.

1. Graph the projectile path on the coordinate axes below.

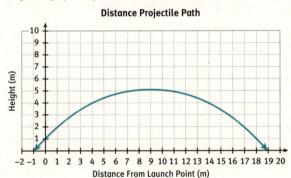

Distance Projectile Path

> **CONNECT TO AP**
>
> In AP Calculus, you will represent projectile motion with special types of equations called parametric equations.

TRY THESE A

b.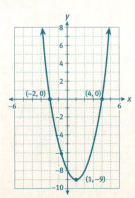

c.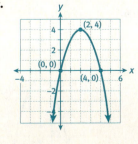

d.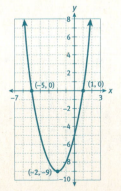

ACTIVITY 31 Continued

Check Your Understanding
Debrief students' answers to these items by asking students to describe the steps they followed to graph each quadratic function. Follow up with questions to check students' general understanding of parabolas, such as "What is true about a parabola if its vertex is below the x-intercepts?" (The parabola opens upward.)

Answers
3.
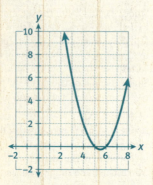

4.
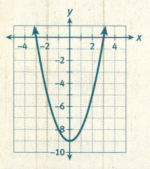

ASSESS

Students' answers to Lesson Practice problems will provide you with a formative assessment of their understanding of the lesson concepts and their ability to apply their learning.

See the Activity Practice for additional problems for this lesson. You may assign the problems here or use them as a culmination for the activity.

ADAPT

If students' graphs are incorrect, have them explain how they determined the vertex and zeros. Help them analyze their work to find and correct the errors. If students have difficulty with functions that contain $-x^2$, encourage them to write the coefficient of -1 and recall the common factor.

ACTIVITY 31 continued

Lesson 31-3
Graphing a Quadratic Function

2. **Make sense of problems.** Last year's winning trebuchet launched a projectile a horizontal distance of 19.5 m. How does the team's trebuchet compare to last year's winner?
   Answers may vary. The team's trebuchet traveled 19 meters from its launch point. This is 0.5 m short of last year's winning distance.

Check Your Understanding

Graph each quadratic function.
3. $y = x^2 - 11x + 30$
4. $y = x^2 - 9$

LESSON 31-3 PRACTICE

Graph the quadratic functions. Label the vertex, axis of symmetry, and zeros on each graph.

5. $y = x^2 + 5x$
6. $y = -x^2 + 2x$
7. $y = -x^2 + \frac{1}{4}$
8. $y = x^2 - 3x - 4$
9. $y = -x^2 + 3x + 4$
10. **Attend to precision.** Describe the similarities and differences between the graphs of the functions in Items 8 and 9. How are these similarities and differences indicated by the functions themselves?

LESSON 31-3 PRACTICE

5.
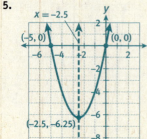

6.
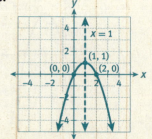

7.
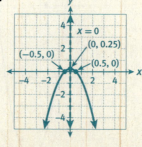

464 SpringBoard® Mathematics Algebra 1, Unit 5 • Quadratic Functions

Solving Quadratic Equations by Graphing and Factoring
Trebuchet Trials

ACTIVITY 31 PRACTICE
Write your answers on notebook paper.
Show your work.

Lesson 31-1

Use the graphs to determine the zeros of the quadratic functions.

1. $y = 0.5x^2$

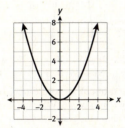

2. $y = x^2 - 4$

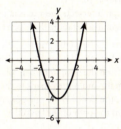

3. $y = -x^2 + 4x - 3$

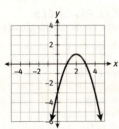

4. Identify the zeros of the quadratic function $y = (x - a)(x - b)$.

5. Which of the following functions has zeros at $x = 2$ and $x = -3$?
   A. $y = x^2 + x - 6$ B. $y = -x^2 + 5x + 6$
   C. $y = 2x^2 + x - 3$ D. $y = x^2 + 2x - 3$

6. What are the solutions to the equation $x^2 - 6x = -5$?
   A. $-1, 5$ B. $2, 3$
   C. $1, 5$ D. $-2, 3$

Solve by factoring.

7. $0 = x^2 - 25$
8. $0 = x^2 + 17x$
9. $0 = x^2 - 5x - 66$
10. $0 = x^2 + 99x - 100$
11. $0 = -x^2 - 4x + 32$
12. $0 = x^2 + 10x + 25$
13. $0 = x^2 + 8x + 7$

14. A fountain at a city park shoots a stream of water vertically from the ground. The function $y = -8x^2 + 16x$ models the height of the stream of water in feet after x seconds.
    a. What is the maximum height of the stream of water?
    b. At what time does the stream of water reach its maximum height?
    c. For how many seconds does the stream of water appear above ground?
    d. Identify a reasonable domain and range for this function.

15. Write a quadratic equation whose roots are -4 and 8.

16. Write a quadratic function whose zeros are 2 and 9.

ACTIVITY 31 Continued
ACTIVITY PRACTICE
1. 0
2. -2 and 2
3. 1 and 3
4. a and b
5. A
6. C
7. $x = -5$ or $x = 5$
8. $x = -17$ or $x = 0$
9. $x = -6$ or $x = 11$
10. $x = -100$ or $x = 1$
11. $x = -4$ or $x = 8$
12. $x = -5$
13. $x = -7$ or $x = -1$
14. a. 8 ft
    b. 1 sec
    c. 2 sec
    d. Domain: $0 \le x \le 2$;
       range: $0 \le y \le 8$
15. Sample answer: $0 = x^2 - 4x - 32$
16. Sample answer: $y = x^2 - 11x + 18$

LESSON 31-3 PRACTICE (cont.)

8.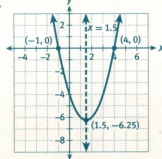

9.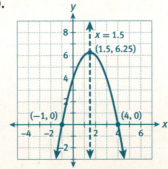

10. The graphs have the same axis of symmetry and intersect the x-axis at the same points, while the vertices are reflections of each other across the x-axis. This is because the second function is -1 times the first function.

ACTIVITY 31 Continued

17. 1
18. $(-4, -1)$
19. $x = -1$
20.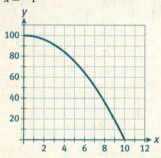
    a. $(0, 100)$
    b. $(10, 0)$ (Note: $x = -10$ is not in the domain based on the problem statement.)
21. $x = -10$ or $x = 10$
22. The positive solution, 10, is the same as the positive x-intercept.
23. 100 meters
24. 10 meters
25. A
26. C
27. B
28. C
29. Using zeros of 5 and 20,
    $y = -(x - 5)(x - 20) = -x^2 + 25x - 100$
30. a. No; $x^2 - x - 3$ cannot be factored.
    b. Answers may vary; -1.5 and 2.5
    c. -1.3 and 2.3

ADDITIONAL PRACTICE

If students need more practice on the concepts in this activity, see the eBook Teacher Resources for additional practice problems.

ACTIVITY 31 continued

Solving Quadratic Equations by Graphing and Factoring
Trebuchet Trials

Lesson 31-2

17. If a quadratic function has zeros at $x = -4$ and $x = 6$, what is the x-coordinate of the vertex?

18. What is the vertex of the quadratic function $y = x^2 + 8x + 15$?

19. Write the equation for the axis of symmetry of the graph of the quadratic function $y = 2x^2 + 4x - 1$.

Use the following information for Items 20–24.

A diver in Acapulco jumps from a cliff. His height y, in meters, as a function of x, his distance from the cliff base in meters, is given by the quadratic function $y = 100 - x^2$, for $x \geq 0$.

20. Graph the function representing the cliff diver's height.
    a. Identify the vertex of the graph.
    b. Identify the x-intercepts of the graph.

21. Determine the solutions of the equation $100 - x^2 = 0$.

22. How do the solutions to the equation in Item 21 relate to the x-intercepts of the graph in Item 20?

23. How high is the cliff from which the diver jumps?

24. How far from the base of the cliff does the diver hit the water?

25. Which of these functions has a graph with the axis of symmetry $x = -2$?
    A. $y = x^2 + 4x - 2$
    B. $y = x^2 - 4x + 2$
    C. $y = 2x^2 + 2x - 3$
    D. $y = 2x^2 - 2x + 3$

Lesson 31-3

26. Lisa correctly graphed a quadratic function and found that its vertex was in Quadrant I. Which function could she have graphed?
    A. $y = x^2 + 4x + 2$
    B. $y = x^2 - 4x + 2$
    C. $y = x^2 - 4x + 6$
    D. $y = x^2 + 4x + 6$

27. Which of these is the equation of the parabola graphed below?

    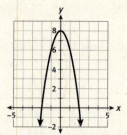

    A. $y = 2x^2 - 2x + 8$ B. $y = -2x^2 + 8$
    C. $y = -x^2 - 8x - 15$ D. $y = -x^2 - 8$

28. DeShawn's textbook shows the graph of the function $y = x^2 + x - 6$. Which of these is a true statement about the graph?
    A. The axis of symmetry is the y-axis.
    B. The vertex is $(0, -6)$.
    C. The graph intersects the x-axis at $(-3, 0)$.
    D. The graph is a parabola that opens downward.

29. Kim throws her basketball up from the ground toward the basketball hoop from a distance of 20 feet away from the hoop. The ball follows a parabolic path and returns back to the gym floor 5 feet from the hoop. Write one possible equation to represent the path of the basketball. Explain your answer.

MATHEMATICAL PRACTICES
Use Appropriate Tools Strategically

30. Consider the quadratic function $y = x^2 - x - 3$.
    a. Is it possible to find the zeros of the function by factoring? Explain.
    b. Use your calculator to graph the function. Based on the graph, what are the approximate zeros of the function?
    c. Use the zero function of your calculator to find more accurate approximations for the zeros. Round to the nearest tenth.

Algebraic Methods of Solving Quadratic Equations   ACTIVITY 32

Keeping it Quadratic
Lesson 32-1 The Square Root Method

Learning Targets:
- Solve quadratic equations by the square root method.
- Provide examples of quadratic equations having a given number of real solutions.

> **SUGGESTED LEARNING STRATEGIES:** Guess and Check, Simplify the Problem, Think-Pair-Share, Create Representations

Nguyen is trying to build a square deck around his new hot tub. To decide how large a deck he should build, he needs to determine the side length, x, of different sized decks given the possible area of each deck. He knows that the area of a square is equal to the length of a side squared. Using this information, Nguyen writes the following equations to represent each of the decks he is considering.

1. Solve each equation. Be prepared to discuss your solution methods with your classmates.

   a. $x^2 = 49$
   $x = -7$ or $x = 7$

   b. $x^2 = 100$
   $x = -10$ or $x = 10$

   c. $x^2 = 15$
   $x = -\sqrt{15}$ or $x = \sqrt{15}$

   d. $2x^2 = 18$
   $x = -3$ or $x = 3$

   e. $x^2 - 4 = 0$
   $x = -2$ or $x = 2$

   f. $x^2 + 2 = 0$
   no real solution

   g. $x^2 + 3 = 3$
   $x = 0$

> **MATH TIP**
> A solution of an equation makes the equation true. For example, $x + 5 = 7$ has the solution $x = 2$ because $2 + 5 = 7$.

2. Refer to the equations in Item 1 and their solutions.
   a. What do the equations have in common?
   They are all quadratic equations.

   b. What types of numbers are represented by the solutions of these equations?
   Real numbers, except for 1f.

   c. How many solutions do the equations have?
   Two solutions, except for 1f and 1g.

   d. **Reason quantitatively.** Which solutions are reasonable for side lengths of the squares? Explain.
   Only the positive solutions are reasonable, because the side length of a square cannot be negative or 0.

Texas Essential Knowledge and Skills for Activity 32

(8)(A) Solve quadratic equations having real solutions by factoring, taking square roots, completing the square, and applying the quadratic formula.
  (i) solve quadratic equations having real solutions by factoring
  (iv) solve quadratic equations having real solutions by applying the quadratic formula

(8)(B) Write, using technology, quadratic functions that provide a reasonable fit to data to estimate solutions and make predictions for real-world problems.
  (i) write, using technology, quadratic functions that provide a reasonable fit to data to estimate solutions
  (ii) write, using technology, quadratic functions that provide a reasonable fit to data to make predictions for real-world problems

---

ACTIVITY 32
Directed

Activity Standards Focus
In Activity 32, students solve quadratic equations by using square roots, completing the square and the quadratic formula. Throughout this activity, be sure to emphasize when each of these solution methods might be appropriate. Students also use the discriminant of a quadratic equation to analyze its solutions, and solve quadratic equations with complex solutions.

Lesson 32-1

PLAN

Materials
- graphing calculators (optional)

Pacing: 1 class period

Chunking the Lesson
#1–2   Example A   #3
#4   Example B, #5
Check Your Understanding
Lesson Practice

TEACH

Bell-Ringer Activity
Ask students to determine or simplify each square root.
1. $\sqrt{100}$ [10]   2. $\sqrt{225}$ [15]
3. $\sqrt{80}$ [$4\sqrt{5}$]   4. $\sqrt{54}$ [$3\sqrt{6}$]

1–2 Shared Reading, Guess and Check, Group Presentation, Look for a Pattern, Quickwrite In these items, students develop methods to solve quadratic equations that can be written in the form $x^2 = c$. They also make some generalizations about this type of equation. In Item 1, you may want to assign each group a different equation to solve. Have groups present their solutions, and allow other groups to respond to the solutions. It is acceptable for students not to use formal language at this point. The goal is to get them thinking about equations that have more than one solution and to make sure they understand that not all quadratic equations have exactly two real solutions. Some conclusions students should draw about equations of the form $x^2 = c$ include:

- When there are two real solutions, the solutions consist of a number and its opposite.
- There are ways to solve this type of equation besides factoring.

Be sure to ask students why the equation in Item 1f has no real solution.

ACTIVITY 32 Continued

Example A Close Reading, Note Taking, Identify a Subtask In this example, students are introduced to solving quadratic equations that can be written in the form $ax^2 + c = 0$ by using the square root method. Take the time to review the basic steps of solving linear equations if needed. Students who cannot successfully isolate x^2 will have difficulty with problems of this type. The examples are beginning to leave out some steps when showing how to solve equations. If needed, reinsert those steps as you work through the examples with your class. For instance, in Step 1, you may want to insert the equation $3x^2 - 6 + 6 = 0 + 6$. In Step 3, be sure students understand what the symbol $\pm$ represents and why it is needed when taking the square root of both sides of the equation.

3 Graphic Organizer, Think-Pair-Share, Sharing and Responding In this item, students give examples of quadratic equations with two, one and no real solutions. The graphic organizer helps students recognize the different numbers of solutions possible for equations of the form $x^2 = c$ and when they occur. As students share responses, be sure all have recorded the correct equation(s) in the example column. If students have questions about the last row in the table, discuss the information in the Teacher to Teacher box.

TEACHER to TEACHER

Because the square of a real number is always positive or zero, it should present a dilemma to students when c is a negative number in the equation $x^2 = c$. How can something nonnegative equal a negative quantity? In the real number system, the answer is that it cannot. Hence, there is no real solution. However, later in this activity, students will be introduced to the imaginary unit $i = \sqrt{-1}$, and they will learn that equations of the form $x^2 = c$ where $c < 0$ have imaginary solutions.

ACTIVITY 32 continued

My Notes

MATH TIP
Every positive number has two square roots, the principal square root and its opposite. For example, $\sqrt{5}$ is the principal square root of 5 and $-\sqrt{5}$ is its opposite.

READING MATH
The $\pm$ symbol is read "plus or minus."

MATH TIP
The square root of a negative number is not a real number, so equations of the form $x^2 = c$ where $c < 0$ have no real solutions.

Lesson 32-1
The Square Root Method

To solve a quadratic equation of the form $ax^2 + c = 0$, isolate the x^2-term and then take the square root of both sides.

Example A
Solve $3x^2 - 6 = 0$ using square roots.

Step 1: Add 6 to both sides. $\quad 3x^2 - 6 = 0$
$\qquad\qquad\qquad\qquad\qquad\qquad\quad 3x^2 = 6$

Step 2: Divide both sides by 3. $\quad \dfrac{3x^2}{3} = \dfrac{6}{3}$
$\qquad\qquad\qquad\qquad\qquad\qquad\quad x^2 = 2$

Step 3: Take the square root of both sides. $\quad \sqrt{x^2} = \pm\sqrt{2}$
$\qquad\qquad\qquad\qquad\qquad\qquad\qquad\qquad x = +\sqrt{2} \text{ or } -\sqrt{2}$

Solution: $x = +\sqrt{2}$ or $x = -\sqrt{2}$

Try These A
Solve each equation using square roots.

a. $x^2 - 10 = 1$
   $\pm\sqrt{11}$

b. $\dfrac{x^2}{4} = 1$
   ± 2

c. $4x^2 - 6 = 14$
   $\pm\sqrt{5}$

3. Quadratic equations can have 0, 1, or 2 real solutions. Fill in the table below with equations from the first page that represent the possible numbers of solutions.
   Answers may vary.

Number of Solutions	Result When x^2 is Isolated	Example(s)
Two	x^2 = positive number	$x^2 = 49$ $x^2 = 15$
One	$x^2 = 0$	$x^2 + 3 = 3$
No real solutions	x^2 = negative number	$x^2 + 2 = 0$

Common Core State Standards for Activity 32 (continued)

HSA-REI.B.4b   Solve quadratic equations by inspection (e.g., for $x^2 = 49$), taking square roots, completing the square, the quadratic formula and factoring, as appropriate to the initial form of the equation. Recognize when the quadratic formula gives complex solutions and write them as $a \pm bi$ for real numbers a and b.

HSF-IF.C.8   Write a function defined by an expression in different but equivalent forms to reveal and explain different properties of the function.

HSF-IF.C.8a   Use the process of factoring and completing the square in a quadratic function to show zeros, extreme values, and symmetry of the graph, and interpret these in terms of a context.

Lesson 32-1
The Square Root Method

4. **Reason abstractly.** A square frame has a 2-in. border along two sides as shown in the diagram. The total area is 66 in^2. Answer the questions to help you write an equation to find the area of the unshaded square.

   a. Label the sides of the unshaded square x.
   b. Fill in the boxes to write an equation for the total area in terms of x.

Area in terms of x	=	Area in square in.
$(x+2)^2$		66

You can solve quadratic equations like the one you wrote in Item 4 by isolating the variable.

Example B
Solve $(x+2)^2 = 66$ using square roots. Approximate the solutions to the nearest hundredth.

Step 1: Take the square root of both sides.
$$(x+2)^2 = 66$$
$$\sqrt{(x+2)^2} = \pm\sqrt{66}$$
$$x+2 = \pm\sqrt{66}$$

Step 2: Subtract 2 from both sides.
$$x = -2 \pm \sqrt{66}$$
$$x = -2 + \sqrt{66} \text{ or } x = -2 - \sqrt{66}$$

Step 3: Use a calculator to approximate the solutions.

Solution: $x \approx -10.12$ or $x \approx 6.12$

5. Are both solutions to this equation valid in the context of Item 4? Explain your response.
   No. The variable x represents the length of one side of a square. It has to be a positive number. The only valid solution is the positive solution, 6.12 inches.

Try These B
Solve each equation using square roots.
a. $(x-5)^2 = 121$
   $x = 16$ or -6
b. $(2x-1)^2 = 6$
   $x = \dfrac{1 \pm \sqrt{6}}{2}$
c. $x^2 - 12x + 36 = 2$
   $x = 6 \pm \sqrt{2}$

ACTIVITY 32 Continued

Differentiating Instruction

Support students who have difficulty determining what step to do first when solving the Try These problems by reinforcing the idea that solving an equation for *x* involves undoing the operations performed on *x* in reverse order. Ask students to solve the equations below. For Items b and c, ask students to explain what they must do before taking the square root of both sides of the equation.

a. $(x-1)^2 = 8$ $[x = 1 \pm 2\sqrt{2}]$
b. $(x-3)^2 + 2 = 5$ $[x = 3 \pm \sqrt{3}]$
c. $2(x+1)^2 = 8$ $[x = -3 \text{ or } x = 1]$

Check Your Understanding

Debrief students' answers to these items to make sure they can solve quadratic equations by using square roots and can give examples of quadratic equations with a given number of solutions. For Items 6–8, ask students to explain each step of their work.

Answers
6. $x = \pm 1$
7. $x = 3$ or $x = 5$
8. $x = \pm \sqrt{7}$
9. Sample answers:
   a. $x^2 - 2 = -2$
   b. $x^2 + 5 = 3$
   c. $x^2 - 16 = 0$

ASSESS

Students' answers to lesson practice problems will provide you with a formative assessment of their understanding of the lesson concepts and their ability to apply their learning.

See the Activity Practice for additional problems for this lesson. You may assign the problems here or use them as a culmination for the activity.

ADAPT

Debrief students' answers to the Lesson Practice to ensure that students understand concepts related to solving quadratic equations by using square roots. Support struggling students by asking them to begin several problems, showing only the steps necessary to isolate the squared expression. For many, this is the most difficult part of solving using square roots. Also remind students that they must consider both the positive and negative square root when taking the square root of both sides of an equation.

ACTIVITY 32 continued

Lesson 32-1
The Square Root Method

Check Your Understanding

Solve each equation using square roots.

6. $x^2 + 12 = 13$
7. $(x-4)^2 = 1$
8. $3x^2 - 6 = 15$

9. Give an example of a quadratic equation that has
   a. one real solution.
   b. no real solutions.
   c. two real solutions.

LESSON 32-1 PRACTICE

10. If the length of a square is decreased by 1 unit, the area will be 8 square units. Write an equation for the area of the square.
11. Calculate the side length of the square in Item 10.

Solve each equation.

12. $x^2 - 22 = 0$
13. $(x+5)^2 - 4 = 0$
14. $x^2 - 4x + 4 = 0$
15. $(x+1)^2 = 12$

16. **Model with mathematics.** Alaysha has a square picture with an area of 100 square inches, including the frame. The width of the frame is *x* inches.

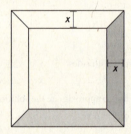

a. Write an equation in terms of *x* for the area *A* of the picture inside the frame.
b. If the area of the picture inside the frame is 64 square inches, what are the possible values for *x*?
c. If the area of the picture inside the frame is 64 square inches, how wide is the picture frame? Justify your response.

LESSON 32-1 PRACTICE

10. $(x-1)^2 = 8$
11. $\left(1 + 2\sqrt{2}\right)$ units
12. $x = \pm\sqrt{22}$
13. $x = -3$ or $x = -7$
14. $x = 2$
15. $x = -1 \pm 2\sqrt{3}$
16. a. $A = (10 - 2x)^2$
    b. $x = 1$ or $x = 9$
    c. 1 in.; If the frame were 9 inches wide, then opposite sides of the frame would overlap and there would be no room for the picture.

470 SpringBoard® Mathematics **Algebra 1, Unit 5** • Quadratic Functions

Lesson 32-2
Completing the Square

Learning Targets:
- Solve quadratic equations by completing the square.
- Complete the square to analyze a quadratic function.

SUGGESTED LEARNING STRATEGIES: Note Taking, Graphic Organizer, Identify a Subtask

As shown in Example B in Lesson 32-1, quadratic equations are more easily solved with square roots when the side with the variable is a perfect square. When a quadratic equation is written in the form $x^2 + bx + c = 0$, you can complete the square to transform the equation into one that can be solved using square roots. **Completing the square** is the process of adding a term to the variable side of a quadratic equation to transform it into a perfect square trinomial.

Example A
Solve $x^2 + 10x - 6 = 0$ by completing the square.

Step 1:	Isolate the variable terms.	$x^2 + 10x - 6 = 0$
	Add 6 to both sides.	$x^2 + 10x = 6$
Step 2:	Transform the left side into a perfect trinomial.	
	Divide the coefficient of the x-term by 2.	$10 \div 2 = 5$
	Square the 5 to determine the constant.	$5^2 = 25$
	Complete the square by adding 25 to both sides of the equation.	$x^2 + 10x + \square = 6 + \square$ $x^2 + 10x + \boxed{25} = 6 + \boxed{25}$ $x^2 + 10x + 25 = 31$
Step 3:	Solve the equation. Write the trinomial in factored form.	$(x + 5)(x + 5) = 31$
	Write the left side as a square of a binomial.	$(x + 5)^2 = 31$
	Take the square root of both sides.	$\sqrt{(x+5)^2} = \pm\sqrt{31}$
	Solve for x.	$(x + 5) = \pm\sqrt{31}$
	Leave the solutions in $\pm$ form.	$x = -5 \pm \sqrt{31}$

Solution: $x = -5 \pm \sqrt{31}$

MATH TIP

Use a graphic organizer to help you complete the square.

	x	5
x	x^2	$5x$
5	$5x$	25

$x^2 + 5x + 5x + 25 = (x+5)(x+5)$
$x^2 + 10x + 25 = (x+5)^2$

Try These A
Make use of structure. Solve each quadratic equation by completing the square.

a. $x^2 - 8x + 3 = 11$
   $x = 4 \pm 2\sqrt{6}$

b. $x^2 + 7 = 2x + 8$
   $x = 1 \pm \sqrt{2}$

MINI-LESSON: Using a Graphic Organizer to Complete the Square

Visual and kinesthetic learners may benefit from using graphic organizers of the same type used in Unit 4 for factoring trinomials to help them complete the square. A mini-lesson is available to provide practice with this process.

See SpringBoard's eBook Teacher Resources for a student page for this mini-lesson.

ACTIVITY 32 Continued

Example B Shared Reading, Note Taking, Identify a Subtask In this example, students write the equation of a quadratic function in the form $y = a(x - h)^2 + k$ by completing the square. They then use this form to analyze characteristics of the function. In Step 1, have students discuss why 9 was added to each side of the equation. Be sure students are able to identify the values of a, h, and k once the equation has been written in the form $y = a(x - h)^2 + k$. Ask students to discuss what they can learn about the quadratic function from the values of a and k. Finally, ask students to explain how the graph verifies that the function has a minimum value of 4 and no x-intercepts.

TEACHER to TEACHER

When the equation of a quadratic function is written in the form $y = a(x - h)^2 + k$, it is said to be in *vertex form*. The vertex of the graph of the function is the point (h, k), and the axis of symmetry is the line $x = h$. If a is negative, k is the maximum value of the function, and if a is positive, k is the minimum value of the function.

Differentiating Instruction

Support students who are visual learners by suggesting that they use a quick sketch to determine whether a quadratic function has a minimum or maximum value. If a is positive, the parabola opens upward, and a sketch shows that the function has a minimum value. Similar steps can be used when a is negative.

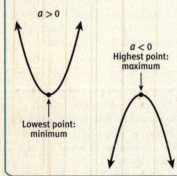

ACTIVITY 32 continued

My Notes

CONNECT TO PHYSICS

Quadratic functions can describe the *trajectory*, or path, of a moving object, such as a ball that has been thrown.

Lesson 32-2
Completing the Square

Completing the square is useful to help analyze specific features of quadratic functions, such as the maximum or minimum value and the possible number of zeros.

When a quadratic equation is written in the form $y = a(x - h)^2 + k$, you can determine whether the function has a maximum or minimum value based on a and what that value is based on k. This information can also help you determine the number of x-intercepts.

Example B

Analyze the quadratic function $y = x^2 - 6x + 13$ by completing the square.

Step 1: Complete the square on the right side of the equation.

$y = x^2 - 6x + 13$

Isolate the variable terms. $y - 13 = x^2 - 6x$

Transform the right side into a perfect square trinomial. $y - 13 + 9 = x^2 - 6x + 9$

Factor and simplify. $y - 4 = (x - 3)^2$

Write the equation in the form $y = a(x - h)^2 + k$. $y = (x - 3)^2 + 4$

Step 2: Identify the direction of opening and whether the vertex represents a maximum or minimum.

Since the value of a is positive, the parabola opens upward and the vertex represents a minimum.

Opens upward
Vertex is a minimum.

Step 3: Determine the maximum or minimum value. This parabola is the graph of the parent function $y = x^2$ translated up 4 units. So the parent function's minimum value of 0 is increased to 4.

Minimum value: 4

Step 4: Determine the number of x-intercepts. Since the minimum value of the function is $y = 4$ and the parabola opens upward, the function will never have a y-value less than 4. The graph will never intersect the x-axis, so there are no x-intercepts.

no x-intercepts

Step 5: Verify by graphing.

Solution: The graph of the quadratic function opens upward, the minimum value is 4, and there are no x-intercepts.

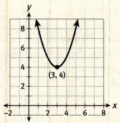

472 SpringBoard® Mathematics **Algebra 1, Unit 5** • Quadratic Functions

Lesson 32-2
Completing the Square

Try These B

Write each of the following quadratic functions in the form $y = a(x - h)^2 + k$. Identify the direction of opening, vertex, maximum or minimum value, and number of x-intercepts.

a. $y = x^2 - 4x + 9$

b. $y = -x^2 - 6x - 8$

c. $y = x^2 + 8x + 15$

Check Your Understanding

Solve by completing the square.

1. $x^2 + 2x + 3 = 0$
2. $x^2 + 6x + 4 = 0$

LESSON 32-2 PRACTICE

Solve by completing the square.

3. $2 = x^2 - 10x$
4. $4x = x^2 - 4x - 32$
5. $-2x^2 + 4 = -x^2 + x - 7$
6. $x + 1 = 6x - x^2$

Complete the square to determine the vertex and maximum or minimum value. Determine the number of x-intercepts.

7. $y = x^2 - 2x + 2$
8. $y = -x^2 + 8x - 6$

9. **Make sense of problems.** In a model railroad, the track is supported by an arch that is represented by $y = -x^2 + 10x - 16$, where y represents the height of the arch in inches and x represents the distance in inches from a cliff. Complete the square to answer the following questions.
   a. How far is the center of the arch from the cliff?
   b. What is the maximum height of the arch?

10. The bubbler is the part of a drinking fountain that produces a stream of water. The water in a drinking fountain follows a path given by $y = -x^2 + 6x + 4.5$, where y is the height of the water in centimeters above the basin, and x is the distance of the water from the bubbler. What is the maximum height of the water above the basin?

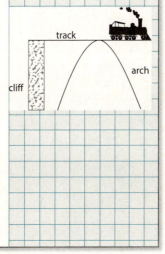

ACTIVITY 32 Continued

Lesson 32-3

PLAN

Materials
- graphing calculators (optional)

Pacing: 1 class period

Chunking the Lesson
Introduction   Example A
Check Your Understanding
Lesson Practice

TEACH

Bell-Ringer Activity

Ask students to solve the following equations by completing the square. Ask volunteers to explain the steps they used to solve the equations.

1. $x^2 + 4x + 1 = 0 \quad x = -2 \pm \sqrt{3}$
2. $x^2 - 8x - 4 = 0 \quad x = 4 \pm 2\sqrt{5}$

Quadratic Formula This page shows a derivation of the quadratic formula by completing the square on the general quadratic equation $ax^2 + bx + c = 0$. Discuss each step as you guide students through the derivation. The basic steps are these:

1. Isolate the variable terms, and rewrite the equation so that the coefficient of x^2 is 1.
2. Set up for completing the square.
3. Complete the square.
4. Factor the perfect square trinomial on the left, and write the terms on the right with a common denominator.
5. Add the terms on the right, and then take the square root of both sides.
6. Isolate x.
7. Simplify the right side.

ACTIVITY 32
continued

Lesson 32-3
The Quadratic Formula

My Notes

Learning Targets:
- Derive the quadratic formula.
- Solve quadratic equations using the quadratic formula.

SUGGESTED LEARNING STRATEGIES: Close Reading, Note Taking, Identify a Subtask

Generalizing a solution method into a formula provides an efficient way to perform complicated procedures. You can complete the square on the general form of a quadratic equation $ax^2 + bx + c = 0$ to find a formula for solving all quadratic equations.

$$ax^2 + bx + c = 0$$

$$x^2 + \frac{b}{a}x = -\frac{c}{a}$$

$$x^2 + \frac{b}{a}x + \Box = -\frac{c}{a} + \Box$$

$$x^2 + \frac{b}{a}x + \left(\frac{b}{2a}\right)^2 = -\frac{c}{a} + \left(\frac{b}{2a}\right)^2$$

$$\left(x + \frac{b}{2a}\right)^2 = -\frac{4ac}{4a^2} + \frac{b^2}{4a^2}$$

$$x + \frac{b}{2a} = \pm\sqrt{\frac{b^2 - 4ac}{4a^2}}$$

$$x = -\frac{b}{2a} \pm \frac{\sqrt{b^2 - 4ac}}{2a}$$

$$x = \frac{-b \pm \sqrt{b^2 - 4ac}}{2a}$$

Quadratic Formula
When $a \neq 0$, the solutions of $ax^2 + bx + c = 0$ are
$$x = \frac{-b \pm \sqrt{b^2 - 4ac}}{2a}.$$

Lesson 32-3
The Quadratic Formula

To apply the quadratic formula, make sure the equation is in standard form $ax^2 + bx + c = 0$. Identify the values of a, b, and c in the equation and then substitute these values into the quadratic formula. If the expression under the radical sign is not a perfect square, write the solutions in simplest radical form or use a calculator to approximate the solutions.

Example A
Solve $x^2 + 3 = 6x$ using the quadratic formula.

Step 1: Write the equation in standard form.
$$x^2 + 3 = 6x$$
$$x^2 - 6x + 3 = 0$$

Step 2: Identify a, b, and c.
$$a = 1, b = -6, c = 3$$

Step 3: Substitute these values into the quadratic formula.
$$x = \frac{-(-6) \pm \sqrt{(-6)^2 - 4(1)(3)}}{2(1)}$$

Step 4: Simplify using the order of operations.
$$x = \frac{6 \pm \sqrt{36 - 12}}{2} = \frac{6 \pm \sqrt{24}}{2}$$

Step 5: Write as two solutions.
$$x = \frac{6 + \sqrt{24}}{2} \text{ or } x = \frac{6 - \sqrt{24}}{2}$$

Solution: Use a calculator to approximate the two solutions.
$$x \approx 5.45 \text{ or } x \approx 0.55$$

If you do not have a calculator, write your solution in simplest radical form. To write the solution in simplest form, simplify the radicand and then divide out any common factors.

$$x = \frac{6 \pm \sqrt{24}}{2} = \frac{6 \pm \sqrt{4 \cdot 6}}{2} = \frac{6 \pm 2\sqrt{6}}{2} = \frac{2(3 \pm \sqrt{6})}{2} = 3 \pm \sqrt{6}$$

Try These A
Solve using the quadratic formula.

a. $3x^2 = 4x + 3$

$x = \dfrac{2 \pm \sqrt{13}}{3}$

b. $x^2 + 4x = -2$

$x = -2 \pm \sqrt{2}$

ACTIVITY 32 Continued

Check Your Understanding

Debrief students' answers to these items to ensure that they understand how to solve quadratic equations by using the quadratic formula. You may also want to have students discuss the advantages and disadvantages of using the quadratic formula to solve quadratic equations.

Answers

1. $x = \dfrac{5 \pm \sqrt{13}}{6}$
2. $x = -4 \pm \sqrt{22}$

ASSESS

Students' answers to lesson practice problems will provide you with a formative assessment of their understanding of the lesson concepts and their ability to apply their learning.

See the Activity Practice for additional problems for this lesson. You may assign the problems here or use them as a culmination for the activity.

LESSON 32-3 PRACTICE

3. $x = \dfrac{-5 \pm \sqrt{29}}{2}$
4. $x = \dfrac{-1 \pm \sqrt{33}}{4}$
5. $x = \dfrac{5 \pm \sqrt{73}}{8}$
6. $x = -2 \pm \sqrt{5}$
7. $x = \dfrac{-3 \pm \sqrt{21}}{3}$
8. a. $15 = -16x^2 + 30x + 5$
   b. $x \approx 0.4$ or $x \approx 1.4$; The ball has a height of 15 ft at 0.4 s and at 1.4 s.
   c. two solutions; The ball is at a height of 15 feet once on the way up and once on the way down.
9. Marta is correct; Jose should have rewritten the equation as $x^2 + 4x + 3 = 0$ and used the value $c = 3$ in the quadratic formula.

ADAPT

Check students' answers to the Lesson Practice questions to make sure that they understand how to apply the quadratic formula. If students need more practice, they can use the quadratic formula to solve the quadratic equations that appeared in Lessons 31-1 and 32-2.

ACTIVITY 32 continued

My Notes

Lesson 32-3
The Quadratic Formula

Check Your Understanding

Solve using the quadratic formula.

1. $3x^2 - 5x + 1 = 0$
2. $x^2 + 6 = -8x + 12$

LESSON 32-3 PRACTICE

Solve using the quadratic formula.

3. $x^2 + 5x - 1 = 0$
4. $-2x^2 - x + 4 = 0$
5. $4x^2 - 5x - 2 = 1$
6. $x^2 + 3x = -x + 1$
7. $3x^2 = -6x + 4$
8. A baseball player tosses a ball straight up into the air. The function $y = -16x^2 + 30x + 5$ models the motion of the ball, where x is the time in seconds and y is the height of the ball, in feet.
   a. Write an equation you can solve to find out when the ball is at a height of 15 feet.
   b. Use the quadratic formula to solve the equation. Round to the nearest tenth.
   c. How many solutions did you find for Part (b)? Explain why this makes sense.
9. **Critique the reasoning of others.** José and Marta each solved $x^2 + 4x = -3$ using two different methods. Who is correct and what is the error in the other student's work?

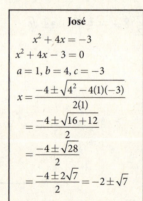

José	Marta
$x^2 + 4x = -3$	$x^2 + 4x = -3$
$x^2 + 4x - 3 = 0$	$x^2 + 4x + \square = -3 + \square$
$a = 1, b = 4, c = -3$	$x^2 + 4x + 4 = -3 + 4$
$x = \dfrac{-4 \pm \sqrt{4^2 - 4(1)(-3)}}{2(1)}$	$(x+2)^2 = 1$
$= \dfrac{-4 \pm \sqrt{16 + 12}}{2}$	$x + 2 = \pm\sqrt{1}$
$= \dfrac{-4 \pm \sqrt{28}}{2}$	$x + 2 = 1$ or $x + 2 = -1$
$= \dfrac{-4 \pm 2\sqrt{7}}{2} = -2 \pm \sqrt{7}$	$x = -1$ or $x = -3$

MINI-LESSON: Simplifying Radicals

If students need additional practice on writing radicals in simplest form, a mini-lesson is available. Many state assessments do not allow students to use a calculator, so students will need to be able to simplify their answers completely to perform well on these exams.

See SpringBoard's eBook Teacher Resources for a student page for this mini-lesson.

Lesson 32-4
Choosing a Method and Using the Discriminant

ACTIVITY 32 continued

Learning Targets:
- Choose a method to solve a quadratic equation.
- Use the discriminant to determine the number of real solutions of a quadratic equation.

SUGGESTED LEARNING STRATEGIES: Think-Pair-Share, Graphic Organizer, Look for a Pattern, Create a Plan, Quickwrite

There are several methods for solving a quadratic equation. They include factoring, using square roots, completing the square, and using the quadratic formula. Each of these techniques has different advantages and disadvantages. Learning how and why to use each method is an important skill.

1. Solve each equation below using a different method. State the method used.
   a. $x^2 + 5x - 24 = 0$
      Solve by factoring. $x = -8$ or $x = 3$

   b. $x^2 - 6x + 2 = 0$
      Solve by completing the square. $x = 3 \pm \sqrt{7}$

   c. $2x^2 + 3x - 5 = 0$
      Solve by quadratic formula. $x = 1$ or $x = -2.5$

   d. $x^2 - 100 = 0$
      Solve by taking a square root. $x = 10$ or $x = -10$

2. How did you decide which method to use for each equation in Item 1?
   Answers may vary. Check students' work.

ACTIVITY 32 Continued

Developing Math Language
This lesson includes the vocabulary word *discriminant*. This word is important for students to know as they develop their understanding of quadratic equations and functions. Encourage students to mark the text by adding notes explaining the new term in their own words. Add the term to the classroom Word Wall.

3–4 Graphic Organizer, Create Representations, Look for a Pattern, Debriefing In these items, students explore the relationship between the discriminant of a quadratic equation, the number of real solutions of the equation, and the number of x-intercepts of the related quadratic function. Ask students to explain how they know that the last equation in the table has no real solutions. As this lesson is debriefed, be sure students recognize the fact that the sign of the discriminant can be used to determine the number of real solutions of a quadratic equation and the number of x-intercepts of the related function.

Differentiating Instruction

Support students who are having trouble graphing the functions by reminding them that they can start by graphing the points containing the x-intercepts (if any). They can identify additional points on the graph by using a table of values or reflecting known points across the axis of symmetry. Students could also use a graphing calculator to graph the functions.

ACTIVITY 32 continued

**Lesson 32-4
Choosing a Method and Using the Discriminant**

My Notes

The expression $\sqrt{b^2 - 4ac}$ in the quadratic formula helps you understand the nature of the quadratic equation. The **discriminant**, $b^2 - 4ac$, of a quadratic equation gives information about the number of real solutions, as well as the number of x-intercepts of the related quadratic function.

3. Solve each equation using any appropriate solution method. Then complete the rest of the table.

Equation	Discriminant	Solutions	Number of Real Solutions	Number of x-Intercepts	Graph of Related Quadratic Function
$x^2 + 2x - 8 = 0$	36	$x = 2$ or $x = -4$	2	2	
$x^2 + 2x + 1 = 0$	0	$x = -1$	1	1	
$x^2 + 2x + 5 = 0$	-16	$x = \dfrac{-2 \pm \sqrt{-16}}{2}$ no real solutions	0	0	

Lesson 32-4
Choosing a Method and Using the Discriminant

4. **Express regularity in repeated reasoning.** Complete each statement below using the information from the table in Item 3.

   - If $b^2 - 4ac > 0$, the equation has __2__ real solution(s) and the graph of the related function has __2__ x-intercept(s).

   - If $b^2 - 4ac = 0$, the equation has __1__ real solution(s) and the graph of the related function has __1__ x-intercept(s).

   - If $b^2 - 4ac < 0$, the equation has __no__ real solution(s) and the graph of the related function has __no__ x-intercept(s).

Check Your Understanding

Use the discriminant to determine the number of real solutions.

5. $4x^2 + 2x - 12 = 0$
6. $x^2 - 7x + 14 = 0$
7. $x^2 - 10x + 25 = 0$

LESSON 32-4 PRACTICE

For each equation, use the discriminant to determine the number of real solutions. Then solve the equation.

8. $x^2 - 1 = 0$
9. $x^2 - 4x + 4 = 0$
10. $-4x^2 + 3x = -2$
11. $x^2 - 2 = 12x$
12. $x^2 + 5x - 1 = 0$
13. $-x^2 - 2x - 10 = 0$
14. **Model with mathematics.** Lin launches a model rocket that follows a path given by the function $y = -0.4t^2 + 3t + 0.5$, where y is the height in meters and t is the time in seconds.
    a. Explain how you can write an equation and then use the discriminant to determine whether Lin's rocket ever reaches a height of 5 meters.
    b. If Lin's rocket reaches a height of 5 meters, at approximately what time(s) does it do so? If not, what is the rocket's maximum height?

ACTIVITY 32 Continued

Lesson 32-5

PLAN

Pacing: 1 class period
Chunking the Lesson
#1–2
Example A   Example B
Check Your Understanding
Lesson Practice

TEACH

Bell-Ringer Activity
Ask students to use the discriminant to determine the number of real solutions of each quadratic equation.

1. $x^2 - 6x + 11 = 0$ [0]
2. $x^2 + 2x - 4 = 0$ [2]
3. $4x^2 - 12x + 9 = 0$ [1]

1–2 Shared Reading, Create Representations, Think-Pair-Share, Construct an Argument In these items, students graph a quadratic function to determine that it has no x-intercepts. They should conclude that the related equation has no real solutions and a negative discriminant. These items are designed to lead into a discussion of nonreal solutions of quadratic equations.

TEACHER to TEACHER

Every quadratic equation of the form $ax^2 + bx + c = 0$, where a, b and c are real numbers and $a \neq 0$, has two roots, including repeated roots. When the discriminant is positive, the equation has two distinct real roots. When the discriminant is zero, the equation has a repeated, or double, real root, which corresponds to one distinct real root. When the discriminant is negative, the equation has two complex roots. The complex roots are always complex conjugates, which means that their real parts are the same and their imaginary parts are opposites.

ACTIVITY 32 continued

Lesson 32-5
Complex Solutions

My Notes

Learning Targets:
- Use the imaginary unit i to write complex numbers.
- Solve a quadratic equation that has complex solutions.

SUGGESTED LEARNING STRATEGIES: Think-Pair-Share, Close Reading, Note Taking, Construct an Argument, Identify a Subtask

When solving quadratic equations, there are always one, two, or no real solutions. Graphically, the number of x-intercepts is helpful for determining the number of real solutions.

- When there is one real solution, the graph of the related quadratic function touches the x-axis once, and the vertex of the parabola is on the x-axis.
- When there are two real solutions, the graph crosses the x-axis twice.
- When there are no real solutions, the graph never crosses the x-axis.

1. Graph the function $y = x^2 - 6x + 13$. Use the graph to determine the number of real solutions to the equation $x^2 - 6x + 13 = 0$.
   No real solutions.

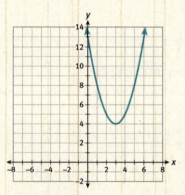

2. **Construct viable arguments.** What does the number of real solutions to the equation in Item 1 indicate about the value of the discriminant of the equation? Explain.
   There are no real solutions so the discriminant is negative.

When the value of the discriminant is less than zero, there are no *real* solutions. This is different from stating there are *no* solutions. In cases where the discriminant is negative, there are two solutions that are *not* real numbers. **Imaginary numbers** offer a way to determine these non-real solutions. The **imaginary unit**, i, equals $\sqrt{-1}$. Imaginary numbers are used to represent square roots of negative numbers, such as $\sqrt{-4}$.

Lesson 32-5
Complex Solutions

Example A
Simplify $\sqrt{-4}$.

Step 1: Write the radical as a product involving $\sqrt{-1}$. $\quad \sqrt{-4} = \sqrt{-1} \cdot \sqrt{4}$

Step 2: Replace $\sqrt{-1}$ with the imaginary unit, i. $\quad = i\sqrt{4}$

Step 3: Simplify the radical. The principal square root of 4 is 2. $\quad = 2i$

Solution: $\sqrt{-4} = 2i$

Try These A
Simplify.
a. $\sqrt{-16}$
$4i$

b. $-\sqrt{-9}$
$-3i$

c. $\sqrt{-8}$
$2i\sqrt{2}$

Problems involving imaginary numbers can also result in **complex numbers**, $a + bi$, where a and b are real numbers. In this form, a is the real part and b is the imaginary part.

Example B
Solve $x^2 - 6x + 12 = 0$.

Step 1: Identify a, b, and c. $\quad a = 1, b = -6, c = 12$

Step 2: Substitute these values into the quadratic formula. $\quad x = \dfrac{-(-6) \pm \sqrt{(-6)^2 - 4(1)(12)}}{2(1)}$

Step 3: Simplify using the order of operations. $\quad x = \dfrac{6 \pm \sqrt{36 - 48}}{2} = \dfrac{6 \pm \sqrt{-12}}{2}$

Step 4: Simplify using the imaginary unit, i. $\quad x = \dfrac{6 \pm \sqrt{-1}\sqrt{12}}{2} = \dfrac{6 \pm i\sqrt{12}}{2}$

Step 5: Simplify the radical and the fraction. $\quad x = \dfrac{6 \pm 2i\sqrt{3}}{2} = 3 \pm i\sqrt{3}$

Solution: $x = 3 \pm i\sqrt{3}$

Try These B
Solve each equation.
a. $x^2 + 100 = 0$
$x = \pm 10i$

b. $x^2 - 4x = -11$
$x = 2 \pm i\sqrt{7}$

WRITING MATH

Notice that the solution to Example A is written $2i$ and not $i2$. However, when a value includes a radical and i, i is written in front of the radical, as in $i\sqrt{3}$.

ACTIVITY 32 Continued

Check Your Understanding

Debrief students' answers to these items to make sure that they can simplify square roots of negative numbers and determine complex solutions of quadratic equations.

Answers
3. $3i\sqrt{3}$
4. $x = -2 \pm i\sqrt{2}$

ASSESS

Students' answers to lesson practice problems will provide you with a formative assessment of their understanding of the lesson concepts and their ability to apply their learning.

See the Activity Practice for additional problems for this lesson. You may assign the problems here or use them as a culmination for the activity.

LESSON 32-5 PRACTICE

5. $-i\sqrt{11}$
6. $i\sqrt{42}$
7. $\pm 9i$
8. $x = \dfrac{1 \pm i\sqrt{14}}{5}$
9. $x = \pm i\sqrt{6}$
10. $x = 1 \pm i\sqrt{3}$
11. a. $4 - 4c$
    b. If both zeros are imaginary numbers, then the discriminant is negative, so solve $4 - 4c < 0$. The solution is $c > 1$.
    c. Since $c > 1$ in this function, both zeros are imaginary numbers.

ADAPT

Use the Lesson Practice to ensure that students understand concepts related to imaginary numbers and quadratic equations with complex solutions. If students need more help simplifying square roots of negative numbers, be sure to emphasize the difference between a negative square root and the square root of a negative number. A negative square root is indicated by a negative sign *outside* the square root symbol. For example, $-\sqrt{4}$ is the negative square root of 4, which is -2. By contrast, the square root of a negative number is indicated by a negative sign *inside* the square root symbol. For example, $\sqrt{-4}$ is the principal square root of -4, which is $2i$.

Lesson 32-5
Complex Solutions

Check Your Understanding

3. Simplify $\sqrt{-27}$.
4. Solve $x^2 + 4x + 6 = 0$.

LESSON 32-5 PRACTICE

Simplify.

5. $-\sqrt{-11}$
6. $\sqrt{-42}$
7. $\pm\sqrt{-81}$

Solve.

8. $5x^2 - 2x + 3 = 0$
9. $-x^2 - 6 = 0$
10. $(x - 1)^2 + 3 = 0$
11. **Make use of structure.** Consider the quadratic function $y = x^2 + 2x + c$, where c is a real number.
    a. Write and simplify an expression for the discriminant.
    b. Explain how you can use your result from Part (a) to write and solve an inequality that tells you when the function will have two zeros that involve imaginary numbers.
    c. Use your results to describe the zeros of the function $y = x^2 + 2x + 3$.

Algebraic Methods of Solving Quadratic Equations
Keeping it Quadratic

ACTIVITY 32 PRACTICE
Write your answers on notebook paper. Show your work.

Lesson 32-1

Solve each equation using square roots.

1. $x^2 + 7 = 43$
2. $(x - 5)^2 + 2 = 11$
3. $x^2 - 8x + 16 = 3$
4. Antonio drops a rock from a cliff that is 400 feet high. The function $y = -16t^2 + 400$ gives the height of the rock in feet after t seconds. Write and solve an equation to determine how long it takes the rock to land at the base of the cliff. (*Hint*: At the base of the cliff, the height y is 0.)
5. Maya wants to use square roots to solve the equation $x^2 - 6x + 9 = k$, where k is a positive real number. Which of these is the best representation of the solution?
   A. $x = 3 \pm \sqrt{k}$ B. $x = -3 \pm \sqrt{k}$
   C. $x = \pm\sqrt{k+3}$ D. $x = \pm\sqrt{k-3}$

Lesson 32-2

6. Given the equation $x^2 - 8x = 3$, what number should be added to both sides to complete the square?
   A. -4 B. 8
   C. 16   D. 64

Write each of the following equations in the form $y = a(x - h)^2 + k$. Then identify the direction of opening, vertex, maximum or minimum value, and x-intercepts.

7. $y = x^2 - 4x + 11$
8. $y = -x^2 - 6x - 8$
9. $y = x^2 + 2x - 8$

10. A golfer stands on a platform 16 feet above a driving range. Once the golf ball is hit, the function $y = -16t^2 + 64t + 16$ represents the height of the ball in feet after t seconds.
    a. Write an equation you can solve to determine the number of seconds it takes for the ball to land on the driving range.
    b. Solve the equation by completing the square. Leave your answer in radical form.
    c. Use a calculator to find the number of seconds, to the nearest tenth, that it takes the ball to land on the driving range.

11. Which of the following is a true statement about the graph of the quadratic function $y = x^2 - 2x + 3$?
    A. The vertex of the graph is $(-1, 2)$.
    B. The graph intersects the x-axis at $x = 1$.
    C. The graph is a parabola that opens upward.
    D. There is exactly one x-intercept.

Solve by completing the square.

12. $x^2 - 4x = 12$
13. $x^2 + 10x + 21 = 0$
14. $2x^2 - 4x - 4 = 0$
15. $x^2 + 6x = -10$

16. A climbing structure at a playground is represented by the function $y = -x^2 + 4x + 1$, where y is the height of the structure in feet and x is the distance in feet from a wall. What is the maximum height of the structure?
    A. 1 foot   B. 2 feet
    C. 4 feet   D. 5 feet

ACTIVITY 32 Continued

ACTIVITY PRACTICE

1. $x = \pm 6$
2. $x = 2, x = 8$
3. $x = 4 \pm \sqrt{3}$
4. $-16t^2 + 400 = 0$; 5 seconds
5. A
6. C
7. $y = (x - 2)^2 + 7$; opens upward; vertex is $(2, 7)$; minimum is 7; no x-intercepts
8. $y = -(x + 3)^2 + 1$; opens downward; vertex is $(-3, 1)$; maximum is 1; x-intercepts are -2 and -4
9. $y = (x + 1)^2 - 9$; opens upward; vertex is $(-1, -9)$; x-intercepts are -4 and 2
10. a. $0 = -16t^2 + 64t + 16$
    b. $t = 2 \pm \sqrt{5}$
    c. 4.2 seconds
11. C
12. $x = -2, x = 6$
13. $x = -7, x = -3$
14. $x = 1 \pm \sqrt{3}$
15. no real solutions
16. D

ACTIVITY 32 Continued

17. $x = -\frac{1}{2}, x = \frac{3}{2}$
18. $x = 2, x = -\frac{1}{5}$
19. $x = 1 \pm \sqrt{5}$
20. a. $11 = -16t^2 + 32t + 3$
    b. $t \approx 0.3$ or $t \approx 1.7$; 0.3 seconds and 1.7 seconds
21. No; in the fourth line of the solution, the expression inside the square root sign should be $36 + 8$, not $36 - 8$. The correct solution is $x = \frac{-3 \pm \sqrt{11}}{2}$.
22. no real solutions
23. one real solution
24. D
25. The dolphin would reach a height of 7 feet when $-16t^2 + 20t = 7$. The equation $-16t^2 + 20t - 7 = 0$ has a discriminant of $20^2 - 4(-16)(-7) = -48$. Because the discriminant is negative, the equation has no real solutions, and therefore the dolphin is never at a height of 7 feet.
26. $\pm 2i$
27. $-5i$
28. $2i\sqrt{2}$
29. $-11i$
30. $12 - 12i$
31. $\pm 4i\sqrt{2}$
32. $x = \frac{5 \pm i\sqrt{15}}{4}$
33. $x = \frac{-1 \pm i\sqrt{11}}{2}$
34. $x = \frac{-3 \pm i\sqrt{3}}{6}$
35. $x = \frac{-1 \pm i\sqrt{7}}{2}$
36. Sample answer: If $y = x^2 + 4x + p$ has two real zeros, then the discriminant of $0 = x^2 + 4x + p$ is positive. Therefore, $4^2 - 4(1)p > 0$ or $16 - 4p > 0$. Solving shows that $p < 4$. So, the function has two real zeros when $p < 4$.

ADDITIONAL PRACTICE

If students need more practice on the concepts in this activity, see the eBook Teacher Resources for additional practice problems.

ACTIVITY 32 continued

Algebraic Methods of Solving Quadratic Equations
Keeping it Quadratic

Lesson 32-3

Solve using the quadratic formula.

17. $4x^2 - 4x = 3$
18. $5x^2 - 9x - 2 = 0$
19. $x^2 = 2x + 4$
20. A football player kicks a ball. The function $y = -16t^2 + 32t + 3$ models the motion of the ball, where t is the time in seconds and y is the height of the ball in feet.
    a. Write an equation you can solve to find out when the ball is at a height of 11 feet.
    b. Use the quadratic formula to solve the equation. Round to the nearest tenth.
21. Kyla was asked to solve the equation $2x^2 + 6x - 1 = 0$. Her work is shown below. Is her solution correct? If not, describe the error and give the correct solution.

    $2x^2 + 6x - 1 = 0$
    $a = 2, b = 6, c = -1$

    $x = \frac{-6 \pm \sqrt{6^2 - 4(2)(-1)}}{2(2)}$

    $= \frac{-6 \pm \sqrt{36 - 8}}{4}$

    $= \frac{-6 \pm \sqrt{28}}{4}$

    $= \frac{-6 \pm 2\sqrt{7}}{4}$

    $= \frac{-3 \pm \sqrt{7}}{2}$

Lesson 32-4

Use the discriminant to determine the number of real solutions.

22. $x^2 + 3x + 5 = 0$
23. $4x^2 - 4x + 1 = 0$
24. The discriminant of a quadratic equation is -1. Which of the following must be a true statement about the graph of the related quadratic function?
    A. The graph intersects the x-axis in exactly two points.
    B. The graph lies entirely above the x-axis.
    C. The graph intersects the x-axis at $x = -1$.
    D. The graph has no x-intercepts.
25. A dolphin jumps straight up from the water. The quadratic function $y = -16t^2 + 20t$ models the motion of the dolphin, where t is the time in seconds and y is the height of the dolphin, in feet. Use the discriminant to explain why the dolphin does not reach a height of 7 feet.

Lesson 32-5

Simplify.

26. $\pm\sqrt{-2}$
27. $-\sqrt{-25}$
28. $\sqrt{-8}$
29. $-\sqrt{-121}$
30. $12 - \sqrt{-144}$
31. $\pm\sqrt{-32}$

Solve.

32. $2x^2 - 5x + 5 = 0$
33. $x^2 + x + 3 = 0$
34. $-3x^2 - 3x - 1 = 0$
35. $-x^2 - x - 2 = 0$

MATHEMATICAL PRACTICES
Construct Viable Arguments and Critique the Reasoning of Others

36. For what values of p does the quadratic function $y = x^2 + 4x + p$ have two real zeros? Justify your answer.

---

484 SpringBoard® Mathematics Algebra 1, Unit 5 • Quadratic Functions

Applying Quadratic Equations
Rockets in Flight
Lesson 33-1 Fitting Data with a Quadratic Function

ACTIVITY 33

Learning Targets:
- Write a quadratic function to fit data.
- Use a quadratic model to solve problems.

> **SUGGESTED LEARNING STRATEGIES:** Create Representations, Look for a Pattern, Summarizing, Predict and Confirm, Discussion Groups

Cooper is a model rocket fan. Cooper's model rockets have single engines and, when launched, can rise as high as 1000 ft depending upon the engine size. After the engine is ignited, it burns for 3–5 seconds, and the rocket accelerates upward. The rocket has a parachute that will open as the rocket begins to fall back to Earth.

Cooper wanted to investigate the flight of a rocket from the time the engine burns out until the rocket lands. He set a device in a rocket, named *Spirit*, to begin collecting data the moment the engine shut off. Unfortunately, the parachute failed to open. When the rocket began to descend, it was in *free fall*.

The table shows the data that was collected.

My Notes

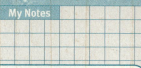

CONNECT TO SCIENCE

A *free falling* object is an object that is falling under the sole influence of gravity. The approximate value of acceleration due to gravity for an object in free fall on Earth is 32 ft/s² or 9.8 m/s².

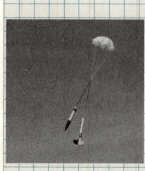

The *Spirit*	
Time Since the Engine Burned Out (s)	Height (ft)
0	512
1	560
2	576
3	560
4	512
5	432
6	320

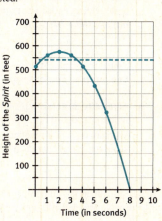

1. **a.** Use the table to determine whether the height of the *Spirit* can be modeled by a linear function.

   The data in the table for the *Spirit* are not linear. There is a constant increase in the time, but not a constant increase or decrease in the height.

   b. Graph the data for the height of the *Spirit* as a function of time on the grid.

   See graph. The curve on the graph is the answer to Item 3b below. The dashed line is the answer to Item 2 of Lesson 33-2.

Common Core State Standards for Activity 33

HSN-Q.A.3	Choose a level of accuracy appropriate to limitations on measurement when reporting quantities.
HSA-CED.A.1	Create equations and inequalities in one variable and use them to solve problems. *Include equations arising from linear and quadratic functions and simple rational and exponential functions.*
HSF-IF.B.5	Relate the domain of a function to its graph and, where applicable, to the quantitative relationship it describes.*
HSF-BF.A.1	Write a function that describes a relationship between two quantities.*
HSF-BF.A.1a	Determine an explicit expression, a recursive process, or steps for calculation from a context.

ACTIVITY 33
Investigative

Activity Standards Focus
In Activity 33, students write quadratic functions to fit data. They then apply the quadratic models to solve problems. Students also interpret solutions of quadratic equations in real-world contexts. Throughout this activity, be sure to emphasize that the meaning of the variables in a function helps to determine its reasonable domain and range.

Lesson 33-1

PLAN

Materials
- graphing calculators

Pacing: 1 class period

Chunking the Lesson
#1 #2
#3–4 #5
Check Your Understanding
Lesson Practice

TEACH

Bell-Ringer Activity
Ask students to discuss how they can determine whether a linear function is a good model for a set of data presented in a table.

1 Shared Reading, Summarizing, Look for a Pattern, Create Representations, Think-Pair Share
In this item, students analyze a set of data to determine that a linear model is not a good fit, both by analyzing a table of values and by creating a graph. Some students may question why the rocket initially continues to increase in height after the engine cuts off. Remind students that the rocket's momentum will tend to move it in the same direction it was initially traveling. After a time, however, the force of gravity will reduce the rocket's upward speed to 0 ft/s, and only then will the rocket begin to fall.

ACTIVITY 33 Continued

2 Predict and Confirm, Discussion Groups In this item, students use the data set from Item 1 to solve problems. They also use the data set to make an estimate, which they will later confirm when they develop a functional model for the data. Students may use the table of values as well as the graph to answer these questions. For part c, various estimates may be given and should be accepted. Some students may give $2 < t \leq 6$ as the time of the free fall because this is the time range shown in the table. Other students may give $2 < t < 8$ because a curve that approximates the data points in the graph will intersect the horizontal axis at about $t = 8$. Still others may give $t > 2$.

3–4 Create Representations, Discussion Groups In these items, students use a graphing calculator to develop a quadratic function that models the data. If students are not already familiar with finding a regression equation on a graphing calculator, you will need to guide them through this skill prior to these items. See the mini-lesson below. For Item 3b, students can graph the function on their graphing calculator first, and then use that display to help them sketch the graph on the grid in Item 1. In Item 4, students confirm the prediction they made in Item 2 concerning the height of the rocket when the engine burned out.

TEACHER to TEACHER

If graphing calculators are not available, give students the function $h(t) = -16t^2 + 64t + 512$. Students will need this function to answer subsequent questions throughout this activity. An alternative is to have students substitute three ordered pairs from the table into the function $h(t) = at^2 + bt + c$ to form a system of three equations. Students can solve the system to find the values of a, b, and c. For example, using the ordered pairs $(0, 512)$, $(1, 560)$, and $(2, 576)$ results in the system
$\begin{cases} 512 = c \\ 560 = a + b + c \\ 576 = 4a + 2b + c \end{cases}$.
Solving this system shows that $a = -16$, $b = 64$, and $c = 512$.

ACTIVITY 33 continued

My Notes

CONNECT TO TECHNOLOGY

For Item 3a, enter the data from the table above Item 1 into a graphing calculator. Use the calculator's quadratic regression feature to find a representative function.

Lesson 33-1
Fitting Data with a Quadratic Function

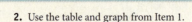

2. Use the table and graph from Item 1.

   a. How high was the *Spirit* when the engine burned out?
      512 feet

   b. How long did it take the rocket to reach its maximum height after the engine cut out?
      2 seconds

   c. Estimate the time the rocket was in free fall before it reached the earth.
      Answers may vary. See notes on Item 2c to the left.

3. Use the table and graph from Item 1.

   a. **Use appropriate tools strategically.** Use a graphing calculator to determine a quadratic function $h(t)$ for the data.
      $h(t) = -16t^2 + 64t + 512$

   b. Sketch the graph of the function on the grid in Item 1.
      See graph on previous page.

   c. **Attend to precision.** Give a reasonable domain and range for $h(t)$ within the context of the problem. Be sure to include units.
      Domain: $0 \leq t \leq 8$ seconds
      Range: $0 \leq h(t) \leq 576$ feet

4. Use the function found in Item 3 to verify the height of the *Spirit* when the engine burned out.
   $h(0) = -16(0)^2 + 64(0) + 512 = 512$

5. **Construct viable arguments.** Use the graph of $h(t)$ to approximate the time interval in which the *Spirit* was in free fall. Explain how you determined your answer.
   The *Spirit* was in free fall for about 6 seconds. Explanations may vary. From the graph, I can see that the maximum value appears to occur when t is 2. The *Spirit* hit the ground at 0 feet when t is 8. So the interval between these times was about 6 seconds.

MINI-LESSON: Quadratic Regression

If students need help performing quadratic regressions on a graphing calculator, a mini-lesson is available to provide practice. Check the manual for the calculators your class uses to adapt the steps if necessary.

See SpringBoard's eBook Teacher Resources for a student page for this mini-lesson.

Lesson 33-1
Fitting Data with a Quadratic Function

Check Your Understanding

Use the data in the table for Items 6–8.

x	0	1	2	3
f(x)	5	4.5	3	0.5

6. Graph the data in the table. Do the data appear to be linear? Explain.
7. Use a graphing calculator to write a function to model $f(x)$.
8. Use the function to determine the value of $f(x)$ when $x = 1.5$.

LESSON 33-1 PRACTICE

9. **Model with mathematics.** Cooper wanted to track another one of his rockets, the *Eagle*, so that he could investigate its time and height while in flight. He installed a device into the nose of the *Eagle* to measure the time and height of the rocket as it fell back to Earth. The device started measuring when the parachute opened. The data for one flight of the *Eagle* is shown in the table below.

The *Eagle*										
Time Since Parachute Opened (s)	0	1	2	3	4	5	6	7	8	9
Height(ft)	625	618	597	562	513	450	373	282	177	58

a. Graph the data from the table.
b. Use a graphing calculator to determine a quadratic function $h(t)$ that models the data.

10. Use the function from Item 9b to find the time when the rocket's height is 450 ft. Verify that your result is the same as the data in the table.

11. After the parachute opened, how long did it take for the rocket to hit Earth?

LESSON 33-1 PRACTICE

9. a.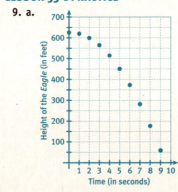

b. $h(t) = -7t^2 + 625$
10. $t = 5$ seconds
11. $t \approx 9.449$ seconds

ACTIVITY 33 Continued

5 Predict and Confirm, Discussion Groups, Construct an Argument, Debriefing Students use their quadratic models to better estimate the time interval during which the rocket was in free fall. Have students discuss what characteristics of the graph they used to determine their estimates. Ask students to identify the x and y intercepts and the vertex of the parabola used to model the flight of the rocket and to relate their meaning within the problem situation.

Check Your Understanding
Debrief students' answers to these items to be sure they can write and apply a quadratic function to model a set of data. Ask students to determine whether their model has a minimum or maximum value and what that value is.

Answers
6. The data do not appear to be linear, because the points do not lie on a straight line.

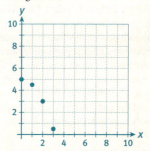

7. $f(x) = -0.5x^2 + 5$
8. 3.875

ASSESS

Students' answers to lesson practice problems will provide you with a formative assessment of their understanding of the lesson concepts and their ability to apply their learning.

See the Activity Practice for additional problems for this lesson. You may assign the problems here or use them as a culmination for the activity.

ADAPT

Ensure that students understand concepts related to fitting data with quadratic functions. Discuss ways students can check that their model is a good fit for the data set. One way is by graphing the model and the data set on the same coordinate grid to check that the data points lie on or close to the parabola given by the model. Another way is by substituting input values from the data set into the equation of the function to check that the function rule gives the appropriate output values.

Activity 33 • Applying Quadratic Equations 487

ACTIVITY 33 Continued

Lesson 33-2

PLAN

Materials
- graphing calculators

Pacing: 1 class period

Chunking the Lesson
#1 #2–3 #4
Check Your Understanding
Lesson Practice

TEACH

Bell-Ringer Activity

Review the methods of solving quadratic equations by asking students to solve each of the following equations. Then ask students to describe which methods they used and why.

1. $8x^2 + 16x - 24 = 0$ $[x = -3, x = 1]$
2. $x^2 - 4x - 1 = 0$ $[x = 2 \pm \sqrt{5}]$

1 Activating Prior Knowledge, Identify a Subtask, Think-Pair-Share
Students use factoring to solve an equation based on the quadratic model they developed in the previous lesson. Be sure students recall what t and $h(t)$ represent in this situation. As students share responses, be sure they have reached the conclusion that 8 seconds is the only reasonable response for the situation.

2–3 Predict and Confirm, Activating Prior Knowledge, Discussion Groups In these items, students use a graph to estimate the times at which the rocket was at a height of 544 feet. Drawing the line $y = 544$ on the graph will help students see that the rocket was at a height of 544 feet between 0 seconds and 1 second and again between 3 seconds and 4 seconds.

ACTIVITY 33 continued

Lesson 33-2
Interpreting Solutions of Quadratic Equations

My Notes

Learning Targets:
- Solve quadratic equations.
- Interpret the solutions of a quadratic equation in a real-world context.

SUGGESTED LEARNING STRATEGIES: Create Representations, Predict and Confirm, Think-Pair-Share, Discussion Groups

1. The total time that the *Spirit* was in the air after the engine burned out is determined by finding the values of t that make $h(t) = 0$.

   a. Rewrite the equation identified in Item 3a in Lesson 33-1 and set it equal to 0.
   $-16t^2 + 64t + 512 = 0$

   b. Completely factor the equation.
   $-16(t - 8)(t + 4) = 0$

   c. **Make use of structure.** Identify and use the appropriate property to find the time that the *Spirit* took to strike Earth after the engine burned out.
   The Zero Product Property can be used.
   $(t - 8) = 0$ $(t + 4) = 0$
   $t = 8$ $t = -4$
   The solution $t = 8$ seconds is the only solution in the reasonable domain of the function.

2. Draw a horizontal line on the graph in Item 1 in Lesson 33-1 to indicate a height of 544 ft above Earth. Estimate the approximate time(s) that the *Spirit* was 544 ft above Earth.
   Estimates may vary, but should be approximately 0.5 seconds and 3.5 seconds.

488 SpringBoard® Mathematics **Algebra 1, Unit 5** • Quadratic Functions

Lesson 33-2
Interpreting Solutions of Quadratic Equations

3. The time(s) that the *Spirit* was 544 ft above Earth can be determined exactly by finding the values of t that make $h(t) = 544$.
   a. Rewrite the equation from Item 3a in Lesson 33-1 and set it equal to 544.
   $-16t^2 + 64t + 512 = 544$

   b. Is the method of factoring effective for solving this equation? Justify your response.
   Answers may vary. Factoring is not an effective method for solving this problem, because the quadratic that remains after factoring out -16 cannot be factored easily.

   c. Is the quadratic formula effective for solving this problem? Justify your response.
   Answers may vary. The quadratic formula is a more effective method for solving this problem, because the quadratic formula can be used to solve any quadratic equation.

   d. Determine the time(s) that the rocket was 544 ft above Earth. Round your answer to the nearest thousandth of a second. Verify that this solution is reasonable compared to the estimated times from the graph in Item 2.
   $t = \frac{4 \pm \sqrt{8}}{2} = 2 \pm \sqrt{2} \approx 2 \pm 1.414$ $t \approx 0.586$ seconds and $t \approx 3.414$ seconds.

   e. **Attend to precision.** Explain why it is more appropriate in this context to round to the thousandths place rather than to use the exact answer or an approximation to the nearest whole number.
   Answers may vary. Approximate decimal values give a measure that can be understood and used to evaluate the data. Rounding to the nearest whole number does not give an accurate enough measure.

4. Cooper could not see the *Spirit* when it was higher than 528 ft above Earth.
   a. Calculate the values of t for which $h(t) = 528$.
   $t = 2 + \sqrt{3} \approx 3.732$ seconds and $t = 2 - \sqrt{3} \approx 0.268$ seconds

   b. **Reason quantitatively.** Write an inequality to represent the values of t for which the rocket was not within Cooper's sight.
   $2 - \sqrt{3} < t < 2 + \sqrt{3}$ or $0.268 < t < 3.732$

CONNECT TO AP
In AP Calculus, calculations are approximated to the nearest thousandth.

MINI-LESSON: Solving a Quadratic Equation by Graphing

Students can also solve the quadratic equation in Item 3 by using a graphing calculator. A mini-lesson is available to demonstrate this process.

See SpringBoard's eBook Teacher Resources for a student page for this mini-lesson.

ACTIVITY 33 Continued

2–3 (continued) Students calculate more accurate times by using the quadratic model from the previous lesson. In Items 3b and 3c, students must justify their answers regarding effective methods of solving the quadratic equation.

TEACHER to TEACHER
When solving equations using the quadratic formula, it is acceptable to divide out a common factor of the terms to make the computations easier. In Item 3, for instance, after students subtract 544 from both sides to get $-16t^2 + 64t - 32 = 0$, they can divide both sides by -16 to get $t^2 - 4t + 2 = 0$. Applying the quadratic formula to the last equation gives the same results as the original equation and is easier to compute.

CONNECT TO AP
As stated on the student page, decimal answers on the AP Calculus Exam must be given to three decimal places unless otherwise stated in a problem. Caution students not to round intermediate steps when using a calculator, but to round only the final answer.

TEACHER to TEACHER
If your class does not have access to technology, you can either have students leave their answers in simplest radical form in this lesson or have students use guess and check to estimate square roots to the nearest tenth.

4 Create Representations, Think-Pair-Share, Debriefing This item can be used as formative assessment as students again solve a quadratic equation to determine when the rocket was at a given height. They then write an inequality to express the values of t for which the rocket was above the given height. As this portion of the lesson is debriefed, ask students to explain why they chose the method they did to solve the quadratic equation. Is it always appropriate to use the quadratic formula?

Activity 33 • Applying Quadratic Equations

ACTIVITY 33 Continued

Check Your Understanding

Debrief students' answers to these items to ensure that they understand how to solve real-world problems involving quadratic models. For Item 5, have students justify their choice of method for solving the resulting quadratic equation and explain why they eliminated one of the solutions.

Answers

5. $t \approx 5.372$ seconds
6. $0 < t < 2.8125$

ASSESS

Students' answers to lesson practice problems will provide you with a formative assessment of their understanding of the lesson concepts and their ability to apply their learning.

See the Activity Practice for additional problems for this lesson. You may assign the problems here or use them as a culmination for the activity.

LESSON 33-2 PRACTICE

7. $t \approx 8.558$ or $t \approx -2.308$
8. $t \approx 7.948$ or $t \approx -1.698$
9. 30 ft; $h(0) = 30$
10. $t = 0.25$ second; 31 ft
11. 1.642 seconds
12. $0 < t < 1.079$; Solve $20 = -16t^2 + 8t + 30$ to determine when the diver is at a height of 20 ft. She is at a height greater than 20 ft at all times before this.

ADAPT

Check students' answers to the Lesson Practice questions to make sure that they can solve problems involving quadratic equations and functions and interpret the solutions. For students having difficulty relating the quadratic functions to the problem situations, encourage the use of visualization. Ask students to describe the path of the object in question. When is the object "going up" and when is it "going down"? At what point does this movement change? Where on the graph is "the ground"?

ACTIVITY 33 continued

Lesson 33-2
Interpreting Solutions of Quadratic Equations

Check Your Understanding

The path of a rocket is modeled by $h(t) = -16t^2 + 45t + 220$, where h is the height in feet and t is the time in seconds.

5. Determine the time t when the rocket was on the ground. Round to the nearest thousandth.
6. Identify the times, t, when the height $h(t)$ was greater than 220 ft.

LESSON 33-2 PRACTICE

Solve the quadratic equations. Round to the nearest thousandth.

7. $-16t^2 + 100t + 316 = 0$
8. $-16t^2 + 100t + 316 = 100$

Make sense of problems. For Items 9–12, use the function $h(t) = -16t^2 + 8t + 30$, which represents the height of a diver above the surface of a swimming pool t seconds after she dives.

9. The diver begins her dive on a platform. What is the height of the platform above the surface of the swimming pool? How do you know?
10. At what time does the diver reach her maximum height? What is the maximum height?
11. How long does it take the diver to reach the water? Round to the nearest thousandth.
12. Determine the times at which the diver is at a height greater than 20 ft. Explain how you arrived at your solution.

Applying Quadratic Equations
Rockets in Flight

ACTIVITY 33 continued

ACTIVITY 33 PRACTICE
Write your answers on notebook paper.
Show your work.

Lesson 33-1

A model rocket is launched from the ground. Its height at different times after the launch is recorded in the table below. Use the table for Items 1–7.

Time Since Launch of Rocket (s)	Height of Rocket (ft)
1	144
2	256
3	336
4	384
5	400
6	384
7	336

1. Are the data in the table linear? Justify your response.
2. Use a graphing calculator to determine a quadratic function $h(t)$ for the data.
3. Identify a reasonable domain for the function within this context.
4. How high is the rocket after 8 seconds?
5. After how many seconds does the rocket come back to the ground?
6. What is the maximum height the rocket reached?
7. **a.** At what time(s) t will the height of the rocket be equal to 300 ft?
   b. How many times did you find in Part (a)? Explain why this makes sense in the context of the problem.

8. Which of the following functions best models the data in the table below?

x	f(x)
1	29
2	66
3	101
4	134
5	165
6	194

  A. $f(x) = x + 28$
  B. $f(x) = -x^2 + 40x - 10$
  C. $f(x) = -16x^2 + 10x + 100$
  D. $f(x) = -|x + 3| + 42$

9. As part of a fireworks display, a pyrotechnics team launches a fireworks shell from a platform and collects the following data about the shell's height.

Time Since Launch of Shell (s)	Height of Shell (ft)
1	68
2	100
3	100
4	68
5	4

Which of the following is a true statement about this situation?
  A. The launch platform is 68 ft above the ground.
  B. The maximum height of the shell is 100 ft.
  C. The shell hits the ground after 6 seconds.
  D. The shell starts to descend 2.5 seconds after it is launched.

ACTIVITY 33 Continued
ACTIVITY PRACTICE

1. No; the data are not linear since the change in time is constant while the change in height is not constant.
2. $h(t) = -16t^2 + 160t$
3. domain: $0 \leq t \leq 10$
4. 256 ft
5. 10 seconds
6. 400 ft
7. **a.** $t = 2.5$ seconds and $t = 7.5$ seconds
   b. two times; The rocket is at a height of 300 ft once on the way up and once on the way down.
8. B
9. D

ACTIVITY 33 Continued

10. 4 ft; $h(0) = 4$
11. $h(t)$

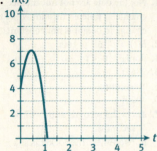

12. $t = 0.4375$ second
13. 7.0625 ft
14. $t \approx 1.102$ seconds
15. Domain: $0 \leq t \leq 1.102$; range: $0 \leq h \leq 7.0625$
16. $0.180 < t < 0.695$; Solving $-16t^2 + 14t + 4 = 6$ gives the times when the ball is at a height of exactly 6 ft; it is higher than 6 ft between these times.
17. ≈ 3.449 seconds
18. No; the equation $-16x^2 + 32x + 80 = 100$ has no real solutions because the discriminant of the equation $-16x^2 + 32x - 20 = 0$ is negative.
19. C
20. B
21. always
22. always
23. never
24. sometimes
25. sometimes
26. never
27. Sample answer: Free fall motion is not modeled by linear functions because of the acceleration due to gravity. This causes the rate of change of the height to not be constant.

ADDITIONAL PRACTICE

If students need more practice on the concepts in this activity, see the eBook Teacher Resources for additional practice problems.

ACTIVITY 33 continued

Applying Quadratic Equations
Rockets in Flight

Lesson 33-2

A soccer player passes the ball to a teammate, and the teammate kicks the ball. The function $h(t) = -16t^2 + 14t + 4$ represents the height of the ball, in feet, t seconds after it is kicked. Use this information for Items 10–16.

10. What is the height of the ball at the moment it is kicked? Justify your answer.
11. Graph the function.
12. Calculate the time at which the ball reaches its maximum height.
13. What is the maximum height of the ball?
14. Assuming no one touches the ball after it is kicked, determine the time when the ball falls to the ground.
15. Identify a reasonable domain and range for the function.
16. Determine the times when the ball is higher than 6 ft. Explain how you arrived at your solution.

Casey is standing on the roof of a building. She tosses a ball into the air so that her friend Leon, who is standing on the sidewalk, can catch it. The function $y = -16x^2 + 32x + 80$ models the height of the ball in feet, where x is the time in seconds. Use this information for Items 17–20.

17. Leon lets the ball hit the sidewalk. Determine the total time the ball is in the air until it hits the sidewalk.
18. Is the ball ever at a height of 100 ft? Justify your answer.
19. How many times is the ball at a height of exactly 92 ft?
    A. never
    B. one time
    C. two times
    D. three times
20. Casey solves the equation shown below. What does the solution of the equation represent?

    $$10 = -16x^2 + 32x + 80$$

    A. The height of the ball after 10 seconds
    B. The time when the ball is at a height of 10 ft
    C. The time when the ball has traveled a total distance of 10 ft
    D. The time it takes the ball to rise vertically 10 ft from the rooftop

The function $h(t) = -16t^2 + 50t + k$, where $k > 0$, gives the height, in feet, of a marble t seconds after it is shot into the air from a slingshot. Determine whether each statement is always, sometimes, or never true.

21. The initial height of the marble is k feet.
22. There is some value of t for which the height of the marble is 0.
23. The graph of the function is a straight line.
24. The marble reaches a height of 50 ft.
25. The marble reaches a height of 65 ft.
26. The maximum height of the marble occurs at $t = 1$ second.

MATHEMATICAL PRACTICES
Reason Abstractly and Quantitatively

27. Why do you think quadratic functions are used to model free-fall motion instead of linear functions?

Solving Quadratic Equations
EGG DROP

Embedded Assessment 2
Use after Activity 33

Every fall the Physics Club hosts an annual egg-drop contest. The goal of the egg-drop contest is to construct an egg-protecting package capable of providing a safe landing upon falling from a fifth-floor window.

During the egg-drop contest, each contestant drops an egg about 64 ft to a target placed at the foot of a building. The area of the target is about 10 square feet. Points are given for targeting, egg survival, and time to reach the target.

Colin wanted to win the egg-drop contest, so he tested one of his models with three different ways of dropping the package. These equations represent each method.

Method A	$h(t) = -16t^2 + 64$
Method B	$h(t) = -16t^2 - 8t + 64$
Method C	$h(t) = -16t^2 - 48t + 64$

1. Quadratic equations can be solved by using square roots, by factoring, or by using the quadratic formula. Solve the three equations above to find t when $h(t) = 0$. Use a different solution method for each equation. Show your work, and explain your reasoning for choosing the method you used.

2. Colin found that the egg would not break if it took longer than 1.5 seconds to hit the ground. Which method(s)—A, B, or C—will result in the egg not breaking?

3. Colin tried another method. This time he recorded the height of the egg at different times after it was dropped, as shown in the table below.

Elapsed Time Since Drop of Egg (s)	Height of Egg (ft)
0	64
0.25	62
0.5	58
0.75	52
1.00	44
1.25	34

   a. Use a graphing calculator to determine a function $h(t)$ that models the quadratic data.
   b. After how many seconds will the egg hit the ground? Show the work that justifies your answer mathematically.
   c. At what time, t, will the egg be 47 feet above the ground? Show the work that justifies your answer mathematically.

Embedded Assessment 2
Assessment Focus

- Solving quadratic equations by factoring
- Solving quadratic equations by the square root method
- Solving quadratic equations using the quadratic formula
- Choosing a method to solve a quadratic equation
- Writing the equation of a quadratic function to fit data
- Using a quadratic model to solve problems
- Interpreting solutions of a quadratic equation

Answer Key

1. Method A: $t = 2$ seconds; Method B: $t \approx 1.766$ seconds; Method C: $t = 1$ second
2. Methods A and B will result in the egg not breaking.
3. a. $h(t) = -16t^2 - 4t + 64$
   b. $t \approx 1.879$ seconds
   c. $t \approx 0.913$ second

Common Core State Standards for Embedded Assessment 2

HSN-Q.A.3	Choose a level of accuracy appropriate to limitations on measurement when reporting quantities.
HSA-SSE.B.3	Choose and produce an equivalent form of an expression to reveal and explain properties of the quantity represented by the expression.*
HSA-SSE.B.3a	Factor a quadratic expression to reveal the zeros of the function it defines.
HSA-CED.A.1	Create equations and inequalities in one variable and use them to solve problems. Include equations arising from linear and quadratic functions and simple rational and exponential functions.
HSA-REI.B.4	Solve quadratic equations in one variable.
HSA-REI.B.4b	Solve quadratic equations by inspection (e.g., for $x^2 = 49$), taking square roots, completing the square, the quadratic formula and factoring, as appropriate to the initial form of the

Embedded Assessment 2

TEACHER to TEACHER

You may wish to read through the scoring guide with students and discuss the differences in the expectations at each level. Check that students understand the terms used.

Unpacking Embedded Assessment 3

Once students have completed this Embedded Assessment, turn to Embedded Assessment 3 and unpack it with them. Use a graphic organizer to help students understand the concepts they will need to know to be successful on Embedded Assessment 3.

Embedded Assessment 2
Use after Activity 33

Solving Quadratic Equations
EGG DROP

Scoring Guide	Exemplary	Proficient	Emerging	Incomplete
	The solution demonstrates the following characteristics:			
Mathematics Knowledge and Thinking (Item 1)	• Effective understanding of and accuracy in solving quadratic equations	• Adequate understanding of how to solve quadratic equations, leading to solutions that are usually correct	• Difficulty solving quadratic equations	• Inaccurate or incomplete understanding of how to solve quadratic equations
Problem Solving (Item 2)	• Appropriate and efficient strategy that results in a correct answer	• Strategy that may include unnecessary steps but results in a correct answer	• Strategy that results in some incorrect answers	• No clear strategy when solving problems
Mathematical Modeling/ Representations (Item 3)	• Clear and accurate understanding of how to use technology to model real-world data and how to use a graph to solve a real-world problem	• Some difficulty understanding how to use technology to model real-world data and/or how to use a graph to solve a real-world problem, but a correct answer is present	• Partial understanding of how to use technology to model real-world data and/or how to use a graph to solve a real-world problem that results in some incorrect answers	• Little or no understanding of how to use technology to model real-world data and/or how to use a graph to solve a real-world problem
Reasoning and Communication (Items 1, 3b, 3c)	• Precise use of appropriate math terms and language to explain a choice of solution method • Clear and accurate use of mathematical work to justify an answer	• Adequate explanation of choice of solution method • Correct use of mathematical work to justify an answer	• Misleading or confusing explanation of choice of solution method • Partially correct justification of an answer using mathematical work	• Incomplete or inaccurate explanation of choice of solution method • Incorrect or incomplete justification of an answer using mathematical work

Common Core State Standards for Embedded Assessment 2 (cont.)

	equation. Recognize when the quadratic formula gives complex solutions and write them as $a \pm bi$ for real numbers a and b.
HSF-IF.B.5	Relate the domain of a function to its graph and, where applicable, to the quantitative relationship it describes.*
HSF-IF.C.8	Write a function defined by an expression in different but equivalent forms to reveal and explain different properties of the function.
HSF-IF.C.8a	Use the process of factoring and completing the square in a quadratic function to show zeros, extreme values, and symmetry of the graph, and interpret these in terms of a context.
HSF-BF.A.1	Write a function that describes a relationship between two quantities.*

Modeling with Functions

Photo App
Lesson 34-1 Constructing Models

Learning Targets:
- Construct linear, quadratic, and exponential models for data.
- Graph and interpret linear, quadratic, and exponential functions.

> **SUGGESTED LEARNING STRATEGIES:** Discussion Groups, Look for a Pattern, Create Representations, Sharing and Responding, Construct an Argument

Jenna, Cheyenne, and Kim all use a photo app on their smartphones to share their photos online and to track how many people follow their photo stories. The people who choose to follow the girls' photo stories can also stop following or "unsubscribe" at any time. After the first eight months, the data for each of the girl's monthly number of followers was compiled and is shown in the table below.

Months Since Account Was Opened	Jenna's Total Followers	Cheyenne's Total Followers	Kim's Total Followers
1	37	2	16
2	42	5	29
3	46	9	38
4	52	16	42
5	56	33	43
6	63	65	39
7	66	128	31
8	72	251	20

1. Describe any patterns you observe in the table. As you share your ideas with your group, be sure to use mathematical terms and academic vocabulary precisely. Make notes to help you remember the meaning of new words and how they are used to describe mathematical concepts.

   Answers may vary. The total number of Jenna's followers increases by 3 to 6 people every month. The total number of Cheyenne's followers comes close to doubling every month. The total number of Kim's followers increases and then decreases.

Common Core State Standards for Activity 34

- **HSF-IF.C.7:** Graph functions expressed symbolically and show key features of the graph, by hand in simple cases and using technology for more complicated cases.
- **HSF-IF.C.7a:** Graph linear and quadratic functions and show intercepts, maxima, and minima.
- **HSF-IF.C.7b:** Graph square root, cube root, and piecewise-defined functions, including step functions and absolute value functions.
- **HSF-IF.C.7e:** Graph exponential and logarithmic functions, showing intercepts and end behavior, and trigonometric functions, showing period, midline, and amplitude.

ACTIVITY 34
Investigative

Activity Standards Focus
In this activity, students will connect what they know about linear, quadratic, exponential and piecewise functions by comparing, contrasting, writing and interpreting appropriate models.

Lesson 34-1

PLAN

Pacing: 1 class period
Chunking the Lesson
#1–3   #4–5   #6
Check Your Understanding
Lesson Practice

TEACH

Bell-Ringer Activity
Ask students to create a table with three columns. They should use one column to list attributes of linear functions, one column to list the attributes of quadratic functions, and one column to list attributes of exponential functions.

1–3 Discussion Groups, Look for a Pattern, Create Representations, Sharing and Responding Focus students' thinking by asking questions that relate to reading the table. For example: which girl had the greatest total of followers for the most months? (Jenna, for 5 months) There are many possible patterns in the table. As students share, record all the patterns that are mentioned for students to refer to as they work through the activity.

ACTIVITY 34 Continued

1–3 (continued) As students continue to share, be sure they are able to justify their reasoning and the reasonableness of their solutions when choosing a function to model each girl's data. Remind students to use specific details and precise mathematical language in their justifications.

If necessary, review how to use the regression function on a graphing calculator. Writing correct functions is crucial for the remaining questions in the lesson.

4–5 Discussion Groups, Construct an Argument Monitor student discussions carefully to be sure students are using the information from Items 1–3 to respond appropriately. Students should recognize slope as the growth rate in the linear function modeling Jenna's followers and 1.964 as the growth factor in Cheyenne's function.

Differentiating Instruction

Support students who have a difficult time responding to Items 4 and 5. Review prior concepts, including the meaning of slope in a real-world context and the meaning of the base in an exponential function. Remind students that they can refer back to the data in the table to help them answer the Items. Students should have an understanding of how to use both the tables *and* the equations to answer the items.

Developing Math Language

As students respond to questions or discuss possible solutions to problems, monitor their use of new terms and descriptions of applying math concepts to ensure that they understand and are able to use language correctly and precisely.

ACTIVITY 34 continued

Lesson 34-1
Constructing Models

My Notes

2. **Reason quantitatively.** Use the terms *linear*, *quadratic*, or *exponential* to identify the type of function that would best model each girl's total number of followers during the first eight months.
   a. Jenna: linear
   b. Cheyenne: exponential
   c. Kim: quadratic

3. Use the regression function of a graphing calculator to write a function that could be used to model each girl's total number of followers over the first eight months.
   a. Jenna: $y = 5x + 31.75$
   b. Cheyenne: $y = 1.136(1.964)^x$
   c. Kim: $y = -2.048x^2 + 18.929x - 0.714$

4. Approximately how many followers does Jenna gain each month? Justify your response.
   Jenna gains about 5 followers each month. This equals the slope of the function $y = 5x + 31.75$.

5. **Critique the reasoning of others.** Cheyenne tells the other girls she thinks her following is almost doubling each month. Is she correct? Justify your response using Cheyenne's function or the data in the table.
   Yes; the base of the function $y = 1.136(1.964)^x$ is 1.964, which is approximately 2. This matches the doubling amount in the data.

6. **Model with mathematics.** Use the functions from Item 3 to create graphs to represent each of the girl's total number of followers over the first eight months. Label at least three points on each graph.

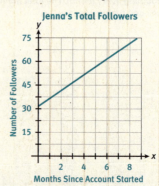

Jenna's Total Followers

Common Core State Standards for Activity 34 *(continued)*

HSF-IF.C.8:	Write a function defined by an expression in different but equivalent forms to reveal and explain different properties of the function.
HSF-IF.C.8a:	Use the process of factoring and completing the square in a quadratic function to show zeros, extreme values, and symmetry of the graph, and interpret these in terms of a context.
HSF-BF.A.1:	Write a function that describes a relationship between two quantities.
HSF-BF.A.1b:	Combine standard function types using arithmetic operations.
HSF-LE.A.3:	Observe using graphs and tables that a quantity increasing exponentially eventually exceeds a quantity increasing linearly, quadratically, or (more generally) as a polynomial function.

496 SpringBoard® Mathematics **Algebra 1, Unit 5** • Quadratic Functions

Lesson 34-1
Constructing Models

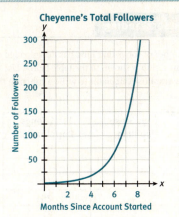

Cheyenne's Total Followers

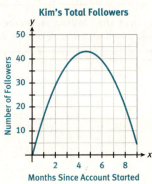

Kim's Total Followers

a. Describe the similarities and differences between the reasonable domains and ranges of each of the functions represented by the graphs.
   The domains of all three functions are the same: $x \geq 0$ (months).
   The ranges are not all the same:
   Jenna: $y \geq 0$ (people)
   Cheyenne: $y \geq 0$ (people)
   Kim: $0 \leq y \leq 43$ (people)

b. Identify the maximum values, if they exist, of each of the functions represented by the graphs.
   Jenna: no maximum value
   Cheyenne: no maximum value
   Kim: maximum value of 43 people

ACTIVITY 34 Continued

6 Discussion Groups, Create Representations, Debriefing

Students can draw the graphs on grid paper or use a calculator. Students using grid paper will be able to easily compare and contrast the functions. Students should recognize that while the domains of all three functions are the same, the ranges are not. While the data suggest that Jenna and Cheyenne could have an unlimited number of followers, Kim is restricted to 43 or fewer followers.

Students should understand that the domain represents the number of months and the range represents the number of people. For advanced learners, consider asking if the domain and the range values must be restricted to the positive integers. Debrief this portion of the lesson by asking students to use precise mathematical vocabulary to describe the real-world situations modeled here.

TEACHER to TEACHER

Words and terms are added to the classroom Word Wall frequently. To remind students to refer to the Word Wall, ask them to choose words to add to any of the descriptions. Another way to reinforce language acquisition is to have each student choose a word from the Word Wall and then pair-share to discuss its meaning and use.

Activity 34 • Modeling with Functions 497

ACTIVITY 34 Continued

Check Your Understanding

Debrief with students by asking them to name the type of function that models the data. Ask how they know the number of songs roughly triples each day. Discuss why the data do not model a linear or a quadratic function.

Answers

7. $y = 0.332(3.011)^x$
8. The number of songs Ping downloads roughly triples each day.

Lesson 34-1
Constructing Models

Check Your Understanding

7. The total number of songs that Ping has downloaded since he joined an online music club is shown in the table below. Use the regression function of a graphing calculator to write a function that best models the data.

Days Since Joining	Total Songs Downloaded
1	1
2	3
3	9
4	28
5	81

8. Use your function from Item 7 to describe how the total number of songs that Ping has downloaded is changing each day.

Lesson 34-1
Constructing Models

LESSON 34-1 PRACTICE

Alice and Ahmad have started an in-home pet grooming business. They each recorded the number of clients they visited each month for the first seven months. Their data are shown in the table below.

Months in Business	Number of Alice's Clients	Number of Ahmad's Clients
1	5	0
2	7	1
3	10	2
4	11	4
5	13	8
6	15	15
7	18	32

9. Examine the data in the table to determine the type of function that would best model the number of each groomer's clients.

10. Use the regression function of a graphing calculator to find a function to represent the number of Alice's clients.

11. Use the regression function of a graphing calculator to find a function to represent the number of Ahmad's clients.

12. Graph the functions from Items 10 and 11 on separate graphs.

13. **Attend to precision.** Describe the similarities and differences between the two functions. Compare and contrast the following features:
    a. domain and range
    b. maximum and minimum values
    c. rate of increase per day
    d. groomer with the most clients during different months

ACTIVITY 34 Continued

ASSESS

Students' answers to lesson practice problems will provide you with a formative assessment of their understanding of the lesson concepts and their ability to apply their learning.

See the Activity Practice for additional problems for this lesson. You may assign the problems here or use them as a culmination for the activity.

LESSON 34-1 PRACTICE
9. Alice: linear
   Ahmad: exponential
10. $y = 2.071x + 3$
11. $y = 0.375(1.848)^x$
12.
    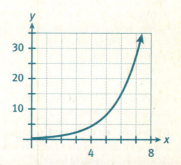

13. Answers will vary. The domain is the same for both of the groomers. The range for Alice's function is $3 \leq y$ and for Ahmad's function is $1 \leq y$. Ahmad's function has a smaller minimum value than Alice's, because he started with fewer clients. However, Ahmad eventually gained more clients per month than Alice did. Alice had more clients until the sixth month in which they were tied, but then Ahmad quickly gained more clients.

ADAPT

Check students' answers to the Lesson Practice to ensure that they can identify data that represent linear, quadratic, and exponential models. For students who are having difficulty identifying types of functions by examining data, graphing the data can help identify what kind of function models the data. Ask students to explain how they determined the answers to Item 13. Discuss whether the answers can be found algebraically, graphically, or in both ways.

ACTIVITY 34 Continued

Lesson 34-2

PLAN

Pacing: 1 class period
Chunking the Lesson
#1–5   #6   #7–9
Check Your Understanding
Lesson Practice

TEACH

Bell-Ringer Activity
Ask students to identify each function as linear, quadratic, exponential, or none of these.

a. $y = 5x^2 + 3$ [quadratic]
b. $y = 2x + 3^2$ [linear]
c. $y = 3x^2 + x - 1$ [quadratic]
d. $2^4(1.2)^x$ [exponential]
e. $y = 1$ [linear]
f. $y = \dfrac{3}{x}$ [none]

1–5 Visualization, Discussion Groups, Construct an Argument
Remind students they have three representations they can use to respond to these items—the data in the table, equations, and graphs. Students should understand how to use each representation to answer the items on this page and the next. If necessary, have a student re-read aloud the opening paragraph in Lesson 34-1.

6 Construct an Argument Students use what they know about the rates of growth of the different types of functions to determine whose followers will eventually outnumber the rest. It is important that students understand what *grew twice as quickly as initially thought* means in terms of the data before beginning this item.

Differentiating Instruction

To **support** students in using proper math language, especially as discussion focuses on increasingly more challenging concepts, provide linguistic support through translations of key terms and other language. Group students carefully to ensure participation of all group members in class discussions.

ACTIVITY 34 continued

Lesson 34-2
Comparing Models

My Notes

Learning Targets:
- Identify characteristics of linear, quadratic, and exponential functions.
- Compare linear, quadratic, and exponential functions.

SUGGESTED LEARNING STRATEGIES: Visualization, Discussion Groups, Construct an Argument, Create Representations

1. Rewrite the functions you found for Jenna, Kim, and Cheyenne in Lesson 34-1.

   Jenna: $y = 5x + 31.75$
   Cheyenne: $y = 1.136(1.964)^x$
   Kim: $y = -2.048x^2 + 18.929x - 0.714$

2. Which of the three girls—Jenna, Kim, or Cheyenne—has a constant rate of change in her number of followers per month? Explain.
   Jenna's data is the only data represented by a linear function, and therefore has a constant rate of change in the number of followers. She gains 5 followers every month.

3. Which girl had the greatest number of followers initially? Justify your response using both the functions and the graphs.
   Jenna started with the greatest number of followers. This corresponds to the y-intercept on the graph. And this is the value of the function when $x = 0$.

4. Did any of the girls experience followers who "unsubscribed" from, or stopped following, their photo story? Explain how you know.
   Yes; Kim had followers who unsubscribed. Kim's number of followers declines after the fifth month. Kim's data is modeled by a quadratic function that first increases and then decreases.

5. Which girl's photo story gained the most followers over the eight months? Justify your response using the functions or the graphs.
   Cheyenne gained the most followers over the entire eight months even though she started with the fewest followers.

6. **Critique the reasoning of others.** Kim states that even if the number of Jenna's followers had grown twice as quickly as it did, Cheyenne's followers would still eventually outnumber Jenna's followers. Is this assumption reasonable? Justify your response.
   Yes; the value of an exponential function will always eventually exceed the value of a linear function.

500 SpringBoard® Mathematics **Algebra 1, Unit 5** • Quadratic Functions

Lesson 34-2
Comparing Models

ACTIVITY 34
continued

My Notes

7. Write a new function that describes the total number of Jenna's and Kim's followers combined.
   $y = -2.048x^2 + 23.929x + 31.036$

8. Determine the reasonable domain and range, as well as any maximum or minimum values, of the function you wrote in Item 7.
   Domain: $x \geq 0$
   Range: $0 \leq y \leq 101$
   Maximum value: 101

9. **Construct viable arguments.** Will the number of Cheyenne's followers ever exceed the total number of Jenna and Kim's followers combined? Justify your response.
   Yes; the value of an exponential function will always eventually exceed the value of a quadratic function, especially because this function opens downward, and therefore eventually decreases to zero.

Check Your Understanding

Write the letter of the description that matches the given function.

10. $f(x) = -x^2 - 4x - 3$ A. has a constant rate of change
11. $f(x) = 4x - 3$ B. has a maximum value
12. $f(x) = 3(4)^x$ C. increases very quickly at an ever increasing rate

ACTIVITY 34 Continued

ASSESS

Students' answers to Lesson Practice problems will provide you with a formative assessment of their understanding of the lesson concepts and their ability to apply their learning.

See the Activity Practice for additional problems for this lesson. You may assign the problems here or use them as a culmination for the activity.

LESSON 34-2 PRACTICE
13. Roberto's Books
14. Roberto's Books; this Web site recommended *My Story* 4 times each day
15. It is possible for 1465 recommendations for a book to be made during one year.
16. Yes; since Tyler's Time to Read model is exponential, those number of recommendations will eventually exceed those of Roberto's Books.
17. $f(x) = -4x^2 + 3x + 14$

ADAPT

Check students' answers to the Lesson Practice to ensure that students can answer Items 13–16 using only a graph. Have them classify the functions shown on the graph as linear, quadratic, or exponential. For students having difficulty identifying the type of function, create a graphic organizer summarizing the representations of each function and the information generated from each representation. Students will get additional practice comparing and combining linear, quadratic, and exponential models in the Activity Practice.

ACTIVITY 34 continued

My Notes

MATH TIP

You can combine algebraic representations for two different functions by adding, subtracting, multiplying, or dividing the expressions.

Lesson 34-2
Comparing Models

LESSON 34-2 PRACTICE

The graphs show the number of times two online retailers—Roberto's Books and Tyler's Time to Read—recommended the bestselling book *My Story* to their customers after x days. Use the graphs for Items 13–16.

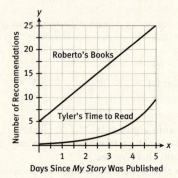

13. Who had given more recommendations of *My Story* after three days?
14. Who recommended *My Story* the same number of times each day? How many times was it recommended each day?
15. If this model continues, Roberto's Books will have recommended *My Story* approximately 1465 times after one year. Is this a reasonable amount? Justify your response.
16. Will the number of times Tyler's Time to Read recommends *My Story* ever exceed the number of times Roberto's Books recommends *My Story*? Explain your reasoning.
17. **Make use of structure.** The total number of times that Roberto's Books recommended the bestseller *A Fatal Memory* after x days is modeled by the function $f(x) = 3x + 2$. The total number of times that Penny's Place recommended the same book after x days is modeled by the function $g(x) = -4x^2 + 12$. Write a function to model the combined number of times that Roberto's Books and Penny's Place recommended *A Fatal Memory*.

Lesson 34-3
Extending Models

Learning Targets:
- Compare piecewise-defined, linear, quadratic, and exponential functions.
- Write a verbal description that matches a given graph.

SUGGESTED LEARNING STRATEGIES: Look for a Pattern, Create Representations, Critique Reasoning, Think-Pair-Share, Marking the Text

Rosa begins using the photo app to create a photo story and track her number of followers. The table below shows her results.

Months Since Account Was Started	Rosa's Total Followers
1	25
2	25
3	25
4	25
5	50
6	50
7	50
8	50

1. Describe any patterns you observe in the table.
   Answers may vary. Rosa has a constant number of followers for the first four months, then her followers double in the fifth month, and then remain constant again.

2. Graph the data in the table.

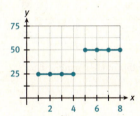

3. Write a function to model Rosa's number of followers for the first four months.
   f(*x*) = 25

4. Write a function to model Rosa's number of followers for months five through eight.
   f(*x*) = 50

ACTIVITY 34 Continued

3–6 Create Representations, Critique Reasoning Monitor student discussions carefully to be sure students are defining the domain and range of the piecewise function appropriately. It is especially important for students to indicate an understanding of the difference between *constant* and *constant rate of change*.

Differentiating Instruction

To **extend** learning, ask students to investigate a special kind of piecewise function called a step function. Ask them to determine whether the piecewise function in this lesson is a step function.

7–10 Discussion Groups, Create Representations, Look for a Pattern Some students may recognize that the coordinates of the vertex of the absolute value function (4, 45) are related to the numbers in the equation of the function, much like the coordinates of the vertex in a quadratic function. It is not necessary for students to know or understand that this is a transformation of the absolute value parent function $y = |x|$, but it can be discussed if students are curious about the relationship between the numbers.

ACTIVITY 34 continued

My Notes

MATH TERMS

A **piecewise-defined function** is a function that is defined differently for different disjoint intervals in its domain. For example, for the piecewise function
$$f(x) = \begin{cases} x \text{ when } x < 0 \\ 2 \text{ when } x \geq 0 \end{cases}, f(x) = x$$
when $x < 0$ and $f(x) = 2$ when $x \geq 0$.

**Lesson 34-3
Extending Models**

5. Write a *piecewise-defined function* to represent the number of followers Rosa has in any given month.

$$f(x) = \begin{cases} 25 \text{ when } 0 \leq x < 5 \\ 50 \text{ when } 5 \leq x \leq 8 \end{cases}$$

6. **Critique the reasoning of others.** Rosa says that her total number of followers is changing at a constant rate of 25 followers per month for the first four months. Is Rosa's statement correct? Explain your reasoning.
   Rosa is incorrect. The *number* of her followers is constant at 25 but the *rate of change* is 0.

Juanita has recorded the number of followers for her latest photo story over the last seven days. She finds that the function $f(x) = -5|x - 4| + 45$ represents the number of followers after x days.

7. Complete the table for the number of followers each day.

Days Since Photo Story Was Posted	Number of Followers
1	30
2	35
3	40
4	45
5	40
6	35
7	30

8. Graph the function that models Juanita's data.

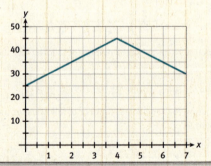

504 SpringBoard® Mathematics Algebra 1, Unit 5 • Quadratic Functions

Lesson 34-3
Extending Models

9. Determine the reasonable domain and range, as well as the maximum and minimum values of the function within the context of the problem.
   Domain: $x \geq 0$; range: $0 \leq y \leq 45$; maximum value: 45; minimum value: 0

10. **Reason abstractly.** Describe the similarities and differences between the function $f(x) = -5|x - 4| + 45$ and a quadratic function.
    This function is similar to a quadratic function in that it increases on some intervals and decreases on others, and it has a maximum value at its vertex. It is different in that the rate of change when it increases and decreases is constant.

11. The graphs below show the number of followers for two photo stories over one week. Describe how the number of followers changed over time.

    a.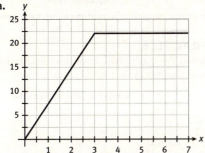
    The number of followers increased at a constant rate for the first three days and then remained constant for the rest of the week (no more followers subscribed).

    b.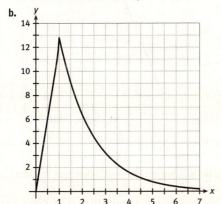
    The number of followers increased linearly for the first day and then decreased exponentially for the rest of the week.

ACTIVITY 34 Continued

Check Your Understanding
Debrief the lesson by asking them how to graph a quadratic equation. Students should describe different ways to help graph a function including making a table, setting the function equal to zero to find the x-intercepts, and using a calculator. Have students describe how to use a graph to determine the domain, range, and maximum or minimum of a quadratic function.

Answers
12. Up until hour 4
13. The number tripled each hour starting in hour 4.

ASSESS

Students' answers to Lesson Practice problems will provide you with a formative assessment of their understanding of the lesson concepts and their ability to apply their learning.

See the Activity Practice for additional problems for this lesson. You may assign the problems here or use them as a culmination for the activity.

LESSON 34-3 PRACTICE
14. Days 2 to 5
15. Days 2 to 7
16. The number of tickets that Mike sold remained constant from days 0 to 2, increased at a constant rate from days 2 to 5, and then remained constant from days 5 to 7.
17. The number of tickets Mike sold did not change for the first two days but then increased at a constant rate from days 2 through 5. After day 5, Mike did not sell any more tickets. Ryan started out the week selling tickets at a constant rate until day 2, after which he did not sell any more tickets.

ADAPT

Check students' answers to the Lesson Practice to ensure that students can identify all the properties of a quadratic function. Encourage students who are having difficulty examining entire functions or graphs to cover all but one "piece" of the graph or function and examine the function or graph one "piece" at a time. Students will get additional practice writing, graphing, and interpreting quadratic functions in the Activity Practice.

ACTIVITY 34 continued

My Notes

Lesson 34-3
Extending Models

Check Your Understanding

Use the graph for Items 12 and 13.

12. During which time period was the number of bacteria constant?

13. Describe the type of change in the number of bacteria for hours four through seven.

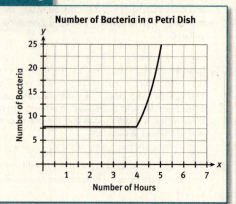

LESSON 34-3 PRACTICE

Make sense of problems. The graphs show the number of raffle tickets Mike and Ryan each sold to raise money for a school group.

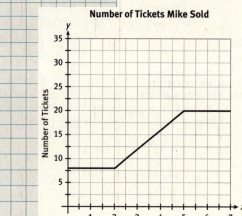 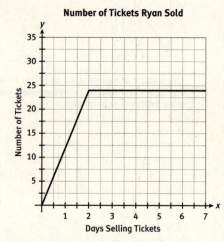

14. For which days did the number of raffle tickets that Mike sold increase at a constant rate?

15. For which days did the number of raffle tickets that Ryan sold not increase?

16. Describe any patterns you see in the number of tickets that Mike sold over the seven days.

17. Compare and contrast any patterns you observe in the number of tickets that Mike sold and the number of tickets Ryan sold over the seven days.

Modeling with Functions
Photo App

ACTIVITY 34
continued

ACTIVITY 34 PRACTICE
Write your answers on notebook paper.
Show your work.

Li and Alfonso have both opened accounts with an online music store. The data in the table below show the total number of songs each of them has downloaded since opening his account. Use the table for Items 1–12.

Days Since Account Opened	Total Number of Songs Downloaded by Li	Total Number of Songs Downloaded by Alfonso
1	1	5
2	2	15
3	5	26
4	8	34
5	16	45
6	33	56
7	65	67

Lesson 34-1

1. What type of function would best model the number of songs that Li has downloaded after x days?
   A. linear
   B. quadratic
   C. exponential
   D. absolute value

2. What type of function would best model the number of songs that Alfonso has downloaded after x days?
   A. linear
   B. quadratic
   C. exponential
   D. absolute value

3. Use the regression function of a graphing calculator to write a function that models the number of songs that Li has downloaded after x days.

4. Use the regression function of a graphing calculator to write a function that models the number of songs that Alfonso has downloaded after x days.

5. Determine the reasonable domain and range for each function.

Lesson 34-2

6. How would the number of songs downloaded by Alfonso change if the rate of change of Alfonso's downloads remained the same, but he had not downloaded any songs on day 1?

7. Describe the similarities and differences between the rates of change in the number of songs downloaded by the two boys.

8. If the models continue to represent the number of songs downloaded, who do you predict will have downloaded more songs after 30 days? Explain your reasoning.

9. If the rate of change of the number of songs downloaded by Alfonso tripled, how many songs will he have downloaded after 30 days? Is this a reasonable number?

10. How many songs will Li have downloaded after 30 days if his model continues? Is this a reasonable number?

ACTIVITY 34 Continued
ACTIVITY PRACTICE
1. C
2. A
3. $y = 0.528(1.992)^x$
4. $y = 10.25x - 5.571$
5. The domain of each is $x \geq 0$ (days); the range of each is $y \geq 0$ ("likes").
6. All of his daily totals, including the final total, would decrease by 5.
7. Both boys' "likes" increased, but Alfonso's rate of increase was approximately constant, and Li's was approximately exponential.
8. Li will win the competition, since the values of an exponential function will always eventually exceed a linear function, and after only 7 days they are almost equal, meaning that Li's total will soon exceed Alfonso's.
9. Approximately 917 downloads; this number is reasonable
10. Approximately 502,705,857 downloads; this number is unreasonable.

ACTIVITY 34 Continued

11. She downloads 10 songs on the first day, 20 more on the fifteenth day, and 15 additional songs on the twenty-fifth day.
12. The number of songs Alfonso downloaded increased by a constant number every day; Caily ordered additional songs only three particular days during this time.
13. $k(x) = -0.5x^2 + 14x + 32$
14. The maximum value is 130 fish.

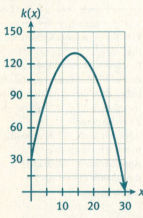

15. Since she received a rapid, constant increase in votes over days 0 to 10 and again over days 20 to 30, she may argue that the constant value (or no additional votes) during days 10 to 20 is not in line with the rest of her data; and this may mean that there was something wrong with the recording tool used.

ADDITIONAL PRACTICE

If students need more practice on the concepts in this activity, see the eBook Teacher Resources for additional practice problems.

ACTIVITY 34 continued

Modeling with Functions
Photo App

Lesson 34-3

Caily opened an account at the same time with the same online music store. The following piecewise function represents the total number of songs she has downloaded over the first x days.

$$f(x) = \begin{cases} 10 \text{ when } 1 \leq x < 15 \\ 30 \text{ when } 15 \leq x < 25 \\ 45 \text{ when } 25 \leq x \leq 30 \end{cases}$$

Use this function for Items 11 and 12.

11. Describe the number of songs Caily has downloaded during this time.
12. Describe the difference between the number of songs downloaded by Caily and by Alfonso over the first 30 days.
13. The functions $g(x) = 10x + 2$ and $h(x) = -0.5x^2 + 4x + 30$ represent the total numbers of two different types of fish in a pond over x weeks. Write a function $k(x)$ that represents the combined number of fish during the same period of time.
14. Graph the function $k(x)$ from Item 13. What is the maximum value of the function?

MATHEMATICAL PRACTICES
Construct Viable Arguments and Critique the Reasoning of Others

15. The graph shows the number of "likes" a blogger received for a blog post t days after it was posted.

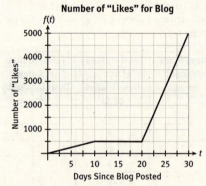

The blogger believes that there may have been something wrong with how the "likes" were recorded between days 10 and 20. Why might she believe this?

Systems of Equations
Population Explosion
Lesson 35-1 Solving a System Graphically

ACTIVITY 35

Learning Targets:
- Write a function to model a real-world situation.
- Solve a system of equations by graphing.

SUGGESTED LEARNING STRATEGIES: Marking the Text, Create Representations, Sharing and Responding, Discussion Groups, Visualization

Professor Hearst is studying different types of bacteria in order to determine new ways to prevent their population overgrowth. Each bacterium in the first culture that she examines divides to produce another bacterium once each minute. In the second culture, she observes that the number of bacteria increases by 10 bacteria each second.

1. Each population began with 10 bacteria. Complete the table for the population of each bacteria sample.

Elapsed Time (minutes)	Population of Sample A	Population of Sample B
0	10	10
1	20	610
2	40	1210
3	80	1810
4	160	2410
5	320	3010

CONNECT TO BIOLOGY
Bacteria can be harmful or helpful to other organisms, including humans. Bacteria are commonly used for food fermenting, waste processing, and pest control.

2. **Reason quantitatively.** Describe the type of function that would best model each population.
   Sample A: exponential
   Sample B: linear

ACTIVITY 35 Guided

Activity Standards Focus
In In this activity, students learn to solve linear/exponential and linear/quadratic systems by graphing and by using algebraic methods.

Lesson 35-1

PLAN
Materials
- graphing calculators

Pacing: 1 class period

Chunking the Lesson
#1–4   #5–6   #7
Check Your Understanding
Lesson Practice

TEACH
Bell-Ringer Activity
Instruct students to draw an example and give a basic equation for a linear function. Ask them to do the same for an exponential function and a quadratic function. Encourage students to refer to their math notebooks for assistance, if needed.

1–4 Shared Reading, Summarizing, Marking the Text, Sharing and Responding, Create Representations, Discussion Groups Students should mark important information as it is read in order to complete the table. Monitor group discussions carefully to be sure students are using the change in populations correctly. In order to compare the functions, students should write both with time in terms of minutes.

Common Core State Standards for Activity 35

HSA-REI.C.7: Solve a simple system consisting of a linear equation and a quadratic equation in two variables algebraically and graphically.

HSA-REI.D.11: Explain why the x-coordinates of the points where the graphs of the equations $y = f(x)$ and $y = g(x)$ intersect are the solutions of the equation $f(x) = g(x)$; find the solutions approximately, e.g., using technology to graph the functions, make tables of values, or find successive approximations. Include cases where $f(x)$ and/or $g(x)$ are linear, polynomial, rational, absolute value, exponential, and logarithmic functions. *

HSF-IF.C.9: Compare properties of two functions each represented in a different way (algebraically, graphically, numerically in tables, or by verbal descriptions).

HSF-LE.A.3: Observe using graphs and tables that a quantity increasing exponentially eventually exceeds a quantity increasing linearly, quadratically, or (more generally) as a polynomial function.

Activity 35 • Systems of Equations

ACTIVITY 35 Continued

1–4 (continued) Encourage students to check their functions by checking that they generate the same data as in the table. If they do not, student groups should determine whether the error is in the table or in the function. As students share responses, ask them to explain how they determined the number of bacteria in Sample B after each minute, given only seconds.

5–6 Discussion Groups, Create Representations, Sharing and Responding As students work through these problems, they draw conclusions about the solutions and explain what they mean in mathematical terms as well as in the context of the situation. Have students explain why it was necessary to determine the amount of growth of both populations over the same unit of time.

> **CONNECT TO TECHNOLOGY**
>
> Although students can adjust the window to see the points of intersection, it is difficult to get accurate approximations of the coordinates visually. Demonstrate how to use the trace function or the table feature to get a good estimation of the solutions of the system.

ACTIVITY 35 continued

Lesson 35-1
Solving a System Graphically

3. Write a function $A(t)$ to model the number of bacteria present in Sample A after t minutes.
   $A(t) = 10(2)^t$

4. Write a function $B(t)$ to model the number of bacteria present in Sample B after t minutes.
   $B(t) = 600t + 10$

5. **Use appropriate tools strategically.** Use a graphing calculator to graph $A(t)$ and $B(t)$ on the same coordinate plane.
   a. Sketch the graph below and label several points on each graph.

   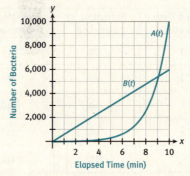

   b. Determine the points of intersection of the two graphs. Round non-integer values to the nearest tenth.
   (0, 10); (9.1, 5466.7)

   c. **Make use of structure.** What do the points of intersection indicate about the two graphs? Explain.
   At least one graph is not a line. Two distinct lines can intersect in at most one point, and these graphs intersect at two points.

   d. Interpret the meaning of the points of intersection within the context of the bacteria samples.
   The points of intersection represent the time t when the two populations of bacteria are equal. The populations of Samples A and B are equal at the beginning of Professor Hearst's observations, and then again after about 9.1 minutes.

6. Which bacteria population contains more bacteria? Explain.
   Between the beginning of Professor Hearst's observations and at about 9.1 minutes, Sample B had more bacteria. After this time, Sample A had more bacteria.

> **CONNECT TO TECHNOLOGY**
>
> You can use a graphing calculator to determine the point of intersection in several ways. You may choose to graph the functions and determine the point of intersection. You also may use the table feature to identify the value of x when y_1 and y_2 are equal.

Lesson 35-1
Solving a System Graphically

The solutions you found in Item 5b are solutions to the **nonlinear system of equations** $A(t) = 10(2)^t$ and $B(t) = 600t + 10$.

Just as with linear systems, you can solve nonlinear systems by graphing each equation and determining the intersection point(s).

7. Solve each system of equations by graphing.

   a. $y = 2x + 1$
      $y = x^2 + 1$
      (0, 0) and (2, 5)

   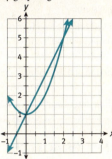

   b. $y = x - 1$
      $y = 3^x - 2$
      (0, −1)

   c. $y = 2(5)^x$
      $y = -x^2 + 2x - 2$
      No solution

   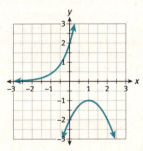

MATH TERMS

A **nonlinear system of equations** is a system in which at least one of the equations is not linear.

ACTIVITY 35 Continued

Developing Math Language
Have students apply their knowledge of the terms *nonlinear* and *system of equation* to grasp the meaning of a *nonlinear system of equations*. Have them write the definition in their math notebooks. Point out that a linear/exponential system is only one type of a nonlinear system.

7 Visualization, Create Representations, Activating Prior Knowledge, Debriefing
Before students graph each system, have them identify each equation as linear, quadratic, or exponential. Then have them visualize and sketch possible ways they could intersect. After students have solved the systems by graphing, compare the results with their sketches. As you debrief this lesson, focus on the number of times any two types of nonlinear functions might intersect. Ask students to give pictorial examples of linear/exponential, linear/quadratic, and exponential/quadratic intersections.

Differentiating Learning

Support students who have difficulty with the vocabulary they need to describe the graphs. Provide students with a word bank to use to describe the functions they are examining. Some words to include are *line*, *linear function*, *parabola*, *quadratic function*, and *exponential function*. Sketch the graph of a line that intersects the parabola at two points. Ask students to verbally describe the diagram using proper mathematical terminology. Provide other examples as needed for practice.

Activity 35 • Systems of Equations 511

ACTIVITY 35 Continued

Check Your Understanding

Students must be able to visualize different types of systems to answer Item 8. As you debrief students' answers to Items 9 and 10, make sure they have written $g(t)$ correctly before they find the points of intersection.

Answers

8. 0, 1, or 2 solutions are possible. This is different from a linear system because a linear system can have at most one solution.
9. $g(t) = 10(2)^t$
10. $t \approx 1.9$ minutes

ASSESS

Students' answers to lesson practice problems will provide you with a formative assessment of their understanding of the lesson concepts and their ability to apply their learning.

See the Activity Practice for additional problems for this lesson. You may assign the problems here or use them as a culmination for the activity.

LESSON 35-1 PRACTICE

11. (1, 2)
12. (1.4, 6.1) and (−1.4, −8.1)
13. no solution
14. Answers will vary. Students' graphs should show a line and a parabola that do not intersect.
15. Yes; graphs will vary. Students' graphs should show a line and an exponential curve that intersect in two points.
16. Fred is correct. The population is increasing at a constant rate and is therefore linear. The model for the population is $P(t) = 200 + 100t$. Francis has written an exponential function representing a population that is multiplied by 100 every minute.

ADAPT

If students cannot visualize various types of intersections, allow the use of graphing calculators to graph multiple examples of each type of system. Students should sketch and label each graph with the type of system it shows.

ACTIVITY 35 continued

Lesson 35-1
Solving a System Graphically

Check Your Understanding

8. Examine the graphs in Item 7. How many solutions are possible for a nonlinear system of linear, quadratic, and/or exponential equations? Describe how this is different from the number of possible solutions for a linear system.

A population of bacteria is given by $f(t) = t^2 + 2t + 30$, where t is in minutes. Another population begins with 10 bacteria and doubles every minute.

9. Write a function $g(t)$ to model the population of the second sample of bacteria.
10. Use a graph to determine at what time the two populations are equal.

LESSON 35-1 PRACTICE

For Items 11–13, solve each system of equations by graphing.

11. $y = -2x + 4$
    $y = -x^2 + 3$
12. $y = x^2 + 5x - 3$
    $y = -x^2 + 5x + 1$
13. $y = -4x^2 - 1$
    $y = 4(0.5)^x$

14. A nonlinear system contains one linear equation and one quadratic equation. The system has no solutions. Sketch a possible graph of this system.

15. Is it possible for a system with one linear equation and one exponential equation to have two solutions? If so, sketch a graph that could represent such a system. If not, explain why not.

16. **Critique the reasoning of others.** A population of 200 bacteria begins increasing at a constant rate of 100 bacteria per minute. Francis writes the function $P(t) = 200(100)^t$, where t represents the time in minutes, to model this population. Fred disagrees. He writes the function $P(t) = 200 + 100t$ to model this population. Who is correct? Justify your response.

Lesson 35-2
Solving a System Algebraically

Learning Targets:
- Write a system of equations to model a real-world situation.
- Solve a system of equations algebraically.

SUGGESTED LEARNING STRATEGIES: Create a Plan, Identify a Subtask, Construct an Argument, Close Reading, Note Taking, Visualization, Think-Pair-Share, Discussion Groups

Just as with linear systems, nonlinear systems of equations can be solved algebraically.

Example A
Solve the system of equations algebraically.

$$y = -x + 3$$
$$y = x^2 - 2x - 3$$

Step 1: The first equation is solved for y. Substitute the expression for y into the second equation.
$$y = x^2 - 2x - 3$$
$$(-x + 3) = x^2 - 2x - 3$$

Step 2: Solve the resulting equation using any method. In this case, solve by factoring.
$$0 = x^2 - x - 6$$
$$0 = (x + 2)(x - 3)$$

Step 3: Apply the Zero Product Property. $x + 2 = 0$ or $x - 3 = 0$

Step 4: Solve each equation for x. $x = -2$ or $x = 3$

Step 5: Calculate the corresponding y-values by substituting the x-values from Step 4 into one of the original equations.
When $x = -2$, $y = -(-2) + 3 = 5$
When $x = 3$, $y = -(3) + 3 = 0$

Step 6: Check the solution by substituting into the other original equation.
When $x = -2$, $y = (-2)^2 - 2(-2) - 3 = 5$
When $x = 3$, $y = (3)^2 - 2(3) - 3 = 0$

Solution: $(-2, 5)$ and $(3, 0)$

Try These A
Solve each system algebraically.

a. $y = -x + 2$
   $y = x^2 - x + 2$
   $(0, 2)$

b. $y = 2x^2 - 7$
   $y = 7x - 3$
   $(4, 25)$ and $(-0.5, -6.5)$

c. $y = -x + 3$
   $y = x^2 - 2x - 4$
   $(-2.2, 5.2)$ and $(3.2, -0.2)$

ACTIVITY 35 Continued

1 Construct an Argument, Activating Prior Knowledge As students develop their arguments, they need to consider many possibilities. When they cannot factor the equation using real numbers, they must remember that sometimes the roots are irrational. They need to apply what they know about discriminants to analyze the number of solutions. When they study the graph of the system, they must recall that the graphs of the functions extend infinitely and consider how they behave outside of the window shown.

Differentiating Instruction

Support student understanding of linear/quadratic systems. Ask them to visualize systems with different numbers of solutions. Ask them to sketch, if possible, the graphs of a line and a parabola that
- do not intersect
- intersect at one point
- intersect at two points
- intersect at three points.

If any sketches are not possible, have students explain why that is the case.

2 Visualization, Think-Pair-Share, Create a Plan, Identify a Subtask This item guides students through the steps necessary to solve the problem. Some students may benefit by sketching a visual representation of the situation. Students should determine that although there are two solutions to this system of equations, only one makes sense in the context of the problem.

ACTIVITY 35 continued

My Notes

Lesson 35-2
Solving a System Algebraically

1. Lauren solved the following system of equations algebraically and found two solutions.

$$y = -x + 3$$
$$y = x^2 - 3x - 4$$

Will solved the system by graphing and said that there is only one solution. Who is correct? Justify your response both algebraically and graphically.

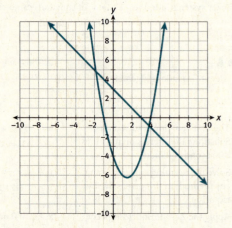

Lauren is correct. The graph shows two points of intersection. The equation $-x + 3 = x^2 - 3x - 4$ written in standard form is $x^2 - 2x - 7 = 0$. For this equation, the discriminant is positive, which means there are two real solutions.

2. **Model with mathematics.** Deshawn drops a ball from the 520-foot-high observation deck of a tower. The height of the ball in feet after t seconds is given by $f(t) = -16t^2 + 520$. At the moment the ball is dropped, Zoe begins traveling up the tower in an elevator that starts at the ground floor. The elevator travels at a rate of 12 feet per second. At what time will Zoe and the ball pass by each other?

   a. Write a function $g(t)$ to model Zoe's height above the ground after t seconds.

   $g(t) = 12t$

   b. Write a system of equations using the function modeling the height of the ball and the function you wrote in Part (a).

   $y = -16t^2 + 520$
   $y = 12t$

   c. Solve the system that you wrote in Part (b) algebraically. Round to the nearest hundredth, if necessary.

   $$t = \frac{-3 \pm \sqrt{9 - 16(-130)}}{8} \approx 5.34 \text{ or } -6.09$$

514 SpringBoard® Mathematics Algebra 1, Unit 5 • Quadratic Functions

Lesson 35-2
Solving a System Algebraically

   d. Interpret the meaning of the solution in the context of the problem. Does the solution you found in Part (c) make sense? Explain.
   The solution represents the time that Zoe and the ball pass by each other, so only the positive solution makes sense.

   e. Determine the height at which Zoe and the ball pass by each other. Explain how you found your answer.
   Evaluate either of the functions at $t = 5.34$; about 64 feet above the ground.

3. At the same moment that Deshawn drops the ball, Joey begins traveling down the tower in another elevator that starts at the observation deck. This elevator also travels at a rate of 12 feet per second.

   a. Write a function $h(t)$ to model Joey's height above the ground after t seconds. Explain any similarities or differences between this function and the function in Item 2a.
   $h(t) = -12t + 520$; the slope is now negative because the height is decreasing, and the constant is now 520 because that is Joey's initial height above the ground.

   b. Solve the system of equations algebraically.
   $t = 0$ or $t = 0.75$

   c. Interpret the solution in the context of the problem.
   This means that Joey and the ball are at the same height when the ball is first dropped, and again after 0.75 seconds.

   d. Construct viable arguments. Determine whether Joey or the ball reaches the ground first. Justify your response.
   The ball reaches the ground first. Students may justify this by setting each equation equal to zero and solving for t. Students may also justify their answer in this way: Between 0 and 0.75 seconds, Joey is closer to the ground. However, after 0.75 seconds, the ball is not only closer to the ground than Joey is, but the ball is also approaching the ground at an increasing rate, while Joey approaches the ground at a constant rate.

ACTIVITY 35 Continued

TEACHER to TEACHER
If students graph this system of equations, their graphs will not look the same as the sketches: their sketches show the paths taken by Zoe and the ball, while the graphs show the heights of Zoe and the ball relative to time.

3 Activating Prior Knowledge, Visualization, Discussion Groups, Create a Plan, Identify a Subtask, Construct an Argument, Debriefing
Students should be able to explain how the scenario differs from the previous one. Visual learners should make a new sketch to show the difference and then attempt to explain the difference algebraically. As students interpret the solution, they must apply algebraic thinking to a real-world situation. Have students explain why the two solutions to this system of equations make sense in the context of this problem, compared to the solutions for Item 2. When you debrief this lesson, it is important to discuss context in determining reasonable solutions to systems that model real-world situations.

Activity 35 • Systems of Equations

ACTIVITY 35 Continued

Check Your Understanding
As you debrief the answers, encourage students to share the methods they used to solve the system of equations in Item 4. Discuss the advantages of each method.

Answers
4. 0, 1, or 2
5. (4, 12) and (7, 15)

ASSESS

Students' answers to lesson practice problems will provide you with a formative assessment of their understanding of the lesson concepts and their ability to apply their learning.

See the Activity Practice for additional problems for this lesson. You may assign the problems here or use them as a culmination for the activity.

LESSON 35-2 PRACTICE
6. (2, 19)
7. no solution
8. $\left(\dfrac{-1+\sqrt{17}}{2}, \dfrac{-1+\sqrt{17}}{2}\right)$ and $\left(\dfrac{-1-\sqrt{17}}{2}, \dfrac{-1-\sqrt{17}}{2}\right)$

9. The solutions, or intercepts, do not fall on integers, which are easily read from the graph. The algebraic method of solving would provide a more precise answer instead of just an estimate.
10. 3 feet

ADAPT

If students have difficulty with these problems, ask them to begin by outlining a plan to determine the steps necessary to solve the problem at hand. Do they have all the information they need? Do they need to create equations? Would a graph or visual representation be helpful? What types of functions are necessary? What methods could they choose from to determine a solution? Students can refer to their math notebooks along the way to review the process of solving systems of equations algebraically. Once they solve the quadratic equations, discuss what one solution, two solutions, and no solution mean in terms of the graphs.

Lesson 35-2
Solving a System Algebraically

Check Your Understanding

4. How many solutions could the following system of equations have?
$$y = x + 8$$
$$y = x^2 - 10x + 36$$

5. Solve the system of equations in Item 4 using any appropriate algebraic method.

LESSON 35-2 PRACTICE
Solve each system of equations algebraically.

6. $y = 16x - 13$
   $y = 4x^2 + 3$

7. $y = 5$
   $y = -x^2 - x + 1$

8. $y = x$
   $y = x^2 + 2x - 4$

9. Jessica has decided to solve a system of equations by graphing. Her graph is shown. Why might she prefer to solve this system algebraically?

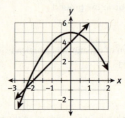

10. **Make sense of problems.** A competitive diver dives from a 33-foot high diving board. The height of the diver in feet after t seconds is given by $u(t) = -16t^2 + 4t + 33$. At the moment the diver begins her dive, another diver begins climbing the diving board ladder at a rate of 2 feet per second. At what height above the pool deck do the two divers pass each other?

Systems of Equations
Population Explosion

ACTIVITY 35 PRACTICE
Write your answers on notebook paper.
Show your work.

Lesson 35-1

1. Which function models the size of a neighborhood that begins with one home and doubles in size every year?
   A. $P(t) = t + 2$
   B. $P(t) = 2t + 1$
   C. $P(t) = 2(1)^t$
   D. $P(t) = (2)^t$

2. Which function models the size of a neighborhood that begins with 4 homes and increases by 6 homes every year?
   A. $P(t) = 4t + 6$
   B. $P(t) = 6t + 4$
   C. $P(t) = 6^t + 4$
   D. $P(t) = 4(6)^t$

3. When will the number of homes in Items 1 and 2 be equal?

For Items 4 and 5, sketch the graph of a system that matches the description. If no such system exists, write *not possible*.

4. The system contains a linear equation and an exponential equation. There are no solutions.

5. The system contains two exponential equations. There is one solution.

For Items 6–8, solve each system of equations by graphing.

6. $y = -x^2 + 3$
   $y = x^2 + 4$

7. $y = 0.5x^2 + x - 2$
   $y = x - 2$

8. $y = 2x^2 + 5$
   $y = -2x + 5$

9. Twin sisters Tamara and Sandra each receive $50 as a birthday present.

   Tamara puts her money into an account that pays 3% interest annually. The amount of money in Tamara's account after x years is given by the function $t(x) = 50(1.03)^x$.

   Sandra puts her money into a checking account that does not pay interest, but she plans to deposit $50 per year into the account. The amount of money in Sandra's account after x years is given by the function $s(x) = 50x + 50$.

   Use a graphing calculator to graph this system of equations. When will Tamara and Sandra have an equal amount of money in their accounts?

10. Josie and Jamal sold granola bars as a fund raiser, and they each started with 128 granola bars to sell.

    Josie sold 30 granola bars every day. The number of granola bars that Josie sold after x days is given by the function $y = -30x + 128$.

    Every day, Jamal sold half the number of the granola bars than he sold the day before. The number of granola bars that Jamal sold after x days is given by the function $y = 128(0.5)^x$.

    Use a graphing calculator to graph this system of equations. After how many days did Josie and Jamal have the same number of unsold granola bars?

ACTIVITY PRACTICE

1. D
2. B
3. about 5.1 years
4. Answers will vary. Students' graphs should show a line and an exponential curve that do not intersect.

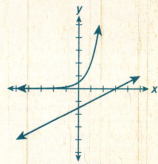

5. Answers will vary. Students' graphs should show two exponential curves that intersect in one point.

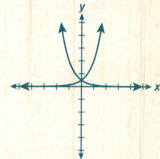

6. no solution
7. (0, 2)
8. (−1, 7) and (0, 5)
9. at the beginning and after 175 years
10. at the beginning and after 4 days

ACTIVITY 35 Continued

11. after 2 minutes
12. at the beginning and after raking 6 bags
13. $\frac{3 \pm \sqrt{69}}{6}, \frac{12 \pm \sqrt{69}}{3}$
14. (8, −17) and (10, 19)
15. (2, 8)
16. (1, 6) and (−1, 0)
17. (0, 0) and $\left(\frac{3}{2}, \frac{9}{2}\right)$
18. after 16 seconds
19. at about 0.7 second and 2.7 seconds
20. Answers may vary. The quadratic formula will work for systems that contain a linear equation and a quadratic equation, or two quadratic equations. But the quadratic formula will not work for systems that contain other nonlinear equations, such as exponential equations.

ADDITIONAL PRACTICE

If students need more practice on the concepts in this activity, see the eBook Teacher Resources for additional practice problems.

ACTIVITY 35 continued

Systems of Equations
Population Explosion

For Items 11 and 12, write a system of equations to model the situation. Then solve the system by graphing.

11. A sample of bacteria starts with 2 bacteria and doubles every minute. Another sample starts with 4 bacteria and increases at a constant rate of 2 bacteria every minute. When will the populations be equal?

12. Jennie and James plan to save money by raking leaves. Jennie already has 1 penny. With each bag of leaves she rakes, she doubles the amount of money she has. James earns 10 cents per bag. When will Jennie have more money than James?

Lesson 35-2

For Items 13–17, solve each system of equations algebraically.

13. $y = 3x^2 - x - 2$
    $y = 2x + 3$

14. $y = x^2 - 81$
    $y = 18x - 161$

15. $y = x^2 + 4$
    $y = 4x$

16. $y = 3x + 3$
    $y = x^2 + 3x + 2$

17. $y = 3x$
    $y = 2x^2$

For Items 18 and 19, write a system of equations to model the scenarios. Then solve the system of equations algebraically.

18. Simone is driving at a rate of 60 mi/h on the highway. She passes Jethro just as he begins accelerating onto the highway from a complete stop. The distance that Jethro has traveled in feet after t seconds is given by the function $f(t) = 5.5t^2$. When will Jethro catch up to Simone? (*Hint:* Use the fact that 60 mi/h is equivalent to 88 ft/s to write a function that gives the distance Simone has traveled.)

19. Jermaine is playing soccer next to his apartment building. He kicks the ball such that the height of the ball in feet after t seconds is given by the function $g(t) = -16t^2 + 48t + 2$. At the same moment that he kicks the ball, Jermaine's father begins descending in the elevator from his apartment at a rate of 5 ft/s. The apartment is 30 feet above the ground floor. At what time(s) are Jermaine's father and the soccer ball at the same height?

MATHEMATICAL PRACTICES
Construct Viable Arguments and Critique the Reasoning of Others

20. Rachel is solving systems of equations and has concluded that the quadratic formula is always an appropriate solution method when solving a nonlinear system algebraically. Do you agree with Rachel's conclusion? Use examples to support your reasoning.

Solving Systems of Equations
SPORTS COLLECTOR

Embedded Assessment 3
Use after Activity 35

Emilio loves sports and sports memorabilia. He has collected many different items over the years, but his favorite items are a signed baseball card, the catcher's mitt of his favorite catcher from 1979, and a vintage pennant from his favorite team. Emilio enjoys keeping track of how much his items are worth and has tracked their values for the last 10 years. The table below shows the values he has recorded thus far.

Years Since Item Acquired	Signed Baseball Card	Catcher's Mitt	Vintage Pennant
1	$50.00	$30.00	$6.00
2	$53.00	$37.00	$13.00
3	$56.00	$45.00	$22.00
4	$59.00	$55.00	$33.00
5	$62.00	$68.00	$46.00
6	$65.00	$83.00	$61.00
7	$68.00	$101.00	$78.00
8	$71.00	$124.00	$97.00
9	$74.00	$152.00	$118.00
10	$77.00	$186.00	$141.00

1. Identify the type of function that can be used to represent the value of each of the items shown in the table. Use the regression functions of a graphing calculator to determine functions $S(t)$, $C(t)$, and $P(t)$ that model the value of the signed baseball card, the catcher's mitt, and the vintage pennant, respectively.

2. In addition to the items in the table, Emilio also has a baseball that he caught during his favorite game of all time. The function $B(t) = -100|t-5| + 1300$ can be used to model the value of the ball. Graph this function and each of the other three functions on separate coordinate planes.

3. Identify the appropriate domain for each function. Justify your responses.

4. Which item reached the greatest value during the last 10 years? What is that maximum value? Justify your response.

5. Which item's value is changing at a constant rate? Support your response using a graph or the table.

6. Which item or items have a decreasing value? For what values of t do the function(s) decrease? Justify your response.

7. Which item's value is increasing the fastest? Explain your reasoning.

8. After how many years will the signed baseball card be worth more than the catcher's mitt? Use graphs to justify your response.

9. Create a system of equations to represent the relationship between the values of the signed baseball card and the vintage pennant. Solve the system algebraically. Interpret the solution within the context of the sports memorabilia. Describe how the values of the two items compare before and after the time represented by the system's solution.

Common Core State Standards for Embedded Assessment 3

HSN-Q.A.3	Choose a level of accuracy appropriate to limitations on measurement when reporting quantities.
HSA-REI.C.7	Solve a simple system consisting of a linear equation and a quadratic equation in two variables algebraically and graphically.
HSF-IF.B.5	Relate the domain of a function to its graph and, where applicable, to the quantitative relationship it describes.*
HSF-BF.A.1	Write a function that describes a relationship between two quantities.*
HSF-LE.A.3	Observe using graphs and tables that a quantity increasing exponentially eventually exceeds a quantity increasing linearly, quadratically, or (more generally) as a polynomial function.

Embedded Assessment 3

Assessment Focus
- Identifying the type of function necessary to represent the value of items in a table
- Graphing linear, quadratic, and exponential functions
- Identifying the domain of a function
- Identifying increasing and decreasing functions
- Identifying the function with the greatest maximum value
- Solving systems of equations

Answer Key

1. Baseball card: linear; $S(t) = 3t + 47$
   Catcher's mitt: exponential; $C(t) = 24.57(1.224)^t$
   Pennant: quadratic; $P(t) = t^2 + 4t + 1$

2. $S(t)$
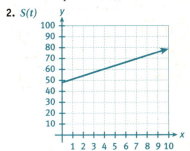

$C(t)$
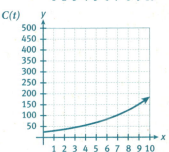

$P(t)$
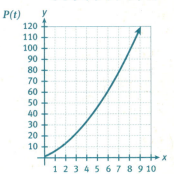

$B(t)$
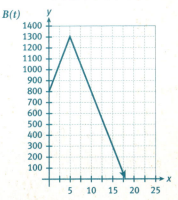

Unit 5 • Quadratic Functions

Embedded Assessment 3

TEACHER to TEACHER

You may wish to read through the scoring guide with students and discuss the differences in the expectations at each level. Check that students understand the terms used.

3. $t \geq 0$; because t represents the number of years since Emilio acquired each item, negative values for t do not make sense.
4. The ball; $B(t)$ reaches the greatest maximum value, $1300. The ball has a much greater value than the other items to begin with, and that value increased more rapidly than the other functions' values.
5. The baseball card; the table shows that its value has a constant rate of change and the graph of $S(t)$ is a line. The graph of $B(t)$ shows that it has a constant rate of change for the first 5 years but a different constant rate of change for the second 5 years.
6. The ball; $B(t)$ decreases over [5, 10]: The value of $B(t)$ is greatest at $t = 5$, but then decreases after this.
7. Over the first 5 years, the ball's value increased the fastest ($100/yr). However, once the value of the ball started to decrease, the value of the catcher's mitt, which is represented by an increasing exponential function, increased the fastest.
8. After 4 years; the graphs of these two functions intersect between $t = 4$ and $t = 5$.
9. $S(t) = 3t + 47$ and $P(t) = t^2 + 4t + 1$; the solution is (6.3, 65.9) and it represents the time (6.3 years) when the two items have an equal value ($65.90). For up to 6 years, the value of the signed baseball card is greater than the value of the vintage pennant, but after 6 years the value of the vintage pennant is greater than the value of the signed baseball card.

Embedded Assessment 3
Use after Activity 35

Solving Systems of Equations
SPORTS COLLECTOR

Scoring Guide	Exemplary	Proficient	Emerging	Incomplete
	The solution demonstrates the following characteristics:			
Mathematics Knowledge and Thinking (Items 1–9)	• Clear and accurate understanding of piecewise-defined, linear, quadratic, and exponential functions and the key features of their graphs • Effective understanding of how to solve systems of equations graphically and algebraically	• Largely correct understanding of piecewise-defined, linear, quadratic, and exponential functions and the key features of their graphs • Adequate understanding of how to solve systems of equations graphically and algebraically	• Partial understanding of piecewise-defined, linear, quadratic, and exponential functions and the key features of their graphs • Some difficulty solving systems of equations graphically and/or algebraically	• Inaccurate or incomplete understanding of piecewise-defined, linear, quadratic, and exponential functions and the key features of their graphs • Little or no understanding of how to solve systems of equations graphically and/or algebraically
Problem Solving (Items 4, 8)	• Appropriate and efficient strategy that results in a correct answer	• Strategy that may include unnecessary steps but results in a correct answer	• Strategy that results in a partially incorrect answer	• No clear strategy when solving problems
Mathematical Modeling / Representations (Items 1–9)	• Clear and accurate understanding of how to find, graph, interpret, and compare regression functions that model real-world data • Effective understanding of how to write and interpret the solution of a system of equations that represents a real-world scenario	• Largely correct understanding of how to find, graph, interpret, and compare regression functions that model real-world data • Adequate understanding of how to write and interpret the solution of a system of equations that represents a real-world scenario	• Partial understanding of how to find, graph, interpret, and/or compare regression functions that model real-world data • Some difficulty writing and/or interpreting the solution of a system of equations that represents a real-world scenario	• Inaccurate or incomplete understanding of how to find, graph, interpret, and/or compare regression functions that model real-world data • Little or no understanding of how to write and/or interpret the solution of a system of equations that represents a real-world scenario
Reasoning and Communication (Items 3–9)	• Precise use of appropriate math terms and language to describe and compare characteristics of several functions	• Adequate description and comparison of characteristics of several functions	• Misleading or confusing description and comparison of characteristics of several functions	• Incomplete or inaccurate description and comparison of characteristics of several functions

Unit 6 — Planning the Unit

In this unit, students study univariate data, using statistics and graphs to compare different distributions. They use two-way tables to summarize bivariate categorical data. Technology is used to calculate a measure of strength and direction for relationships in bivariate data that are linear in form, and distinguish between correlation/association and causation.

Vocabulary Development

The key terms for this unit can be found on the Unit Opener page. These terms are divided into Academic Vocabulary and Math Terms. Academic Vocabulary includes terms that have additional meaning outside of math. These terms are listed separately to help students transition from their current understanding of a term to its meaning as a mathematics term. To help students learn new vocabulary:

- Have students discuss meaning and use graphic organizers to record their understanding of new words.
- Remind students to place their graphic organizers in their math notebooks and revisit their notes as their understanding of vocabulary grows.
- As needed, pronounce new words and place pronunciation guides and definitions on the class Word Wall.

Embedded Assessments

Embedded Assessments allow students to do the following:

- Demonstrate their understanding of new concepts.
- Integrate previous and new knowledge by solving real-world problems presented in new settings.

They also provide formative information to help you adjust instruction to meet your students' learning needs.

Prior to beginning instruction, have students unpack the first Embedded Assessment in the unit to identify the skills and knowledge necessary for successful completion of that assessment. Help students create a visual display of the unpacked assessment and post it in your class. As students learn new knowledge and skills, remind them that they will be expected to apply that knowledge to the assessment. After students complete each Embedded Assessment, turn to the next one in the unit and repeat the process of unpacking that assessment with students.

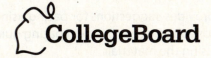

AP / College Readiness

This unit develops students' understanding of statistics by:

- Investigating applications of univariate and bivariate data.
- Communicating mathematical relationships graphically, visually and verbally.
- Using technology to experiment, analyze and interpret results, and support conclusions.
- Developing an understanding of and using the vocabulary of statistics.

Unpacking the Embedded Assessments

The following are the key skills and knowledge students will need to know for each assessment.

Embedded Assessment 1

Comparing Univariate Distributions, *Splitting the Bill*

- Visual comparison of univariate graphical displays
- Computational comparisons of center and spread
- Determining outliers and creating modified box plots
- Determining appropriate measures of variability

Embedded Assessment 2

Bivariate Distributions, *Dear Traveling Tooth*

- Describing a bivariate numerical relationship and associating that description with a correlation coefficient
- Developing a linear model, interpreting its components, using the model for prediction, and recognizing its limitations
- Analyzing row percentages and segmented bar graphs to investigate association

Planning the Unit continued

Suggested Pacing

The following table provides suggestions for pacing using a 45-minute class period. Space is left for you to write your own pacing guidelines based on your experiences in using the materials

	45-Minute Period	Your Comments on Pacing
Unit Overview/ Getting Ready	1	
Activity 36	3	
Activity 37	4	
Embedded Assessment 1	1	
Activity 38	3	
Activity 39	5	
Activity 40	3	
Embedded Assessment 2	1	
Total 45-Minute Periods	**21**	

Additional Resources

Additional resources that you may find helpful for your instruction include the following, which may be found in the eBook Teacher Resources.

- Unit Practice (additional problems for each activity)
- Getting Ready Practice (additional lessons and practice problems for the prerequisite skills)
- Mini-Lessons (instructional support for concepts related to lesson content)

Probability and Statistics

Unit Overview
Ask students to read the unit overview and to relate examples of previous graphs and statistics they have used. Ask them to recall methods they have used to present and analyze univariate and bivariate data in previous units.

Materials
- calculator

Key Terms
As students encounter new terms in this unit, help them to choose an appropriate graphic organizer for their word study. As they complete a graphic organizer, have them place it in their math notebooks and revisit as needed as they gain additional knowledge about each word or concept.

Essential Questions
Read the essential questions with students and ask them to share possible answers. As they complete the unit, revisit the essential questions to help them adjust their initial answers as needed.

Unpacking Embedded Assessments
Prior to beginning the first activity in this unit, turn to Embedded Assessment 1 and have students unpack the assessment by identifying the skills and knowledge they will need to complete the assessments successfully. Guide students through a close reading of the assignment, and use a graphic organizer or other means to capture their identification of the skills and knowledge. Repeat the process for each Embedded Assessment in the unit.

Unit Overview
In this unit you will investigate univariate data, using statistics and graphs to compare different distributions and to comment on similarities and differences among them. You will also use two-way tables to summarize bivariate categorical data and find a "best-fit line" to summarize bivariate numerical data. You will use technology to calculate a measure of strength and direction for relationships that are linear in form. Finally, you will learn to distinguish between correlation/association and causation.

Key Terms
As you study this unit, add these and other terms to your math notebook. Include in your notes your prior knowledge of each word, as well as your experiences in using the word in different mathematical examples. If needed, ask for help in pronouncing new words and add information on pronunciation to your math notebook. It is important that you learn new terms and use them correctly in your class discussions and in your problem solutions.

Academic Vocabulary
- cluster
- associate

Math Terms
- sample
- sampling error
- measurement error
- standard deviation
- outlier
- normal distribution
- z score
- correlate
- correlation coefficient
- residual
- best-fit line
- segmented bar graph
- row percentages

ESSENTIAL QUESTIONS
- How are dot plots, histograms, and box plots used to learn about distributions of numerical data?
- How can the scatter plot, best-fit line, and correlation coefficient be used to learn about linear relationships in bivariate numerical data?
- How can a two-way table be used to learn about associations between two categorical variables?
- When is it reasonable to interpret associations as evidence for causation?

EMBEDDED ASSESSMENTS
These assessments, following Activities 37 and 40, will give you an opportunity to demonstrate your understanding of how graphical displays and numerical summaries are used to compare distributions and of methods for summarizing and describing relationships in bivariate data.

Embedded Assessment 1:	
Comparing Univariate Distributions	p. 557
Embedded Assessment 2:	
Bivariate Distributions	p. 609

Developing Math Language
As this unit progresses, help students make the transition from general words they may already know (the Academic Vocabulary) to the meanings of those words in mathematics. You may want students to work in pairs or small groups to facilitate discussion and to build confidence and fluency as they internalize new language. Ask students to discuss new academic and mathematics terms as they are introduced, identifying meaning as well as pronunciation and common usage. Remind students to use their math notebooks to record their understanding of new terms and concepts.

As needed, pronounce new terms clearly and monitor students' use of words in their discussions to ensure that they are using terms correctly. Encourage students to practice fluency with new words as they gain greater understanding of mathematical and other terms.

UNIT 6
Getting Ready

You may wish to assign some or all of these exercises to gauge students' readiness for Unit 6 topics.

Prerequisite Skills

- Developing a trend line (Item 1) 8.SP.A.2
- Interpreting slope in context (Item 2) 8.F.B.4, 8.SP.A.3
- Determining missing values in a two-way table (Item 3) 8.SP.A.4
- Developing row percentages from two-way tables (Item 4) 8.SP.A.4
- Computing summary measures of center for univariate data (Items 5a, b) 6.SP.A.3, 6.SP.B.5c
- Developing a graph for univariate data (Item 5c) 6.SP.B.4
- Describing the shape of a univariate distribution (Item 5d) 6.SP.A.2, 6.SP.5B.d

Answer Key

1. Answers may vary significantly, but sample equations are:
   a. $y = 1 + \dfrac{5}{6}x$
   b. $y = -5 + \dfrac{10}{3}x$
   c. $y = 50 - 2x$
2. a. Cost of flight increases by about 19 cents per mile on average.
   b. Test score decreases by about 1.2 points per hour of TV watched on average.
3. b. Percentage of seventh graders voting for Greg = $\left(\dfrac{63}{150}\right) \cdot 100\% = 42\%$;
   percentage of eighth graders voting for Greg = $\left(\dfrac{166}{250}\right) \cdot 100\% = 66.4\%$
4. a. 66.5 inches
   b. 66.5 inches
   c.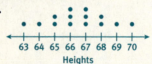
   d. Symmetric

UNIT 6
Getting Ready

Write your answers on notebook paper.
Show your work.

1. Each scatter plot below shows a set of (x, y) coordinate pairs with an approximate linear trend. Estimate the equations of the trend lines for the graphs below.

   a.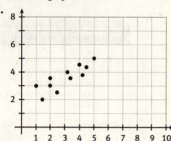

   b.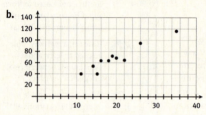

   c.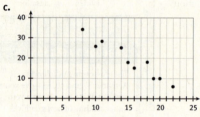

2. For each trend line below, interpret the slope of the trend line in relation to the variable quantities.
   a. $y = 75 + 0.19x$, where x is the number of miles traveled and y is the cost of an airline flight in dollars
   b. $y = 100 - 1.2x$, where x is the number of hours of TV watched per week and y is the test score on last week's test

3. An election was held at Greg's school; Greg and his friend Mary were both nominated. The table below shows the results of the voting.

Grade	Voted for Greg	Voted for Mary	Voted for Another Candidate	Total
Seventh	63	81		150
Eighth		71	13	250
Total	229		19	400

   a. Fill in the three remaining cells of the table above.
   b. What percentage of seventh graders voted for Greg? What percentage of eighth graders voted for Greg?

4. The heights (to the nearest inch) of 14 students are given below. Use these data for parts a–d.

68	66	67	70	66	68	67
69	65	67	64	66	63	65

   a. Compute the mean height of these students.
   b. Compute the median height of these students.
   c. Construct a dot plot of the heights of the students.
   d. Describe the shape of the distribution shown in the dot plot.

Getting Ready Practice

For students who may need additional instruction on one or more of the prerequisite skills for this unit, Getting Ready practice pages are available in the eBook Teacher Resources. These practice pages include worked-out examples as well as multiple opportunities for students to apply concepts learned.

3. a.

Grade	Voted for Greg	Voted for Mary	Voted for Another Candidate	Total
Seventh	63	81	6	150
Eighth	166	71	13	250
Total	229	152	19	400

Measures of Center and Spread

To Text, or Not to Text
Lesson 36-1 Mean, Median, Mode, and MAD

ACTIVITY 36

Learning Targets:
- Interpret differences in center and spread of data in context.
- Compare center and spread of two or more data sets.
- Determine the mean absolute deviation of a set of data.

> **SUGGESTED LEARNING STRATEGIES:** Summarizing, Interactive Word Wall, Create Representations, Look for a Pattern, Think-Pair-Share

Zach is a high school student who enjoys texting with friends after school. Recently, Zach's parents have become concerned about the amount of time that he spends text messaging on school nights.

Zach decides to compare the amount of time he spends text messaging to that of his good friend Olivia. Both of them record the number of minutes they spend text messaging on school nights for one week.

	Sunday	Monday	Tuesday	Wednesday	Thursday
Zach	10 min	60 min	20 min	135 min	75 min
Olivia	60 min	60 min	60 min	60 min	60 min

One way to describe a set of data is by explaining how the data *cluster* around a value, or its center. The measures of center include the **mean**, the **median**, and the **mode**.

1. Find the mean amount of time that Zach spends text messaging each night. Show how you determined your answer.

   60 hours; $\frac{(10 + 60 + 20 + 135 + 75)}{5} = 60$

2. Find the mean amount of time that Olivia spends text messaging each night. Show how you determined your answer.

   60 hours; $\frac{(60 + 60 + 60 + 60 + 60)}{5} = 60$

ACADEMIC VOCABULARY
To *cluster* means to group around.

MATH TIP
The **mean** of a set of data is found by adding the values and dividing the sum of the values by the number of values.

The **median** of a set of data is found by listing the values in order and finding the value in the middle. If there is an even number of values, the median is the mean of the two values in the middle.

The **mode** of a set of data is the value that appears most often. If there are two or more values that appear most often, then each of these values is the mode. If no data value repeats, then there is no mode.

Common Core State Standards for Activity 36

HSS-ID.A.2 Use statistics appropriate to the shape of the data distribution to compare center (median, mean) and spread (interquartile range, standard deviation) of two or more different data sets.

ACTIVITY 36
Investigative

Activity Standards Focus
Measures of center and spread will be developed more extensively using the formulas. The mean will be used to find the mean absolute deviation and standard deviation, which are measures of variability. Skills include calculating and interpreting the standard deviation of a numerical data set, as well as selecting appropriate measures of spread by examining the shape of a distribution.

Lesson 36-1

PLAN

Pacing: 1 class period
Chunking the Lesson
Introduction   #1–3   #4–6
#7–8   #9   #10–11
Check Your Understanding
#14   #15–16   #17–18   #19
Check Your Understanding
Lesson Practice

TEACH

Bell-Ringer Activity
The word *average* is used often in daily life. List two places where you have seen the word used and describe what one of them means in your own words.

Introduction Shared Reading, Paraphrasing The scenario is a text messaging context. After allowing students the opportunity to read the material and bring up any words whose meanings they do not know, ask a student to put in his or her own words what the context of the activity will address.

1–3 Activating Prior Knowledge, Interactive Word Wall Ask students to share answers to Items 1 and 2 on whiteboards or with partners to recall how to find the mean. Although it is a review term, add *mean* to the class Word Wall. Allow students to work Item 3 by themselves before debriefing.

ACTIVITY 36 Continued

4–6 Activating Prior Knowledge, Create Representations, Look for a Pattern, Debriefing These items allow students to review a prior skill by finding the range and to note that the mean and range do not provide enough information to distinguish between Zach's and Trey's data.

ACTIVITY 36 continued

My Notes

MATH TIP
Another word for spread is **variability**.

DISCUSSION GROUP TIPS
As you share your ideas, be sure to use mathematical terms and academic vocabulary precisely. Make notes to help you remember the meaning of new words and how they are used to describe mathematical concepts.

Lesson 36-1
Mean, Median, Mode, and MAD

3. **Reason quantitatively.** Compare the amounts of time that Zach and Olivia spend text messaging. Describe similarities and differences.
   Answers may vary. The mean amount of time that Zach and Olivia spend text messaging is the same. Zach's times varied, while Olivia's times were consistent.

Zach knows that data can be described by center and also by spread. **Spread** indicates how far apart the data values are in the set. Measures of spread include the **range** and the **mean absolute deviation**.

Zach asks his friend Trey to record the amount of time he spends text messaging on school nights. To measure spread, Zach chooses the *range*.

4. Find the mean and range of Trey's data.

	Sunday	Monday	Tuesday	Wednesday	Thursday
Zach	10 min	60 min	20 min	135 min	75 min
Trey	10 min	10 min	135 min	135 min	10 min

mean: 60; $\frac{(10 + 10 + 135 + 135 + 10)}{5} = 60$

range: 125; (max − min) = (135 − 10) = 125

5. Complete the table below. How do the mean and range of Trey's data compare to those of Zach's?

	Mean	Range
Trey	60	125
Zach	60	125

The mean and range of Trey's data are the same as those of Zach's data.

6. **Construct viable arguments.** Describe how the two data sets are different. Did the mean and range help you to identify these differences? Explain.
   Answers may vary. The individual data values in the two sets varied except for their maximum and minimum times, which were the same. Since the mean and the range of the two sets are the same, they do not help identify ways that the two data sets are different.

Lesson 36-1
Mean, Median, Mode, and MAD

Because the range is based on only two values, it does not reflect any variation in the data between the greatest and least values. The range is greatly influenced by extreme values.

Another measure of spread that is not as influenced by extremes is the mean absolute deviation, which is computed using all the data values. The **mean absolute deviation** is the mean (average) of the absolute values of the deviations of the data. The **deviation** is a measure of how far a data value is from the mean.

7. To find the mean absolute deviation of Zach's data, begin by completing the table. Use the mean for Zach's data that you calculated in Item 1.

Zach	Deviation	Absolute Deviation
Time x	Time − Mean $= (x - \bar{x})$	|Time − Mean| $= \|x - \bar{x}\|$
10	$10 - 60 = -50$	$\|10 - 60\| = 50$
60	$60 - 60 = 0$	$\|60 - 60\| = 0$
20	$20 - 60 = -40$	$\|20 - 60\| = 40$
135	$135 - 60 = 75$	$\|135 - 60\| = 75$
75	$75 - 60 = 15$	$\|75 - 60\| = 15$

8. To finish calculating the mean absolute deviation of Zach's data, find the mean of the numbers in the third column. Determine the sum of the numbers in the third column and then divide by the number of data values (the number of items in the first column).

 mean absolute deviation: 36;
 $$\frac{(50 + 0 + 40 + 75 + 15)}{5} = \frac{180}{5} = 36$$

WRITING MATH

The symbol $\bar{x}$ is used to represent the mean of a set of values.

TEACHER to TEACHER

This section of this activity acts as the foundation for calculating the standard deviation. The mean absolute deviation (MAD) statistic is reintroduced, and the application of absolute value notation is reinforced in that context.

Students will be asked to compute deviations and the absolute value of deviations. The contents of this chunk are very straightforward, but students may require additional guidance with the notation.

7–8 Create Representations By filling in the values in the chart, students set up the calculation of the mean absolute deviation in Item 8.

ACTIVITY 36 Continued

9 Debriefing Since the Deviation column entries add up to zero, the mean of the deviations is zero and does not provide any useful information about the relative spread of the data.

10–11 Discussion Groups, Quickwrite, Debriefing These items ask students to describe the differences between Zach's, Olivia's, and Trey's text messaging data in terms of the mean absolute deviations. Students should verify the information in the Math Tip with their answers to these items.

ACTIVITY 36 continued

My Notes

Lesson 36-1
Mean, Median, Mode, and MAD

9. **Reason abstractly.** Why would statisticians use the mean absolute deviation rather than the mean of the deviations (in the second column)?

   Answers may vary. Statisticians are looking for the average of the deviations. The deviations need to be positive numbers. Thus, they use the absolute value of the deviations. The mean of the deviations is 0 and does not represent the spread of times.

	Sunday	Monday	Tuesday	Wednesday	Thursday
Zach	10 min	60 min	20 min	135 min	75 min
Olivia	60 min	60 min	60 min	60 min	60 min
Trey	10 min	10 min	135 min	135 min	10 min

10. Trey's text messaging minutes are shown in the table above.
    a. Find the mean absolute deviation for the amount of time that Trey spends text messaging.

    mean absolute deviation: 60;
    $$\frac{|10-60|+|10-60|+|135-60|+|135-60|+|10-60|}{5} = \frac{300}{5} = 60$$

    b. Why is the mean absolute deviation for Trey's data set greater than the mean absolute deviation for Zach's data set?

    Answers may vary. Trey's data values are farther from the mean than Zach's.

11. Olivia's text messaging minutes are also given in the table above Item 10.
    a. Find the mean absolute deviation for the amount of time that Olivia spends text messaging.

    mean absolute deviation: 0; $\frac{(0+0+0+0+0)}{5} = 0$

    b. **Make sense of problems.** Explain why Olivia's mean absolute deviation is descriptive of her data.

    Explanations may vary. Olivia's minutes were the same each day, and thus were equal to the mean. There was no deviation.

Lesson 36-1
Mean, Median, Mode, and MAD

Check Your Understanding

12. During the annual food drive, Mr. Binford's homeroom collected canned goods for a month. The numbers of cans collected are given below.

    Boys

12	42	69	91	97	61
15	37	104	38	82	90
51	96	19	66	8	24

    Girls

20	63	18	89	67	19
66	108	96	24	16	44

    Compare and contrast the results for the boys and girls using the mean and the range.

13. Remi recorded data on her car's fuel efficiency for five trips in the table below.

Trip	1	2	3	4	5
Miles per Gallon	23.7	25.5	25.2	24.8	25.4

    a. Calculate the absolute deviation for each trip if the average number of miles per gallon for the trips was 24.9.
    b. Find the mean absolute deviation for the five trips.

Zach would like to assure his parents that his text messaging time is not unusual for a high school student. His average text messaging time is the same as Olivia's and Trey's average times, but the variability differs for each set of data.

Before reporting to his parents, Zach decides to gather additional information. He considers using a *census* of the 1250 students in his school.

14. Conducting a census is often difficult. What difficulties might Zach encounter if he proceeds with his census?
    Answers may vary. Difficulties could include getting responses from all members of the population, and the time it will take to compile all the data.

ACADEMIC VOCABULARY

A *census* gathers information about every member of the population.

ACTIVITY 36 Continued

15–16 Interactive Word Wall, Think-Pair-Share Have students share answers with a partner, then with their groups. Monitor students' discussions to ensure that they are using precise mathematical language.

Lesson 36-1
Mean, Median, Mode, and MAD

Instead of gathering information from all 1250 students in the school, Olivia suggests that Zach conduct a survey of 40 students.

15. What is the population of Zach's survey? What is the *sample*?
    The population of Zach's survey is the 1250 students at his school. The sample is the 40 students he will survey.

Olivia warns Zach to choose his sample wisely. A selection method that produces samples that are not representative of the total population will introduce a *sampling error* into the process.

16. Explain why each sampling method produces sampling error when surveying the school's population.
    Answers will vary.

    a. Zach asks the 30 students in his algebra class how many minutes they spend per school night text messaging with friends.
       The sample represents only one class of algebra students in his age group, not the whole school.

    b. Zach leaves questionnaires on a table in the main hall. One of the questions asks students to indicate the number of minutes they spend per school night text messaging with friends.
       It is possible that the responding group could be made up of one small subgroup, such as the girls' basketball team, and no one else.

MATH TERMS

A **sample** is a portion of the population. A good sample looks like the entire population and provides useful information about the population.

Sampling error occurs because particular subgroups of the population are missing from the sample or are over-represented. Sampling error is also called **sample selection bias**.

Lesson 36-1
Mean, Median, Mode, and MAD

Only sampling methods that incorporate random chance into the sample selection method can hope to avoid sampling error.

17. Zach obtains a list of the names of every student in the school. Explain how Zach could use the list to randomly select a sample of 40 students.
    Answers may vary. Zach could write the name of each student on a slip of paper and place the slips in a box. Then he could thoroughly mix the slips and select 40 names without looking into the box.

18. Error can also occur when the method for obtaining a response is flawed. Suppose that Zach has properly selected a random sample of 40 students. Determine why each method produces *measurement error*.

    a. Zach gives each student a questionnaire. One question states: "It is very important for young people to have time to socialize with each other and today's method of communication seems to be text messaging. How many minutes do you typically spend text messaging with friends on school nights?"
    Answers may vary. The sentence before the question gives an opinion that may influence the students answering the question and produce inaccurate data and measurement error.

    b. Zach verbally surveys 40 students. Zach feels that the student responses are too low, so he reminds them that he is trying to convince his parents that 60 minutes per night is not too much time to spend text messaging.
    Answers may vary. Zach's reminder to students about what he hopes or feels is true could influence students to give false data in support of Zach's intended result.

ACTIVITY 36
continued

My Notes

MATH TERMS

Measurement error occurs when incorrect or misleading data are collected that can be ascribed to the interviewer, the respondent, the survey instrument, or the method used for recording the data.

CONNECT TO AP

In AP Statistics, it is important to understand whether a particular sample or method for gathering information contains the potential for error or bias.

ACTIVITY 36 Continued

17–18 Think-Pair-Share, Interactive Word Wall The answer to Item 17 gives teachers formative information regarding whether students have been exposed to using random chance to select a sample.

CONNECT TO AP

In AP Statistics, students will need to understand whether a particular sample or method for gathering information contains the potential for error or bias. The study below comes from a multiple-choice question found in the AP Statistics Course Description (p. 26).

George and Michelle each claimed to have the better recipe for chocolate chip cookies. They decided to conduct a study to determine whose cookies were really better. They each backed a batch of cookies using their own recipes. George asked a random sample of his friends to taste his cookies and to complete a questionnaire on their quality. Michelle asked a random sample of her friends to complete the same questionnaire for her cookies. They then compared the results.

After completing this lesson, ask your students to list some of the things George and Michelle did incorrectly in designing their study and some of the things they did correctly.

ACTIVITY 36 Continued

19 Debriefing Students' answers to this item provide you with formative information regarding whether they understand measurement error and sampling error.

Check Your Understanding

Debrief students' answers to these items to ensure that they understand the difference between a sample and a census.

Answers
20. sample
21. census
22. sample

---

continued

Lesson 36-1
Mean, Median, Mode, and MAD

19. **Attend to precision.** Gathering good data involves avoiding error in the process.

   a. Write a question that Zach can use to gather data about text messaging habits and avoid measurement error.
   Answers may vary; sample answer: How many minutes a night do you spend text messaging?

   b. Describe a method that Zach can use to collect answers from a group of 40 students and avoid sampling error.
   Answers may vary. Write the names of the 1250 students in the school on slips of paper, one name per slip. Mix up the slips of paper thoroughly and select 40 slips without looking.

Check Your Understanding

Classify each as a sample or a census.

20. Surveying every ninth-grade student regarding the new dress code policy for the students in all grades in the school to determine the student opinion

21. Asking every eighth-grade student whom they will vote for in the upcoming eighth-grade student election

22. Selecting every fourth student from an alphabetical list of students to gather data on absenteeism for the school

530 SpringBoard® Mathematics Algebra 1, Unit 6 • Probability and Statistics

Lesson 36-1
Mean, Median, Mode, and MAD

LESSON 36-1 PRACTICE

Work with your group to answer Items 23–27. As you discuss your solutions, speak clearly and use precise mathematical language. Remember to use complete sentences and words such as *and, or, since, for example, therefore, because of* to make connections between your thoughts.

23. Vernice asked 12 classmates to record the number of hours they spent watching television during one week. The table shows the data she collected.

10	11	22	7	17	17
20	31	0	12	19	23

   a. Calculate the mean and mean absolute deviation for the data.
   b. What statements could Vernice make about the viewing habits of these classmates?

24. Rewrite each question to avoid measurement error.
   a. Don't you agree that seniors should be dismissed early on Fridays at least once each semester?
   b. Drinking beverages with sugar promotes tooth decay and obesity. How many soft drinks with sugar did you drink in the past week?

25. Scores from the same benchmark test were collected from two algebra classes, each with 30 students enrolled. One class had a mean score of 79 with a mean absolute deviation of 5, and the other had a mean score of 81 with a mean absolute deviation of 10. What can be said about the distribution of scores on this test for the two classes?

26. Mitch and a group of his friends have estimated how long it will take each of them to run 400 meters around the track. Their estimates in seconds are 115, 76, 94, 81, 78, 99, 68, and 84.
   a. Calculate the mean and the range.
   b. Which estimate stands out as unusual?
   c. What might be a reason for such an unusual estimate?

27. **Construct viable arguments.** What are the advantages of taking a census of the population instead of a sample? Describe some examples when it would be worth the time and effort to conduct a census.

ACTIVITY 36 Continued

Lesson 36-2

PLAN

Materials
- calculator

Pacing: 1 class period

Chunking the Lesson
Introduction   #1–3
#4   #5   #6
Check Your Understanding
Lesson Practice

TEACH

Bell-Ringer Activity
Describe with words or an example how to find the mean absolute deviation.

Introduction Shared Reading, Summarizing, KWL Chart Use Reading strategies to be sure students understand variability in data and how to calculate the MAD.

1–3 Create Representations, Activating Prior Knowledge While most of the activity is scaffolded using data sets with no specific context, some analysis is performed in the activity in the context of the amount of talk time a cell phone battery offers. You may want to consider asking students what they feel an acceptable (or expected) value for talk time might be for a standard cell phone battery on a full charge.

Remind students that absolute value is used so that the sum of the MAD does not equal 0, which would not provide useful information.

ACTIVITY 36 continued

Lesson 36-2
Another Measure of Variability

Learning Targets:
- Use summation and subscript notation.
- Calculate and interpret the standard deviation of a numerical data set.
- Select appropriate measures of spread by examining the shape of a distribution.

SUGGESTED LEARNING STRATEGIES: Summarizing, KWL Chart, Create Representations, Self Revision/Peer Revision, Think-Pair-Share

In the previous lesson, you studied one way to describe the spread, or variability, in a data set—the mean absolute deviation (MAD). The mean absolute deviation is the mean (or average) difference of the data values from the mean of a numerical data set.

Consider the following data from C|NET (www.cnet.com), a tech media website that publishes information about technology and consumer electronics. These data are taken from a review of cell phone battery lifetimes. The table below gives the talk times (in hours) for the top 10 brands of batteries.

Talk Time (hours)	
19.78	11
14.55	10.7
13.4	10.6
12.75	10.6
12	10.3

1. Calculate the mean for these 10 talk times.
   12.568 hours

2. Complete the table below by finding the absolute values of the deviations (the absolute value of the difference between each data value and the mean calculated in Item 1).

Talk Time and Deviations from the Mean			
Value	|Deviation|	Value	|Deviation|
19.78	7.212	11.00	1.568
14.55	**1.982**	10.70	**1.868**
13.40	**0.832**	10.60	**1.968**
12.75	**0.182**	10.60	**1.968**
12.00	**0.568**	10.30	**2.268**

3. **Attend to precision.** The mean absolute deviation for this data set is the mean of the 10 absolute deviations. Calculate the mean absolute deviation.
   2.0416

Lesson 36-2
Another Measure of Variability

Another common measure of variability is the **standard deviation**. Before calculating the standard deviation, let's introduce some notation.

The Greek letter Σ (sigma) is used to indicate a sum of several values. For example, $\sum_{i=1}^{5} i$ means "the sum of the values of i as i goes from 1 to 5":

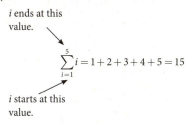

$$\sum_{i=1}^{5} i = 1 + 2 + 3 + 4 + 5 = 15$$

In the summation above, i can be replaced by any expression. Also, the values of i can start and end at any number:

$$\sum_{i=3}^{6} (i+2) = (3+2) + (4+2) + (5+2) + (6+2) = 26$$

4. Determine each sum.

a. $\sum_{i=1}^{4} i^2$ b. $\sum_{i=0}^{2} 2^i$ c. $\sum_{i=2}^{7} |5-i|$

   30   7   9

The standard deviation is similar to the MAD in that it is based on deviations from the mean. The formulas below show the similarities between the MAD and the standard deviation s.

$$\text{MAD} = \frac{\sum_{i=1}^{n} |x_i - \bar{x}|}{n} \qquad s = \sqrt{\frac{\sum_{i=1}^{n} (x_i - \bar{x})^2}{n-1}}$$

As you can see, the calculation of the standard deviation is a little more involved than MAD. Why this additional complexity? The basic answer is that this measure has some advantages in more advanced statistical settings.

You will calculate both the MAD and standard deviation using a data set consisting of four battery talk times (in hours):

$$x_1 = 7.00, \ x_2 = 10.00, \ x_3 = 8.10, \ x_4 = 9.32$$

The mean of these 4 observations is 8.605.

MATH TERMS

The **standard deviation** is a measure of variability in a data set.

DISCUSSION GROUP TIPS

As needed, refer to the Glossary to review definitions and pronunciations of key terms. Incorporate your understanding into group discussions to confirm your knowledge and use of key mathematical language.

MATH TIP

In sigma notation, the letter i is called the **index of summation**.

MATH TIP

In a data set with n values, the individual values are referred to as $x_1, x_2, x_3, \ldots, x_n$. In the formulas for the MAD and the standard deviation, x_i, as i goes from 1 to n, refers to these values.

ACTIVITY 36 Continued

5 Create Representations, Self Revision/Peer Revision, Debriefing Have students use whiteboards to complete the table to recognize the values that are used in the formula. Debrief by discussing squared values, sum, and square root. Squared values do not make sense in the context of the data but are necessary due to the sum of the deviations adding to 0. By taking the average, students are finding the sum of the deviation squared and dividing by $(n-1)$. This allows the data to make sense in terms of the context.

Standard deviation is an important statistic that is used for calculating and analyzing data.

Using Technology

The calculation of standard deviation can be cumbersome for many data sets. Using a calculator helps to find the standard deviation quickly.
Steps for the TI-nSpire:
Open list and spreadsheet
Name the list
Enter data
Control I to insert a new page
Menu
Statistics
1 variable

6 Think-Pair-Share Students extend their knowledge of standard deviation. After they have calculated the standard deviation by hand, if technology is available show students the steps on the calculator and recalculate the standard deviation for Item 4.

ACTIVITY 36 continued

Lesson 36-2
Another Measure of Variability

5. Complete the table. Then compute the MAD and the standard deviation.

x_i	$\bar{x}$	$|x_i - \bar{x}|$	$(x_i - \bar{x})^2$
$x_1 = 7.00$	8.605	1.605	2.576025
$x_2 = 10.00$	8.605	1.395	1.946025
$x_3 = 8.10$	8.605	0.505	0.255025
$x_4 = 9.32$	8.605	0.715	0.511225
$\sum_{i=1}^{4} x_i = 34.42$		$\sum_{i=1}^{4}|x_i - \bar{x}| = 4.22$	$\sum_{i=1}^{4}(x_i - \bar{x})^2 = 5.2883$

$$\text{mean absolute deviation} = \frac{\sum_{i=1}^{4}|x_i - \bar{x}|}{n} = 1.055$$

$$\text{standard deviation} = \sqrt{\frac{\sum_{i=1}^{4}(x_i - \bar{x})^2}{n-1}} = 1.328$$

6. The weights (in ounces) of three newborns are:

$$w_1 = 120; \quad w_2 = 115; \quad w_3 = 125$$

a. Compute the mean of these three weights.
   $\bar{w} = 120$ ounces

b. Complete the table below and calculate both the MAD and the standard deviation.

w_i	$\bar{w}$	$|w_i - \bar{w}|$	$(w_i - \bar{w})^2$
120	120	0	0
115	120	5	25
125	120	5	25
$\sum_{i=1}^{3} w_i = 360$		$\sum_{i=1}^{3}|w_i - \bar{w}| = 10$	$\sum_{i=1}^{3}(w_i - \bar{w})^2 = 50$

$$\text{mean absolute deviation} = \frac{\sum_{i=1}^{3}|w_i - \bar{w}|}{n} = 3.333$$

$$\text{standard deviation} = \sqrt{\frac{\sum_{i=1}^{3}(w_i - \bar{w})^2}{n-1}} = 5$$

You can see that calculating the standard deviation can be a lot of work. Fortunately, calculators and computers can be used to do the calculations.

Lesson 36-2
Another Measure of Variability

ACTIVITY 36
continued

My Notes

Check Your Understanding

7. Calculate the mean and standard deviation of the original data for the cell phone battery talk times (shown below) using a calculator or computer software.

Talk Times (hours)	
19.78	11.00
14.55	10.70
13.40	10.60
12.75	10.60
12.00	10.30

LESSON 36-2 PRACTICE

The table shows the speeds of the 10 fastest roller coasters in the United States. Use the table for Items 8–11.

Fastest Roller Coasters (mi/h)	
76	81
85	67
74	72
63	59
73	80

8. Find $|x - \bar{x}|$ for each data value.

Value	76	81	85	67	74	72	63	59	73	80
$\lvert x - \bar{x} \rvert$										

9. Find the mean absolute deviation using the values you calculated in Item 8.

10. Using technology, calculate the standard deviation for the data set.

11. **Make sense of problems.** A new roller coaster is being built, scheduled to be complete in the spring of next year. It is projected to reach speeds of 90 mi/h. This new value will cause the lowest speed to be removed from the list. Calculate the new standard deviation and describe the changes.

ACTIVITY 36 Continued

Check Your Understanding

Debrief the lesson by asking students to explain how to calculate the mean, MAD, and standard deviation.

Answers

7. mean: 12.568; standard deviation: 2.903

ASSESS

Students' answers to Lesson Practice problems will provide you with a formative assessment of their understanding of the lesson concepts and their ability to apply their learning.

See the Activity Practice for additional exercises for this lesson. You may assign the items here or use them as a culmination for the activity.

LESSON 36-2 PRACTICE

8.

Value	$\lvert x - \bar{x} \rvert$
76	3
81	8
85	12
67	6
74	1
72	1
63	10
59	14
73	0
80	7

9. 6.2

10. 8.165

11. The new standard deviation is 8.144. The standard deviation decreased slightly.

ADAPT

Check students' answers to the Lesson Practice to ensure that they understand how to calculate the standard deviation and have a basic understanding of how the standard deviation is affected by changes to the data. Students who may need more practice with the concepts in this lesson may benefit from an explanation using test scores. The standard deviation can give information on the diversity of the scores. If you are talking about the end-of-course exam for Algebra 1 and a school in your district has a higher average than your school, some might say that their school is smarter, but there is more to the picture than just the mean; maybe there is a life skills/special education program at your school or a gifted program at their school. Standard deviation with the real-life explanations helps to explain these differences.

ACTIVITY 36 Continued

ACTIVITY PRACTICE

1. **a.** Yes, this is a statistical question.
   b. No, this is not a statistical question because it is a "yes/no" question. What is your favorite ice cream flavor?
   c. No, this is not a statistical question because it is a "yes/no" question. How many hours of TV do you watch at night?
   d. Yes, this is a statistical question.
2. Answers will vary.
3. There will be little variability. This is a "yes/no" question and predictable. Only a few students will want more homework.
4. Sample answer: The people in my class come from around the world.
5. Answers will vary.
6. 4
7. 8
8. mean: 33.2
   MAD: 6.4
   standard deviation: 8.6
9. The mean would decrease, and the standard deviation would increase.

ADDITIONAL PRACTICE

If students need more practice on the concepts in this activity, see the eBook Teacher Resources for additional practice problems.

ACTIVITY 36
continued

Measures of Center and Spread
To Text, or Not to Text

ACTIVITY 36 PRACTICE

Write your answers on notebook paper.
Show your work.

1. For each question, decide:
   - Is it a statistical question?
   - If not, explain why not and rewrite the question so that it is a statistical question.

   a. How many states have you visited?
   b. Do you like chocolate ice cream?
   c. Do you watch TV at night?
   d. How many sports do you play?

2. Write three examples of statistical questions whose responses would show different levels of variability. Include at least one question whose responses would show "lots" of variability and at least one question whose responses would show "little" variability.

3. Describe the variability associated with the question "Do students in my school need more homework each night?" Do you think there will be a lot or a little variability? Explain.

4. Suppose the results of a survey about where your classmates were born showed great variability. What would these results tell you about the people in your class?

5. The following question was asked of your classmates today before school started:

   "How many states have you visited in the last year?"

   The responses to this question showed little variability. How could you change this question to gather answers that show greater variability?

For Items 6 and 7, use the data set
$x_1 = 2.3$, $x_2 = 4.1$, $x_3 = 1.6$, and $x_4 = 2.0$.

6. Compute the value of $\frac{x_1 + x_2}{x_3}$.

7. Compute the value of $\sum_{i=1}^{3} x_i$.

Use the information below for Items 8 and 9.

The amount of caffeine in beverages presents an important health concern, especially for women of childbearing age. In a recent study of carbonated sodas, the numbers of milligrams of caffeine detected were as follows:

29.5, 38.2, 39.6, 29.5, 31.7, 27.4, 45.4, 48.2, 36.0, 33.8, 19.4, 18.0, 34.6

8. Calculate the mean, mean absolute deviation, and standard deviation for these data.

MATHEMATICAL PRACTICES
Reason Abstractly and Quantitatively

9. In the report of the caffeine levels, five sodas were left out because no caffeine was detected. If these sodas were given a value of 0 mg of caffeine and added to the data above, how would the mean and standard deviation change?

Dot and Box Plots and the Normal Distribution
Disturbing Coyotes
Lesson 37-1 Dot Plots and Box Plots

ACTIVITY 37

Learning Targets:
- Construct representations of univariate data in a real-world context.
- Describe characteristics of a data distribution, such as center, shape, and spread, using graphs and numerical summaries.
- Compare distributions, commenting on similarities and differences among them.

SUGGESTED LEARNING STRATEGIES: Summarizing, Paraphrasing, Look for a Pattern, Discussion Groups, Quickwrite

Professional wildlife managers and the public are concerned with the impact of human activity on wildlife. One measure studied is animals' "home range," the typical area in which an animal spends its time.

Researchers were concerned that the home ranges of some coyotes in a portion of Colorado were affected by military maneuvers involving jeeps, tanks, helicopters, and jet fighter flyovers. To evaluate these potential effects, several coyotes were collared with radio transmitters. The researchers used the transmitters to track the movement of the coyotes. Coyotes were monitored before, during, and after the military maneuvers.

Home Range Before Maneuvers (km²)	Home Range During Maneuvers (km²)	Home Range After Maneuvers (km²)
3.9	7.5	3.1
5.4	32.6	5.4
5.7	3.2	4.5
4.8	2.7	4.1
5.3	7.3	8.6
5.3	9.1	8.0
5.5	18.0	6.4
11.4	6.5	7.8
3.6	2.1	1.0
3.5	5.2	3.7
6.3	4.3	10.7

A **dot plot** is an effective method for representing univariate (one-variable) data when dealing with small data sets.

1. A dot plot of the "Before" data is shown below. Make dot plots of the "During" and "After" data sets.

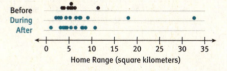

MATH TIP

To make a dot plot, draw a number line with an appropriate scale for the data. Then mark a dot for each data value above the appropriate number on the number line.

Common Core State Standards for Activity 37

HSS-ID.A.1 Represent data with plots on the real number line (dot plots, histograms, and box plots).

HSS-ID.A.2 Use statistics appropriate to the shape of the data distribution to compare center (median, mean) and spread (interquartile range, standard deviation) of two or more different data sets.

HSS-ID.A.3 Interpret differences in shape, center, and spread in the context of the data sets, accounting for possible effects of extreme data points (outliers).

HSS-ID.A.4 Use the mean and standard deviation of a data set to fit it to a normal distribution and to estimate population percentages. Recognize that there are data sets for which such a procedure is not appropriate. Use calculators, spreadsheets, and tables to estimate areas under the normal curve.

ACTIVITY 37
Investigative

Activity Standards Focus
Students will construct representations of univariate data in a real context, and describe characteristics of the data distribution such as center, shape, and spread using graphs and numerical summaries. They will compare distributions, commenting on similarities and differences among them. Students will learn to create a five-number summary and create a modified box plot.

Lesson 37-1

PLAN

Materials
- calculator

Pacing: 1 class period

Chunking the Lesson
Introduction
#1–2   #3–5   #6–8
Check Your Understanding
Lesson Practice

TEACH

Bell-Ringer Activity
Display the following data set for students to consider.
15 18 12 9 21 15 18 14 17
These numbers represent the points scored in each of the first nine games this season by the basketball team's high scorer. Compute the five-number summary for this data set.
[Answers: minimum: 9; first quartile: 13; median: 15; third quartile: 18; maximum: 21]

TEACHER to TEACHER
To contextualize the lesson, ask students what they know about coyotes and coyote behavior. You may want to provide some coyote pictures and information. You can also ask students to predict whether they think that coyotes (or similar animals) would be disturbed or affected by noise, and how such animals may react or adapt.

1–2 Create Representations, Summarizing, Paraphrasing Students will reconnect with the processes involved with developing a dot plot to organize data. Students will be asked to make comparisons of univariate distributions in terms of center spread.

Technology Tip
The TI-Nspire can be used to make dot plots and box plots.

Activity 37 • Dot and Box Plots and the Normal Distribution 537

ACTIVITY 37 Continued

1–2 (continued) As students are asked to record comparisons, encourage them to use contextual terms and clear, precise language. For example, in Item 2, if a student comments "one data set is wider than the other," encourage the student to be more specific (e.g., "The 'During' data set may have a slightly higher center, and it certainly has a greater spread than…").

3–5 Look for a Pattern, Quickwrite, Debriefing Students will compute a five-number summary and create a box plot. Students may need additional review of the process for computing the values of a five-number summary. Students should discuss and analyze the data by looking at the shape, center, and spread.

ACTIVITY 37 continued

My Notes

MATH TIP

The first quartile (Q_1) is the median of the data values to the left of the overall median, and the third quartile (Q_3) is the median of the data values to the right of the overall median. A box plot (sometimes called a box-and-whisker plot) is a graph of the five-number summary that consists of a central box from Q_1 to Q_3 that has a vertical line segment at the median. Horizontal line segments ("whiskers") extend from the box to the minimum and maximum data values.

Lesson 37-1
Dot Plots and Box Plots

2. Compare and contrast the centers and spreads of the three data sets.
   The "Before" and "After" data sets appear to have similar centers and spreads. The "During" data set may have a slightly greater center than the other two data sets, but it certainly has a greater spread due to two very extreme data values.

A five-number summary provides a numerical summary of a set of data. It is used to construct a **box plot**. The five-number summary for the "Before" home ranges is shown below, together with the resulting box plot.

Home Ranges: Before Maneuvers	
Minimum:	3.5
First quartile (Q_1):	3.9
Median:	5.3
Third quartile (Q_3):	5.7
Maximum:	11.4

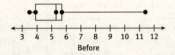

3. Create five-number summaries of the "During" and "After" data.

Home Ranges: During Maneuvers	
Minimum:	2.1
First quartile:	3.2
Median:	6.5
Third quartile:	9.1
Maximum:	32.6

Home Ranges: After Maneuvers	
Minimum:	1.0
First quartile:	3.7
Median:	5.4
Third quartile:	8.0
Maximum:	10.7

4. Use the summaries in Item 3 to construct box plots of the "During" and "After" data sets in the space below.

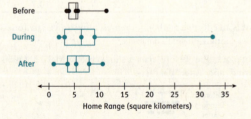

538 SpringBoard® Mathematics **Algebra 1, Unit 6** • Probability and Statistics

Lesson 37-1
Dot Plots and Box Plots

5. Based on the box plots and five-number summaries in Items 3 and 4:
   a. Which data set seems to have the least overall spread? Which data set seems to have the greatest overall spread?
   The "Before" and "After" groups have similar, low overall spread, but the "During" group's overall spread is much greater.

   b. Which data set seems to have the least spread in its "middle 50%" box? Which data set seems to have the greatest spread in its "middle 50%" box?
   The "Before" group has the least spread in its "middle 50%" box, and the "During" group has the greatest spread in its "middle 50%" box.

Remember that the initial concern before the data gathering was that the home ranges of the local coyotes might change during the military maneuvers. To investigate this concern, you will use the graphs, numerical summaries, and comparisons you have developed as a starting point for your analysis.

6. **Reason abstractly and quantitatively.** Based on the data and graphs, does it appear that there was a substantial change in the coyotes' home ranges during the military maneuvers? Write a few sentences specifically comparing the "Before" and "During" data sets. Use numerical values where possible.
   Answers will vary. The "During" and "Before" data sets seem to be very different. The medians are quite different (6.5 and 5.3, respectively) and the median home range is 1.2 square kilometers greater for the coyotes during the maneuvers. Also, the overall spread of the "During" data set is much greater than the overall spread of the "Before" data set, and the same can be said for the spreads of the respective "middle 50%" values. The "During" data set also has two data values that are quite high (18.0 and 32.6)—in fact, those two values are greater than *any* "Before" values. The coyotes appear to have been disturbed during the maneuvers in terms of their home ranges, but further analysis may be needed.

7. Do there appear to be any substantial permanent changes to the coyotes' home ranges after the military maneuvers? Write a few sentences specifically comparing the "Before" and "After" data sets. Use numerical values where possible.
   Answers will vary. The "After" and "Before" data sets seem to be relatively similar. The medians are very close (5.3 and 5.4, respectively) and the overall spreads appear to be similar from both the box plots and the dot plots. The coyotes may have been disturbed during the maneuvers, but they seem to be returning to normal after the military maneuvers.

ACTIVITY 37 Continued

6–8 Quickwrite, Discussion Group
These questions contain several opportunities for writing. The first few questions are scaffolded in a way that is similar to previous questions. Items 6 and 7 are open-ended writing exercises that are intended to produce student work appropriate for a presentation, publication, etc., for an audience with minimal understanding of statistics. Encourage students to work with one another as a means of developing the strongest writing possible. If you wish, students could work on these exercises collectively (e.g., multiple students could each contribute a sentence or two, or a shared wall or document could be developed).

ACTIVITY 37 Continued

6–8 (continued) Item 8 encourages students to recognize that different types of plots have different strengths and weaknesses. The choice as to which is the most effective plot to use to illustrate a viewpoint can be influenced by characteristics of the data.

Encourage students to share their responses to the questions with one another. If possible, facilitate some discussion as to the characteristics that make an answer "good," such as concise language, use of measurement values, comparative terms, etc.

Check Your Understanding

After allowing work time, permit students to compare answers. Debrief student answers to reveal their level of understanding regarding dot plots, box plots, and center.

Answers

9. a.

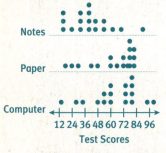

Dot Plot of Notes, Paper, Computer

b. The distribution of the Notes distribution is skewed to the right, and the Paper strategy distribution is skewed to the left. The Computer distribution is, in comparison, relatively symmetric. The median of the Notes strategy is smaller than the other two (39 vs. 71 and 66.5). The variability in the distributions for the three strategies is similar.

10.

	Student Notes	Paper Help	Computer Help
Minimum	15	18	13
First quartile	31	47	53
Median	39	71	66.5
Third quartile	54	77	80
Maximum	89	82	100

ACTIVITY 37 continued

Lesson 37-1
Dot Plots and Box Plots

8. The "During" data set contains two values that are far away from the rest of the data. The "Before" data set contains one such value also. Suppose you wish to call attention to the fact there are such far-away data values. Which type of plot—the dot plot or the box plot—would be your choice? Why?

   The dot plot would be the choice. Explanations may vary. The dot plot shows the actual values of the far-away data values relative to the other values. Also, for the "During" data, it is not clear from the box plot that there is more than one data value that is far from the majority of the data.

Check Your Understanding

9. A teacher in a statistics class allows her students to use notes about statistical procedures on tests. She believes that a teacher-made study sheet will be more effective in helping students recall the procedures. In each of her three classes she used one of three helping strategies: (a) student-made notes with information about the procedures, (b) teacher-made information printed on paper in the form of a flowchart, and (c) teacher-made information delivered by computer access during the exam. Each of her classes has 18 students. The test scores for her students are given as percent correct.

Student Notes	89	15	39	15	31	69	39	54	31	62	46	39	54	39	15	46	23	31
Paper Help	76	24	77	71	18	29	59	77	41	77	77	47	71	82	82	82	59	65
Computer Help	100	13	73	73	33	53	60	60	27	80	80	47	73	80	80	93	60	53

   a. In order to compare the results for these three groups, construct a dot plot for each of the three data sets.
   b. Describe the three data sets with specific attention to center and spread.

10. For each group, create a five-number summary for these data.

	Student Notes	Paper Help	Computer Help
Minimum			
First quartile			
Median			
Third quartile			
Maximum			

11. Use the summaries in Item 10 to construct box plots of the three data sets: "Notes," "Paper," and "Computer."

540 SpringBoard® Mathematics Algebra 1, Unit 6 • Probability and Statistics

Lesson 37-1
Dot Plots and Box Plots

12. a. For these data, what advantages do you see in using the dot plots to display the data sets?
    b. For these data, what advantages do you see in using the box plots to display the data sets?
13. Comparing the test scores for these groups and using specific information from the five-number summary and/or your dot plots and box plots, answer the following in a few sentences.
    a. Which group appears to have done the best on the exam?
    b. Which group appears to have done the worst on the exam?

LESSON 37-1 PRACTICE

Some researchers believed that one reason students often have unhealthy sleeping habits is that they don't adequately manage their time. The researchers wanted to test whether providing information to students about time management could help. Eighteen students were divided into three groups. Students in Group 1 were taught to use a planner for time management and asked to sleep 7–8 hours daily. Students in Group 2 were taught to use a planner but not given any instruction on how many hours to sleep daily. Students in Group 3 were simply instructed to sleep as they usually do. At the conclusion of the study the participants were given a questionnaire to measure their anxiety levels. (A high score on the questionnaire indicates low levels of anxiety.)

Dot plots of the data from the questionnaire are shown below. Use the dot plots for Items 14 and 15.

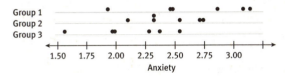

14. In a few sentences, compare the centers of these data sets. Do the three groups have approximately equal centers? If not, how do they differ?
15. In a few sentences, compare the spreads of these data sets. Do the three groups have very similar spreads? If not, how do they differ?

ACTIVITY 37 Continued

17.

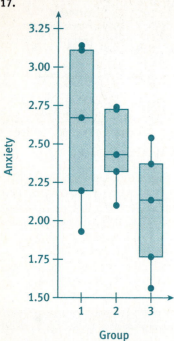

18. Answers may vary. The box plot makes it easier to compare the centers; the spread can be compared with either plot, although the dot plot has more detail.

ADAPT

If some students need additional work on these topics, you may want to access activities from previous grade levels (middle school SpringBoard material, for example) that address the visualization of distributions and the computation of summary measures. Peer review of writing may also benefit some students.

ACTIVITY 37 continued

Lesson 37-1
Dot Plots and Box Plots

A table of the data from the questionnaires is shown below. Use the table for Item 16.

Group 1: Planner & Sleep Instruction	Group 2: Planner Only	Group 3: Neither Planner nor Sleep Instruction
2.86	2.32	1.56
3.14	2.71	1.97
2.48	2.32	2.28
1.93	2.54	2.54
2.46	2.74	1.99
3.08	2.10	2.37

16. Create a five-number summary for each group.

	Group 1: Planner & Sleep Instruction	Group 2: Planner Only	Group 3: Neither Planner nor Sleep Instruction
Minimum	1.93	2.10	1.56
First quartile		2.32	
Median		2.43	2.135
Third quartile			2.37
Maximum	3.14	2.74	2.54

17. Use the five-number summaries you created in Item 16 to draw box plots for the three groups.

18. Use appropriate tools strategically. You now have two different graphs for these data. Which graph—dot plots or box plots—do you feel makes it easier to compare centers and spreads? Explain your reasoning.

16.

	Group 1: Planner & Sleep Instruction	Group 2: Planner Only	Group 3: Neither Planner nor Sleep Instruction
Minimum	1.93	2.10	1.56
First quartile	2.195	2.32	1.765
Median	2.67	2.43	2.135
Third quartile	3.11	2.725	2.37
Maximum	3.14	2.74	2.54

Lesson 37-2
Modified Box Plots

ACTIVITY 37
continued

Learning Targets:
- Use modified box plots to summarize data in a way that shows outliers.
- Compare distributions, commenting on similarities and differences among them.

SUGGESTED LEARNING STRATEGIES: Summarizing, Paraphrasing, Think-Pair-Share, Create Representations, Quickwrite

You already are familiar with many ways to summarize data graphically and numerically. In this lesson, you will see a new type of plot called a **modified box plot**.

Below are the dot plots of the data about the coyotes from Lesson 37-1. Refer back to Lesson 37-1 for the actual data values.

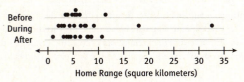

Notice that there are two unusually large values in the "During" data set and one in the "Before" data set.

Outliers are values that differ so much from the rest of a one-variable data set that attention is drawn to them. Outliers may arise for many reasons, including measurement errors or recording errors. It may also be the case that there actually are unusual values in a data set.

Outliers can also occur in bivariate (two-variable) data. It is important to consider the impact of outliers when summarizing and analyzing data.

1. Which, if any, of the data values shown in the dot plots do you consider to be outliers? List each data value that you think might be an outlier and the data set from which it came.
   Answers may vary. For the "Before" data set, the value 11.4 might be an outlier; for "During," 18.0 and 32.6.

2. Describe the method you used in Item 1 to determine whether a data value is an outlier. Is it based on distance? Is it based on the concentration of other data values? Compare your method with those of your classmates.
   Answers may vary, but answers should include some informal system of visualization and/or quantification to identify an outlier.

MATH TERMS

An **outlier** is a data point that is unusual enough that it draws attention during the data analysis.

Activity 37 • Dot and Box Plots and the Normal Distribution   543

Lesson 37-2
Modified Box Plots

MATH TIP

Notation Review

Q_1 First quartile

Q_3 Third quartile

$IQR = Q_3 - Q_1$

It is not surprising that people do not always agree about how different a value must be in order to be called an outlier. For consistency, a data value is considered to be an outlier if it is more than $1.5 \times (IQR)$ from the nearest quartile. Recall that *IQR* stands for **interquartile range** and is the difference between the third quartile (Q_3) and the first quartile (Q_1): $IQR = Q_3 - Q_1$.

Mathematically, this means that a data value x is an outlier if:

$$x > Q_3 + 1.5(IQR)$$
or
$$x < Q_1 - 1.5(IQR)$$

We will use this definition of an outlier to draw a *modified* box plot. The idea behind modifying the box plot is to create a plot that shows outliers.

When creating a *modified* box plot, the procedure for determining the length of the whiskers is different. The whiskers extend only to the least data value that is not an outlier and to the greatest data value that is not an outlier. Outliers are then shown by adding dots to the box plot to indicate their locations.

Let's walk through the steps together using the data from the "Before" data set. Here are the data, arranged in order, as well as the five-number summary and box plot of the data from Lesson 37-1:

| 3.5 | 3.6 | 3.9 | 4.8 | 5.3 | 5.3 | 5.4 | 5.5 | 5.7 | 6.3 | 11.4 |

Home Ranges: Before Maneuvers	
Minimum:	3.5
First quartile:	3.9
Median:	5.3
Third quartile:	5.7
Maximum:	11.4

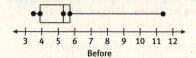

The maximum value, 11.4, is much greater than the other data values. In fact, it is more than 5 units away from the next-greatest value of 6.3. The remaining data (from the minimum value of 3.5 to 6.3) have a range of only 2.8.

Now let's modify the box plot to show outliers.

Lesson 37-2
Modified Box Plots

3. For the "Before" data set, how great or how small must a data value be to be identified as an outlier? Perform the calculations below to find the upper and lower boundary values that separate outliers from the rest of the data.

$$Q_3 + 1.5(IQR) = 5.7 + 1.5(5.7 - 3.9) = 8.4$$

$$Q_1 - 1.5(IQR) = 3.9 + 1.5(5.7 - 3.9) = 1.2$$

4. Are any data values less than the lower boundary for outliers identified in Item 3?
   No; none are less than 1.2.

5. Are any data values greater than the upper boundary for outliers identified in Item 3?
   Yes; 11.4 is greater than 8.4.

Next, identify the least and greatest values in the data set that are not outliers. These values will determine the endpoints of the whiskers in the modified box plot.

6. What is the least data value that is not an outlier?
   the minimum number, 3.5

7. What is the greatest data value that is not an outlier?
   6.3

ACTIVITY 37 Continued

8 Create Representations Students will use the numbers they have determined to draw a modified box plot. Discuss with the students why there is a space between the whisker and dot representing the outlier.

TEACHER to TEACHER

It is important that students pay close attention to SCALE in developing their box plots, as the positioning with respect to the number line of the box, the whisker endpoints, and the outliers are critical for effectively communicating the variability in the distribution. Address any errors or any common confusion.

Check Your Understanding

Debrief students' answers to these items to ensure their understanding of five-number summary, outliers, and upper and lower boundaries, including how to find each of the values. Students should be able to draw the modified box plot.

Answers

9. upper boundary = 116.5
   lower boundary = −23.5
10. Increasing the third quartile value by ten would change your upper boundary to 141.5, because the interquartile range would also increase.
11. Disagree. The IQR is 55. $1.5(55) = 82.5$; $56 + 82.5 = 138.5$; therefore, any number larger than 138.5 is considered an outlier.

ACTIVITY 37 continued

Lesson 37-2
Modified Box Plots

Now you have everything you need to draw the modified box plot. The modified box plot is constructed as follows:

Step 1. Draw the box as usual.

Step 2. Extend the whiskers to the least and greatest data values that are not outliers.

Step 3. Place dots above the scale to indicate the outliers.

8. Follow the directions above to draw the modified box plot for the "Before" data:

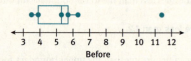

Check Your Understanding

9. A data set has a third quartile of 64 and a first quartile of 29. What are the upper and lower boundary values that separate outliers from the rest of the data set?

10. **Make sense of problems.** If the third quartile of the data set in Item 9 were increased by 10, how would this change the upper boundary for outliers? Explain your answer.

11. **Critique the reasoning of others.** In a survey of 21 teenage girls about their text message usage for one month, the five-number summary is minimum = 0, $Q_1 = 1$, median = 31, $Q_3 = 56$, and maximum = 1305. In analyzing these data, Michelle determined that there are no outliers. Do you agree or disagree? Explain.

Lesson 37-2
Modified Box Plots

LESSON 37-2 PRACTICE

12. Describe the effects an outlier can have on a set of data.
13. **Construct viable arguments.** In a data set, the third quartile is 36 and the first quartile is 12. Would a value of 52 be considered an outlier? Why or why not?
14. Lisa recorded the heights of her classmates. Her data are shown in the table below.

Heights of Classmates in Inches				
61	58	62	60	57
67	68	61	64	70
72	64	63	59	69

Calculate the upper and lower boundaries for outliers for the heights of Lisa's classmates.

15. List two values that would be considered outliers in the data set in Item 14. Include one value that is less than the lower boundary for outliers and one that is greater than the upper boundary for outliers.

ACTIVITY 37 Continued

Lesson 37-3

PLAN

Materials
- calculator

Pacing: 1 class period

Chunking the Lesson
Introduction
#1–4   #5–7   #8–10
#11   #12   #13–14   #15–16
Check Your Understanding
Lesson Practice

TEACH

Bell-Ringer Activity
Calculate the mean and standard deviation for the following data:
10, 11, 12, 14, 14, 16. [mean = 12.83, standard deviation = 2.23]

Developing Math Language
There are a great number of math terms in this lesson that may be new to students. Be sure to have students discuss and use the class Interactive Word Wall to assist with completing this lesson.

1–4 Shared Reading, Create Representations, Look for a Pattern
Students will engage in a discussion and gather data on the number of hours that a student slept the previous night. The reason behind this is so that the students can look at a dot plot of their class data and, using the statistics learned in the previous activities, make statements about the shape of the distribution.

Differentiating Instruction

Students who need extra help with the terminology addressed here may understand better through an exercise in which the students line up from shortest to tallest. They should be able to see that many of the students are close to the center height. There are only a few who are much shorter or much taller than the others.

ACTIVITY 37 continued

Lesson 37-3
Normally Distributed

My Notes

Learning Targets:
- Use the mean and standard deviation to fit a normal distribution.
- Develop an understanding of the normal distribution.
- Use technology to estimate the percentages under the normal curve.

SUGGESTED LEARNING STRATEGIES: Visualization, Think-Pair-Share, Create Representations, Look for a Pattern, Quickwrite

There is a relationship between sleep quality and health. Many studies have concluded that good-quality rest helps to relieve health issues such as high blood pressure, depression, weight gain or loss, and fatigue.

Answers to Items 1–4 will vary.

1. How many hours did you spend sleeping in the last 24 hours?

2. Gather all of your classmates' answers to Item 1. In the *My Notes* section of this page, create a dot plot of the number of hours that students in your class spent sleeping.

3. Use the dot plot to describe the data. Identify the mean, standard deviation, maximum, and minimum.

4. What do you consider to be a normal amount of time spent sleeping in a 24-hour period?

Data can be distributed in different ways.

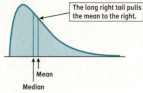 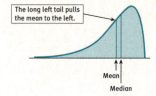

The majority of the data can be grouped to the left (skewed right).   The majority of the data can be grouped to the right (skewed left).

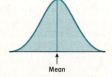

The majority of the data can be grouped symmetrically around the center.

This last type of distribution is called a **normal distribution**.

MATH TERMS

In a **normal distribution**, the data values are grouped symmetrically about the mean, with most of the data values occurring near the mean. Because of its shape, a normal distribution is sometimes called a bell curve.

Lesson 37-3
Normally Distributed

Many real-world data are approximately normally distributed—for instance, height, weight, grades, blood pressure, and time people spend sleeping. Because so many data sets can be modeled by a normal distribution, a table of values is used to analyze and make decisions about normally distributed data. For this to be accomplished, the data values are converted to "scores" that can be easily compared. This is called *standardizing*. A standard score means that the data have a mean of 0 and a standard deviation of 1.

5. Below is a list of the hours slept by the students in Mr. Trent's class. Calculate the mean number of hours these students slept.
   2, 4, 5, 6, 7, 7, 7, 8, 9, 10
   6.5 hours

6. Find the probability that a student in Mr. Trent's class slept more than 7 hours.
   3 out of 10, or 30%

Below is a graph of the hours slept by Mr. Trent's students. Dashed vertical lines have been drawn at one standard deviation above and below the mean.

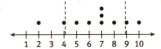

7. Calculate the percent of students whose time spent sleeping is within one standard deviation above or below the mean.
   5, 6, 7, 7, 7, and 8, so 6 out of 10, or 60%

The first step in standardizing the data values from Mr. Trent's class is to calculate each value's deviation from the mean.

8. For each data value, calculate the deviation from the mean, $(x - \bar{x})$, and fill in the second column of the table. Leave the third column blank for now.

Hours Slept and Deviation from the Mean		
Hours	$(x - \bar{x})$	$\dfrac{x - \bar{x}}{s}$
2	−4.5	−1.875
4	−2.5	−1.042
5	−1.5	−0.625
6	−0.5	−0.208
7	0.5	0.208
7	0.5	0.208
7	0.5	0.208
8	1.5	0.625
9	2.5	1.042
10	3.5	1.458

ACTIVITY 37 Continued

8–10 Activating Prior Knowledge
Students are completing the deviation score, which is a review of previous work. They will use this skill in the rest of the activity.

11 Create Representations, Look for a Pattern Students will complete the table using the given z formula. The process is to divide the deviation by 2.4, which can be completed with a calculator. If calculators are not available, the teacher may want to give them the answers. The patterns that develop are important in helping to visualize the area.

12 Visualization, Use Manipulatives Lead a discussion with the students, visually demonstrating the grouping of the data, including mean and upper and lower boundaries, based on the context of the number of hours slept.

ACTIVITY 37 continued

My Notes

MATH TERMS

A **z score** is a standard score that indicates by how many standard deviations a data value is above or below the mean.

Lesson 37-3
Normally Distributed

9. Calculate the sum of the deviations from the mean.
   0

10. Explain the numerical value that you calculated in Item 9.
    Because you are subtracting the mean from each amount of time slept, the sum equals 0.

The next step is to divide by the standard deviation.

11. The standard deviation s for Mr. Trent's class is 2.4. Use this to complete the last column of the table in Item 8. Round values to the nearest thousandth, if necessary.

    See table in Item 8.

The numbers in the last column of the table represent the standardized scores for each data value. A standardized score is called a **z score**: $z = \frac{x - \bar{x}}{s}$. The set of z scores is a data set with mean 0 and standard deviation 1. This represents a standardized version of the original data set.

12. Specific z scores of ±1 are indicated with vertical lines on the graph below. Draw vertical lines to indicate z scores of ±2 and ±3.

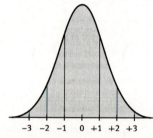

The graph above shows a standardized normal distribution, which can be used to find probabilities. For example, the probability that a data value lies between −1 and +1 standard deviation from the mean is about 68%.

Use a normal distribution and a calculator to estimate the probability that a randomly selected student from Mr. Trent's class slept between 7 and 10 hours. The steps for one type of graphing calculator are:

Step 1. Press 2nd VARS.

Step 2. Select normalcdf(.

Step 3. Determine the lower boundary and enter that number, 7, followed by a comma.

Step 4. Determine the upper boundary and enter that number, 10, followed by a comma.

Step 5. Enter the mean, 6.5, followed by a comma.

Step 6. Enter the standard deviation, 2.4, followed by).

Step 7. Press Enter.

---

550 SpringBoard® Mathematics **Algebra 1, Unit 6** • Probability and Statistics

Lesson 37-3
Normally Distributed

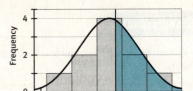

13. Using a calculator, find this probability.
    0.345

14. Compare your answer to Item 13 with your answer to Item 6.
    The two values should be close.

15. **Make sense of problems.** Use a normal distribution and a graphing calculator to estimate the probability that a student in Mr. Trent's class slept fewer than 6 hours. Does the answer that the calculator gives make sense when you look at the data? Explain.

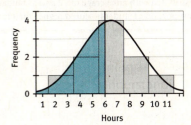

normalcdf(0, 6, 6.5, 2.4) = 0.414. The actual probability is $\frac{3}{10} = 0.3$, but by looking at the graph we see that the normal curve includes values that could be, for example, sleeping 1 hour.

16. Estimate the probability that a student in Mr. Trent's class slept fewer hours than you did in the last 24 hours.
    Answers will vary.

ACTIVITY 37 Continued

Check Your Understanding

Debrief students' answers to these items to ensure that students understand the normal distribution and the probability information it provides.

17. Answers will vary.
18. Answers will vary. Use a calculator to determine the percent.
19. Answers will vary. Sample answer: how many times have students in your class have traveled to China? Most students will answer zero and there will be little variation.
20. normalcdf(1.76, 2, 1.69, 0.05) = 0.081; there is about an 8% probability of this happening.
21. normalcdf(0, 1.62, 1.69, 0.05) = 0.081; there is about an 8% probability of this happening.
22. normalcdf(78, 100, 72,5) = 0.115 = 11.5% of the class

ACTIVITY 37 continued

Lesson 37-3
Normally Distributed

My Notes

Check Your Understanding

17. How many students from your class slept fewer hours than you did in the last 24 hours?

18. What percent of the students in your class slept between 6 and 8 hours in the last 24 hours? Explain how you would find this value and then calculate it.

19. You have discussed several examples of data sets that are normally distributed. Give an example of a real-world data set that would not be normally distributed and explain why.

20. The weights of 1.69-oz bags of M&Ms are normally distributed with a mean of 1.69 oz and a standard deviation of 0.05 oz. What is the probability that a bag selected at random weighs between 1.76 oz and 2 oz?

21. What is the probability that a bag of M&Ms selected at random weighs less than 1.62 oz?

22. Ian scored a 78 on last week's algebra test. The scores for the class were normally distributed with an average of 72 and a standard deviation of 5. What proportion of students scored higher than Ian?

Lesson 37-3
Normally Distributed

LESSON 37-3 PRACTICE

Model with mathematics. The times between eruptions of Old Faithful, a geyser at Yellowstone National Park, vary from 44 to 122 minutes. The average time between eruptions is 91 minutes. The table below lists the times between eruptions for January 1, 2011.

Time in Minutes	
85	85
85	92
100	88
85	90
99	100
91	101
86	96

23. Draw a dot plot and calculate the mean and standard deviation.
24. Determine the interval that represents 1 standard deviation on either side of the mean. Calculate the proportion of data values that lie in this interval, and show this on your dot plot.
25. Use a normal distribution to estimate the proportion of data values that lie between 85 and 92 minutes.
26. Compare your answers to Items 24 and 25.
27. Write a statement that explains how a normal distribution is related to the data on eruptions of Old Faithful.

ACTIVITY 37 Continued

ACTIVITY PRACTICE

1. The median of the distribution of the Cost/Serving amount is more apparent in the box plot than in the dot plot.
2. The dot plot shows that the data falls into two clusters.
3. The greatest number of Cooper's hawks weighed between 400 g and 480 g. A smaller subgroup weighed between 270 g and 290 g. Explanations may vary. This may be due to the time of year the birds breed and eggs hatch.
4.

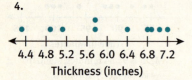

5. Yes; all locations randomly selected on the surface of the lake had an ice thickness of more than 4 inches.
6. No; although most randomly selected locations on the surface of the lake had an ice thickness of more than 5 inches, I would not snowmobile on the surface of the lake.
7. 6 inches
8. 0.95 inches
9. Less worried because the lake is most likely of uniform thickness.
10. More worried because the thickness of the ice is less uniform. It is likely to have more areas where the ice is too thin.

ACTIVITY 37 continued
Dot and Box Plots and the Normal Distribution
Disturbing Coyotes

ACTIVITY 37 PRACTICE
Write your answers on notebook paper.
Show your work.

A restaurant specializing in Mexican food offers nine different choices of enchiladas. They vary in cost and in sodium content. Data for the nine options are given in the table below.

Option	Cost/Serving (dollars)	Sodium Content (mg)
1	3.03	780
2	1.07	1570
3	1.28	1500
4	1.53	1370
5	1.05	1700
6	1.27	1330
7	2.34	440
8	2.47	520
9	2.09	660

1. A dot plot and a box plot of the Cost/Serving amounts are shown below. What feature of the data set is apparent in the box plot but not particularly apparent in the dot plot?

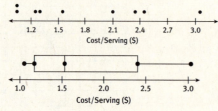

2. A dot plot and a box plot of the Sodium Content amounts are shown below. What feature of the data set is hidden in the box plot but apparent in the dot plot?

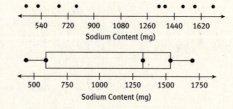

3. In a study of raptors in the western United States, 110 Cooper's hawks were trapped and their weights (in grams) recorded. A dot plot of these weights is shown below. What interesting feature do you notice about this data set? What do you think might be the reason for this interesting feature?

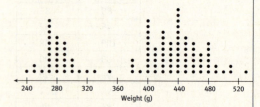

For winter sports enthusiasts, the thickness of ice is a significant safety issue. The Minnesota Department of Natural Resources recommends that ice thickness be at least 4 inches for walking or skating on the ice, and at least 5 inches for operating a snowmobile or all-terrain vehicle on the ice. Ice thicknesses (in inches) were measured at 10 randomly selected locations on the surface of a lake. The thicknesses were as follows:

5.8, 6.4, 6.9, 7.2, 5.1, 4.9, 4.3, 5.8, 7.0, 6.8

4. Construct a dot plot of the ice thicknesses.
5. On the basis of your dot plot, do you think it is safe to play hockey on this lake? Explain why or why not.
6. On the basis of your dot plot, do you think it is safe to operate a snowmobile on this lake? Explain why or why not.
7. Calculate the mean ice thickness for the locations in this sample.
8. Calculate the standard deviation of the ice thicknesses.
9. If the mean of the thicknesses were greater and the standard deviation were the same, would you be more worried or less worried about operating a snowmobile on the ice on this lake? Explain.
10. If the mean of the thicknesses were the same and the standard deviation were greater, would you be more worried or less worried about walking or skating on the ice on this lake? Explain.

Dot and Box Plots and the Normal Distribution
Disturbing Coyotes

ACTIVITY 37
continued

Distributors of soft drinks are aware that end-aisle displays in stores are effective for increasing sales. A distributor is testing new designs of displays, where the image on the display is varied. Each image pictures one or two smiling people holding an open container of the soft drink. The distributor would like to know which images increase sales the most. The three different images are one man, one woman, and a pair of individuals (one man and one woman). Each image was used in 11 stores for 1 month, and the percent increases in sales compared to the same month of the previous year were recorded. Data from the 33 stores are shown in the table below.

Percent Increase in Sales

Image: One Man	Image: One Woman	Image: Man and Woman
4.79	5.71	8.18
5.71	6.29	9.14
5.74	7.44	9.70
5.54	6.03	9.25
4.43	5.54	7.40
6.42	5.90	9.25
6.07	5.23	8.42
3.97	7.96	8.12
5.84	4.75	9.07
5.55	4.68	7.84
6.76	5.90	8.09
5.62	5.71	9.18

11. For each of these three data sets, calculate the five-number summary and the upper and lower boundaries for outliers.

	Man	Woman	Both
Minimum			
First quartile			
Median			
Third quartile			
Maximum			
Lower outlier boundary			
Upper outlier boundary			

12. Use the information from the table in Item 11 to sketch modified box plots of these three data sets. Be sure to indicate any outliers.

13. In a few sentences, describe the similarities and differences among the three data sets.

14. Which image would you recommend that the distributor use and why?

ACTIVITY 37 Continued

11.

	Man	Woman	Both
Minimum	3.97	4.68	7.4
Q_1	4.9775	5.3075	8.0975
Median	5.665	5.805	8.745
Q_3	6.0125	6.225	9.2325
Maximum	6.76	7.96	9.7
Lower outlier boundary	3.425	3.9313	6.395
Upper outlier boundary	7.565	7.6013	10.935

13. Sample answer: The median percentage increases for the images with one man and one woman are about the same (5.7% and 5.8%), but the image with both a man and a woman indicates a significantly higher median percentage increase (8.7%). The variability of the distribution for all three images is about the same, as measured by the IQR. The distributions are roughly symmetric for the image with one man and the image with both individuals; the distribution for the image with one woman is skewed to the right and has two outliers (7.96).

14. Sample answer: I would recommend the image with both a man and a woman. The increases in sales for this image were greater than for the other two images, and the increases didn't vary as much from store to store.

12.

Box plots showing Man, Woman, and Both along a Percent Increase axis from 3 to 10.

ACTIVITY 37 Continued

15.

	Mail	Phone	Direct
Minimum	1000	900	900
Q_1	1100	1500	1000
Median	1300	1750	1200
Q_3	1500	1850	1350
Maximum	1800	2000	1550
Lower outlier boundary	500	975	475
Upper outlier boundary	2100	2375	1875

16.

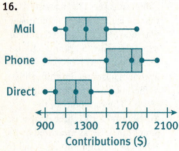

17. The Phone method appears to give the best results on average.
18. A standard box plot of a data set that contained any outliers would have noticeably long whiskers.
19. The interquartile range is 13. The value 45 is considered an outlier because it falls outside the upper boundary of 40.5.
20. Sample answer: Forty-five is an outlier. It represents that someone has visited 45 states. Yes, this makes sense as an outlier because it is very uncommon for someone to have visited 45 of the 50 states.

21.

22. Sample answer: Adding the two new students will decrease the lower boundary and increase the upper boundary. The upper boundary is now 43.75; therefore, 45 is still considered an outlier.
23. Answers will vary.

ADDITIONAL PRACTICE

If students need more practice on drawing box plots and modified box plots and finding outliers, see the eBook Teacher Resources for additional practice problems.

ACTIVITY 37 continued

Dot and Box Plots and the Normal Distribution
Disturbing Coyotes

A regional symphony orchestra needs money to repair their theater, which was seriously damaged by flooding. They have tested three different methods of asking for donations: mail, phone, and direct appeal at social gatherings. Each method was used with 11 potential donors, and the amounts donated for each method are shown below. Use the table for Items 15–17.

Contributions ($)

Mail	Phone	Direct
1000	1700	900
1500	1800	1000
1200	1900	1200
1800	1750	1500
1600	2000	1200
1100	1700	1550
1000	1800	1000
1250	1850	1100
1400	1500	1250
1300	900	1250
1400	1400	1350

15. Complete the table below.

Data Summary Table ($)

	Mail	Phone	Direct
Minimum			
First quartile	1100	1500	1000
Median	1300	1750	1200
Third quartile	1500	1850	1350
Maximum			
Lower outlier boundary			
Upper outlier boundary			

16. Construct modified box plots for the different methods.
17. Based on the data and your box plots, which method would you recommend and why?
18. Sometimes it is not clear whether a box plot is a modified box plot or a standard box plot. If you were looking at a box plot and outliers were not visible, what characteristics of the plot would lead you to believe it was standard rather than modified?

The following values represent the number of states visited by students in a class:

3, 12, 17, 2, 21, 14, 14, 8, 45, 29

Use these data for Items 19–22.

19. Find the interquartile range and any outliers for the data set.
20. If you found an outlier in Item 19, what does this number represent? Does it make sense that this number would be an outlier in this context? Explain your answer.
21. Create a modified box plot for the data.
22. Two new students joined the class, both of whom have visited only two states each. What effects, if any, does this have on the upper and lower boundaries for outliers?

MATHEMATICAL PRACTICES
Use Appropriate Tools Strategically

23. In Item 12 of Lesson 37-1, you were asked about advantages of using box plots and dot plots to describe and compare distributions of scores. Do you think the advantages you found would exist not only for these data, but for numerical data in general? Explain.

Comparing Univariate Distributions
SPLITTING THE BILL

Embedded Assessment 1
Use after Activity 37

In restaurant settings, groups of diners are faced with the problem of how to pay the bill. Three common methods are (a) each pays his or her own (PYO), (b) the bill is split evenly (SE), and (c) someone else pays the entire bill (FREE).

In a recent experiment, diners were told which method would be used to see if this would have an effect on what people ordered. Twelve diners were told that they would each pay their own share of the bill. The cost of the food and drink ordered by each individual was determined. This was repeated with a different group of 12 people who were told the bill would be split evenly and a third group of 12 diners who were told that the experimenter would pay for everyone. The data from the first two groups (PYO and SE) are summarized in the table and the box plots below.

	Cost of Food and Drink (dollars)		
Statistics	Pay Your Own (PYO)	Split Evenly (SE)	Free Meal (FREE)
Minimum:	12.00	22.00	
First quartile:	31.00	40.00	
Median:	39.50	46.50	
Third quartile:	45.75	61.50	
Maximum:	59.00	81.00	
IQR:	14.75	21.50	
Mean:	37.29	50.92	
Standard deviation:	12.54	14.33	

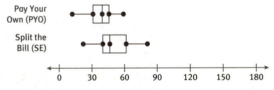

1. Using only the box plots, comment on similarities and differences in the two data sets.
2. Based on the summary statistics in the table, comment on the center and spread of the data sets for those in the PYO group and those in the SE group.

The cost data for the FREE group are shown in the table below.

168	123	101	94	81	75
69	61	59	57	51	49

Embedded Assessment 1

Assessment Focus
- Visual comparison of univariate graphical displays
- Computational comparisons of center and spread
- Computing specific measures of center and spread (including five-number summary)
- Determining outliers
- Creating modified box plots
- Determining appropriate measures of variability

Materials
- calculator

Answer Key
1. Answers will vary. The median is higher and the spread is greater for the SE group.
2. All the statistics increase! The Split the Bill treatment results in higher and more variable costs to the individuals.

Common Core State Standards for Embedded Assessment 1

HSS-ID.A.1 Represent data with plots on the real number line (dot plots, histograms, and box plots).

HSS-ID.A.2 Use statistics appropriate to the shape of the data distribution to compare center (median, mean) and spread (interquartile range, standard deviation) of two or more different data sets.

HSS-ID.A.3 Interpret differences in shape, center, and spread in the context of the data sets, accounting for possible effects of extreme data points (outliers).

Embedded Assessment 1

3.

Statistics	Cost of Food and Drink ($)		
	Pay Your Own	Split The Bill	Free Meal
Minimum:	12.00	22.00	49.00
First Quartile:	31.00	40.00	57.50 58
Median:	39.50	46.50	72.00
Third Quartile:	45.75	61.50	99.25 97.5
Maximum:	59.00	81.00	168.00
IQR:	14.75	21.50	41.75 39.5
Mean:	37.29	50.92	82.33
Standard Deviation:	12.54	14.33	34.96

4.

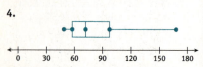

5. The presence of the outlier suggests that the *IQR* would be better. It is less affected by the outlier.

> **TEACHER to TEACHER**
>
> You may wish to read through the scoring guide with students and discuss the differences in the expectations at each level. Check that students understand the terms used.

Unpacking Embedded Assessment 2

Once students have completed this Embedded Assessment, turn to Embedded Assessment 2 and unpack it with them. Use a graphic organizer to help students understand the concepts they will need to know to be successful on Embedded Assessment 2.

Embedded Assessment 1
Use after Activity 37

Comparing Univariate Distributions
SPLITTING THE BILL

3. Complete the FREE column in the table from the previous page.

4. On the scale below, sketch the box plot for the FREE group.

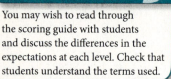

5. If you wanted to describe variability in the FREE group, would you use the *IQR* or the standard deviation? Explain your choice.

Scoring Guide	Exemplary	Proficient	Emerging	Incomplete
	The solution demonstrates the following characteristics:			
Mathematics Knowledge and Thinking (Items 1–5)	• Clear and accurate understanding of univariate data distributions, including center, shape, and spread	• Adequate understanding of univariate data distributions, including center, shape, and spread	• Partial understanding of univariate data distributions, center, shape, and spread	• Little or no understanding of univariate data distributions, center, shape, or spread
Problem Solving (Item 4)	• An appropriate and efficient strategy that results in a correct answer	• A strategy that may include unnecessary steps but results in a correct answer	• A strategy that results in some incorrect answers	• No clear strategy when solving problems
Mathematical Modeling / Representations (Items 1–5)	• Clear and accurate understanding of representations of univariate data • Clear and accurate understanding of how to display data in box plots	• Adequate understanding of representations of univariate data • Mostly accurate displays of data in box plots	• Partial understanding of representations of univariate data • Partial understanding of how to display data in box plots	• Little or no understanding of representations of univariate data • Inaccurate or incomplete understanding of how to display data in box plots
Reasoning and Communication (Items 1, 2, 5)	• Precise use of appropriate math terms and language to characterize univariate data distributions, including center, spread, and variability	• Correct characterization of univariate data distributions, including center, spread, and variability	• Misleading or confusing characterization of univariate data distributions	• Incomplete or inaccurate characterization of univariate data distributions

Correlation
What's the Relationship?
Lesson 38-1 Scatter Plots

ACTIVITY 38

Learning Targets:
- Describe a linear relationship between two numerical variables in terms of direction and strength.
- Use the correlation coefficient to describe the strength and direction of a linear relationship between two numerical variables.

> **SUGGESTED LEARNING STRATEGIES:** Graphic Organizer, Think-Pair-Share, Create Representations, Predict and Confirm, Quickwrite

Scatter plots are used to visualize the relationship between two numerical variables. When you look at a scatter plot, determine whether there appears to be a relationship (pattern) between the two variables.

For example, consider the following three scatter plots.

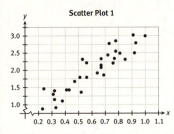 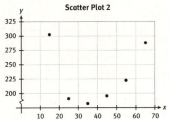 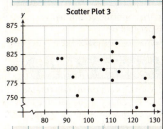

For Scatter Plot 1, there does appear to be a relationship between x and y because greater values of x tend to be paired with greater values of y. Notice that the pattern in the plot looks roughly linear, so you would say that there is a linear relationship between these two variables.

Two numerical variables are related if they tend to vary together in a predictable way.

1. For Scatter Plot 2 above, does there appear to be a relationship between x and y? If so, describe the pattern.
   Yes, there does appear to be a pattern. The pattern is curved.

2. For Scatter Plot 3 above, does there appear to be a relationship between x and y? If so, describe the pattern.
   No, there isn't really a pattern in the scatter plot.

Common Core State Standards for Activity 38

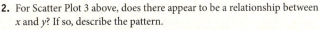

HSS-ID.C.8 Compute (using technology) and interpret the correlation coefficient of a linear fit.
HSS-ID.C.9 Distinguish between correlation and causation.

Activity 38 • Correlation 559

ACTIVITY 38 Continued

Developing Math Language
This lesson develops the term *correlation* as a way to describe the strength and direction of a linear relationship and the use of the value *r* to numerically describe the data.

3 Predict and Confirm, Think-Pair-Share
Students will participate in a matching exercise to reinforce concepts that are key both for describing bivariate numerical data and for properly using and interpreting the correlation coefficient. (Note: This could be run as a group or all-class exercise.) Students are then asked to consider and to record which characteristics of the graphs led them to describe the bivariate relationships as linear, weak, etc. Positive and negative association concepts are presented in "math terms."

ACTIVITY 38 continued

Lesson 38-1
Scatter Plots

MATH TERMS
Two numerical variables are **correlated** if one variable tends to increase (or decrease) as the other variable increases.

MATH TERMS
The **correlation coefficient** is a measure of the strength and direction of a linear relationship.

The correlation coefficient is denoted by *r*.

MATH TIP
Variables are **positively related** if lesser values of one variable tend to occur with lesser values of the other variable.

Variables are **negatively related** if lesser values of one variable tend to occur with greater values of the other variable.

DISCUSSION GROUP TIP
As you read and define new terms, discuss their meanings with other group members and make connections to prior learning.

Just as the mean and standard deviation are used to describe center and variability in a data set, there is a summary statistic to describe the strength (how close the points are to a line) and direction (positive or negative) of a linear relationship. This statistic is called the **correlation coefficient** and is denoted by *r*.

3. Given below are seven scatter plots and seven verbal descriptions of relationships. Match each scatter plot with the appropriate description. (Each scatter plot goes with one and only one description.)

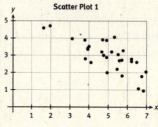
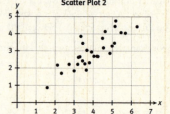
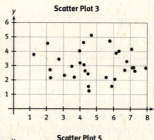
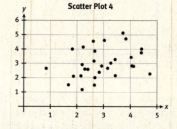
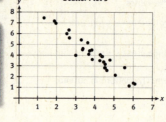
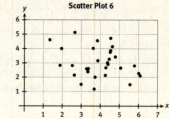
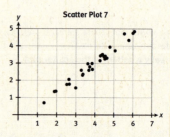

560 SpringBoard® Mathematics **Algebra 1, Unit 6** • Probability and Statistics

Lesson 38-1
Scatter Plots

A. Very strong positive linear relationship ($r = 0.981$) ___7___

B. Relatively strong positive linear relationship ($r = 0.828$) ___2___

C. Relatively weak positive linear relationship ($r = 0.310$) ___4___

D. Very slight or no linear relationship ($r = 0.043$) ___3___

E. Relatively weak negative linear relationship ($r = -0.238$) ___6___

F. Relatively strong negative linear relationship ($r = 0.772$) ___1___

G. Very strong negative linear relationship ($r = -0.95$) ___5___

4. What feature(s) of the scatter plots did you consider when deciding whether a relationship was positive or negative?
   Answers may vary; whether or not the greater values of one variable tend to be paired with the greater (for positive) or lesser (for negative) values of the other variable

5. What feature(s) of the scatter plots did you consider when deciding whether a relationship was relatively weak, relatively strong, or very strong?
   Answers may vary. If the points were "close" to lying on a line, the relationship was judged to be strong.

6. **Make sense of problems.** Examine the values of r for each relationship in Item 3. How does the value of r relate to the scatter plots? What makes r increase or decrease?
   Answers may vary. The weaker the relationship, the closer the value of r is to 0. The value of r is greater than 0 when the relationship is positive; the stronger the positive relationship, the closer r is to 1. The value of r is less than 0 when the relationship is negative; the stronger the negative relationship, the closer r is to -1.

Here is a summary of important characteristics of r:
- The value of r quantifies the strength of a linear relationship.
- The sign of r describes the direction of the relationship: positive or negative.
- r ranges in value between -1 (perfect negative linear relationship) and $+1$ (perfect positive linear relationship).

ACTIVITY 38 Continued

Check Your Understanding

Debrief students' answers to these items to be sure that they know the difference in strength and direction of linear relationships.

Answers

7. **a.** positive linear relationship of moderate strength
   b. very weak or no relationship
   c. perfect (very strong) linear relationship

ACTIVITY 38
continued

Lesson 38-1
Scatter Plots

Check Your Understanding

The following table displays costs to travel, round-trip, to various cities from Cedar Rapids, Iowa. The costs are calculated assuming a June 1 departure and a 3-day stay. Driving costs were calculated based on $0.20 per mile.

Travel Cost (dollars)			
Destination	**Train**	**Plane**	**Car**
New York City	268	391	204
Chicago	74	453	49
Atlanta	483	703	168
Washington, D.C.	254	577	186
New Orleans	338	342	189
Denver	221	384	160
Albuquerque	354	486	222
Seattle	510	647	367
San Francisco	290	435	385
Los Angeles	390	299	362
Kansas City	184	523	64

Scatter plots of cost versus distance for each of the three travel methods are shown below.

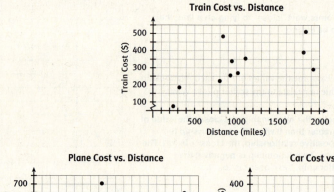

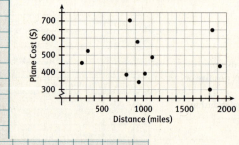

Lesson 38-1
Scatter Plots

7. **Reason abstractly.** How would you describe the relationship between cost and distance for each method of transportation? Be sure to indicate whether you think the relationship is linear and to comment on the strength and direction of the relationship.
   a. Train
   b. Plane
   c. Car

LESSON 38-1 PRACTICE

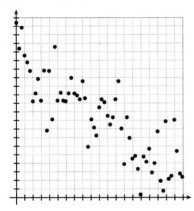

8. Describe the relationship shown in the scatter plot above.
9. **Reason abstractly.** In your own words, describe the similarities and differences between a scatter plot that shows a strong positive relationship and a scatter plot that shows a weak positive relationship.
10. What type of relationship would you expect to see between height and age? Explain your answer.
11. Describe two real-world quantities that would have a strong negative relationship.
12. Describe two real-world quantities that would have no correlation.
13. For positive linear relationships, as the value of r increases, is the linear relationship getting stronger or weaker?

ACTIVITY 38 Continued

Lesson 38-2

PLAN

Materials
- calculator

Pacing: 1 class period

Chunking the Lesson
Introduction #1–3 #4
Check Your Understanding
Lesson Practice

> **TEACHER to TEACHER**
>
> Prior to the students attempting this next chunk, it is assumed that the instructor will have demonstrated how students can use available classroom technology to compute the correlation coefficient (r) of a data set using technology. If no access to technology is available, consider simply providing students with the values of the correlation coefficients for Items 1 and 2.
>
> The data sets used for Items 1–3 are taken from *Consumer Reports*, a source for information about products. As a framework for these questions, consider asking the students if they think that more expensive items necessarily provide better service, quality, etc.

TEACH

Bell-Ringer Activity
Have students list an example of two items that they believe would have a positive correlation and two items they believe would have a negative correlation. Discuss students' examples with the class.

Technology Tip
Remember to turn the diagnostic on to calculate the correlation coefficient.

Introduction, 1–3 Make a Prediction, Activating Prior Knowledge, Quickwrite Students make a prediction based on the given information. They apply the skill previously learned about linear relationships in a real-life context.

Lesson 38-2
Correlation Coefficient

Learning Targets:
- Calculate correlation.
- Distinguish between correlation and causation.

SUGGESTED LEARNING STRATEGIES: Think-Pair-Share, Vocabulary Organizer, Quickwrite

The calculation of r gives additional information that helps to describe the data.

This data set shows price (in dollars) and quality ratings for 12 different brands of bike helmets. The quality rating is a number from 0 (worst) to 100 (best) that measures various factors such as how well the helmet absorbed the force of an impact, the strength and ventilation of the helmet, and its ease of use.

Bicycle Helmets	Price (dollars)	35	20	30	40	50	23	30	18	40	28	20	25
	Quality Rating	65	61	60	55	54	47	47	43	42	41	40	32

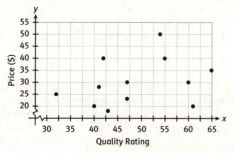

1. **a.** How would you describe the relationship between price and quality rating?

   It appears that generally quality tends to increase as price increases.

   b. Make a prediction of what you think the correlation coefficient might be.

   Answers will vary.

2. Using a graphing calculator, enter the prices as one list and the quality ratings as another list. What is the value of the correlation coefficient for these two variables?

   0.3034

Lesson 38-2
Correlation Coefficient

3. **Reason abstractly.** How would you interpret the value of the correlation coefficient in the context of this problem?
   There is a positive linear relationship between price and quality, but the relationship is not very strong.

At this point you have used scatter plots to visually represent the relationship between two numerical variables and you have used a numerical measure to describe the strength and direction of a linear relationship. When a relationship is uncovered by statistics, the next task is to explain its meaning.

Sometimes the interpretation of a relationship may not be obvious. For example, across European countries there is a positive linear relationship between the number of storks and the number of newborn babies. Do storks bring babies? Are storks attracted by babies? Are both babies and storks brought by the tooth fairy? Do parents with newborns have warmer houses, and therefore their chimneys attract storks looking for warm places to nest?

Humans want to make sense of their world, and sometimes leap too quickly from seeing a correlation to inferring *causation* (a cause-and-effect relationship between two variables). This tendency should be resisted! There are many reasons why two variables might be related other than cause and effect.

Here are some common examples where a correlation should not be interpreted as a cause-and-effect relationship:
- The number of fire engines responding to a fire is positively correlated with the total damage. (Should fewer fire engines—perhaps 0—be sent to fires to reduce damage?)
- The number of people drowning at beaches is positively correlated with ice cream sales. (Is ice cream dangerous?)
- Shoe size is strongly correlated with reading ability. (Should parents start their children off with size 12?)
- The number of doctors per 1000 people is positively correlated with the rate of serious disease. (Are doctors spreading disease?)

4. **Make sense of problems.** For each of the correlations above, what do you think is the correct explanation for the correlation?
   Fire engines: More engines are sent to larger fires.
   Drowning: Both are the result of increased temperatures.
   Shoe size: Both increase with age.
   Doctors: Doctors and patients tend to occur together in hospitals.

Be sure to use common sense when determining causation.

ACTIVITY 38 Continued

Check Your Understanding
Debrief students' answers to these items to ensure that they understand how to determine if there is a linear relationship and also the strength and direction of that relationship. They should be able to make these decisions based on words and/or graphically.

Answers
5. a. negative correlation
   b. positive correlation
   c. positive correlation
   d. correlation close to zero
6.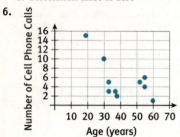

ACTIVITY 38 continued

My Notes

Lesson 38-2
Correlation Coefficient

Check Your Understanding

5. For each of the following pairs of variables, indicate whether you would expect a positive correlation, a negative correlation, or a correlation close to 0. Explain your choice.
   a. Weight of a car and gas mileage
   b. Size and selling price of a house
   c. Height and weight
   d. Height and number of siblings

The table below gives data on age and number of cell phone calls made in a typical day for each person in a random sample of 10 people. Use the table for Items 6–8.

Age (years)	Number of Cell Phone Calls
55	6
33	5
60	1
38	2
55	4
19	15
30	10
33	3
37	3
52	5

6. Sketch a scatter plot of these data using the grid below.

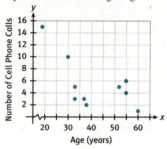

MINI-LESSON: How the correlation coefficient works

Students can gain some insight into how the correlation coefficient, symbolized as r, measures the strength and direction of a linear relation by examining the formula for it:

$$r = \frac{1}{n-1}\sum\left(\frac{x_i - \bar{x}}{s_x}\right)\left(\frac{y_i - \bar{y}}{s_y}\right)$$

In the formula, x_i and y_i are the numbers in the ith ordered pair, (x_i, y_i). The s_x and s_y are the standard deviations of the variables—finding them is one reason the calculations are tedious. The $n-1$ and the standard deviations are constants in this formula, and they act together to force r to have a value between -1 and $+1$, no matter how many points are in the data set. Using some advanced algebra, r can be rewritten like this:

566 SpringBoard® Mathematics **Algebra 1, Unit 6** • Probability and Statistics

Lesson 38-2
Correlation Coefficient

7. Describe the direction and strength of the relationship between these two variables.

8. Calculate the value of the correlation coefficient. Do the sign and the magnitude of the correlation coefficient agree with your answer in Item 7? Explain.

9. Suppose that you shot an arrow into the air and kept track of how high it was every 1.0 second. If you made a scatter plot of the data (time, height), the resulting pattern of points would be in the shape of a parabola. Do you feel the correlation coefficient should be used to describe the strength of the relationship between time and height? Why or why not?

LESSON 38-2 PRACTICE
Model with mathematics. Consider tablet computers with 9- to 10-inch screens.

10. For tablet computers, do you think there is a relationship between price and battery life? If so, do you think the relationship is positive or negative?

11. For tablet computers, do you think there is a relationship between price and weight? If so, do you think the relationship is positive or negative?

MINI-LESSON: How the correlation coefficient works *(continued)*

$$r = \frac{1}{n-1}\left(\frac{1}{s_x}\frac{1}{s_y}\right)\sum(x_i - \bar{x})(y_i - \bar{y})$$

Written in this form, one can see how the calculation of r reacts to a positive or negative relationship. If the two variables, x and y, are positively associated, above average values for the x variable tend to be associated with numbers that are above average for the y variable. In that case, the products, $(x_i - \bar{x})(y_i - \bar{y})$, tend to be more positive than negative and their sum tends to be positive. In a similar manner, for variables that are negatively associated, the products, $(x_i - \bar{x})(y_i - \bar{y})$, tend to be more negative than positive, and their sum tends to be negative. These tendencies for the sums to be positive and negative show up as positive and negative values for r. In the case of no relation, the sum of the products will tend to cancel out, giving a value close to 0.

ACTIVITY 38 Continued

LESSON 38-2 PRACTICE (cont.)

12. 0.3638
13. −0.1267
14. Answers will vary, but the correlation coefficients should be consistent with the direction of the anticipated relationship.

ADAPT

Check students' answers to the Lesson Practice to ensure that they understand how to describe the strength and direction of a linear relationship. Students who may need more practice with the concepts in this lesson may benefit from other prepared scatterplots. Have students speculate as to the data's *r*-value strictly from visual inspection. Also, have students consider numerical relationships that may be strongly correlated but do not have a direct cause-and-effect relation. A famous example: reading comprehension tends to increase with shoe size in elementary school children, but better comprehension won't make for bigger feet!

If some students need additional work on these topics, you may want to access bivariate numerical data sets in previous SpringBoard material or from the Internet (such as lib.stat.cmu.edu/DASL/) and ask students to estimate and then compute the corresponding *r* statistic. Some students may also wish to know more about the formula behind the correlation coefficient and why the statistic "behaves" as it does. Consider sharing some of the Teacher to Teacher/Mini-Lesson addressing the algebra behind the statistic.

Lesson 38-2
Correlation Coefficient

Data for tablet computers with 9- to 10-inch screens are shown in the table and scatter plots below.

9- to 10-inch Tablets		
Price (dollars)	Battery Life (hours)	Weight (pounds)
730	11.6	1.3
570	8.4	1.0
600	11.6	1.3
600	9.3	1.2
600	9.1	1.3
800	8.9	1.3
600	10.5	1.6
850	11.5	1.6
500	11.0	1.6
470	9.0	1.5
500	8.6	1.4
500	8.4	1.6
480	7.4	1.7
500	8.6	1.7
570	7.7	1.7
780	8.1	1.7
580	9.5	2.1

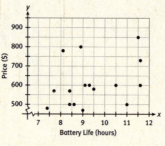
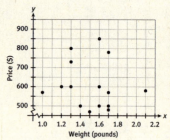

12. Calculate the correlation coefficient for Price and Battery Life.

13. Calculate the correlation coefficient for Price and Weight.

14. Are the values of the correlation coefficients consistent with your predictions in Items 10 and 11? Explain.

Correlation
What's the Relationship?

ACTIVITY 38 continued

ACTIVITY 38 PRACTICE

Write your answers on notebook paper.
Show your work.

1. Describe as precisely as you can how the appearance of a scatter plot showing a positive linear relationship between two quantitative variables differs from the appearance of a scatter plot showing a negative relationship between two quantitative variables.

2. Describe as precisely as you can how the appearance of a scatter plot showing a strong linear relationship between two quantitative variables differs from the appearance of a scatter plot showing a weak linear relationship between two quantitative variables.

The basking shark is the second-largest fish (after the whale shark) swimming in the oceans today. In a study of these creatures, their length and average swimming speed were measured from a safe distance. The results are shown in the table below. Use the table for Items 3–5.

Body Length (meters)	Average Speed (meters/sec)
4.0	0.89
4.5	0.83
4.0	0.76
6.5	0.94
5.5	0.94

3. Sketch a scatter plot of these data.

4. Calculate the correlation coefficient for these data using your available technology.

5. How would you describe this relationship in terms of strength and direction? Support your description with specific references to the scatter plot and/or the correlation coefficient.

One danger of premature human birth is low birth weight. It is thought that low birth weight results in small hippocampus volume, which might be cause for concern because the hippocampus is important in later brain functioning. The scatter plot below displays data from a study of the relationship between hippocampus volume and birth weight in premature infants. The correlation coefficient for these data is $r = 0.51$.

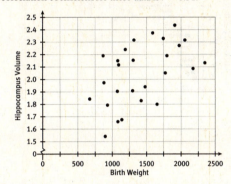

6. Describe the strength and direction of this relationship.

7. Does this relationship appear to be reasonably described as linear? Explain.

ACTIVITY 38 Continued

ACTIVITY PRACTICE

1. The points in the scatterplot of a positive linear relationship are generally oriented from the lower left to the upper right. Scatterplots with a negative linear relationship generally have points oriented from upper left to lower right.

2. In a strong linear relationship, the points tend to cluster tightly around a straight line; in a weak linear relationship, the points are not tightly clustered.

3.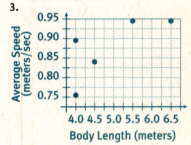

4. $r = 0.749$

5. The relationship as indicated by the scatterplot and the correlation coefficient is positive; there is a lower left to upper right orientation of the points and $r > 0$. The relationship does not appear to be particularly strong—the points do not cluster tightly about any straight line, although the value of the correlation coefficient is 0.749.

6. The direction is positive and the strength is moderate at best.

7. Yes—there is no clear indication of curvature.

When young children are prepared for surgery, a tracheal tube is inserted to allow the unconscious child to breathe. It is very important to get the correct insertion depth. Researchers investigated the relationship between best insertion depth and the weight of the child in a large sample of children, and a scatter plot of their data is shown. The correlation coefficient is $r = 0.878$.

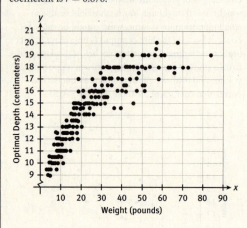

8. Describe the strength and direction of this relationship.

MATHEMATICAL PRACTICES
Reason Abstractly and Quantitatively

9. Does this relationship appear to be reasonably described as linear? Why or why not?

The Best-Fit Line
Regressing Linearly
Lesson 39-1 Line of Best Fit

ACTIVITY 39

Learning Targets:
- Describe the linear relationship between two numerical variables using the best-fit line.
- Use the equation of the best-fit line to make predictions and compare the predictions to actual values.

> **SUGGESTED LEARNING STRATEGIES:** Look for a Pattern, Interactive Word Wall, Predict and Confirm, Graphic Organizer, Discussion Groups

In recent activities, you created scatter plots as a way to graphically summarize bivariate numerical data. In addition, you learned how to use the correlation coefficient as a numerical summary of the strength and direction of a linear relationship.

In this activity, you will see a way to summarize bivariate numerical data called the "best-fit line." You will use technology to determine the slope and y-intercept of the best-fit line for a data set.

1. The scatter plots below show linear relationships of different strengths and directions. For each scatter plot, use your judgment to draw a line that you feel best represents the linear relationship.
   Answers will vary.

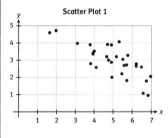
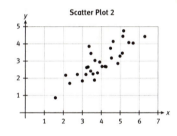
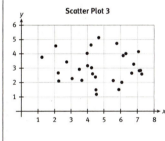
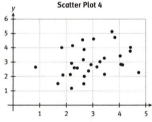

Common Core State Standards for Activity 39

HSS-ID.B.6 Represent data on two quantitative variables on a scatter plot, and describe how the variables are related:
  a. Fit a function to the data; use functions fitted to data to solve problems in the context of the data. Use given functions or choose a function suggested by the context. Emphasize linear, quadratic, and exponential models.
  b. Informally assess the fit of a function by plotting and analyzing residuals.
  c. Fit a linear function for a scatter plot that suggests a linear association.

HSS-ID.C.7 Interpret the slope (rate of change) and the intercept (constant term) of a linear model in the context of the data.

ACTIVITY 39
Investigative

Activity Standards Focus
Students will estimate best-fit lines visually, make predictions from best-fit lines, determine the difference between actual variable value and predicted value (i.e., residual), determine the best-fit line (using technology), interpret slope and intercept of a linear model in context, recognize cases where the use of a best-fit line is not recommended, recognize situations that do not fit a linear model, and use a residual plot to decide if a scatterplot has a linear relationship.

Lesson 39-1

PLAN

Materials
- graphing calculator

Pacing: 1 class period

Chunking the Lesson
#1–2   #3–5   #6–9
Check Your Understanding
Lesson Practice

TEACH

Bell-Ringer Activity
Determine the slope of the line that connects the points (1, 3), (3, 6), (5, 9). Describe what it means to find the slope.

1–2 Activating Prior Knowledge, Look for a Pattern
Students will draw a variety of different lines, discuss the reasons they chose for the placement of the line and compare the lines. Ask the students, "Did you all have the same graph or were there differences? Why do you think that happened?"

Activity 39 • The Best-Fit Line 571

ACTIVITY 39 Continued

Paragraphs Shared Reading Have students read the text following Item 2. Have the class come to a consensus on what "best" means in terms of using the best-fit line for data.

Lesson 39-1
Line of Best Fit

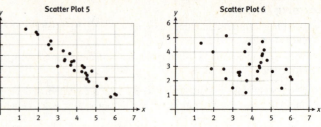

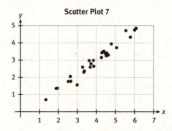

2. Compare the lines you drew with the lines drawn by another student in your class. Did you draw identical lines? Were your lines more similar for scatter plots where the linear relationship was strong or where the linear relationship was weak?

   Answers will vary, but students should see that they don't always agree. Typically, agreement is good when the linear relationship is strong and not as good when the relationship is weak.

Because informal assessments of what line might best describe a linear relationship don't always agree, we need to come to some agreement about what "best" means.

Before we look at how to define the best-fit line, let's first consider how the best-fit line might be used.

One reason for finding a best-fit line to describe the relationship between two variables is so that you can use the line to make predictions. For example, you might want to predict the age (in years) of a black bear from its weight (in pounds). This would be helpful to wildlife biologists, because it is a lot easier to weigh a bear than to ask a bear its age!

572 SpringBoard® Mathematics **Algebra 1, Unit 6** • Probability and Statistics

Lesson 39-1
Line of Best Fit

Suppose you know that for adult black bears, the relationship between age and weight can be approximately described by the line

$$y = -3.69 + 0.115x$$

where y = age in years and x = weight in pounds. You can use this equation to predict the age of a bear that weighs 100 pounds:

predicted age = $-3.69 + 0.115(100) = -3.69 + 11.5 = 7.81$ years

3. Using the equation $y = -3.69 + 0.115x$, what is the predicted age of a bear that weighs 115 pounds?

   9.54 years

The line $y = -3.69 + 0.115x$ is the best-fit line for the following data. These data are from a study in which nine black bears of known age were weighed.

Bear	Weight (x)	Age (y)
1	88.2	6.5
2	88.2	7.5
3	92.6	5.5
4	110.3	8.0
5	112.5	10.5
6	112.5	9.5
7	119.1	10.5
8	121.3	9.0
9	130.1	11.5

4. Construct a scatter plot for the bear data.

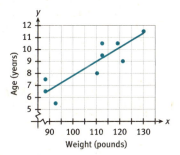

5. Add the best-fit line to your scatter plot.

   (*Hint*: Find two points on the line by picking two x values and using the equation of the best-fit line to find the corresponding predicted ages. Then plot these two (x, predicted age) pairs on the scatter plot and draw the line that goes through those two points.)

   Line shown on scatter plot above.

Lesson 39-1
Line of Best Fit

Below is a scatter plot of the bear data, the best-fit line, and a line that is not the best-fit line.

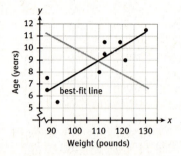

6. Why is the best-fit line a better description of the relationship between age and weight than the other line graphed above?

   Answers may vary. The points in the scatter plot tend to be closer to the best-fit line than to the line graphed above. Also, the overall trend is positive, and the line graphed above has a negative slope.

7. One bear in the data set (Bear 3) was 5.5 years old and weighed 92.6 pounds. If you used the best-fit line ($y = -3.69 + 0.115x$) to predict the age of this bear based on its weight, how far off would you be from the bear's actual age?

   predicted age = $-3.69 + 0.115(92.6) = 6.96$ years

   The actual age was 5.5 years, so the predicted age is off by 1.46 years.

8. If you used the line graphed above to predict the age of this bear, do you think your prediction would be closer to or further from the bear's actual age? What feature(s) of the scatter plot shown above supports your answer?

   It would be further from the actual age because the data point is farther from the line graphed above than from the best-fit line.

9. **Attend to precision.** For the bear that was 5.5 years old and weighed 92.6 pounds, the best-fit line led to a predicted age that was greater than the bear's actual age. Will age predictions based on the best-fit line be greater than the actual age for *all* of the bears in the data set? If so, explain why. If not, give an example of a bear in the data set for which the predicted age is less than the bear's actual age.

   No, some predictions will be less than the bear's actual age. For example, the predicted age of Bear 2 is less than the bear's actual age.

Lesson 39-1
Line of Best Fit

Check Your Understanding

10. The scatter plot below shows two lines, Line 1 and Line 2. One of these lines is the best-fit line. Which one is it?

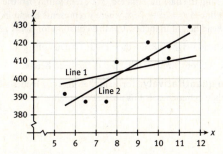

11. Suppose that for students taking a statistics class, the best-fit line for a data set where y is a student's test score (out of 100 points) and x is the number of hours spent studying for the test is $y = 43 + 12x$.

    a. What is the predicted test score for a student who studied for one hour?

    b. What is the predicted test score for a student who studied for three hours?

ACTIVITY 39 Continued

ASSESS

Students' answers to the lesson practice will allow you to assess their understanding of the concept of a line of best fit as well as their ability to make an accurate prediction using a model.

Teacher to Teacher

The lesson practice requires students to change 5 foot 3 inches to 63 inches and also gives the y-value, and the students have to predict the x-value. Students will need to be careful about labeling what x and y represent.

LESSON 39-1 PRACTICE
12. 64.75 inches
13. 8.07 inches
14. Sample answer: The equation predicts that Tori is 62.5 inches tall, and since this is just a prediction, it is possible that she is 64 inches tall.

ADAPT

Check students' answers to the Lesson Practice to ensure that they understand how to make a prediction and also can determine which values represent x and which y. Students who may need more practice with the concepts in this lesson may need to complete the Activity Practice for additional exercises for this lesson. You may assign the items here or use them as a culmination for the activity.

Lesson 39-1
Line of Best Fit

LESSON 39-1 PRACTICE

Mr. Trent examined some data on head height and a person's actual height and found that a person's height is about 7.5 times his or her head height. "Head height" refers to the distance from the top of the head to the bottom of the chin. Using the data he gathered, Mr. Trent found that the equation of the best-fit line is $y = 2.5 + 7.5x$, where y represents height in inches and x represents head height in inches. Use this equation for Items 12–14.

12. Xavier's head height is 8.3 inches. Predict Xavier's height.
13. Tamisha is 5 foot 3 inches tall. Predict her head height.
14. **Reason quantitatively.** Tori said that she is 64 inches tall and her head height is 8 inches. Is this possible? Explain.

Lesson 39-2
Residuals

Learning Targets:
- Use technology to determine the equation of the best-fit line.
- Describe the linear relationship between two numerical variables using the best-fit line.
- Use residuals to investigate whether a given line is an appropriate model of the relationship between numerical variables.

SUGGESTED LEARNING STRATEGIES: Predict and Confirm, Look for a Pattern

Below is a scatter plot of the bear data with the best-fit line and the points in the scatter plot labeled according to which bear the data point represents.

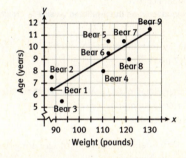

1. For which bears does the best-fit line predict an age that is less than the bear's actual age?
   Bears 1, 2, 5, 6, 7, and 9

2. Look at the points in the scatter plot that correspond to the bears whose predicted ages are less than their actual ages. What do the points all have in common relative to the best-fit line?
   They are all above the best-fit line.

Lesson 39-2
Residuals

For Bear 3, the actual age was 5.5 years and the predicted age from the best-fit line was 6.96 years. The difference between the actual age and the predicted age is

$$5.5 - 6.96 = -1.46$$

3. Look at the scatter plot and locate the point corresponding to Bear 3. What does 1.46 represent in terms of the scatter plot?
   1.46 is the vertical distance from the data point that represents Bear 3 to the best-fit line. The point is 1.46 y-units (years) below the best-fit line.

The difference between an actual y-value and a predicted y-value is called a **residual**. A residual is positive when the actual y-value is greater than the predicted y-value.

4. When is a residual negative?
   A residual is negative when the predicted y-value is greater than the actual y-value.

5. For which of the bears is the residual positive?
   Bears 1, 2, 5, 6, 7, and 9

6. Look at the scatter plot. Do data points that fall above the best-fit line have positive or negative residuals?
   They have positive residuals.

The table below shows the actual ages, predicted ages using the best-fit line, and residuals for the nine bears.

Bear	Age (years)	Predicted Age (years)	Residual
1	6.5	6.45	0.05
2	7.5	6.45	1.05
3	5.5	6.96	−1.46
4	8.0	9.00	−1.00
5	10.5	9.25	1.25
6	9.5	9.25	0.25
7	10.5	10.01	0.49
8	9.0	10.26	−1.26
9	11.5	11.27	0.23

7. **Make sense of problems.** What is the sum of all nine residuals? Does this value surprise you? Explain why or why not.
   −0.40. It is not surprising because some residuals are positive and some are negative.

Note: The sum of the residuals for the best-fit line is equal to zero. Here, because of rounding in the calculation of the slope and y-intercept of the best-fit line and rounding in calculating the predicted values, the sum of the residuals is not exactly 0.

MATH TERMS

A **residual** is a difference between an actual y-value and a predicted y-value.

Residual = actual y − predicted y

Lesson 39-2
Residuals

8. The scatter plot below shows the best-fit line and another line. If you ignore the sign of the residuals, which line has greater residuals overall? (*Hint*: Look at the distances of points to each of the two lines.)
   The other line has greater residuals overall.

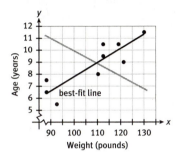

A line is a good description of a bivariate data set if the residuals tend to be small overall. To measure the overall "goodness" of a line, you might think about adding all of the residuals. The problem with this is that some residuals are positive and some are negative, and so you can get a sum that is zero (or close to zero) even for lines that are not good descriptions of the data. So, instead of judging the "goodness" of a line by looking at the sum of the residuals, we look at the **sum of the squared residuals** (**SSR**, for short). The squared residuals are all positive, so positive and negative values don't offset one another.

9. Look again at the scatter plot above that shows the bear data and the two different lines. Which line do you think has the lesser SSR? Explain your reasoning.
   The best-fit line has the lesser SSR because the points tend to be closer to this line.

The *best-fit line* for a particular data set is the line that has the least sum of squared residuals (least SSR). In the scatter plot with the two lines, not only does the best-fit line have an SSR less than that of the other line shown, it has an SSR less than that of *any* other line.

Calculating the equation of the best-fit line by hand is very time-consuming, especially if there are a lot of values in the data set. Because of this, you will use a graphing calculator or computer software to do the calculations.

10. Enter the bear data and use technology to verify that the equation of the best-fit line is $y = -3.69 + 0.115x$.

MATH TIP

SSR stands for the sum of the squared residuals.

MATH TERMS

The **best-fit line** is the line for which the sum of squared residuals (SSR) is less than that of any other line.

ACTIVITY 39 Continued

8–10 Look for a Pattern Students may also ask why the sum of the *squared* residuals is used (rather than the sum of the residuals) to determine the "best" line. Mathematically, the residuals of *any* line will always sum to zero (or near zero with rounding), as generally the sum of any linear model's negative residual values and the sum of its positive residual values will be the same in terms of absolute value.

Differentiating Instruction

If access to the TI-nSpire is available, it can be used to show the boxes of the sum of the squared residuals, or those boxes can be drawn onto a graph. This visual representation can help students to see the process.

MINI-LESSON: Using a TI-83 to Determine a Best-Fit (Least-Squares Regression) Line

1. Press (STAT) and choose the EDIT option.
2. Enter the *x*-values (such as "weight" in Item 8) in L1 and the corresponding *y*-values (such as "age" in Item 8) in L2.
3. Press (STAT), choose the CALC option, the "LinReg" option, and (ENTER).

Activity 39 • The Best-Fit Line

ACTIVITY 39 Continued

Check Your Understanding
Debrief this lesson by having students describe the differences in positive and negative residuals. Ask them to explain how you find a residual.

Answers
11.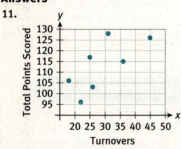

12. There is a moderate to strong positive linear relationship between x and y.
13. $y = 86.4 + 0.917x$
14. 113.91

---

continued

Lesson 39-2
Residuals

Check Your Understanding

The men's basketball coach at Grinnell College employs a style of basketball known as "system ball." The idea behind system ball is that forcing turnovers on defense leads to more shots, especially 3-point shots, on offense, and thus a higher point total. Data on the number of turnovers committed by the opposing team and the total points scored by Grinnell for a sample of seven games are given below.

Turnovers (x)	Total Points Scored (y)
36	115
45	126
26	103
18	106
25	117
31	128
22	96

11. Construct a scatter plot for this data set.

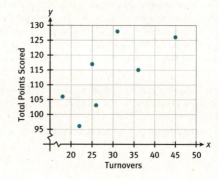

12. Based on the scatter plot, how would you describe the relationship between x and y?
13. Use technology to find the equation of the best-fit line.
14. Use the best-fit line to predict the total points scored in a game with 30 turnovers.

Lesson 39-2
Residuals

LESSON 39-2 PRACTICE

The table below shows the historical minimum wage (in dollars per hour) for the State of New York. Use the table for Items 15–18.

Year (x)	Wage (y)
1962	1.15
1968	1.60
1974	2.00
1978	2.65
1981	3.35
1990	3.80
1991	4.25
2000	5.15
2005	6.00
2012	7.25

15. Construct a scatter plot of the data.
16. Does there appear to be a relationship between the year and the minimum wage? If so, describe the relationship.
17. **Use appropriate tools strategically.** Find the equation for the best-fit line and the correlation coefficient.
18. **Construct viable arguments.** Do the equation you found and the correlation coefficient support your answer in Item 16? Explain.

ACTIVITY 39 Continued

ASSESS

Use the lesson practice to determine if students know how to use technology to find the line of best fit and are able to accurately find a prediction.

See the Activity Practice for additional problems for this lesson. You may assign the problems here or use them as a culmination for the activity.

LESSON 39-2 PRACTICE

15.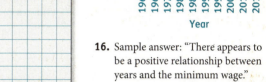
 Historical Minimum Wage for NY

16. Sample answer: "There appears to be a positive relationship between years and the minimum wage."
17. $y = 0.12x - 238.14$ (using full year); $r = 0.991$
18. Sample answer: "Yes, my line supports my answer, because the line of best fit has a positive slope."

ADAPT

Check students' answers to the Lesson Practice to ensure that they understand how to draw a scatterplot and find the line of best fit. Students who may need more practice with the concepts in this lesson may benefit from extra practice using technology.

Activity 39 • The Best-Fit Line 581

ACTIVITY 39 Continued

Lesson 39-3

PLAN

Pacing: 1 class period
Chunking the Lesson
#1–3   #4–5   #6–9
Check Your Understanding
Lesson Practice

TEACH

Bell-Ringer Activity
Ask students to write the formula for slope and give the definition of slope. Lead a discussion to be sure that students know slope is a ratio of change in y to change in x.

> **TEACHER to TEACHER**
>
> Students should feel comfortable with the notion of slope as the change in the value of the predicted y variable that is associated with a change of one unit in the value of the x variable. Stated another way, the units of slope would be "y units per x units." Consider reminding students of this before they perform additional slope interpretations.

1–3 Quickwrite Make sure that students are developing their interpretation of slope with proper units and proper context. Consider using examples to explain to students that a model developed over a certain span of an explanatory variable may not be applicable for other spans of that variable.

> **CONNECT TO AP**
>
> In Statistics students are expected to be able to interpret the slope in a real-life context and use the line of best fit to make predictions.

ACTIVITY 39 continued

Lesson 39-3
Interpreting the Slope and Intercept of the Best-Fit Line

Learning Targets:
- Interpret the slope of the best-fit line in the context of the data.
- Distinguish between scatter plots that show a linear relationship and those where the relationship is not linear.

SUGGESTED LEARNING STRATEGIES: Quickwrite, Look for a Pattern, Visualization, Create Representations, Guess and Check

Once you have determined the equation of the best-fit line, it is possible to interpret the slope of the line. For the bear data, the slope of the best-fit line is 0.115. This means that for each additional pound of weight, the predicted age increases by 0.115 years.

1. The best-fit line for a data set where y is the time to complete a task (in seconds) and x is the room temperature (in degrees Fahrenheit) is $y = 128 + 2x$. The slope of this line is 2. Interpret this value in the context of this problem.
   The predicted time to complete the task increases by 2 seconds with each additional degree increase in temperature.

> **MATH TIP**
>
> The slope of the best-fit line can be interpreted as the change in the value of the predicted y variable that is associated with a change of 1 in the value of the x variable.

2. What does it mean when the slope of the best-fit line is negative?
   The predicted y-value decreases when x increases.

Lesson 39-3
Interpreting the Slope and Intercept of the Best-Fit Line

3. The best-fit line for a data set where y is the fuel efficiency of a car (in miles per gallon) and x is the weight of the car (in pounds) is $y = 40 - 0.005x$. The slope of this line is -0.005. Interpret this value in the context of this problem.

   The predicted fuel efficiency decreases by 0.005 mpg for each additional pound of car weight.

It is sometimes also possible to interpret the y-intercept of the best-fit line, but this is not always a sensible thing to do. The y-intercept of the line is the point whose x-coordinate is 0. But it doesn't make sense to predict the age of a bear whose weight is 0 or the fuel efficiency of a car that weighs 0 pounds. So, in most cases, you won't want to interpret the y-intercept.

When is it NOT okay to use the best-fit line to make predictions?

There are three situations when it doesn't make sense to use the best-fit line to make predictions.

You should not use the best-fit line
- to predict a value that is far outside the range of values in the data set used to find the best-fit line.
- when the linear relationship between x and y is very weak, which means that the residuals are very large overall.
- when the relationship between x and y is not linear and would be better described by a curve.

4. **Make sense of problems.** Why do you think it is not a good idea to predict a value that is far outside the range of the data used to find the best-fit line? For example, why would it not be a good idea to predict the fuel efficiency of a 500-pound car if the data set had only cars that weighed between 2000 and 3500 pounds?

   The linear relationship might not continue outside this range. We know what the relationship looks like only over the range where we have data.

Lesson 39-3
Interpreting the Slope and Intercept of the Best-Fit Line

5. Even though the best-fit line has an SSR that is less than the SSR of any other line, the residuals might still be large. That is, the points in the data set might still tend to fall far from the line. If this is the case, do you think predictions based on the line will tend to be close to actual y-values? Explain your reasoning.

If the residuals are great, points in the scatter plot aren't close to the line and predicted values based on the line may not be close to the actual values.

Sometimes there is a relationship between two numerical variables, but the relationship is not linear. For example, consider the scatter plot below. This scatter plot was constructed using data on y, the time to finish a marathon (in minutes), and x, the age (in years), for six women.

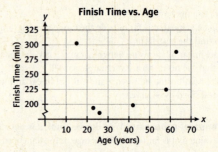

This scatter plot shows a nonlinear relationship between finish time and age.

Lesson 39-3
Interpreting the Slope and Intercept of the Best-Fit Line

6. Below are scatter plots that show the best-fit line and the best-fit quadratic curve. To predict finish time based on age, would you recommend using the best-fit line or the best-fit quadratic curve? Explain why you made this choice.

   I would recommend the quadratic curve. The points in the scatter plot tend to be much closer to the best-fit quadratic curve than to the best-fit line.

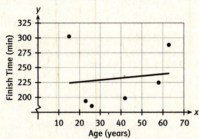

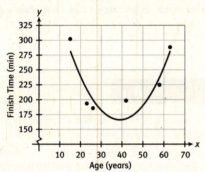

ACTIVITY 39 Continued

Check Your Understanding
Debrief students' answers to these items to ensure that students are able to interpret slope in context.

Answers
10. The weights of the bears in the data set ranged from 88.2 pounds to 130.1 pounds. 40 pounds is way outside this range. We don't know if the linear relationship continues for bear weights less than 88.2 pounds.
11. Yes, because 110.5 is within the range used to make the best-fit line, so it would result in a good prediction.
12. $y = 10.56x$

    A tennis player burns approximately 10.56 calories per minute playing singles.

---

continued

Lesson 39-3
Interpreting the Slope and Intercept of the Best-Fit Line

My Notes

7. What do you think it means to say a quadratic curve is the best-fit quadratic curve?

    The best-fit quadratic curve has an SSR less than that of any other quadratic curve.

8. The equation of the best-fit quadratic curve for the marathon data is

    $$\text{finish time} = 473.3 - 15.80(\text{age}) + 0.2030(\text{age})^2$$

    What is the predicted finish time for a 30-year-old woman?
    predicted finish time = 473.3 − 15.80(30) + 0.2030(30)² = 473.3 − 474.0 + 82.70 = 182 minutes

9. Would you recommend using the best-fit quadratic curve to predict the finish time for a woman who is 80 years old? Explain why or why not.
    No. The oldest woman in the data set is about 65 years old.

Check Your Understanding

10. Explain why you should not use the best-fit line for the bear data to predict the age of a 40-pound bear.
11. Suppose a bear weighs 110.5 pounds. Would you recommend using the best-fit line to predict the age of this bear? Why or why not?
12. The Chestnut High School tennis team recorded the amount of time that each player's last tennis match lasted as well as the number of calories the player burned during the match. Dakota plays number 1 singles and her match lasted 3 hours; she burned 1900 calories. Briana is number 2 singles; she beat her opponent with no problem and her match lasted only 45 minutes. She burned 475 calories. Maria beat her opponent in 1.5 hours and she burned 950 calories. Determine the equation of the best-fit line and interpret the slope in the context of the problem.

Lesson 39-3
Interpreting the Slope and Intercept of the Best-Fit Line

LESSON 39-3 PRACTICE

Use the following information for Items 13–16.

At a recent football game you noticed that students tend to be near the same height as their parent of the same gender. You surveyed several students and their parents and recorded the data below.

Student's Height (in.)	Parent's Height (in.)
61	56
62	57
63	62
65	66
65	60
66	66
67	68
68	63
68	69
70	71
72	68
71	73
73	67
74	71

13. **Model with mathematics.** Draw a scatter plot and describe the relationship between the students' heights and their parents' heights.

14. Using a graphing calculator, determine the equation of the best-fit line and calculate the correlation coefficient.

15. Use the best-fit line to predict the height of your classmate Tyler's father. Tyler is 70 inches tall.

16. **Attend to precision.** Tyler's father came to pick him up and you asked his height. He is 6 feet tall. What is his residual and where does the data point lie in relation to the best-fit line?

Lesson 39-4
Plotting Residuals

Learning Targets:
- Create a residual plot given a set of data and the equation of the best-fit line.
- Use residuals to investigate whether a line is an appropriate description of the relationship between numerical variables.

SUGGESTED LEARNING STRATEGIES: Create Representations, Look for a Pattern, Quickwrite

In the marathon data scatter plot from the previous lesson, it is obvious that the relationship between finish time and age is not linear. But sometimes, the nonlinearity of the relationship between two variables isn't always this obvious. One way to decide whether a curve describes a relationship better than a line is to look at a residual plot.

A *residual plot* is a scatter plot of the (x, residual) pairs.

To see how to make a residual plot, let's return to the bear data. For the bear data set, the data and the residuals for the best-fit line are shown in the table below.

Bear	Weight (pounds)	Age (years)	Predicted Age (years)	Residual
1	88.2	6.5	6.45	0.05
2	88.2	7.5	6.45	1.05
3	92.6	5.5	6.96	−1.46
4	110.3	8.0	9.00	−1.00
5	112.5	10.5	9.25	1.25
6	112.5	9.5	9.25	0.25
7	119.1	10.5	10.01	0.49
8	121.3	9.0	10.26	−1.26
9	130.1	11.5	11.27	0.23

The first (x, residual) pair is (88.2, 0.05). This point has been graphed below.

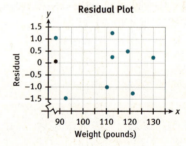

1. Complete the residual plot by adding the other eight (x, residual) points to the plot.

   See scatter plot above.

MATH TERMS

A **residual plot** is a scatter plot of the (x, residual) pairs.

Lesson 39-4
Plotting Residuals

Notice that there is no pattern in the residual plot for the bear data best-fit line. The points appear to be scattered at random in this plot.

Now look at a residual plot for the best-fit line for the marathon data set.

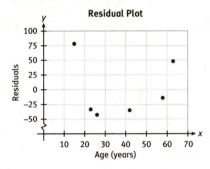

Residual Plot

Here there is a very strong curved pattern. It is this pattern in the residual plot that confirms that a line is **not** the best way to describe the relationship between finish time and age.

Now let's look at another example where biologists were interested in predicting the ages of lobsters. The data below are the shell lengths and ages for 12 lobsters.

Shell Length (mm)	Age (years)
63	1.00
83	1.42
109	1.82
111	1.80
118	2.17
143	3.70
90	1.40
125	2.51
136	2.92
75	1.10
142	3.17
148	3.75

MATH TIP

A strong pattern in the residual plot for the best-fit line indicates that a line is not the best way to describe the relationship.

Lesson 39-4
Plotting Residuals

A scatter plot of the data is shown below.

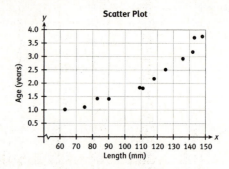

2. The equation of the best-fit line for this data set is $y = -1.41 + 0.0325x$, where y = age (in years) and x = shell length (in mm). Use this equation to complete the following table. Round values to the nearest hundredth if necessary.

Shell Length (mm)	Age (years)	Predicted Age (years)	Residual
63	1.00	0.64	0.36
83	1.42	1.29	0.13
109	1.82	2.13	−0.31
111	1.80	2.20	−0.40
118	2.17	2.43	−0.26
143	3.70	3.24	0.46
90	1.40	1.52	−0.12
125	2.51	2.65	−0.14
136	2.92	3.01	−0.09
75	1.10	1.03	0.07
142	3.17	3.21	−0.04
148	3.75	3.40	0.35

A residual plot (scatter plot of the (x, residual) pairs) is shown here:

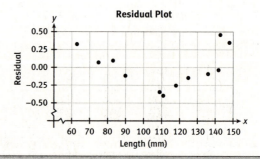

Lesson 39-4
Plotting Residuals

3. Describe the pattern in the residual plot.
   There is a definite curved pattern in the plot.

4. **Critique the reasoning of others.** The biologists who collected these data decided to use an exponential curve to describe the relationship between age and shell length. Do you think this was a reasonable choice? Explain why or why not.
   Yes. Even though there was not an obvious curved pattern in the scatter plot, once the residual plot indicates that the relationship is not linear, we should take another careful look at the scatter plot. Looking for a curved relationship in the scatter plot, we can see that there is a curved pattern that looks like an exponential curve.

Check Your Understanding

No tortilla chip lover likes soggy chips, so chip makers want to know what makes them crispy. The data below are from an experiment to see how the moisture content of tortilla chips is related to the frying time.

Here, x = frying time (in seconds) and y = moisture content (in percent).

Frying Time (x)	Moisture Content (y)
5	16.3
10	9.7
15	8.1
20	4.2
25	3.4
30	2.9
45	1.9
60	1.3

5. Construct a scatter plot for this data set.
6. Find the equation of the best-fit line.

ACTIVITY 39 Continued

TEACHER to TEACHER

If a linear model generates a low SSR, and if the variables have a strong correlation coefficient value, these alone are NOT sufficient for determining if a linear model is truly appropriate. A pattern in the residual plot is a problem that lets a student realize that a linear fit may not be appropriate. A thorough analysis should include investigating the residual plot and looking beyond the number values of SSR and r, no matter how "good" the fit is.

Check Your Understanding

Debrief students' answers to these items to ensure that they understand how to use a scatterplot and line of best fit to create a residual plot. Once they have created the residual plot, be sure that they understand how to identify whether the data should be represented by a linear relationship.

Answers

5.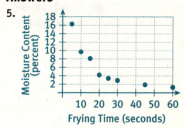

6. $y = 11.9 - 0.224x$

Activity 39 • The Best-Fit Line 591

ACTIVITY 39 Continued

7. These numbers were calculated using the entire line of best fit from the calculator. If using the rounded equation from Item 2, the answers will be slightly different; for example, the predicted y for 5 is 10.78.

x	y	Predicted y	Residual
5	16.3	10.74	5.56
10	9.7	9.62	0.08
15	8.1	8.50	−0.40
20	4.2	7.38	−3.18
25	3.4	6.26	−2.86
30	2.9	5.13	−2.23
45	1.9	1.77	0.13
60	1.3	−1.59	2.89

8.
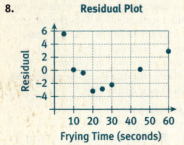

9. No; there is a curved pattern in both the scatterplot and the residual plot.

Lesson 39-4
Plotting Residuals

7. Compute the predicted values and the residuals.

x	y	Predicted y	Residual
5	16.3		
10	9.7		
15	8.1		
20	4.2		
25	3.4		
30	2.9		
45	1.9		
60	1.3		

8. Construct a residual plot.

9. **Construct viable arguments.** Based on the scatter plot and the residual plot, would you recommend using the best-fit line to describe the relationship between x and y? Explain.

592 SpringBoard® Mathematics Algebra 1, Unit 6 • Probability and Statistics

Lesson 39-4
Plotting Residuals

LESSON 39-4 PRACTICE

10. Why is it important to look at the scatter plot and the residual plot before deciding whether it is appropriate to describe the relationship between two numerical variables using the best-fit line?

11. A recent study of birds of prey resulted in data on x = wing length and y = total weight for 16 different species. The scatter plot for these data is shown below. The correlation coefficient is $r = 0.897$. Describe the relationship between these variables in terms of strength, direction, and shape.

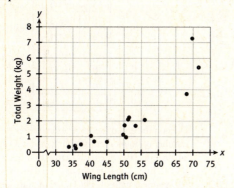

12. The residual plot for the data in Item 11 is shown below. Does the residual plot support your description of the relationship between these two variables? Explain, referring to specific characteristics of the residual plot.

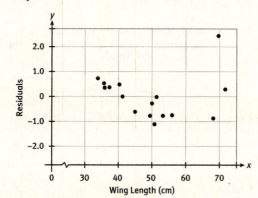

ACTIVITY 39 Continued

ACTIVITY PRACTICE

1.
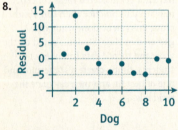

2. The predicted weight increases by 31.7 pounds for each additional pound of food.
3. It does not make sense to interpret the intercept since an x-value of 0 (no food) is not reasonable. The smallest amount of food in the data set is 0.5 pound, so 0 is outside the range of x-values in the data set.
4. predicted $y = 24.6 + 31.7x = 24.6 + 31.7(0.8) = 49.5$
   Predicted value is off by $50 - 49.5 = 0.5$.
5. The residual is for Dog 2.
6. Points with positive residuals would be above the line; points with negative residuals would be below the line.
7. If the residual turned out to be 2.0, the point would be closer to the best-fit line but still above it. That movement of the point would make the points more clustered around the best-fit line, and therefore r would increase.
8.

No, there is no curved pattern in the residual plot. The most unusual feature of the residual plot is the one data value with the very large residual.
9. The predicted end-of-semester exam score increases by 11 points with each additional hour studied.

ADDITIONAL PRACTICE
If students need more practice on the concepts in this activity, see the eBook Teacher Resources for additional practice problems.

ACTIVITY 39
continued

ACTIVITY 39 PRACTICE
Write your answers on notebook paper.
Show your work.

A veterinarian is studying the relationship between the weight of one-year-old golden retrievers in pounds (y) and the amount of dog food the dog is fed each day in pounds (x). A random sample of 10 one-year-old golden retrievers yielded the data in the table below. Use the table for Items 1–8.

Dog	1	2	3	4	5	6	7	8	9	10
x	0.5	1.0	0.6	1.0	1.3	1.5	0.5	0.7	0.8	0.6
y	42	67	47	55	62	71	36	42	50	43

1. Construct a scatter plot of the dog food data.

The equation of the best-fit line is $y = 24.6 + 31.7x$; the correlation coefficient is $r = 0.92$.

2. Interpret the value of the slope of the best-fit line.
3. Does it make sense to interpret the meaning of the y-intercept of the best-fit line?
4. One dog in this data set is fed 0.8 pound of food per day. If you used the best-fit line to predict the weight of this dog, how far off would you be from the dog's actual weight?
5. A box plot of the residuals is shown below. One of the residuals is an outlier. To which dog does this residual belong?

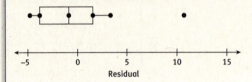

6. Suppose that you graphed the best-fit line on your scatter plot in Item 1. Looking at the scatter plot with the best-fit line, how would you know whether a point had a positive or a negative residual?
7. Suppose that the point with the greatest residual was the result of an error in data recording, and that the actual residual for that point is 2.0. Would the correlation increase or decrease? Explain your reasoning in a few sentences.

10. The predicted score of a student who did not study for the end-of-semester exam is 42.
11. Ten hours of study time is greater than any of the study times used to find the best-fit line. We don't have any reason to believe that the linear relationship continues for study times greater than 6 hours.

The Best-Fit Line
Regressing Linearly

8. Below are the 10 residuals for the best-fit line. Construct a residual plot. Are there any patterns in the residual plot that indicate the relationship between x and y is not linear?

Dog	Residual
1	1.55
2	10.7
3	3.38
4	−1.3
5	−3.81
6	−1.15
7	−4.45
8	−4.79
9	0.04
10	−0.62

Data on the end of semester exam scores (y) and the number of hours spent studying (x) for 34 students in an algebra class were used to find the equation of the best-fit line. A scatter plot of these data showed a strong linear pattern. The equation of the best-fit line was $y = 42 + 11x$. Use this information for Items 9–11.

9. Interpret the value of the slope of the best-fit line.
10. Interpret the meaning of the y-intercept of the best-fit line.

MATHEMATICAL PRACTICES
Construct Viable Arguments and Critique the Reasoning of Others

11. Study times for these 34 students ranged from 0 to 6 hours. Explain why it is not reasonable to use the best-fit line to predict the test score of a student who studied for 10 hours.

Bivariate Data
Categorically Speaking
Lesson 40-1 Bivariate Categorical Data

ACTIVITY 40

Learning Targets:
- Summarize bivariate categorical data in a two-way frequency table.
- Interpret frequencies and relative frequencies in two-way tables.

SUGGESTED LEARNING STRATEGIES: Summarizing, Paraphrasing, Create Representations

In previous activities, you analyzed bivariate numerical data. You measured the strength of association between two numerical variables and learned how to predict the value of one variable given the value of another, using the best-fit line. In this activity, you will work with bivariate categorical data.

The first example you will consider involves a famous data set from a tragic historical event: survival data from the luxury ship SS *Titanic*. The *Titanic* set sail on her maiden voyage on April 10, 1912. On a moonless night, April 12, she struck an iceberg and sank, and many people died. Hundreds of books and scholarly studies have tried to answer the question of how a ship so well constructed could have come to such a sad end, and government inquiries resulted in recommendations for future ship construction and safety regulations. This accident has been the topic of many popular movies, plays, and televisions specials.

Table 1 is taken from the United States Senate report dated May 28, 1912: Investigation into Loss of S. S. "Titanic". The format of the table has been preserved for historical accuracy.

Table 1: Original Senate Data

	On board.			Saved.			Lost.			Percent saved.
	Women and children.	Men.	Total.	Women and children.	Men.	Total.	Women and children.	Men.	Total.	
Passengers										
First class	156	173	329	145	54	199	11	119	130	60
Second class . . .	128	157	285	104	15	119	24	142	166	42
Third class	224	486	710	105	69	174	119	417	536	25
Total passengers	508	816	1,324	354	138	492	154	678	832	. . .
Crew	23	876	899	20	194	214	3	682	685	24
Total	531	1,692	2,223	374	332	706	157	1,360	1,517	32

There are quite a few categorical variables summarized in this table: Person Type (men, women/children); Ticket Class (first, second, third); Role (passengers, crew); and Fate (saved, lost). You will use numbers from the Senate report table to form your own tables.

Common Core State Standards for Activity 40

HSS-ID.B.5 Summarize categorical data for two categories in two-way frequency tables. Interpret relative frequencies in the context of the data (including joint, marginal, and conditional relative frequencies). Recognize possible associations and trends in the data.

ACTIVITY 40
Investigative

Activity Standards Focus

Students will learn to summarize categorical data in a two-way frequency table. They will use the table to interpret frequency and relative frequency. By using these tables they will learn to recognize and describe patterns of association. They will also learn about creating row percentages, developing a segmented bar graph, and analyzing row percentages and segmented bar graphs to investigate association.

Lesson 40-1

PLAN

Pacing: 1 class period
Chunking the Lesson
#1–4   #5–6
Check Your Understanding
Lesson Practice

TEACH

Bell-Ringer Activity

Have the students describe the difference in these two items, the number of desks in the classroom and the color shirt that each student is wearing. Discuss the difference between numerical values and attributes (categorical).

TEACHER to TEACHER

The initial section of this activity uses survival statistics from the *Titanic* to reintroduce students to two-way tables, and the *Titanic* data are then used as a case throughout the remainder of the activity. Thus there are several media available to consider for contextualizing the lesson. Sports examples are also used in some activities, and the hope is that students will see that analysis of bivariate categorical data is quite common and applicable to many areas.

Activity 40 • Bivariate Data   595

Lesson 40-1
Bivariate Categorical Data

1. Use the Senate report table to answer the following questions:
   a. How many first-class passengers were there?
   329

   b. How many of the first-class passengers were men?
   173

   c. How many male first-class passengers were saved?
   54

2. **Reason quantitatively.** Compare the number of male second-class passengers saved with the number of male third-class passengers saved.
   The number of male second-class passengers saved (15) was less than the number of male third-class passengers saved (69).

3. Was the percentage of second-class passengers saved greater than or less than the percentage of third-class passengers saved?
   The percentage of second-class passengers saved (42) was greater than the percentage of third-class passengers saved (25).

4. Based on your answers to Items 2 and 3, does it seem like the second-class or the third-class passengers were at greater risk of death?
   It appears that the third-class passengers were at greater risk.

To analyze these data we can consider two variables at a time. Bivariate categorical data are often summarized and arranged in a *two-way frequency table*. A two-way frequency table has rows and columns corresponding to the different possible values of the two categorical variables. For example, suppose you were interested in the variables Ticket Class and Fate. There are three values for Ticket Class and two values for Fate. The data can be summarized using a table with three rows and two columns. You can label the rows with the values for Ticket Class and the columns with the values for Fate.

5. Use the information from the Senate report table to complete the two-way table Ticket Class vs. Fate.

Ticket Class vs. Fate

	Saved	Lost
First class	199	130
Second class	119	166
Third class	174	536

MATH TERMS

A **two-way frequency table** summarizes the distribution of values for bivariate categorical data.

Lesson 40-1
Bivariate Categorical Data

Row and column totals are usually also included in a two-way table, as shown below. These totals are called *marginal* totals. The marginal totals describe the univariate distribution for each of the variables. For example, there were a total of 329 passengers with first-class tickets, and a total of 492 passengers were saved. The total number of passengers is shown in the bottom right column, the *grand total*.

Ticket Class vs. Fate Frequencies

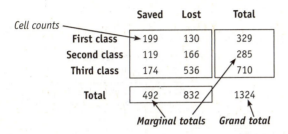

MATH TIP

Frequency refers to the number of times a particular value occurs in a data set.

Relative frequency is the proportion of the time that a particular value occurs in a data set.

Marginal total is the sum of frequencies or relative frequencies in a row or column of a two-way table.

It is common to convert these frequencies into *relative frequencies* by dividing by the grand total. These values are expressed in decimal form. As an example, consider the second-class passengers who were lost. The table in Item 5 shows that there were 119 second-class passengers who were saved. The grand total was 1324. Dividing 119 by 1324 gives 0.090 as the relative frequency.

6. Calculate the remaining relative frequencies and complete the two-way table below.

Ticket Class vs. Fate

	Saved	Lost
First class	0.150	0.098
Second class	0.090	0.125
Third class	0.131	0.405

ACTIVITY 40 Continued

Check Your Understanding

Debrief students' answers to these items to ensure that they have a good understanding of table completion and relative frequency calculations. Encourage students to compare answers and verify that students have a good understanding of these skills.

Answers

7. See table below.
8. See table below.
9. 0.144
10. 0.192
11. 0.244

ACTIVITY 40 continued

My Notes

Lesson 40-1
Bivariate Categorical Data

Check Your Understanding

In recent years, there has been a great deal of controversy over the use of Native American-related team names and mascots for sports teams in the United States. *Sports Illustrated* magazine commissioned a survey of Native Americans living on reservations, Native Americans living off reservations, and non-Native-American sports fans to determine their feelings. One question asked about the "tomahawk chop" chant used in home games of the Atlanta Braves baseball team. Results are shown in the frequency table below.

7. Complete the frequency table below.

Person vs. Attitude Toward "Tomahawk Chop" Frequency Table

	Non-NA Fans	NA on Res	NA off Res	Total
Like it	208	24	44	
Don't care	379		66	545
Is objectionable	156	85	24	265
Total	743	209		

8. Use the frequencies from the table above to complete the table below.

Person vs. Attitude Toward "Tomahawk Chop" Relative Frequency Table

	Non-NA Fans	NA on Res	NA off Res	Total
Like it	0.192	0.022		0.254
Don't care		0.092	0.061	0.502
Is objectionable	0.144	0.078	0.022	
Total	0.684	0.192	0.123	

9. What is the relative frequency of non-Native-American fans who find the "tomahawk chop" objectionable?

10. What is the marginal frequency of Native Americans living on a reservation?

11. What is the marginal relative frequency of those who find the "tomahawk chop" objectionable?

Answers

7. **Person vs. Attitude Toward "Tomahawk Chop" Frequency Table**

	Fans	NA on Res	NA off Res	Total
Like it	208	24	44	276
Don't care	379	100	66	545
Is objectionable	156	85	24	265
Total	743	209	134	1086

8. **Person vs. Attitude Toward "Tomahawk Chop" Relative Frequency Table**

	Fans	NA on Res	NA off Res	Total
Like it	0.192	0.022	0.041	0.254
Don't care	0.349	0.092	0.061	0.502
Is objectionable	0.144	0.078	0.022	0.244
Total	0.684	0.192	0.123	1.000

Lesson 40-1
Bivariate Categorical Data

LESSON 40-1 PRACTICE

12. Complete the following table on United States usage of multimedia devices in minutes per day.

United States Usage of Multimedia in Minutes Per Day				
Year	Web Browsing	Mobile Applications	Television	Total
2010	70	66	162	
2011	72	94	168	
2012	70	127	168	
Total				

13. Give the frequency of mobile application usage in 2012.
14. Determine the relative frequency of total time spent Web Browsing in 2010 for all usage of multimedia.
15. **Critique the reasoning of others.** Jayson states that people use multimedia devices of some type for 30% of the day. Do you agree with this statement? Why or why not?

12.

United States Usage of Multimedia in Minutes Per Day				
	Web Browsing	Mobile Applications	Television	Total
2010	70	66	162	298
2011	72	94	168	334
2012	70	127	168	365
Total	212	287	498	997

ACTIVITY 40 Continued

Lesson 40-2

PLAN

Pacing: 1 class period
Chunking the Lesson
#1–2   #3–5   #6–8
Check Your Understanding
Lesson Practice

TEACH

Bell-Ringer Activity
Have students look around the room and determine two things that are categorical and that they might want to compare percentages on, and describe them—e.g., the number of boys wearing a sweatshirt compared to the number of girls.

Developing Math Language
Students use segmented bar graphs for displaying categorical data. It involves the use of a bar to represent the entire category, but different sections of the bar are based on the area of a particular category.

1–2 Think-Pair-Share In this chunk, students will begin to compile relative frequencies for individual rows of a table. Students should begin to note that if the distribution of percentages for one subgroup is very different from the distribution of percentages in another subgroup, then there may in fact be an association between the two categorical variables.

As students study and complete the tables in this lesson, familiarize them with some of the terminology associated with two-way tables.

- The percent or decimal entries can represent *joint relative frequencies*, which are found by dividing the number in each entry by a total number of values.
- *Marginal relative frequencies* are the sums of the joint relative frequencies in each row or column.
- *Conditional relative frequencies* are the quotients of joint relative frequencies and marginal relative frequencies.

ACTIVITY 40 continued

Lesson 40-2
Presenting Relative Frequency Data Graphically

My Notes

MATH TERMS

A **segmented bar graph** summarizes categorical data.

The total data set is represented by a bar, and the different possible categories are represented by sections of the bar. The area of the section for a particular category is proportional to the relative frequency of that category.

MATH TERMS

Row percentages are calculated by dividing a number by the corresponding row total and then multiplying by 100.

Learning Targets:
- Interpret frequencies and relative frequencies in two-way tables.
- Recognize and describe patterns of association in two-way tables.

SUGGESTED LEARNING STRATEGIES: Think-Pair-Share, Create Representations, Look for a Pattern

For most people, a visual presentation of data leads to faster understanding of the relationships between pairs of variables. A **segmented bar graph** is an effective way to present bivariate categorical data so that these relationships can be easily seen.

To see how a segmented bar graph is constructed, consider data from Major League Baseball.

Major League players are categorized as relief pitchers (RP), starting pitchers (SP), catchers (C), infielders (In), and outfielders (Out). The frequency table below summarizes data on position and team for the 75 players on three different teams (Cubs, Reds, and Pirates).

Frequencies
Positions for Three Different Baseball Teams

	RP	SP	C	In	Out	Total
Cubs	6	6	2	7	4	25
Reds	7	5	2	6	5	25
Pirates	7	4	2	8	4	25
Total	20	15	6	21	13	75

A segmented bar graph can be constructed by first calculating percentages within a row of data. Because we are interested in the percentages in each of the position categories for the different teams, we treat each individual row as a "whole." For each row, compute the percentages by first dividing each number in that row by the corresponding row total and then multiplying by 100.

The *row percentages* for the positions on the Cubs team are shown in the table below. Notice that the percentages in the Cubs row add to 100% because we are considering each row separately.

1. Use the frequencies in the table above to complete the following table.

Row Percentages
Positions for Three Different Baseball Teams

	RP	SP	C	In	Out	Total
Cubs	24%	24%	8%	28%	16%	100%
Reds	28%	20%	8%	24%	20%	100%
Pirates	28%	16%	8%	32%	16%	100%

600 SpringBoard® Mathematics Algebra 1, Unit 6 • Probability and Statistics

© 2014 College Board. All rights reserved.

Lesson 40-2
Presenting Relative Frequency Data Graphically

The segmented bar graph below shows the percentages of different positions (Outfield, Infield, etc.) for each of the three teams.

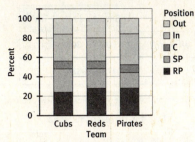

Positions for Three Different Baseball Teams

2. **Reason quantitatively.** Does it appear from the segmented bar graph that the distributions of positions are very similar for the three teams, or are they noticeably different? How are the distributions similar or different?
   The distributions are very similar. The percents for each position category do not differ by more than about 5% across the different teams. The greatest difference appears to be in infield positions; the least difference appears to be in catchers.

Two categorical variables are said to be *associated* if knowing the value of one of the variables gives you information about the value of the other variable. With categorical data we will ask whether the distribution of values for one variable is similar to or different from the distribution of values for the other variable. A segmented bar graph presents these distributions of values graphically. For the baseball example, the distributions of the position categories (distribution across the columns in each row) are very similar for all three teams (rows). We would say that the categorical variables of position and team are **not** associated. That is, knowing the percentage of pitchers on a team does not provide information about which team it is.

The table below is a two-way frequency table for the variables Person Type and Fate on the *Titanic*.

Person Type vs. Fate

	Saved	Lost	Total
Woman/Child	354	154	508
Man	138	678	816
Total	492	832	1324

ACADEMIC VOCABULARY
To **associate** means to connect or to link.

TEACHER to TEACHER
With the segmented bar graphs, students should note that when there is an association between row and column variables, the segmented bar graph components for the corresponding categories should be noticeably different since the corresponding percentages are noticeably different. Student writing should discuss association between *variables*, not categories.

ACTIVITY 40 Continued

3–5 Create Representations, Debriefing Ask students to compare and contrast the row percentages and segmented bar graphs obtained for the cases of association (e.g., baseball player positions or man/woman vs. fate on the *Titanic*). Emphasize how drastically different percentages across subgroups lead to striking, visible differences in the corresponding segmented bar graphs and consequently make a strong case for association between the categorical variables of interest. Make sure that students' writing discusses association between *variables*, not categories.

Lesson 40-2
Presenting Relative Frequency Data Graphically

3. Calculate the percentages needed to construct a segmented bar graph by completing the table below.

Person Type vs. Fate

	Saved	Lost	Total
Woman/Child	70%	30%	100%
Man	17%	83%	100%

A segmented bar graph for these variables is shown below.

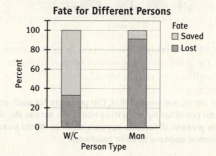

4. About what percent of the area of the bar for women and children (W/C) is ▭? Does this correspond to the percentage of women and children who were saved?
   About 70%; yes; this corresponds to the percentage of women and children saved.

5. About what percentage of the area of the bar for men is ▭? Does this correspond to the percentage of men who were lost?
   About 80%; yes; this corresponds to the percentage of men lost.

Compare this segmented bar graph with the segmented bar graph for the baseball players. The segmented bars representing baseball positions were very similar for the three teams, indicating that the position and team are not associated. The segmented bars above are quite different in terms of the percentages of Lost and Saved. There is an association between Fate and Person Type because knowing the person type (for example, women and children) does provide useful information about fate.

Lesson 40-2
Presenting Relative Frequency Data Graphically

The table below is a two-way frequency table for the two variables Role and Person Type for the *Titanic* data.

	Passenger	Crew	Total
Woman/Child	508	23	531
Man	816	876	1692
Total	1324	899	2223

To construct a segmented bar graph, one variable is placed on the horizontal axis, and the vertical axis is labeled and scaled with percentages from 0% to 100%.

The two-way frequency table above shows that 508 of the 531 women and children (96%) were passengers and 23 of 531 (4%) were crew members. The vertical bar for women and children is "segmented" to reflect this distribution of percentages.

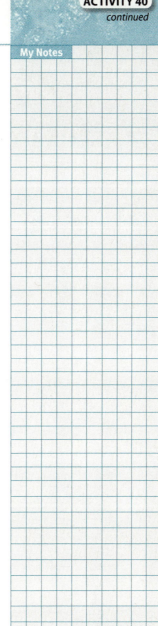

6. Complete the table below, rounding percentages to the nearest whole number.

Role vs. Person Type

	Passenger	Crew	Total
Woman/Child	96%	4%	100%
Man	48%	52%	100%

ACTIVITY 40 Continued

6–8 Look for a Pattern, Create Representations Students will complete the segmented bar graph and look for association.

ACTIVITY 40 continued

Lesson 40-2
Presenting Relative Frequency Data Graphically

My Notes

7. Add the missing bar to the segmented bar graph below.

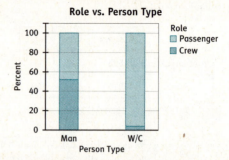

8. **Make sense of problems.** Comment on the association, if any, between these variables.
   The percentages of men vs. women/children differ greatly for the passengers and crew. These variables are associated.

Check Your Understanding

Kidney stones are solid lumps of crystals that separate from urine and build up on the inner surface of the kidney. If left untreated they can lead to kidney failure. Is there an association between sleep position and the location of kidney stones? Researchers asked patients with kidney stones to identify their preferred sleep positions. The table below classifies the responses by sleep position and kidney stone location.

Frequencies of Kidney Stone Locations and Preferred Sleep Positions

	Left Kidney	Right Kidney	Total
Right Side Down	31	13	44
No Preference	8	9	17
Left Side Down	9	40	49
Total	48	62	110

604 SpringBoard® Mathematics **Algebra 1, Unit 6** • Probability and Statistics

Lesson 40-2
Presenting Relative Frequency Data Graphically

9. From the data summarized in the table above, find the row percentages that would be used to construct a segmented bar graph. Round the percentages to the nearest whole number and use them to complete the following table.

Kidney Stone Locations and Preferred Sleep Positions

	Left Side	Right Side	Total
Right Side Down	70%		100%
No Preference			100%
Left Side Down		82%	100%

A segmented bar graph for the kidney stone data is shown below.

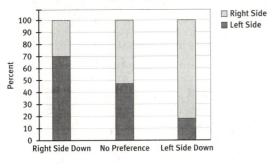

10. **Make sense of problems.** Does there appear to be an association between sleep position and kidney stone location? What feature(s) of the segmented bar graph support your answer?

11. How is sleep position related to the location of kidney stones?

ACTIVITY 40 Continued

Check Your Understanding
Debrief students' answers to these items to ensure that they understand how to create a segmented bar graph.

Answers

9. **Kidney Stone Locations and Preferred Sleep Positions**

	Left Side	Right Side	Total
Right Side Down	70%	30%	100%
No Preference	47%	53%	100%
Left Side Down	18%	82%	100%

10. There does appear to be an association. The percentages for the different location categories differ quite a bit for the three different preferred sleep positions.

11. The kidney stones tend to occur on the side that is not down in the sleeping position.

Activity 40 • Bivariate Data   605

ACTIVITY 40 Continued

ASSESS

Students' answers to these questions will provide you with formative information about their ability to complete and read two-way tables, to assess those tables for possible association between row and column variables, and to develop graphs that display/support the association claim.

See the Activity Practice for additional problems for this lesson. You may assign the problems here or use them as a culmination for the activity.

LESSON 40-2 PRACTICE

12. **Treatment vs. Cold Percentages Across Rows**

	Cold	No Cold	Total
Placebo	22%	78%	100%
Vitamin C	12%	88%	100%

13.
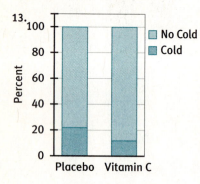

14. It appears that Vitamin C might help. The percentages of "No Cold" are quite a bit larger for the Vitamin C group than for the placebo group.

ADAPT

If some students need further work or have further interest, consider having them conduct a survey at school regarding a hypothesis of interest, collecting data for at least two variables. Students might also use data from established websites that list bivariate or multivariate categorical data (such as through Pew Research's public website or the "Census at School" project).

My Notes

Lesson 40-2
Presenting Relative Frequency Data Graphically

LESSON 40-2 PRACTICE

Does Vitamin C help prevent colds? In a study of 279 French skiers, 140 of them were given a placebo (a sham treatment with no active ingredients) and 139 of them were given Vitamin C. They were followed for one week and whether or not they caught a cold was recorded. The data from this study are shown below.

Treatment vs. Cold Frequency Table

	Cold	No Cold	Total
Placebo	31	109	140
Vitamin C	17	122	139
Total	48	231	279

12. Find the row percentages that would be used to construct a segmented bar graph. Round the percentages to the nearest whole number and use them to complete the table below.

Treatment vs. Cold Percentages Across Rows

	Cold	No Cold	Total
Placebo			100%
Vitamin C			100%

13. Sketch a segmented bar graph using the axes shown below.

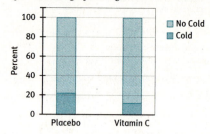

14. **Make use of structure.** It appears that there is an association between these two variables. Does this association suggest that Vitamin C might help prevent colds? What features of the data and the segmented bar graph support your answer?

Bivariate Data
Categorically Speaking

ACTIVITY 40 PRACTICE
Write your answers on notebook paper.
Show your work.

As part of the United States Census, data are collected on the number of persons in each household. The census data for four decades are summarized below.

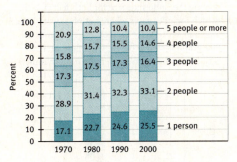

Households by Size: Selected Years, 1970 to 2000

Source: U.S. Census Bureau, Current Population Survey, March Supplements: 1970 to 2000.

1. Which size household increased the most on a percentage basis between 1970 and 2000?
2. Which size household decreased the most on a percentage basis between 1970 and 2000?

Myopia (nearsightedness) is a condition in which a person's eye is slightly longer than it should be, resulting in blurry images of objects far away. Hyperopia (farsightedness) is a condition in which a person's eye is slightly shorter than it should be, resulting in blurry images of near objects. There is some evidence that nighttime light exposure during sleep before the age of two years may be associated with myopia. Data from a survey of parents of children aged 2 to 16 seen at an eye clinic are summarized in the table below.

Vision Condition vs. Light History Frequencies

	Darkness	Night Light	Room Light	Total
Hyperopia	23	17	16	56
Vision OK	66	50	29	145
Myopia	10	34	55	99
Total	99	101	100	300

3. Use the information given in the table above to complete the table of relative frequencies below.

Vision Condition vs. Light History Relative Frequencies

	Darkness	Night Light	Room Light	Total
Hyperopia	0.077	0.057	0.053	
Vision OK	0.220		0.097	0.483
Myopia		0.113	0.183	0.330
Total	0.330	0.337		

ACTIVITY 40 Continued
ACTIVITY PRACTICE
1. The percentage of 1-person households increased the most, by 8.1 percentage points.
2. The percentage of 5-person households decreased the most, by 10.5 percentage points.
3. See table below.

3.

	Darkness	Night Light	Room Light	Total
Hyperopia	0.077	0.057	0.053	0.187
Vision OK	0.220	0.167	0.097	0.483
Myopia	0.033	0.113	0.183	0.330
Total	0.330	0.337	0.333	1.000

ACTIVITY 40 Continued

ACTIVITY PRACTICE

4. Answers will vary. The data suggest that there is an association between vision condition and light history. The percentage with myopia increases with increasing amount of light, but the percentage with hyperopia is similar for the three light conditions.

5. The percentages saved are clearly different for the different ticket classes. A higher percentage of first-class passengers were saved than second-class passengers, and the percentage of second-class passengers saved was higher than that of third-class passengers.

6. Answers will vary. The percentages for the foot asymmetry categories are clearly different for right-handed men and women. A much larger percentage of women than men have a left foot that is larger than the right foot. A much larger percentage of men than women have a left foot that is smaller than the right foot. About the same percentage of men and women have feet that are about the same size.

ADDITIONAL PRACTICE

If students need more practice on the concepts in this activity, see the eBook Teacher Resources for additional practice problems.

ACTIVITY 40 continued

Bivariate Data
Categorically Speaking

4. A segmented bar graph of Vision Condition vs. Light History is shown below. In a few sentences, describe the association between Vision Condition and Light History. (You may assume that room light has more light than a night light.)

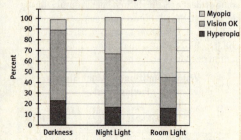

5. The segmented bar graph below shows the relationship between Ticket Class and Fate for those on the *Titanic*. Comment on the association between these two variables.

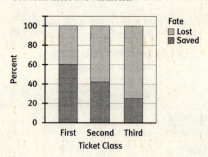

MATHEMATICAL PRACTICES
Make Sense of Problems and Persevere in Solving Them

6. In a study of right-handed men and women, data were gathered on gender and foot asymmetry. An individual was classified as having a left foot more than half a shoe size larger than the right foot (L > R), having a left foot more than half a shoe size smaller than the right foot (L < R) or having the same shoe size for both feet (L = R, which includes cases where both feet were within one half shoe size of each other). A segmented bar graph is shown below. Comment on the association between gender and foot asymmetry.

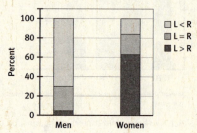

Bivariate Distributions
DEAR TRAVELING TOOTH

Embedded Assessment 2
Use after Activity 40

A vital tooth is one that has a functioning nerve. A nonvital tooth is one where the nerve has died. The chances of restoring a nonvital tooth to health are slim. A team of orthodontists wondered if there was a relationship between vital and nonvital tooth temperatures. They measured vital and nonvital tooth temperatures of 32 patients.

A scatter plot of the data and the best-fit line are shown below.

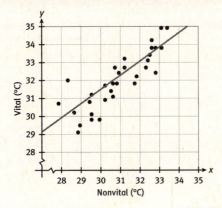

The best-fit line for predicting the vital tooth temperature from the nonvital tooth temperature is given by the equation $y = 6.0 + 0.84x$, where y is the vital tooth temperature and x is the nonvital tooth temperature. The correlation coefficient is $r = 0.856$.

1. Describe the direction of the relationship between the vital and nonvital tooth temperatures. What feature(s) of the graph support your description?

2. Describe the strength of the relationship between the vital and nonvital tooth temperatures. What feature(s) of the graph support your description?

3. One patient has a nonvital tooth temperature of 28.3°C and a vital tooth temperature of 32°C. If you used the best-fit line to predict the vital tooth temperature, how far off would you be from the actual temperature? Is the residual positive or negative?

4. The slope of the best-fit line is 0.84. Interpret this value in the context of this problem.

5. Is it sensible to interpret the y-intercept of the best-fit line for these data? Why or why not?

Embedded Assessment 2

Assessment Focus
- Describing a bivariate numerical relationship and associating that description with a correlation coefficient
- Developing a linear model, interpreting its components, using the model for prediction, and recognizing its limitations
- Reading a two-way table
- Creating row percentages
- Developing a segmented bar graph
- Analyzing row percentages and segmented bar graphs to investigate association

Answer Key
1. The direction of the relationship is positive, as indicated by both the positive slope of the best-fit line and the positive correlation coefficient.
2. The relationship appears to be strong, as indicated by the high correlation coefficient, 0.856.
3. $y = 6.0 + 0.84x$
   $= 6.0 + 0.84(28.3)$
   $= 29.77$
   The residual is: $y -$ predicted y
   $= 32 - 29.77$
   $= 2.23$
   The prediction would be off by 2.23°C. The residual is positive.
4. The predicted vital temperature increases by 0.84°C with each additional degree of nonvital tooth temperature.
5. No. Tooth temperatures near 0°C would be near freezing! The lowest temperature in the data set was about 27°C.

Common Core State Standards for Embedded Assessment 2

HSS-ID.B.5 Summarize categorical data for two categories in two-way frequency tables. Interpret relative frequencies in the context of the data (including joint, marginal, and conditional relative frequencies). Recognize possible associations and trends in the data.

HSS-ID.B.6 Represent data on two quantitative variables on a scatter plot, and describe how the variables are related:
a. Fit a function to the data; use functions fitted to data to solve problems in the context of the data. Use given functions or choose a function suggested by the context. Emphasize linear, quadratic, and exponential models.
b. Informally assess the fit of a function by plotting and analyzing residuals.
c. Fit a linear function for a scatter plot that suggests a linear association.

Embedded Assessment 2

6. The direction of the relationship is negative, and ordinarily a correlation of –0.80 would indicate a strong relationship. However, the scatter plot and especially the residual plot show that the relationship is not linear. The best-fit line is not an appropriate way to describe the the relationship.

Embedded Assessment 2
Use after Activity 40

Bivariate Distributions
DEAR TRAVELING TOOTH

The social behavior of deer can be affected by their environment. Biologists observed the average group size of deer and the amount of woodland in the surrounding area. They found that as the percentage of woodland increased, deer were seen in smaller groups. Their data and a best-fit line are shown below. A residual plot is also shown. The equation of the best-fit line is $y = 7.8 - 0.069x$, where y is the average group size and x is the percentage of the environment that is woodland. The correlation coefficient is $r = -0.80$.

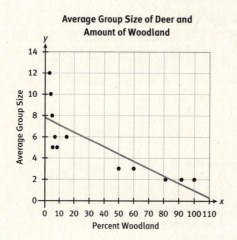
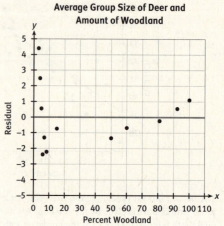

6. Comment on the direction and strength of the relationship between average group size and the percentage of woodland. Is the best-fit line an appropriate way to describe this relationship? Why or why not?

Common Core State Standards for Embedded Assessment 2 (cont.)

HSS-ID.C.7 Interpret the slope (rate of change) and the intercept (constant term) of a linear model in the context of the data.

HSS-ID.C.8 Compute (using technology) and interpret the correlation coefficient of a linear fit.

Bivariate Distributions
DEAR TRAVELING TOOTH

Embedded Assessment 2
Use after Activity 40

A marketing consultant for a travel agency surveyed 400 travelers about their reasons for traveling and their most important travel concerns. The data are summarized in the table below.

Reason for Travel	Most Important Travel Concern		
	Hotel Room	On-Time Arrival	Rental Car Cost
Business	25	150	25
Leisure	50	50	100

7. Sketch a segmented bar graph for business travelers and leisure travelers.
8. About what percent of the travelers in this sample are most concerned about the rental car cost?
9. What is the biggest difference between the business and leisure travelers? Why do you think this is the case?
10. Is there an association between the reason for travel and the most important travel concern? What feature(s) of the segmented bar graph support your answer?

Embedded Assessment 2

7.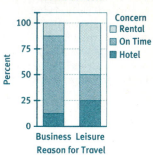
Most Important Travel Concern

8. The percentage of travelers in this sample who felt rental car cost was their greatest concern was $100 \times \left(\frac{125}{400}\right)$, approximately 31.25%.

9. The concern for on-time arrival is the most different. This might be because the business traveler typically has meetings to get to, whereas the leisure traveler can be more flexible with his or her time.

10. There appears to be an association between reason for travel and most important travel concern. The percentages for the Most Important Concern categories are noticeably different for the two types of travelers.

Embedded Assessment 2

TEACHER to TEACHER

You may wish to read through the scoring guide with students and discuss the differences in the expectations at each level. Check that students understand the terms used.

Embedded Assessment 2
Use after Activity 40

Bivariate Distributions
DEAR TRAVELING TOOTH

Scoring Guide	Exemplary	Proficient	Emerging	Incomplete
	The solution demonstrates the following characteristics:			
Mathematics Knowledge and Thinking (Items 1–10)	• Clear and accurate understanding of relationships and associations in bivariate data • Clear and accurate understanding of lines of best fit, correlation coefficients, and residuals	• Recognition of relationships and associations in bivariate data • Adequate understanding of lines of best fit, correlation coefficients, and residuals	• Partial recognition of relationships and associations in bivariate data • Partial understanding of lines of best fit, correlation coefficients, and residuals	• Little or no understanding of relationships and associations in bivariate data • Little or no understanding of lines of best fit, correlation coefficients, and residuals
Problem Solving (Item 3)	• Clear and accurate interpretation of bivariate data to make a prediction	• Interpreting bivariate data to make a prediction	• Difficulty making an accurate prediction from bivariate data	• Inaccurate interpretation of bivariate data
Mathematical Modeling / Representations (Items 1–7, 10)	• Clear and accurate understanding of scatter plots and lines of best fit • Clear and accurate understanding of two-way tables, relative frequency, and segmented bar graphs	• A functional understanding of scatter plots and lines of best fit • A functional understanding of two-way tables, relative frequency, and segmented bar graphs	• Partial understanding of scatter plots and lines of best fit • Partial understanding of two-way tables, relative frequency, and segmented bar graphs	• Little or no understanding of scatter plots and lines of best fit • Inaccurate understanding of two-way tables, relative frequency, and segmented bar graphs
Reasoning and Communication (Items 1, 2, 4–6, 9, 10)	• Precise use of appropriate math terms and language to characterize relationships and associations in bivariate data	• Correct characterization of relationships and associations in bivariate data	• Misleading or confusing characterization of relationships and associations in bivariate data	• Incomplete or inaccurate characterization of relationships and associations in bivariate data

Symbols

<	is less than
>	is greater than
≤	is less than or equal to
≥	is greater than or equal to
=	is equal to
≠	is not equal to
≈	is approximately equal to
$\lvert a \rvert$	absolute value: $\lvert 3 \rvert = 3$; $\lvert -3 \rvert = 3$
$\sqrt{}$	square root
%	percent
⊥	perpendicular
∥	parallel
(x, y)	ordered pair
$\overset{\frown}{AB}$	arc AB
$\overleftrightarrow{AB}$	line AB
$\overrightarrow{AB}$	ray AB
$\overline{AB}$	line segment AB
∠A	angle A
m∠A	measure of angle A
△ABC	triangle ABC
π	pi; $\pi \approx 3.14$; $\pi \approx \frac{22}{7}$

Formulas

Perimeter	
P	= sum of the lengths of the sides
Rectangle	$P = 2l + 2w$
Square	$P = 4s$
Circumference	$C = 2\pi r$

Area	
Circle	$A = \pi r^2$
Parallelogram	$A = bh$
Rectangle	$A = lw$
Square	$A = s^2$
Triangle	$A = \frac{1}{2}bh$
Trapezoid	$A = \frac{1}{2}h(b_1 + b_2)$

Surface Area	
Cube	$SA = 6e^2$
Rectangular Prism	$SA = 2lw + 2lh + 2wh$
Cylinder	$SA = 2\pi r^2 + 2\pi rh$
Cone	$SA = \pi r^2 + \pi rl$
Regular Pyramid	$SA = B + \frac{1}{2}pl$
Sphere	$SA = 4\pi r^2$

Volume	
Cylinder	$V = Bh, B = \pi r^2$
Rectangular Prism	$V = lwh$
Triangular Prism	$V = Bh, B = \frac{1}{2}bh$
Pyramid	$V = \frac{1}{3}Bh$
Cone	$V = \frac{1}{3}\pi r^2 h$
Sphere	$V = \frac{4}{3}\pi r^3$

Linear function	
Slope	$m = \frac{y_2 - y_1}{x_2 - x_1}$
Slope-intercept form	$y = mx + b$
Point-slope form	$y - y_1 = m(x - x_1)$
Standard form	$Ax + By = C$

Quadratic Equations	
Standard Form	$ax^2 + bx + c = 0$
Quadratic Formula	$x = \frac{-b \pm \sqrt{b^2 - 4ac}}{2a}$

Other Formulas	
Pythagorean Theorem	$a^2 + b^2 = c^2$, where c is the hypotenuse of a right triangle
Distance	$d = \sqrt{(x_2 - x_1)^2 + (y_2 - y_1)^2}$
Direct variation	$y = kx$
Inverse variation	$y = \frac{k}{x}$

Temperature	
Celsius	$C = \frac{5}{9}(F - 32)$
Fahrenheit	$F = \frac{9}{5}C + 32$

Properties of Real Numbers

Reflexive Property of Equality	For all real numbers a, $a = a$.
Symmetric Property of Equality	For all real numbers a and b, if $a = b$, then $b = a$.
Transitive Property of Equality	For all real numbers a, b, and c, if $a = b$ and $b = c$, then $a = c$.
Substitution Property of Equality	For all real numbers a and b, if $a = b$, then a may be replaced by b.
Additive Identity	For all real numbers a, $a + 0 = 0 + a = a$.
Multiplicative Identity	For all real numbers a, $a \cdot 1 = 1 \cdot a = a$.
Commutative Property of Addition	For all real numbers a and b, $a + b = b + a$.
Commutative Property of Multiplication	For all real numbers a and b, $a \cdot b = b \cdot a$.
Associative Property of Addition	For all real numbers a, b, and c, $(a + b) + c = a + (b + c)$.
Associative Property of Multiplication	For all real numbers a, b, and c, $(a \cdot b) \cdot c = a \cdot (b \cdot c)$.
Distributive Property of Multiplication over Addition	For all real numbers a, b, and c, $a(b + c) = a \cdot b + a \cdot c$.
Additive Inverse	For all real numbers a, there is exactly one real number $-a$ such that $a + (-a) = 0$ and $(-a) + a = 0$.
Multiplicative Inverse	For all real numbers a and b where $a \neq 0$, $b \neq 0$, there is exactly one number $\frac{b}{a}$ such that $\frac{b}{a} \cdot \frac{a}{b} = 1$ and $\frac{a}{b} \cdot \frac{b}{a} = 1$.
Multiplication Property of Zero	For all real numbers a, $a \cdot 0 = 0$ and $0 \cdot a = 0$.
Addition Property of Equality	For all real numbers a, b, and c, if $a = b$, then $a + c = b + c$.
Subtraction Property of Equality	For all real numbers a, b, and c, if $a = b$, then $a - c = b - c$.
Multiplication Property of Equality	For all real numbers a, b, and c, if $a = b$, then $a \cdot c = b \cdot c$.
Division Property of Equality	For all real numbers a, b, and c, $c \neq 0$ if $a = b$, then $\frac{a}{c} = \frac{b}{c}$.
Zero Product Property of Equality	For all real numbers a and b, if $a \cdot b = 0$ then $a = 0$ or $b = 0$ or both a and b equal 0.
Addition Property of Inequality*	For all real numbers a, b, and c, if $a > b$, then $a + c > b + c$.
Subtraction Property of Inequality*	For all real numbers a, b, and c, if $a > b$, then $a - c > b - c$.
Multiplication Property of Inequality *	For all real numbers a, b, and c, $c > 0$, if $a > b$, then $a \cdot c > b \cdot c$. For all real numbers a, b, and c, $c < 0$, if $a > b$, then $a \cdot c < b \cdot c$.
Division Property of Inequality*	For all real numbers a, b, and c, $c > 0$ if $a > b$, then $\frac{a}{c} > \frac{b}{c}$. For all real numbers a, b, and c, $c < 0$ if $a > b$, then $\frac{a}{c} < \frac{b}{c}$.

*These properties are also true for $<$, $\leq$, $\geq$.

Properties of Exponents

For any numbers a and b and all integers m and n,

$a^m \cdot a^n = a^{m+n}$

$(a^m)^n = a^{mn}$

$(ab)^m = a^m b^m$

$\dfrac{a^m}{a^n} = a^{m-n}, a \neq 0$

$\left(\dfrac{a}{b}\right)^m = \dfrac{a^m}{b^m}, b \neq 0$

$a^{-n} = \dfrac{1}{a^n}, a \neq 0$ and $\dfrac{1}{a^{-n}} = a^n, a \neq 0$

$a^0 = 1, a \neq 0$

Properties of Radicals

In the expression $\sqrt[n]{a}$,

a is the radicand, $\sqrt{}$ is the radical symbol, and n is the root index.

$\sqrt[n]{a} = b$, if $b^n = a$ b is the nth root of a.

$a\sqrt{b} \pm c\sqrt{b} = (a \pm c)\sqrt{b}$, where $b \geq 0$.

$(a\sqrt{b})(c\sqrt{b}) = ac\sqrt{bd}$, where $b \geq 0, d \geq 0$.

$\dfrac{a\sqrt{b}}{c\sqrt{d}} = \dfrac{a}{c}\sqrt{\dfrac{b}{d}}$, where $b \geq 0, c \neq 0, d > 0$.

Table of Measures

Customary	Metric
Distance/Length 1 foot (ft) = 12 inches (in.) 1 yard (yd) = 3 feet (ft) = 36 inches (in.) 1 mile (mi) = 5280 feet (ft)	1 centimeter (cm) = 10 millimeters (mm) 1 meter (m) = 100 centimeters (cm) 1 kilometer (km) = 1000 meters (m)
Volume 1 cup (c) = 8 fluid ounces (fl oz) 1 pint (pt) = 2 cups (c) 1 quart (qt) = 2 pints (pt) 1 gallon (gal) = 4 quarts (qt)	1 liter (L) = 1000 milliliters (mL)
Weight/Mass 1 pound (lb) = 16 ounces (oz)	1 gram (g) = 1000 milligrams (mg) 1 kilogram (kg) = 1000 grams (g)
Time 1 minute (min) = 60 seconds (sec) 1 hour (hr) = 60 minutes (min) 1 day (d) = 24 hours (hr) 1 week (wk) = 7 days (d)	1 year (yr) = 365 days (d) 1 year (yr) = 52 weeks (wk) 1 year (yr) = 12 months (mo)

SpringBoard Learning Strategies

READING STRATEGIES

STRATEGY	DEFINITION	PURPOSE
Activating Prior Knowledge	Recalling what is known about a concept and using that information to make a connection to a new concept	Helps students establish connections between what they already know and how that knowledge is related to new learning
Chunking the Activity	Grouping a set of items/questions for specific purposes	Provides an opportunity to relate concepts and assess student understanding before moving on to a new concept or grouping
Close Reading	Reading text word for word, sentence by sentence, and line by line to make a detailed analysis of meaning	Assists in developing a comprehensive understanding of the text
Graphic Organizer	Arranging information into maps and charts	Builds comprehension and facilitates discussion by representing information in visual form
Interactive Word Wall	Visually displaying vocabulary words to serve as a classroom reference of words and groups of words as they are introduced, used, and mastered over the course of a year	Provides a visual reference for new concepts, aids understanding for reading and writing, and builds word knowledge and awareness
KWL Chart (Know, Want to Know, Learn)	Activating prior knowledge by identifying what students know, determining what they want to learn, and having them reflect on what they learned	Assists in organizing information and reflecting on learning to build content knowledge and increase comprehension
Marking the Text	Highlighting, underlining, and/or annotating text to focus on key information to help understand the text or solve the problem	Helps the reader identify important information in the text and make notes about the interpretation of tasks required and concepts to apply to reach a solution
Predict and Confirm	Making conjectures about what results will develop in an activity; confirming or modifying the conjectures based on outcomes	Stimulates thinking by making, checking, and correcting predictions based on evidence from the outcome
Levels of Questions	Developing literal, interpretive, and universal questions about the text while reading the text	Focuses reading, helps in gaining insight into the text by seeking answers, and prepares one for group and class discussions
Paraphrasing	Restating in your own words the essential information in a text or problem description	Assists with comprehension, recall of information, and problem solving
Role Play	Assuming the role of a character in a scenario	Helps interpret and visualize information in a problem
Shared Reading	Reading the text aloud (usually by the teacher) as students follow along silently, or reading a text aloud by the teacher and students	Helps auditory learners do decode, interpret, and analyze challenging text
Summarizing	Giving a brief statement of the main points in a text	Assists with comprehension and provides practice with identifying and restating key information
Think Aloud	Talking through a difficult text or problem by describing what the text means	Helps in comprehending the text, understanding the components of a problem, and thinking about possible paths to a solution
Visualization	Picturing (mentally and/or literally) what is read in the text	Increases reading comprehension and promotes active engagement with the text
Vocabulary Organizer	Using a graphic organizer to keep an ongoing record of vocabulary words with definitions, pictures, notes, and connections between words	Supports a systematic process of learning vocabulary

SpringBoard Learning Strategies

COLLABORATIVE STRATEGIES

STRATEGY	DEFINITION	PURPOSE
Critique Reasoning	Through collaborative discussion, respond to the arguments of others; question the use of mathematical terminology, assumptions, and conjectures to improve understanding and to justify and communicate conclusions	Helps students learn from each other as they make connections between mathematical concepts and learn to verbalize their understanding and support their arguments with reasoning and data that make sense to peers
Debriefing T	Discussing the understanding of a concept to lead to consensus on its meaning	Helps clarify misconceptions and deepen understanding of content
Discussion Groups	Working within groups to discuss content, to create problem solutions, and to explain and justify a solution	Aids understanding through the sharing of ideas, interpretation of concepts, and analysis of problem scenarios
Group Presentation	Presenting information as a collaborative group	Allows opportunities to present collaborative solutions and to share responsibility for delivering information to an audience
Jigsaw	Reading different texts or passages, students become "experts" and then move to a new group to share their information; after sharing, students go back to the original group to share new knowledge	Provides opportunities to summarize and present information to others in a way that facilitates understanding of a text or passage (or multiple texts or passages) without having each student read all texts
Sharing and Responding	Communicating with another person or a small group of peers who respond to a piece of writing or proposed problem solution	Gives students the opportunity to discuss their work with peers, to make suggestions for improvement to the work of others, and/or to receive appropriate and relevant feedback on their own work
Think-Pair-Share	Thinking through a problem alone, pairing with a partner to share ideas, and concluding by sharing results with the class	Enables the development of initial ideas that are then tested with a partner in preparation for revising ideas and sharing them with a larger group

WRITING STRATEGIES

STRATEGY	DEFINITION	PURPOSE
Drafting	Writing a text in an initial form	Assists in getting first thoughts in written form and ready for revising and refining
Note Taking	Creating a record of information while reading a text or listening to a speaker	Helps in organizing ideas and processing information
Prewriting	Brainstorming, either alone or in groups, and refining thoughts and organizing ideas prior to writing	Provides a tool for beginning the writing process and determining the focus of the writing
Quickwrite	Writing for a short, specific amount of time about a designated topic	Helps generate ideas in a short time
RAFT (Role of Writer, Audience, Format, and Topic)	Writing a text by consciously choosing a viewpoint (role of the writer), identifying an audience, choosing a format for the writing, and choosing a topic	Provides a framework for communicating in writing and helps focus the writer's ideas for specific points of communication
Self Revision / Peer Revision	Working alone or with a partner to examine a piece of writing for accuracy and clarity	Provides an opportunity to review work and to edit it for clarity of the ideas presented as well as accuracy of grammar, punctuation, and spelling

SpringBoard Learning Strategies
PROBLEM-SOLVING STRATEGIES

Strategy	Definition	Purpose
Construct an Argument	Use mathematical reasoning to present assumptions about mathematical situations, support conjectures with mathematically relevant and accurate data, and provide a logical progression of ideas leading to a conclusion that makes sense	Helps develop the process of evaluating mathematical information, developing reasoning skills, and enhancing communication skills in supporting conjectures and conclusions
Create a Plan	Analyzing the tasks in a problem and creating a process for completing the tasks by finding information needed for the tasks, interpreting data, choosing how to solve a problem, communicating the results, and verifying accuracy	Assists in breaking tasks into smaller parts and identifying the steps needed to complete the entire task
Create Representations	Creating pictures, tables, graphs, lists, equations, models, and /or verbal expressions to interpret text or data	Helps organize information using multiple ways to present data and to answer a question or show a problem solution
Guess and Check	Guessing the solution to a problem, and then checking that the guess fits the information in the problem and is an accurate solution	Allows exploration of different ways to solve a problem; guess and check may be used when other strategies for solving are not obvious
Identify a Subtask	Breaking a problem into smaller pieces whose outcomes lead to a solution	Helps to organize the pieces of a complex problem and reach a complete solution
Look for a Pattern	Observing information or creating visual representations to find a trend	Helps to identify patterns that may be used to make predictions
Simplify the Problem	Using "friendlier" numbers to solve a problem	Provides insight into the problem or the strategies needed to solve the problem
Work Backward	Tracing a possible answer back through the solution process to the starting point	Provides another way to check possible answers for accuracy
Use Manipulatives	Using objects to examine relationships between the information given	Provides a visual representation of data that supports comprehension of information in a problem

Glossary
Glosario

A

absolute value (p. 49) The distance from a number, n, to zero on a number line, written $|n|$.
valor absoluto (pág. 49) Distancia entre un número y el cero en una recta numérica, escrita como $|n|$.

absolute value equation (p. 49) An equation involving the absolute value of a variable expression.
ecuación con valor absoluto (pág. 49) Ecuación que involucra el valor absoluto de una expresión con variables.

absolute value inequality (p. 54) An inequality involving the absolute value of a variable expression.
desigualdad con valor absoluto (pág. 54) Desigualdad que involucra el valor absoluto de una expresión con variables.

arthithmetic sequence (p. 160) A sequence in which the difference of consecutive terms is constant.
progresión aritmética (pág. 160) Sucesión en la que la diferencia entre términos consecutivos es constante.

axis of symmetry of a parabola (p. 433) A line that passes through the vertex, dividing the parabola into two symmetrical halves.
eje de simetría de una parábola (pág. 433) Recta que pasa a través del vértice, dividiendo la parábola en dos mitades simétricas.

B

base (p. 287) The factor in an exponential expression that is being raised to a power.
basa (pág. 287) El factor en una expresión exponencial que esta siendo elevado a una potencia.

binomial (p. 385) A a sum or difference of two monomials.
binomio (pág. 385) Una suma o diferencia de dos monomios.

boundary line (p. 242) A line that divides the coordinate plane into two regions, called half-planes.
frontera (pág. 242) Recta que divide el plano de coordenadas en dos regiones, llamadas semiplanos.

C

causation (p. 197) The relation between two events, where the second event is a consequence of the first.
causalidad (pág. 197) La relación entre dos eventos, donde el segundo evento es una consecuencia del primer evento.

census (p. 527) A study that gathers information about every member of the population.
censo (pág. 527) Estudio que reúne información acerca de cada miembro de la población.

coefficient (p. 356) A number by which a variable is multiplied. For example, in the term $6x$, 6 is the coefficient.
coeficiente (pág. 356) Número por el cual se multiplica una variable. Por ejemplo, en el término $6x$, 6 es el coeficiente.

coincident lines (p. 264) Lines that occupy the same space or location in the plane and pass through the same set of points.
líneas coincidentes (pág. 264) Líneas que ocupan el mismo espacio o ubicación en el plano y pasan por el mismo conjunto de puntos.

common difference (p. 160) The difference between consecutive terms in an arithmetic sequence.
diferencia común (pág. 160) La diferencia entre los términos consecutivos de una progresión aritmética.

common ratio (p. 314) The ratio, typically denoted by the letter r, between consecutive terms in a geometric sequence.
proporción común (pág. 314) La proporcion, normalmente denotado por la letra r, entre los terminos consecutivos de una progresión geométrica.

compound inequality (p. 44) Two inequalities joined by the word *and* or by the word *or*.
desigualdad compuesta (pág. 44) Dos desigualdades unidas por la palabra *y* o por la palabra *o*.

compound interest (p. 341) Interest calculated on the total principal plus the interest earned or owed during the previous time period.
interés compuesto (pág. 341) Interés calculado sobre el capital total más el interés devengado o adeudado durante el período anterior.

conjunction (p. 45) Two statements joined by the word *and*.
conjunción (pág. 45) Dos enunciados unidos por la palabra *y*.

consecutive (p. 5) Refers to items that follow each other in order.
consecutivo (pág. 5) Se refiere a los elementos que se suceden en el orden.

constant term (p. 356) A term in an expression that does not change in value because it does not contain a variable. For example, the constant term in the expression $3n + 6$ is 6.
término constante (pág. 356) Término cuyo valor no cambia en una expresión, pues no contiene variables. Por ejemplo, el término constante en la expresión $3n + 6$ es 6.

correlation (p. 197) A collection of data points has a *positive correlation* if it has the property that y tends to increase as x increases. It has a *negative correlation* if y tends to decrease as x increases. A correlation is also known as an **association.**
correlación (pág. 197) Un conjunto de datos tiene una *correlación positiva* si tiene la propiedad de que y tiende a aumentar a medida que aumenta x. Tiene una *correlación negativa* si y tiende a disminuir a medida que aumenta x. Una correlación se conoce también como **asociación.**

cube root (p. 301) The cube root of a number, n, is the number which when used as a factor three times results in the product n.
raíz cúbica (pág. 301) La raíz cúbica de un número n, es el número que usado tres veces como factor da un producto de n.

D

degree of a polynomial (p. 357) The largest degree of any term in the polynomial.
grado de un polinomio (pág. 357) El grado mayor entre todos los términos del polinomio.

degree of a term (p. 356) The sum of the exponents of the variables contained in the term.
grado de un termino (pág. 356) Suma de los exponentes de las variables contenidas en el término.

dependent variable (p. 81) The variable whose value is determined by the input or value of the independent variable.
variable dependiente (pág. 81) Variable cuyo valor queda determinado por la entrada o valor de la variable independiente.

difference of two squares (p. 376) A polynomial of the form $a^2 - b^2$, which is the product of binomials of the form $(a + b)(a - b)$.
diferencia de cuadrados (pág. 376) Polinomio de la forma $a^2 - b^2$, que es el producto de binomios de la forma $(a + b)(a - b)$.

descending order (p. 357) Terms are arranged in order from largest to smallest.
orden descendente (pág. 357) Los términos se ordenan de mayor a menor.

discrete data (p. 82) A set of data with a finite number of data values; a graph of discrete data appears as individual points on a number line or coordinate plane.
datos discretos (pág. 82) Conjunto de datos con un número finito de valores; una gráfica de datos discretos aparece como puntos individuales en una recta numérica o plano de coordenadas.

disjunction (p. 43) Two statements joined by the word *or*.
disyunción (pág. 43) Dos enunciados unidos por la palabra *o*.

E

elimination method (p. 263) A method for solving a system of equations that involves eliminating variables. Also called the **linear combination method**.
método de eliminación (pág. 263) Método para resolver un sistema de ecuaciones que involucra eliminar variables. También llamado **método de combinación lineal**.

equation (p. 15) A mathematical statement that shows that two expressions are equal.
ecuación (pág. 15) Enunciado matemático que muestra que dos expresiones son iguales.

equilateral (p. 11) A polygon with all sides congruent.
equilátero (pág. 11) Polígono con todos sus lados congruentes.

explicit formula (p. 162) Describes any term in a sequence.
fórmula explícita (pág. 162) Describe cualquier término de una progresión.

exponent (p. 287) The number in an exponential expression that tells how many times to use the base as a factor.
exponente (pág. 287) El número en una expresión exponencial que indica el número de veces para usar la base como un factor.

exponential decay (p. 331) A decrease in a quantity due to multiplying by the same factor during each time period. In a decay function, the constant factor is greater than zero but less than 1.
disminución exponencial (pág. 331) Disminución en una cantidad debido a la multiplicación por el mismo factor durante cada período de tiempo. En una función de disminución, el factor constante es mayor que cero pero menor que 1.

exponential function (p. 326) A function of the form $f(x) = a \cdot b^x$, where a and b are constants, x is the domain, $f(x)$ is the range, and $a \neq 0, b > 0, b \neq 1$.
función exponencial (pág. 326) Función de la forma $f(x) = a \cdot b^x$, donde a y b son constantes, x es el dominio, $f(x)$ es el rango y $a \neq 0, b > 0, b \neq 1$.

exponential growth (p. 326) A increase in a quantity due to multiplying by the same factor during each time period. In a growth function, the constant factor is greater than 1.
crecimiento exponencial (pág. 326) Incremento en una cantidad debido a la multiplicación por el mismo factor durante cada período de tiempo. En una función de crecimiento, el factor constante es mayor que 1.

exponential regression (p. 347) A method used to find an exponential function that models a set of data.
regresión exponencial (pág. 347) Un método para encontrar una función exponencial que modela un conjunto de datos.

expression (p. 5) A mathematical phrase that uses numbers, or variables, or both.
expresión (pág. 5) Frase matemática que usa números, o variables, o ambos.

extrema (p. 83) Refers to all maximum and minimum values.
extremidad (pág. 83) Se refiere a todos los valores máximos y mínimos.

F

factor (p. 347) Any of the numbers or symbols that when multiplied together form a product; or the process of finding the factors that form a product.
factor (pág. 347) Cualquiera de los números o símbolos que al multiplicarse entre sí forman un producto; o el proceso de hallar los factores que forman un producto.

factor (p. 386) The process of finding the factors that form a product.
factorizar (pág. 386) Proceso de hallar los factores que forman un producto.

G

graph of an inequality (p. 35) All the points on a number line or half-plane that make the inequality true.
grafica de una desigualdad (pág. 35) Todos los puntos de una recta numérica o semiplano que hacen que la desigualdad sea verdadera.

geometric sequence (p. 314) A sequence in which the ratio of consecutive terms is a constant.
progresión geométrica (pág. 314) Sucesión en que la razón de los términos consecutivos es constante.

H

half-plane (p. 242) One of the two regions of a coordinate plane created by a line in the plane. A half-plane is closed if its boundary line is included in the region. A half-plane is open if its boundary line is not included in the region.
semiplano (pág. 242) Una de las dos regiones de un plano de coordenadas creadas por una recta sobre el plano. Un semiplano es cerrado si su frontera está incluida en la región. Un semiplano es abierto si su frontera no está incluida en la región.

I

independent variable (p. 81) The variable for which input values are substituted in a function.
variable independiente (pág. 81) Variable que es reemplazada por los valores de entrada en una función.

L

leading coefficient (p. 357) The coefficient of the term with the highest degree in a polynomial.
coeficiente líder (pág. 357) Coeficiente del término que tiene el mayor grado en un polinomio.

least common multiple (p. 413) The smallest multiple that two or more numbers have in common.
mínimo común múltiplo (mcm) (pág. 413) El menor múltiplo que dos o más números tienen en común.

like terms (p. 361) Terms that have the same variable(s) raised to the same power(s).
términos semejantes (pág. 361) Términos que tienen las mismas variables elevadas a las mismas potencias.

line of best fit (p. 198) A line on a graph showing the general direction of a group of points.
recta de major ajuste (pág. 198) Una línea de un gráfico que muestra la dirección general de un grupo de puntos.

linear equation (p. 175) An equation in which all the terms have a degree of one or zero.
ecuación lineal (pág. 175) Ecuación en la cual todos los términos tienen grado uno o cero.

linear inequality (p. 240) An inequality in which all the terms have a degree of one or zero.
desigualdad lineal (pág. 240) Desigualdad en la que todos los términos tienen grado uno o cero.

linear regression (p. 198) A method used to find the line of best fit.
regresión lineal (pág. 198) Un método para encontrar la línea de major ajuste.

literal equation (p. 28) A equation containing several different variables.
ecuación literal (pág. 28) Ecuación que contiene varias variables diferentes.

M

monomial (p. 385) A number, a variable, or a product of numbers and variables with whole-number exponents.
monomio (pág. 385) Un número, una variable, o un producto de números y variables con exponents de números enteros.

N

negative correlation (p. 197) Two variables are related in a negative association when values for one variable tend to decrease as values for the other variable increase; an increase in x corresponds with a decrease in y.
correlación negativa (pág. 197) Dos variables tienen una correlación negativa cuando los valores de una variable tienden a decrecer cuando los valores de la otra variable aumentan; un incremento en x corresponde a un decremento en y.

O

opposite (p. 364) A number's additive inverse.
opuesto (pág. 364) El inverso aditivo de un número.

outliers (p. 543) Data points in a set of data that do not fit the overall pattern of the data set.
valores atípicos (pág. 543) Puntos de un conjunto de datos que no calzan en el patrón general del conjunto de datos.

P

parabola (p. 427) The graph of a quadratic function.
parábola (pág. 427) Gráfica de una función cuadrática.

parallel lines (p. 264) Lines in the same plane that do not intersect. Parallel lines on a coordinate plane have the same slope.
rectas paralelas (pág. 264) Rectas que están en el mismo plano, pero no se intersecan. Las rectas paralelas sobre un plano de coordenadas tienen la misma pendiente.

parent function (p. 111) The most basic function of a particular type, such as $f(x) = x$ (linear); $f(x) = x^2$ (quadratic); $f(x) = |x|$ (absolute value); and $f(x) = b^x$ (exponential).
función básica (pág. 111) La función más simple de un tipo en particular, como $f(x) = x$ (lineal), $f(x) = x^2$ (cuadrática), $f(x) = |x|$ (valor absoluto) y $f(x) = b^x$ (exponencial).

perfect square trinomials (p. 389) Trinomials that have the form $a^2 + 2ab + b^2$ or $a^2 - 2ab + b^2$ and are the result of squaring binomials of the form $(a + b)^2$ and $(a - b)^2$, respectively.
trinomios de cuadrados perfectos (pág. 389) Trinomios de la forma $a^2 + 2ab + b^2$ o $a^2 - 2ab + b^2$ que resultan de elevar al cuadrado binomios de la forma $(a + b)^2$ y $(a - b)^2$, respectivamente.

piecewise defined function (p. 217) A function that is defined differently for different disjoint intervals in its domain.
función por tramos (pág. 217) Función que se define de manera diferente para diferentes intervalos de su dominio.

point-slope form (p. 179) A linear equation with the form $y - y_1 = m(x - x_1)$, where m is the slope of the line and (x_1, y_1) is a point on the line.
forma punto-pendiente (pág. 179) Ecuación lineal expresada de la forma $y - y_1 = m(x - x_1)$, donde m es la pendiente de la recta y (x_1, y_1) es un punto sobre la recta.

polynomial (p. 356) A single term or the sum of two or more terms.
polinomio (pág. 356) Un único término o la suma de dos o más términos.

positive correlation (p. 197) Two variables are related in a positive association when values for one variable tend to increase as values for the other variable also increase; an increase in x corresponds with an increase in y.
correlación positiva (pág. 197) Dos variables tienen una correlación positiva cuando los valores de una variable tienden a aumentar cuando los valores de la otra variable también aumentan; un aumento en x corresponde a un aumento en y.

power (p. 287) A mathematical expression with two parts, a base and an exponent. For example, in the power 5^3, 5 is the base and 3 is the exponent.
potencia (pág. 287) Expresión matemática con dos partes, una base y un exponente. Por ejemplo, en la potencia 5^3, 5 es la base y 3 es el exponente.

principal square root (p. 300) The positive square root of a number.
raíz cuadrada primaria (pág. 300) La raíz cuadrada positiva de un número.

Q

quadratic function (p. 425) A function in one variable with the form $f(x) = ax^2 + bx + c$, where a, b, and c are real numbers and $a \neq 0$.
función cuadrática (pág. 425) Función en una variable de la forma $f(x) = ax^2 + bx + c$, donde a, b y c son números reales y $a \neq 0$.

Resources **623**

quadratic regression (p. 202)   A method used to find a quadratic function that models a set of data.
regresion cuadrático (pág. 202)   Un método para encontrar una función cuadrática que modela un conjunto de datos.

R
radical expression (p. 300)   An expression of the form $\sqrt[n]{a}$.
expresión radical (pág. 300)   Expresión de la forma $\sqrt[n]{a}$.

rate of change (p. 128)   The ratio of the change in y to the change in x.
proporcion de cambio (pág. 128)   La relación entre el cambio en y para el cambio en x.

rational expression (p. 403)   An expression that can be written as the ratio of two polynomials.
expresión racional (pág. 403)   Expresión que puede escribirse como razón de dos polinomios.

rationalize the denominator (p. 309)   To make the denominator of a fraction rational by multiplying the fraction by an appropriate form of 1.
racionalizar el denominador (pág. 309)   Hacer que el denominador de una fracción sea racional, multiplicando la fracción por una forma apropiada de 1.

residual (p. 578)   The difference between an actual value and a predicted value.
residual (pág. 578)   La diferencia entre un valor real y un valor predicho.

S
scatter plot (p. 194)   A graphic display of bivariate data on a coordinate plane that may be used to show a relationship between two variables.
diagrama de dispersión (pág. 194)   Representación gráfica de datos bivariados sobre un plano de coordenadas, que puede usarse para mostrar una relación entre dos variables.

sequence (p. 160)   A list of items or numbers.
sucesión (pág. 160)   Lista de elementos o números.

slope (p. 124)   The ratio of the vertical change of a line to the horizontal change of the line.
pendiente (pág. 124)   La razón del cambio vertical de una recta al cambio horizontal de la recta.

Two angles whose measures Dos ángulos cuyas

slope-intercept form (p. 175)   A linear equation of the form $y = mx + b$ where m is the slope and b is the y-intercept.
forma pendiente-intercepto (pág. 175)   Ecuación lineal de la forma $y = mx + b$, donde m es la pendiente y b es el intercepto en el eje de las y.

solution (p. 15)   Any value that makes an equation or inequality true when substituted for the variable.
solución (pág. 15)   Cualquier valor que hace verdadera una ecuación o desigualdad al reemplazar la variable.

solution region (p. 275)   The part of the coordinate plane in which the ordered pairs are solutions to all inequalities in a system.
región solución (pág. 275)   La parte del plano de coordenadas en el que los pares ordenados son soluciones a las desigualdades en un sistema.

solution of an inequality (p. 35)   An ordered pair or set of ordered pairs that makes an inequality true when substituted for the variables.
solución de una desigualdad (pág. 35)   Par ordenado o conjunto de pares ordenados que hacen verdadera una desigualdad al reemplazar las variables.

solutions of the linear inequality (p. 240)   An ordered pair or set of ordered pairs that makes an inequality true when substituted for the variables.
solución de una desigualdad lineal (pág. 240)   Par ordenado o conjunto de pares ordenados que hacen verdadera una desigualdad al reemplazar las variables.

standard form of a linear equation (p. 183)   A linear equation of the form $Ax + By = C$, where $A \geq 0$, A and B cannot both be zero, and A, B, and C are integers whose greatest common factor is one.
forma estándar de una ecuación lineal (pág. 183)   Ecuación lineal de la forma $Ax + By = C$, donde $A \geq 0$, A y B no pueden ser ambos cero y A, B y C son enteros cuyo máximo común divisor es uno.

standard form of a polynomial (p. 357)   A way of writing a polynomial so that the terms are in descending order of degree.
forma estándar de un polinomio (pág. 357)   Manera de escribir un polinomio de modo que los términos estén en orden decreciente de grado.

standard form of a quadratic function (p. 425)   A function of the form $f(x) = ax^2 + bx + c$, where a, b, and c are real numbers and $a \neq 0$
forma estándar de una función cuadrática (pág. 425)   Función de la forma $f(x) = ax^2 + bx + c$, donde a, b y c son números reales y $a \neq 0$.

substitution method for solving a system of linear equations (p. 258)   A method that involves solving one of the equations for one of the variables and then substituting that value in the other equation(s).
método de sustitución para resolver un sistema de ecuaciones lineales (pág. 258)   Método que involucra resolver una de las ecuaciones para una de las variables y luego sustituir ese valor en la otra ecuación o ecuaciones.

system of linear equations (p. 253)   Two or more linear equations using the same variables.
sistema de ecuaciones lineales (pág. 253)   Dos o más ecuaciones lineales que usan las mismas variables.

system of linear inequalities (p. 274)   Two or more linear inequalities using the same variables.
sistema de desigualdades lineales (pág. 274)   Dos o más desigualdades lineales que usan las mismas variables.

T
term (p. 355)   A number, a variable, or the product of a number and variable(s).
término (pág. 355)   Número, variable, o producto de un número por una o más variables.

transformation (p. 113)   A change in the position, size, or shape of a parent graph.
transformación (pág. 113)   Cambio en la posición, tamaño o forma de una gráfica original.

translation (p. 113)   A transformation that moves each point of a figure the same distance and in the same direction.
traslación (pág. 113)   Transformación que mueve cada punto de una figura la misma distancia y en la misma dirección.

tree diagram (p. 313)   A graphic organizer for listing the possible outcomes of an experiment.

diagrama de árbol (pág. 313)   Organizador gráfico para registrar los resultados posibles de un experimento.

trend line (p. 194)   A line drawn on a scatterplot to show the general direction of the association or correlation between two sets of data.

línea de tendencia (pág. 194)   Línea que se dibuja en un diagrama de dispersión para mostrar la dirección general de la asociación o correlación entre dos conjuntos de datos.

V

variable (p. 5)   A letter or symbol used to represent one or more numbers.

variable (pág. 5)   Letra o símbolo que se usa para representar uno o más números.

vertex of a parabola (p. 427)   The point at which a maximum or minimum value of a function occurs.

vértice de una parábola (pág. 427)   Punto en el que ocurre un valor máximo o mínimo de una función.

vertical line test (p. 81)   A visual inspection for checking whether or not a graph represents a function.

prueba de la recta vertical (pág. 81)   Inspección visual para comprobar si una gráfica representa o no una función.

X

x-intercept (p. 102)   The point where a line crosses the x-axis. Its coordinates will be of the form $(a,0)$, where a is a real number.

intercepto en x (pág. 102)   Punto donde una línea cruza el eje de las x. Sus coordenadas serán de la forma $(a,0)$, donde a es un número real.

Y

y-intercept (p. 103)   The point where a line crosses the y-axis. Its coordinates will be of the form $(0,a)$, where a is a real number.

intercepto en y (pág. 103)   Punto donde una línea cruza el eje de las y. Sus coordenadas serán de la forma $(0,a)$, donde a es un número real.

Verbal & Visual Word Association

Definition in Your Own Words	**Important Elements**
Visual Representation	**Personal Association**

Academic Vocabulary Word

Word Map

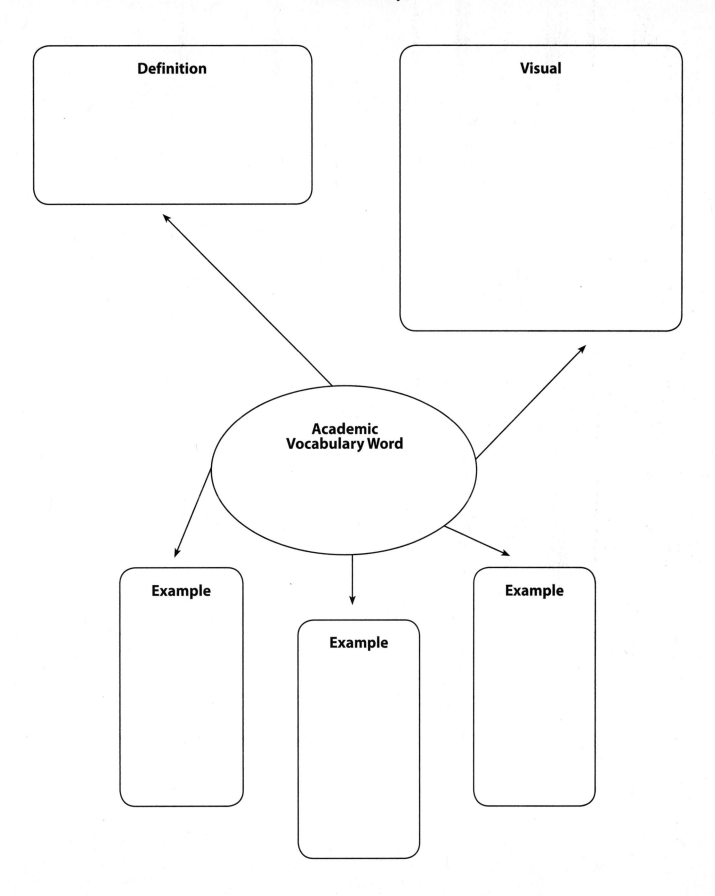

Eight Circle Spider

Venn Diagram

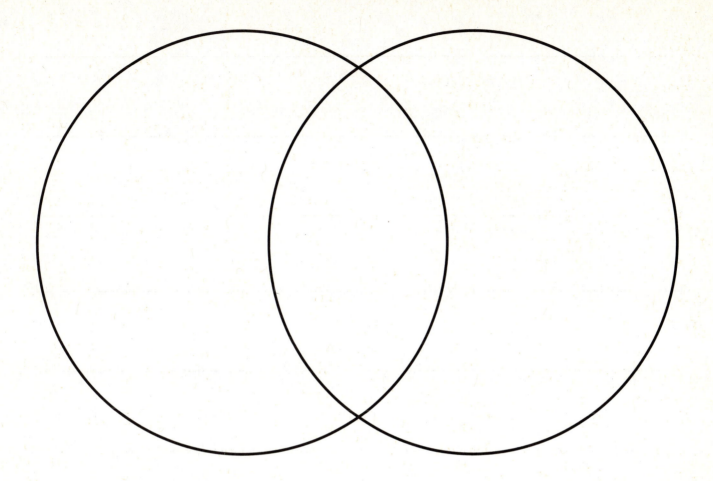

Number Lines

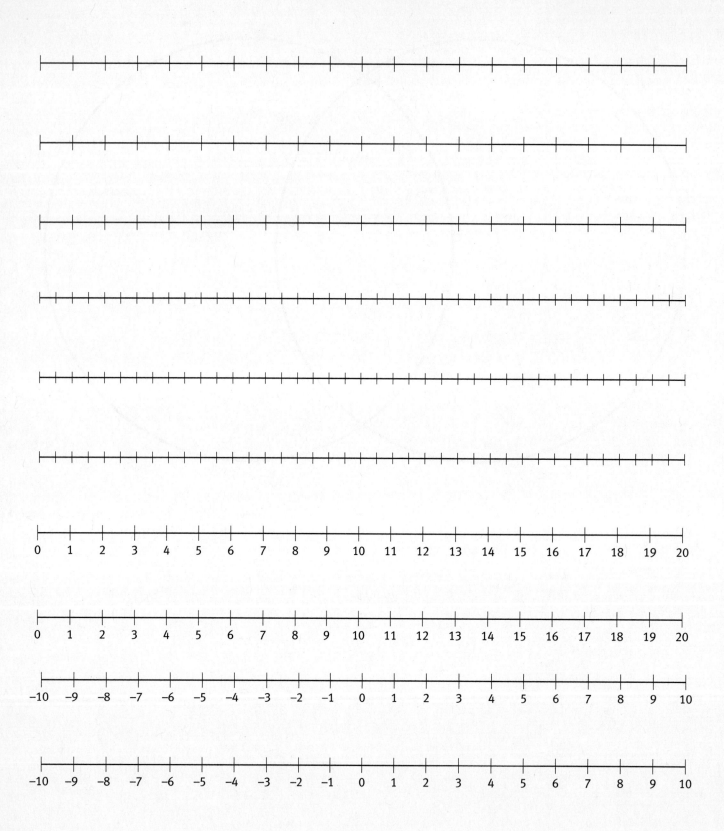

Algebra Tiles

Resources 631

Name: _____ Date: _____ Class: _____

Quarter Inch Grid Paper

Name: _____ Date: _____ Class: _____

Name: _____ Date: _____ Class: _____

Centimeter Grid Paper

Tables and Coordinate Grids

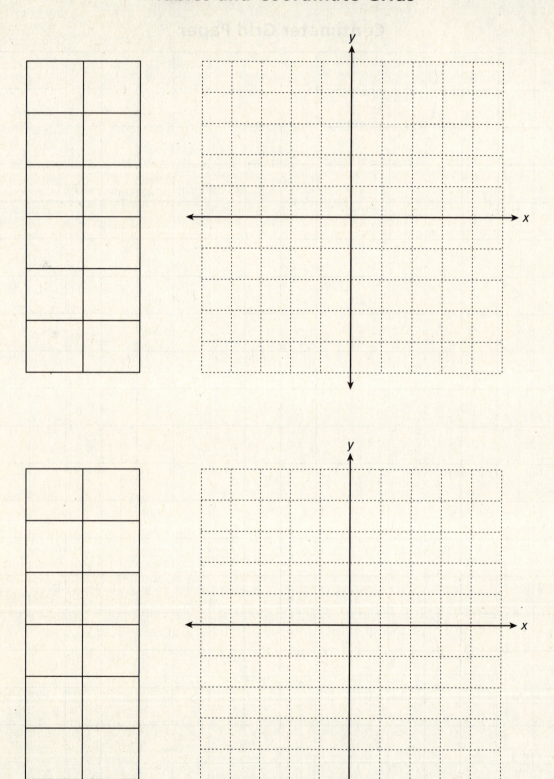

Vertical/Horizontal T-Table

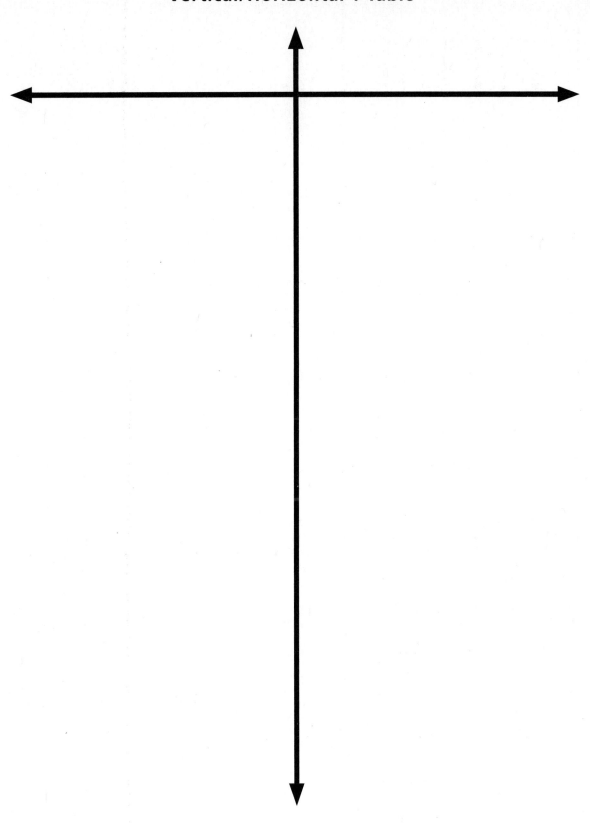

1st Quadrant Grids

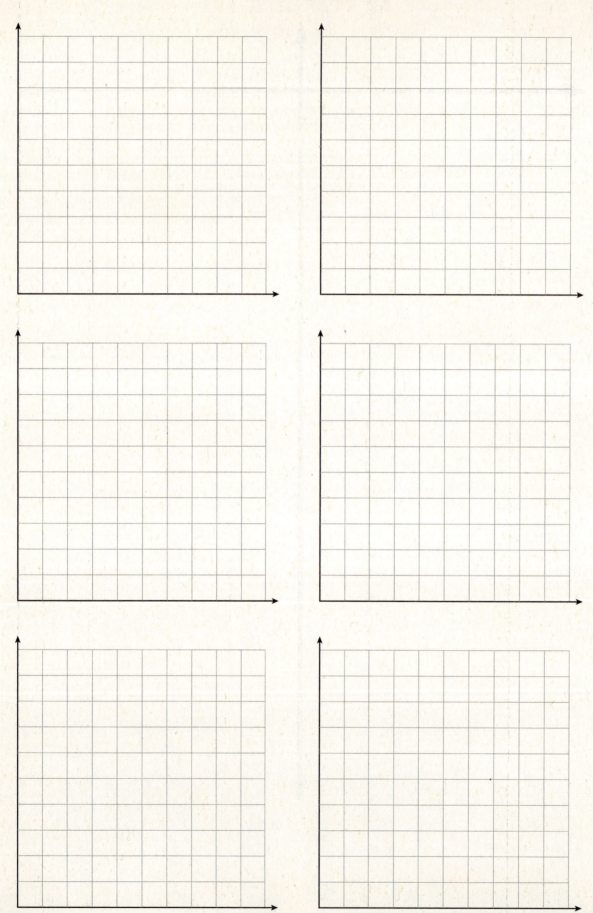

5 by 5 Coordinate Grids

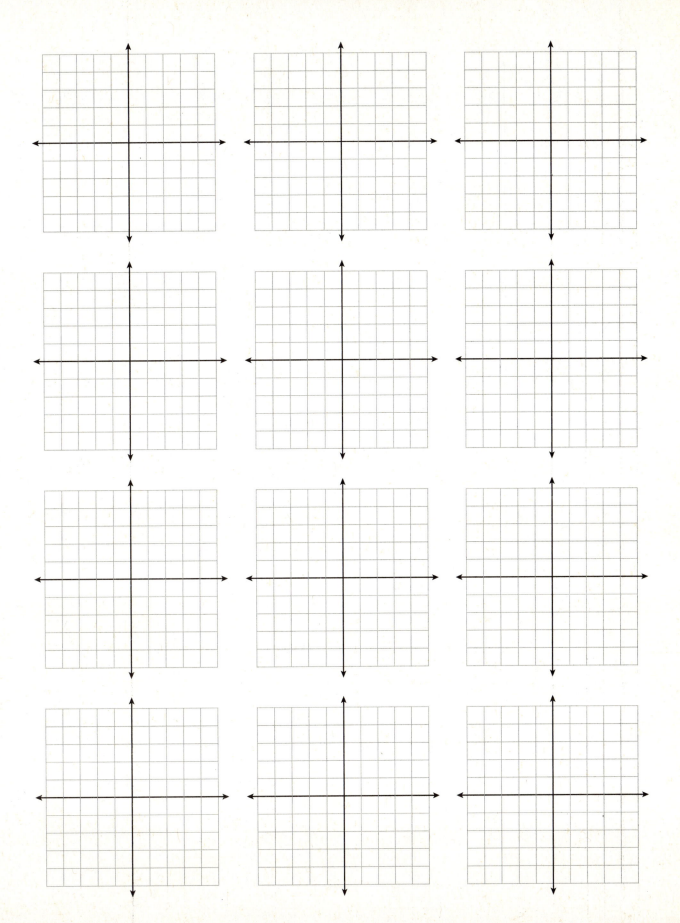

20 × 20 Coordinate Grids

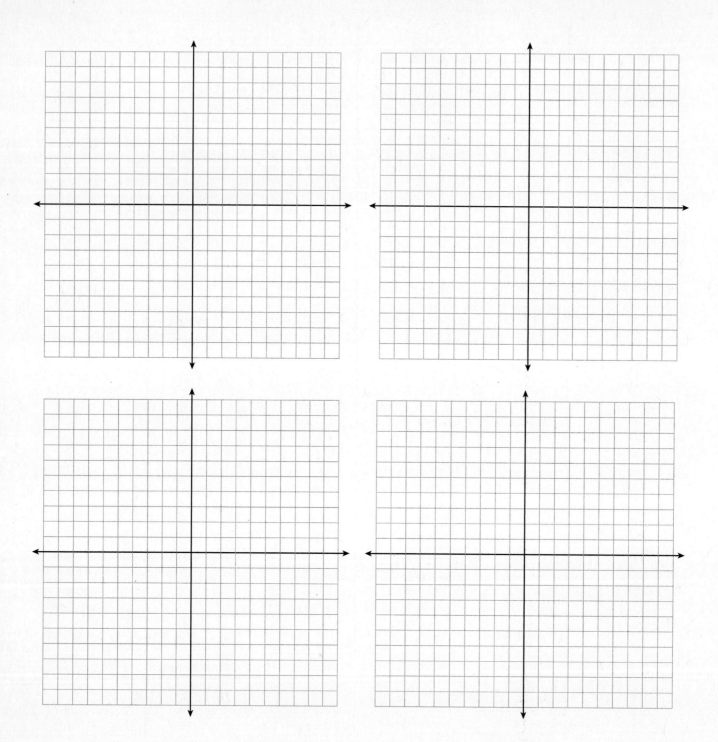

Index

A

Absolute maximum, 82–91
Absolute minimum, 82–91
Absolute value
  defined, 49
  equations, 49–53
  inequalities, 54–58
  notation for, 49
Acceleration, 29
Addition
  Addition Property of Radicals, 304
  Associative Property, 16, 362
  Commutative Property, 16, 362
  polynomials, 359–363
  Property of Equality, 16, 17
  radical expressions, 304–306
  rational expressions, 413–416
  square roots, 304–306
Addition Property of Radicals, 304
Additive inverse, 364
Algebra-tile method
  adding polynomials, 359–360
  factoring trinomials, 393
Americans with Disabilities Act (ADA), 123
Arithmetic sequences
  common difference, 160–161, 315
  constant difference, 160
  defined, 160, 315
  explicit formula for, 162–165
  as function, 166–167
  identifying, 159–161
  recursive formula, 168–170
Associate, 601
Associative Property
  of Addition, 16, 362
  of Multiplication, 16
Atmospheric pressure, 175
Axis of symmetry, parabola, 433, 459

B

Bar graphs, segmented bar graph, 600–606
Base, 287
Bell curve, 548
Best-fit line, 571–592
  defined, 579
  introduction to, 571–576
  residuals and, 577–581
  slope of and intercept of, 582–587
Binomial
  defined, 357, 385
  difference of two squares, 376–378
  Distributive Property, 373–374
  multiplication, 369–378
  square of, 377–378
Bivariate data, 595–606
Boundary line, 242
Box-and-whisker plot, 538
Box plot, 538–547
  defined, 538
  modified, 543–547
Boyle's Law, 30
Break-even point, 20

C

Calories, 61
Causation, 197
  compared to correlation, 561
Census, 527–528
Closed half-plane, 242
Closed numbers, 304, 308
Cluster, 523
Coefficient
  leading, 357
  of polynomials, 356
  of radical expression, 308
Coincident lines, 264–266
Comets, 453
Common difference
  of arithmetic sequence, 160
  defined, 160, 315
Common ratio, 314
Commutative Property
  of Addition, 16, 362
  of Multiplication, 16
Completing the square, 471–473
Complex numbers, 308, 481
Compound inequalities
  conjunctions, 43–46
  defined, 43
  disjunctions, 43–46
  solving, 43–46
Compound interest, 341–346
Conjunctions, 43–46
Consecutive, 5
Consistent, 267–270
Constant difference, 5
  of arithmetic sequence, 160
  defined, 160
Constant of variation, 140–143
Constant rate of change, linear function, 133–136
Constant ratio, 314
Constant term, of polynomials, 356
Continuous data, 82
Coordinate plane
  boundary line, 242
  closed half-plane, 242
  half-planes, 242
  open half-plane, 242
Correlated, 560
Correlation
  compared to causation, 561
  defined, 195
  negative, 197
  no, 197
  positive, 197
Correlation coefficient, 559–568
  defined, 560
  negatively related, 560–561
  positively related, 560–561
Costs, 20
Cubed root, as radical expression, 301
Curie, Marie, 101
Cylinder, surface area of, 387

D

Data
  continuous, 82
  discrete, 82
  equations from, 193–204
  negative correlation, 197
  no correlation, 197
  numeric and graphic representation of, 3–7
  positive correlation, 197
Data analysis. *See* Probability and statistics
Degree of a polynomial, 357
Degree of a term, 356
Denominator, rationalize, 309
Density, 288
Dependent system of linear equation, 267–270
Dependent variable, 81
Descending order of degree, 357
Deviation
  defined, 525
  mean absolute deviation, 524–526, 534
  standard deviation, 532–535
Dietary calories, 61
Difference of two squares, 376–378
  factoring polynomials, 390
Direct proportion, 140
Direct variation, 139–143, 140
Discrete data, 82
Discriminant, 477–479
Disjunctions, 43–46
Distance formula, 30
Distribution, normal distribution, 548–553
Distributive Property, 16, 22, 373–374
  multiplication, 414
  point-slope form, 179
Division
  Division Property of Radicals, 308–309
  exponents, 288–289
  polynomials, 406–410
  Property of Equality, 16, 29
  radical expression, 307–310
  rational expression, 403–405, 411–412
  square root, 307–310
Division Property of Radicals, 308–309
Domain
  defined, 71
  function notation for, 76
  of functions, 71–72, 75
  independent variable, 81–86
  of linear function, 215–218
Dot plot, 537–542

E

Elimination method, 261–263
Equality, Properties of, 16, 29
Equations. *See also* Linear equations
    absolute value, 49–53
    from data, 193–204
    defined, 15
    literal, 28–30
    solving
        algebraic method, 15–18
        with infinitely many solutions, 25–27
        literal equations for a variable, 28–30
        with multiple steps, 22–24
        with no solution, 25–27
        with variable on both sides, 19–21
    writing, 15–18
Equilateral triangle, 11
Explicit formula, for arithmetic sequence, 162–165
Exponential decay, 329–332
Exponential function, 325–338
    compound interest, 341–346
    construct and compare model of, 495–506
    defined, 203, 326
    exponential decay, 329–332
    exponential growth, 325–328
    exponential regression, 347–348
    graphs of, 333–338
    half-life, 329–332
    population growth, 347–350
Exponential growth, 325–328
Exponential regression, 203, 347–348
Exponents, 287–296
    base, 287
    defined, 287
    division, 288–289
    multiplication, 287–288
    Negative Power Property, 291
    power, 287
    Power of a Power Property, 294
    Power of a Product Property, 295
    Power of a Quotient Property, 295
    Product of Powers Property, 288
    Quotient of Powers Property, 289
    simplifying expressions and, 289–290, 292–293, 295–296
    terms of, 287
    Zero Power Property, 292
Expression
    defined, 5
    patterns to write, 8–13
Extrema, 83

F

Factor, 386
Factored form of polynomial, 456
Factoring
    polynomials
        difference of two squares, 390
        monomial from polynomial, 385–387
        perfect square of trinomial, 389–390
        prime polynomial, 396
        trinomials, 393–400
    quadratic equations, 455–458

Falling object experiment, 101–104
Fibonacci sequence, 169
Finite set, 73
First quartile, in five-number summary, 538
Five-number summary, 538
Formulas
    for arithmetic sequences, 162–165
    Boyle's Law, 30
    defined, 28
    density, 287
    distance, 30
    explicit, 162–165
    gravitational energy, 30
    interest, 263
    kinetic energy, 30
    pressure, 30
    quadratic, 474–476
    recursive, 168–170, 316–320
    velocity, 29
Free falling object, 485–486
Frequency
    defined, 597
    marginal total, 597
    relative, 597
    two-way frequency tables, 595–606
Function machine, 73–74
Functions. *See also* Linear function; Quadratic functions
    absolute maximum, 82–91
    absolute minimum, 82–91
    arithmetic sequences as, 166–167
    comparing properties of, 221–224
    comparing with inequalities, 231–234
    continuous data, 82
    defined, 68, 314
    dependent variable, 81
    discrete data, 82
    domain of, 71–72, 75
    exponential, 203, 325–338, 341–350
    finite set, 73
    function machine, 73–74
    graphs of
        falling object experiment, 101–104
        horizontal translation, 116
        key features, 227–230
        key features of, 81–91
        radioactive decay experiment, 105–108
        real-world situations, 92–94
        spring experiment, 97–100
        transformations, 111–118
        vertical translation, 113
    independent variable, 81
    infinite set, 73
    modeling with, 495–506
    notation for, 76–78, 211–214
    one-to-one, 155
    ordered pairs, 66
    parent, 111, 433
    piecewise defined function, 217–220
    range of, 71–72, 75
    relations and, 65–70
    relative maximum, 82–91
    relative minimum, 82–91
    vertical line test for, 81
    y-intercept, 82–91
    zero of, 458

G

Galileo, 101
Geometric sequence
    common ratio, 314
    defined, 314
    identify, 313–315
    recursive formula, 316–320
    tree diagram, 313
Graphs
    boundary line, 242
    closed half-plane, 242
    direct variation, 139–143
    of functions
        exponential, 333–338
        falling object experiment, 101–104
        key features of, 81–91, 227–230
        radioactive decay experiment, 105–108
        real-world situations, 92–94
        spring experiment, 97–100
    graphing method for systems of linear equations, 251–255
    horizontal translation, 116
    indirect variation, 144–147
    inequalities, 35–37, 43–46
        in two variables, 239–246
    open half-plane, 242
    piecewise defined function, 219–220
    quadratic equations, 455–458
        axis of symmetry and vertex, 459–462
        real roots of equation, 458
        zeros of the function, 458
    quadratic functions
        key features of, 427–430
        multiple transformations of, 444–450
        parent function, 433–439
        reflection, 444
        translations, 436–439
        with vertex and zeros, 462–464
        vertical shrink, 440–443
        vertical stretch, 440–443
    segmented bar graph, 600–606
    system of equations, 509–512
    systems of linear inequalities, 273–280
    transformation, 113
    vertical translation, 113
    write linear equation from, 227–230
Gravitational energy formula, 30
Greatest common factor (GCF), 385
    of polynomial, 385–387
Growler, 287

H

Half-life, 105, 329–332
Half-plane, 242
Hooke's Law, 98
Horizontal translation, 116

I

Icebergs, 287
Imaginary number, 480–482
Imaginary unit, 480
Inconsistent, 267–270
Independent systems of linear equations, 267–270

Independent variable, 81
Index, 300, 308
Index of summation, 533
Indirect variation, 144–147
Inequalities
   absolute value, 54–58
   comparing functions with, 231–234
   compound, 43–46
   domain and range of functions as, 215–218
   graphing, 35–37, 43–46
   multiplication by negative number, 40
   multi-step, 38–42
   solution of, 35–37
   solving, 35–46
   systems of linear inequalities, 273–280
   in two variables, 239–246
Infinite set, 73
Input, 66–67
   function machine, 73–74
   mapping, 66
Interest
   compound, 341–346
   formula, 263
   simple, 263
Inverse function, of linear function, 152–156
Inverse variation, 144–147
Irrational number, 299

K

Kinetic energy formula, 30

L

Leading coefficient, 357
Least common multiple (LCM), 413
Like terms
   addition of polynomials, 361–362
   defined, 361
   subtraction of polynomials, 364–366
Linear combination method, 261–263
Linear equations
   boundary line, 240
   from data, 193–199
   defined, 175
   linear regression, 197–199
   slope
      of parallel lines, 187–190
      of perpendicular lines, 187–190
   slope-intercept form, 175–178, 267
   standard form of, 183–186
   systems of, 251–270
   write
      from graph or table, 227–230
      from verbal description, 235–236
Linear function
   constant rate of change, 133–136
   construct and compare model of, 495–506
   direct variation, 139–143
   find domain and range of, 215–218
   indirect variation, 144–147
   inverse function for, 152–156
   inverse variation, 144–147
   point-slope form, 179–182
   rate of change, 211–214
   slope-intercept form, 175–178
   standard form of linear equation, 183–186
   writing, 215–218
Linear inequalities
   defined, 240
   graphing, 239–246
   solution of, 240
   systems of, 273–280
   in two variables, 239–246
   write, 239–241
Linear model
   equation from data, 193–199
   linear regression, 197–199
Linear regression, 197–199
   line of best fit, 198
Line of best fit, 198
Lines
   best-fit line, 571–592
   coincident, 264–266
   dependent systems, 267–270
   independent systems, 267–270
   parallel, 187–190, 264–266, 267–270
   perpendicular, 187–190
Literal equation, 28

M

Mapping
   defined, 66
   to identify function, 66–69
Marginal total, 597
Maximum, in five-number summary, 538
Mean, 523
Mean absolute deviation, 524–526, 534
Measurement error, 529
Measures of center, 523
   mean, 523
   median, 523
   mode, 523
Median, 523
   in five-number summary, 538
Mesa Verde National Park, 8
Metric measurement, 207
Minimum, in five-number summary, 538
Mode, 523
Modified box plot, 543–547
   outliers and, 543–547
Monomial
   defined, 357, 385
   factoring from polynomial, 385–387
Multiplication
   Associative Property, 16
   binomial, 369–378
   Commutative Property, 16
   Distributive Property, 414
   exponents, 287–288
   inequalities and negative numbers, 40
   Multiplication Property of Radicals, 308
   polynomials, 369–380
   Property of Equality, 16
   radical expressions, 307–310
   rational expressions, 411–412
   square root, 307–310
Multiplication Property of Radicals, 308
Multiple steps, solving equations with, 22–24

N

Negative correlation, 197
Negatively related, 560–561
Negative Power Property, 291
Negative square root, 300
Newton, Isaac, 453
No correlation, 197
Nonlinear system of equations, 511
Normal distribution, 548–553
Number line
   absolute value, 49–58
   graphing inequalities on, 35–37, 43–46
Numbers
   absolute value, 49
   complex, 308, 481
   imaginary, 480–482
   opposite of, 364

O

One-to-one functions, 155
Open half-plane, 242
Opposite, 364
Ordered pair
   defined, 66
   function machine, 73–74
   functions, 69
   relation, 66
Outlier
   defined, 543
   modified box plot and, 543–547
Output, 66–67
   function machine, 73–74
   mapping, 66

P

Parabola
   axis of symmetry, 433, 459
   defined, 427
   maximum, 427
   minimum, 427
   vertex of, 427
Parallel lines
   slope of, 187–190
   systems of linear equations, 264–266
Parent function
   defined, 111
   quadratic functions, 433
Patterns
   expressions and, 8–13
   investigating, 3–13
Perfect square of trinomial, 389–390
Perpendicular lines, slope of, 187–190
Photovoltaic panels, 355
Piecewise-defined function, 217–220
   construct and compare, 504–506
   defined, 217
   graphing, 219–220
Point-slope form, 179–182
Polynomials
   addition, 359–363
   Associative Property, 362
   classification of, 357
   coefficients of, 356

Commutative Property, 362
constant term of, 356
defined, 356
degree of a term, 356
degree of polynomial, 357
descending order of degree, 357
difference of two squares, 376–378
Distributive Property, 373–374
division, 406–410
factored form of, 456
factoring
   difference of two squares, 390
   greatest common factor of, 385–387
   monomial from polynomial, 285–390
   perfect square of trinomial, 389–390
   prime polynomial, 396
   trinomials, 393–400
leading coefficient, 357
like terms, 361–362, 364–365
multiplication, 369–380
rational expressions
   addition, 413–416
   defined, 403
   division, 403–405, 411–412
   multiplication, 411–412
   simplifying, 403–405
   subtraction, 413–416
square of binomials, 377–378
standard form of, 357
subtraction, 364–366
terminology for, 355–358
terms of, 356
Population growth, 347–350
Positive correlation, 197
Positively related, 560–561
Power
   defined, 287
   Power of a Power Property, 294
   Power of a Product Property, 295
   Power of a Quotient Property, 295
Power of a Power Property, 294
Power of a Product Property, 295
Power of a Quotient Property, 295
Pressure
   atmospheric, 175
   defined, 175
Pressure formula, 30
Prime number, 397
Prime polynomial, 396
Principal square root, 300
Probability and statistics
   best-fit line, 571–592
   bivariate data, 595–606
   box-and-whisker plot, 538
   box plot, 538–547
   census, 527–528
   correlation coefficient, 559–568
   deviation, 525
   dot plots, 537–542
   five-number summary, 538
   mean, 523
   mean absolute deviation, 524–526, 534
   measurement error, 529
   measure of center, 523
   median, 523
   mode, 523
   modified box plot, 543–547

   normal distribution, 548–553
   outliers, 543–547
   range of data set, 524
   residual plot, 588–592
   residuals and, 577–580
   sample, 528
   sample selection bias, 528
   sampling error, 528
   segmented bar graph, 600–606
   spread, 524
   standard deviation, 532–535
   sum of the squared residuals (SSR), 579
   two-way frequency tables, 595–606
   variability, 524
   z score, 550
Product of Powers Property, 288
Profit, 20
Projectile motion, 463
Properties
   Addition Property of Radicals, 304
   Associative, 16
   Distributive, 16, 22, 373–374
   Division Property of Radicals, 308–309
   of Equality, 16, 29
   Multiplication Property of Radicals, 308
   Negative Power Property, 291
   Power of a Power Property, 294
   Power of a Product Property, 295
   Power of a Quotient Property, 295
   Product of Powers Property, 288
   Quotient of Powers Property, 289, 403
   Symmetric Property of Equality, 16
   Zero Power Property, 292
   Zero Product Property, 456–457, 513
Pythagorean Theorem, 299

Q

Quadratic equations
   axis of symmetry, 433, 459–461
   discriminant, 477–479
   graphing
      axis of symmetry and vertex, 459–462
      with vertex and zeros, 462–464
   rocket application of, 485–487
   solving
      algebraic methods of, 467–482
      choosing method for, 477–479
      completing the square, 471–473
      complex numbers and, 480–482
      discriminant, 477–479
      factoring, 455–458
      imaginary numbers and, 480–482
      interpreting solutions of, 488–490
      quadratic formula, 474–476
      square root method, 467–470
      Zero Product Property, 456–457
Quadratic formula, 474–476
Quadratic functions
   construct and compare models
      of, 495–506
   defined, 202, 425
   graphing
      key features of, 427–430
      multiple transformations of, 444–450
      parent function, 433–439
      real roots of equation, 458

      reflection, 444
      translations, 436–439
      vertical shrink, 440–443
      vertical stretch, 440–443
      zeros of the function, 458
   introduction to, 423–426
   parent function, 433
   standard form of, 425
Quadratic regression, 202
Quotient of Powers Property, 289, 403

R

Radical expression, 299–310
   addition, 304–306
   Addition Property of Radicals, 304
   components of, 300, 308
   defined, 300
   division, 307–310
   Division Property of Radicals, 308–309
   multiplication, 307–310
   Multiplication Property of Radicals, 308
   negative square root, 300
   principal square root, 300
   rationalize the denominator, 309
   simplify, 299–303
   subtraction, 304–306
   write, 299–303
Radicand, 300, 308
Radioactive decay experiment, 105–108
Radon, 329–332
Range
   of data set, 524
   defined, 71
   dependent variable, 81
   function notation for, 76
   of functions, 71–72, 75
   of linear function, 215–218
Rate of change
   defined, 128
   linear function, 133–136, 211–214
   slope and, 128–132
Rational expression
   addition, 413–416
   defined, 403
   division, 403–405, 411–412
   multiplication, 411–412
   simplifying, 403–405
   subtraction, 413–416
Rationalized, 309
Rationalize the denominator, 309
Reading Math, 29, 49, 154, 160, 300, 358, 468
Real roots of equation, 458
Rectangular prism
   defined, 149
   volume, 144
Recursive formula
   for arithmetic sequence, 168–170
   defined, 168
   for geometric sequence, 316–320
Reflection, of quadratic functions, 444
Relation
   defined, 66
   identifying, 65–70
Relative frequency, 597
Relative maximum, 82–91

Relative minimum, 82–91
Residual plot, 588–592
Residuals, 577–580
　defined, 578
　residual plot, 588–592
　sum of the squared residuals (SSR), 579
Revenue, 20
Rockets, quadratic equations and, 485–487
Root index, 300, 308
Row percentages, 600

S

Sample, 528
Sample selection bias, 528
Sampling error, 528
Scatter plot
　best-fit line, 571–576
　correlation coefficient, 559–563
　defined, 194
　line of best fit, 198
　negative correlation, 197
　no correlation, 197
　positive correlation, 197
　trend line, 194
Segmented bar graph, 600–606
Sequence. *See also* Arithmetic sequences
　arithmetic, 159–170, 315
　defined, 4, 160
　Fibonacci sequence, 169
　geometric, 313–320
　terms of, 160
Sigma notation, 533
Simple interest, 263
Slope
　of best-fit line, 582–587
　change in y/change in x, 124–127
　defined, 124
　finding
　　point-slope form, 179–182
　　slope-intercept form, 175–178
　　standard form of linear equation, 183–186
　negative, 133–136
　of parallel lines, 187–190
　of perpendicular lines, 187–190
　point-slope form, 179–182
　positive, 133–136
　rate of change and, 128–132
　rise/run, 123–125
　slope-intercept form, 175–178
　undefined, 133–136
　vertical change/horizontal change, 124–127
　zero, 133–136
Slope-intercept form, 175–178
　classifying systems of linear equations, 267–270
Solar panels, 355
Solution
　defined, 15
　equations with infinitely many solutions, 25–27
　equations with no solution, 25–27
　of inequality, 35–37
　of linear inequality, 240
　systems of linear inequalities, 273–280

Solution region, 275–280
Solving equations
　with algebraic method, 15–18
　inequalities, 35–46
　with infinitely many solutions, 25–27
　literal equations for a variable, 28–30
　multiple steps, 22–24
　with no solution, 25–27
　quadratic equations
　　algebraic methods of, 467–482
　　choosing method for solving, 477–479
　　completing the square, 471–473
　　complex numbers and, 480–482
　　discriminant, 477–479
　　factoring, 455–458
　　imaginary numbers and, 480–482
　　interpreting solutions of, 488–490
　　quadratic formula, 474–476
　　square root method, 467–470
　　Zero Product Property, 456–457
　system of equations, 509–516
　systems of linear equations
　　elimination method, 261–263
　　graphing method, 251–255
　　linear combination method, 261–263
　　substitution method, 258–260
　　using tables, 256–260
　systems of linear inequalities, 273–280
　with variable on both sides, 19–21
Spread, 524
Spring experiment, 97–100
Square of binomial, 377–378
Square root
　addition, 304–306
　Addition Property of Radicals, 304
　division, 307–310
　multiplication, 307–310
　Multiplication Property of Radicals, 308
　negative square root, 300
　principal square root, 300
　as radical expression, 300
　rationalize the denominator, 309
　simplify, 300
　subtraction, 304–306
Square root method, 467–470
Standard deviation
　calculating, 532–535
　defined, 533
　normal distribution and, 549–551
Standard form
　of linear equation, 183–186
　of polynomial, 357
　of quadratic functions, 425
Statistics. *See* Probability and statistics
Subscript, 29
Substitution method, 258–260
Subtraction
　polynomials, 364–366
　Property of Equality, 16, 17, 29
　radical expression, 304–306
　rational expression, 413–416
　square root, 304–306
Sum of the squared residuals (SSR), 579
Surface area, of cylinder, 387
Symmetric Property of Equality, 16

Systems of equations
　nonlinear, 511
　solve algebraically, 513–516
　solve by graphing, 509–512
Systems of linear equations, 251–270
　classification
　　consistent, 267–270
　　dependent, 267–270
　　inconsistent, 267–270
　　independent, 267–270
　　slope-intercept form and, 267
　coincident lines, 264–266
　defined, 253
　parallel lines, 264–266
　solving
　　elimination method, 261–263
　　graphing method, 251–255
　　linear combination method, 261–263
　　substitution method, 258–260
　　using tables, 256–260
　　without unique solution, 264–266
　write, 261–263
Systems of linear inequalities, 273–280
　defined, 274
　solution region, 275–279

T

Tables
　to solve systems of linear equations, 256–260
　two-way frequency tables, 595–606
　write linear equation from, 227–230
Terms
　constant, 356
　defined, 355
　degree of, 356
　like terms, 361
　of sequence, 160
Third quartile, in five-number summary, 538
Trajectory, 473
Transformation
　defined, 113, 442
　of functions, 111–118
　horizontal translation, 116
　multiple, 444–450
　reflection, 444
　translation, 436–439
　vertical shrink, 442
　vertical stretch, 442
　vertical translation, 113
Translations
　horizontal, 116
　of quadratic functions, 436–439
Trebuchet, 455
Tree diagram, 313
Trend line
　defined, 194
　line of best fit, 198
Triangles, equilateral, 11
Trinomial
　defined, 357
　factoring, 393–400
　perfect square of trinomial, 389–390
Two-way frequency tables, 595–606

V

Variability
- defined, 524
- standard deviation, 532–535

Variables
- defined, 5
- dependent, 81
- equations with variables on both sides, 19–21
- independent, 81
- inequalities in two, 239–246
- negatively related, 560
- positively related, 560
- solving literal equations for specified variable, 28–30

Variation
- constant of, 140
- direct, 139–143
- indirect, 144–147
- inverse, 144–147

Velocity, 29
Vertical line test for, 81
Vertical shrink, 442
Vertical stretch, 442
Vertical translation, 113
Volume, rectangular prism, 144

W

Write
- equations, 15–18
- inequalities in two variables, 239–241
- linear equation
  - from graph or table, 227–230
  - from verbal description, 235–236
- linear function, 215–218
- radical expression, 299–303
- systems of linear equations, 261–263

Writing Math, 35, 37, 44, 71, 126, 228, 232, 481, 525

X

x-intercept
- defined, 102
- finding, standard form of linear equation, 183–186
- zero of function, 458

Y

Yellowstone National Park, 3

y-intercept
- defined, 175
- finding
  - slope-intercept form, 175–178
  - standard form of linear equation, 183–186
- of function, 82–91
- residuals and, 577–580

Z

Zero
- graphing quadratic equation with, 462–464
- slope, 133–136
- as subscript, 29
- Zero Power Property, 292
- Zero Product Property, 456–457, 513
- zeros of the function, 458

Zero Power Property, 292
Zero Product Property, 456–457, 513
Zeros of the function, 458
z score, 550